Lecture Notes in Artificial Intelligence   7499

Subseries of Lecture Notes in Computer Science

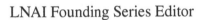
T0217439

Petr Sojka   Aleš Horák
Ivan Kopeček   Karel Pala (Eds.)

Text, Speech and Dialogue

15th International Conference, TSD 2012
Brno, Czech Republic, September 3-7, 2012
Proceedings

 Springer

Series Editors

Randy Goebel, University of Alberta, Edmonton, Canada
Jörg Siekmann, University of Saarland, Saarbrücken, Germany
Wolfgang Wahlster, DFKI and University of Saarland, Saarbrücken, Germany

Volume Editors

Petr Sojka
Masaryk University
Faculty of Informatics
Department of Computer Graphics and Design
Botanická 68a, 602 00, Brno, Czech Republic
E-mail: sojka@fi.muni.cz

Aleš Horák
Ivan Kopeček
Karel Pala
Masaryk University
Faculty of Informatics
Department of Information Technologies
Botanická 68a, 602 00 Brno, Czech Republic
E-mail: {hales, kopecek, pala}@fi.muni.cz

ISSN 0302-9743   e-ISSN 1611-3349
ISBN 978-3-642-32789-6   e-ISBN 978-3-642-32790-2
DOI 10.1007/978-3-642-32790-2
Springer Heidelberg Dordrecht London New York

Library of Congress Control Number: 2012944853

CR Subject Classification (1998): I.2, H.3-5, J.1, H.2, I.5, F.1

LNCS Sublibrary: SL 7 – Artificial Intelligence

Typesetting: Camera-ready by author, data conversion by Scientific Publishing Services, Chennai, India

Printed on acid-free paper

Springer is part of Springer Science+Business Media (www.springer.com)

Preface

The annual Text, Speech and Dialogue Conference (TSD), which originated in 1998, is in the middle of its second decade. So far more than 1,000 authors from 45 countries have contributed to the proceedings. TSD constitutes a recognized platform for the presentation and discussion of state-of-the-art technology and recent achievements in the field of natural language processing. It has become an interdisciplinary forum, interweaving the themes of speech technology and language processing. The conference attracts researchers not only from Central and Eastern Europe but also from other parts of the world. Indeed, one of its goals has always been to bring together NLP researchers with different interests from different parts of the world and to promote their mutual cooperation. One of the ambitions of the conference is, as its title says, not only to deal with dialogue systems as such, but also to contribute to improving dialogue between researchers in the two areas of NLP, i.e., between text and speech people. In our view, the TSD Conference was successful in this respect in 2012 again.

This volume contains the proceedings of the 15th TSD Conference, held in Brno, Czech Republic, in September 2012. In the review process, 82 papers were accepted out of 173 submitted, an acceptance rate of 47%.

We would like to thank all the authors for the efforts they put into their submissions and the members of the Program Committee and reviewers who did a wonderful job in helping us to select the most appropriate papers. We are also grateful to the invited speakers for their contributions. Their talks provide insight into important current issues, applications, and techniques related to the conference topics.

Last year the workshop on Natural Language Processing of Baltic and Slavonic Languages took place in the frame of the TSD Conference. This year a workshop on Hybrid Machine Translation was organized together with the conference. In part it was related to the EU project PRESEMT (FP7/2007-2013, ICT 248307) in which organizers participate.

Special thanks are due to the members of Local Organizing Committee for their tireless effort in organizing the conference.

The TeXpertise of Petr Sojka resulted in the production of the volume that you are holding in your hands.

We hope that the readers will benefit from the results of this event and disseminate the ideas of the TSD Conference all over the world. Enjoy the proceedings!

July 2012

Aleš Horák
Ivan Kopeček
Karel Pala
Petr Sojka

Organization

TSD 2012 was organized by the Faculty of Informatics, Masaryk University, in co-operation with the Faculty of Applied Sciences, University of West Bohemia in Plzeň. The conference webpage is located at http://www.tsdconference.org/tsd2012/

Program Committee

Hermansky, Hynek (USA),
    General Chair
Agirre, Eneko (Spain)
Černocký, Jan (Czech Republic)
Ferencz, Attila (Romania)
Fišer, Darja (Slovenia)
Garabík, Radovan (Slovakia)
Gelbukh, Alexander (Mexico)
Guthrie, Louise, (UK)
Hajič, Jan (Czech Republic)
Hajičová, Eva (Czech Republic)
Hanks, Patrick (UK)
Hitzenberger, Ludwig (Germany)
Hlaváčová, Jaroslava
    (Czech Republic)
Horák, Aleš (Czech Republic)
Hovy, Eduard (USA)
Kopeček, Ivan (Czech Republic)
Krauwer, Steven (The Netherlands)
Kunzmann, Siegfried (Germany)
Loukachevitch, Natalija (Russia)
Matoušek, Václav (Czech Republic)
McCarthy, Diana (UK)
Ney, Hermann (Germany)

Nöth, Elmar (Germany)
Oliva, Karel (Czech Republic)
Pala, Karel (Czech Republic)
Pavešić, Nikola (Slovenia)
Petkevič, Vladimír (Czech Republic)
Pianesi, Fabio (Italy)
Piasecki, Maciej (Poland)
Przepiorkowski, Adam (Poland)
Psutka, Josef (Czech Republic)
Pustejovsky, James (USA)
Rothkrantz, Leon (The Netherlands)
Rusko, Milan (Slovakia)
Skrelin, Pavel (Russia)
Smrž, Pavel (Czech Republic)
Sojka, Petr (Czech Republic)
Steidl, Stefan (Germany)
Stemmer, Georg (Germany)
Tadić, Marko (Croatia)
Varadi, Tamas (Hungary)
Vetulani, Zygmunt (Poland)
Vintsiuk, Taras (Ukraine)
Wiggers, Pascal (The Netherlands)
Wilks, Yorick (UK)
Zakharov, Victor (Russia)

Referees

Alegria, Iñaki
Arregi Uriarte, Olatz
Baisa, Vít
Beňuš, Štefan
Broda, Bartosz
Cerňak, Miloš

Diaz de Ilarraza, Arantza
Evdokimova, Vera
Evgrafova, Karina
Fellbaum, Christiane
Grézl, František
Guthrie, Joe

Hannemann, Mirko
Hlaváčková, Dana
Holub, Martin
Héja, Enikő
Kamshilova, Olga
Karafiát, Martin
Khokhlova, Maria
Kocharov, Daniil
Lopez de Lacalle, Oier
Marcińczuk, Michał
Mareček, David
Materna, Jiří
Matějka, Pavel
Maziarz, Marek
Mihelič, France
Miháltz, Márton
Mikolov, Tomáš
Mitrofanova, Olga

Mráková, Eva
Nedoluzhko, Anna
Němčík, Václav
Nevěřilová, Zuzana
Oravecz, Csaba
Otegi, Arantxa
Peterek, Nino
Popel, Martin
Radziszewski, Adam
Rigau Claramunt, German
Růžička, Michal
Sass, Bálint
Sazhok, Mykola
Stemmer, Georg
Szöke, Igor
Veselý, Karel
Wardyński, Adam

Organizing Committee

Dana Hlaváčková *(Administrative Contact)*, Aleš Horák *(Co-chairs)*,
Dana Komárková *(Secretary)*, Ivan Kopeček, Karel Pala *(Co-chair)*,
Adam Rambousek *(Web System)*, Pavel Rychlý, Petr Sojka *(Proceedings)*

Sponsors and Support

The TSD conference is regularly supported by International Speech Communication Association (ISCA). We would like to express our thanks to the Lexical Computing Ltd., for their kind sponsoring contribution to TSD 2012.

Table of Contents

Part III: Speech

Part IV: Dialogue

Part I

Invited Papers

Getting to Know Your Corpus

Adam Kilgarriff

Lexical Computing Ltd., Brighton, UK
http://www.sketchengine.co.uk

Abstract. Corpora are not easy to get a handle on. The usual way of getting to grips with text is to read it, but corpora are mostly too big to read (and not designed to be read). We show, with examples, how keyword lists (of one corpus *vs.* another) are a direct, practical and fascinating way to explore the characteristics of corpora, and of text types. Our method is to classify the top one hundred keywords of corpus1 *vs.* corpus2, and corpus2 *vs.* corpus1. This promptly reveals a range of contrasts between all the pairs of corpora we apply it to. We also present improved maths for keywords, and briefly discuss quantitative comparisons between corpora. All the methods discussed (and almost all of the corpora) are available in the Sketch Engine, a leading corpus query tool.

Keywords: corpora, corpus similarity, keywords, keyword lists, Sketch Engine.

1 Spot the Difference

accord actually amendment among bad because behavior believe bill blog ca center citizen color defense determine do dollar earth effort election even evil fact faculty favor favorite federal foreign forth guess guy he her him himself his honor human kid kill kind know labor law let liberal like man maybe me military movie my nation never nor not nothing official oh organization percent political post president pretty professor program realize recognize say shall she sin soul speak state suppose tell terrorist that thing think thou thy toward true truth unto upon violation vote voter war what while why woman yes

accommodation achieve advice aim area assessment available band behaviour building centre charity click client club colour consultation contact council delivery detail develop development disabled email enable enquiry ensure event excellent facility favourite full further garden guidance guide holiday improve information insurance join link local main manage management match mm nd offer opportunity organisation organise page partnership please pm poker pp programme project pub pupil quality range rd realise recognise road route scheme sector service shop site skill specialist st staff stage suitable telephone th top tour training transport uk undertake venue village visit visitor website welcome whilst wide workshop www

These two lists are the keywords we see when we compare one web-crawled English corpus (UKWaC, [1]) with another (enTenTen, [2]).

It does not take long to spot recurring themes: one classification (with each word assigned to one and only one category) is shown in Table 1.

P. Sojka et al. (Eds.): TSD 2012, LNCS 7499, pp. 3–15, 2012.

Table 1. Keywords; enTenTen *vs.* UKWaC

enTenTen keywords	UKWaC keywords
American spellings: among behavior center color defense favor favorite honor labor organization program realize recognize toward while	**British spelling:** behaviour centre colour favourite organise organisation programme realise recognise whilst
American politics: amendment bill blog citizen election federal law liberal nation official president state vote voter	**Schools, training:** assessment council guidance local pupil scheme skill training workshop
Bible: believe evil forth nor sin soul speak thou thy true truth unto upon	**Business:** achieve advice aim building client consultation develop development facility improve information manage management offer opportunity partnership project quality sector service specialist staff undertake
Informal: guy kid oh pretty yes	
Core verbs: be determine do guess know let say shall suppose tell think	
War and terrorists: foreign military kill terrorist violation war	**Furniture of web pages:** available contact click enquiry detail email further guide join link page please site telephone visit visitor website welcome www
Pronouns: he her him his me my she	
Negatives: never nor not nothing	
Other adverbs: even actually even maybe like pretty	**British lexical variants:** garden holiday shop transport (American equivalents: yard vacation store transportation)
Other grammatical words: because that what why	**British culture:** pub village
Academic: faculty professor	**Music:** band event stage tour venue
Core nouns: kind thing fact effort man woman human	**Addresses:** rd road route st
	Nonwords: th pm uk nd mm pp
Other: accord bad ca dollar earth movie percent post	**Adjectives:** main suitable excellent full wide
	Other: accommodation area charity club disabled enable ensure insurance main match poker range top

So: enTenTen has more American, more politics, more informal material, more war and terrorism, plus seams of biblical and academic material. UKWaC has a corresponding set of Britishisms and more on schools and training, business and music.

At all times we should note that the terms on one list were not missing from the other – American spellings are found in large numbers in UKWaC too – just that the balance is different. Needless to say many words might belong in multiple categories (shouldn't *believe* go with verbs?) and the classification requires some guessing about dominant meanings and word classes of polysemous words: I think it will be the adverbial *pretty* ("a pretty good idea"), not the adjectival ("a pretty dress"). There is nothing intrinsically business-y about words like *development, project, opportunity* but it is my hunch, on seeing them all in the same list, that their prevalence is due to their appearance in texts where companies are giving an account of all the good work they do.

We can attempt to check which words belong where by looking at concordances, but this turns out to be hard. Typically many patterns of use will be common in both corpora and, other than the plain statistic, there is no obvious way to summarise contrasts. For cases like *pretty*, where different meanings are for different word classes, we can see how the word classes differ (and the evidence from automatic pos-tagging confirms my guess: the adverb is an order of magnitude more frequent than the adjective (in both corpora) and dominates the frequencies). However that only helps in a limited range of cases (and part-of-speech tagging makes many mistakes with ambiguous words).

The lists are sorted by ratio of normalised frequencies (after addition of the simplemaths parameter, see below) and the lists are then simply the top 100 items, with no manual editing. Other settings were:

- Lemmas (as opposed to word forms).
  - This does some generalising over, for example, singular and plural form of the same noun.
  - It is only possible when the same lemmatisation procedure has been applied to both corpora. Otherwise, even if the differences seem minimal, the top of the keyword list will be dominated by the cases that were handled differently.
- Simplemaths parameter: 100 (see below)
- Only items containing exclusively lowercase a-to-z characters were included
- A minimum of two characters

Varying the setting gives other perspectives. Looking at capitalised items, reducing simplemaths parameter to ten and setting minimum length to five, we find the top items in enTenTen are *Obama Clinton Hillary McCain*, and for UKWaC *Centre Leeds Manchester Edinburgh*. (I changed the parameters to get longer and potentially lower-frequency items. Otherwise the lists had many acronyms and abbreviations: I wanted to see names.) enTenTen was collected in the run-up to the US Presidential election, which also explains the 'political' cluster. In UKWaC we have many places. *rd* and *st* are abbreviations, as usually used, in addresses, for 'road' and 'street'.

The settings I most often use are: simplemaths 100, exclusively lowercase words, of at least three characters. This usually gives a set of core-vocabulary words with minimal noise. Shorter items (one and two characters) are often not words For the lists above I used two characters, thereby including the two-letter words *be do he me my* and non-words *oh ca rd st th nd pm uk mm pp*. While my usual preference is for words, these all tell their story too, with, for example, *oh* vouching for informality and *mm* for the preference for the metric system in the UK (the USA more often uses inches).

If there are differences in how the data was prepared, they tend to dominate keyword lists. Between these two corpora there were not many differences – but the technology available for 'cleaning up' and removing duplicated material from web datasets has improved (thanks particularly to the work of Jan Pomikálek, whose tools were used for enTenTen). The 'furniture of web pages' cluster in UKWaC is probably there because, in 2009, we removed repeated material from web pages more effectively than in 2006.

1.1 Formality

Whereas lower-frequency items will support an understanding of the differences of content between the two corpora, as linguists we are also interested in differences of register. As Biber shows, the dominant dimension according to which text varies, across a wide range of text types and also languages, is from formal to informal, or to use his more specific terms, from interactional (for which everyday conversation is the prototype) to informational (with an academic paper as an extreme case) [3,4]. There are many features of text that vary according to where it sits on this dimension. Ones that are easily counted include word class: interactional language uses more verbs, personal pronouns and adverbs, informational uses more nouns, articles and adjectives [5].

We can see that there is a higher proportion of less formal material in enTenTen from the categories pronouns, core verbs, adverbs as well as the one marked informal, and the adjectives category in UKWaC is perhaps an indicator of higher formality. Both corpora have been tagged by TreeTagger[1]

and we can investigate further by looking at a keyword list, not of words or lemmas, but of word classes.

The word classes with a ratio between relative frequencies of 1.2 or greater are shown in Table 2.

Table 2. Key word classes, enTenTen *vs.* UKWaC

enTenTen key word classes	*UKWaC key word classes*
PP, PP$ personal pronoun (regular, possessive)	**POS** Possessive ending
VVD, VVP lexical verb (past, present tense)	**NP** Proper noun
IN/that that as subordinator	
WP, WDT wh-pronoun, wh-determiner	
UH interjection	
VHD, VH the verb have, base form and past tense	
RB adverb	

This confirms the greater formality (on average) of UKWaC.

2   Simple Maths for Keywords

The statistics used here for identifying keywords improve on those used elsewhere.

"This word is twice as common here as there." This is the simplest way to make a comparison between a word's frequency in one text type and its frequency in another. "Twice as common" means the word's frequency (per thousand or million words) in the first corpus is twice its frequency in the second. We count occurrences in each corpus, divide each number by the number of words in that corpus, optionally multiply by 1,000 or 1,000,000 to give frequencies per thousand or million, and divide the first number by the second to give a ratio. (Since the thousands or millions cancel out when we carry out the division, it makes no difference whether we use thousands or millions.)

If we find the ratio for all words, and sort by the ratio, we find the words that are most associated with each corpus as against the other. This will give a first pass at two

---

[1] See http://www.ims.uni-stuttgart.de/projekte/corplex/TreeTagger/; for the tagset used, see https://trac.sketchengine.co.uk/wiki/tagsets/penn

keyword lists, one (taken from the top of the sorted list) of corpus1 *vs.* corpus2, and the other, taken from the bottom of the list (with scores below 1 and getting close to 0), for corpus2 *vs.* corpus1. (In the discussion below we will refer to the two corpora as the focus corpus or *fc*, for which we want to find keywords, and the reference corpus or *rc*. We divide relative frequency in the *fc* by relative frequency in the *rc* and are interested in the high-scoring words.)

One problem with preparing keyword lists in this way is that you can not divide by zero, so it is not clear what to do about words which are present in the *fc* but absent in the *rc*.

A second problem is that, even setting aside the cases of zero occurrences, the list will be dominated by words with very few occurrences in the *rc*. There is nothing very surprising about a contrast between 10 in *fc* and 1 in *rc*, giving a ratio of 10, and we expect to find many such cases; but we would be very surprised to find words with 10,000 hits in *fc* and only 1,000 in *rc*, even though that also gives a ratio of 10. Simple ratios will give a list dominated by rare words.

A common solution to the zeros problem is 'add one'. If we add one to all the frequencies, including those for words which were present in *fc* but absent in *rc*, then we have no zeros and can compute a ratio for all words. A word with 10 hits in *fc* and none in *rc* gets a ratio of 11:1 (as we add 1 to 10 and to 0) or 11. "Add one" is widely used as a solution to a range of problems associated with low and zero frequency counts, in language technology and elsewhere [6].

This suggests a solution to the second problem. Consider what happens when we add 1, 100, or 1000 to all counts from both corpora. The results, for the three words *obscurish, middling,* and *common,* in two hypothetical corpora, are presented in Figure 1.

word	fc	rc	Add 1			Add 100			Add 1000		
	freq	freq	AdjFs	R1	R2	AdjFs	R1	R2	AdjFs	R1	R2
obscurish	10	0	11, 1	11.0	1	110, 100	1.1	3	1010, 1000	1.01	3
middling	200	100	201, 101	1.99	2	300, 200	1.50	1	1200, 1100	1.09	2
common	12000	10000	12001, 10001	1.20	3	12100, 10100	1.20	2	13000, 11000	1.18	1

Fig. 1. Frequencies, adjusted frequencies (AdjFs), ratio (R1), and keyword rank (R2), for three Simplemaths parameter settings, for rare, medium, and common words

All three words are notably more common in *fc* than *rc*, so all are candidates for the keyword list, but they are in different frequency ranges.

- When we add 1, *obscurish* comes highest on the keyword list, with *middling* second, and *common* last.
- When we add 100, the order is *middling, common, obscurish*.
- When we add 1000, it is *common, middling, obscurish*.

Different values for the 'add-N' or 'simplemaths' parameter give prominence to different frequency ranges. For some purposes a keyword list with commoner words is desirable; for others, we would want more rarer words. Our model lets the user specify the keyword list they want by adjusting the parameter. The model provides a way of identifying keywords without unwarranted mathematical sophistication, and reflects the fact that there is no one-size-fits-all list and different lists are wanted for different research questions.

The model is called 'simple maths' in contrast to other methods for keyword extraction, several of which use a hypothesis testing approach to see by what margin the null hypthesis is disproved. Such approaches both have much more complex maths, and are built on a flawed analysis of corpus statistics: the case is presented in full in [7].

3 Comparing Corpora of Known, Different Genres

Our first test case –UKWaC *vs.* enTenTen– was one in which we did not, at the outset, know what the differences were between the two corpora. The same method can be used where we know the differences of text type, which are there by design, and then we can use the keyword lists to find out more about the distinctions between the two text types, and also to find other, possibly unintended, contrasts between the two corpora.

We compared BAWE (British Academic Written English [8]) with SiBol/Port (comprising British broadsheet national newspapers [9]) and classified the top hundred words (word forms, with simplemaths 100, at least three letters, all lowercase) as follows.

A side-effect of using word forms rather than lemmas is that we see, in many cases, multiple forms of the same lemma (*factor factors, theory theories, use used using, played player players playing* etc.) While in one way this means we have had to waste time on multiple copies of the same word, in another it is reassuring: it shows how systematic the process is, where, of all the tens of thousands of English words that could have appeared in these top-100 lists, the words that do are so often different forms of the same lemma.

Much could be said about the analyses above, and what they tell us about academic writing, journalism, and the contrasts between them. A few brief comments:

1. Academic writing is more formal. The BAWE list is mostly nouns, with some adjectives and prepositions. The verbs that do appear are mostly past or past participle forms, with some (*associated, cited*) that rarely occur except in the passive. By contrast the SiBol/Port list has many pronouns and verbs.
2. Discourse structure is a central theme for academic prose, and discourse markers appear in the BAWE list.
3. The nouns listed under 'theory' for BAWE are a set of highly general and polysemous words, most of which have concrete meanings as well as abstract ones, so defy easy classification: a *solution* can be a solution in water as well as a solution of a problem, *development* can be what a plant does, or what a society does.

Table 3. Keywords; BAWE *vs.* SiBol/Port

BAWE keywords	SIBOL/Port keywords
Nouns:	**Time:** ago day days former last latest minutes
Experimental method: analysis control data error equation factor factors graph model method output sample variables	month months next never night now season summer week weekend year years yesterday
Theory: behaviour characteristics concept context development differences effect effects extent function information individual individuals knowledge nature states social systems process product products results theories theory type value values	**Money, numbers:** billion cent five million per pounds shares six
	Bosses: chairman chief director executive head minister secretary spokesman
	Sport: ball club football game games hit manager match played player players playing team top victory win won
Not-quite-so-general: cell cells communication environment gender human labour language learning meaning protein species temperature	**Verbs:** announced came come get going got had say says said think told took want went
Academic process: eds essay program project research section study	**Pronouns:** him his you your she who (there may well have been more but for the three-letter minimum)
Verbs: associated cited considered defined increase increased occur required shown use used using	**Prepositions/particles:** about ahead back down like round off
Adjectives: different important negative significant specific various	**News/politics:** cut died election news party police
Prepositions: between upon within	**Adjectives:** big young
Discourse connectives: due *(to)* hence therefore these thus whilst	**Non-time adverbs:** just really
	Other: bit com home house music thing television www

4. Newspapers are very interested in time (and money).
5. Sport forms a substantial component of SiBol/Port.

Both journalism and academic writing have been objects of extended study, with corpus work including, for journalism, [9,10], and for academic writing, [11,12,13]. Our current goal is simply to show how keyword methods can very quickly and efficiently contribute to those areas of research, as well as highlighting aspects of contrasting datasets that researchers might not have considered before.

4 Designed Corpora and Crawled Corpora

Two contrasting approaches to corpus-building are:

Design: Start from a design specification and select what goes into the corpus accordingly
Crawl: Crawl the web, and put whatever you find into the corpus.

The British National Corpus is a model designed corpus. UKWaC and enTenTen are both crawled.

The relative merit of the two approaches is a live topic [14,15,16,1]. Crawling is very appealing, since it involves no expert linguist input, is fast and cheap, and can be used to prepare vast corpora. But can we trust a crawled corpus? How do we know what is in it, or if it does a good job of representing the language?

Table 4. Keywords; enTenTen *vs.* BNC

enTenTen keywords	*BNC keywords*
Pronouns: our your	**Pronouns:** he herself her
Encoding: don percent	**Encoding**
Web: com site email request server internet comments click website online posted web list access data search www files file blog address page	**Speech transcription:** cos cent erm gon per pound pounds
University: article campus faculty graduate information project projects read research science student students	**Numbers:** eight fifty five forty four half hundred nine nineteen seven six ten thorty three twenty two
American spelling : behavior center color defense favor favorite labor organizations program programs toward	**British spelling:** behaviour centre colour defence favour labour programme round towards
Bible: believe evil faith forth sin soul thee thou thy unto upon	**British lexical variants:** bloody pupils shop
Politics: current federal global laws nation president security world	**Past tense verbs:** got felt turned smiled sat looked stood was said been seemed had went were knew put thought
Creative industries: author content create digital film game images media movie review story technology	**Particles:** away back down off
Informal: folks guess guy guys kids	**Local government:** council firm hospital local industrial police social speaker
Language change: issues	**Household nouns:** bed car door eyes face garden girl hair house kitchen mother room tea
Other: code efforts entire focus human include including human located mission persons prior provides	**Informal:** alright mean quite perhaps sort yeah yes
	Language change: chairman
	Other: although club considerable could head know main manager night there studio yesterday

4.1 BNC *vs.* enTenTen

Here there is no simple story to tell regarding formality. Both lists include pronouns: the BNC has three third person singular feminine pronouns, whereas enTenTen has a first and a second person one. This, along with the 'informal' cluster, suggests enTenTen has more interactional material. It is the BNC that has the verbs but they are all in the past tense. Biber shows that narrative is a central dimension of variation in language. The cluster of features associated with narrative includes past tense verbs and third person pronouns. The BNC has 16% fiction, and also a large quantity of newspaper, where the 'story' is central, so it seems that these two components place the BNC further along the narrative dimension than enTenTen. The daily newspaper material accounts for an abundance of *yesterday*, and the fiction, the 'household nouns' cluster.

don (in enTenTen) and *gon* (in BNC) arise from tokenisation issues: *don't* and *gonna* ('going to') both have different possible tokenisations, and different choices were made in the processing of the two corpora. Also there are different conventions on spelling out '%'.

10% of the BNC is transcribed speech. The BNC transcription manual specified that *erm* (pause filler), *cos* (spoken variant of *because*) and *gonna* (again) should be transcribed as *erm, cos* and *gonna*, and that *pound(s)* should be spelt out. So should numbers: hence the numbers cluster.

Whereas enTenTen has a biblical seam, the BNC has a local government one.

Language has changed in the two decades separating the two corpora, with *chairman* becoming less politically acceptable and *issues* acquiring a popular new sense, as in "we have some issues with that". And the world has changed: the web was unknown outside academia at the time of the BNC. Hence the web cluster.

We now have two comparisons involving enTenTen that we can compare. Some of the clusters (bible, American spellings) are much the same in both cases but most are quite different. Both tell us about enTenTen, but from different vantage points. The more corpora we compare our corpus with, the better we will get to know it.

4.2 Czech: CNC *vs.* czTenTen

The Czech National Corpus, as used in this study, comprises three 'balanced' 100m-word components (from 1990–99, 2000–04 and 2005–09) and one billion words of newspapers and magazines (1989–2007) [17]. czTenTen is a web corpus crawled in 2011.

Here there were no constraints on case or item-length, the simplemaths parameter was again 100, and there was a little manual editing to remove tokenization anomalies, words with missing diacritics, and Slovak words, and to merge multiple forms of the same lemma.

As with enTenTen *vs.* BNC, the web corpus is more interactional, with many first and second person forms of verbs, and 2nd person personal pronoun. As for English, there is a web cluster. With a large part of CNC being newspaper, it shares narrative characteristics like past tense reporting verbs with the BNC but also with SiBol/Port, with which it also shares politics, economics, sport and time.

5 Quantitative Approaches: Measuring Distances between Corpora

In this paper we have presented a keywords-based method of comparing corpora. This is just one method, and a qualitative one, empoying skills typically taught in humanities departments. enTenTen is more similar to UKWaC than the BNC, but this fact has not been foregrounded in the keyword-list analysis. A complementary approach is a quantitative one, in which we measure distances between corpora. [18] makes the case for corpus distance measures (and the closely related case for homogeneity/heterogeneity measures) and makes some proposals. Using a variant of

Table 5. Keywords; czTenTen *vs.* CNC

CczTenTen keywords	CNC keywords
informal: taky, teda, moc, sem, fakt, dneska, taky, sme, zas, dost, můžu, tak, takže, nějak, prostě, ahoj, tohle, super, jinak, fotky, jak, takže, holky, takhle, fajn, doufám **verbs in 1st or 2nd person:** jsi, můžete, najdete, mám, děkuji, bych, budete, máte, máš, nevím, prosím, děkuji, myslím, jsem, budu, díky, chci, naleznete, nemám, ráda, budeš, nejsem, vím, chcete **pronouns** (half are forms of second person plural): Vám, Vás, Vaše, vám, moje, ten, Váš, něco, nějaký, tebe, ono, vás, toto, nějaké, toho **adverbs:** trošku, naprosto, opravdu, akorát, docela, bohužel, trochu, krásně, jinak, pěkně, tam **web, computing:** web, aplikace, stránky, Windows, verze, video, online, odkaz, server, nastavení **other:** dobrý, dle, jestli, článek, pokud, zeptat, použití, nachází, pomocí, snad, jelikož, napsat, odpověď, den, nebo, přeci, týče	**politics/functions/institutions:** policie, starosta, ODS, unie, radnice, ČSSD, ředitel, úřad, policisté, policejní, nemocnice, klub, předseda, šéf, vedoucí, USA, ministr, prezident, banka, vláda **economics/mostly numerals:** koruna, tisíc, procento, milion, pět, čtyři, miliarda, tři, deset, šest, sto, osm, sedm, dvacet, dolar, padesát **spokesman-related words** (told, stated, said, spokesman, explained): uvedl, řekl, mluvčí, dodal, tvrdí, uvedla, prohlásil, říká, vysvětlil, řekla, sdělil **sports:** trenér, utkání, domácí, liga, kouč, vítězství, soutěž **names:** Jiří, Josef, Jan, Jaroslav, Vladimír, Pavel, Petr, Miroslav, Václav, Zdeněk, František, Karel, Milan **places:** Praha, ulici, náměstí, Brno, Ústí, Plzeň, Králové, město, České, Ostrava, Hradec, Liberec **time:** včera, letos, hodina, sobota, loni, pondělí, neděle, dosud, nyní, zatím, víkend, úterý **other:** výstava, expozice, však, totiž, Právo, muž, například, zřejmě, zhruba, lidé, uskuteční

Table 6. Distances between English corpora

	BNC	enTenTen	SiBol/Port	UKWaC
BAWE	2.15	1.98	2.39	1.92
BNC		1.51	1.64	1.63
enTenTen			1.75	1.42
SiBol/Port				1.74

the method found to work best there, we computed the distances between the five English corpora.

The most similar two corpora are indeed enTenTen and UKWaC, although enTenTen and BNC are only slightly further apart. Of the five, BAWE, comprising exclusively academic prose, is the outlier.

A careful comparison between two corpora generally requires both quantitative and qualitative approaches.

6   Functionality in the Sketch Engine

The Sketch Engine is a leading corpus query tool, in use for lexicography at Oxford University Press, Cambridge University Press, Collins, Macmillan, Cornelsen, Le Robert, and ten national language institutes (including those for Czech and Slovak), and for teaching and research at over one hundred universities worldwide. The Sketch Engine website has, already installed in the Sketch Engine and accessible to all users, large corpora for over sixty languages. For English there are many others as well. Users can install their own corpora and make comparisons between it and any other corpus (or subcorpus) of the same language.

The Sketch Engine provides functions for generating a range of lists, including all the keyword lists used in this paper. The interface for specifying a list (which may be a simple list, or a contrastive 'keyword' one) is shown in Figure 2.

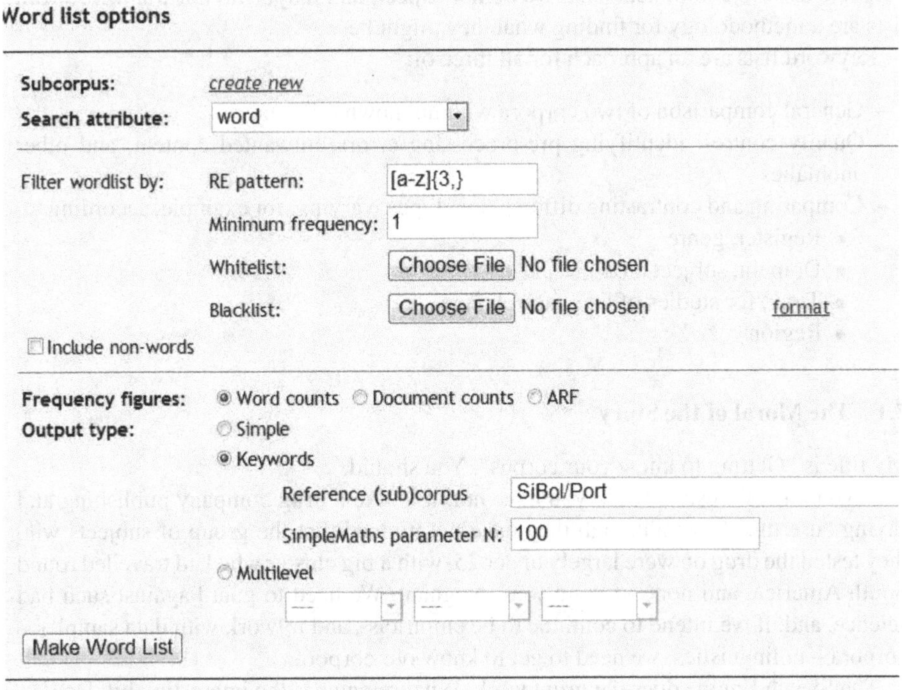

Word list options

Subcorpus:	*create new*	
Search attribute:	word ▾	

Filter wordlist by:	RE pattern:	[a-z]{3,}
	Minimum frequency:	1
	Whitelist:	Choose File No file chosen
	Blacklist:	Choose File No file chosen format
☐ Include non-words		

Frequency figures:	◉ Word counts ○ Document counts ○ ARF
Output type:	○ Simple
	◉ Keywords
	Reference (sub)corpus SiBol/Port
	SimpleMaths parameter N: 100
	○ Multilevel
	[--- ▾] [--- ▾] [--- ▾]

[Make Word List]

Fig. 2. Sketch Engine's form for specifying a word list, including specifying whether the list should be of word forms, lemmas, word classes etc., any pattern that should be matched, and whether the list should be a simple list or a keyword list. For English there are twenty corpora, installed and available, that one might choose to make comparisons with.

Until recently one might have argued that, while the procedures outlined in this paper for getting to know your corpus were sensible and desirable, they were hard to do, and unreasonable to expect of busy researchers, particularly those without programming skills. As it is now straightforward to use the Sketch Engine to prepare the lists, this argument is no longer valid.

7   The Bigger Picture

Corpora are not easy to get a handle on. The usual way of engaging with text is to read it, but corpora are mostly too big to read (and are not designed to be read). So, to get to grips with a corpus, we need some other strategy: perhaps a summary. A summary in isolation is unlikely to be helpful, because we do not know what we expect a corpus summary to look like. The summary only becomes useful when we can compare it with a summary for another corpus. A keyword list does this in the most straightforward way: it takes frequency lists as summaries of the two corpora, and shows us the most contrasting items.

Corpora are usually mixtures, and any two corpora vary in a multitude of ways, according to what their components are, and in what proportions. Any large, general corpus will have components that we do not expect, and maybe do not want. Keyword lists are a methodology for finding what they might be.

Keyword lists are an approach for all three of:

- General comparison of two corpora with unknown differences
- Quality control: identifying pre-processing errors, unwanted content, and other anomalies
- Comparing and contrasting different text types, varying, for example, according to:
  - Register, genre
  - Domain, subject area
  - Time, for studies of language change
  - Region

7.1   The Moral of the Story

My title is "Getting to know your corpus". You should.

If you publish results when you have not, it is like a drug company publishing and saying "use this drug" although they have not noticed that the group of subjects who they tested the drug on were largely under 25, with a big cluster who had travelled round South America, and none of them were pregnant. We need to guard against such bad science, and, if we intend to continue to be empiricist, and to work with data samples – corpora – in linguistics, we need to get to know our corpora.

The Sketch Engine does the grunt work. What remains is the interesting bit. Do it.

Acknowledgements. With thanks to Vít Suchomel for the analysis for Czech.

References

1. Baroni, M., Bernardini, S., Ferraresi, A., Zanchetta, E.: The wacky wide web: a collection of very large linguistically processed web-crawled corpora. Language Resources and Evaluation Journal 43, 209–226 (2009)
2. Pomikàlek, J., Rychlý, P., Kilgarriff, A.: Scaling to billion-plus word corpora. Advances in Computational Linguistics. Special Issue or Research in Computer Science 41 (2009)
3. Biber, D.: Variation across speech and writing. Cambridge University Press (1988)
4. Biber, D.: Dimensions of Register Variation: a cross-linguistic study. Cambridge University Press (2006)
5. Heylighen, F., Dewaele, J.M.: Formality of language: definition, measurement and behavioral determinants. Technical report, Free University of Brussels (1999)
6. Manning, C., Schütze, H.: Foundations of Statistical Natural Language Processing. MIT Press (1999)
7. Kilgarriff, A.: Language is never ever ever random. Corpus Linguistics and Linguistic Theory 1, 263–276 (2005)
8. Heuboeck, A., Holmes, J., Nesi, H.: The BAWE corpus manual. Technical report, Universities of Warwick, Coventry and Reading (2007)
9. Partington, A.: Modern diachronic corpus-assisted discourse studies MD-CADS on UK newspapers: an overview of the project. Corpora 5, 83–108 (2010)
10. Baker, P., Gabrielatos, C., McEnery, T.: Discourse Analysis and Media Bias: The representation of Islam in the British Press. Cambridge University Press (2012)
11. Biber, D.: University Language: A corpus-based study of spoken and written registers. John Benjamins (2006)
12. Paquot, M.: Academic Vocabulary in Learner Writing. Continuum (2010)
13. Kosem, I.: Designing a model for a corpus-driven dictionary of Academic English. Ph.D. thesis, Aston University, UK (2010)
14. Keller, F., Lapata, M.: Using the web to obtain frequencies for unseen bigrams. Computational Linguistics 29, 459–484 (2003)
15. Sharoff, S.: Creating general-purpose corpora using automated search engine queries. In: Baroni, M., Bernardini, S. (eds.) WaCky! Working papers on the Web as Corpus, Gedit, Bologna (2006)
16. Leech, G.: New resources, or just better old ones? the holy grail of representativeness. In: Hundt, M., Nesselehauf, N., Biewer, C. (eds.) Corpus Linguistics and the Web, pp. 133–149. Rodopi, Amsterdam (2007)
17. Čermák, F., Schmiedtová, V., Křen, M.: Czech national corpus – syn. Technical report, Institute of the Czech National Corpus (Prague, Czech Republic) http://www.korpus.cz (accessed on June 08, 2012)
18. Kilgarriff, A.: Comparing corpora. Int. Jnl. Corpus Linguistics 6, 263–276 (2001)

Coreference Resolution:
To What Extent Does It Help NLP Applications?

Ruslan Mitkov, Richard Evans, Constantin Orăsan, Iustin Dornescu, and Miguel Rios

Research Institute in Information and Language Processing
University of Wolverhampton, United Kingdom
{R.Mitkov,R.J.Evans,C.Orasan,I.Dornescu2,M.Rios}@wlv.ac.uk

Abstract. This paper describes a study of the impact of coreference resolution on NLP applications. Further to our previous study [1], in which we investigated whether anaphora resolution could be beneficial to NLP applications, we now seek to establish whether a different, but related task — that of coreference resolution, could improve the performance of three NLP applications: text summarisation, recognising textual entailment and text classification. The study discusses experiments in which the aforementioned applications were implemented in two versions, one in which the BART coreference resolution system was integrated and one in which it was not, and then tested in processing input text. The paper discusses the results obtained.

Keywords: coreference resolution, text summarisation, recognising textual entailment, text classification, extrinsic evaluation.

1 Introduction

In [1], we conducted the first extensive study into whether NLP applications could benefit from anaphora resolution. In this work we conducted extrinsic evaluation of our anaphora resolution system MARS [2] by seeking to establish whether and to what extent anaphora resolution can improve the performance of three NLP applications: text summarisation, term extraction and text categorisation. On the basis of the results we concluded that the deployment of anaphora resolution has a positive albeit limited impact. More specifically, the deployment of anaphora resolution increased the performance rates of these applications but the difference was not statistically significant.

In this study we revisit this topic but this time we have opted for seeking to establish the impact that coreference resolution could have on NLP applications. While some authors use the terms coreference (resolution) and anaphora (resolution) interchangeably, it is worth noting that they are completely distinct terms or tasks [3]. Anaphora is cohesion which points back to some previous item, with the 'pointing back' word or phrase called an anaphor, and the entity to which it refers, or for which it stands, its antecedent. Coreference is the act of picking out the same referent in the real world. A specific anaphor and more than one of the preceding (or following) noun phrases may be coreferential, thus forming a coreferential chain of entities which have the same referent.

P. Sojka et al. (Eds.): TSD 2012, LNCS 7499, pp. 16–27, 2012.

Coreference is typical of anaphora realised by pronouns and non-pronominal definite noun phrases, but does not apply to varieties of anaphora that are not based on referring expressions, such as verb anaphora. However, not every noun phrase triggers coreference. Bound anaphors which have as their antecedent quantifying noun phrases such as *every man*, *most computational linguistics*, *nobody*, etc. are another example where the anaphor and the antecedent do not corefer. As an illustration, the relation in '*Every man* has *his* own agenda' is only anaphoric, whereas in '*John* has *his* own agenda' is both anaphoric and coreferential. In addition, while identity-of-reference nominal anaphora involves coreference by virtue of the anaphor and its antecedent having the same real-world referent, identity-of-sense anaphora (e.g. 'The man who gave his paycheck to his wife was wiser than the man who gave it to his mistress') does not. Finally, there may be cases where two items are coreferential without being anaphoric. Cross-document coreference is an obvious example: two mentions of the same person in two different documents will be coreferential, but will not stand in anaphoric relation.

Having explained the difference between the terms/phenomena *anaphora* and *coreference*, we should point out that the tasks *anaphora resolution* and *coreference resolution* are not identical either. Whereas the task of anaphora resolution has to do with tracking down an antecedent of an anaphor, coreference resolution seeks to identify all coreference classes (chains).

In this study we seek to establish whether the employment of coreference resolution to NLP applications is beneficial. The investigation has been undertaken by means of experiments involving three applications: text summarisation, textual entailment and text classification. It differs from our 2007 study, not only in the employment of a specific NLP task (coreference resolution as opposed to anaphora resolution) and in the applications covered (recognising textual entailment is a new NLP application), but also in the data selected for the current experiments. Since 2007 there have been significant developments in the construction and sharing of large-scale resources and this has been an ongoing trend in Natural Language Processing. By way of example, research in Textual Entailment is supported by the availability of several annotated datasets. These resources typically consist of sets of T-H pairs manually annotated with a Boolean value to indicate whether or not H is entailed by T. In the current paper, datasets RTE1 [4], RTE-2 [5], and RTE-3 [6] are used to evaluate the impact of coreference resolution on automatic RTE. We have opted to use such publicly available resources in spite of the fact that, as a result, we had to resort to the exploitation of different data for every evaluation/application in contrast to our previous experiments where we benefited from a common corpus.

The development of automatic coreference resolution systems began in earnest in 1996 in response to the MUC-6 competition organised by NRAD with the support of DARPA [7]. Since then, numerous coreference resolution systems have been developed, typically using machine learning, and exploiting supervised [8]; [9]; [10] and unsupervised (clustering) methods [11]. Current approaches continue to exploit machine learning, seeking improved models for the resolution process, based on various linguistic and contextual features [12].

In the research described in the current paper, the publicly available BART toolkit was exploited [13]. The coreference resolution system distributed with BART is reported to offer state of the art performance in coreference resolution, particularly with regard to the resolution of pronominal mentions. With reported recall in pronoun resolution at the level of 73.4%, BART's performance is close to that of specialised pronoun resolution systems.

BART works by first preprocessing input documents in order to detect potential mentions such as pronouns, noun chunks, base noun phrases, and named entities. For each detected anaphor, the system extracts each pair consisting of the anaphor and a potential antecedent for that anaphor. Pairs are represented using a feature set that combines features exploited in the system developed by Soon et al. [8] with features encoding the syntactic relation between anaphors and their potential antecedents [14], and features based on knowledge extracted from Wikipedia. The coreferentiality of each pair of mentions is then determined using machine learning.

The automatic identification of coreference chains enables practical NLP applications to substitute semantically ambiguous references to entities (such as pronouns) with more informative phrases, before subsequent processing.

The rest of the paper is organised as follows. In Section 2 we review related research. In Sections 3, 4 and 5 we present the experimental settings related to the application of BART to text summarisation, recognising textual entailment and text classification respectively. In Section 6 we discuss the evaluation results and finally in the concluding Section 7 we summarise the results of this study.

2 Related Research

The research described in the current paper is motivated by previous research in the fields of automatic summarisation, recognition of textual entailment, and text classification. This work highlights various challenges to be addressed in each area, and describes different attempts to ameliorate them.

2.1 Automatic Summarisation

One of the drawbacks of most implementations of keyword-based summarisation is that they consider words in isolation. This means that most implementations will fail to recognise when two words are in an anaphoric relation or they are part of the same coreferential chain. For this reason, it was argued that it is possible to improve the results of an automatic summarisation system that relies only on keywords by obtaining better frequency counts using the information from an anaphora or coreference resolver.

Previous experiments using a pronominal anaphora resolver showed limited impact. Orăsan [15] shows that an automatic pronoun resolver does not really improve the results of an automatic summarisation method for scientific documents. However, he uses the annotated data to simulate a pronoun resolver and shows that a high accuracy pronoun resolution is useful in the summarisation process. Experiments on newswire texts show similar results [1] and lead to the conclusion that it may be possible to improve the results of the automatic summariser by using a coreference resolver instead of just a pronoun resolver. This conclusion is also supported by the research presented in [16] where the results on an LSA-based summariser are improved when a coreference resolver is used. In his thesis [17] and similar to our previous findings [1], Kabadjov establishes that the employment of his anaphora resolution system GUITAR to summarisation through substitution leads to limited (statistically insignificant) improvement. On the other hand, anaphora resolution leads to statistically significant improvements, when lexical and anaphoric knowledge is integrated into an LSA-based summariser.

2.2   Textual Entailment

The exploitation of coreference information in tasks related to RTE is motivated by previous work which has demonstrated encouraging results. In the context of RTE competitions exploiting the RTE-3 dataset, systems exploiting coreference information have not been the highest ranking ones in terms of performance, but results have been encouraging in paraphrase recognition [18] and in the identification of sentences entailed by input queries [19]. Research reported in [20] suggests that discourse information can improve RTE in the context of the Search Task, but that such information should be integrated into the inference engine as opposed to serving as a preprocessing step or a feature exploited by an ML algorithm.

2.3   Text Classification

One challenge for TC systems is caused by synonymy, e.g. when new terms are synonyms of observed terms but are ignored by the classifiers, and by polysemy, e.g. when a new sense of a known term is used. Previous work studied the use of WordNet in addressing the first problem, or WSD systems in addressing the second. Coreference is another natural language phenomenon which affects TC in a similar way to synonymy and polysemy: on the one hand the same entities or concepts are mentioned multiple times but using different words; on the other hand the same words can be used to refer to different concepts, as is usually the case with pronominal anaphoric expressions.

Coreference resolution provides discourse level information which could help classifiers alleviate some of these issues, but its usefulness for TC has received little attention in the literature. Incorporating coreference information in TC is usually achieved by changing the weights of those terms which occur in coreference chains [21].

2.4   Other Related Work

Hendrickx et al. [22] investigate the effect resulting from the deployment of a coreference resolution system for Dutch [23] in relation to information extraction and question answering. Their findings point to some increase in performance of information extraction after incorporating coreference resolution. However, this increase is not statistically significant. When incorporating their coreference system for Question Answering, while the number of extracted facts increases by 50%, overall performance decreases significantly. Overall, due to the number of additional facts retrieved, this leads to a 5% improvement in performance on the QA@CLEF 2005 test set.

3   Automatic Summarisation: Experimental Settings

This section presents the settings for an experiment where a keyword-based summariser is enhanced with information from a coreference resolver. Section 6.1 describes the evaluation results and discussion.

For the experiments presented in this section, we reimplemented the keyword-based summariser described in [24] and used only the best performing setting identified there. Therefore, for our experiment, words are scored on the basis of their frequency in the document and stopwords are filtered out.[1] On the basis of previous research, we decided to count words as they appear in the text and not to do any morphological processing. The final score of a sentence is calculated by adding up the scores of the words contained in the sentence. The summary is produced by extracting the sentences with the highest scores until the desired length is reached.

In order to obtain better frequency counts, we used the information from a coreference resolver to boost the scores of the sentences which contain coreferential chains by the scores of the chains. For this purpose we used two settings: In the first, we increased the score of a chain by the score of the longest mention which occurs in the chain. When several mentions of the same length are found in the text, the first one is used. This setting is used in order to have a setting similar to the other experiments presented in this paper. In the second setting, the score of the chain is given by the greatest score of the mentions it contains. This is to reflect the fact that the importance of a mention is given by its content, not by its length. In both experiments, the chains containing only one element are discarded.

The evaluation was carried out using 89 randomly selected texts from the CAST corpus [25]. The CAST corpus is a corpus of newswire texts annotated with information about the importance of the sentences with regards to the topic of the document. Annotators were asked to manually annotate 15% of the most important sentences as *ESSENTIAL* and a further 15% as *IMPORTANT*. In this way, it is possible to evaluate summarisation methods which produce summaries of 15% and 30% compression rates. All the texts selected for this experiment were annotated with coreference information using BART [13].

For the evaluation (Section 6.1), we compared the sets of sentences selected by the program with the set of sentences annotated by humans. On the basis of this comparison, we calculated precision and recall and we report the results using F-measure.

4   Recognising Textual Entailment: Experimental Settings

RTE can be regarded as a binary classification task in which each pair of text and hypothesis is classified according to whether or not the text is entailed by the hypothesis. In this context, RTE benchmark datasets are used to train a classifier [26]. We followed the methodology used by Castillo [19] to process coreference chains in which each mention in a chain is substituted by the longest (most informative) mention. In contrast

---

[1] One of the differences between the current implementation and the implementation reported in [24,15] comes from the fact that the current implementation is in Python and uses NLTK for the stoplist and processing.

to that approach, we used the two-way benchmark datasets (i.e. text and hypothesis pairs that have been manually classified as true/false) for training and testing. We appended each T-H pair as one piece of text, and processed each pair using the BART[2] coreference resolver. Then, for each coreference chain we selected the mention with the greatest word length and replace all other mentions in the chain with this most informative mention.

The RTE system is based on a supervised Machine Learning algorithm. The algorithm is trained to classify T-H pairs by means of metrics that assess the similarity between T and H. These include *lexical metrics* (precision, recall, and F-score), used with a bag-of-words representation of the T-H pairs; and metrics such as *BLEU* [27]; *METEOR* [28]; and *TINE* [29].

With these metrics we built a vector of similarity scores used as features to train a Machine Learning algorithm. We used the development datasets from the RTE 1 to 3 benchmark to train a Support Vector Machine algorithm distributed with Weka[3] with no parameter optimisation. Then, we tested the models using 10-fold-cross-validation over the development datasets and we compared them against the test datasets.

5   Text Classification: Experimental Settings

A TC system has three processing stages: document processing, classifier learning and evaluation. During document processing, or document indexing, textual documents are analysed and represented in a compact form as a weighted term vector, where each term corresponds to a feature and the weight quantifies its importance for a particular document. Terms often correspond to words mentioned in the document, but usually stop-words are removed and stemming can be applied. The weight of each term is usually computed using either statistical or probabilistic techniques, with tf · idf being one of the most popular methods. As unseen documents are likely to use vocabulary terms which did not occur in training, classifiers tend to perform better on training data than on test data. To reduce overfitting, a dimensionality reduction step can be employed which also reduces the computational complexity for building classifier models. Dimensionality reduction in TC usually involves a term selection method in which only the most relevant terms are used to represent documents.

A standard BOW approach was used in this study: punctuation and stop-words have been removed, all words have been converted to lower-case characters and Porter's stemmer was applied. Both single words and bigrams were used as terms [30].

Several studies [31,32] found that feature selection methods based on χ^2 statistics consistently outperformed those based on other criteria (including information gain) for the most popular classifiers used in TC. The terms with a document frequency less than 5 were also removed, as χ^2 is known to be less reliable for rare words [31]. Both methods were applied and 10% of the terms were selected for the vector space representation.

Length-normalised feature vectors were built using the standard $tf \cdot$idf function using log smoothing: $\text{tfidf}(t_k, d_j) = \text{tf}(t_k, d_j) \cdot \log \frac{|D|}{|D_k|}$, where $\text{tf}(t_k, d_j) = 1 + \log(\text{occ}(t_k, d_j))$

---

[2] http://www.bart-coref.org/

[3] http://www.cs.waikato.ac.nz/ml/weka/

for terms t_k with at least one occurrence in document d_j and 0 otherwise, $|D|$ is the collection size and $|D_k|$ is the document frequency of term t_k.

The BART [13] coreference resolution system was run on the original plain text version of the documents in the R(10) corpus to identify coreference chains. This information was used to boost the weights of terms included in the chains, by using a modified term frequency function: $\text{tf}^{\text{coref}}(t_k, d_j) = \sum_{c \in C_{k,j}} \text{len}(c)$, where $\text{len}(c)$ is the length of chain c, $C_{k,j}$ is the set of chains in document d_j containing at least one mention of term t_k. Essentially this function acts as if a term occurs in all mentions of a chain, as long as it occurs in at least one of them.

The SVM classifier was used in a binary mode: a different model was built for each of the 10 classes, including the term selection step, also known as local-selection [32]. The average precision of the individual classifiers is used for evaluation (Section 6.3).

Some of the most popular collections used to compare different approaches to text categorisation are 20-newsgroups[4], Reuters-21578[5] and Reuters Corpus Volume 1 [33]. A study of the impact of class distribution on the performance of automatic TC systems [32] showed that the relative ranking of several approaches depends on which subset of the Reuters-21578 corpus is used. The study also revealed that the SVM classifier usually outranks other learners and that χ^2 usually achieves better results than other selection methods such as information gain, information ratio and mutual information.

In this paper, TC performance was assessed using a subset of the ApteMod dataset. ApteMod[6] is a collection of 10,788 documents from the Reuters-21578 corpus, partitioned into a training set with 7,769 documents and a test set with 3,019 documents. The subset exploited in the current work consists of the 10 categories with the highest number of positive training examples, also known as R(10) in the literature. This subset has also been exploited in research presented in [34,35,36].

6   Results and Discussion

This section presents an evaluation of the impact that automatically obtained coreference information [13] has on the three NLP applications described in Sections 3–5. In each case, a comparison is made between the efficacy of systems exploiting such information and those that do not.

6.1   Automatic Summarisation

Table 1 presents the results of the evaluation. Column *Without BART* shows the results of the system which does not use any coreference information. Columns *With BART len* and *With BART weight* show the results when information from BART is used and correspond to the two settings presented described in Section 3. To our surprise, the results of the summarisation process decrease when coreference information is added. For both experiments, the decrease is statistically significant at 15% compression rate, but not at 30% compression rate.

---

[4] http://people.csail.mit.edu/jrennie/20Newsgroups/

[5] http://www.daviddlewis.com/resources/testcollections/reuters21578

[6] http://www.cpan.org/authors/Ken_Williams/data/reuters-21578.readme

Table 1. Evaluation results of the automatic summarisation method

Compression rate	Without BART	With BART len	With BART weight
15%	32.88%	28.62%	27.14%
30%	46.34%	45.88%	45.19%

The results presented in the table were obtained by giving a weight of 1 to the contribution from the coreference resolver. In order to find out whether it is possible to obtain better results using a different weight for this contribution, we run an experiment where the contribution increased from 0 to 10 in 0.25 increments. Figure 1 and 2 show that as the contribution of coreference resolver increases, the results of the summariser decrease. This is the case for both experiments.

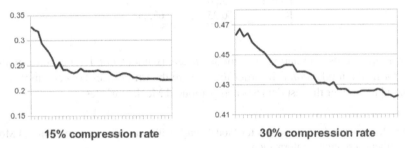

Fig. 1. The results for the first setting of the summarisation experiment when the contribution of the coreference resolver increases

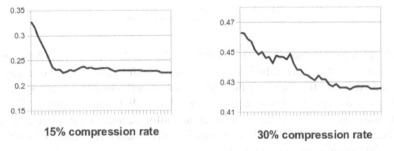

Fig. 2. The results for the second setting of the summarisation experiment when the contribution of the coreference resolver increases

On the basis of the experiment presented in this section, it can be concluded that using information from a coreference resolver in such a simple way is not beneficial for automatic summarisation. The main reason for this is the errors introduced by the coreference resolver. As future research, we plan to employ the approach proposed by [15] and use a gold standard to simulate a coreference resolver to find out what level of accuracy is necessary in order to improve the results of an automatic summariser.

6.2   Recognising Textual Entailment

Two different models for RTE were trained and tested, one of which exploits coreference information and one of which does not. The models use the same features, but
with different preprocessed input data, where model *coref* denotes data processed with
coreference information and model *token* denotes data processed without coreference
information. Table 2 shows the comparison of both models' accuracy via 10-fold cross-
validation over the development datasets.

Table 2. Results of 10-cross-fold-validation for Model *coref*: with coreference information and
Model *token*: without coreference information

Dataset	Model coref	Model token
RTE-1	54.14	56.61
RTE-2	58.50	60
RTE-3	60.25	67.25

For 10-fold cross-validation, the model *token* (without coreference information)
outperforms the model *coref*. In order to measure the differences between models we
compared them over the test datasets and computed McNemar's test.

Table 3. Results over the test datasets for Model *coref*: with coreference information and Model
token: without coreference information

Dataset	Model coref	Model token
RTE-1	56.87	56.87
RTE-2	57.12	59.12
RTE-3	60.25	61.75

Table 3 shows the results of both models over the test datasets. The models in which
coreferential mentions are substituted show worse performance than those which do
not make such substitutions, but the differences in performance are not statistically
significant. Furthermore, when assessed over the RTE-1 dataset, this RTE system
outperforms the system exploiting coreference information for paraphrase detection
described in [18], but the models show similar performance regardless of whether or
not coreference information is exploited.

We analysed the datasets in order to investigate cases in which the quality of
the coreference resolver is decisive in affecting the performance of the method.
For example, in the RTE-1 test dataset with 800 T-H pairs the average number of
coreference chains per document is 2.1, the average number of words is 38.16 and the
number of pairs with no chain is 60. In the RTE-3 dataset, over which the model obtains
the best result, with 800 T-H pairs and a similar average number of chains (1.80), the
number of pairs without chains increases to 104. Thus, the method reduces the amount
of errors with fewer coreference-enhanced T-H pairs. However, the appended T-H pairs
do not differ from one dataset to another in terms of the number of words. Therefore

the number of entities is insufficient to make a significant difference to the result. More conclusive results may be achieved in the context of the Search Task.

6.3   Text Categorisation

In text categorisation, the two term weighting functions yield two experimental settings: *run-bow* using the standard pipeline, and *run-bart* which boosts the weight of terms occurring in coreference chains, in proportion to the chain length. The results of the experiments show that the difference between the two settings is small: the macro-averaged precision for *run-bow* is 95.6% and for *run-bart* it is 95.7%. The performance difference between the corresponding binary classifiers is also small, suggesting that the state-of-the-art approach using the bag-of-words representation does not take advantage of coreference information. This result confirms that of [21] who used a different coreference system and a slightly different weighting function.

This result can be partially explained by errors in the coreference chains produced by the resolver, but also suggests that a more explicit way of employing this information is necessary. The intuition is that the presence or absence of a particular entity or term better indicates the topic of the document than the actual number of times it is mentioned. Future investigations should consider ways in which coreference information can be used to enhance TC systems using semantic features to represent documents, which can make use of what entities represent, instead of just using entity names. A TC system employing a semantic representation using external knowledge could exploit coreference information directly, e.g. it could know that a document is about sport based on the number of mentions of sportspeople instead of their actual names, which could be very sparse and occur in too few documents.

7   Conclusions

This study sought to establish whether or not coreference resolution could have a positive impact on NLP applications, in particular on text summarisation, recognising textual entailment, and text categorisation. The evaluation results presented in Section 6 are in line with previous experiments conducted both by the present authors and other researchers: there is no statistically significant benefit brought by automatic coreference resolution to these applications. In this specific study, the employment of the coreference resolution system distributed in the BART toolkit generally evokes slight but not significant increases in performance and in some cases it even evokes a slight deterioration in the performance results of these applications. We conjecture that the lack of a positive impact is due to the success rate of the BART coreference resolution system which appears to be insufficient to boost performance of the aforementioned applications.

References

1. Mitkov, R., Evans, R., Orăsan, C., Ha, L.A., Pekar, V.: Anaphora Resolution: To What Extent Does It Help NLP Applications? In: Branco, A. (ed.) DAARC 2007. LNCS (LNAI), vol. 4410, pp. 179–190. Springer, Heidelberg (2007)

2. Mitkov, R., Evans, R., Orăsan, C.: A New, Fully Automatic Version of Mitkov's Knowledge-Poor Pronoun Resolution Method. In: Gelbukh, A. (ed.) CICLing 2002. LNCS, vol. 2276, pp. 168–187. Springer, Heidelberg (2002)
3. Mitkov, R.: Anaphora Resolution. Longman, Cambridge (2002)
4. Sekine, S., Inui, K., Dagan, I., Dolan, B., Giampiccolo, D., Magnini, B. (eds.): Proceedings of the ACL-PASCAL Workshop on Textual Entailment and Paraphrasing. Association for Computational Linguistics, Prague (2007)
5. Ido, R.B.H., Dagan, I., Dolan, B., Ferro, L., Giampiccolo, D., Magnini, B., Szpektor, I.: The second pascal recognising textual entailment challenge (2006)
6. Dagan, I., Glickman, O.: The PASCAL recognising textual entailment challenge. In: Proceedings of the PASCAL Challenges Workshop on Recognising Textual Entailment (2005)
7. Grishman, R., Sundheim, B.: Message understanding conference-6: A brief history. In: Proceedings of the 16th International Conference on Computational Linguistics, COLING 1996, Copenhagen, Denmark (1996)
8. Soon, W.M., Ng, H.T., Lim, D.C.Y.: A machine learning approach to coreference resolution of noun phrases. Computational Linguistics 27, 521–544 (2001)
9. Ng, V., Cardie, C.: Improving machine learning approaches to coreference resolution. In: Proceedings of ACL 2002. Association for Computational Linguistics (2002)
10. Uryupina, O.: Coreference resolution with and without linguistic knowledge. In: Proceedings of LREC 2006, Genoa, Italy, pp. 893–898 (2006)
11. Cardie, C., Wagstaff, K.: Noun phrase coreference as clustering. In: Proceedings of the 1999 Joint SIGDAT Conference on Empirical Methods in Natural Language Processing and Very Large Corpora, pp. 82–89. Association for Computational Linguistics, College Park (1999)
12. Ng, V.: Supervised noun phrase coreference research: The first fifteen years. In: Proceedings of ACL 2010. Association for Computational Linguistics (2010)
13. Versley, Y., Ponzetto, S.P., Poesio, M., Eidelman, V., Jern, A., Smith, J., Yang, X., Moschitti, A.: Bart: A modular toolkit for coreference resolution. In: Proceedings of LREC 2008 (2008)
14. Yang, X., Su, J., Tan, C.L.: Kernel-based pronoun resolution with structured syntactic knowledge. In: Proceedings of CoLing/ACL 2006. Association for Computational Linguistics (2006)
15. Orăsan, C.: The Influence of Pronominal Anaphora Resolution on Term-based Summarisation. In: Nicolov, N., Angelova, G., Mitkov, R. (eds.) Recent Advances in Natural Language Processing. Current Issues in Linguistic Theory, vol. 309, pp. 291–300. John Benjamins, Amsterdam (2009)
16. Steinberger, J., Kabadjov, M.A., Poesio, M., Sanchez-Graillet, O.: Improving LSA-based summarization with anaphora resolution. In: Proceedings of Human Language Technology Conference and Conference on Empirical Methods in Natural Language Processing (HLT/EMNLP), Vancouver, Canada, pp. 1–8 (2005)
17. Kabadjov, M.: A Comprehensive Evaluation of Anaphora Resolution and Discourse-new Classification. Ph.D. thesis, Department of Computer Science, University of Essex (2007)
18. Andreevskaia, A., Li, Z., Bergler, S.: Can shallow predicate argument structures determine entailment? In: Proceedings of the PASCAL Challenges Workshop on Recognising Textual Entailment (2005)
19. Castillo, J.J.: Textual entailment search task: An initial approach based on coreference resolution. In: International Conference on Intelligent Computing and Cognitive Informatics, pp. 388–391 (2010)
20. Mirkin, S., Dagan, I., Padó, S.: Assessing the role of discourse references in entailment inference. In: ACL, pp. 1209–1219 (2010)

21. Li, Z., Zhou, M.: Use semantic meaning of coreference to improve classification text representation. In: The 2nd IEEE International Conference on Information Management and Engineering (ICIME), pp. 416–420 (2010)
22. Hendrickx, I., Bouma, G., Coppens, F., Daelemans, W., Hoste, V., Kloosterman, G., Mineur, A.M., Vloet, J.V.D., Verschelde, J.L.: A coreference corpus and resolution system for Dutch. In: Proceedings of the Sixth International Conference on Language Resources and Evaluation (LREC 2008), pp. 144–149 (2008)
23. Hendrickx, I., Hoste, V., Daelemans, W.: Semantic and Syntactic Features for Dutch Coreference Resolution. In: Gelbukh, A. (ed.) CICLing 2008. LNCS, vol. 4919, pp. 351–361. Springer, Heidelberg (2008)
24. Orăsan, C.: Comparative evaluation of term-weighting methods for automatic summarization. Journal of Quantitative Linguistics 16, 67–95 (2009)
25. Hasler, L., Orăsan, C., Mitkov, R.: Building better corpora for summarisation. In: Proceedings of Corpus Linguistics 2003, Lancaster, UK, pp. 309–319 (2003)
26. Dagan, I., Dolan, B., Magnini, B., Roth, D.: Recognizing textual entailment: Rational, evaluation and approaches – erratum. Natural Language Engineering 16, 105 (2010)
27. Papineni, K., Roukos, S., Ward, T., Zhu, W.J.: Bleu: a method for automatic evaluation of machine translation. In: Proceedings of the 40th Annual Meeting on Association for Computational Linguistics (ACL 2002), Stroudsburg, PA, USA, pp. 311–318 (2002)
28. Banerjee, S., Lavie, A.: METEOR: An automatic metric for MT evaluation with improved correlation with human judgments. In: Proceedings of the ACL Workshop on Intrinsic and Extrinsic Evaluation Measures for Machine Translation and/or Summarization, pp. 65–72. Ann Arbor, Michigan (2005)
29. Rios, M., Aziz, W., Specia, L.: Tine: A metric to assess mt adequacy. In: Proceedings of the Sixth Workshop on Statistical Machine Translation, pp. 116–122. Association for Computational Linguistics, Edinburgh (2011)
30. Tan, C.M., Wang, Y.F., Lee, C.D.: The use of bigrams to enhance text categorization. Inf. Process. Manage. 38, 529–546 (2002)
31. Rogati, M., Yang, Y.: High-performing feature selection for text classification. In: Proceedings of the Eleventh International Conference on Information and Knowledge Management, CIKM 2002, pp. 659–661. ACM, New York (2002)
32. Debole, F., Sebastiani, F.: An analysis of the relative hardness of Reuters-21578 subsets: Research articles. J. Am. Soc. Inf. Sci. Technol. 56, 584–596 (2005)
33. Lewis, D.D., Yang, Y., Rose, T.G., Li, F.: Rcv1: A new benchmark collection for text categorization research. J. Mach. Learn. Res. 5, 361–397 (2004)
34. Bennett, P.N.: Using asymmetric distributions to improve text classifier probability estimates. In: Proceedings of the 26th Annual International ACM SIGIR Conference on Research and Development in Informaion Retrieval, SIGIR 2003, pp. 111–118. ACM, New York (2003)
35. Bennett, P.N., Dumais, S.T., Horvitz, E.: Probabilistic combination of text classifiers using reliability indicators: models and results. In: Proceedings of the 25th Annual International ACM SIGIR Conference on Research and Development in Information Retrieval (SIGIR 2002), pp. 207–214. ACM, New York (2002)
36. Nigam, K., McCallum, A.K., Thrun, S., Mitchell, T.: Text classification from labeled and unlabeled documents using em. Mach. Learn. 39, 103–134 (2000)

Part II

Text

"**Text:** A book or other written or printed work, regarded in terms of its content rather than its physical form: a text which explores pain and grief."
NODE (The New Oxford Dictionary of English), Oxford, OUP, 1998, page 1998, meaning 1.

Semantic Similarity Functions
in Word Sense Disambiguation

Łukasz Kobyliński and Mateusz Kopeć

Institute of Computer Science, Polish Academy of Sciences
ul. Jana Kazimierza 5, 01-248 Warszawa, Poland
{lkobylinski,m.kopec}@ipipan.waw.pl

Abstract. This paper presents a method of improving the results of automatic Word Sense Disambiguation by generalizing nouns appearing in a disambiguated context to concepts. A corpus-based semantic similarity function is used for that purpose, by substituting appearances of particular nouns with a set of the most closely related similar words. We show that this approach may be applied to both supervised and unsupervised WSD methods and in both cases leads to an improvement in disambiguation accuracy. We evaluate the proposed approach by conducting a series of lexical sample WSD experiments on both domain-restricted dataset and a general, balanced Polish-language text corpus.

1 Introduction

Word Sense Disambiguation (WSD) is now a well known task of computational linguistics, for which many automated methods have already been proposed. It is a problem that consists of assigning the meaning to a given instance of a polysemous word, based on the context, in which it has been used.

In this article we propose a method of improving existing approaches to word sense disambiguation by using available linguistic knowledge resources to generalize information included in contexts of the disambiguated words. We argue that using either a corpus- or relation-based Semantic Similarity Function (SSF, discussed in Section 3) to find lexemes closely related to the words appearing in disambiguated contexts may significantly increase disambiguation accuracy. By using an SSF we include important semantic information in the purely statistical process of selecting the correct sense for a particular word. This benefits both the unsupervised, knowledge-based approaches to WSD (as described in Section 4) by increasing the chances of matching a particular context with a sense definition and supervised methods, in which case the contexts extended with semantically related words translate to richer training material for machine learning methods, as we describe in Section 5. In Section 6 we provide results of experiments validating the proposed approach by comparing original WSD methods and methods extended with an SSF.

2 Previous Work

As reported by the organizers of public evaluations of WSD methods (e.g. [1]), supervised learning approaches currently achieve the highest accuracy in the task of

P. Sojka et al. (Eds.): TSD 2012, LNCS 7499, pp. 31–38, 2012.

WSD. This class of approaches requires that a training corpus is available, annotated with information about the sense in which each or some of the words appear in the text.

Unsupervised methods, which use external knowledge sources, such as WordNet [2] or Wikipedia and unsupervised learning approaches, can be used in situations where very little or no training data in the form of annotated corpus is available. For example in [3] a graph-based approach has been presented, where WordNet has been used as a lexical knowledge base containing hierarchical information about relationships between ambiguous words and other elements of the language. In the context of Polish language, an approach to WSD, which involved 106 polysemous target words and a large corpus of more than 30,000 instances has been described in [4]. A WSD method based on mining class association rules has been presented in [5].

An idea related to the one presented in this contribution, which concerns the expansion of training data with WordNet parents, has been proposed in [6]. Lesk method has been modified to use WordNet relations in [7], while in [8] a distributional similarity has been used to expand the sense definitions.

3   Semantic Similarity Functions

Semantic similarity function (SSF) is defined as a mapping from pairs of words (or lemmas) into real numbers: $W \times W \to \mathbb{R}$. The value of this function for a given pair of semantically strongly related words should be greater than the value for another pair of words, between which the relation is weaker. For example, $SSF(book, page)$ should be greater than $SSF(book, pen)$, because although both pairs show some kind of relatedness, the former is arguably stronger. If we take the type of the linguistic resource (used to extract the similarity of two words) into consideration, semantic similarity functions can be broadly divided into two types: distribution- (or corpus-) based and relation-based (see for example [9] and [10], respectively).

Corpus-based similarity functions rely on the following idea: the more often two words occur in similar context (and are used in similar way), the more semantically similar they are. Based on their frequency or more sophisticated statistical features of the contexts, in which they occur or co-occur, a real number representing the similarity may be calculated. The second type of semantic similarity functions relies on the existence of structural linguistic resources such as WordNet. They traverse the semantic network (or any taxonomy) between words to calculate the value of the function. A large number of WordNet SSFs have been developed, but their requirement is the availability of such a large structured language resource, which makes these methods less applicable for languages other than English.

For the Polish language, there were only a few attempts to create similarity functions. In this paper we use a corpus-based Rank Weight Function (RWF, [11]), to find nouns most closely related to the ones appearing in disambiguated contexts. Further in this paper, we are going to refer to this function simply as semantic similarity function (or SSF).

4 Extending the Knowledge-Based Approach to WSD

Although most often the supervised WSD techniques achieve the highest accuracy, the need to investigate unsupervised methods still exists, because of one main reason: lack of sufficient amount of training data. As every disambiguated word needs its own set of training examples, the *all-words* WSD task is most often best conducted by the knowledge-based methods, instead of the machine learning algorithms. In this paper we evaluate an extension of the simplified Lesk algorithm, with and without the help of the semantic similarity function.

The proposed extension builds on the idea of comparing the coincidence of sense definition with the context (*Coincidence(i, c)*, where c – context, i – i-th sense definition) and choosing the sense for which the value of the function is the highest. In the Extended Lesk approach the coincidence function is calculated in a more complex way than intersecting the sets of words (as in original Lesk algorithm). Individual steps of the algorithm are presented below.

1. Create two empty maps: W_i (for sense i) and W_c (for context c). They will store pairs: *(lemma, weight)*, *lemma* being the base form of a word, *weight* being a real number.
2. Choose the size of context window (i.e. number of tokens before and after polysemous word to take into account). Tested sizes are: $1, 2, 5, 10, 30, 50$.
3. Insert the base forms of words from the dictionary definition of sense i and from the context window into W_i and W_c respectively. Each entry should have a weight equal to its number of occurrences in definition/context.
4. Multiply each weight of each lemma by its Inverse Document Frequency (IDF). In the case of the words from context, frequency is based on the corpus, treating the corpus text as a document. In the case of words from sense definition, IDF is calculated treating all possible sense definitions of currently disambiguated word as the set of documents.
5. Remove from both maps entries with outlying weights, high or low (defined as having higher/lower value than a chosen percentage of the highest one). Tested threshold can be 0% or 1% in case of low outliers, 99% and 100% in case of high ones.
6. Normalize definition of sense i, by dividing each weight in W_i by the number of words in this definition. It prevents bias to longer definitions.
7. Normalize context by dividing the weights from W_c by values dependent on the textual distance of word from the disambiguated one. Three options can be tested – no normalization, division by the distance, division by the squared distance.
8. Extend both maps with related words extracted using Semantic Similarity Function by choosing the words with the highest score with a chosen similarity threshold. For example, if there is a word w in W_c, we extend W_c with 20 words most similar to w, regarding SSF. If a threshold is set, we only add these words from the top 20, which acquire result higher than the threshold. We have tested 4 threshold values: $0.0, 0.1, 0.2, 0.3$ and a version, in which we do not extend maps with related words.
9. Compare the maps using product measure or the Jaccard coefficient.

In this way we acquire a single real number, representing the coincidence between the sense i and the context c. The sense with the maximum value is chosen as the correct one. The *Coincidence* function can have a large number of variants, depending on the choice of parameters[1]. Each of these versions produces an individual disambiguation method.

5 Extending the Supervised Learning Approach to WSD

In the supervised learning approach to WSD we use an annotated text corpus to train state-of-the-art machine learning methods and then measure their classification accuracy using the ten-fold cross-validation approach. The fundamental problem we face using machine learning methods for WSD is the selection of an adequate feature representation method, which allows us to express the knowledge about an ambiguous word and its context in the form of a feature vector. We thus transform the WSD task into a classification problem and represent the textual data in the form of fixed-length number vectors.

We have chosen the following representation, implemented as feature generators in the WSD Development Environment [12]. Thematic Feature Generator (TFG): represents the existence of a word in a window around the disambiguated lexeme with window size: 5–25, lemmatization: on/off, generation of related words using a semantic similarity function: on/off and SSF threshold value: 0.1, 0.2, 0.3, or 0.4. Structural Feature Generator 1 (SFG1): existence of a word on a particular position in a small window relative to the disambiguated lexeme with window size: 1–5 and lemmatization: on/off. Structural Feature Generator 2 (SFG2): existence of a part-of-speech on a particular position in a small window relative to the disambiguated lexeme with window size: 1–5 and tagset: full or simplified. Keyword Feature Generator (KFG): grammatical form of the disambiguated lexeme with tagset: full or simplified.

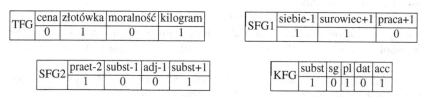

TFG	cena	złotówka	moralność	kilogram
	0	1	0	1

SFG1	siebie-1	surowiec+1	praca+1
	1	1	0

SFG2	praet-2	subst-1	adj-1	subst+1
	1	0	0	1

KFG	subst	sg	pl	dat	acc
	1	0	1	0	1

Fig. 1. Examples of feature vectors used with supervised learning WSD methods

Examples of feature vectors created by the generators described above are presented on Figure 1. We follow the common approach of representing the context of a particular polysemous word with a variant of the bag-of-words representation. The TFG generator captures the information about the existence of a particular word or lemma in the context, while the SFG1 generator analyzes a smaller window around the disambiguated lexeme and adds information about the position of the word in context relative to the

---

[1] During the development of the Extended Lesk method more extensions were tested than are presented in this paper. We describe only the parameter values, which were found to be the most successful.

lexeme. We also use the SFG2 generator, which is analogous to SFG1, but takes parts-of-speech appearing in context into account and KFG, which notes the grammatical form of the disambiguated lexeme.

We have used the SSF to extend the TFG generator and to include the information about the words most similar to the words appearing in context in the final feature vector. This allows us to train a more general classifier, which is not closely tied to a particular word, but rather to a general concept. For example, in case of the word "kilogram", the used SSF returns such closely related words as: "kg" (similarity rating: 0.299), "kilo" (0.287), "tona" ("ton", 0.241) and "gram" (0.206). All such words, having the similarity rating above a selected threshold are appended as additional attributes to the feature vector.

The bag-of-words approach to text representation produces a very large number of attributes, which is impractical in the subsequent classifier learning phase. In order to reduce the size of the feature vectors we employ an attribute selection method (still using the training data set), which chooses between 50 to 400 most important attributes, by calculating their information gain with respect to the class.

6 Experimental Results

6.1 Evaluation Corpora

Evaluation of the proposed improvement to automatic WSD has been performed on two corpora, each having its own dictionary of polysemous words (nouns, verbs and adjectives). The larger of the corpora comes from the National Corpus of Polish (NCP) project, described in [13]. It contains 1,215,513 tokens, including 34,114 polysemous ones, in 3,889 texts. It is a balanced corpus, spanning multiple types of textual sources and thematic domains. To verify the performance of the proposed approach on a domain-restricted collection of documents, we have used the Econo corpus, presented in [5]. It consists of 370,182 tokens with 22,520 polysemous ones in 1,861 texts from the domain of economy. Each of the corpora has been manually sense-annotated by qualified linguists (each example was annotated independently by two annotators and an additional third annotator in case of a disagreement) to serve as a verification data set and training material for supervised learning methods. In case of the NCP corpus the sense inventory contained 106 polysemous words and 2.85 sense definitions per word on average. The smaller Econo corpus was annotated using a dictionary of 52 polysemous words and containing 3.62 sense definitions per word on average.

6.2 Unsupervised Methods

Based on a combination of parameters of *Coincidence* function (see Table 1), two sets of methods were defined. The first set (EL) was formed by creating all combinations of possible parameter values without the usage of Semantic Similarity Function, which resulted in total number of 288 methods. The second set (EL-SS) consisted of 1,152 methods, as the usage of SSF with different parameter values was included. The only difference between these two sets lies in the use of SSF and a chosen similarity threshold, while all the other parameters remain unchanged.

We have used the following experimental framework for the available data. Each corpus was split into two parts of similar size (on the level of texts). The first part was designated as the development part, while the other as the evaluation part. All methods from both sets were tested on the development part. Based on the results from that test, the best methods from EL and EL-SS sets for each lexeme were chosen, as well as single methods, which performed best when used on all of the lexemes. Final results are calculated on the evaluation part using best methods from the previous step.

Table 1. Unsupervised methods achieving the highest accuracy on all lexemes from the inventory (as measured on the development part)

Parameter	Possible values	NCP	Econo
Context size	1,2,5,10,30,50	50	10
Low threshold	0.01,0.00	0.01	0.01
High threshold	1.00,0.99	1.00	0.99
Definition normalization	yes, no	yes	no
Context normalization	none, linear, square	square	none
Comparison measure	product, jaccard	jaccard	product
SSF	yes, no	yes	yes
SSF threshold	0.0, 0.1, 0.2, 0.3	0.1	0.2

6.3 Supervised Methods

In the case of supervised methods, a best performing method has been chosen for each of the lexemes in the sense inventory. The selection of the most accurate classification method has been done by searching through the space of feature representation methods, their parameters (as described in Section 5), machine learning algorithms (e.g. NaiveBayes, C4.5, RandomForest) and the number of selected attributes. All the experiments have been performed using the ten-fold cross-validation approach. The accuracy figures given in the following section are calculated by choosing the best supervised method for each of the 52 (in case of Econo corpus) or 106 (in case of the NCP corpus) disambiguated words.

6.4 Results

To assess the improvement in disambiguation accuracy gained from the proposed approach, we have experimented individually with each of the words found in the Econo and NCP sense inventories. The improvement varies greatly between the lexemes, ranging from no improvement to an increase of more than 60 percentage points. Disambiguation accuracy of the polysemous lexemes from the Econo corpus has been presented on Figure 2. Context generalization using the SSF proved to increase the accuracy of disambiguation for 33 polysemous words in case of the unsupervised approach and 19 in case of the supervised methods.

The improvement is also considerable in case of the NCP corpus, as presented in Table 2. The parameters of single unsupervised methods that performed best on all the

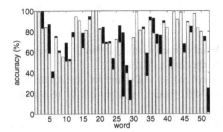 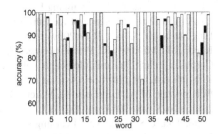

Fig. 2. Disambiguation accuracy of the Econo corpus using the extended Lesk method (left) and supervised learning approach (right) and the increase in accuracy gained using the semantic similarity function (black bars)

lexemes have been presented in Table 1. In case of the supervised methods the use of a SSF has influenced the disambiguation accuracy of lexemes, which appear less often in the dataset and for that reason the improvement of overall results for this group of approaches is less significant than for the Extended Lesk method.

Table 2. Overall results of supervised and unsupervised methods on Econo and NCP corpora

Econo corpus			NCP corpus		
Method	Method set	Accuracy	Method	Method set	Accuracy
Single best	EL	60.04%	Single best	EL	62.46%
	EL-SS	61.70%		EL-SS	65.64%
Best per lexeme	EL	77.27%	Best per lexeme	EL	75.09%
	EL-SS	80.92%		EL-SS	80.03%
	BestSupervised	97.29%		BestSupervised	91.47%
	BestSupervised-SS	97.52%		BestSupervised-SS	91.94%

7   Conclusions and Future Work

In this contribution we have described our experiments concerning Word Sense Disambiguation performed on two Polish language corpora: a general, balanced NCP corpus and domain-restricted Econo corpus. We have presented the results of a knowledge-based and supervised learning approaches to WSD in these corpora and proposed and improvement applicable to any method relying on context to perform the disambiguation. In future we would like to explore the possibilities of combining the proposed SSF extension of WSD methods with the knowledge available in WordNet-like resources, to fine-tune the generalization of words in disambiguated contexts.

References

1. Pradhan, S., Loper, E., Dligach, D., Palmer, M.: Semeval-2007 task-17: English lexical sample srl and all words. In: Proceedings of SemEval 2007 (2007)
2. Fellbaum, C.: WordNet: An Electronic Lexical Database. Bradford Books (1998)

3. Agirre, E., Soroa, A.: Personalizing PageRank for word sense disambiguation. In: Proceedings of the 12th Conference of the European Chapter of the Association for Computational Linguistics, pp. 33–41. ACL (2009)
4. Kopeć, M., Młodzki, R., Przepiórkowski, A.: Word Sense Disambiguation in the National Corpus of Polish. Prace Filologiczne, vol. LX (forthcoming, 2012)
5. Kobyliński, Ł.: Mining Class Association Rules for Word Sense Disambiguation. In: Bouvry, P., Kłopotek, M.A., Leprévost, F., Marciniak, M., Mykowiecka, A., Rybiński, H. (eds.) SIIS 2011. LNCS, vol. 7053, pp. 307–317. Springer, Heidelberg (2012)
6. Kohomban, U.S., Lee, W.S.: Learning semantic classes for word sense disambiguation. In: Proceedings of the 43rd Annual Meeting on Association for Computational Linguistics, pp. 34–41. ACL (2005)
7. Banerjee, S., Pedersen, T.: An Adapted Lesk Algorithm for Word Sense Disambiguation Using WordNet. In: Gelbukh, A. (ed.) CICLing 2002. LNCS, vol. 2276, pp. 136–145. Springer, Heidelberg (2002)
8. Iida, R., McCarthy, D., Koeling, R.: Gloss-based semantic similarity metrics for predominant sense acquisition. In: Proceedings of the Third International Joint Conference on Natural Language Processing, pp. 561–568 (2008)
9. Lin, D.: Automatic retrieval and clustering of similar words. In: COLING-ACL, pp. 768–774 (1998)
10. Budanitsky, A., Hirst, G.: Evaluating WordNet-based measures of lexical semantic relatedness. Computational Linguistics 32, 13–47 (2006)
11. Piasecki, M., Szpakowicz, S., Broda, B.: Automatic Selection of Heterogeneous Syntactic Features in Semantic Similarity of Polish Nouns. In: Matoušek, V., Mautner, P. (eds.) TSD 2007. LNCS (LNAI), vol. 4629, pp. 99–106. Springer, Heidelberg (2007)
12. Młodzki, R., Przepiórkowski, A.: The WSD development environment. In: Proceedings of the 4th Language and Technology Conference (2009)
13. Przepiórkowski, A., Bańko, M., Górski, R.L., Lewandowska-Tomaszczyk, B. (eds.): Narodowy Korpus Języka Polskiego. Wydawnictwo Naukowe PWN, Warsaw (forthcoming)

Opinion Mining on a German Corpus
of a Media Response Analysis

Thomas Scholz[1], Stefan Conrad[1], and Lutz Hillekamps[2]

[1] Heinrich-Heine-University, Institute of Computer Science, Düsseldorf, Germany
{scholz,conrad}@cs.uni-duesseldorf.de
[2] pressrelations, Editorial Department & Media Analysis, Düsseldorf, Germany
lutz.hillekamps@pressrelations.de

Abstract. This contribution introduces a new corpus of a German Media Response Analysis called the pressrelations dataset which can be used in several tasks of Opinion Mining: Sentiment Analysis, Opinion Extraction and the determination of viewpoints. Professional Media Analysts created a corpus of 617 documents which contains 1,521 statements. The statements are annotated with a tonality (positive, neutral, negative) and two different viewpoints. In our experiments, we perform sentiment classifications by machine learning techniques which are based on different methods to calculate tonality.

Keywords: Corpora and Language Resources, Media Response Analysis, Sentiment Analysis, Opinion Extraction, Viewpoint Determination.

1 Motivation

Opinion Mining for a Media Response Analysis (MRA) [14] presents a major challenge and a high benefit at the same time. Public relations departments of companies, organisations (such as political parties) and even distinguished public figures need an analysis of their public image (for advertising, PR campaigns, or elections campaigns). Therefore, it represents an own business segment for media monitoring companies. So, Opinion Mining is very interesting in the context of news articles because it would save time and human effort. One reason why this is urgently necessary is that the number of relevant articles is growing continuously and at the same time analyses in more detail are requested.

In a MRA, the most important key performance indicators are the media reach and the sentiment (we speak also of the tonality). The media reach shows how much the article is distributed in the media (how many sources like online portals, for example, publish this article). The calculation of the media reach is rather simple and automatically realised. An automated calculation of the tonality is much more difficult. In the simplest case, an article can be positive, neutral or negative as a whole. In the more interesting case for a media analysis, it contains statements which are positive, neutral or negative. In addition, the statements belong to a viewpoint, which shows that this statement is positive for that viewpoint, for example:

(1) The SPD man Klaus Wowereit is elected governing mayor of Berlin. (**Code:** positive, SPD (Note: The SPD is a German political party.))

P. Sojka et al. (Eds.): TSD 2012, LNCS 7499, pp. 39–46, 2012.

Today, crawler systems collect news articles which contain a given search string (the string is usually defined by the customer of a MRA and can be the company's name or the name of a product). But it still requires a big human effort for media analysts to read the articles and than select relevant statements. Furthermore, the analysts have to set the polarity of the sentiment of the statement. In many tasks, they code the statement for a certain group. This represents a viewpoint for the tonality. A positive statement for one viewpoint might be negative or neutral for another. So, the tonality cannot merely determined by assigning tonality values to single words (like in dictionary-based approaches).

To design new automated approaches, there is a lack of resource, especially in the newspaper context and for other languages than English. Moreover, the existing resources do not fulfil the requirements of Opinion Mining tasks for a MRA, because they do not include the concept of relevant statements nor a sentiment value for a single statement. Also, the tonality does not belong to a certain viewpoint.

Therefore, we introduce the pressrelations dataset which is a new corpus for Opinion Mining in newspaper articles. The corpus is created by professional specialists. The corpus can be used for several tasks in Opinion Mining: Sentiment Classification [2,12,16], Subjectivity Analysis [8,16], Opinion Extraction [12], the determination of viewpoints [3], the identification of argumentation stands [10,13], and the creation of sentiment dictionaries [2,11].

The remainder of this paper describes the following: In section 2 we characterise related work which addresses primarily corpora and language resources for Opinion Mining. In the third section, we explain our dataset and in particular the annotation scheme and differences to other resources. Then we apply different methods to classify tonality by machine learning techniques in section 4, before we summarise in the last section.

2   Related Work

Opinion Mining and Sentiment Analysis [9] is a far-reaching subject. Corpora and resources have been designed for many different tasks.

In the context of film and customer reviews, the dataset of [5] of 322 reviews, which have been increased [2] to 445 documents, is a benchmark dataset [2] for sentiment analysis in product reviews. The products are two digital cameras, two cellular phones, a MP3 player, a DVD player, a router and one anti-virus software. The dataset [8] of Pang and Lee contains subjective and objective sentences which are extracted from film reviews and plots. They collect 5,000 subjective sentence and sentence fragments (from www.rottentomatoes.com) as subjective examples and 5,000 sentences from plot summaries (www.imdb.com) as objective examples. The sentences are at least 10 words long.

For Opinion Mining tasks in news articles, the MPQA Corpus [15] contains a word- and phrase-based annotated corpus which consists of 535 English news documents (11,112 sentences and 19,962 subjective expressions). The tasks evaluated on this dataset cover contextual polarity [16]. In contrast to our corpus, here single words and phrases are annotated with sentiments and the strength of a sentiment is given. It is not designed as a MRA, because it does not contain relevance areas or viewpoints.

In German, resources are limited. SentiWS [11] is the first public available dictionary for Sentiment Analysis. It includes 1,686 positive and 1,686 negative words in lemma, which cover 16,406 positive and 16,328 negative word forms in the German Language. Momtazi [7] introduces the first German corpus for Opinion Mining in social media. It contains 500 short texts about celebrities.

The main difference is that all these corpora do not include marked areas which represent relevant statements and viewpoints for the tonality of statements. But these components are necessary to develop, to train, and to evaluate approaches which tackle the extraction of relevant statements, the calculation of their tonality and the determination of their viewpoint for a MRA.

3   The Corpus

3.1   The Task of the MRA

Two media analysts (professional experts in field of media monitoring and analysis, hereinafter called the annotators) annotate news articles about the two biggest political parties in Germany: The CDU (the party of the German Christian Democrats) is governing party under its chairwoman chancellor Merkel and the SPD (the party of the German Social Democrats) is the strongest opposition party. Web crawler systems have collected press releases because those have a huge media reach and must not be licensed or have archiving costs (what is a big problem for the provision of a public available MRA corpus). The annotators collect 617 relevant articles (consisting of 15,089 sentences) and annotate them as follows, what took 8 person days.

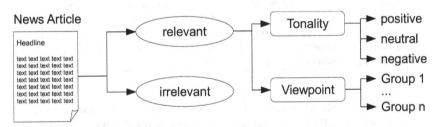

Fig. 1. Hierarchical relationships between the categories of relevance, tonality and viewpoint

3.2   The Annotation Scheme

There are four different categories of text in a MRA: Positive statements, negative statements, neutral statements and not relevant text areas. The text passages of statements (positive, negative, neutral) are also called relevant. Figure 1 shows the hierarchical relationships between these categories.

In the annotation process, the media analysts extract relevant statements. The statements have these attributes:

- **News no.:** This is the identification number of the news article which contains this statement.

- **Statement text:** The complete text of the statement is usually between one and four sentences long.
- **Tonality value:** The tonality can have one of three values (positive: one (1), neutral: zero (0), and negative: minus one (-1)).
- **Codes:** The codes specify the viewpoint. The statement is relevant and have the given tonality for the indicated company or organisation.

The corresponding news articles have the attributes **news no.**, **headline** and **text**. The crawlers have collected these articles because they contain the string 'CDU' or 'SPD'. Then the annotators have rejected articles which are no press releases or do not contain relevant statements. The inter annotators' agreement is 88,06% using Cohen's kappa. The quality assurance of a MRA ensures a annotation quality of at least 80% of correct annotations.

3.3   Comparison with Other Resources

Table 1 compares different corpora for Opinion Mining. The most similar corpus is the MPQA corpus, because it also deals with news articles. In the MPQA, the annotations cover single words and phrases, whereas our annotations belong to statements.

Table 1. Comparison of ressources for Opinion Mining

Aspect	Pang [8]	MPQA [15]	Momtazi [7]	our corpus
Language	English	English	German	German
Area	reviews	news	social media	news
Texts	-	523	500	617
Sentences	10,000	11,112	890	15,089
containing Statements	no	no	no	yes (1,521)
containing Viewpoints	no	no	no	yes (2)

While it is possible to perform fine-grained word classifications about sentiments [16] with the MPQA, our intention is to perform different tasks of a MRA. This requires statements which are annotated with tonality and a viewpoint. Our corpus contains 1521 of these annotations (consisting of 3,283 sentences and 55,174 words). Table 2 shows the distribution of the statements on the three tonality classes and the two viewpoints. The complete pressrelations dataset is available at http://www.pressrelations.de/research/.

Table 2. Distribution of tonality and viewpoints

Viewpoint	Positive	Neutral	Negative	All
CDU	257	265	470	992
SPD	189	227	113	529
All	446	492	583	1,521

4 Experiments

4.1 Creation of Sentiment Dictionaries

For our sentiment dictionaries, we use 420 of our statements (140 positive, 140 negative, and 140 neutral statements) to calculate a tonality value t for one word in lemma by the **chi-square** method, the **PMI** method, **association rule mining**, the **SentiWS** and **TF-IDF**.

The statistical measure **chi-square** value [6] can be used as a polarity value. The higher probability sets the polarity of the score value:

$$t\chi^2(w) = \begin{cases} \chi^2(w) & \text{if } P(w|neg) < P(w|pos) \\ -\chi^2(w) & \text{otherwise} \end{cases} \tag{1}$$

$P(w|y)$ is the probability for word w to appear with tonality y and the null hypothesis suppose that this is equal for both: positive and negative tonality.

$$\chi^2(w) = \sum_{x\in\{w,\neg w\}} \sum_{y\in\{pos,neg\}} \frac{(f(x,y) - \hat{f}(x,y))^2}{\hat{f}(x,y)} \tag{2}$$

$f(x,y)$ is the frequency of x in statements with tonality y and $\hat{f}(x,y)$ is the expected value of $f(x,y)$ when the null hypothesis is supposed.

The **pointwise mutual information** [1,6] uses the strength of the association between the word w and positive and negative statements, respectively.

$$t_{PMI}(w) = PMI(w, pos) - PMI(w, neg) = \log_2 \frac{P(w|pos)}{P(w|neg)} \tag{3}$$

Association rule mining for polarity [4] searches for rules: the word determines polarity. It needs minimum support (word w appears in x_1 cases with the polarity y) and a minimum confidence (word w appears in $x_2\%$ of all cases with a positive polarity).

$$\text{support}(w, y) = \frac{f(w, y)}{N} \geq x_1 \qquad \text{confidence}(w \to y) = \frac{f(w, y) * 100}{f(w)} \geq x_2 \tag{4}$$

N is the number of all statements. $f(w)$ is the number of statements which contain w. If the word w fulfils both conditions, the word w gets the value $+1$ and -1, respectively.

Also, we use a **TF-IDF** matrix $\omega_{w,y}$ to weigh the terms.

$$\omega_{w,y} = tf_{w,y} * idf_w = f(w, y) * \log \frac{N}{f(w)} \tag{5}$$

$$t_{tf-idf}(w) = \begin{cases} \omega_{w,pos} & \text{if } \omega_{w,pos} \geq \omega_{w,neg} \\ -\omega_{w,neg} & \text{otherwise} \end{cases} \tag{6}$$

All these methods are used to differentiate between positive and negative words and to calculate four values for the polarity classification (one value for each word category: adverbs, adjectives, verbs and nouns). In order to distinguish between subjective (positive and negative) statements and objective (neutral) statements, we use the same methods by changing the positive class to the subjective class and the negative to the objective class. So, we get four additional values for subjectivity analysis. As another

Table 3. Results of the classification of the tonality

Method	SVM	Naive Bayes	Neural Net	Decision Tree	k-means
Chi-square	53.23%	**54.71%**	46.31%	43.13%	43.51%
PMI	46.31%	49.26%	43.02%	45.29%	46.23%
Association Rule Mining	33.48%	34.05%	39.95%	40.86%	39.78%
TF-IDF	48.92%	47.90%	52.44%	48.92%	38.96%
SentiWS	40.75%	34.62%	39.73%	42.00%	35.15%

baseline, we use the values (polarity classification) and the absolute values (subjectivity analysis) of the **SentiWS** dictionary [11].

4.2   Classification and Results

To classify the statements, we use different machine learning techniques[1] (see Table 3). We select randomly 220 statements for training and 881 for testing. The accuracies of the tonality classification are shown in Table 3. For the SVM, we choose a two-way classification: First we differentiate between subjective and objective statements, before we classify the subjective statements as positive or negative. Naive Bayes and the chi-square method achieve the best result (54.71%).

On average, the Naive Bayes classification performs better than all the other techniques for the subjective analysis (Table 4 top) and the polarity classification (Table 4 bottom). So the results are based on this technique for these two classification problems. We use the related four values as input for each task. The chi-square method can differentiate most suitable between subjective and objective examples (71.74%) and also produces the best values for the polarity classification (68.78%).

Table 4. Results of the subjectivity analysis and the polarity classification

Method	Accuracy	Subjective		Objective	
		Precision	Recall	Precision	Recall
Chi-square	**71.74%**	73.51%	90.90%	62.76%	31.82%
PMI	69.35%	69.70%	12.59%	64.29%	96.64%
Association Rule Mining	65.72%	67.76%	6.99%	35.71%	93.95%
TF-IDF	67.99%	69.34%	94.29%	52.78%	13.29%
SentiWS	36.55%	80.00%	8.07%	33.37%	95.80%

Method	Accuracy	Positive		Negative	
		Precision	Recall	Precision	Recall
Chi-square	**68.78%**	59.23%	65.53%	76.11%	70.88%
PMI	65.28%	56.52%	49.79%	69.90%	75.27%
Association Rule Mining	58.60%	38.18%	8.94%	60.66%	90.66%
TF-IDF	59.27%	48.79%	47.53%	76.55%	77.45%
SentiWS	55.26%	45.82%	41.21%	73.53%	77.02%

---

[1] All techniques use the rapidminer standard implementation with default parameters (http://rapid-i.com).

5   Conclusion

The comparison of related work shows clearly that this corpus is required to develop and evaluate new approaches of Opinion Mining for a MRA and, in addition, for other tasks of sentiment analysis in German. Beyond the improvement of tonality classification the important tasks in further research are especially the extraction of relevant statements and the identification of different viewpoints of statements. Nevertheless, this corpus can be used to create large sentiment dictionaries including positive, negative and also neutral examples, too.

Acknowledgments. This work is funded by the German Federal Ministry of Economics and Technology under the ZIM-program (Grant No. KF2846501ED1). The authors want to thank Sonja Hansen for annotating the dataset.

References

1. Church, K.W., Hanks, P.: Word association norms, mutual information, and lexicography. In: Proc. of the 27th Annual Meeting of the ACL, ACL 1989, pp. 76–83 (1989)
2. Ding, X., Liu, B., Yu, P.S.: A holistic lexicon-based approach to opinion mining. In: Proc. of the Intl. Conf. on Web Search and Web Data Mining, WSDM 2008, pp. 231–240 (2008)
3. Greene, S., Resnik, P.: More than words: syntactic packaging and implicit sentiment. In: Proc. of Human Language Technologies: The 2009 Annual Conf. of the North American Chapter of the ACL, NAACL 2009, pp. 503–511 (2009)
4. Harb, A., Plantié, M., Dray, G., Roche, M., Trousset, F., Poncelet, P.: Web opinion mining: how to extract opinions from blogs? In: Proc. of the 5th Intl. Conf. on Soft Computing as Transdisciplinary Science and Technology, CSTST 2008, pp. 211–217 (2008)
5. Hu, M., Liu, B.: Mining and summarizing customer reviews. In: Proc. of the 10th ACM SIGKDD Intl. Conf. on Knowledge Discovery and Data Mining, KDD 2004, pp. 168–177 (2004)
6. Kaji, N., Kitsuregawa, M.: Building lexicon for sentiment analysis from massive collection of html documents. In: Proc. of the 2007 Joint Conf. on Empirical Methods in Natural Language Processing and Computational Natural Language Learning, EMNLP-CoNLL (2007)
7. Momtazi, S.: Fine-grained german sentiment analysis on social media. In: Proc. of the 9th Intl. Conf. on Language Resources and Evaluation, LREC 2012 (2012)
8. Pang, B., Lee, L.: A sentimental education: Sentiment analysis using subjectivity summarization based on minimum cuts. In: Proc. of the 42nd Meeting of the ACL, pp. 271–278 (2004)
9. Pang, B., Lee, L.: Opinion mining and sentiment analysis. Foundations and Trends in Information Retrieval 2(1-2), 1–135 (2008)
10. Park, S., Lee, K., Song, J.: Contrasting opposing views of news articles on contentious issues. In: Proc. of the 49th Annual Meeting of the ACL: Human Language Technologies, HLT 2011, vol. 1, pp. 340–349 (2011)
11. Remus, R., Quasthoff, U., Heyer, G.: Sentiws – a publicly available German-language resource for sentiment analysis. In: Proc. of the 7th Intl. Conf. on Language Resources and Evaluation, LREC 2010 (2010)
12. Sarvabhotla, K., Pingali, P., Varma, V.: Sentiment classification: a lexical similarity based approach for extracting subjectivity in documents. Inf. Retr. 14(3), 337–353 (2011)

13. Somasundaran, S., Wiebe, J.: Recognizing stances in ideological on-line debates. In: Proc. of the NAACL HLT 2010 Workshop on Computational Approaches to Analysis and Generation of Emotion in Text, CAAGET 2010, pp. 116–124 (2010)
14. Watson, T., Noble, P.: Evaluating public relations: a best practice guide to public relations planning, research & evaluation. PR in practice series, ch. 6, pp. 107–138. Kogan Page (2007)
15. Wiebe, J., Wilson, T., Cardie, C.: Annotating expressions of opinions and emotions in language. Language Resources and Evaluation 39(2-3), 165–210 (2005)
16. Wilson, T., Wiebe, J., Hoffmann, P.: Recognizing contextual polarity: An exploration of features for phrase-level sentiment analysis. Computational Linguistics 35(3), 399–433 (2009)

The Soundex Phonetic Algorithm Revisited for SMS Text Representation[*]

David Pinto[1], Darnes Vilariño[1], Yuridiana Alemán[1],
Helena Gómez[1], Nahun Loya[1], and Héctor Jiménez-Salazar[2]

[1] Faculty of Computer Science
Benemérita Universidad Autónoma de Puebla, Mexico
{dpinto,darnes}@cs.buap.mx,
{yuridiana.aleman,helena.adorno}@gmail.com,
israel_loya@hotmail.com
[2] Information Technologies Department
Universidad Autónoma Metropolitana, DF, Mexico
hgimenezs@gmail.com

Abstract. The growing use of information technologies such as mobile devices has had a major social and technological impact such as the growing use of Short Message Services (SMS), a communication system broadly used by cellular phone users. In 2011, it was estimated over 5.6 billion of mobile phones sending between 30 and 40 SMS at month. Hence the great importance of analyzing representation and normalization techniques for this kind of texts. In this paper we show an adaptation of the Soundex phonetic algorithm for representing SMS texts. We use the modified version of the Soundex algorithm for codifying SMS, and we evaluate the presented algorithm by measuring the similarity degree between two codified texts: one originally written in natural language, and the other one originally written in SMS "sub-language". Our main contribution is basically an improvement of the Soundex algorithm which allows to raise the level of similarity between the texts in SMS and their corresponding text in English or Spanish language.

1 Introduction

SMS is a very popular short message-based communication service among mobile phone users. However, SMS is also synonym of the short message itself which can contain up to 160 characters. The length limitation of an SMS has lead to create a sort of "sub-language" which includes a vocabulary of words, phonetically similar to that of the original natural language, but that regularly omit grammatical forms, punctuation marks and vowels.

In this paper we present a study based on lexical similarity, for different adaptations to the Soundex phonetic algorithm in order to represent SMS texts. The input is an SMS text and the output is a code or a set of codes. The aim is to have a family of words that matches with the same code, and to use the codified text version in

---

[*] This project has been partially supported by projects CONACYT #106625, VIEP #VIAD-ING11-II y #PIAD-ING11-II.

P. Sojka et al. (Eds.): TSD 2012, LNCS 7499, pp. 47–55, 2012.

some natural language tasks, such as information extraction or question answering. The presented algorithms have been evaluated in three different datasets (two languages, English and Spanish) by comparing the lexical similarity between pairs of texts (SMS, natural language) already codified with the purpose of having an overview of the level of generality obtained with the different codifications.

The remainder of the paper is organized as follows. In section 2 we summarize different works reported in the literature that are related to the one presented in this paper. In Section 3 we discuss the Soundex phonetic algorithm. In Section 4 we present different modifications we have done to the Soundex algorithm in order to have a proper codification for SMS texts, together with a preliminary study that indicates the degree of similarity between pairs of texts which express exactly the same meaning, having different textual representation, in one case they are written using the standard vocabulary of the language (Spanish or English) and in the other case they are written in the SMS sub-language. Finally we conclude this paper by resuming the strengths of our contributions and sketching future research issues.

2   Related Work

The Soundex code has often been applied in the information retrieval task, particularly when it is based on transcriptions of spoken language, because it is known that speech recognition produce transcription errors that are phonetically similar but ortographically dissimilar. In [1], however, it is claimed that the use of this codification does not improve regular string-matching based IR. The purpose of this paper is to study phonetic-based representations for SMS messages. Therefore, we are interested in those works reported in litereature dealing with the task of SMS analysis, basically by considering normalization of SMS. In [2], for instance, the authors provide a brief description on their input pre-processing work for an English to Chinese SMS translation system using a word group model. The same authors provide an excellent work for SMS normalization in [3]. They prepared a training corpus of 5,000 SMS aligned with reference messages manually prepared by two persons which are then introduced to a phrase-based statistical method to normalize SMS messages. In the context of SMS-based FAQ retrieval, the most salient works are the ones presented in [4] and [5], where authors formulate a similarity criterion of the search process as a combinatorial problem in which the search space is conformed of all the different combinations for the vocabulary of the query terms and their N best translations. Unfortunately, the corpus used in these experiments is not available and, therefore, it is not possible to use it in our experiments. To the best of our knowledge, in the literature there is not a particular phonetic algorithm particularly adapted for representing SMS and, therefore, we consider that the approach presented in this paper would be of high benefit.

3   Phonetic Representation

The phonetic representation has several applications. It allows to search concepts based on pronunciation rather than on the spelling, as it is traditionally done. There exist different algorithms for codifying text according to its phonetic pronunciation. Some of the

Table 1. Soundex phonetic codes for the English language

Numeric code	Letter
0	a,e,i,o,u,y,h,w
1	b,p,f,v
2	c,g,j,k,q,s,x,z
3	d,t
4	l
5	m,n
6	r

most known and used phonetic algorithms are: Soundex [6,7], NYSIIS [8], Metaphone and Double Metaphone [9]. For the purposes of these preliminar experiments, we have started by considering the Soundex algorithm, which is better described as follows:

The Soundex phonetic algorithm was mainly used in applications involving searching of people's names like air reservation systems, censuses, and other tasks presenting typing errors due to phonetic similarity [10]. As shown in [11], the Soundex algorithm evaluates each letter in the input word and assigns a numeric value. The main function of this algorithm is to convert each word into a code made up of four elements.

Soundex uses numeric codes (see Table 1) for each letter of the string to be codified. The Soundex algorithm can be depicted as follows:

1. Replace all but the first letter of the string by its phonetic code
2. Eliminate any adjacent reptitions of codes
3. Eliminate all occurrences of code 0 (that is, eliminate vowels)
4. Return the first four characters of the resulting string

The Soundex algorithm has the following features:

- It is intuitive in terms of operation.
- The simplicity of the code allows to implement changes according to the objective.
- The processing time is relatively short.
- It has a high tolerance for variations in words that sound very similar or are exactly the same.

3.1 Different Adaptations to the Soundex Algorithm

We have observed the SMS representation in some languages like Spanish and English. That is why we propose some improvements to the basic Soundex algorithm with the purpose of obtaining the same code for a word in the SMS representation and its corresponding normal way of writing (standard vocabulary). We have done different adaptations of the two languages studied in this paper: Spanish and English. Below we resume the changes made to the Soundex algorithm for the Spanish language:

1. To keep the four digits of the Soundex code, but replace the first letter with its numeric representation.

2. If the letter "X" appears aside to a consonant, then change it for the letters "PR", and immediately code this two letters to its numeric representation. The rationale of this proposal is that the "x" letter is often used for expressing the word "por (because)".

3. To replace all symbols for its corresponding name, i.e. $ = pesos (Mexican pesos), % = "por ciento (percent)", hs = horas (hours), among others using the SMS dictionaries latterly introduced.

4. To replace all numbers in the original text for its textual representation, and then compute its Soundex code.

For the experiments carried out in this paper, we have tested four different approaches. The first one is when no phonetic codification is applied, which we have named *Uncodified* version. The second approach is named *Soundex* because it uses de basic Soundex algorithm. The following two approaches: *NumericSoundex* and *SoundexMod* are summarized as follows:

1. *NumericSoundex:* To keep the four digits of the Soundex code, replacing the first letter with its numeric representation.

2. *SoundexMod:* Before applying *NumericSoundex*, we use a dictionary of common SMS acronyms and phrases abbreviations for normalizing the SMS texts. A freely SMS dictionary available online[1] was used.

Similar changes to the Soundex algorithm was done in order to deal with the English language, obtaining four approches to be compared in the experiments carried out (*Uncodified, Soundex, NumericSoundex* and *SoundexMod*). For the English *SoundexMod* version, the freely SMS dictionary available online[2] was used.

4   Evaluating the Different Soundex Adaptations for SMS Text Representation

In this section we study the behavior of the different phonetic representations proposed when considering lexical similarity. The description of the corpus characteristics is done in the following SubSection. The metric used for determining the performance of the different phonetic algorithms together with the lexical similarity values found are shown in SubSection 4.2.

4.1   Datasets

In order to evaluate the adaptations made to the Soundex Algorithm for the Spanish language, we have constructed a parallel SMS corpus on the basis of the book named *"En Patera y haciendo agua"* [12] which is freely available online[3]. The salient features of this parallel corpus are shown in Table 2.

---

[1] http://www.diccionariosms.com

[2] http://smsdictionary.co.uk/

[3] http://www.adiccionesdigitales.es/libro

In order to evaluate the phonetic modification for the English language, we have used two different corpora. The first one is a parallel SMS corpus of 5,000 SMS aligned with reference messages manually prepared by two persons prepared by [3] as a training dataset. The salient features of this corpus are shown in Table 3.

Table 2. A Spanish parallel corpus of SMS

Feature	SMS	Original text
Number of messages	316	316
Number of tokens	30,195	30,270
Vocabulary size	7,061	6,448
Average message length in words	95.55	95.79
Average message length in characters	454.11	552.70
Average characters per word	4.75	5.77

Table 3. An English parallel corpus of SMS [3]

Feature	SMS	Original text
Number of messages	5,000	5,000
Number of tokens	68,666	69,521
Vocabulary size	6,814	5,746
Average message length in words	13.73	13.9
Average message length in characters	57.95	62.44
Average characters per word	4.22	4.49

The second corpus used in this experiment was the corpus of the SMS-based FAQ Retrieval task of the FIRE 2011 competition (Forum for Information Retrieval Evaluation)[4]. The salient features of this comparable corpus were already shown in Table 4.

4.2 Evaluation Based on Lexical Similarity

We are using a similarity measure in order to observe how well the different approaches codify correctly the SMS. We basically, applied a similarity measure for verifying in which percentage the Soundex-like codes in Spanish or English original texts are equal/similar to the Soundex-like codes for the text written in SMS for both languages. In particular, we use the Jaccard coefficient [13] to measure the similarity between the sample sets. Let SMS' be the SMS codified and T' be the original text codified, both with one of the above presented Soundex-like phonetic representation, then in Eq. 1 it is shown the Jaccard coefficient between SMS' and T'.

$$Jaccard(SMS', T') = \frac{|SMS' \cap T'|}{|SMS' \cup T'|} \tag{1}$$

---

[4] http://www.isical.ac.in/%7Efire/

Table 4. An English comparable corpus of SMS

Feature	SMS	Original text
Number of messages	721	721
Number of tokens	5,573	7,337
Vocabulary size	2,121	2,034
Average message length in words	7.74	10.14
Average message length in characters	37.28	56.62
Average characters per word	4.82	5.58

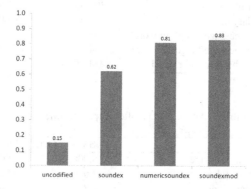

Fig. 1. Average Jaccard similarity for the different SMS text representations with the Spanish SMS parallel corpus

The intersection represents the number of matches between the SMS codified and the original text in Spanish or English also codified with some of the proposed Soundex-like algorithm. The union represents the total number of words in the data set. We apply the Jaccard coefficient between the SMS and the correct translation (or associated text in the case of the comparable corpus). The greater the value of the Jaccard coefficient, the better, the matching between the pair of codified texts. In Figure 1 we show the average Jaccard coefficient values obtained after comparing each pair (text,SMS) for the Spanish SMS parallel corpus. A 0.15 of average similarity in the *Uncodified* corpus show the great difference that exist between SMS and the corresponding translation. By applying the *Soundex* algorithm we obtain a similarity average value of 0.62. However, by just modifying the Soundex algorithm considering the first element of the code to be a numeric value (*NumericSoundex*), we improve the Soundex representation obtaining a similarity value of 0.81. Finally, the *SoundexMod* approach obtains the best value of similarity (0.83).

In Figure 2 we show the average Jaccard coefficient values obtained after comparing each pair (text,SMS) for the English SMS parallel corpus. A 0.652 of average similarity in the *Uncodified* corpus indicates that there exist a more or less stable way of writing SMS in this corpus, using a small number of acronyms, contractions and elimination of vowels. By applying the *Soundex* algorithm we obtain a similarity average value

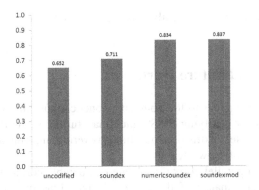

Fig. 2. Average Jaccard similarity for the different SMS text representations with the English SMS parallel corpus

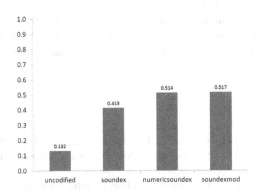

Fig. 3. Average Jaccard similarity for the different SMS text representations with the English SMS comparable corpus (FIRE)

of 0.711. However, by just modifying the Soundex algorithm considering the first element of the code to be a numeric value (*NumericSoundex*), we improve the Soundex representation obtaining a similarity value of 0.834. Finally, the *SoundexMod* approach obtains the best value of similarity (0.837).

In Figure 3 we show the average Jaccard coefficient values obtained after comparing each pair (text,SMS) for the English SMS comparable corpus. A 0.131 of average similarity in the *Uncodified* corpus show the great difference that exist between the query written in SMS format and the corresponding FAQ question associated. By applying the *Soundex* algorithm we obtain a similarity average value of 0.413. However, by just modifying the Soundex algorithm considering the first element of the code to be a numeric value (*NumericSoundex*), we improve the Soundex representation obtaining a similarity value of 0.514. Finally, the *SoundexMod* approach obtains the best value of similarity (0.517).

In summary, we have observed that the Soundex algorithm is useful for codifying SMS, but the simple modification *NumericSoundex* greatly improves the similarity between SMS codified and original texts codified. Ad-hoc modifications to the Soundex method for a particular language slightly improves the results, but generates a language dependent algorithm.

5   Conclusions and Future Work

We have proposed adaptations to the Soundex phonetic algorithm in order to provide a suitable algorithm for codifying SMS which may further be used in other natural language tasks such as text extraction, question answering, information retrieval, etc., which use SMS as part of the written texts.

We have used two parallel corpora and one comparable corpus with the purpose of evaluating the performance of the proposed algorithms in a more challenging environment. The Soundex method greatly improves the matching between SMS and their corresponding associated words, but the modifications proposed in this paper improve also the Soundex method in all the cases.

As future work, we would like to study the familiy of words clustered around a single SMS word in order to determine the existence of semantic groups. Also we would like to evaluate the performance of the presented phonetic codification in real natural language tasks.

References

1. Reyes-Barragán, A., Villaseñor Pineda, L., Montes-y-Gómez, M.: INAOE at QAst 2009: Evaluating the usefulness of a phonetic codification of transcriptions. In: Proceedings of CLEF 2009 Workshop. Springer (2009)
2. Aiti, A., Min, Z., Pohkhim, Y., Zhenzhen, F., Jian, S.: Input normalization for an english-to-chinese sms translation system. In: MT Summit 2005 (2005)
3. Aw, A., Zhang, M., Xiao, J., Su, J.: A phrase-based statistical model for sms text normalization. In: Proceedings of the COLING/ACL on Main Conference Poster Sessions, COLING-ACL 2006, pp. 33–40. Association for Computational Linguistics, Stroudsburg (2006)
4. Kothari, G., Negi, S., Faruquie, T.A., Chakaravarthy, V.T., Subramaniam, L.V.: SMS based interface for FAQ retrieval. In: Proceedings of the Joint Conference of the 47th Annual Meeting of the ACL and the 4th International Joint Conference on Natural Language Processing of the AFNLP, ACL-IJCNLP 2009, vol. 2, pp. 852–860. Association for Computational Linguistics, Morristown (2009)
5. Contractor, D., Kothari, G., Faruquie, T.A., Subramaniam, L.V., Negi, S.: Handling noisy queries in cross language faq retrieval. In: Proceedings of the 2010 Conference on Empirical Methods in Natural Language Processing, EMNLP 2010, pp. 87–96. Association for Computational Linguistics, Stroudsburg (2010)
6. Hall, P.A.V., Dowling, G.R.: Approximate string matching. ACM Comput. Surv. 12, 381–402 (1980)
7. Rajkovic, P., Jankovic, D.: Adaptation and application of daitch-mokotoff soundex algorithm on Serbian names. In: XVII Conference on Applied Mathematics (2007)
8. Taft, R.: Name search techniques. Special report. Bureau of Systems Development, New York State Identification and Intelligence System (1970)
9. Philips, L.: Hanging on the metaphone. Computer Language Magazine 7, 38–44 (1990)

10. Knuth, D.E.: The art of computer programming, vol. 3: sorting and searching, 2nd edn. Addison Wesley Longman Publishing Co., Inc., Redwood City (1998)
11. Manning, C.D., Raghavan, P., Schtze, H.: Introduction to Information Retrieval. Cambridge University Press, New York (2008)
12. Romero, J., et al.: En patera y haciendo agua. Adicciones Digitales (2011), http://www.adiccionesdigitales.es/libro
13. Jaccard, P.: Etude comparative de la distribution florale dans une portion des Alpes et du Jura. Bulletin de Société vaudoise des Sciences Naturelles 37, 547–579 (1901)

Sentence Modality Assignment
in the Prague Dependency Treebank

Magda Ševčíková and Jiří Mírovský

Charles University in Prague, Faculty of Mathematics and Physics
Institute of Formal and Applied Linguistics

Abstract. The paper focuses on the annotation of sentence modality in the Prague Dependency Treebank (PDT). Sentence modality (as the contrast between declarative, imperative, interrogative etc. sentences) is expressed by a combination of several means in Czech, from which the category of verbal mood and the final punctuation of the sentence are the most important ones. In PDT 2.0, sentence modality was assigned semi-automatically to the root node of each sentence (tree) and further to the roots of parenthesis and direct speech subtrees. As this approach was too simple to adequately represent the linguistic phenomenon in question, the method for assigning the sentence modality has been revised and elaborated for the forthcoming version of the treebank (PDT 3.0).

Keywords: sentence modality, Prague Dependency Treebank, dependency tree, coordination root, coordinated clause.

1 Introduction

Recognition of the contrast between declarative, imperative, interrogative, and possibly other types of sentences, which is referred to as sentence modality in the present paper, is a salient subtask needed for NLP applications in the domain of question answering, machine translation etc.: for instance, without distinguishing assertions vs. questions, it is not possible to choose the right verb form and place the words in the right order during the Czech-to-English translation. The present paper focuses on the assignment of sentence modality in the Prague Dependency Treebank (PDT), which is a richly annotated collection of Czech newspaper texts.

In Section 2, we explain how the term 'sentence modality' is used in the paper. Section 3 introduces the original annotation of sentence modality, released as a part of PDT 2.0 [3] in 2006 and yet not described in any published paper. As this annotation proved to be too simple to adequately represent the linguistic phenomenon in question, the sentence modality assignment has been revised and extended for the forthcoming version of PDT (PDT 3.0; Sect. 4).[1]

---

[1] PDT 3.0 is planned as a treebank consisting of the data of PDT 2.0 with corrections and revisions of several types, and of new data annotated in a comparable way. If we refer to the PDT 3.0 in the present paper, only the corrected and revised PDT 2.0 data are meant (i.e., PDT 2.0 and PDT 3.0 consist of the same texts and have the same size).

P. Sojka et al. (Eds.): TSD 2012, LNCS 7499, pp. 56–63, 2012.

Table 1. Means used for expressing sentence modality in Czech written texts

particle/wh-word	verbal mood	final punct. mark	sentence modality
Ø	indicative/conditional	. / Ø	declarative
Ø	indicative/conditional	?	interrogative (polar (yes/no) question)
wh-word	indicative/conditional	?	interrogative (non-polar (wh-)question)
Ø	imperative	! / .	imperative
Ø/ at', kéž, necht'	indicative/conditional	! / .	desiderative
Ø	indicative	!	exclamative

2 Sentence Modality as a Modal Meaning of the Sentence

In PDT as well as in the linguistic framework of Functional Generative Description (FGD; [12]), which the PDT annotation scenario is based on, sentence modality is understood as a modal meaning of the sentence; it is the function of the sentence to assert a content, ask a question, require that someone performs something etc.[2] In Czech written texts, these functions are conventionally expressed by combinations of formal means of different types, namely by the mood of the verb form, by the final punctuation mark, by the word order, and by modal particles *at', kéž, necht'*.[3]

Five types of sentence modality are distinguished in PDT and FGD according to the Czech linguistic tradition (e.g. [13,2]):

– declarative modality (e.g. *Ekonomika jde do vzestupu už letos.* 'The economy rises already this year.'),
– interrogative modality (*Jaká je nezaměstnanost v této zemi?* 'How big is the unemployment in this country?'),
– imperative modality (*Podívej se na mě!* 'Look at me!'),
– desiderative modality (*At' si provincie konečně oddychne.* 'Let the province finally relax.') or
– exclamative modality (*To nejsou špatně rozdané karty!* 'The cards have been dealt not at all badly!').

Although the sentence modality and thus the choice of formal means mirror the speaker's intention to state something or to learn a piece of information etc. (cf.

---

[2] The terminology is far from uniform. Portner [8] speaks about sentential force, or simply about clause types or sentence types; all these terms are subsumed under discourse modality whereas the term sentential modality is used for isolated linguistic means operating "at the level of the whole sentence". Zaefferer [14] makes a terminological distinction between sentence mood (close to our usage of sentence modality) and sentential modality (underlying intention of the speaker).

[3] In spoken texts, prosodic features (esp. intonation) are reckoned for the most important means for conveying sentence modality. However, these features are not available in written texts, which we are concerned with.

illocution in the Speech Act Theory by Austin [1] and Searle [11]), neither the classification nor the annotation aim at capturing this intention since extra-linguistic factors (politeness conventions etc.) can play a crucial role in how the intention is expressed.[4] The theoretical delimitation of the five modality types as well as our annotation approach are based on linguistic means explicitly coded in the sentence; see Table 1 for the respective combinations.[5]

The relatively transparent relations between the sentence modalities and the formal markers seem to be a solid base for annotating the sentence modality in real language data. However, an essential question, namely which parts of the sentence are to be assigned the sentence modality, must be answered before any annotation starts. The sentence modality is often defined as a modal meaning of the whole sentence (cf. footnote 2), however, there are sentences with a more complicated structure, for which this definition is not satisfactory.

In FGD, the sentence modality is supposed to be a characteristic of the sentence as a whole if it involves just one main (syntactically independent) clause (see ex. (1)), but in a coordination structure, each of the syntactically independent clauses can have a different modality (ex. (2)). Similarly, an embedded but syntactically independent structure, such as direct speech, expresses its 'own' sentence modality, which may differ from the modality of the respective matrix clause (ex. (3)). The annotation of sentence modality in PDT 2.0 did not meet all these requirements (Sect. 3), they are reflected in the more advanced approach introduced in Section 4.

(1) *Neptejte*.imper *se mě, proč jsem přijel do Prahy.* 'Do not ask.imper me why I came to Prague.' (the modality is marked with the head of the respective structure, see Sect. 3.1 for the explanation of the values used)

(2) *Poprvé jste nastoupil*.enunc *v závěru zápasu v Benešově, jaké to bylo*.inter? 'For the first time you entered.enunc the game before the end of the match in Benešov, what was.inter it like?'

(3) *Kam se poděla*.inter *má bojovnost? ptala*.enunc *se sama sebe po utkání Martinezová.* 'Where did my fighting spirit disappear.inter? Martinezová asked.enunc herself after the match.'

3 Annotation of Sentence Modality in PDT 2.0

3.1 Sentence Modality as a Part of the Deep-syntactic Annotation

PDT 2.0 is a treebank of Czech written texts enriched with a complex annotation of three types (at three layers): the morphological layer (where each token was assigned a lemma and a POS tag), the so-called analytical layer, at which the surface-syntactic structure of the sentence (subject, object etc. relations) is represented as a dependency tree, and the tectogrammatical layer, at which the linguistic meaning of the sentence is represented. Nodes of the tectogrammatical tree represent auto-semantic words whereas

---

[4] Cf. the examples of asking *Is there any salt?* or stating *It is cold here*, which both can be meant as a request (to pass over the salt and close the window, respectively).

[5] Some of the means for expressing sentence modality (esp. verbal mood, modal particles) are used, in combination with further ones, to recognize the so-called factuality of events in FactBank [10] or the factuality of conditions within the annotation of discourse relations in the Penn Discourse TreeBank 2.0 [9].

functional words (such as prepositions, auxiliaries, subordinating conjunctions) and punctuation marks have no node of their own in the tree.[6] The nodes are labeled with a tectogrammatical lemma, with a functor (dependency relation; e.g. Actor ACT, Patient PAT, Location LOC) and other attributes; see [4]. One of the node attributes is the attribute sentmod, capturing the sentence modality of the respective syntactic structure. For this attribute, five values were defined: enunc for declarative modality (enunciation), inter for interrogative modality,[7] imper for imperative modality, excl for exclamative modality, and desid for desiderative modality.

Annotation at all three layers is available for 3,168 documents, containing altogether 49,442 sentences with 833,357 tokens (word forms and punctuation marks). The statistics reported in this paper have been measured on the training set of these data (2,533 documents, 38,727 sent., 652,544 tokens).[8]

3.2 Semi-automatic Assignment of Sentence Modality

Due to the large amount of data and a limited amount of time, a simplified approach to the sentence modality was carried out in PDT 2.0. The simplification consisted in that only one sentence modality value was determined for the whole syntactic structure; the fact that coordinated clauses can have different sentence modalities was intentionally omitted. Two types of embedded syntactic structures (direct speech and parenthesis) were assigned a separate sentmod value.[9]

As the first step of the sentence modality assignment in the PDT 2.0 data, the set of candidate nodes to be assigned a sentmod value was delimited as follows:

(a) child nodes of the technical root node, i.e. nodes representing the main verb or noun and the root nodes of coordination structures (corresponding to a conjunction or punctuation; 'coordination roots' in the sequel);

(b) root nodes of subtrees representing a direct speech; these nodes were identified on the basis of the node attribute is_dsp_root, which had been assigned before the sentmod annotation was carried out;

(c) root nodes of parenthesis subtrees (labeled with the functor PAR).

With the nodes identified as (a), (b) or (c), the value of the sentmod attribute was filled in according to the following 'algorithm', taking advantage of the links between the tectogrammatical, analytical and morphological annotation:

---

[6] There are certain, rather technical exceptions, e.g. coordinating conjunctions used for representation of coordination constructions are present in the tree structure.

[7] The difference between polar (yes/no) questions and non-polar (wh-)questions is not captured by the sentmod value but by the non/presence of the wh-word in the tree.

[8] For searching in the PDT 2.0 data and for data manipulation, we used the Netgraph query language [5] and the PMLTQ extension [7] to the Tree Editor TrEd [6].

[9] It means, a sentence with an embedded direct speech or parenthesis was assigned two sentmod values: one value was specified for the sentence as a whole (and assigned to the child node of the technical root node; see under (a) bellow), one value for the direct speech or parenthesis as a whole (see under (b) and (c)). The inner structure of the direct speech subtrees and parenthesis subtrees was not analyzed here.

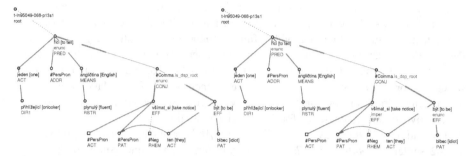

Fig. 1. The tectogrammatical tree for the sentence *"Nevšímejte si jich, jsou to blbci," řekl mi plynulou angličtinou jeden z přihlížejících.* ("Do not take notice of them, they are idiots," told me one of the onlookers in fluent English.), in which two clauses with different sentence modalities are coordinated within a direct speech (the coordination root is assigned the functor CONJ and the attribute is_dsp_root). Within the original PDT 2.0 annotation (on the left side), the CONJ node was assigned the enunc value, the imperative modality of the first clause of the direct speech was omitted. The new annotation specifying the modality for each clause of the direct speech (imper & enunc) as well as for the matrix clause (enunc) is on the right side.

1. if the node represented an imperative verb form (i.e., technically, if one of the morphological tokens which the node was interlinked with was assigned the tag Vi.* (imperative verb form)), the node was assigned the sentmod value imper;
2. if the syntactic structure to which the node belonged ended with a question mark (technically, if the node corresponded to an analytical node that had a question mark among its child nodes), the sentmod value inter was filled in;
3. from the rest of the nodes, nodes that were a part of a sentence introduced by the particles *ať, kéž, nechť* and/or ended with an exclamation mark were identified (92 occurrences in the training data of PDT 2.0) and assigned manually one of the sentmod values desid, excl or imper;
4. the remaining nodes were assigned the sentmod value enunc.

A tectogrammatical tree with sentmod values assigned according to this algorithm is displayed in Figure 1 (on the left side). The distribution of the sentmod values in PDT 2.0 is given in Table 2.

4   An Extended Approach to Sentence Modality in PDT 3.0

4.1   Identification of Weak Points of the Annotation

The main motivation for revision of the annotation of sentence modality was the insufficient treatment of coordination structures. However, at the very beginning of the revision, we wanted to find out whether there are, in addition to direct speech and parenthesis, further types of embedded structures which express a sentence modality on their own and thus require a separate sentmod value. A simple test, based on the direct interconnection between imperative mood and imperative sentence modality,[10] has pointed out to one more type of such structures, namely to sentence-like titles assigned the functor ID in the annotation (see ex. (4): the title *Pohlad'te si králíčka* 'Stroke a bunny rabbit' expressing the imperative modality is embedded in a matrix clause with declarative modality).

(4) *Zítra bude u příležitosti III. výročí české a slovenské edice Playboy otevřena*.enunc *výstava Pohlad'te*.imper *si králíčka sestavená z ilustrací pro časopis Playboy.* 'An exhibition Stroke.imper a bunny rabbit consisting of illustrations for the magazine Playboy will be opened.enunc tomorrow on the occasion of the 3rd anniversary of the Czech and Slovak editions of Playboy.'

4.2 Redesigning the Assignment Process

When considering the relation between the sentence modality annotation in PDT 2.0 (which concerned the nodes listed under (a) to (c) in Sect 3.2) and the new aim to specify a sentmod value for each clause in coordinations as well as for the title structures, it was not possible to preserve the current annotation and just to add sentmod values to the new candidates, since the decision to deal with coordinations affects all current subgroups (a) to (c). Another reason in favor of repeating the annotation was the fact that errors of several types were corrected during a systematic revision of the PDT 2.0 annotation carried out in the recent two years. Therefore, the sentmod values available in the PDT 2.0 data were canceled and the assignment process has been redesigned for PDT 3.0 and applied to the data from the scratch.

First of all, the set containing the candidate nodes (a) to (c) was extended by (d) the root nodes of title subtrees (functor ID). Secondly, from all these candidates, coordination roots were extracted and handled separately (see Sect. 4.3). Thirdly, for the remaining (non-coordination) nodes the steps described under 1 to 4 in Sect. 3.2 were applied; manual annotation (step 3) was needed for 82 nodes (in the training data).

4.3 Assigning Coordinated Clauses with Sentence Modality

Coordinations were handled as a homogeneous group, regardless which of the subgroups (a) to (d) they belonged to. On the basis of the extracted list of coordination roots, the set of root nodes of coordinated clauses which were to be assigned a sentmod value was delimited: 17,320 roots of coordinated clauses (governed by 7,598 coordination roots) were identified in the training data.

For the sake of specification of the sentmod value for the root of each coordinated clause, the step 1 of the algorithm could be applied "locally", i.e. just for the particular clause of the coordination structure, not for all the clauses in a coordination: 64 root nodes of the individual coordinated clauses (in the training data) were assigned the value imper since they represented an imperative form.

---

[10] In Czech, the imperative mood occurs exclusively in sentences with the imperative sentence modality and, the other way round, the imperative modality is mostly expressed by sentences with an imperative verb form.

Table 2. The sentmod values in the PDT 2.0 and PDT 3.0 training data. The values of the last column are involved in the values of the third column.

sentmod value	frequency in PDT 2.0	frequency in PDT 3.0	with coordinated clauses
enunc	41,949	57,608	17,106
inter	777	828	130
imper	175	271	64
desid	17	13	2
excl	84	62	18
total	43,002	58,782	17,320

Those non-imperative clauses which were coordinated with the imperative ones were extracted to be assigned a sentmod value manually. The second portion for manual annotation were roots of coordinated clauses that were part of a coordination structure ending with a question mark. Our assumption that the question mark occurring as the final punctuation mark of the whole coordination structure is to be interpreted as a signal of the sentence modality just for the final clause of the coordination structure (i.e. it does not mirror the sentence modality of the non-final clauses) proved to be true during the annotation. Roots of coordinated clauses which were part of a coordination structure ending with an exclamation mark or involving the particles *at'*, *kéž* and *necht'* were the third portion for manual annotation. The manual annotation thus concerned 268 roots of coordinated clauses in total. It was carried out by two annotators in parallel, with the inter-annotator agreement of 93.7% (Cohen's Kappa 0.89).

All the remaining coordination structures ended with a period (or without punctuation etc.) and involved only clauses with an indicative or conditional verb form. As in 100 coordination structures randomly selected from this group, only coordinated clauses with declarative modality were found, clauses in these coordination structures were automatically assigned the sentmod value enunc.

The distribution of the sentmod values in the training data of PDT 3.0 is listed in Table 2, besides the overall statistics (3rd column of the Table), the frequency of the values with the coordinated clauses is given as well (4th column). Lower frequency of the values desid and excl in PDT 3.0 in contrast to PDT 2.0 is due to some recent theoretical clarifications and corrections which were reflected in the manual annotation to be included in the PDT 3.0. The substantial increase with the other values is connected with the assignment of coordinated clauses. The differences between the sentence modality assignment in PDT 2.0 vs. 3.0 are illustrated in Figure 1.

5   Conclusions

In the Prague Dependency Treebank, sentence modality is understood as a modal meaning of the sentence, or of each of its syntactically independent parts, and represented by a special node attribute sentmod in the tectogrammatical annotation. Within the original, simplified annotation of sentence modality, which was implemented in the PDT 2.0 data, sentence modality was assigned to the root node of each sentence and to the roots of parentheses and direct speech. Within the recent months, the sentence modality assignment has been elaborated with coordinated clauses and extended to embedded titles. The resulting annotation is thus more consistent and theoretically adequate, it will be released as a part of the PDT 3.0.

Acknowledgments. The research reported on in the present paper was supported by the grants GA ČR P406/2010/0875 and GA ČR P406/12/P175. This work has been using language resources developed and/or stored and/or distributed by the LINDAT-Clarin project of the Ministry of Education of the Czech Republic (project LM2010013).

References

1. Austin, J.L.: How to Do Things with Words. Harvard (2005)
2. Daneš, F., Hlavsa, Z., Grepl, M., et al.: Mluvnice češtiny 3, Praha (1987)
3. Hajič, J., et al.: Prague Dependency Treebank 2.0. CD-ROM, Philadelphia (2006)
4. Mikulová, M., et al.: Annotation on the tectogrammatical level in the Prague Dependency Treebank. Annotation manual. Tech. Rep. Nr. 2006/30, Prague (2006)
5. Mírovský, J.: Searching in the Prague Dependency Treebank, Prague (2009)
6. Pajas, P., Štěpánek, J.: Recent Advances in a Feature-rich Framework for Treebank Annotation. In: Proceedings of Coling 2008, Manchester, pp. 673–680 (2008)
7. Pajas, P., Štěpánek, J.: System for Querying Syntactically Annotated Corpora. In: Proceedings of the ACL-IJCNLP 2009 Software Demonstrations, Singapore, pp. 33–36 (2009)
8. Portner, P.: Modality, Oxford (2009)
9. Prasad, R., et al.: The Penn Discourse Treebank 2.0. In: Proceedings of LREC 2008, Marrakech, pp. 2961–2968 (2008)
10. Saurí, R., Pustejovsky, J.: FactBank: a corpus annotated with event factuality. Language Resources and Evaluation 43, 227–268 (2009)
11. Searle, J.: Speech Acts, Cambridge (1969)
12. Sgall, P., Hajičová, E., Panevová, J.: The Meaning of the Sentence in Its Semantic and Pragmatic Aspects, Dordrecht, Praha (1986)
13. Šmilauer, V.: Novočeská skladba. 3rd edn., Praha (1969)
14. Zaefferer, D.: On the coding of sentential modality. In: Bechert, J., et al. (eds.) Towards a Typology of European Languages, Berlin, New York, pp. 215–237 (1990)

Literacy Demands and Information to Cancer Patients

Dimitrios Kokkinakis, Markus Forsberg,
Sofie Johansson Kokkinakis, Frida Smith, and Joakim Öhlen

University of Gothenburg, P.O. Box 100
SE-405 30 Gothenburg, Sweden
{dimitrios.kokkinakis,markus.forsberg,
sofie.johansson.kokkinakis,frida.smith,joakim.ohlen}@gu.se
http://spraakbanken.gu.se/
http://www.gpcc.gu.se/

Abstract. This study examines language complexity of written health information materials for patients undergoing colorectal cancer surgery. Written and printed patient information from 28 Swedish clinics are automatically analyzed by means of language technology. The analysis reveals different problematic issues that might have impact on readability. The study is a first step, and part of a larger project about patients' health information seeking behavior in relation to written information material. Our study aims to provide support for producing more individualized, person centered information materials according to preferences for complex and detailed or legible texts and thus enhance a movement from *receiving* information and instructions to *participating* in knowing. In the near future the study will continue by integrating focus groups with patients that may provide valuable feedback and enhance our knowledge about patients' use and preferences of different information material.

Keywords: Health literacy, Readability, Natural Language Processing.

1 Introduction

This preliminary study examines language complexity of printed and written health information materials given to patients undergoing colorectal cancer (CRC) surgery in Sweden. The overall aim of this research is to investigate whether the implementation of adapted, person-centered information and communication for patients diagnosed with CRC undergoing elective surgery, can enhance the patients' self-care beliefs and well-being during recovery in the phase following diagnosis and initial treatment (i.e., three months). Several explorative, qualitative studies are planned and will be function both as a basis for the proposed interventions and provide explanations for the actual processes leading to the desired outcomes. Patients' knowledge enablement will be reached by several interrelated intervention strategies. The goal of one of these strategies is to provide patients with written information materials according to preferences for complex and detailed or legible texts.

2 Background

There is a strong belief in the scientific community that misunderstandings in health information increases the risk of making unwise health decisions, leading to poorer

P. Sojka et al. (Eds.): TSD 2012, LNCS 7499, pp. 64–71, 2012.

health and higher health care costs. There is also a consensus that it is of benefit to health care consumers if the provided information is easier to read and comprehend by, e.g., avoiding unknown vocabulary and difficult grammatical structures [1,2,3,4,5,6,7,8]. The knowledge gained from such endeavor can empower health consumers to ask more informed questions when seeing their caregiver and it lessens their fear of the unknown. A large group of such health care individuals are cancer patients with documented "unmet communication needs" [4,5,6,7,8]. Therefore, readability and suitability of health information materials is considered important for patient empowerment. Readability is a quantitative measure that estimates the difficulty in reading text [9,10], without considering layout, familiarity of the subject or subject complexity, and there is a large number of readability indices described in literature. The majority of these metrics are based on the statistical properties of the text, e.g., the number of words in a sentence, while other measures try to overcome the limitations of traditional metrics by taking into account other less shallow characteristics such as text coherence features through an entity grid representation of discourse [11,12].

3   Material

Written and printed patient information material (hereafter called the CRC-corpus) from 28 Swedish clinics for patients diagnosed with CRC undergoing elective surgery were selected for analysis by means of standard metrics and more elaborate language technology techniques [13]. The CRC-corpus comprises 185 documents, divided into booklets (79), information sheets (70), brochures (29) and letters (7). All material was converted from PDF to UTF-8 text format, tokenized and separated into sentences. For some of the experiments we also annotated the CRC-corpus with part-of-speech information, controlled vocabularies and syntactically analyzed with a shallow parser [14]. The total number of all tokens in the corpus is 179,300 (153,800 without punctuation, numerical information, email addresses and URLs); the total number of types (unique tokens) is 22,218, while the total amount of sentences was 16,000. The average length of the documents was 919 tokens. The brochures are the longest documents (2,540 tokens/doc), the letters on average the shortest (290 tokens/doc). Previous studies assessing the health literacy of written materials have relied on readability formulas alone. We also start with such approach but continue by applying other, more elaborate techniques in order to get a comprehensive view of the data. For each document we initially applied a number of readability-related scores, namely the LIX, a readability index commonly used in Scandinavia [15] and as complementary methods we measured the word variation index (OVIX) and the nominal ratio (NR). We then move onto calculating the percentage of word overlap of the information material against a newspaper corpus. This way we can calculate the frequency profile of the data which is an important instrument for assessing vocabulary knowledge [16]. We also calculate the terminology load of the corpus based on the content of two official controlled vocabularies; performed shallow syntactic analysis and also computed other features, i.e., function words, and very long words, that we believe provide substantial support with respect to how popularized some of the CRC-corpus are. Table 1 summarizes some of characteristics of the data and some of the investigated variables and metrics.

Table 1. Complexity indices for the CRC-corpus (*excl. punctuation and numerals)

	booklets n=79	inf. sheets n=70	brosch. n=29	letters n=7
All tokens	64,017	39,562	73,661	2,036
All words	54,920	33,424	63,830	1,620
Types*	7,030	5,697	8,780	711
Avg word/sent.	10,6	11,4	11,8	9,8
Words over 6	14,925 (27.17%)	9,006 (26.94%)	16,946 (26.54%)	552 (34.1%)
Words over 13	1,827 (3.32%)	937 (2.8%)	1,612 (2.52%)	65 (4.01%)
LIX	37	37,1	37,5	42,1
OVIX	63	66,47	67,6	71,1
NR	1,29	1,29	1,25	1,33

4   Method

For the readability assessment we applied a number of formulas commonly used on Swedish textual data [17,18]. The LIX index measures the number of words per sentence and also the number of long words (over 6 chars) in the text:

$$\text{LIX} = n(w)/n(s) + (n(words > 6chars) \times 100)/n(w) \qquad (1)$$

Here, n(w) denoted the number of words and n(s) the number of sentences. A high LIX indicates greater difficulty, a LIX score below 30 indicates very easy to read text, between 30–40 indicates normal text or fiction, between 40–50 usually indicates newspaper language or factual language and 50–60 professional language. Lexical variation or OVIX measures the ratio of unique tokens in a text and is used to indicate the ideal density, usually in conjunction with the nominal ratio (see below).

$$\text{OVIX} = \log(n(w))/\log(2 - (\log(n(nw))/\log(n(w)))) \qquad (2)$$

Here, n(w) denotes the number of words and n(uw) the number of unique words. Nominal ratio (NR) is calculated by dividing the number of nouns, prepositions and adjectives/participles with the number of pronouns, adverbs and verbs:

$$\text{NR} = n(noun) + n(prep) + n(part)/n(pron) + n(adv) + n(v) \qquad (3)$$

Here $n(noun)$ denotes the number of nouns, $n(prep)$ the number of prepositions, $n(part)$ the number of adjectives and participles, $n(pron)$ the number of pronouns, $n(adv)$ the number of adverbs, and $n(v)$ the number of verbs. A higher nominal ratio indicates a more professional and stylistically developed text, while a lower value indicate informal language, a value around 1 is considered normal (daily newspaper level); cf. Table 1. Note that LIX as well as the OVIX index do not depend on text length since they only consider ratios, e.g., proportion of long words in LIX.

4.1   Frequency Bands

Using frequency bands we can calculate the percentage of word overlap in the information material with various types of corpora (i.e., frequency profile). In our experiment we chose a lemmatized daily newspaper corpus to compare with. The 10,000 most frequent lemmas in the newspaper corpus were extracted and compared with the vocabulary in the information material. Table 2 shows that the "outsiders" (out-of-vocabulary words) word types, that are not in the 10,000 most frequent lemmas, are high in all material, the booklets are almost fifteen percent more in the letters (54.77% compared to 39.03%). This experiment resulted into finding uncommon words, which would presumably have a highly technical content, words that do not appear in common language and therefore require special attention by the author of the text. Such support can take the form of better explanations in the form of e.g., a list of defining indexed words. Obviously, not all technical words can be explained, for instance, there were a number of medication names, sometimes long lists, that cannot be explained further since they are easily understandable as such in context, e.g., in one of the texts one could read: *Mediciner som du skall sluta använda 7 dagar innan operationen är: Albyl minor, Alganex, Alindrin, Alka-Seltzer, Alpoxen, Ardinex, Arthrotec, Asasantin, Aspirin, Bamyl [. . .]*, i.e. "Medications that you must stop using seven days before the surgery are: . . . ". The results with respect to the newspaper corpus were also uneven at the top-1000 lemma level, which means that 14.77%–36.51% of the lemmas in the CRC-corpus could be found among the most frequent, top-1000 lemmas, of the newspaper corpus. In Table 2 the figures in parenthesis denote the percentage of lemmas not found among the 10,000 newspaper lemma list (39.03–54.77%).

Table 2. CRC-corpus frequency profile based on lemmas, compared with a newspaper corpus

	news/top1000	outsider examples
booklets	14.77% / 54.77%	*kolostomi* colostomy
broschures	13.21% / 53.4%	*bandage* bandage
inf. sheets	17.1% / 49.41%	*gasbesvär* flatulence
letters	36.51% / 39.03%	*narkosläkare* anesthetist

4.2 Vocabulary Assessment

Terminology. In order to assess the medical content of the corpora we performed a series of experiments by automatically identifying medical terms. This way we could evaluate an important aspect of the vocabulary style used in the different types of the corpus by comparing them with existing official, professional controlled vocabularies that are medical in nature [9,10]. The two vocabularies tested were the Swedish Systematized Nomenclature of Medicine Clinical Terms, SNOMED CT (290,000 terms), and the 2011 version of the Swedish Medical Subject Headings, MeSH (27,000 terms). The content of these vocabularies can be considered as an important style marker since complex medical terminology might be a negative factor in understanding the textual content. Several solutions are possible for instance the use of explanations; some already provided in some texts, usually as parenthesized alternatives; e.g. *agraffer (metallklämmor)* 'surgical staples (metal clips); or by defining context using verbs such as *betyda* 'mean' or *innebära* 'imply', for instance *ordet stomi kommer fran grekiskan och betyder 'öppning'* i.e. 'the word stoma comes from the Greek and means 'opening".

Table 3. The CRC-corpus terminology profile

	amount	most frequent values
letters	149(SNOMED) 232(MeSH)	qualifier (54) - M01/person(36)
booklets	7,595(SNOMED) 8,504(MeSH)	qualifier (2695) - E04/surg. procedure(823)
inf. sheets	4,503(SNOMED) 5,099(MeSH)	qualifier (1502) - E04/surg. procedure(810)
broschures	9,153(SNOMED) 9,639(MeSH)	qualifier (2908) - E04/surg. procedure(1958)

The amount of complex words (terms) in the corpus varied considerably between the different genres. With respect to the SNOMED CT, all four types of material were rich in qualifier values (which are usually easy to comprehend modifiers) such as *möjligt* 'possible'. Letters and brochures were also rich in (everyday) physical objects, such as *telefon* 'telephone' and *bandage*; booklets were rich in findings such as *feber* 'fever' and information sheets in body structures such as *tarm* 'intestines'. With respect to MeSH, letters contained most terms in category M01+, such as patient and L01+ such as e-mail, both characteristic of the written style of letters as a communication tool. The category E04+ was a frequent category in the rest, such as *operation* 'surgical procedure'. Table 3 summarizes the results.

Function words. Popularized texts are expected to contain more personal pronouns than in the scientific documents. In the case of the CRC-material, there is a relatively high frequency of occurrence of the second person singular subject pronoun du 'you' and the object form dig 'you'. First person singular, jag 'I', and plural, vi 'we', are also frequent. These pronouns are strong indicators that we are dealing with a popularized text. Not, that the second person, plural form *ni* 'you' was fairly uncommon, 28 occurrences in total. Table 4 illustrates these findings. The total number of function words was also measured, as function words we counted adverbs, conjunctions, prepositions, pronouns, auxiliary verbs, articles and particles. The use of function words, and in particular pronouns, allows the creator of the document to make the content more personal, while the scientific and expert literature remains abstract. In this way, patients should feel more directly addressed by the documents, and may be more involved in the process of communication with experts. If so, they can feel more involved in the caregiving process as well.

Table 4. Function words in the CRC-corpus

	du 'you'	*dig* 'your'	*jag* 'I'	*vi* 'we'	function words
booklets	2,170	465	137	200	28,149(51.25%)
broschures	1,395	376	162	113	32,618(51.1%)
inf. sheets	1,143	271	52	112	17,258(51.63%)
letters	68	21	-	20	654(40.37%)

Very long words. This is a characteristic of texts that are hard to comprehend. The limit discussed in different studies slightly varies, here we use the figure also applied by [19] in which very long words are considered those with over 13 characters long, note also that Swedish is a compounding language so the proportion of long words is expected to be relatively high. The percentage of very long words lies between 2.52–4.01%, and in certain cases it might have been advantageous to paraphrase with simpler constructions. Examples of very long words in the corpus are: *kolorektalsjuksköterskemottagning* 'colorectal nurse reception', *sociallagsstiftningsfraga* 'social service act case' and *hormonersättningsmedel* 'hormone replacement drug'.

Table 5. Simplex (NP-1-3) and complex NPs (NP>3) in the CRC-corpus

	all NPs	Np1	Np2	Np3	Np over 3
booklets	15,362	10767 (70.1%)	2994 (19.49%)	1045 (6.89%)	556 (3.62%)
broschures	16,557	10690 (64.56%)	3630 (21.92%)	1531 (9.24%)	798 (4.82%)
inf. sheets	8,964	6046 (67.44%)	1874 (20.9%)	650 (7.25%)	394 (4.4%)
letters	522	362 (69.34%)	109 (20.88%)	26 (4.98%)	25 (4.79%)

Grammatical Analysis. There are very few investigations focusing on the syntactical level of medical documents. For the grammatical analysis, we evaluated the use of noun phrases (NP) in the corpus. We applied a Swedish shallow parser [14] and counted the number of simplex noun phrases (phrases containing 1–3 tokens) and complex noun phrases (longer phrases containing over 3 tokens, noun phrases that take prepositional phrases as an argument as well as coordinated constructions). Examples of simplex noun phrases are: *np[ytterligare information]* 'more information' and *np[en välbalanserad kost]* 'a well balanced diet'. Examples of complex noun phrases are: *np[ditt psykisk och fysisk tillstäand]* 'your mental and physical condition' and *np[kontaktsjuksköterska, stomisköterska, dietist, kurator och arbetsterapeut]* 'contact nurse, stoma nurse, dietitian, social worker and occupational therapist'. We also count as complex nouns phrases the ones that take a syntactic construction, a prepositional phrases, as obligatory argument; for instance, *np[orsak pp[till [din operation]]]* 'reason for your surgery' and *np[konkret räad pp[om [kost]]]* 'practical advice on diet'. Some of the long noun phrases were not as complex as one might have thought since the were mostly enumerative ones, like the previously given example with the 'contact nurse, stoma nurse, …'. On the contrary, simplex NPs can be problematic mainly due to the use of figurative language, such as *np[fantomsmärta]* 'phantom pain' or ad-hoc combinations of abbreviations in compound words such as *np[postopkost]* 'post-op diet' or technical vocabulary such as *np[titthäalskirurgi]* 'keyhole surgery'.

5   Discussion and Conclusions

The corpus study presented in this paper represent a first attempt at characterizing the language of Swedish written and printed information material for CRC-patients. Our aim has been to develop a comprehensive and balanced toolkit of metrics and analytic approaches that will provide an indication on how difficult documents in each genre are to read and understand. We examined various text parameters such as lexical variation, frequency bands and the use of terminology. The findings from this study indicate that there is space for improvement in many aspects of the written material. For instance, an unexpectedly finding was that letters were scored higher with respect to most of the measures tested, such as LIX, OVIX, NR and had less function words than the rest of the material. Although sentences are not generally longer in letters, they still contain more noun phrases and they differ in the number of long and very long words compared with the other three text types. Printed and written health information materials for health consumers place certain demands on their readers since it is implied that the reader to have the ability to read, comprehend, interpret, analyze and apply the information they gain from the materials. As a matter of fact, it has been discussed [1] that uncertainty in illness can be due to inconsistent information in situations characterized by ambiguous, complex and unpredictable language. In our study we did not take into account other aspects of readability, such as layout of the page, illustrations or font size of the text, even though these features play a significant role in readability and suitability of written or printed documents. The vision we have for the future is to both to be able to personalize or customize health-related information material to the individual patient's level of reading and cognitive skills and also to apply the features and text characteristics in such a way that can predict readability in a better and more reliable way since language technology techniques can produce a large number of relevant linguistic features [21]. Thus, we envisage a scenario that such printed information material can be dynamically generated on demand, using natural language generation techniques [20], based on individualized parameters, such as education level, linguistic, ethnic and religious background, explanation level of complex concepts, demand for illustrations and pictures etc. To a certain degree, such techniques have been already applied for Web-based tailored interventions for various patient groups [22].

References

1. Mishel, M.H., Clayton, M.F.: Uncertainty in Illness Theory. In: Smith, M.J., Liehr, P. (eds.) Middle Range Theory for Nursing. Spring Publications (2003)
2. Shieh, C., Hosei, B.: Printed Health Information Materials: Evaluation of Readability and Suitability. J. Community Health Nurs. 25(2), 73–90 (2008)
3. Weintraub, D., et al.: Suitability of prostate cancer education materials: applying a standardized assessment tool to currently available materials. J PatEduCouns 55(2), 275–280 (2004)
4. Weert, J.C., et al.: Tailored information for cancer patients on the Internet: Effects of visual cues and language complexity on information recall & satisfaction. J. PatEduCouns (2011)
5. Helitzer, D., et al.: Health literacy demands of written health information materials: an assessment of cervical cancer prevention materials. Cancer Control 16(1), 70–78 (2009)
6. Doak, C.C., Doak, L.G., Root, J.H.: Assessing Suitability of Materials. In: Teaching Patients with Low Literacy Skills. Lippincott Williams & Wilkins (1996)
7. Ownby, R.L.: Influence of vocabulary & sentence complexity & passive voice on the readability of consumer-oriented mental health info on the internet. In: AMIA Symp., pp. 585–588 (2005)
8. Hack, T.F., Degner, L.F., Parker, P.A.: The communication goals and needs of cancer patients: a review. Psychooncology 14(10), 831–845 (2005)
9. Leroy, G., Eryilmaz, E., Laroya, B.T.: Health Information Text Characteristics. In: AMIA Symp., pp. 479–483 (2006)

10. Leroy, G., Helmreich, S., Cowie, J.R., Miller, T., Zheng, W.: Evaluating Online Health Information: Beyond Readability Formulas. In: AMIA Symp., pp. 394–398 (2008)

11. Heilman, M., et al.: Combining Lexical and Grammatical Features to Improve Readability Measures for 1st and 2nd Language Texts. In: HLT-NAACL, Rochester, NY, pp. 460–467 (2007)

12. Barzilay, R., Lapata, M.: Modeling local coherence: an entity-based approach. In: 43rd Annual Meeting on Association for Comp. Ling, USA, pp. 141–148 (2005)

13. Borin, L., et al.: Empowering the patient with language technology. SemanticMining NoE 507505: D27.2 (2007),
    `http://demo.spraakdata.gu.se/svedk/pbl/WP27-02.pdf`

14. Kokkinakis, D., Johansson Kokkinakis, S.: A Cascaded Finite-State Parser for Syntactic Analysis of Swedish. In: The 9th EACL, Norway, pp. 245–248 (1999)

15. Björnsson, C.-H.: Läsbarhet. Stockholm. Liber (1968)

16. Laufer, B., Nation, P.: Vocabulary size and use: Lexical richness in L2 written production. Applied Linguistics 16(3), 307–329 (1995)

17. Mühlenbock, K., Johansson Kokkinakis, S.: LIX 68 revisited – An extended readability measure. Corpus Linguistics. U. of Liverpool, UK (2009)

18. Johansson Kokkinakis, S., Magnusson, U.: Computer based quantitative methods applied to 1st & 2nd language student writing. In: Göteborgsstudier i nordisk spräakvet, GNS (2011)

19. Melin, L., Lange, S.: Att analysera text. Studentlitteratur AB (2000) (in Swedish)

20. Miller, T., Leroy, G.: Dynamic generation of a Health Topics Overview from consumer health information documents. J. Biomed. Eng. and Tech. 1(4), 395–414 (2008)

21. Oosten, P., Tanghe, D., Hoste, V.: Towards an Improved Methodology for Automated Readability Prediction. In: Conf. on Lang. Resources and Eval., LREC, Malta, pp. 775–782 (2010)

22. Lustria, M.L., et al.: Computer-tailored health interventions delivered over the web: Review and analysis of key components. Patient Edu. and Counseling 74(2), 156–173 (2009)

Expanding Opinion Attribute Lexicons

Aleksander Wawer and Konrad Gołuchowski

Institute of Computer Science, Polish Academy of Science
ul. Jana Kazimierza 5, 01-238 Warszawa, Poland
axw@ipipan.waw.pl, kodieg@gmail.com

Abstract. The article focuses on acquiring new vocabulary used to express opinion attributes. We apply two automated expansion techniques to a manually annotated corpus of attribute-level opinions. The first method extracts opinion attribute words using patterns. It has been augmented by the second, wordnet and similarity-based expansion procedure. We examine the types of errors and shortcomings of both methods and end up proposing two hybrid, machine learning approaches that utilise all the available information: rules, lexical and distributional. One of them proves highly successful.

1 Introduction and Existing Work

A large amount of work has been done on document-level [4] and word-level [10] sentiment and opinion recognition. Due to limited space and because of the subject of our paper, we describe here only research on feature-based opinion mining. Features of a product are its attributes, sometimes called also aspects or components of the object under review (opinion target). For example, a camera may have *good lens, nice zoom* but *poor battery*.

In their seminal work, Hu and Liu [2] observed that attributes are typically expressed by nouns and opinion words by adjectives. They introduced a method of product feature extraction based on association rule mining. In this approach frequent nouns are assumed to be the most likely candidates for features. The technique is constrained to single nouns only and does not take into account the possibility of expressing the same feature using different words. A similar work [7] proposes to discover noun phrase (noun noun) features using PMI between the phrase and a set of discriminators.

More recently, a number of methods has been proposed to jointly bootstrap feature and opinion vocabulary, such as [11]. In [13] authors introduce a dependency-based approach to analyse movie reviews. Perhaps more closely related to our work is the pattern mining method described in [3]. Unlike in our paper though, their patterns are relations between features and opinion words.

Another interesting study [9] describes double propagation: an unsupervised, iterative algorithm based on syntactic relations of opinion words and features. It extracts noun features and suffers from low precision on large corpora.

Contrary to the mentioned papers, our approach assumes a small seed of manually annotated examples of attributes, as described in section 2, and focuses on its automated expansion by comparing and finally merging various techniques: lexicon-based, rule-based and distributional. There are two key aspects which distinguish our

P. Sojka et al. (Eds.): TSD 2012, LNCS 7499, pp. 72–80, 2012.

methods. First, we assume that expression of the same opinion attributes is possible using different words, not necessarily constrained to synonyms. Second, unlike some bootstrapping algorithms discussed above, we do not assume interdependence between opinion words and attribute words.

The paper is organized as follows. In Section 2 we briefly describe the manual corpus, which is the subject of automated rule-based expansion, as described in Section 3, and lexical-based expansion presented in Section 4. Sections 5 and 6 discuss two hybrid expansion methods. We discuss the results in Section 7.

2 Manual Annotation

In our experiments we used the corpus of product reviews obtained from one of the leading Polish review aggregator websites. Random samples of perfumes and women underwear reviews were manually annotated by a group of four annotators. They started their work independently and finally met to discuss annotation differences and create single, the most objective set of annotations. The annotation process and corpus design have been described in detail in [12]. The work focused on reviews of two product types, perfumes and women underwear, as listed in Table 1.

Table 1. Manual annotation statistics of the corpus

	Perfumes	Women underwear
# of words	4323	2446
# of attributes	31	20
# of attribute instances	343	305

In our experiments we treated all attributes separately and did not use the relations between attributes.

In the following sections we describe methods of automated expansion of the words identified in the two manually annotated subcorpora to all available, unannotated reviews of each product type. The set consisted of 5,498 perfume reviews and 10,177 women underwear reviews.

3 Expansion through Extraction Rules

This section describes the generation of candidates for opinion attribute words using rule-based approach. The hypothesis that underlies this method is that there exists a class of expressions (or patterns) which indicate presence of an attribute.

For example, in the below sentence:

```
Cena była jego jedynym plusem.
(The price was its only advantage).
```

one may expect that the `price` denotes an attribute. It seems intuitive that finding candidates for attribute words could be facilitated using similar patterns, when applied in the corpus. To test this approach, a small set of rules was created manually using the corpus described in [12]. The set consisted of 41 patterns designed to identify candidates for opinion attributes and yielded 12 different perfume attributes (out of 31) and 15 underwear attributes (out of 20).

Because of these promising initial results, methods of automated generation of such patterns from the review corpus and expansion of attribute vocabulary using these patterns have been devised. To increase the number of both extraction patterns and attribute word candidates, an iterative, bootstrapping technique has been proposed, which starts with nine, high quality, manually created rules to choose best words that may denote attributes. It seeks new patterns in the large unannotated corpus. These new patterns, together with the ones created previously (also manually), are used to find attribute word candidates.

3.1   Extraction Rule Generation

To represent patterns, Spejd's formalism and a shallow parser [8] have been applied. Each pattern is represented as a rule that may refer to lexemes in orthographic (inflected) forms, part of speech or word base forms. For instance, a rule that could be created for the example sentence above could look like:

```
Match: [pos~"subst"] [base~"była"] [orth~"jego"] [orth~"jednym"]
(Match: [pos~"noun"] [base~"be"] [orth~"it's"] [orth~"only"])
```

To generate such rules we use an approach similar to the one used in Brill tager [1]. Namely, a set of pattern templates that contain placeholders for PoS, lexemes or word base forms and the position of an attribute word has been created manually. For example, a template from which the above rule has been created might look like the one below. A placeholder (marked with the _ sign) may be filled with either word directly encountered from text (orth), word base form (base) or its part of speech (pos) depending on the type of clause it appears in. So, a placeholder in the clause [pos~"_"] expects a part of speech, whereas placeholder in the clause [base~"_"] expects to be filled with a word base form.

```
Match: [pos~"subst"] [base~"_"] [orth~"_"] [orth~"_"]
Attribute is first clause ([pos~"subst"])
```

Then, if we encounter this sentence where the word *price* is an attribute, the template is filled.

```
The        price         was           it's          only
           [pos~"noun"]  [base~"_"]   [orth~"_"]    [orth~"_"]
Match:     [pos~"noun"]  [base~"be"]  [orth~"it's"] [orth~"only"]
```

Of course, not all lexemes indicated by rules are correct attributes. Therefore, one has to add a filtering method, described in section 5 and remove inefficient patterns.

This leads to the problem of choosing the best rules from a set of possible candidates. For both rule and attribute candidates evaluation two basic scoring methods have been proposed. Intuitively, good rules are the ones that identify many correct word attribute candidates. Moreover, this has been augmented by a manually compiled a list of frequent incorrect words which are known *not* to denote any attribute. A good rule should not extract words from this list. Therefore, rules are scored using a linear combination of three components:

– number of correctly recognized words of known attributes,
– number of words that are on the manually compiled list of incorrect words,
– number of unseen words, potentially correct or incorrect, yet not on the above list.

The final step is the selection of a subset of rules with the highest scores. Most likely, correct attribute words are those pointed to by multiple rules. Intuitively, the number of various rules which extract a candidate word seems a plausible indicator that the word under consideration is really an attribute. Thus, one chooses words with the highest rules support.

3.2 Template Pattern Analysis

This section describes how a rule templates design affects recall. As it was described in section 3.1, rule templates are used to create new rule candidates. Results of the attribute extraction algorithm are compared on different sets of templates. The focus here is on maximizing recall because it is assumed that high precision is to be achieved in the second step, described in Section 5.

Templates were categorized using two criteria: length (from 3 to 5 elements) and whether a template contains PoS placeholders. Therefore, templates were divided into 6 categories: 4 lexeme placeholders, 3 lexeme placeholders, 2 lexeme placeholders, 3 lexeme placeholders and a PoS placeholder, 2 lexeme placeholders and a PoS placeholder, 1 lexeme placeholder and a PoS placeholder.

For each template category, templates have been generated with all possible locations of an argument (attribute candidate word). In other words, a template that expects four lexemes and then an attribute, then one that expects three lexemes followed by an attribute, and so on. To decrease the number of possible rule templates, the position of a PoS placeholder was fixed and set to either one word before the attribute placeholder or one word after it.

The rule templates have been evaluated mostly according to their recall defined by the percentage of correct opinion attribute words, obtained from the manually annotated corpus after its expansion as described in Section 4.

The first evaluation compared templates for each category separately. The second involved the most specific templates (longest without PoS) and extended the template set with less detailed ones. The third evaluation focused on using only templates that contain a PoS placeholder. The number of iterations was limited to 75, because initial experiments indicated that it is sufficient to obtain maximum possible recall. For each set of templates, precision has been measured – defined as the ratio of known correct attribute words to all extracted words. This was done only to evaluate how various sets of templates affect precision, nevertheless the main purpose of this evaluation was to select rule templates with the highest recall.

Additionally, for each set of templates, we recorded iteration number of the bootstrap algorithm, where the maximum recall was reached, as well as the number of rules in that iteration. It is worth noticing that because each iteration expands set of rules, recall may only rise in subsequent iterations. Results are shown in Table 2.

Table 2. Results for templates sensitivity analysis for reviews of perfumes

Templates	Iteration	# of rules	P	R	F
Individual templates					
(1) 4 x orth	0	9	0.213	0.262	0.235
(2) 3 x orth	43	113	0.181	0.423	0.253
(3) 2 x orth	74	287	0.122	0.624	0.204
(4) 3 x orth + pos	5	49	0.211	0.272	0.237
(5) 2 x orth + pos	28	209	0.197	0.380	0.260
(6) 1 x orth + pos	70	569	0.108	0.688	0.187
All categories					
(1) – (2)	16	137	0.209	0.290	0.243
(1) – (3)	33	262	0.184	0.464	0.264
(1) – (4)	32	261	0.182	0.471	0.263
(1) – (5)	12	105	0.191	0.452	0.268
(1) – (6)	70	569	0.125	0.642	0.209
Containing PoS placeholder					
(4) – (5)	26	208	0.200	0.376	0.261
(4) – (6)	59	481	0.118	0.649	0.200

Use of 4 lexeme placeholders in template reaches maximum recall in the first iteration, when only hand-written rules are considered. Surprisingly, using $1 \times$ **orth + pos** set of templates alone achieves the highest recall, which holds true in both domains. Therefore, this set of templates has been applied in further experiments.

4   Expansion through Lexical Resources

The method described previously has several shortcomings. It is designed to facilitate selection of attribute word candidates, but without identifying associated attributes. This is because extraction rules are generic and not related to any specific attribute. The example rule mentioned earlier in Section 3.1 may identify words such as price, quality, and whatever else may be an advantage of a product under review. Moreover, many extracted words are incorrect and additional steps are needed to remove them. Therefore, in this section another technique of attribute lexicon expansion has been proposed. Basing on words that are already in the annotated corpus with known, assigned attributes, it generates candidates for words denoting attributes using:

- Synonymy and hyponymy, as in the Polish Wordnet [6] – for each word annotated in the corpus as an attribute, it extracts its synonyms and all possible hyponyms of its hypernyms.
- Words with high MSR similarity as in [5] that exceed certain similarity threshold. The method involves construction of a matrix of co-occurrences, its transformation and subsequent calculations of similarity.

This expansion results in a significantly larger set of words that could denote attributes. For example, in the perfume domain a set of 83 source words generated as many as 881 word candidates. For each obtained word human annotators verified whether it describes an attribute and whether the attribute is identical as for the source word. From all candidates, 310 were verified as correct attribute descriptors and from those 310 words, 276 were present in the corpus.

The analysis revealed several sources of errors:

1. Spelling errors.
2. Errors in morphological disambiguation: incorrectly identified word base forms of originally inflected words.
3. Incorrect word-sense and incorrect hypernyms.
4. No word can replace the source word because of linguistic constraints such as collocations.
5. The hypernym is generally correct, but the meaning of the specific hyponym is too distant. This could be caused by hypernyms with wide meaning so that the method generates words which do not fit into the general context.
6. Candidate word does not fit into attribute's context. Example: from the source word *dress*, through *clothes*, the candidate word *sweter* will not be correct for such an attribute as "appropriateness for special occasion".

The last two items are in fact closely related and the last item is a special case of the former.

5   From Lexicon to Rules

This section describes a machine learning approach to automatically identify attribute words obtained as in Section 4. The purpose is to recognize correct attribute lexemes. The method does not directly distinguish word attributes, but by the definition, correct attribute lexemes denote the same attribute as the original, source word from the annotated corpus. Therefore, the expanded lexicon also carries attribute information.

Features were binary and indicated if a candidate lexeme has ever been extracted by a given extraction rule. Features were selected using the recursive feature elimination algorithm (RFE) with support vector machine (SVM) classifier. The RFE method selects features by recursively considering smaller and smaller sets of features and at each step prunes features with smallest weights (the least relevant). Figure 1 presents the total number of misclassified lexemes in 3-fold cross validation using the expanded set of perfume words and rules as features.

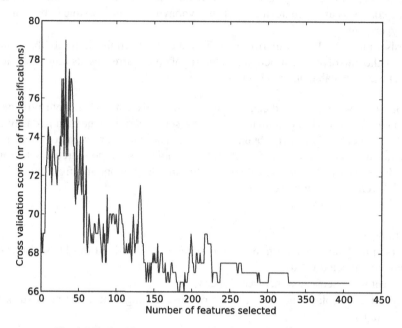

Fig. 1. Misclassifications and the number of features (RFM+SVM)

Classifiers trained on the optimal subset of extraction rules reached total average precision of 0.87 (3-fold cross validation). This result should be considered highly promising and also intuitive, because it confirms that extraction rules are a good indicator of whether a lexeme is a correct attribute denotation or not.

6   Attribute Recognition

The technique described in the last sections begins with WordNet expansion to filter the candidate words using extraction rules. The question that naturally arises at this point is whether it is possible to reverse the process: identify the candidate words using extraction rules, drop non-attribute words and then identify correct attributes for the remaining ones. One potential relative advantage of the reversed approach is vocabulary size: in the perfume domain, extraction rules identify 1,738 candidates for lexemes which denote attributes.

The problem of recognizing the correct attribute for a candidate word can be formulated as that of multiclass classification. Such classifiers have been trained on the feature space of 5,640 lexemes seen in contexts (concordances) of each of the manually annotated perfume attribute words. Context was defined as a window spanning two words left and right from a given attribute word, around each occurrence in the large unannotated corpus. The features were binary – word presence rather than frequency.

Table 3 presents average accuracies on 3-fold cross validation obtained using two classifiers: K-nearest neighbours (KNN) with K=5 and an SVM in one-vs-rest

Table 3. Average attribute classification accuracy; 3-fold cross validation

	KNN	SVM
all	0.142	0.176
correct	0.202	0.246

paradigm. The classifiers were tested in two settings: on all word candidates (including incorrect ones as one of the classes) and only on correct ones. Random baseline was at around 0.04. Both classifiers performed over the baseline, but the most likely explanation of otherwise low precision is that contexts are not informative enough to distinguish attributes.

Many opinion words, which typically precede attribute words in product reviews, are independent of the following attribute. For instance, common opinionated adjectives like *good*, *nice*, *poor* could describe almost any aspect of any product. Adjectives such as *comfortable*, which hint at specific attributes, are not common enough to facilitate an efficient classification. The same observation applies to extraction rules, which also appear in contexts of attribute words: they are not attribute specific.

7   Conclusions and Future Work

In this section we describe two methods of automated expansion of a corpus annotated with opinion attributes. The first approach is based on extraction rules and the second on lexicons and distributional similarity. None of the two methods can be successfully applied on its own due to low precision. Therefore two hybrid, two-step approaches to the problem have been proposed.

The first hybrid method, outlined in Section 5 demonstrates that using extraction rules to filter word candidates generated by lexical-based expansion produces high-quality classifiers for the problem of identifying opinion attribute words. The main advantage is a high precision and information about attributes. The minor disadvantage is the fact that the acquired lexicon is smaller by about a half than the list of potential word candidates for attributes obtained using extraction rules.

The reversed approach discussed in Section 6, where one begins with words obtained by extraction rules and then classifies them as denoting one of the attributes, is not satisfactory. Classifiers trained on corpus contexts of manually annotated words are better than the random baseline but the precision is low. Likely causes of this finding have been discussed.

An interesting challenge for the future work would be to merge both methods into one semi-supervised scenario. Probably the most attractive possibility is to train attribute classifiers, as in Section 6, on corpus contexts of words identified as correct by the method described in Section 5.

Acknowledgements. This research is supported by the POIG.01.01.02-14-013/09 project which is co-financed by the European Union under the European Regional Development Fund.

References

1. Brill, E.: A simple rule-based part of speech tagger. In: Proceedings of the Third Conference on Applied Natural Language Processing, pp. 152–155 (1992)
2. Hu, M., Liu, B.: Mining opinion features in customer reviews. In: Proceedings of the 19th National Conference on Artifical Intelligenc, AAAI 2004, pp. 755–760. AAAI Press (2004)
3. Kobayashi, N., Inui, K., Matsumoto, Y.: Extracting aspect-evaluation and aspect-of relations in opinion mining. In: Proceedings of EMNLP (2007).
4. Pang, B., Lee, L., Vaithyanathan, S.: Thumbs up? sentiment classification using machine learning techniques. In: Proceedings of EMNLP (2002)
5. Piasecki, M., Broda, B.: Semantic Similarity Measure of Polish Nouns Based on Linguistic Features. In: Abramowicz, W. (ed.) BIS 2007. LNCS, vol. 4439, pp. 381–390. Springer, Heidelberg (2007), http://dl.acm.org/citation.cfm?id=1759779.1759813
6. Piasecki, M., Szpakowicz, S., Broda, B.: A Wordnet from the Ground Up. Oficyna Wydawnicza Politechniki Wroclawskiej (2009)
7. Popescu, A.M., Etzioni, O.: Extracting product features and opinions from reviews. In: Proceedings of EMNLP (2005)
8. Przepiorkowski, A., Buczynski, A.: Spade: Shallow parsing and disambiguation engine. In: Proceedings of the 3rd Language and Technology Conference, Poznan (2007)
9. Qiu, G., Liu, B., Bu, J., Chen, C.: Expanding domain sentiment lexicon through double propagation. In: Proceedings of IJCAI (2009)
10. Turney, P., Littman, M.: Measuring praise and criticism: Inference of semantic orientation from association. ACM Transactions on Information Systems 21, 315–346 (2003)
11. Wang, B., Wang, H.: Bootstrapping both product features and opinion words from chinese customer reviews with cross-inducing. In: Proceedings of IJCNLP 2008 (2008)
12. Wawer, A., Sakwerda, K.: How Opinion Annotations and Ontologies Become Objective? In: Bouvry, P., Kłopotek, M.A., Leprévost, F., Marciniak, M., Mykowiecka, A., Rybiński, H. (eds.) SIIS 2011. LNCS, vol. 7053, pp. 391–400. Springer, Heidelberg (2012)
13. Zhuang, L., Feng, J., Xiao-yan, Z.: Movie review mining and summarization. In: Proceedings of CIKM (2006)

Taggers Gonna Tag: An Argument against Evaluating Disambiguation Capacities of Morphosyntactic Taggers*

Adam Radziszewski[1] and Szymon Acedański[2]

[1] Institute of Informatics, Wrocław University of Technology
[2] Institute of Computer Science, Polish Academy of Sciences

Abstract. Usually tagging of inflectional languages is performed in two stages: morphological analysis and morphosyntactic disambiguation. A number of papers have been published where the evaluation is limited to the second part, without asking the question of what a tagger is supposed to do. In this article we highlight this important question and discuss possible answers. We also argue that a fair evaluation requires assessment of the whole system, which is very rarely the case in the literature. Finally we show results of the full evaluation of three Polish morphosyntactic taggers. The discrepancy between our results and those published earlier is striking, showing that these issues do make a practical difference.

Keywords: morphosyntactic tagging, morphological analysis, tokenisation, evaluation.

1 Introduction

Part-of-speech (POS) tagging is a well-researched Natural Language Processing (NLP) task. Taggers assign POS tags to words and word-like units (*tokens*) in text. In languages with rich inflection the tags usually include significantly more information that just parts-of-speech, e.g. nouns may be specified for values of number, gender and case, adverbs may be specified for degree. In such a setting, the tags are often called morphosyntactic tags and the task is referred to as morphosyntactic tagging.

It has been noted that the tagging accuracy has significant impact on performance of other NLP tasks, such as parsing [1]. A reliable tagger evaluation procedure is therefore vital for unbiased selection of the best tagger for a particular application. What is more, using an inaccurate evaluation procedure may bring about long-standing consequences: as long as the generally used procedure neglects pratical aspects related to actual tagger mispredictions, those shortcomings are unlikely to be overcome. In this paper we show that the latter is often the case. We also offer a simple alternative that allows avoiding unjustified bias. Our discussions are grounded on Polish background, although the observations are also applicable at least to other inflectional languages. We also perform a new evaluation of two state-of-the art Polish taggers along the lines of the proposed methodology and present our results, which are strikingly different from those previously published.

* Work financed by Innovative Economy Programme, POIG.01.01.02-14-013/09.

P. Sojka et al. (Eds.): TSD 2012, LNCS 7499, pp. 81–87, 2012.

2 Common Practice

A common practice in tagging of inflectional languages is to decompose the process into two stages:

1. morphological analysis (dictionary look-up), resulting in sets of possible tags assigned to each token;
2. morphosyntactic disambiguation, that is ruling out of contextually inappropriate tags assigned during the previous stage.

This practice has become so common[1] that it has already started to influence the perception of what the task of a morphosyntactic tagger is. This consideration has important implications for tagger evaluation: depending on which stage of an evaluated tagger we take as a black box, we will assess different parts of the whole system and get different results. The problem is that the very question of what a tagger is supposed to do is almost never asked, while different answers are implicitly assumed, rendering comparisons of published results impossible. There are at least three possible answers, all of them at least occasionally assumed:

1. The tagger tags running (plain) text, that is a sequence of characters; assumed in [3].
2. The tagger tags a sequence of unlabelled tokens, possibly given sentence boundaries; assumed in [4,5,6].
3. The tagger's task is only that of morphosyntactic disambiguation; assumed in [7,8,9,10,2].

The second approach assumes that the tokenisation performed by the tagger is perfect (tokenisation is taken from the reference corpus). The third approach is most controversial, since it neglects both tokenisation errors and deficiencies of morphological analysis. In spite of that, this is actually the most popular approach, at least to evaluation of Polish taggers.

3 Proposed Methodology

We argue that taggers are best evaluated on plain text, while the two other approaches (as outlined in the previous section) are biased and should be avoided:

1. In a typical scenario, the user has access to plain text and is interested in obtaining reasonable tokenisation and accurate labelling of those tokens. No reference morphological analysis is normally available.
2. One of the symptoms is that separate figures for tagging known and unknown words are not reported in the Polish literature, which is otherwise a common practice. This is because in such a setting there are virtually no unknown words — the reference tag is always there to be chosen. This may lead to an absurd situation, when proper tagging of out-of-vocabulary words is the easiest task for the black box[2].

---

[1] E.g. [2] state that "given the nature of inflectional languages (…) it is necessary to employ morphological analysis before the tagging proper".

[2] Such a situation indeed occurs in [8,9,10] as the corpus employed for evaluation assigns exactly two possible tags for all the unknown words: the proper tag and a special *out-of-vocabulary* tag [11] — a sufficient winning strategy is to never choose the *out-of-vocabulary* tag.

3. Such an approach makes it impossible to assess the impact of different morphological analysers on the overall tagging performance. This is a serious consideration: multiple analysers are available, while the two-stage implementation of tagging facilitates integration with different analysers. At the same time, the choice of morphological analyser has an obvious impact on the final tagging accurracy. Hence, publishing results of only the disambiguation part silently ignores the influence of the choice and quality of the analyser.
4. Such evaluation procedures show only differences in disambiguation strategy, while the tagger's guessing capabilities are not assessed. This discourages the development of better strategies for handling unknown words.

It is also worth noting that a fair evaluation has already been performed for two Polish taggers [3], although not mentioned[3]. Note that the following publications involving evaluation of Polish taggers continue to use the 'old' disambiguation-based approach: [8,9,10]. A similar situation happened for tagging Czech: [12] states in a footnote on p. 3 that they had "been simply ignoring the unknown word problem altogether in the past". We hope that this paper will make these issues explicit and encourage fair tagger evaluation in the future.

3.1 Recommendations

We recommend the following methodology, which we also used for experiments reported in the next section:

1. Ten-fold cross-validation is employed, using a manually annotated corpus (gold standard). We split on the basis of paragraphs to account for taggers which may use context information that crosses sentence boundaries.
2. The main statistic calculated we call *accuracy lower bound*. It is the measure we advocate for general-purpose tagger evaluation, where all discrepancies in segmentation are penalised. We expect the segmentation to match the gold standard exactly to promote authors to create consistent segmenters instead of tweaking evaluation methods to match particular taggers. We calculate this statistic as a percentage of tokens in the gold standard which have a lexically matching segment and it is correctly tagged.
3. We also calculate an additional statistic designed to show the possible influence of segmentation errors on tagging quality. *Accuracy upper bound* is a hypothetical upper limit of tagger performance, treating all the tokens subjected to segmentation changes as correctly tagged. It is a percentage of gold standard segments, which either have a lexically corresponding segment with the correct tag, or have no lexically corresponding segment.

4 Experiments for Polish Taggers

It is worth noting that in Polish, tags are composed of a part-of-speech label and a number of labels for grammatical categories. For example, `subst:sg:nom:f` is

---

[3] It was confirmed by Danuta Karwańska (personal communication, 6 October 2011).

a feminine, singular noun (substantive) in the nominative case. Some attributes are defined as optional and during evaluation we expand them to multiple tags, following recommendation in [3]. As the gold standard we used the published 1 million token manually annotated subcorpus of the National Corpus of Polish, version 1.0 [13].

4.1 PANTERA

PANTERA [8] is a morphosyntactic tagger based on Brill's Algorithm [14] adapted for morphologically rich languages, targeted at Polish. The tagger automatically generates limited-context rules which are then applied in order to the text, pre-tagged using an unigram tagger. PANTERA is also a 2-tier tagger, which first disambiguates the part of speech, case and person, and then the rest of grammatical categories. In our experiments we train PANTERA with the threshold rule quality of 6, using morphological analyser Morfeusz[4] [15] with the TaKIPI guesser module [7] enabled. The authors report 92.68% accuracy on the NCP.

4.2 WMBT

WMBT [10] is a simple memory-based tagger that operates on as many tiers as there are attributes in the tagset. WMBT itself is actually a disambiguation engine — it should be run after performing tokenisation and morphological analysis. The authors recommend using Maca [16] software for that purpose. Thus, the tests presented here were performed this way, using a Maca configuration recommended for this scenario, namely `morfeusz-nkjp-official-guesser`. The authors report 92.98% accuracy on the NCP.

4.3 MBT

To make the comparison more insightful, we decided to also include a tagger which is not targeted specifically at Polish. MBT [5] is a generic memory-based tagger that may be used for various languages. MBT is trained with a corpus that must contain a sequence of tokens and their corresponding tags. Optionally, external features may also be included. A particularly interesting feature of MBT is that it creates two models: one for known words and one for unknown words (the latter one is obtained by analysing forms that were infrequent in the training data).

A drawback of MBT when applied to inflectional languages is that it treats feature values atomically, hence it cannot reason using the attribute values inferred from tags. This can be altered by introducing additional features directly to the input before running MBT. We decided to exploit this possibility by testing three variants:

1. A *naive* variant (*Features 0*): each input token is described by its wordform.
2. Simplified feature set (*Features 1*): each token is described by its wordform, but also possible part-of-speech labels and values of three gramamtical categories: number, gender and case taken from a window $(-3, \ldots, +2)$ surrounding the token.

---

[4] For all the experiments described in this paper we use a 64-bit version of Morfeusz SGJP, ver. 0.82 (code timestamp 22/02/2010, linguistic data from 15/04/2011).

3. Rich feature set (*Features 2*) is exactly the same feature set as used internally by WMBT, that is *Features 2* extended with some tests for morphosyntactic agreement [10], generated using the WCCL toolkit [17].

4.4 Evaluation Results

We performed two experiments:

1. Evaluation of disambiguation capabilities only (Acc_{dis}). The morphological data and tokenisation were taken directly from the reference corpus.
2. Evaluation according to our proposal: the testing material was turned to plain text and the taggers were to produce valid output. We report the values of accuracy lower and upper bound (Acc_{lower}, Acc_{upper}), but also, accuracy lower bound measured separately for words known and unknown to the morphological analyser (Acc_{lower}^{K} and Acc_{lower}^{U}, respectively).

Table 1. Accuracy measures obtained using the proposed methodology

Tagger	Acc_{dis}	Acc_{lower}	Acc_{upper}	Acc_{lower}^{K}	Acc_{lower}^{U}
PANTERA	92.95%	88.79%	89.09%	91.08%	14.70%
WMBT	93.00%	87.50%	87.82%	89.78%	13.57%
MBT: Features 0	79.31%	79.11%	79.44%	80.30%	40.49%
MBT: Features 1	88.03%	84.14%	84.46%	85.79%	30.74%
MBT: Features 2	87.12%	83.39%	83.72%	85.00%	31.36%

The results are presented in Table 1. The difference between accuracy lower and upper bounds is not very substantial. In the following discussion we focus on the lower bounds then.

The figures for disambiguation accuracy of PANTERA and WMBT are in line with previous publications. But what is striking here is the gap between these figures and the accuracy of the full tagging process (that is, accuracy lower bound). The gap apparently stems from the fact that the disambiguation accuracy neglects the errors made during morphological analysis and tokenisation. Interestingly, the errors neglected make up almost a half of the total tagging error. What is more, PANTERA turns out to outperform WMBT, while the opposite seems to hold if observing disambiguation accuracy alone.

As one could expect, both measures do converge for MBT. This is because MBT is by design not given access to external morphological analyser, hence it is also unable to take advantage of the reference morphological analysis available in the reference corpus. This is another argument against evaluation based on the reference morphological analysis: the taggers that assume two-stage operation are able to peek at the reference annotation and win undeserved points.

While the overall tagging accuracy of MBT is not impressive, a couple of interesting observations could be made. First, the introduction of additional features brought substantial improvement. This confirms our presumption that for best results in tagging inflectional languages, values of grammatical categories should be represented as separate features. Even then MBT is still bound to output whole tags during one run; this is a likely explanation for its performance being much lower than that of WMBT, a tiered tagger. On the other hand, the figures recorded for tagging unknown words are much higher in the case of MBT than those achieved by the two state-of-the-art taggers made specifically for Polish: the best result for MBT is 40.5%, while the figures reported for the Polish taggers are lower than 15%. This is probably due to a separate module for tagging unknown words in MBT. The lower accuracy of other taggers could also be attributed to the prevalence of evaluation based on disambiguation capabilities in Polish NLP community: a problem unnoticed is likely to remain unsolved.

5 Conclusion

In this paper we pointed out and discussed a number of choices which need to be made consciously to ensure comparable evaluation of taggers. We also re-evaluated the taggers described in published papers, now in a comparable way. This brought significantly different results. Therefore we proposed a general-purpose tagger evaluation methodology. By providing a simple enough method, we promote publishing comparable and useful tagging performance values.

References

1. Hajič, J., Krbec, P., Květoň, P., Oliva, K., Petkevič, V.: Serial combination of rules and statistics: A case study in Czech tagging. In: Proceedings of the 39th Annual Meeting on Association for Computational Linguistics, pp. 268–275. Association for Computational Linguistics (2001)
2. Hajič, J., Vidová-Hladká, B.: Tagging inflective languages: Prediction of morphological categories for a rich, structured tagset. In: Proceedings of the COLING - ACL Conference, ACL, pp. 483–490 (1998)
3. Karwańska, D., Przepiórkowski, A.: On the evaluation of two Polish taggers. [18]
4. Schmid, H., Laws, F.: Estimation of conditional probabilities with decision trees and an application to fine-grained POS tagging. In: Proceedings of COLING 2008, vol. 1, pp. 777–784. Association for Computational Linguistics (2008)
5. Daelemans, W., Zavrel, J., Van den Bosch, A., van der Sloot, K.: MBT: Memory-Based Tagger, version 3.2. Technical Report 10-04, ILK (2010)
6. Acedański, S., Przepiárkowski, A.: Towards the adequate evaluation of morphosyntactic taggers. In: Proceedings of the 23rd International Conference on Computational Linguistics (COLING 2010), Poster Session, Beijing, pp. 1–8 (2010)
7. Piasecki, M.: Polish tagger TaKIPI: Rule based construction and optimisation. Task Quarterly 11, 151–167 (2007)
8. Acedański, S.: A Morphosyntactic Brill Tagger for Inflectional Languages. In: Loftsson, H., Rögnvaldsson, E., Helgadóttir, S. (eds.) IceTAL 2010. LNCS, vol. 6233, pp. 3–14. Springer, Heidelberg (2010)
9. Śniatowski, T., Piasecki, M.: Combining Polish Morphosyntactic Taggers. In: Bouvry, P., Kłopotek, M.A., Leprévost, F., Marciniak, M., Mykowiecka, A., Rybiński, H. (eds.) SIIS 2011. LNCS, vol. 7053, pp. 359–369. Springer, Heidelberg (2012)
10. Radziszewski, A., Śniatowski, T.: A memory-based tagger for Polish. In: Proceedings of the 5th Language & Technology Conference, Poznań (2011)

11. Przepiórkowski, A., Murzynowski, G.: Manual annotation of the National Corpus of Polish with Anotatornia. [18]
12. Hajič, J.: Morphological tagging: Data vs. dictionaries. In: Proceedings of the 6th Applied Natural Language Processing and the 1st NAACL Conference, pp. 94–101 (2000)
13. Przepiórkowski, A., Górski, R.L., Łaziński, M., Pęzik, P.: Recent developments in the National Corpus of Polish. In: Proceedings of the Seventh International Conference on Language Resources and Evaluation, LREC 2010, ELRA, Valletta, Malta (2010)
14. Brill, E.: A simple rule-based part of speech tagger. In: Proceedings of the Third Conference on Applied Natural Language Processing, pp. 152–155. Association for Computational Linguistics, Morristown (1992)
15. Woliński, M.: Morfeusz — a Practical Tool for the Morphological Analysis of Polish. In: Intelligent Information Processing and Web Mining, pp. 511–520 (2006)
16. Radziszewski, A., Śniatowski, T.: Maca — a configurable tool to integrate Polish morphological data. In: Proceedings of the Second International Workshop on Free/Open-Source Rule-Based Machine Translation (2011)
17. Radziszewski, A., Wardyński, A., Śniatowski, T.: WCCL: A morpho-syntactic feature toolkit. In: Proceedings of the Balto-Slavonic Natural Language Processing Workshop. Springer (2011)
18. Goźdź-Roszkowski, S. (ed.): The proceedings of Practical Applications in Language and Computers PALC 2009. Frankfurt am Main, Peter Lang (2010)

An Ambiguity Aware Treebank Search Tool

Marcin Woliński and Andrzej Zaborowski

Institute of Computer Science, Polish Academy of Sciences, Warsaw

Abstract. We present a search tool for constituency treebanks with some interesting new features. The tool has been designed for a treebank containing several alternative trees for any given sentence, with one tree marked as the correct one. The tool allows to compare the selected tree with other candidates.

The query language is modelled after TIGER Search, but we extend the use of the negation operator to be able to use a class of universally quantified conditions in queries.

The tool is built on top of an SQL engine, whose indexing facilities provide for efficient searches.

Keywords: treebanks query/search tools, constituency trees, syntactic ambiguity.

1 Introduction

The primary goal of the work being reported is to facilitate the access to a treebank we are building. This treebank — Składnica [10] currently comprises constituency trees for about 8,200 Polish sentences. The treebank is under active development, so we are interested in a search tool that would not only be useful for "end users" but also which would assist in finding errors and inconsistencies in the currently available trees.

The present treebank was built semi-automatically. Sets of candidate parse trees were generated automatically using a parser and then disambiguated by human annotators. In such a setup it is desirable to be able to confront the whole set of solutions proposed by the parser with the single one selected by annotators. Unfortunately, available treebank search tools [4,6,7] assume a single tree for each sentence.

We decided to build our own tool that would be aware of ambiguous syntactic structures present in Składnica. The tool is web based, which provides for an easy access to the treebank from a standard web-browser and without installing specialised programs. We also decided to build our query language on the TIGER Search syntax [4], which looks familiar for the users of Poliqarp [2] commonly used for searching in lower levels of annotation in our corpus.

2 Quick Overview of the TIGER Query Language

A query in the TIGER language [3, Chapter III] describes a set of matching nodes. The construct [] matches any node in a tree. It can be limited by specifying values of features combined with logical operators: negation !, conjunction &, and alternative |. For example, the following query matches nodes of category NP with feature fa equal val1 or feature fb different from val2:

P. Sojka et al. (Eds.): TSD 2012, LNCS 7499, pp. 88–94, 2012.
© Springer-Verlag Berlin Heidelberg 2012

```
[cat="NP" & (fa = "val1" | fb != "val2")]
```

The feature `cat` is commonly used to denote category, i.e., the name of the nonterminal unit. In general, however, feature names and values depend on a given treebank.

Alternatives can also be used on the right hand side of an equation, which provides for compact queries like:

```
[cat=("NP"|"PP")]
```

Nodes in a query can be linked using several node relations. The following query

```
[cat="S"] > [cat="NP"]
```

uses the direct dominance relation >, so the first node is to be the parent of the second.

Other node relations include indirect dominance >* (transitive closure of >), labelled dominance >1bl (if edges in the treebank are labelled), direct precedence ., indirect precedence .*, sibling (common parent) relation $, and some more variations.

The language uses unification variables denoted with #name. A variable can bind the right hand side of an equation, so in the following query we require the two adjacent nodes to be of the same category irrespective which of the two allowed categories it is:

```
[cat=#c:("NP"|"PP")] . [cat=#c]
```

A variable can also bind a single node, so the following query requires the node #n to be the parent both to the NP and VP node:

```
#n:[cat="S"] & #n > [cat="NP"] & #n > [cat="VP"]
```

The third type of variables are "variables for feature constraints", which bind all features of a particular node, e.g., the query:

```
[#f:(pos="prep")] .* [#f]
```

matches trees where a node with exactly the same features appears twice. Assuming `pos="prep"` selects a terminal being a preposition, both matching nodes must be prepositions with the same orthographic form and the same tag. We don't find this behaviour very useful, since it can be used only when all features match. Therefore this type of binding was skipped in our implementation. We prefer to state the features that should unify explicitly:

```
[pos="prep" & orth=#o] .* [pos="prep" & orth=#o]
```

The last element we are going to mention are node predicates. For example the following requires node #n to have at least 2 and at most 4 children:

```
#n:[cat="S"] & arity(#n,2,4)
```

3   Searching in Parse Forests

Składnica uses shared parse forests [1] as a compact representation of all trees generated by the parser for a given sentence. In the shared parse forest each subtree occurring in multiple complete parse trees is represented only once. The forest consists of nodes, each having the name of the nonterminal unit of the grammar and its morpho-syntactic features as attributes. Each combination of a particular span of text, a nonterminal unit, and its attributes corresponds to a separate node.

Moreover, each node carries a set of lists of its possible immediate constituents. Each list represents children of the given node in one of possible trees.

A subset of nodes is marked as comprising the tree selected by annotators.

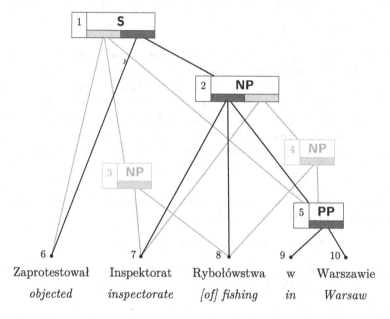

Fig. 1. A simplified shared parse forest for the sentence 'The Inspectorate of Fishing in Warsaw has objected.' The tree selected by the annotators is shown in a darker colour.

For example, Fig. 1 shows the forest for a sentence in which a prepositional phrase 'in Warsaw' can be attached in three places. Node 1, representing the sentence, has two possible lists of children. The first consists of nodes 6 (the verb), 3 (nominal phrase 'Inspectorate of Fishing'), and 5 (prepositional phrase 'in Warsaw'). The second list contains nodes 6 and 2 (nominal phrase 'Inspectorate of Fishing in Warsaw'). This node is part of the selected tree together with its second list of children. Observe that node 2 represents all possible structures of the phrase in question, in particular it represents two attachments of the prepositional phrase: either the Inspectorate is in Warsaw or the Inspectorate deals with fishing in Warsaw. The annotators have selected the first attachment.

The compactness of the representation comes from the fact that each of nodes 5, 7, and 8 is shared by three possible parents and node 6 is shared by two parents. It is worth reminding that a shared parse forest can represent an exponential number of trees in polynomial amount of memory.

When searching in shared forests we need to be able to address particular nodes and check whether they have been selected by the annotators. For that purpose we have equipped the nodes with an extra-grammatical boolean feature sel.

Although searching in all possible trees is important to us as treebank developers, we expect users to mainly search in the disambiguated trees. For that reason we reserve the TIGER notation [ ] for a node selected by the annotators and denote any node in the forest by [[ ]] (which means [...] is a shorthand for [[sel=true & ...]]).

For the forest in Fig. 1 the query [cat="NP"] would return only node 2. On the other hand the query [[cat="NP"]] has three answers: nodes 2, 3, and 4.

When searching among all nodes of a forest it may happen that the matching nodes do not all belong to a single tree. For example the answer to the query

```
#n: [[cat="NP"]] > [[cat="PP"]] & #n > [[cat="NP"]]
```

could consist of nodes 2, 5, and 4. This may be exactly what we are interested in, for example when searching for structures which were disambiguated in some way by the annotators while the parser proposed simultaneously a particular different structure. Sometimes, however, we need to request some nodes with sel=false to come from a single consistent tree. This can be achieved with the use of the predicate same_tree(...), which is true if all nodes enumerated as its arguments appear in one tree in the forest.

Obviously, these considerations do not apply if all nodes in the query are specified with [], since the nodes with sel=true are guaranteed to form a single complete tree.

The following example illustrates a real query we have used to check to what extent the annotators were consistent in their decisions. In Składnica some complements to verbs are labelled as filling the adverbial position while being realised by prepositional-nominal phrases with various prepositions (cf. [8]). The following query matches complements labelled as prepositional by the annotators, while the parser proposed also the adverbial interpretation for the same span of text:

```
[cat="fw" & tfw=/prepnp.*/ & from=#f & to=#t]
  & [[cat="fw" & tfw="advp" & from=#f & to=#t]]
```

In this query we show actual feature values as defined in Składnica. Nonterminal fw denotes a complement. The attribute tfw denotes structural type of the complement. We use the regular expression /prepnp.*/ to allow for prepositional phrases with any preposition. The attributes from= and to= denote positions in the text where the given phrase begins and ends. Their repetition forces the two phrases to span the same text fragment.

4   Searching for the Nonexistent

TIGER Search has a well known limitation with regard to quantification. All node specifications in queries are implicitly existentially quantified. There exists no feature of the query language that would allow to form questions of the form "all nodes such that..." or "there exists no node such that...".

In particular, the authors claim in the manual [3] that the use of the universal quantifier causes computational overhead and so they do not introduce it for the sake of computational simplicity and tractability. However, the TIGER language has subsequently been successfully extended with universal quantification [5].

In the scope of our project it is crucial to be able not only to search for elements present in the trees but also to check for absent elements. For the present version of the search tool we have decided just to extend the possibility of introducing negation in queries. Namely the negation operator ! when applied to nodes has the meaning "there exists no node". For example the query

```
#n: [cat="zdanie"] > [cat="ff"] & (! #n > [tfw="subj"])
```

finds all finite sentences without a subject. Literally it searches for all nodes of category 'zdanie' (sentence) which have a finite verbal phrase ff as a direct descendant and among such direct descendants there is no phrase marked as the subject.

Note that #n in this query is defined outside the scope of the ! operator, so the "there exists no" applies only to the node [tfw="subj"].

We can combine this mechanism with facilities described in the previous section and search, e.g., for sentences which do not have a subject according to the annotators, although a subject is present in some interpretations generated by the parser:

```
#n: [cat="zdanie"] > [cat="ff"] &
(! #n > [tfw="subj"]) & #n > [[tfw="subj"]]
```

5   Searching Trees with SQL

In contrast to the original TIGER Search our tool is built on top of an SQL engine, whose indexing facilities provide for efficient searches.

The query entered by the user is translated into an SQL query using joins in case of multiple node specifications (each node is represented by one row in the node table). Feature specifications are mapped to where clauses in SQL. The most fundamental node attributes like the category and the span of the node are represented with dedicated columns in the node table. Other features, which can differ from category to category, are conveniently stored in a single field of type hstore. In PostgreSQL databases this type can be used to store arbitrary sets of key-value pairs. What is important, PostgreSQL server provides for indexing values of this type, which results in efficient queries.

The user interacts with the search tool via a standard web browser (currently we support Mozilla Firefox and WebKit based browsers including Google Chrome and Safari). The trees are displayed using a JavaScript library developed in the project and used also in our treebank development tool [9].

Current implementation covers a large subset of the TIGER query language including conditions on node features expressed with equality and regular expressions, direct and indirect dominance and precedence (including labelled dominance), node and feature variables and their unification, boolean expressions over node and feature specifications.

```
SELECT filename, tree.id, n0.id, n1.id
FROM node AS n0, node AS n1, tree
WHERE n0.treeid = n1.treeid
  AND (n0.sel AND n0.category = 'zdanie')
  AND (n1.sel AND n1.f::hstore -> 'tfw'='subst')
  AND n1.id = ANY(n0.children)
  AND n0.treeid = tree.id
```

Fig. 2. The result of converting the query [cat="zdanie"] > [tfw="subst"] to SQL.
The feature cat is represented with a designated column, the feature tfw is stored in a hstore
container.

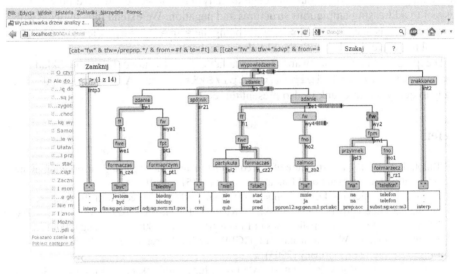

Fig. 3. Results of a query displayed in the web interface

We haven't implemented template definitions nor type definitions. In fact, our tool
does not consider features typed. All values are treated as simple strings. A system of
query templates would help in stating complex queries, so we will probably consider
it in the future. Note, however, that neither of these elements increases the expressive
power of the language.

Our implementation strategy is admittedly simple: we push the problem of effective
finding of matching nodes to the specialised software namely to the database engine. It
seems that PostgreSQL performs very well in this context. This is probably due to its
sophisticated query optimiser.

We have run a few tests comparing performance of the original implementation
of TIGER Search and our tool. The treebank consists of about 8,200 sentences with
varying numbers of trees. The total number of nodes in all parse forests is over
2,500,000. In the experiment, our tool explored the complete set of nodes, while
TIGER Search used only about 300,000 nodes selected by annotators. For queries using
3–5 node specifications and conditions including (indirect) dominance and precedence

our tool gives answers in 5–15 seconds. TIGER Search gives analogous answers in 3–8 seconds. Answer times of our tool are longer, but the data set used is significantly larger. We do not observe any influence of the "not exists" operator on execution time, as queries involving this operator compute in similar times.

6   Conclusions

The tool presented in this paper has been created using a seemingly simple technique, and yet it has proved stable and effective enough to be used as the main search interface of our treebank. Current implementation covers all important features of the TIGER query language. Moreover, our language already has a larger expressive power with respect to addressing ambiguous structures and negation. We extend the language with new operators as need arises, one example being the same_tree predicate. We expect the tool to evolve together with the Składnica treebank.

The program is free software. It is available for download from the Składnica webpage http://zil.ipipan.waw.pl/Składnica. A running installation ready to query our treebank is also present there.

References

1. Billot, S., Lang, B.: The structure of shared forests in ambiguous parsing. In: Meeting of the Association for Computational Linguistics, pp. 143–151 (1989)
2. Janus, D., Przepiórkowski, A.: Poliqarp 1.0: Some technical aspects of a linguistic search engine for large corpora. In: Waliński, J., Kredens, K., Goźdź-Roszkowski, S. (eds.) The Proceedings of Practical Applications of Linguistic Corpora 2005, Peter Lang (2006)
3. König, E., Lezius, W., Voormann, H.: TIGERSearch 2.1 user's manual. Tech. rep., IMS, Universität Stuttgart, Germany (2003)
4. Lezius, W.: TIGERSearch — ein Suchwerkzeug für Baumbanken. In: Busemann, S. (ed.) Proceedings der 6. Konferenz zur Verarbeitung natürlicher Sprache (KONVENS 2002), Saarbrücken (2002)
5. Marek, T., Lundborg, J., Volk, M.: Extending the TIGER query language with universal quantification. In: KONVENS 2008, 9. Konferenz zur Verarbeitung natürlicher Sprache. pp. 5–17 (2008)
6. Maryns, H., Kepser, S.: MonaSearch — a tool for querying linguistic treebanks. In: Van Eynde, F., Frank, A., De Smedt, K. (eds.) Treebanks and Linguistic Theories, pp. 29–40 (2009)
7. Rohde, D.L.T.: Tgrep2 user manual (2005), http://tedlab.mit.edu/~dr/Tgrep2/
8. Świdziński, M., Woliński, M.: Towards a Bank of Constituent Parse Trees for Polish. In: Sojka, P., Horák, A., Kopeček, I., Pala, K. (eds.) TSD 2010. LNCS, vol. 6231, pp. 197–204. Springer, Heidelberg (2010)
9. Woliński, M.: Dendrarium — an open source tool for treebank building. In: Kłopotek, M.A., Marciniak, M., Mykowiecka, A., Penczek, W., Wierzchoń, S.T. (eds.) Intelligent Information Systems, Siedlce, Poland, pp. 193–204 (2010)
10. Woliński, M., Głowińska, K., Świdziński, M.: A preliminary version of Składnica — a treebank of Polish. In: Vetulani, Z. (ed.) Proceedings of the 5th Language & Technology Conference, Poznań, pp. 299–303 (2011)

A New Annotation Tool for Aligned Bilingual Corpora

Georgios Petasis[1] and Mara Tsoumari[2]

[1] Software and Knowledge Engineering Laboratory,
Institute of Informatics and Telecommunications,
National Centre for Scientific Research (N.C.S.R.) "Demokritos",
GR-153 10, P.O. Box 60228, Aghia Paraskevi, Athens, Greece
petasis@iit.demokritos.gr
[2] School of English,
Faculty of Philosophy,
Aristotle University of Thessaloniki,
54 124, P.O. Box 58, Thessaloniki, Greece
mtsoum2@gmail.com, mara@optimum-services.com

Abstract. This paper presents a new annotation tool for aligned bilingual corpora, which allows the annotation of a wide range of information, ranging from information about words (such as part-of-speech tags or named-entities) to quite complex annotation schemas involving links between aligned segments, such as co-reference or translation equivalence between aligned segments in the two languages. The annotation tool is implemented as a component of the Ellogon language engineering platform, exploiting its extensive annotation engine, its cross-platform abilities and its linguistic processing components, if such a need arises. The new annotation tool is distributed with an open source license (LGPL), as part of the Ellogon language engineering platform.

Keywords: Annotation tools, collaborative annotation, adaptable annotation schemas.

1 Introduction

The huge amount of the available information on the Web has created the need of effective information extraction systems that are able to produce meta-data that satisfy user's information needs. The development of such systems, in the majority of cases, depends on the availability of an appropriately annotated corpus in order to learn extraction models. The production of such corpora can be significantly facilitated by annotation tools. While a considerable number of annotation tools can be found in the literature [1,2], they are mostly targeting monolingual documents, lacking any support for aligned bilingual corpora. However, as parallel corpora are often used as linguistic resources in translation, some tools have been developed to facilitate research in translation and multilingual corpus analysis, including the following ones:

GATE [3] is a language engineering platform, which offers support for aligning corpora. Text alignment can be achieved at document, section, paragraph, sentence and word level. "Compound documents" are created by combining existing documents and by aligning various text segments between documents. It is unclear, however, whether the user can annotate information across the participating documents. Another

P. Sojka et al. (Eds.): TSD 2012, LNCS 7499, pp. 95–104, 2012.

popular tool is ParaConc [4] whose main characteristics are an alignment function, concordance search, search for specific words and their possible translations, corpus frequency and collocate frequency. However ParaConc offers no annotating facilities. A fairly recent tool is InterText[1], which is an editor for aligned parallel texts. It has been developed for the project InterCorp, in order to edit and manage alignments of multiple parallel language versions of texts at the level of sentences. However, similar to ParaConc and perhaps GATE, it does not support annotation of documents, only alignment between segments. Another parallel corpus alignment toolbox is Uplug [5], which is a collection of tools for linguistic corpus processing, word alignment and term extraction from parallel corpora. All these tools offer support for aligning segments in bilingual documents, but do not offer other annotation facilities, beyond alignment.

On the other hand, there exist tools that allow any type of linguistic annotation, but provide no special support for bilingual documents. Callisto is a multilingual, multi-platform tool providing a set of "annotation services" [6]. Its standard components are textual annotation view and a configurable table display. Some of the tasks performed are automatic content extraction entity and relation detection, characterization and co-reference, temporal phrase normalization, named entity tagging, event and temporal expression tagging etc. The IAMTC Project combines already existing facilities and newly developed ones and has developed an annotation tool for text manipulation. The Project involves the creation of multilingual parallel corpora with semantic annotation to be used in natural language applications [7]. Annotation includes dependency parsing, associating semantic concepts with lexical units, and assigning theta roles. MULTEXT [8] is a project involving the development of tools on the basis of "software re-usability", and multilingual parallel corpora. It combines NLP and speech, and examines the possibilities for such a combination by harmonizing tools and methods from both areas. The annotation is performed with a segmenter, a morphological analyser, a part of speech disambiguator, an aligner, a prosody tagger, and post-editing tools. Thus, the annotated data provide information about syntax, morphology, prosody and the alignment of parallel texts. Finally, Propbank is a project where a corpus is annotated with semantic roles for verb predicates [9]. Annotation is performed with the help of Jubilee by simultaneously presenting syntactic and semantic information. The process is facilitated by Cornerstone, a user-friendly XML editor, customized to allow frame authors to create and edit frameset files.

The tool presented in this paper is an attempt to narrow the gap between these two types of annotation tools. Allowing the alignment of bilingual, parallel documents at sentence level, the tool allows the annotation of any type of linguistic annotation on document pairs simultaneously, by presenting to the end user a synchronised view of both documents with aligned sentences next to each other.

1.1 Motivation and Features of the New Annotation Tool

The annotation tools presented in the previous section have been developed to cover specific and diverse needs, with each tool exhibiting different characteristics and capabilities, making difficult to find a single tool that concentrates the majority of features. Each of the tools presented in the previous section has significant features and

---

[1] http://wanthalf.saga.cz/intertext

capabilities but none of them is an all-in-one tool. Of course, the desired features of an annotation tool are closely related to the requirements of a specific annotation task, constituting the construction of a generic set of desired features quite difficult. The scope of the research that motivated the creation of this tool combines mainly translation, parallel corpora (original-source texts and translation-target texts), semantics, pragmatics, and discourse. Being developed within the framework of wider research in the analysis of parallel texts from a translation point of view, the new annotation tool concentrates characteristics that cannot be found altogether in a single tool. In particular:

a) it imports aligned texts already processed in an efficient alignment tool, allowing a corpus builder to use an external aligner of one's own choice;

b) each pair (i.e. translation unit) of aligned texts is clearly separated from the other pairs. At the same time, they keep their place in the text manifesting coherence relations and flow of text meaning and discourse in each language;

c) the tool allows the location of possible translation equivalents within a specific context, always keeping the source text item and its target text equivalent in a close, binary relationship. This unfolds the variety of equivalents an item can have that may be either context dependent or context independent, and also reveals translation procedures and strategies;

d) it allows the creation of a comparable profile at sentence level of the source text entry and its target text equivalent by entering accompanying information based on their context (distribution of the entries, collocations, etc.). The source text entry and its equivalent are seen comprehensively as a whole during the annotation process;

e) it displays all the attribute sections and fields for each source text entry and its target text equivalent, providing easy access with one click;

f) it allows the examination of the target text in its own right to identify cases, if any, of linguistic items that are present without being a translation equivalent of a source text entry;

g) it allows the correlation of discourse topics with the frequency of the linguistic items and their translation equivalents in the two languages, and also with their profile. Additionally, it allows both intra-linguistically and inter-linguistically analysis;

h) it provides detailed statistics [10], which allows the grouping of information of the entries for specialized analysis of results; and the tables of statistics are exportable to widely commercial formats such as Microsoft Excel.

2   Reusing Ellogon's Annotation Engine

The Ellogon language engineering platform [11] offers an extensive annotation engine, allowing the construction of a wide range of annotation tools for both plain text and HTML documents. This annotation engine provides a wide range of features, including: *a*) cross-platform graphical user interface, *b*) use of standard formats, *c*) customized annotation schemata, *d*) automatic annotation, and *e*) comparison facilities to identify mismatches among independent annotations of the same document, or calculate inter-annotation agreement. Despite the fact that these features are not unique among the available annotation tools (i.e. most of these features are also supported by tools offered by Callisto, Wordfreak[2], GATE [3], MMAX2 [12], Knowtator [13], and AeroSWARM [14]), reusing an annotation engine allows for rapid and robust development of a new annotation tool, through the re-use of tested components. The annotation engine of the Ellogon language engineering platform is configurable through XML files that define annotation schemas. The tool reads the annotation schema from an XML file, and presents to the annotator a suitable GUI for annotating text segments. The XML annotation schema language provides a variety of types that can be annotated. While most available types, along with their visual representation in the GUI, can be found in [15] and [2], the types that relate to the grouping of several segments and other information in a single annotation to facilitate annotation of co-reference or other types of relations, are shown in the following list: *a*) A **span** or **segment** (fig. 1-A), represented by a textual label (specified by the annotation schema), the text of the segment, its offsets, a button to fill in the segment from the current selection, and a button to clear the segment. *b*) A **description** (fig. 1-B), which the user can fill in with arbitrary text. Represented by a textual label and an entry widget, where arbitrary text can be entered. *c*) A **category** (fig. 1-C), selectable from a set of predefined categories by the annotation schema. Represented by a textual label and a combo-box widget, allowing the user to select a category from a set of predefined categories. *d*) A **boolean value** (fig. 1-D), denoting the presence or absence of an attribute. Represented by a textual label and a check-box widget.

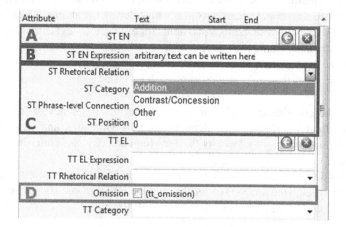

Fig. 1. Example of some annotation types, allowed in annotation schemas

2.1 The Aligned Bilingual Corpora Annotation Tool

After describing the annotation engine, the next step towards the creation of an annotation tool for aligned corpora is: *a*) to define a format for representing aligned bilingual corpora within the Ellogon platform, *b*) to extend the visualisation components to display correctly an aligned document, and *c*) to extend the annotation engine to operate

---

[2] http://wordfreak.sourceforge.net/

on the extended visualisation components. A screenshot of the annotation tool can be seen in fig. 2. Aligned documents are displayed one next to the other, aligned at sentence level. The annotation schema used by the configuration shown in fig. 2 relates to the analysis of parallel texts from a translation point of view. Segments having the same colour between texts in the two languages in fig. 2 denote that they belong to the same annotation (group of features). Clicking on any of them enables the editing/modification of the relevant annotation, through the inputs on the right side of the tool.

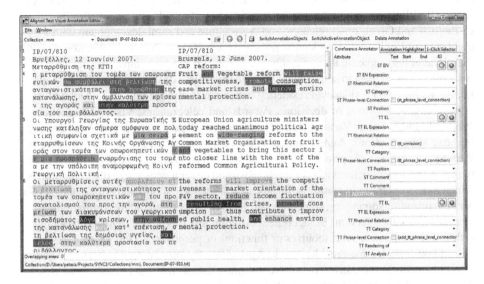

Fig. 2. The New Annotation Tool for Aligned Bilingual Corpora

3   Usage Example: Connectives in Parallel Corpora

The main motivation for the development of this new annotation tool for aligned bilingual corpora has been the need to analyse the role of *connectives* in parallel corpora from both a linguistic and a translation perspective. To start with, the tool allows the researcher to build what we decided to call *parallel comparable corpora*. Parallel comparable corpora can be bilingual or multilingual collections (with more than one target language collections as translation of a source collection); be of the same genre or not; include source texts and their target texts; be grouped according to some textual resemblance – topics, text types etc. – although they cover different general topics; and be compared at various levels. For instance, a collection about the general topic agriculture is divided into subcollections with distinct text types and/or discourses and/or subtopics etc. The subcollections within the same collection can be contrasted with each other at the level of source text/language-intralinguistically or target text/language-intralinguistically. For instance if and how a linguistic form or a feature of context changes profile through different subcollections of the same collection. Also the subcollections can be contrasted at the level of translation-interlinguistically. For instance, if and how the translation of linguistic forms changes through the different subcollections. If more than one collection of a different general topic is included in the corpus, but with, at least to some extent, comparable subcollections, then similar subcollections of different collections can be compared both intralinguistically (compare only source texts among comparable sub-collections or compare only target texts among comparable subcollections) and interlin-guistically (compare how source texts are translated among comparable subcollections). If more than one genre is included then the comparison starts at the level of genres.

The tool is tested on a parallel comparable corpus, a special corpus of press releases of the European Commission, drawn from the electronic text library of all EU press re-leases (RAPID[3]), with two thematic categories-collections, Presidency with fifty pairs of English and Greek press releases, and Agriculture and rural development with fifty eight pairs of English and Greek press releases. Each of these thematic collections is fur-ther divided into separate thematic subcollections that can be comparable between the two collections, at least to some extent, despite the collections being of a different topic, i.e. Agriculture vs Presidency. Presidency collection is divided into six subcollections: Agreements or approval of decisions, Awards and celebrations, Visits and meetings, Proposals and policies, Various, Reports and surveys. Agriculture collection is divided into five subcollections: Agreements or approval of decisions, Proposals and policies, Reports and surveys, Approval of EU countries' plans, Warning and legal action. The names of the subcollections are comprehensive and cover a wide range of similar topics which have something in common, for instance somebody adopts something, the Com-mission proposes something, somebody is related to a meeting/conference/dialogue etc. The thematic subcollections in fact reflect different text types. Different text types may reflect different dominant discourses or functions or purposes. So far, the tool has pro-cessed both collections, Presidency and Agriculture by annotating the entries in ques-tion (adversative/contrastive/concessive discourse connectives and "and" connective). Analysis at this stage is focused on Presidency collection and its subcollections from a translation perspective. Findings show that the contrastive/concessive group of con-nectives keep their role in the Greek text with overwhelming persistence compared to "and". Only "yet" seems to differentiate itself from the rest of its group. Another finding is that the omission of discourse connectives in translation is not necessarily related to the large or small number of these connectives in the source texts. Having higher avail-ability of a specific type of connective in a collection or subcollection is not necessarily related to higher omission rates of that type of connective. Findings from the addition of discourse connectives in the target texts show that contrast/concession is more persistent in the Greek press releases. Relating these findings with findings from a manual contex-tual analysis on a sample corpus of Agriculture collection the conclusion strengthens Sidiropoulou's [16,17] findings about the Greek reader viewing the world from a con-trastive perspective. As to the question whether different text types affect the translation of these two groups of discourse connectives, One-Way ANOVA of "and" and its typi-cal Greek translation equivalent "kai" – this pair was tested due to adequate amount of data – showed that there is a systematic influence of the text type on the frequency of the translation of "and" as και. What determined the outcome of the One-Way Analysis of Variance is a particular subcollection Agreements or approval of decisions which has distinct features from the rest of the subcollections of that collection.

---

[3] http://europa.eu/rapid/searchAction.do

3.1   The Annotation Schema

Annotation is conducted by associating attributes to the linguistic items. The devised annotation schema involves parallel documents in the English (EN) and Greek (EL) languages, and contains three sections of attribute fields: a) The first section is general and the most frequently used. In the first section, the focus is on the *source text entry (ST EN)* and the *target text entry (TT EL)*, where the latter is considered the translation equivalent of the former in that context. The ST EN fields that follow relate to accompanying information of that token based on the particular context. The same goes for the TT EL fields of the TT EL entry. b) The next section, *target text addition (TT Addition)*, involves the addition of the items in question in the target texts, where there exists no connective or discourse marker in the source text. c) The third section, *Context*, involves the context of the texts. The original concept of that section is an attempt to map the differences emerging from the translation process between the two texts.

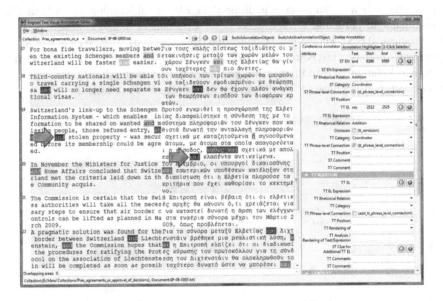

Fig. 3. First Section of Attributes – General Section

First Section of Attributes – General Section. On the right side of the tool (fig. 3), the three sections of attributes defined by the annotation schema are presented. In the first section, the focus is on the *source text entry (ST EN)* and the *target text entry (TT EL)* where the latter is considered the translation equivalent of the former in that context. The ST EN and TT EL fields that follow relate to accompanying information of those annotated segments (tokens) based on the particular context. The fields "ST EN/TT EL Expression" accommodate cases where the ST EN/TT EL entries are part of an expression or form a collocation with the surrounding words. Each entry is also annotated for its rhetorical relation and category in that particular context. The values in these fields have been selected in relation to the connectives and discourse markers of interest. For cases where the discourse marker or connective has another function except for the linking one, the value "0" in the "ST/TT Rhetorical Relation" fields and the value "Other" in the "ST/TT Category" fields have been provided. The check-box of the "ST/TT Phrase-level Connection" provides information about how often the ST and TT markers/connectives in question link predicates or non-predicates (noun phrases, adjectival phrases etc.) in their language respectively. Difference in the type of connection between the ST EN entry and its TT EL equivalent entry manifests different syntactic structures, and perhaps participant roles in the source and target languages. This in turn may reflect translation strategies, i.e. shifts, transpositions, modulations etc.

The "ST/TT Position" fields relate to the distribution of the tokens. When the ST EN entry and its TT EL equivalent are seen in parallel and a change in position is noted, then different thematic and rhematic structures and focus may be reflected in the two languages. Omission of an ST EN entry in the target text is also checked ("ST/TT Omission"). An example can be a token of the additive conjunction "and" (fig. 3): This entry involves the token "and", highlighted with blue colour in the translation unit 19. Based on its attributes, it is a conjunction of addition ("ST Rhetorical Relation" = "Addition"), a coordinator in particular ("ST Category" = "Coordinator"), and connects phrases (non-predicates) ("ST Phrase-level Connection" box checked). The token acting as its equivalent in the target text is και (kae) "and", which is also a conjunction of addition ("TT Rhetorical Relation" = "Addition"), a coordinator ("TT Category" = "Coordinator"), and connects non-predicates ("TT Phrase-level Connection" box checked).

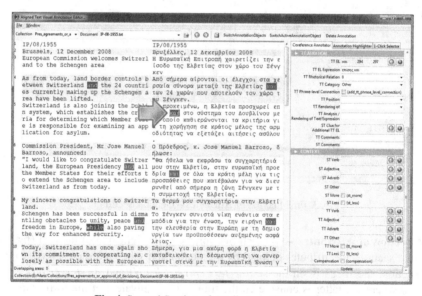

Fig. 4. Second Section of Attributes – TT Addition

Second Section of Attributes – TT Addition. The next section, TT Addition, involves the addition of the items in question in the target texts (fig. 4). There are similar fields as in the first section of attributes. Since in this section of attributes the starting point is the target text, a couple of extra fields of attributes have been added: the "TT Rendering of" field, which attempts to classify the category of the word/phrase in the ST, if any, that motivated the addition of the discourse marker/connective in the TT; the "TT Analysis/Rendering of Text/Expression" field where the ST word/phrase is entered. Finally, there is one more field, "ST Clue for Additional TT EL". Practically, the last two fields have a similar function providing distinct ways to enter information. An example can be found in translation unit 5 of fig. 4: according to the annotation, the TT EL entry και (kae) 'and' that was added in translation unit 5, is not used as a conjunction ("TT Rhetorical Relation" = "0") and performs a different function from coordination in the structure of the sentence ("TT Category" = "Other").

Third Section of Attributes – Context. The third section involves the context of the texts. The motivation of this section is an attempt to map the differences that emerge from the translation process. These differences can be: a) grammatical, i.e. a change in the tense of a verb form; b) semantic, i.e. the choice of a slightly/a lot different semantically TT EL equivalent; c) pragmatic, i.e. the choice of a completely different expression in the TT to render ST meaning; or d) lexical, i.e. the addition or omission of a word/phrase in one of the two texts. The following pairs of fields have been designed: *a)* "ST Verb" (or verb phrase) – "TT Verb" (or verb phrase), *b)* "ST Adjective" (or adjectival phrase) – "TT Adjective" (or adjectival phrase), *c)* "ST Adverb" (or adverbial phrase) – "TT Adverb" (or adverbial phrase), *d)* "ST Other" – "TT Other". The last pair involves differences that do not fall under any of the other pairs. Then the differences recorded can be evaluated compared with each other based on which of the two options – "ST option" or "TT option" – is more or less strong in meaning, more or less informative, more or less appellative, and more or less affective. Some of these differences between the two texts are mandatory driven by language restrictions, for instance, or optional driven by cultural preferences, register, politics etc. Either way, these differences create an effect to the reader. So under the ST fields there are two check-boxes "ST More", "ST Less" and under the TT fields, "TT More" and "TT Less", respectively. For each difference entered the relevant box is checked; ST entry evaluated as "ST More" or "ST Less" and TT equivalent evaluated as "TT More" or "TT Less". Finally, there is a check-box in this section, "Compensation", modelled after the translation strategy. Compensation refers to making up for the loss of meaning or effect in some part of the sentence, in another part of that sentence, or in a contiguous sentence [18]. This box is checked when the difference in context in the two texts is due to the translation strategy of compensation.

4   Conclusions and Future Work

In this paper we present a new annotation tool, which is able to annotate a wide range of information on aligned parallel corpora and parallel comparable corpora, implemented as a plug-in of the Ellogon language engineering platform, and distributed as open source. The annotation tool has been used in the context of analysing parallel/parallel comparable corpora from a translation point of view, concentrating mainly on the role of connectives. The annotation tool presented in this paper was proved extremely user friendly and robust in its operation, for the case studied. It offers to the researcher the advantage of selecting an external alignment tool for aligning a corpus of parallel texts according to his/her needs. In addition, it is very flexible when studying linguistic items and translation issues, and at the same time allows analysis pertaining to discourse, semiotics, ideology, culture etc. [10]. Thus the researcher works with a tool that is easily adjustable to his/her varied needs in relation with the annotation of bilingual data. As future work, we aim to integrate the aligned corpora annotation with the (semi) automatic annotation facilities offered by the Ellogon platform.

Acknowledgments. The authors would like to acknowledge partial support from the FP7–ICT SYNC3 project.

References

1. Uren, V., Cimiano, P., Iria, J., Handschuh, S., Vargas-Vera, M., Motta, E., Ciravegna, F.: Semantic annotation for knowledge management: Requirements and a survey of the state of the art. Web Semant. 4, 14–28 (2006)
2. Fragkou, P., Petasis, G., Theodorakos, A., Karkaletsis, V., Spyropoulos, C.: Boemie ontology-based text annotation tool. In: Proceedings of LREC 2008. ELRA (2008)
3. Cunningham, H., Maynard, D., Bontcheva, K., Tablan, V., Aswani, N., Roberts, I., Gorrell, G., Funk, A., Roberts, A., Damljanovic, D., Heitz, T., Greenwood, M.A., Saggion, H., Petrak, J., Li, Y., Peters, W.: Text Processing with GATE (2011)
4. Barlow, M.: ParaConc: concordance software for multilingual parallel corpora. In: Proceedings of LREC 2002 (2002)
5. Tiedemann, J.: ISA & ICA - two web interfaces for interactive alignment of bitexts. In: Proceedings of LREC 2006 (2006)
6. Day, D., McHenry, C., Kozierok, R., Riek, L.: Callisto: A configurable annotation workbench. In: Proceedings of LREC 2004 (2004)
7. Farwell, D., Helmreich, S., Dorr, B., Green, R., Reeder, F., Miller, K., Levin, L., Mitamura, T., Hovy, E., Rambow, O., Habash, N., Siddharthan, A.: Interlingual Annotation of Multilingual Text Corpora and FrameNet, Berlin (2008)
8. Ide, N., Véronis, J.: Multext: Multilingual text tools and corpora. In: Proceedings of COLING 1994, vol. 1, pp. 588–592. ACL (1994)
9. Choi, J.D., Bonial, C., Palmer, M.: Multilingual propbank annotation tools: Cornerstone and jubilee. In: Proceedings of the NAACL HLT 2010 Demo Session, pp. 13–16. ACL (2010)
10. Tsoumari, M., Petasis, G.: Coreference Annotator - A new annotation tool for aligned bilingual corpora. In: Proceedings of AEPC 2, RANLP 2011 (2011)
11. Petasis, G., Karkaletsis, V., Paliouras, G., Androutsopoulos, I., Spyropoulos, C.D.: Ellogon: A New Text Engineering Platform. In: Proceedings of LREC 2002 (2002)
12. Müller, C., Strube, M.: Multi-level annotation of linguistic data with MMAX2. In: Braun, S., Kohn, K., Mukherjee, J. (eds.) Corpus Technology and Language Pedagogy: New Resources, New Tools, New Methods, Peter Lang, Frankfurt a.M., Germany, pp. 197–214 (2006)
13. Ogren, P.V.: Knowtator: A protégé plug-in for annotated corpus construction. In: Moore, R.C., Bilmes, J.A., Chu-Carroll, J., Sanderson, M. (eds.) HLT-NAACL. The Association for Computational Linguistics (2006)
14. Corcho, O.: Ontology based document annotation: trends and open research problems. Int. J. Metadata Semant. Ontologies 1, 47–57 (2006)
15. Petasis, G.: The SYNC3 Collaborative Annotation Tool. In: Proceedings of LREC 2012 (2012)
16. Sidiropoulou, M.: Contrast in english and greek newspaper reporting: A translation perspective. In: Proceedings of 8th International Symposium on English & Greek: Description and/or Comparison of the Two Languages, Thessaloniki, Greece, School of English, Aristotle University (1994)
17. Sidiropoulou, M.: Linguistic Identities through Translation. Approaches to Translation Studies, vol. 23. Rodopi B.V., Amsterdam (2004)
18. Newmark, P.: A Textbook of Translation. Prentice-Hall International, New York (1988)

Optimizing Sentence Boundary Detection for Croatian

Frane Šarić, Jan Šnajder, and Bojana Dalbelo Bašić

University of Zagreb, Faculty of Electrical Engineering and Computing
Unska 3, 10000 Zagreb, Croatia
{frane.saric,jan.snajder,bojana.dalbelo}@fer.hr

Abstract. A number of natural language processing tasks depend on segmenting text into sentences. Tools that perform sentence boundary detection achieve excellent performance for some languages. We have tried to train a few publicly available language independent tools to perform sentence boundary detection for Croatian. The initial results show that off-the-shelf methods used for English do not work particularly well for Croatian. After performing error analysis, we propose additional features that help in resolving some of the most common boundary detection errors. We use unsupervised methods on a large Croatian corpus to collect likely sentence starters, abbreviations, and honorifics. In addition to some commonly used features, we use these lists of words as features for classifier that is trained on a smaller corpus with manually annotated sentences. The method we propose advances the state-of-the art accuracy for Croatian sentence boundary detection on news corpora to 99.5%.

1 Introduction

Sentence boundary detection (SBD) is an essential building block for many natural language processing systems. Syntax parsing, text summarization, chunking, etc. all depend on correct segmentation. At first glance, the problem is deceptively simple, but naïve solutions quickly prove to be insufficient.

Sentences usually end with a period, but so do abbreviations. A very simple rule based sentence boundary segmentation might use a rule that marks a period as sentence end only if it is not preceded by an abbreviation, but the following examples illustrate that periods can be ambiguous even if abbreviations are detected:

1. a) *...which could be considered by the **U.S.** House of Representatives and Senate as early as next week.*
   b) *...and the United Kingdom all appearing just below the **U.S.** </S> Canada is the...*
2. a) *Nakon teške prometne nesreće **1937.** Tesla je dane provodio...*
   b) *...važenja Zakona o stambenim zadrugama do 31. **XII. 2000.** </S> Za privremenoga upravitelja...*
3. a) *Na skupu je govorio i **dr.** Vilim Herman...*
   b) *...i dalmatinski specijaliteti – sir, pršut, vino i **dr.** </S> Do nedjelje je sajam razgledalo...*

P. Sojka et al. (Eds.): TSD 2012, LNCS 7499, pp. 105–111, 2012.

Although the sentences in examples *1.b*, *2.b*, and *3.b* end with abbreviations, examples *1.a*, *2.a*, and *3.a* demonstrate that the same abbreviations can sometimes be in the middle of the sentence. We could try to add some exceptions to common cases and add more rules, but then the resulting method would be difficult to maintain and adapt to new corpora. Disambiguating periods is even more difficult in Croatian because all ordinal numbers, dates and years end with a period.

Error rates of many SBD systems are relatively low (typical values are less than 2–3%) and depending on the text corpus simple baseline systems can achieve error rates of 5–10%. The reason why it is worth investing more effort in reducing these error rates is that a lot of other stages of a typical natural language processing pipeline depend on accurate sentence boundary detection. It is arguably easier to reduce the error rate of sentence segmentation than, e.g., the error of a syntax parser.

Because the problem is well studied, there are many available tools that use various approaches to solve it. Many existing methods can easily be adapted to multiple languages with minimal effort. In this paper we evaluate some of available methods on texts in Croatian language. We propose some adjustments to existing methods that prove to be useful for Croatian and possibly for other Slavic languages. Our newly designed sentence boundary detection is both more accurate and faster than all other evaluated methods.

2   Related Work

There are roughly three kinds of SBD systems described in the literature: rule-based, supervised, and unsupervised methods. This classification is not crisp because most methods use at least some heuristic rules.

2.1   Rule-based Methods

Rule-based methods are rarely described in the literature, but they are often used in widely available tools. Stanford parser, a part of Stanford CoreNLP tools,[1] provides a heuristic sentence splitter that uses a combination of regular expressions and a list of tokens that cannot start sentences to determine sentence endings. LingPipe library[2] also provides a heuristic sentence boundary detection model that additionally uses a list of impossible penultimate words (usually a list of abbreviations or acronyms).

Boras [1] describes a rule-based method for text segmentation in Croatian that segments text into sentences and sentence clauses. The method is difficult to reproduce because it also uses rule-based morphological analyser and comprehensive lists of abbreviations, names, etc. The result is reported only for the task of sentence clause boundaries detection, i.e., sentence boundary detection was not evaluated separately.

2.2   Supervised Methods

Riley [2] described a system for disambiguating sentence endings that end with a period using a decision tree classifier. The system first calculates probability that the word

---

[1] http://nlp.stanford.edu/software/corenlp.shtml
[2] http://alias-i.com/lingpipe

before a period occurs at the end of the sentence and that the word after a period occurs at the beginning of the sentence. Other features used are lengths of words adjacent to the period, their case class (capitalized, lowercase, all capitals, and numbers), and the abbreviation class of the word before the period. The abbreviations list was manually constructed.

Riley obtained 0.2% error rate on Brown corpus using word probabilities computed from 25 million word AP corpus. Because that corpus was not sentence segmented, the probabilities were calculated only using beginning and end of paragraph marks. It should be noted that the system was evaluated by disambiguating only sentence boundaries after a period (which is the general case for Brown corpus).

OpenNLP[3] is a popular open source library that also provides a tool for sentence segmentation. Just like Riley's method, it also uses supervised learning, but instead of a decision tree, it uses maximum entropy classifier and trains it using generalized iterative scaling. For each token that is a candidate for sentence ending (e.g., for tokens ., ?, !) the system uses the local context around it to determine if it ends the sentence. The default context extractor uses the token containing candidate punctuation and two adjacent tokens as features, along with the information about their case (whether the first letter is capital) and whether they belong to a list of abbreviations. The user of the library provides a list of abbreviations.

MxTerminator [3] is very similar to OpenNLP, but it includes an additional pass in which it induces a list of abbreviations. All words followed by a period that do not end a sentence are added to the list of abbreviations. All supervised methods mentioned so far use only primitive lexical features. Palmer and Hearst [4] show that additional information about tokens can be beneficial. They used part of speech (POS) tagged text to build a POS dictionary and prior probability distribution of POS tags. The best results (1.5% error rate on the WSJ corpus) were achieved using a C4.5 classifier and POS features of three tokens before and three tokens after the period. Abbreviations in the training corpus were manually annotated, i.e., there was a separate POS class used for abbreviations.

2.3 Unsupervised Methods

Mikheev [5] combined the problems of sentence boundary disambiguation and disambiguation of capitalized words in positions where capitalization is expected. A larger unlabeled corpus of texts is used to extract four word lists: (1) common words, (2) common words that are frequent sentence starters, (3) frequent proper names, and (4) abbreviations.

Kiss and Strunk [6] argue that a large number of ambiguities in SBD can be easily solved if a word before the period is known to be an abbreviation. The paper describes an unsupervised system called Punkt that uses statistics obtained from a larger corpus to determine a list of abbreviations and frequent sentence starters. On Brown corpus Punkt achieved 1.02% error rate. Punkt is implemented as a part of NLTK, an NLP library for Python [7].

---

[3] http://incubator.apache.org/opennlp

3   Our Approach

Our primary goal is to reduce the error rate by all means, even if that requires some additional work (e.g., annotating the sentence boundaries). The unsupervised approach can start with low error rates, but in order to reduce them even further, we need to use supervised methods.

3.1   Features

Most supervised approaches use a maximum entropy classifier with binary features. Every token ending with punctuation is classified as either sentence ending or a part of the sentence. Features are extracted from a few words (tokens) surrounding the ambiguous punctuation mark. We have examined the existing tools in order to see what features are commonly used:

- the token itself,
- token is in all caps (capital letters),
- token contains digits,
- token is a single letter,
- token ends with a period,
- token length,
- token is an abbreviation.

The context we have used is "$A\ B\ .\ C$", where A and B are two tokens preceding the punctuation mark (usually a period) and C follows it. Human experts can classify the vast majority of sentence endings in Croatian by considering such a context, which justifies the small size of the context. On the other hand if we remove token A the context is too small even for humans. Although a larger context helps humans resolve some ambiguous cases it reduced the performance of our system probably due to small amount of training data. For each ambiguous punctuation mark we examine its context and extract binary features. The classifier we have used is logistic regression[4], trained using LIBLINEAR [8]. During the classifier training, a weight is assigned to each binary feature and the punctuation is marked as sentence ending if the sum of weights for a particular context is greater than zero.

In Table 1 we list rule-based features used for Croatian SBD. The features are sorted according to the magnitude of the weight assigned by the classifier after the training was complete, so the most important rule based features are listed first. Note that tokens consisting of all digits with length 3 or 4 are usually years. In addition to these features we have also added binary lexical features, i.e., for each word w_i in the dictionary we also add features $A = w_i$, $B = w_i$, and $C = w_i$. Although the number of features is rather large, most of them have zero weights.

Lexical features improve classifier performance, but they are not sufficient because the training set we use is rather small. No matter how large is the supervised training set, the number of different abbreviations it covers is usually too small to generalize

---

[4] We have also experimented with SVM classifiers using nonlinear kernels, but the classification accuracy did not improve.

Table 1. Binary rule-based features used, extracted from context "*A B . C*". The features are listed from most important to least important, based on the magnitude of weights assigned to each feature.

Feature	Indicates sentence boundary
capitalized(C)	yes
length(B) = 1 and capital(B) and capitalized(C)	no
all_digits(C) and $3 \leq$ length(C) ≤ 4	no
all_digits(C)	yes
all_digits(B)	no
all_digits(B) and $3 \leq$ length(B) ≤ 4	yes
capitalized(A)	no

well. Therefore we apply unsupervised methods similar to the ones described in [5] and [6] on a very large unannotated corpus in order to extract some useful word lists. Unlike in [6] we can afford to have lower precision and/or recall of a particular feature because the supervised classifier will automatically assign lower weights to less useful features. For each word class W we add features $A \in W$, $B \in W$ and $C \in W$.

3.2 Unsupervised Word Lists Extraction

By examining some very ambiguous sentence endings from the training set we have concluded that the following word lists might improve the classification accuracy:

1. honorifics (e.g. `prof.`, `mr.`),
2. sentence starters (`Navodeći`, `Sa`),
3. abbreviations (`tzv.`, `tj.`, but also `1997.` with initial lowercase L, which is a result of an OCR error).

In order to extract honorifics we first analyze a large text corpus and find all capitalized words that never occur in lowercase form, most of which are proper nouns. Now honorific extraction is easy – tokens optionally ending with a period that are almost always followed by a proper noun and almost never by other kind of words are honorifics. An example of honorific we do not extract is abbreviation `dr.` (*doktor*) because it is also commonly used in "i dr." (and so on). Kiss and Strunk [6] treat "almost always" and "almost never" more formally. Instead, `dr.` is extracted as a common abbreviation.

Sentence starters are capitalized words that almost always following a period and are almost always followed by a lowercase word (i.e., they do not occur in the middle of the sentence). Abbreviations that are not honorifics are collected by finding all lowercase tokens that almost always end with a period. We have also manually added a list of 20 roman numerals as a separate word class. They proved to be valuable features (see Table 2) because sometimes they end with a period (when they are ordinal numbers) and sometimes they are part of royalty names and they might indicate a sentence boundary.

Table 2. Each token in context "*A B . C*" can belong to a specific word class. The features are listed from most important to least important, based on the magnitude of weights assigned to each feature.

Feature	Indicates sentence boundary
roman_numerals(B)	no
abbreviation(B)	no
sentence_starter(A)	no
is_honorific(B)	no
roman_numerals(C)	no
sentence_starter(C)	yes

4   Results

The method was evaluated on a small set of manually annotated sentences from the Vjesnik newspaper. Honorifics, probable sentence starters and some abbreviations were extracted from a larger part (100 million word) of Vjesnik corpus. A sentence can contain multiple embedded sentences inside quotation marks or inside parentheses. Because all evaluated methods use local context to determine sentence boundaries, we have decided to annotate endings of embedded sentences as endings of normal sentences for the purpose of this evaluation. Superfluous sentence endings can easily be removed using a set of simple rules, for example by tracking open and closed parentheses. We have implemented these rules, but we did not use them for this evaluation because not all tools we compare implement such rules.

Table 3 shows the performance of the method described in this paper. We have tried supplying abbreviations dictionary to OpenNLP, but the results were worse than without it. Perhaps the classifier accuracy could be improved further by using POS features, but probably the easiest way to improve its performance is to increase the training set size.

Table 3. Evaluation using 5-fold (5,600 train, 1,400 test sentences) cross-validation on a set of 7,000 Croatian sentences (156,000 tokens) from the Vjesnik newspaper

System	Error(%)	FP	FN
Punkt (NLTK)	2.5	17	19
OpenNLP	1.5	7	12
MxTerminator	3.2	37	10
Our method	0.5	3	4

Although our focus was not on speed, the gain in accuracy did not slow down the system. The second fastest tool was OpenNLP, which could process approximately 100,000 sentences per second on our computer configuration, and our method can process well over 500,000 sentences per second[5].

---

[5] Unicode-aware UNIX word count utility (wc -w) runs only about 10% faster on the same corpus.

Table 4. Some of the errors the classifier made. Native Croatian speaker can see that these contexts really are too short to resolve false negatives. For example *Bila* is a verb that starts a new sentence, but it could also be a name of a doctor.

Context	Error kind
Člankom 139. Zakona	FP
i dr. Bila	FN
I. kotaciju. (	FN
vrijednosnice d.o.o. Samborski	FN

5 Conclusion

We have tested several publicly available methods for sentence segmentation and applied them to Croatian. The method we described uses unsupervised step to acquire features that are used to train a supervised method. To our best knowledge, the method we propose is the best performing method for Croatian language. The system described in this paper is published as a C++ library with Python bindings under a (liberal) BSD license at http://takelab.fer.hr/sbd. As a future work we will focus on expanding the support for other European languages.

Acknowledgments. We thank the anonymous reviewers for their useful comments. This work has been supported by the Ministry of Science, Education and Sports, Republic of Croatia under the Grant 036-1300646-1986.

References

1. Boras, D.: Teorija i pravila segmentacije teksta na hrvatskom jeziku. Ph.D. thesis, Faculty of Humanities and Social Sciences, Univ. of Zagreb, Zagreb (1998)
2. Riley, M.: Tree-based modelling for speech synthesis. In: The ESCA Workshop on Speech Synthesis (1991)
3. Reynar, J., Ratnaparkhi, A.: A maximum entropy approach to identifying sentence boundaries. In: Proceedings of the Fifth Conference on Applied Natural Language Processing, pp. 16–19. Association for Computational Linguistics (1997)
4. Palmer, D., Hearst, M.: Adaptive multilingual sentence boundary disambiguation. Computational Linguistics 23, 241–267 (1997)
5. Mikheev, A.: Periods, capitalized words, etc. Computational Linguistics 28, 289–318 (2002)
6. Kiss, T., Strunk, J.: Unsupervised multilingual sentence boundary detection. Computational Linguistics 32, 485–525 (2006)
7. Bird, S.: NLTK: The natural language toolkit. In: Proceedings of the COLING/ACL on Interactive presentation sessions, pp. 69–72. Association for Computational Linguistics (2006)
8. Fan, R.E., Chang, K.W., Hsieh, C.J., Wang, X.R., Lin, C.J.: LIBLINEAR: A library for large linear classification. Journal of Machine Learning Research 9, 1871–1874 (2008)

Mining the Web for Idiomatic Expressions Using Metalinguistic Markers

Filip Graliński

Faculty of Mathematics and Computer Science
Adam Mickiewicz University
ul. Umultowska 87, 60–687 Poznań, Poland
`filipg@amu.edu.pl`

Abstract. In this paper, methods for identification and delimitation of idiomatic expressions in large Web corpora are presented. The proposed methods are based on the observation that idiomatic expressions are sometimes accompanied by metalinguistic expressions, e.g. the word "proverbial", the expression "as they say" or quotation marks. Even though the frequency of such idiom-related metalinguistic markers is not very high, it is possible to identify new idiomatic expressions with a sufficiently large corpus (only type identification of idiomatic expressions is discussed here, not the token identification). In this paper, we propose to combine infrequent but reliable idiom-related markers (such as the word "proverbial") with frequent but unreliable markers (such as quotation marks). The former could be used for the identification of idiom candidates, the latter – for their delimitation. The experiments for the estimation of recall upper bound of the proposed methods are also presented in this paper. Even though the paper is concerned with identification and delimitations of Polish idiomatic expressions, the approaches proposed here should also be feasible for other languages with sufficiently large web corpora, English in particular.

Keywords: idiomatic expressions, Web mining.

1 Introduction

Identification of idiomatic expressions (idioms) of a given language is of importance from both theoretical and practical perspectives. First, idiomatic expressions can provide insight into the history and culture of their users [1]. Second, idioms, due to their semantic (and sometimes even syntactic) idiosyncrasy, present a challenge in many NLP applications, especially in machine translation – on the one hand, idiomatic expressions are abundant in informal and semi-formal Internet texts, such as blog entries, message board threads, Facebook posts, on the other hand, idioms are not that frequent in typical parallel corpora[1]. Therefore it would make sense to try to enlist idiomatic expressions and annotate them manually (e.g. in the context of machine translation – translate them manually and create a bilingual lexicon).

---

[1] Parallel corpora derived from film subtitles do contain substantial number of informal idiomatic expressions, the quality of their translations is, however, very poor. Idioms are very often mistranslated or rendered literally (!) by incompetent translators.

P. Sojka et al. (Eds.): TSD 2012, LNCS 7499, pp. 112–118, 2012.

We focus here on *type-based identification*, i.e. the aim is to collect as many idiom types as possible rather than to decide whether a given expression is used idiomatically or literally in a specific context (*token-based classification*). In order to find idiomatic expressions in large Web corpora, idiom-related metalinguistic markers can be used. *Idiom-related metalinguistic markers* are expressions used by speakers of a given language to metatextually mark idiomatic expressions, e.g. words and expressions such as *proverbial, as it is said, as they say* [2]. Obviously, occurrences of idioms marked with idiom-related metalinguistic markers are rather infrequent. Although this makes them rather useless in token-based classification context, such markers can be used in type-based identification provided that a corpus is large enough to comprise a significant number of idiom-related markers.

This paper is concerned mainly with Polish idiomatic expressions (usually referred to as *frazeologizmy* or *związki frazeologiczne* in Polish linguistic literature).

2 Related Work

There is a rich research literature on the automatic identification of multi-word expressions (MWEs), non-compositional expressions, collocations etc. (see e.g. [3]), idiomatic expressions represent, however, only a subset of all MWEs. [4] was specifically concerned with (both type- and token-based) identification of English idiomatic expressions. The method described in [4] was based on distribution of syntactic patterns of an idiom, i.e. variants of an idiom with respect to passivisation, determiners used, pluralisation. Such method is not fully applicable for all languages, for instance in Polish no determiners are used and passivisation is less frequent than in English. Furthermore, [4] focused only on some types of idioms (of the form V+NP).

Using metalinguistic markers for the identification of idiomatic expressions was proposed in [2]. The following Polish markers were considered:

- the adjective *przysłowiowy* (*proverbial*),
- the adverb *przysłowiowo* (*proverbially*),
- quotation marks,
- the phrase *tak zwany* (*so-called*) along with its abbreviated form *tzw.*,
- the phrase *jak to mówią* (*as they say*),
- the phrase *jak to się mówi* (*as it is said*),
- the adverb *dosłownie* (*literally*).

The list is probably exhaustive, with the exception of the markers *frazeologizm, związek frazeologiczny, idiom* (all denoting "idiomatic expression" in Polish) – such markers are used in texts showing more linguistic awareness and are more likely to accompany well-known (and less interesting) idioms, already recorded in dictionaries.

The adjective *przysłowiowy* (along with its adverbial form *przysłowiowo*) appears to be the most distinctive idiom-related marker, other markers were not used in [2]. It should be noted, however, that quotations marks are much more frequent than *przysłowiowy*. The problem is that they are used for many other purposes. Results reported in this paper concern combining *przysłowiowy* (reliable but relatively rare) with quotation marks (frequent but unreliable). In this way, we extend work reported in [2]. Also, some estimations of recall upper bound of methods proposed in [2] and in this paper will be given here.

Exploiting expressions indicative of idiomatic usages (such as *proverbially*, *metaphorically speaking*) and quotation marks was mentioned in [5]. That paper was, however, focusing on token-based classification and, as expected, such indicative terms were not a useful feature in the task reported there.

3   Corpus

The same corpus as in [2] were used (732M words). The corpus contained 29,737 sentences with *przysłowiowy/-o* marker. This sentences were pre-processed in the following way:

- a given sentence was tokenised (PSI-Toolkit[2] was used for the segmentation),
- *przysłowiowy/-o* token was found,
- at most 8 tokens to the left and to the right of the *przysłowiowy/-o* token were taken.

This way, a set of snippets containing the word *przysłowiowy/-o* was obtained.

3.1   Test Set

A random sample of 2000 snippets was selected. Snippets belonging to the development and test set used in [2] were excluded. (The test set of [2] was used as the development set now.)

In each snippet an idiomatic expression referred to by *przysłowiowy/-o* was manually tagged and delimited (if present at all). The number of all idioms marked in the corpus sample was 1008 (50.4%) – the *przysłowiowy/-o* marker is used for other (not idiom-related) purposes by Polish speakers as well.

The following criteria for idiomacity have been used [6]:

- semantic non-compositionality,
- lexicosyntactic fixedness,
- prevalence (in order to reject one-time, ad hoc expressions, phrases not listed in traditional paper dictionaries were checked with Web search engines – 5 independent occurrences were required for an expression to be classified as an idiom),
- "graphicalness" (*obrazowość* in Polish, a criterion traditionally put forth in Polish linguistic literature).

A substantial number (188) of idioms marked in the test set were discontinuous, i.e. split into two or more parts separated by some intervening material (mostly pronouns). For instance, in the snippet (the *przysłowiowy/-o* token is replaced with #):

Początkowo prowadzę uczniów wręcz za # rączkę
= *At the beginning I lead the students just by the # hand*

the idiom *prowadzić za rączkę* (= *lead by the hand*) was split into two parts (not counting discontinuity introduced by the idiom-related marker). Such discontinuous idioms are not likely to be properly delimited by the procedure described in Section 4, they were not, however, removed from the test set.

---

[2] http://psi-toolkit.wmi.amu.edu.pl

4   Identification and Delimitation Procedure

The identification and delimitation proposed in [2] was used as a baseline and starting point. Now we are going to describe briefly this procedure.

The procedure starts with locating the *przysłowiowy/-o* token in a given snippet. Then we attach tokens on the right until some condition is true. This way, the right boundary of an idiom is determined. Next, we switch to the left side and, in a similar manner, attach tokens until some condition is true. The alternative of the following conditions yielded the best results for the right side (as reported in [2]):

- the phrase gathered so far combined with the current token occurs only once in the corpus,
- the association strength between the phrase gathered so far and the current token (measured using pointwise mutual information, PMI [7], on the whole corpus) is below a threshold,
- the current token is a punctuation mark (excluding quotation marks),
- the current token is the second quotation mark.

The conditions for the left side are the same as those for the right side with one exception – the first quotation mark on the left stops the procedure as well.

After the main part of the procedure is finished, an extra filter is applied: words which are unlikely to occur at the beginning (mainly conjunctions) or the end (conjunctions and prepositions) of an idiom are stripped from the delimited expression.

Identification is based on the delimitation: a delimited expression is classified as an idiom if and only if it is composed of two or more tokens. (Note that idioms are, by definition, multi-word expressions).

4.1   Enhancements to the Procedure

The procedure described in [2] is based on the *przysłowiowy/-o* metalinguistic marker. We decided to combine this with the most frequent (though less reliable) idiom-related marker – quotation marks. Note that in the original procedure proposed in [2] quotation marks are used *locally* (i.e. they are taken into account only if they occur in the same snippet as *przysłowiowy/-o*), now we are going to use them *globally*, searching for a given expression enclosed in quotation marks in the *whole corpus*. This enhancement will be applied in both the delimitation and identification step.

Delimitation. As the delimitation step is concerned, we add a new stop condition, namely we stop attaching tokens if:

$$\frac{\Phi(``\, t_{k+1} \ldots t_{i-1} \,'')}{\Phi(t_{k+1} \ldots t_{i-1})} > \theta,$$

where $\Phi(t_{k+1} \ldots t_{i-1})$ is the corpus frequency of the phrase gathered so far ($t_{k+1} \ldots t_{i-1}$ using the notation as in [2]), $\Phi(``t_{k+1} \ldots t_{i-1}'')$ – the corpus frequency of the same phrase enclosed in quotation marks, θ – threshold. In other words, we stop if the phrase gathered so far is used with quotation marks with a sufficient frequency (as evidenced by the corpus). This condition is applied for both the left and the right side of *przysłowiowy/-o* marker. The threshold value of θ (0.07) was tuned on the development set.

For instance, let us consider the following snippet:

że bezpieczeństwo na większości placów budów zaczęło <u>być</u> # <u>oczkiem w głowie</u> *wielu pracodawców*
= *that safety on most constructions sites started to* <u>be</u> # <u>the apple of the eye</u> *[for] many employers*

The snippet contains an idiomatic expression *być oczkiem w głowie* (= *be the apple of the eye*). The original procedure, as presented in [2], would erroneously delimit *oczkiem w głowie wielu* – PMI would be not enough to stop before *wielu* (in turn, *być* would not be attached as the frequency of *być oczkiem w głowie wielu* is zero). With the enhancement proposed here, the idiom would be correctly limited (for the corpus used: Φ(oczkiem w głowie) = 234, Φ("oczkiem w głowie") = 27).

Identification. The only criterion for idiom identification used in [2] was the length of the candidate phrase (at least two tokens after the delimitation step). In order to improve the precision, we decided to add another condition: a phrase will be classified as an idiom only if it is found in the corpus with quotation marks. More precisely, the following expressions are tried for a candidate phrase of the form $t_i \ldots t_{k-1} \# t_{k+1} \ldots t_j$:

- " $t_i \ldots t_{k-1} t_{k+1} \ldots t_j$ "
- $t_i \ldots t_{k-1}$ " $t_{k+1} \ldots t_j$ "
- $t_i \ldots t_{k-1}$ przysłowiowy " $t_{k+1} \ldots t_j$ "

For instance, let us consider the snippet:

I czuje się jak po # *pobycie na wsi i juz teraz rozumiem dlaczego*
= *And I am feeling like after* # *a stay in the countryside and I understand why*

There is no idiomatic expression in this snippet (it is puzzling why the writer used *przysłowiowy* here at all). The original procedure would, however, return *po pobycie na wsi* (= *after a stay in the countryside*) as an idiomatic expression. As none of the following sequences of tokens:

- " *po pobycie na wsi* "
- *po* " *pobycie na wsi* "
- *po przysłowiowym* " *pobycie na wsi* "

are to be found in the corpus, the phrase will be rejected by the modified procedure.

5   Evaluation

The results are given in Table 1. The method denoted as "PMI-based + filter" in [2] was used as baseline. Precision, recall and F-measure are calculated by taking as true positives only those expressions that were correctly identified *and* delimited.

Table 1. Results for the test set. C – number of correctly recognised and delimited idioms, T – number of all reported idiomatic expression, P – precision, R – recall, F – F-measure.

	C	T	P	R	F
baseline	312	1006	0.310	0.307	0.308
baseline + quotes/delim	314	999	0.314	0.309	0.312
baseline + quotes/ident	253	577	0.438	0.249	0.317
baseline + quotes/delim + quotes/ident	258	597	0.432	0.254	**0.320**

"Quotes/delim" denotes using quotation marks in the delimitation step, whereas "quotes/ident" – in the identification step (see Section 4).

Using quotation marks in the delimitation step did not improve the results much. Better results were obtained for quotation marks used in the identification step, the precision was increased without hurting recall too much.

6 Estimations of Recall Upper Bound

What percentage of idiomatic expressions can be retrieved – at best – using idiom-related metalinguistic markers, assuming a corpus of texts indexed by a general Web search engine (such as Google) is available? In order to estimate this[3], we selected a random sample of 301 Polish idioms from *Wielki słownik polsko-niemiecki* (*Great Polish-German dictionary*) [8], the total number of idioms recorded in this dictionary being 7,292. Then, for each idiom from the sample a number of queries were issued to Google search engine. Each query was constructed by taking a given idiom and inserting a form of the adjective *przysłowiowy* or the adverb *przysłowiowo*. It turned out that for 118 idioms (39%) at least one query returned some Web pages containing a given idiomatic expression with the *przysłowiowy* marker.

Similar experiment was carried out for English idiomatic expressions – a random sample of 100 English expressions was selected from over 4,800 idioms listed at English Wiktionary[4]. The queries were generated with words *proverbial*, *proverbially* and *literally*[5]. Web occurrences with metalinguistic markers were found for 88 out of 100 idioms.

Obviously, the bigger the textual haystack, the more proverbial needles – one can resort to other resources than Web texts indexed by a general search engine, for instance Web archives[6] or old newspapers and books digitised for digital libraries. This

---

[3] Note that we estimate the recall against the set of all idiomatic expressions, not against the test set as in Section 5.

[4] https://en.wiktionary.org/wiki/Appendix:English_idioms

[5] Using *literally* might seem paradoxical, as it is supposed to mark non-idiomatic usages. This word *is*, however, sometimes used for idiomatic occurrences [5] (cf. http://xkcd.com/725/). Moreover, *literally* even for a non-idiomatic occurrence suggests that a given expression *can* be used idiomatically anyway, which is enough for type-based identification.

[6] E.g. http://www.archive.org

way, idiomatic expressions can be traced diachronically. For instance, there are 2,834 occurrences of the word *przysłowiowy* in newspapers, magazines and books stored in Polish digital libraries[7] (the number is likely to be underestimated because of the poor quality of the OCR used).

7 Conclusions and Future Work

We showed that looking globally at all occurrences of an expression marked with idiom-related metalinguistic markers improves type-based identification. It would probably make sense to check all corpus occurrences of a candidate expression, not necessarily accompanied with idiom-related markers. Such methods look promising as was shown with estimations of recall upper bound for Polish and English.

Acknowledgements. This paper is based on research funded by the Polish Ministry of Science and Higher Education (Grant No. N N516 480540).

References

1. Liontas, J.I.: Context and idiom understanding in second languages. EUROSLA Yearbook 2(1), 155–185 (2002)
2. Graliński, F.: Looking for proverbial needles in the proverbial haystack. In: Kłopotek, M.A., Marciniak, M., Mykowiecka, A., Penczek, W., Wierzchoń, S.T. (eds.) Intelligent Information Systems. New Approaches. Wydawnictwo Akademii Podlaskiej, Siedlce, Poland, pp. 101–111 (2010)
3. Lin, D.: Automatic identification of non-compositional phrases. In: Proceedings of ACL 1999, pp. 317–324 (1999)
4. Fazly, A., Cook, P., Stevenson, S.: Unsupervised type and token identification of idiomatic expressions. Computational Linguistics 35, 61–103 (2009)
5. Li, L., Sporleder, C.: Linguistic cues for distinguishing literal and non-literal usages. In: Huang, C.R., Jurafsky, D. (eds.) COLING (Posters), pp. 683–691. Chinese Information Processing Society of China (2010)
6. Lewicki, A.M.: Aparat pojęciowy frazeologii. In: Lech Ludorowski, W.M. (ed.): Z badań nad literaturą i językiem. Państwowe Wydawnictwo Naukowe, pp. 135–151 (1974)
7. Gale, W., Church, K., Hanks, P., Hindle, D.: Using statistics in lexical analysis. In: Zernik, U. (ed.) Lexical Acquisition: Exploiting On-Line Resources to Build a Lexicon, pp. 115–164. Lawrence Erlbaum Associates, Hillsdale (1991)
8. Wiktorowicz, J., Frączek, A. (eds.): Wielki słownik polsko-niemiecki. Wydawnictwo Naukowe PWN, Warszawa (2008)

---

[7] Most of them not indexed by Google search engine.

Using Tree Transducers for Detecting Errors in a Treebank of Polish

Katarzyna Krasnowska, Witold Kieraś, Marcin Woliński, and Adam Przepiórkowski

Institute of Computer Science, Polish Academy of Sciences, Warsaw, Poland

Abstract. The paper presents a modification — aimed at highly inflectional languages — of a recently proposed error detection method for syntactically annotated corpora. The technique described below is based on Synchronous Tree Substitution Grammar (STSG), i.e. a kind of tree transducer grammar. The method involves induction of STSG rules from a treebank and application of their subset meeting a certain criterion to the same resource. Obtained results show that the proposed modification can be successfully used in the task of error detection in a treebank of an inflectional language such as Polish.

1 Introduction

Treebanks are an important type of linguistic resource and are currently maintained or developed for numerous languages. Given their crucial role in the task of training probabilistic parsers and, hence, in many natural language processing applications, it is necessary to ensure their high quality. One of the ways to eradicate erroneous structures in a treebank is to develop a method of automated detection of wrongly annotated structures once the resource is created. In this work, we decribe such a method of finding errors in a syntactically annotated corpus and present the results of using it on a treebank of Polish. The paper is divided into two parts: § 2 introduces the method of error detection, and § 3 describes the experiment conducted on the treebank and its results.

2 An STSG-Based Approach to Error Detection

2.1 Synchronous Tree Substitution Grammars

A Synchronous Tree Substitution Grammar [1] is a set of rules which can be seen as a tree transducer. Each rule comprises of a pair of elementary trees $\langle \tau_1, \tau_2 \rangle$ and a one-to-one alignment between their frontier (leaf) nodes. τ_1 and τ_2 are called *source* and *target*. The derivation starts with two tree roots, left and right (initially with no children), and results in a pair of complete trees. Rule application substitutes source and target into aligned nodes in the left and the right tree, respectively. In order to allow a substitution to be performed, the labels of the substitution node and the substituted tree's root must be identical.

Given a treebank with annotation and structural errors, a set of STSG rules can be induced from it and then used for error detection and correction. The following

P. Sojka et al. (Eds.): TSD 2012, LNCS 7499, pp. 119–126, 2012.

subsection describes the basic methods of obtaining an STSG grammar introduced by Cohn and Lapata [2] and its application for treebank correction according to Kato and Matsubara [3]. Proposed modifications and extensions are explained in the subsequent subsection.

2.2 The Basic Procedure

When inducing an STSG from a syntactically annotated corpus, a *pseudo parallel corpus* is first created [2]. Let T be the set of all syntactic trees in the treebank. The pseudo parallel corpus $Para(T)$ is then the set of all pairs of subtrees found in T such that the two trees in the pair have the same root label and yield (i.e. the terminal sequence dominated by the root), but differ in structure. Formally, $Para(T)$ is given by:

$$Para(T) = \{\langle \tau, \tau' \rangle \mid \tau, \tau' \in \bigcup_{\sigma \in T} Sub(\sigma)$$
$$\wedge \tau \neq \tau'$$
$$\wedge yield(\tau) = yield(\tau')$$
$$\wedge root(\tau) = root(\tau')\},$$

where $Sub(\sigma)$ is the set of all subtrees of σ.

From each tree pair $\langle \tau_1, \tau_2 \rangle \in Para(T)$, an STSG rule is extracted as follows. First, an alignment $C(\tau_1, \tau_2)$ between the nodes of the two trees is determined:

$$C(\tau, \tau') = \{\langle \eta, \eta' \rangle \mid \eta \in Nodes(\tau) \wedge \eta' \in Nodes(\tau')$$
$$\wedge \eta \neq root(\tau) \wedge \eta' \neq root(\tau')$$
$$\wedge label(\eta) = label(\eta')$$
$$\wedge yield(\eta) = yield(\eta')\}.$$

Then, for each node pair $\langle \eta_1, \eta_2 \rangle \in C(\tau_1, \tau_2)$, the descendants of both η_1 and η_2 are deleted. If τ_1 and τ_2 still differ after this operation, the rule $\langle \tau_1, \tau_2 \rangle$ is added to the grammar.

Rules extracted this way must be filtered so as to eliminate the ones which would create errors instead of correcting them. For this purpose, we follow Kato and Matsubara [3] in using a *Score* function predicting the soundness of a rule. Given a rule $\langle \tau_1, \tau_2 \rangle$, its *Score* is calculated as:

$$Score(\langle \tau_1, \tau_2 \rangle) = \frac{f(\tau_1)}{f(\tau_1) + f(\tau_2)},$$

where $f(\tau)$ is the frequency of τ in the corpus. Only rules whose *Score* is not lower than a fixed threshold are taken into account.

The resulting set of rules represents structures which are probably erroneous (sources of rules) and their correct counterparts (targets).

2.3   Adaptation to a Treebank of Polish

The treebank we work with is Składnica [4]. It is a bank of constituency trees for Polish currently consisting of about 8,200 trees but still under development. The treebank is being developed in a semi-automatic manner. Sets of candidate parse trees are generated by a parser and subsequently one tree is selected by human annotators. This procedure probably leads to a different type of inconsistencies than in a treebank built fully manually. The structures annotators can choose are limited by the grammar, which should make the trees more uniform than those created manually. On the other hand, when a complete tree is presented to the annotator, it is a considerable piece of information, so he or she can easily overlook problems in the details.

Unlike Kato and Matsubara [3], when constructing the pseudo parallel corpus, we compared the yield of the trees in terms of morphosyntactic tags rather than orthographic forms. There were several reasons for choosing such an approach. First, the treebank is currently relatively small, therefore an insufficient number of sentences with a common subsequence of words can be found in it to produce relevant STSG rules. Moreover, Polish inflection makes it even less probable to have two strictly identical word sequences in a corpus (see Figure 1 for an example). The sparseness of the pseudo parallel corpus resulting from these problems can be overcome by abstracting from the orthographic and base forms of the words, and using morphosyntactic tags instead. This way we gain more material for STSG rule extraction since we are capable to draw a parallel between analogous phrases which use different words (e.g. *very small house* vs. *rather thick book*).

As an exception to this decision, particles are not replaced by their tags, but, instead, represented by their base forms (which often are their only possible forms). This was motivated by the fact that particles play important and at the same time very different role in the structure of sentences; compare, e.g., the negative marker *nie* and the subjunctive marker *by*, both analysed as particles.

Another feature of Składnica which requires special treatment is the fact that nonterminal nodes contain not only syntactic categories, but also morphosyntactic features such as gender, number or case. These must be taken into account while extracting and applying STSG rules in order to reflect modifier node agreement and avoid overgeneralisation of rules. A selected subset of feature values of nonterminal nodes is therefore preserved in the pseudo parallel corpus and in the extracted rules.

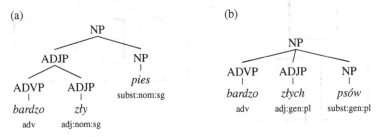

Fig. 1. Two syntactic trees for the *NP* "*very bad dog*" (a) in the nominative singular and (b) in the genitive plural. Even though these sentences clearly represent inconsistent bracketings, no rule would be derived from such trees if orthographic forms of lexemes were taken into consideration.

On the other hand, if rules were kept in such a detailed form, two problems would appear. First, a rule, e.g., for noun phrases in the accusative case, would not be matched by a linguistically relevant structure isomorphic with its source, but occurring, say, in the instrumental. What is more, redundant rules would be created for each combination of morphosyntactic features. As a solution, once a rule is extracted, its feature values are substituted with variables, but retaining any information about agreement, by using the same variable where necessary. Figure 2 shows one example of such an extracted STSG rule, with particular values of number, case, gender and person replaced by variables.

3   Experiments on Składnica

A set of 38 rules with *Score* 0.5 or higher was extracted from the treebank. 323 structures matching the source of some rule were found in 302 trees (in 283 trees one rule was matched, in 17 trees — 2 rules, and in 2 trees — 3 rules). Constructions recognised as errors, as well as proposed corrections (i.e. targets of the matched rules), were manually examined and classified into 5 categories presented and explained in Table 1.

Classification into *ERR* and *INC* is motivated by the question of how many of the structures found by the algorithm are syntactically incorrect, thus belonging to the *ERR* class, and how many represent theoretically possible bracketings which, however, are not compliant with the annotation conventions of the treebank (e.g. binary trees for multiple-modifier NPs detected by the rule in Figure 2); in the latter case the class is *INC*. 185 structures in total were assigned to classes other than *FP*, which means that they were wrongly annotated.

Taking into account the classification introduced in Table 1, we propose the following measures of precision:

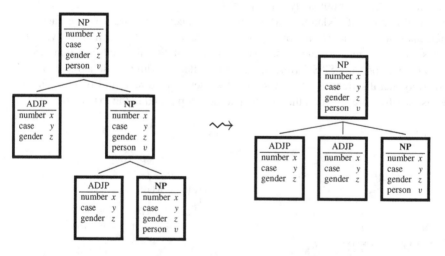

Fig. 2. An example rule flattening the structure of an NP modified by two ADJPs. The source tree is inconsistent with annotation guidelines. The rule matches if the ADJPs agree with the NP in number, case and gender.

Table 1. Classification of structures found using the STSG rules

Category	Occurrences		Comment
ERR_0	74	22.9%	An error in annotation was found and the proposed modification was correct.
INC_0	89	27.6%	An inconsistency in annotation was found and the proposed modification was correct.
ERR_1	18	5.6%	An error in annotation was found, but the proposed modification was incorrect.
INC_1	4	1.2%	An inconsistency in annotation was found, but the proposed modification was incorrect.
FP	134	42.7%	A false positive — correct structure pointed out as erroneous.

$$P_0 = \frac{ERR_0 + INC_0}{ALL},$$

$$P_1 = \frac{ERR_0 + INC_0 + ERR_1 + INC_1}{ALL},$$

$$P_{err} = \frac{ERR_0 + ERR_1}{ALL},$$

where ALL stands for all structures retrieved by the rules. The P_0 measure is more restrictive and only accepts correct modifications, whereas P_1 takes into consideration all correctly recognized wrong structures.

As far as recall is concerned, it is difficult to estimate it without manually inspecting the whole treebank for errors and inconsistencies. In the case of Składnica, such an experiment was performed by its authors [4] on a random sample of 100 sentences and 18 trees were considered wrong. We therefore assume, for the sake of estimating recall, that the whole treebank contains ~18% (1366) erroneous or inconsistent structures.

Table 2 shows the P_0 and P_1 precisions, as well as the estimated recall ($R_\sim$) and F-measure values. For calculating $R_\sim$ and F-measure, all correctly detected wrong trees were treated as true positives (as in P_1). The measures were calculated for the results of applying all 38 rules mentioned above (i.e. *Score* threshold 0.5), as well as only those with *Score* equal or above 0.6, 0.7, 0.8 and 0.9. The results show that P_1 can be increased by raising the threshold, with the obvious trade-off in $R_\sim$. The highest F-measure was achieved for the 0.6 threshold ($R_\sim$ was the same as for 0.5 threshold, i.e. 13.54%, with a slight increase in P_1: 57.63% as compared to 57.28%). Increasing the threshold to 0.9 yielded a very high P_1 precision of 86.62% at the expense of relatively poor $R_\sim$ and F-measure.

22 rules (58% out of 38) achieved a 100% P_1 score, and 18 (47%) — a 100% P_0 score. The score was lower than 50% for 10 rules in case of P_1 and for 12 rules in case of P_0. The average rule's P_1 and P_0 precision were 70.4% and 59.2%, respectively. The plot in Figure 3 shows per-rule P_1 and P_0 results. Figure 4 presents

Table 2. Precision, recall and F-measure for different *Score* thresholds

Score threshold	P_0	P_1	$R_\sim$	F-measure
0.5	50.46%	57.28%	13.54%	21.91%
0.60	**50.78%**	**57.63%**	**13.54%**	**21.93%**
0.7	50.63%	57.59%	13.32%	21.64%
0.8	51.95%	59.09%	13.32%	21.74%
0.9	78.17%	86.62%	9.00%	16.31%

a comparison between P_1 and P_{err} precision (i.e. whether a rule detected more errors or inconsistencies). Rule ordering is the same in both plots.

4 Related Work

Finding errors in manually annotated corpora is a task actively pursued for well over a decade. Initial research was concerned with finding errors at the level of morphosyntactic annotation, e.g. [5,6,7]; while the methodology of these three works differs considerably, they are all based on the same underlying idea: if similar (in some well-defined way) inputs receive different annotations, the less frequent of these annotations are suspected of being erroneous.

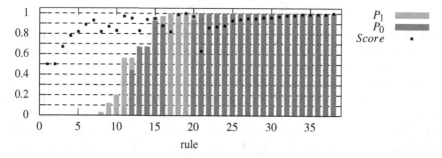

Fig. 3. Per-rule precisions P_1 and P_0

Soon, some of the methods proposed for morphosyntax were generalised to the task of verifying syntactic annotations. In this context, an important line of work is research by Dickinson and Meurers [8,9,10,11]. Before turning to the STSG-based approach described in this work, we had performed some preliminary experiments based on two methods of error detection they proposed: variation n-grams [7] and immediate dominance sets [9]. These did not, however, yield promising results, perhaps because of data sparsity (the combination of relatively small treebank and relatively rich morphology); in fact, both precision and recall were very low.[1]

---

[1] Obviously, these early experiments do not amount to any real evaluation of the methods in question. Rather, the aim of such preliminary experiments was to indicate which method seems to give the quickest returns.

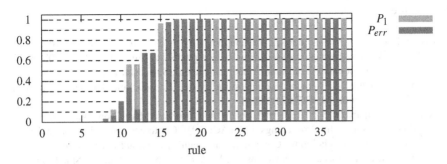

Fig. 4. Per-rule precisions P_1 and P_{err}

5 Conclusion

A method of error detection based on the Synchronous Tree Substitution Grammar formalism was implemented and tested on a treebank of Polish. The system found 185 wrongly annotated structures, achieving a precision of 57.63%. As compared to the work of Kato and Matsubara [3], who achieved a precision of 71.9% with their system, corresponding to a total of 331 structures as erroneous, the present result is worse. Nevertheless, it is worth mentioning that they worked on the Wall Street Journal sections of Penn Treebank, which is not only a bigger, but also was created for English, which has a much simpler inflection. We therefore believe that the precision of the current system is relatively satisfying.

As far as recall is concerned, we do not know of any previous results that ours could be compared to, but it is clear that there is still room for improvement. This could be achieved by finding a way to further generalise the extracted rules without loss of information crucial to retaining reasonable precision. For instance, replacing morphosyntactic features in word-level tags with variables, extending the special treatment of particles to other non-inflected parts of speech, or using base forms during the construction of the pseudo parallel corpus (when searching for subtrees with identical yield) all seem worth an experiment.

Although the method presented above gives relatively satisfying results, we are aware that it recognises only certain classes of inconsistencies. In the current treebank, there exist some serial annotation errors that cannot be found using this technique; wrong classification of verbal dependents as arguments or adjuncts is a typical example. The relatively small size of Składnica is also a problem since there is not enough training data to extract rules that can report some less frequent errors. Hence, the need for further research is clear.

Acknowledgements. The work described in this paper is partially supported by the DG INFSO of the European Commission through the ICT Policy Support Programme, Grant agreement no.: 271022, as well as by the POIG.01.01.02-14-013/09 project co-financed by the European Union under the European Regional Development Fund.

References

1. Eisner, J.: Learning non-isomorphic tree mappings for machine translation. In: Proceedings of the 41st Annual Meeting on Association for Computational Linguistics, ACL 2003, vol. 2, pp. 205–208. Association for Computational Linguistics, Stroudsburg (2003)
2. Cohn, T., Lapata, M.: Sentence compression as tree transduction. Journal of Artificial Intelligence Research 34, 637–674 (2009)
3. Kato, Y., Matsubara, S.: Correcting errors in a treebank based on synchronous tree substitution grammar. In: Proceedings of the ACL 2010 Conference Short Papers, ACLShort 2010, pp. 74–79. Association for Computational Linguistics, Stroudsburg (2010)
4. Woliński, M., Głowińska, K., Świdziński, M.: A preliminary version of Składnica — a treebank of Polish. In: Vetulani, Z. (ed.) Proceedings of the 5th Language & Technology Conference, Poznań, pp. 299–303 (2011)
5. van Halteren, H.: The detection of inconsistency in manually tagged text. In: Proceedings of the 2nd Workshop on Linguistically Interpreted Corpora (LINC 2000) (2000)
6. Eskin, E.: Automatic corpus correction with anomaly detection. In: Proceedings of the 1st Meeting of the North American Chapter of the Association for Computational Linguistics (NAACL 2000), Seattle, WA, pp. 148–153 (2000)
7. Dickinson, M., Meurers, W.D.: Detecting errors in part-of-speech annotation. In: Proceedings of the 10nth Conference of the European Chapter of the Association for Computational Linguistics (EACL 2003), Budapest, pp. 107–114 (2003)
8. Dickinson, M., Meurers, W.D.: Detecting inconsistencies in treebanks. In: Nivre, J., Hinrichs, E. (eds.) Proceedings of the Second Workshop on Treebanks and Linguistic Theories (TLT 2003), Växjö, Norway, pp. 45–56 (2003)
9. Dickinson, M., Meurers, W.D.: Prune diseased branches to get healthy trees! How to find erroneous local trees in a treebank and why it matters. In: Civit, M., Kübler, S., Martí, M.A. (eds.) Proceedings of the Fourth Workshop on Treebanks and Linguistic Theories (TLT 2005), Barcelona, pp. 41–52 (2005)
10. Boyd, A., Dickinson, M., Meurers, D.: On detecting errors in dependency treebanks. Research on Language and Computation 6, 113–137 (2008)
11. Dickinson, M., Lee, C.M.: Detecting errors in semantic annotation. In: Proceedings of the Sixth International Conference on Language Resources and Evaluation, LREC 2008, Marrakech, ELRA (2008)

Combining Manual and Automatic Annotation of a Learner Corpus

Tomáš Jelínek[1], Barbora Štindlová[2], Alexandr Rosen[1], and Jirka Hana[3]

[1] Charles University in Prague, Faculty of Arts
[2] Technical University of Liberec, Faculty of Education
[3] Charles University in Prague, Faculty of Mathematics and Physics

Abstract. We present an approach to building a learner corpus of Czech, manually corrected and annotated with error tags using a complex grammar-based taxonomy of errors in spelling, morphology, morphosyntax, lexicon and style. This grammar-based annotation is supplemented by a formal classification of errors based on surface alternations. To supply additional information about non-standard or ill-formed expressions, we aim at a synergy of manual and automatic annotation, deriving information from the original input and from the manual annotation.

Keywords: learner corpora, error annotation, Czech, morphology, syntax.

1 Introduction

Texts produced by learners of a second or foreign language are a precious source of linguistic evidence for experts in language acquisition, teachers, authors of didactic tools, and students themselves. A corpus of such texts can be annotated by hand or by automatic tools in the usual ways, common in other types of corpora, i.a., by metadata, morphosyntactic categories and syntactic structure [13,2]. However, the value of such texts is mainly in how they differ from the standard language. This information can be extracted by statistical comparisons with texts produced by native speakers, or by an explicit mark-up, offering hypotheses about the writer's intentions in the form of corrections and/or providing a classification of deviations from the standard. Methods and tools for dealing with Czech as a second language in this context have not been explored so far. A part of our learner corpus (about 300K tokens out of 2 million) is now manually (doubly) annotated with error tags and emended (corrected) forms, using a three-level annotation scheme with a complex grammar-based taxonomy of errors in spelling, morphology, morphosyntax, lexicon and style. This type of annotation is supplemented by a formal classification, e.g. an error in morphology can also be specified as being manifested by a missing diacritic or a wrong consonant change.

To assist the annotator and to supply additional information about deviations from the standard, we aim at a synergy of manual and automatic annotation, deriving information from the original input and from the manual annotation. Some methods interact with the annotator (e.g. a spell checker within the annotation editor marks potentially incorrect forms), or use results of manual annotation, including an automatic check for consistency and compliance with the annotation guidelines. After approval by the

P. Sojka et al. (Eds.): TSD 2012, LNCS 7499, pp. 127–134, 2012.

annotator's supervisor, some error tags are specified in more detail and more tags are added automatically. To assist the annotator even further, we experiment with methods of automatic emendation by a mildly context-sensitive spell checker and plan to use a grammar checker or a stochastic model to assign error annotation.

2   Annotating a Learner Corpus of Czech

Our learner corpus includes written texts[1] produced by non-native speakers of Czech at all levels of proficiency and equipped with meta-data about the authors, their learning history and the situation where the text was elicited. The authors are native speakers of Slavic, other Indo-European and some typologically distant languages, such as Chinese, Vietnamese or Arabic. A subcorpus includes texts written by pupils in primary school age with Romani background.

So far, the task of proposing a detailed methodology for teaching Czech as a foreign language has not received enough attention. Especially distinctions related to different target groups are not researched, even those most frequent and obvious between Slavic and non-Slavic students. Our project will help to change this by becoming a resource for research and design of teaching materials. At the same time, it will provide data helping to initiate and develop a systematic research of Czech as a foreign language.

2.1   Annotation Scheme

Since most of the original texts are hand-written, the annotation process starts with their transcription according to detailed rules. A set of codes is used to capture the author's corrections and other properties of the manuscript.

The language of a learner of Czech may deviate from the standard in a number of aspects: spelling, morphology, morphosyntax, semantics, pragmatics or style. To cope with the multi-level options of erring in Czech and to satisfy the goals of the project, our annotation scheme answers the following requirements:

1. Preservation of the original text alongside with the emendations
2. Successive emendation
3. Ability to capture errors in single forms and discontinuous expressions
4. Syntactic relations for errors in agreement, valency, pronominal reference

To meet these requirements, we use a multilevel annotation scheme, supporting successive emendation. As a compromise between several theoretically motivated levels and practical concerns about the process of annotation, the scheme offers two annotation levels. This enables the annotators to register anomalies in isolated forms separately from the annotation of context-based phenomena but saves them from difficult theoretical dilemmas.

Level 0 (L0) includes the transcribed input, where the words represent the original strings of graphemes, with some properties of the hand-written original preserved in the mark-up. Level 1 (L1) gives orthographical and morphological emendation of

---

[1] Spoken data are being collected but not yet transcribed or annotated.

isolated forms; the sentence as a whole can be still incorrect. Level 2 (L2) treats all other deviations, resulting in a grammatically correct sentence. This includes errors in syntax (agreement, government), lexicon, word order, usage, style, reference, negation, or overuse/underuse. The corresponding forms at the neigbouring levels are linked, corrections are assigned error labels, and additional information such as POS tags and lemmas is added.

The whole annotation process proceeds as follows:

1. The transcript is converted into a format where L0 roughly corresponds to the tokenized transcript and L1 is set as equal to L0 by default. Both are encoded in an XML-based format [10].
2. The annotator manually corrects the document and provides some information about errors using the annotation tool *feat* (see Fig. 1 and [5].)
3. Automatic post-processing provides additional information about lemma, POS and morphological categories for emended forms.
4. Error information that can be inferred automatically is added.

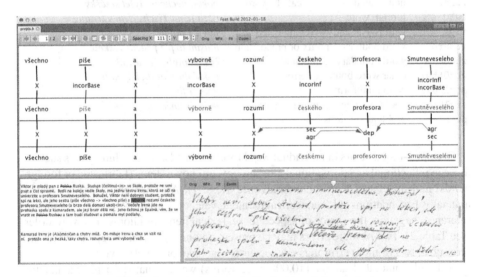

Fig. 1. A sample sentence in the *feat* annotation tool

2.2 Error Taxonomy

Error taxonomies in learner corpora are often based on standard linguistic categories, see [1,3,8]. We use a similar approach, complemented by a classification of surface alternations. A single incorrect form is cross-classified as belonging to one or more types in each of the following two classes:

- grammar-based error types (spelling, morphology, word boundary, agreement, government, lexical issue, style, punctuation)
- formal error types (diacritics, capitalisation, metathesis, missing element)

For some types we identify the locus of the error, e.g. a morphological error is (manually) identified as an error in the stem or in the inflectional ending. Unlike the grammar-based types, the formal errors are more easily detectable by automatic tools. Yet the tools can detect also some of the grammar-based error types. Thus, errors can be identified in the following ways:

- manually
- automatically, by comparing the faulty and the emended forms
- automatically, by specifying a manually assigned error type in more detail, often using the word forms, their morphological tags or lemmas

Table 1. Grammar-based errors at Level 1

Error type	Description	Example
incorInfl	incorrect inflection	*pracovají v továrně*; *bydlím s matkoj*
incorBase	incorrect word base	*lidé jsou mérný*; *musíš to po*světlit
fwFab	non-emendable, „fabricated" word	*pokud nechceš slyšet smášky*
fwNC	foreign word	*váza je na Tisch*; *jsem v truong*
flex	with fwFab and fwNC: inflected	*jdu do shopa*
wbdPre	word boundary: prefix or preposition	*musím to při pravit*; *veškole*
wbdComp	word boundary: compound	*český anglický slovník*
wbdOther	other word boundary error	*mocdobře*; *atak*; *kdy koli*
stylColl	colloquial form	*dobrej film*
stylOther	bookish, dialectal, hypercorrect	*holka s hnědými očimi*
problem	problematic cases	

Grammar-based errors in individual word forms, treated at L1, include errors such as those in inflectional and derivational morphology, unknown stems (fabricated or foreign words) and misplaced word boundaries (see Table 1). All such errors are annotated manually. Emendations at L2 concern agreement, valency, analytical forms, pronominal reference, negation, the choice of aspect, tense, lexical items or idioms, and word order (see Table 2). Two or more errors may be present on one word form. Depending on the error type, two or more error tags may occur at one level or at both levels.

A doubly annotated sample (10,000 word forms) was evaluated for inter-annotator agreement to verify that the annotation scheme and taxonomy are sufficiently robust [12]. Higher agreement was found for formally well-defined categories, with satisfactory results even for those requiring subjective judgment.

3   Automatic Extension of Manual Annotation

Manually emended and error-annotated text can be assigned additional information by automatic tools in the following three ways:

Table 2. Grammar-based errors at L2

Error type	Description	Example
agr	violated agreement rules	*to jsou hezké chlapci*; *Jana čtu*
dep	error in valency	*bojí se pes*; *otázka čas[u]*
ref	error in pronominal reference	*dal jsem to jemu i jejího bratrovi*
vbx	error in analytical or compound verb form	*musíš přijdeš*; *kluci jsou běhali*
rflx	error in reflexive expression	*Pavel si raduje*
neg	error in negation	*žádný to [ne]ví*;
lex	error in lexicon or phraseology	*jsem ruská*; *dopadlo to přírodně*
use	error in the use of a grammar category	*pošta je nejvíc blízko*
sec	secondary error	*stará se o našich holčičkách*
stylColl	colloquial expression	*viděli jsme hezký holky*
stylOther	bookish, dialectal, hypercorrect	*rozbil se mi hadr*
stylMark	redundant discourse marker	*no*; *teda*; *jo*
disr	disrupted construction	*kratka jakost vyborné ženy*
problem	problematic cases	

1. As far as the emended text approximates standard language, at least in grammatical correctness, a tagger/lemmatiser can be used with an error rate similar to that for standard texts [11].
2. Some manually assigned error tags were designed with the intention that they will be specified in more detail by an automatic tool.
3. Yet other tags are only assigned automatically.

3.1   Automatic Addition of Linguistic Information

For practical reasons, especially for corpus searching, words in the corpus should be tagged with their morphological properties, including POS, case, etc. This information is added automatically.

L2 consists of correct Czech sentences only, so we can use standard tools (e.g. [6,7]) to assign each word a lemma and a tag from a standard morphological tagset [4]. L1 consists of correct Czech words, but they might be used with incorrect inflection, word order, etc. Therefore, using standard methods would produce unreliable results. Instead, we combine the result of the morphological analysis with the properties of the word on L2 as follows:

- If the form is the same on both levels we use the tag/lemma from L2.
- If the forms are different, but have the same lemma (for the L1 forms suggested by the morphological analysis), then we use that lemma and the tags appropriate for it. For example, if the L1 form is *má* 'has' or 'my' and the L2 form is *mou* 'my', we assign *má* the lemma *můj* 'my'.
- If the L1 form's lemma is different from the lemma at L2, it receives all possible morphological tags. For example, *má* would be labeled both as a verb with the lemma *mít* 'to have' and as the possessive pronoun *můj* 'my'.

3.2 Automatic Extension and Modification of Error Annotation

Since the start, we assumed that some error types can be identified automatically. This is especially true for formal errors at L1, deducible by a simple comparison of the corresponding L1 and L0 forms, e.g. error in voicing or palatalization. Errors at L2 are more difficult to classify automatically, thus only a limited number of phenomena are tagged this way.

Automatic addition of formal error tags on L1 is based on the comparison of the original L0 form with the corrected L1 form. The manually assigned L1 tags cover the following three types of errors: a wrong form (`incor`), incorrect word boundaries (`wbd`), a neologism or a foreign word (`fw`). The automatically assigned errors are independent of these manual tags. For example, *chrozba/hrozba* 'threat' is manually annotated as `incorBase` (the *h/ch* error is in the stem), and *každécho/každého* 'every$_{masc.sg.gen/acc}$' as `incorInfl` (the *h/ch* error is in the *ého* ending). In both cases the correct h is incorrectly devoiced and the h/ch error is annotated as `formVcd`.[2]

The formal L1 error tags express the way in which an L1 form differs from the original incorrect L0 form. Most of these tags (such as "missing character", "switch error" or even "error in diacritics") only identify surface manifestations. However, a few error types are characterized by linguistic concepts, such as voicing assimilation or palatalization. It is the possibility of their automatic detection that puts them in the same class with the truly formal error types.

Table 3 shows examples of automatically assigned errors on L1. Some errors affect only spelling with no change in pronunciation (capitalization, writing a wedge in *dě/tě/ně*, voicing assimilation, etc.). Other errors always affect pronunciation (vowel quantity, *e* epenthesis). Some errors might affect pronunciation in some contexts, but not others (writing *i/y*, the *c/k* substitution). We list only errors that actually occurred in real texts, using authentic examples, not every possible logical combination.

Table 3. Examples of automatically assigned errors on L1

Error type	Error description	Example
Cap0	capitalization: incor. lower case	*evropě/Evropě*; *štědrý/Štědrý*
Cap1	capitalization: incor. upper case	*Staré/staré*; *Rodině/rodině*
Vcd0	voicing assimilation: incor. voiced	*stratíme/ztratíme*; *nabítku/nabídku*
Vcd1	voicing assimilation: incor. vcless	*zbalit/sbalit*; *nigdo/nikdo*
VcdFin0	word-final voicing: incor. voiceless	*kdyš/když*; *vztach/vztah*
VcdFin1	word-final voicing: incor. voiced	*přez/přes*; *pag/pak*
Vcd	voicing: other errors	*protoše/protože*; *hodili/chodili*
Palat0	missing palatalization (*k,g,h,ch*)	*amerikě/Americe*; *matkě/matce*
Je0	*je/ě*: incorrect *ě*	*ubjehlo/uběhlo*; *Nejvjetší/Největší*
Je1	*je/ě*: incorrect *je*	*vjeděl/věděl*; *vjeci/věci*
ProtV1	protethic *v*: extra *v*	*vosm/osm*; *vopravdu/opravdu*
EpentE0	*e* epenthesis: missing *e*	*domček/domeček*
EpentE1	*e* epenthesis: extra *e*	*rozeběhl/rozběhl*; *účety/účty*

---

[2] In Czech phonology, *h* and *ch* [x] act as voicing counterparts.

Most of the L2 error tags are assigned manually, because the variability of incorrect structures is too high to allow for a reliable automatic error tagging. Thus, only limited amount of information is added automatically:

- The reflexivity error tag (rflx) is added if another type of error concerns a reflexive pronoun.
- Manually assigned error tags for compound verb forms (vbx) are sub-divided as errors in: analytical verb forms (cvf), phase or modal verbs (mod), and copular predicates (vnp). The distinction uses lemmas and morphological tags.
- Tags marking deleted and inserted words are added (odd, miss).
- Word order corrections are tagged (wo). The annotator reorders the words as necessary, but does not tag the change. The label is assigned automatically to one or more misplaced forms using lemmas and tags on L2.

3.3   Automatic Annotation Checking

The system developed for automatic error tagging is also used for evaluating the quality of manual annotation, checking the result for tags that are probably missing or incorrect. For example, if an L0 form is not known by the morphological analyzer, it is likely an incorrect word which should have been emended. Also, if a word was emended and the change affects pronunciation, but no error tag was assigned, an incorr error tag is probably missing. This approach cannot find all problems in emendation and error annotation, but provides a good approximate measure of the quality of annotation and draws annotators' attention to potential errors.

4   Conclusion

We have discussed the schema and process of annotation of a learner corpus of Czech, showing that a combination of manual and automatic annotation can be successful. The corpus will be available soon for on-line queries, both in its error-annotated and merely transcribed parts. We also plan to explore options of (partially) automating the annotation process by presenting emended forms and error labels as suggestions to the annotator or as a raw result. The tools we consider include a context-sensitive spell checker, a grammar checker, and a stochastic model of error corrections. The tools are also tested on a corpus of transcribed speech [9]. Other plans include the use of syntactic annotation, and modifications and development of error taxonomy in response to users' feedback and the experience from annotating larger volumes of data.

Acknowledgments. This research was supported by the Education for Competitiveness programme, funded by the European Structural Fund and the Czech government as a project no. CZ.1.07/2.2.00/07.0259. It was also co-funded by the GACR grant no. P406/10/P328, and by the NAKI programme of the Czech Ministry of Culture, project no. DF11P01OVV013.

References

1. Díaz-Negrillo, A., Fernández-Domínguez, J.: Error tagging systems for learner corpora. Resla 19, 83–102 (2006)
2. Dickinson, M.: Generating learner-like morphological errors in Russian. In: Proceedings of the 23rd International Conference on Computational Linguistics (COLING 2010), Beijing (2010)
3. Granger, S.: Error tagged learner corpora and CALL: A promising synergy. CALICO Journal 20, 465–480 (2003)
4. Hajič, J.: Disambiguation of Rich Inflection: Computational Morphology of Czech. Karolinum, Charles University Press, Praha (2004)
5. Hana, J., Rosen, A., Škodová, S., Štindlová, B.: Error-tagged learner corpus of Czech. In: Proceedings of the Fourth Linguistic Annotation Workshop (LAW IV), Uppsala (2010)
6. Jelínek, T.: Nové značkování v Českém národním korupusu (A new tagging system in the Czech National Corpus). Naše řeč 91, 13–20 (2008)
7. Jelínek, T., Petkevič, V.: Systém jazykového značkování korpusů současné psané češtiny [A system of linguistic markup of corpora of contemporary written Czech]. In: Petkevič, V., Rosen, A. (eds.) Korpusová lingvistika Praha 2011: 3 – Gramatika a značkování korpusů. Studie z korpusové lingvistiky, vol. 16, pp. 154–170. Ústav Českého národního korpusu, Nakladatelství Lidové noviny (2011)
8. Lüdeling, A.: Mehrdeutigkeiten und Kategorisierung: Probleme bei der Annotation von Lernerkorpora. In: Grommes, P., Walter, M. (eds.) Fortgeschrittene Lernervarietäten, Niemeyer, Tübingen, pp. 119–140 (2008)
9. Nouza, J., Blavka, K., Boháč, M., Červa, P., Žďánsky, J., Silovský, J., Pražák, J.: Voice Technology to Enable Sophisticated Access to Historical Audio Archive of the Czech Radio. In: Grana, C., Cucchiara, R. (eds.) MM4CH 2011. CCIS, vol. 247, pp. 27–38. Springer, Heidelberg (2012)
10. Pajas, P., Štěpánek, J.: XML-based representation of multi-layered annotation in the PDT 2.0. In: Hinrichs, R.E., Ide, N., Palmer, M., Pustejovsky, J. (eds.) Proceedings of the LREC Workshop on Merging and Layering Linguistic Information (LREC 2006), Genova, Italy, pp. 40–47 (2006)
11. Spoustová, D., Hajič, J., Votrubec, J., Krbec, P., Květoň, P.: The best of two worlds: Co-operation of statistical and rule-based taggers for Czech. In: Proceedings of the Workshop on Balto-Slavonic Natural Language Processing 2007, pp. 67–74. Association for Computational Linguistics, Praha (2007)
12. Štindlová, B., Škodová, S., Hana, J., Rosen, A.: CzeSL – an error tagged corpus of Czech as a second language. In: Pęzik, P. (ed.) PALC 2011 Practical Applications in Language and Computers, Łódż, April 13-15, Łódź Studies in Language, Peter Lang (to appear, 2012)
13. Van Rooy, B., Schäfer, L.: An evaluation of three POS taggers for the tagging of the Tswana Learner English Corpus. In: Archer, D., Rayson, P., Wilson, A., McEnery, T. (eds.) Proceedings of the Corpus Linguistics 2003 Conference, pp. 835–844. UCREL, Lancaster University (2003)

A Manually Annotated Corpus
of Pharmaceutical Patents*

Márton Kiss[1], Ágoston Nagy[1], Veronika Vincze[2,3],
Attila Almási[1], Zoltán Alexin[4], and János Csirik[2]

[1] University of Szeged, Department of Informatics
6720 Szeged, Árpád tér 2., Hungary
[2] MTA-SZTE Research Group on Artificial Intelligence
6720 Szeged, Tisza Lajos krt. 103., Hungary
[3] Universität Trier, Linguistische Datenverarbeitung
54286 Trier, Universitätsring, Germany
[4] University of Szeged, Department of Software Engineering
6720 Szeged, Árpád tér 2., Hungary
{mkiss,nagyagoston,vinczev,alexin,csirik}@inf.u-szeged.hu,
vizipal@gmail.com

Abstract. The language of patent claims differs from ordinary language to a great extent, which results in the fact that tools especially adapted to patent language are needed in patent processing. In order to evaluate these tools, manually annotated patent corpora are necessary. Thus, we constructed a corpus of English language pharmaceutical patents belonging to the class A61K, on which several layers of manual annotation (such as named entities, keys, NucleusNPs, quantitative expressions, heads and complements, perdurants) were carried out and on which tools for patent processing can be evaluated.

Keywords: patent, corpus, syntactic annotation, named entities.

1 Introduction

For the automatic processing of patents, they are required to be linguistically preprocessed, that is, to be tokenized, POS-tagged and syntactically parsed. However, the language of patent claims differs from ordinary language to a great extent, which results in the fact that tools especially adapted to patent language are needed in patent processing, what is more, manually annotated corpora are desirable to evaluate the performance of these tools.

Thus, we constructed a toolkit that is able to split English language patents into sentences (clauses), to parse them morphologically and to identify key concepts such as named entities or keywords in the texts. In order to evaluate the performance of our tools, we constructed a corpus of pharmaceutical patents belonging to the class A61K, on which several layers of manual annotation were carried out. In this paper, we present our corpus and offer a detailed description of the manual annotations.

---

* This work was supported in part by the National Innovation Office of the Hungarian government within the framework of the project MASZEKER.

P. Sojka et al. (Eds.): TSD 2012, LNCS 7499, pp. 135–142, 2012.

2   Motivation and Related Work

The claims section of patents contains usually the most important information about the topic and the scope of the patents. Among the claims it is the main claim that summarizes the essential content of the patent: all the necessary characteristics of the method, process, tool or product described in the patent have to be listed here. The other claims further detail these characteristics, often with the help of figures, tables and images [1].

2.1   The Linguistic Characteristics of Patent Claims

The linguistic features of patent claims considerably differ from those of ordinary language. The main claim typically consists of one very long sentence with a complex syntactic structure, which is quantitatively supported by the experiments described in [2] and in [3]. There are multiple embedded clauses and noun phrases, lists and coordinated phrases in main claims. Elliptic constructions, anaphoras, post-head modifiers and relative clauses also make the automatic processing of patent claims difficult.

The vocabulary of patents also contains neologisms: it is mostly multiword expressions (noun compounds) that cannot be found in general dictionaries [3]. However, they are compositional, i.e. their meaning can be calculated from the meaning of their parts and from the way they are connected. Sometimes it is also a source of problem that many words acquire a new meaning within the patent since the process or product described is also a novelty, hence it may well be the case that old terms are used in a slightly modified meaning [1].

As authors are required to provide a very detailed description of the subject of the patent, the language used is strict and precise. Still, there is a tendency to generalize over the scope of the patent in order to prevent further abuse [1]. Thus, the scope of the patents can be expanded or other use cases can later be included in the patent. The linguistic strategies applied include the following:

- the use of *etc.* at the end of lists or enumerations;
- the use of *for instance* or *e.g.* at the beginning of lists or enumerations;
- the use of inclusive *or*;
- the use of generalizing adverbs (*usually*, *typically* etc.)

These strategies are comparable to uncertainty cues, in other words, hedges or weasels [4]. Nevertheless, whereas the use of hedges and weasels and other vague and misleading phrases is undesirable in e.g. Wikipedia articles, their frequent occurrence in patent claims is a general phenomenon.

2.2   Related Work

There have been several patent corpora constructed. For instance, the European Patent Corpus contains 130 million sentence pairs from 6 languages, in which sentences are automatically aligned [5]. There is a Japanese–English patent parallel corpus containing 2 million pairs of aligned sentences [6] and a Chinese–English patent parallel corpus has also been constructed with 14 million sentence pairs [7] to name but a few. These corpora can be effectively exploited in cross-lingual information retrieval and machine translation tasks.

The linguistic characteristics of patent claims call for special techniques to be applied when processing the claims. In order to evaluate the tools adapted to patent processing, manually annotated data are needed. However, with the exception of the 100 sentences annotated by [3], we are not aware of any manually annotated patent corpora, which motivated us to build a corpus with several layers of manual annotation on which our processing toolkit can also be evaluated. In the following sections, our corpus and toolkit are presented.

3   The Corpora

We collected 10,000 patents (C10K) (see below) out of which we randomly selected 313 patents (C313) from the class A61K, which includes preparations for medical, dental or toilet purposes and we later narrowed them down to 62 claims (C62). The latter corpus has been chosen to be our benchmark database as it contains all types of manual annotations to be described in this section. Table 1 shows the main characteristics of the main claims of the corpora.

Table 1. Comparison of the corpora

	C62	C313	C10K		C62	C313	C10K
Patent	62	313	8,797	Text	•	•	•
Sentence	62	865	8,793	Key	•		
Token	7,883	59,356	1,771,290	NucleusNP	•	•	
Lemma	1,466	6,010	32,252	Quantity	•	•	
NucleusNP	1,706	14,275		Dependent	•		
Perdurant	664	3,448		Perdurant	•	•	
Quantity	226			Enumeration	•		
Key	415			Headword	•		
NE	825	20 374		Named entity	•	•	

The 10K Corpus. For a start we prepared a corpus of 10,000 patents. The corpus is made of patents chosen randomly from 10 different IPC subclasses (A24F, A61K, A63K, B26D, D21F, E01D, F21K, G10C, G10L, H04M). We downloaded 1,000 patents for each of the above mentioned subclasses from the website of the United States Patent and Trademark Office[1]. We think that this hierarchy level is appropriate for our research, thus we did not carry out further segmentations within the subclasses. Since each patent is downloadable in a well-formatted full-text format from the site of the United States

---

[1] http://patft.uspto.gov/

Patent and Trademark Office, we could easily retrieve the required information, which we converted to XML format. The corpus in XML format was easily manageable in the UIMA[2] system.

The C313 Corpus. The C10K proved too big for the task of annotation, so we filtered it. We chose 313 patents that belong to the subclass A61K. We did initial research on this smaller corpus. On the 313 A61K patents, the following annotations were manually marked: named entities (NEs), NucleusNPs, perdurants, quantitative expressions. When generating the verb frames, the verb frames of the verbs of this C313 corpus were considered.

The C62 Corpus. For marking specific annotations, the C313 corpus seemed enormously big as well. So we constructed a corpus of 62 patents from the 313. We carried out the markings of the enumerations and the keys only on C62.

4 Manual Annotation of the Corpora

In this section, we describe the linguistic phenomena that are manually annotated in our corpora.

4.1 Keys

The main claim of a patent is usually a very complex sentence including many subordinations and coordinations, which is difficult to analyze. To analyze these sentences by the current automated algorithms is hardly possible [3] hence we need to find a solution to break these sentences into sentence fragments that are analyzable by automated algorithms. Therefore we marked the postmodifiers and the beginnings of clauses with keys.

Keys are generally the sections of the processed text where the presence of the modifier-modified noun relation is purely recognizable on formal grounds. Keys consist of a first and a second part. Shinmori et al. [2] apply a similar technique to break Japanese patents into analyzable fragments, however, they only mark the beginning of clauses (i.e. the second part of our keys). **Simple keys** serve to indicate successive keys if no remote second-type key is connected to the first part of the key. E.g.: *substance which, group consisting*. The key is **complex** if the first and the second part of the key are not directly following each other or if more second parts belong to the first part of the key. E.g.: *the **process comprising** the steps of deforming the films (18) to form a multiplicity of recesses (16), **filling** the recesses.*

Keys were marked in C62 by hand to help the development of the automatic key marker module.

---

[2] UIMA means *Unstructured Information Management Architecture*, and is used in this project to give structure to unstructured documents (e.g. plain texts) by different, user-defined or already existing external modules performing annotation tasks and to visualize these annotations in a user-friendly way. http://uima.apache.org/

4.2   NucleusNPs and Their Nominal Heads

In the corpus, it is not the standard NP projection – well-known from generative syntax, e.g. [8] – that is used but another one that we named **NucleusNP**. The main difference between the two types of projection is that a standard NP can have complements and postnominal adjectival adjuncts attached to its head, which is not allowed for NucleusNPs. The nominal head of NucleusNPs marks their end boundary (if it is not followed by a quantitative expression), thus eventual prepositional phrases are not attached to it. Therefore, NPs not having a prepositional complement or a postnominal adjectival phrase coincide with NucleusNPs (e.g. *ascorbic acid*), but the others do not. To sum up, a NucleusNP obligatorily has a nominal head, and optionally prenominal adjectival phrases as well as pre- or postnominal quantitative expressions.

The manual annotation coincides with the above mentioned definition of NucleusNPs. In the following example all NucleusNPs are marked (in italics), with their nominal head (underlined) and quantitative expression (bold).

> A *pharmaceutical composition* comprising [...] in *a ratio* of *paracetamol* to *calcium carbonate* of **3.0:1.0 to 30.0:1.0**, *at least one* binding *agent*, [...] *at least 60% of the paracetamol* is released from *the composition* at **180 seconds** [...] at **40°C.±2°C.** [...]

NucleusNPs are manually marked in C62 and C313.

4.3   Quantitative Expressions

As the corpus describes many chemical compositions, and the ingredients of these have to be detailed as precisely as possible, it possesses many quantitative expressions. As quantitative expressions are important for semantic document indexing, these have to be identified. For that purpose, quantitative expressions were manually annotated in the corpus; however, in the manual annotation phase, these quantitative phrases are marked as a whole, their internal structure is not annotated. The quoted example in the previous subsection showed some different quantitative expressions (marked in bold) annotated in the text.

As it can be seen from the example, quantitative expressions can be (1) intervals (*3.0:1.0 to 30.0:1.0*, *40°C.±2°C.*), (2) numbers with measure units (*180 seconds*), (3) numbers written with letters and eventual measure units (*at least one*), (4) relative expressions (*at least 60%*).

Quantitative expressions are manually annotated in C62 and C313.

4.4   Heads and Complements

The corpus was also annotated in terms of heads and complements, as well. Heads can be (1) finite or non-finite verb forms, (2) adjectival phrases and (3) nominal expressions. (1-3) may have complements introduced by a preposition (e.g. [$_{HEAD}$ *consisting*] [$_{OF-compl}$ *of ascorbic acid*]), only (1) can have complements not introduced by a preposition, which can precede or follow the head.

Heads were marked with bold letters, complements with italics, and of course, character series representing both are in bold and in italics in the same time. Complements are also labelled: this label can be found after the complement introduced by the _ sign. A default complement is attached to the nearest preceding head, and it is an *and*-type coordination; if it is not the preceding head, the number of the heads jumped out is marked between curly brackets { }. Here we show the final result of the annotation on a short text extract:

A blood sugar regulating product *obtainable*_mod1 *from soybean seeds*_from *by* **a process**_by **comprising**_mod1 [...] a) **soaking**_obj *the soybean seeds*_obj [...] b) **drying**{2} *the* [...] *seeds*_obj *to* **reduce**_to *the water content*_obj *of seeds*_of, c) **grinding**{4} *the seeds*_obj [...]

This annotation scheme clearly shows the syntactic relation between heads and complements. For example, *by a process* is the prepositional complement of *obtainable*, and *a process* is the subject of the following verb *comprising*. *drying* and *grinding* are parts of the enumeration starting with *soaking*: these all are the direct objects of *comprising*: therefore, the first element is connected to *comprising*, and the other two to *soaking* as *and*-type coordinations – the numbers in curly brackets show the relative backward position of *soaking* as a head with respect to the actual complement (*soaking* is the second head counted backwards with respect to *drying*, and the fourth to *grinding*).

Heads and their complements are manually annotated in C62.

4.5 Perdurants

According to [9], perdurants – in contrast with endurants – are expressions that designate an event or a state, that is, they can fulfill the same functions as most verbs. So perdurant expressions can manifest themselves by a finite or non-finite verbal form (e.g. *prevents*, *preventing*), a deverbal adjective (e.g. *obtainable*), a noun derived from a verb (e.g. *to access* → *access*).

However, in our analysis, perdurant expressions are defined in another way: they are elements that can have prepositional adjuncts. Adjectives and the other nominal elements can only have prepositional elements that are in their respective valence pattern because it is verbs or perdurant expressions that are more likely to have prepositional adjuncts. Therefore, annotating perdurants facilitates parsing, where heads are connected to their complements because an unattached prepositional phrase is more likely to be linked to a perdurant expression or a verb than to a non-perdurant noun or adjective.

The following annotated text part shows all perdurant words (in italics):

A tablet that readily *disintegrates* in gastric fluid to *give* aspirin crystals *coated* with a polymeric film [...] not *preventing* *access* of gastric fluid to the aspirin [...]

Perdurants are manually marked in the C62 and C313 corpora.

4.6   Other Annotations

Enumeration. In C62 we marked the enumerations by hand. We not only marked the type of the enumerations (discourse or linear) but also the borders of its items.

HeadWord. Pragmatically the subjects of the clause "we claim" (found at the beginning of the main claim) are the headwords. But in most cases this is a word of a very general meaning (*means, composition, method* etc.). A main claim can contain more than one headword. Headwords were marked on C62 as well.

Named entities. On C313 we marked the named entities. Since we examined the A61K subclass, we used the following labels when hand marking: disease (*drunkenness*), special (*succinic acid*), generic (*sugar*).

5   The UIMA Toolkit

During our research we used the UIMA linguistic framework. Now we show how we converted the documents containing manual annotations into the UIMA framework, as well as how we compared manual and automatic results. After that we describe the module that visualized results.

Manual Annotation and Word → UIMA converter. In order to facilitate linguistic annotation we have developed a converter which allows linguists to annotate using certain formatting in Word (e.g.: to change the background color of the marked text). Then we prepared UIMA annotations from the Word text to check and test machine algorithms.

Comparative Module for Annotations. In order to easily compare UIMA annotations with the manual annotations, we prepared a comparative module. The output of the comparison is the recall/precision/F-measure triplet, but we also generated lists for each comparison that contain which annotations were common and which were only included in either one or the other class of comparison.

Visualization. The UIMA annotations can be viewed with a visualizer created by us. This tool is able to visualize annotations and dependency trees as well. This module meant a big relief for the testing and troubleshooting section.

Parsing. After tokenizing and sentence splitting, the program performs the identification of chemical named entities, perdurants, quantitative expressions and NucleusNPs in the UIMA framework. After identifying these units we determine and connect heads with complements.

6   Conclusion

In this paper, we presented our corpora based on pharmaceutical patents belonging to the class A61K developed for the automatic linguistic processing of patents. The corpora contain several layers of manual annotation: keys, quantitative expressions, NucleusNPs, heads and complements, perdurants and named entities.

The development of the corpora contributes to the linguistic processing of patents, which makes it possible to develop applications in the field of information extraction / retrieval. For instance, the potential users of a search engine are usually interested in finding patents related to certain substances, treatments, illnesses etc. Thus, the identification of named entities is essential while further steps like the detection of perdurants are also necessary to extract events from the texts. Finally, for every application, the basic processing steps such as morphological analysis and syntactic parsing should also be carried out.

We would like to further develop our algorithms to recognize the linguistic phenomena annotated in the corpus in the future: its current modules may be ameliorated on the one hand and new modules may be implemented like an enumeration module on the other hand. We are also planning to adapt our toolkit for the processing of other patent classes such as electricity or transporting devices, therefore patent texts from these domains should also be annotated with the same layers.

References

1. Osenga, K.: Linguistics and Patent Claim Construction. Rutgers Law Journal 38, 61–108 (2006)
2. Shinmori, A., Okumura, M., Marukawa, Y., Iwayama, M.: Patent Claim Processing for Readability – Structure Analysis and Term Explanation. In: Proceedings of the ACL Workshop on Patent Corpus Processing, pp. 56–65. Association for Computational Linguistics, Sapporo (2003)
3. Verberne, S., D'hondt, E., Oostdijk, N., Koster, C.H.: Quantifying the Challenges in Parsing Patent Claims. In: Proceedings of the 1st International Workshop on Advances in Patent Information Retrieval (AsPIRe 2010), pp. 14–21 (2010)
4. Farkas, R., Vincze, V., Móra, Gy., Csirik, J., Szarvas, Gy.: The CoNLL 2010 Shared Task: Learning to Detect Hedges and their Scope in Natural Language Text. In: Proceedings of the Fourteenth Conference on Computational Natural Language Learning (CoNLL 2010): Shared Task, pp. 1–12. Association for Computational Linguistics (2010)
5. Täger, W.: The Sentence-Aligned European Patent Corpus. In: Proceedings of the 15th Conference of the European Association for Machine Translation, Leuven, Belgium, pp. 177–184 (2011)
6. Utiyama, M., Isahara, H.: A Japanese-English Patent Parallel Corpus. In: MT Summit XI, pp. 475–482 (2007)
7. Lu, B., Tsou, B.K., Tao, J., Kwong, O.Y., Zhu, J.: Mining Large-scale Parallel Corpora from Multilingual Patents: An English-Chinese example and its application to SMT. In: Proceedings of CIPS-SIGHAN Joint Conference on Chinese Language Processing, pp. 79–86. Chinese Information Processing Society of China, Beijing (2010)
8. Haegeman, L.M.V., Guéron, J.: English grammar: a generative perspective. Blackwell, Oxford (1999)
9. Gangemi, A., Guarino, N., Masolo, C., Oltramari, A., Schneider, L.: Sweetening Ontologies with DOLCE. In: Gómez-Pérez, A., Benjamins, V.R. (eds.) EKAW 2002. LNCS (LNAI), vol. 2473, pp. 166–181. Springer, Heidelberg (2002)

Large-Scale Experiments with NP Chunking of Polish[*]

Adam Radziszewski and Adam Pawlaczek

Institute of Informatics, Wrocław University of Technology

Abstract. The published experiments with shallow parsing for Slavic languages are characterised with small size of the corpora used. With the publication of the National Corpus of Polish (NCP), a new opportunity was opened: to test several chunking algorithms on the 1-million token manually annotated subcorpus of the NCP. We test three Machine Learning techniques: Decision Tree induction, Memory-Based Learning and Conditional Random Fields. We also investigate the influence of tagging errors on the overall chunker performance, which happens to be quite substantial.

Keywords: NP chunking, tagging, Polish, Machine Learning, CRF.

1 Introduction

Text chunking has been proposed as a step towards full parsing [1]. Later, the output of a *chunker* was found useful for many other practical purposes, e.g. Information Extraction tasks. Since then, most of the attention has been devoted to extracting "low-level noun groups" [2], often called *NP chunks*.

A few attempts have been made at chunking of Slavic languages, almost exclusively relying on hand-written grammars, e.g. [3,4,5,6,7]. The only effort involving Machine Learning (ML) algorithms we are aware of is related to the Polish chunking framework *Disaster* [8,9].

All the work mentioned above[1] is characterised by the relatively small corpora involved. We summarise this observation in Tab. 1. The largest dataset employed seems to be the 12 000-token Polish corpus [8]. The F-measure values are given for general orientation as the definitions of NP chunks differ significantly. For instance, the NP chunks assumed in [9] may be quite complex, involving solving PP-attachment problems, while the NP defined in [8] is limited to the extent of morphosyntactic agreement, similarly to [5].

The small size of the available reference corpora could negatively impact the observed performance of the chunkers. This seems especially likely for the systems based on ML techniques. With the recent publication of the National Corpus of Polish (NCP, [12]), the situation suddenly changed: a 1-million token corpus annotated

---

[*] This work was financed by the National Centre for Research and Development (NCBiR) project SP/I/1/77065/10.

[1] There are some other works that present shallow parsing frameworks and rule sets but do not present any experiments/evaluation, e.g. Polish shallow parser *Spejd* [4], Serbo-Croatian formalism for capturing NPs [10].

P. Sojka et al. (Eds.): TSD 2012, LNCS 7499, pp. 143–149, 2012.

Table 1. Comparison of the corpora used for evaluation of Slavic chunkers. The **F** column denotes the reported value of F-measure for NP chunking. Figures marked t include the impact of tagging errors; a denotes that the tagging was partial, resulting in ambiguous input to the chunker; m denotes the usage of reference manual tagging, while u is placed to note that we didn't find the answer.

Language	Corpus size	Algorithm	F	Source
Bulgarian	3500 words	Rule-based	$88.9\%^t$	[3]
Croatian	135 sent.	Rule-based	$96.56\%^u$	[6]
	137 sent.	Rule-based	$92.31\%^a$	[5]
	10,131 tok. (459 sent.)	Rule-based	$88.4\%^t$	[11]
Czech	300 sent.	Deep parser	$93.1\%^m$	[7]
Polish	12,131 tok. (528 sent.)	Hybrid	$85.30\%^a$	[8]
	5,191 tok. (391 sent.)	Mem.-based	$63.0\%^t$	[9]

manually with shallow syntactic structure is available. This allows us to conduct ML-based experiments in chunking of a Slavic language on an unprecedented scale.

We use this opportunity and test an existing chunking framework [8] to perform such experiments with three underlying algorithms: Decision Trees, Memory-Based Learning and Conditional Random Fields. As the NCP also contains manual morphosyntatic annotation, we take an opportunity to inspect the influence of tagger error on the overall chunker performance. The rest of this paper assumes the usage of the 1-million token manually annotated part of the NCP 1.0^2.

2   National Corpus of Polish and Chunks

The syntactic annotation of the NCP is divided into two levels (following [13]):

1. *Syntactic words* are single- or multi-token units behaving syntactically as single words. Typical examples are forms of Polish analytical future tense, e.g. *będę szedł* ('will walk') and multi-word adverbs and prepositions, e.g. *w przeciwieństwie do* ('in contrast with'). Syntactic words constitute a layer of its own, mediating between fine-graind segmentation into tokens and the level of syntactic groups. Each syntactic word is assigned a detailed tag from a tagset specially designated for this layer.

2. *Syntactic groups* (syntactic phrases) are defined in the spirit of shallow parsing [1], i.e. annotating only a partial structure, while avoiding making decisions that seem too hard. The employed NP definition forbids the inclusion of any prepositional phrases to avoid PP-attachment ambiguities. Nevertheless, the set of distinguished phrases is impressive, involving quite fine-grained division of phrase types; e.g., the prepositional phrases are split into prepositional-nominal, prepositional-numeral and prepositional-adjectival. The other feature unusual for a shallow parsing approach is the inclusion of two groups that correspond to whole clauses. For each group, its syntactic and semantic head is marked.

---

^{2} Available at http://nkjp.pl/index.php?page=14&lang=1.

We decided to make a few substantial simplifications to facilitate using the data as material suitable for chunker evaluation. First, we decided against using the whole abundance of available syntactic word and group types. While some interesting data could be obtained from syntactic words (e.g. VP chunks), we omitted the whole layer for simplicity. Next, we joined several types of syntactic groups to form a large type which we labelled simply as noun phrases (NP chunks). Our main motivation was that of simplicity; the other consideration was to obtain a definition closer to those used earlier for Polish and Czech. Our most controversial decision was probably to join actual noun groups, (cardinal) numeral groups and prepositional groups under the name of NP. Similar assumptions are made by [7,9]. The final list of NCP syntactic groups included is as follows:

1. nominal groups (NG, NGadres, NGdata, NGgodz),
2. numeral groups (NumG, NumGb, NumGd, NumGe, NumGk, NumGr, NumGz),
3. prepositional-nominal and prepositional-numeral groups (PrepNG, PrepNumG, PrepNGadres, PrepNGb, PrepNGdata, PrepNGgodz, PrepNGp).

The resulting NPs are larger than the agreement-based phrases defined in [9] (e.g. may include genitive modifiers), but generally smaller that the NPs defined there (e.g. PP modifiers are not allowed).

To cast the groups as chunks we had to deal with two issues: *discontinuity* and *overlapping chunks*. The annotation scheme assumed in the NCP allows for *discontinuous phrases* [13]. Some of the actual instances are cases of genuine discontinuity as accounted for by syntactic theories, other ones stem from the employed shallow parsing approach (e.g. splitting on prepositions). Nevertheless, the average percentage of discontinuous phrases among our 'NP' type was around 2% and hence, we decided for the simplest possible solution: to treat each continuous part as a separate chunk, discarding the information of their connection. The problem of *overlapping chunks* is a conversion artefact; it is caused by the syntactic annotation in the NCP being not-so-shallow. Fortunately, the problem was marginal: out of 240,070 NP chunks, only 17 had to be modified to comply with our task formulation. Again, our solution was crude: given two overlapping chunks, we arbitrarily assigned the conflicting tokens to the left-most one, removing them from the other.

The resulting corpus consists of 1,064,370 tokens, 72,720 sentences and 240,070 NP chunks represented using the IOB2 tags [14].

3 Chunking Algorithms

We employed the *Disaster* chunking framework [8] for the experiments. The framework supports various ML classifiers. We tested three algorithms: Decision Trees (DT), Memory-Based Learning (MBL) classifier and Conditional Random Fields (CRF). To the best of our knowledge, it has been the first attempt at chunking of a Slavic language with CRF. We assume that the input is divided into sentences, each being a sequence of morphosyntactically tagged tokens. The tagset of the NCP is positional, thus each tag consists of the grammatical class (Part-of-Speech) and values of selected grammatical categories applicable for the class. We take advantage of this detailed information: our feature set includes the following items (except for the last point, the features are taken from [9]):

- grammatical class of the token,
- values for the following grammatical categories: number, gender and case,
- the wordform filtered by a frequency list obtained during training (only 800 most frequent forms are retained, the rest replaced with a *rare form* symbol),
- a couple of tests for morphosyntactic agreement on the values of number, gender and case,
- two tests for orthographic form: if it starts with an upper-case letter, if it starts with a lower-case letter.

3.1 Decision Trees

The algorithm assumes processing of each sentence separately. For each token and its local neighbourhood the features are evaluated — a $(-3, -2, \ldots, +2)$ window is used. This way, each token is represented as a fixed-width feature vector. During training, such vectors paired with correct IOB2 tags (class labels) are passed to the decision tree induction module. The performance phase consist in prediction of the IOB2 tag for each unlabelled feature vector by the trained classifier. We use the tree learner from the NLTK suite [15] with default parameter values.

3.2 Memory-Based Learning

Our memory-based chunker is a slight modification of the above DT algorithm: the decision tree classifier is substituted with a memory-based learner, namely the TiMBL module [16]. We re-use the parameters proposed by [9]: 9 neighbours, modified value difference metric, Gain Ratio feature weighting and inverse linear neighbour-distance weighting scheme.

3.3 Conditional Random Fields

Conditional Random Fields (CRF) is a statistical modelling method that has been successfully applied to solve various sequence labelling problems, including NP chunking [17]. The model defines conditional probability distributions of label sequences given input (token) sequences. CRF are inspired by Hidden Markov Models, but thanks to their conditional nature, the undesired independence assumptions are avoided [18].

While the DT and MBL algorithms assumed independent classification of subsequent tokens, the CRF-based chunker predicts tags for a whole sentence (sequence of feature values) at a time. We employed the CRF++ toolkit[3] for the purpose. The feature template used involved the same features as outlined above except the following modifications:

- the orthographic forms were taken unfiltered;
- we also added two form bigrams to the feature set, corresponding to the following windows: $(-1, 0)$ and $(0, 1)$;
- we added grammatical class trigrams: $(-2, -1, 0)$, $(-1, 0, +1)$, $(0, +1, +2)$.

---

[3] http://crfpp.googlecode.com/svn/trunk/doc/index.html

3.4 Spejd as a Chunker

We also compare our results to those obtained using the Spejd parser and the grammar developed manually for the NCP [13]. As Spejd outputs some nested groups, we first leave only those groups that contribute to our NP chunks, and then select the outermost groups as chunks. Note that Spejd is somewhat favoured in this setting, as the very grammar was used during the development of the reference corpus [13].

4 Tagging Errors and Chunker Performance

As the NCP also contains manual morphosyntactic annotation (including tokenisation), we decided to use this information and conduct two types of tests:

1. The chunker operates on the manually tagged material during both training and testing.
2. Re-tagging the test part with a morphosyntactic tagger to estimate the actual error of the tagger–chunker system. Unfortunately, we could not simply re-tag the testing part with an off-the-shelf tagger. This would be unfair, since the taggers available to us that may output in the tagset of the NCP also need to be trained on a manually tagged corpus. The very NCP happens to be the only publicly available corpus suitable for tagger training in our setting. To work around the problem we had to incorporate tagger training into the chunker evaluation procedure. The scheme is as follows: the training part of the corpus is used to train the tagger and chunker. The testing part is converted to plain text, tagged with the trained tagger and then subjected to chunking with the trained chunker. The resulting material will contain occasional differences in tokenisation, some mistaggings and certain amount of chunking errors. This is to mimic the conditions during normal chunker usage: a typical user will not have access to manually tagged input; she will rather be interested in obtaining accurate chunking of plain text documents. We had to prepare special scripts that were able to compare chunks between the original reference test material and the re-tagged output, where tokenisation changes did occur. We employed the most recent version of the WMBT memory-based tiered tagger [19].

5 Results

The results are presented in Tab 2. Figures of standard measures are given: precision (P), recall (R) and F-measure. The column *ref. tagging* shows the performance as measured using the manually tagged and tokenised corpus. The other column show the impact of tagging and tokenisation errors on the overall chunker performance. The difference is quite substantial.

An interesting observation is that the F-measure values are lower than those reported for other languages. This cannot be, however, treated as a definitive judgement, as the chunk definitions employed differ significantly (see Sec. 1).

Another important conclusion is that the CRF chunker outperforms both the DT and ML algorithms, but also the hand-written grammar (Spejd). The F-measure values

Table 2. Performance of the three NP chunking algorithms

	Ref. tagging			Re-tagged		
	P	R	F	P	R	F
CRF	0.9268	0.9194	0.9231	0.8625	0.8613	0.8619
MBL	0.8291	0.8733	0.8506	0.7406	0.8019	0.7701
DT	0.8176	0.8663	0.8413	0.7399	0.8014	0.7694
Spejd	0.8682	0.8909	0.8794	0.7822	0.8148	0.7982

obtained in both types of tests exhibit the same inequalities: CRF > Spejd, Spejd > MBL, Spejd > DT, MBL > DT. The first three differences are significant in both settings (paired t-test with 95% confidence), the fourth one — only in *ref. tagging*. It should be stressed that Spejd outputs more than just chunks, hence it will anyway be a tool of choice for many purposes.

6   Conclusion and Further Work

We presented a pilot study in the usage of the NCP towards creation of chunkers based on ML methods. The results are quite promising, encouraging to follow the ML paradigm. Syntactic groups other than NP are also available in the NCP, making an interesting material for further experiments. Also, information on syntactic and semantic heads could be exploited. It would also be interesting to test the described chunking algorithms and feature sets for other Slavic languages, e.g. on the Croatian CW100 corpus [11].

References

1. Abney, S.: Parsing by chunks. In: Principle-Based Parsing, pp. 257–278. Kluwer Academic Publishers (1991)
2. Ramshaw, L.A., Marcus, M.P.: Text chunking using transformation-based learning. In: Proceedings of the Third ACL Workshop on Very Large Corpora, Cambridge, MA, USA, pp. 82–94 (1995)
3. Osenova, P.: Bulgarian nominal chunks and mapping strategies for deeper syntactic analyses. In: Proceedings of the Workshop on Treebanks and Linguistic Theories, TLT 2002, Sozopol, Bulgaria, September 20–21 (2002)
4. Przepiórkowski, A.: Powierzchniowe przetwarzanie języka polskiego. Akademicka Oficyna Wydawnicza EXIT, Warsaw (2008)
5. Vučković, K., Tadić, M., Dovedan, Z.: Rule-based chunker for croatian. In: Proceedings of the Sixth International Language Resources and Evaluation (LREC 2008). ELRA, Marrakech, Morocco (2008)
6. Vučković, K.: Model parsera za Hrvatski jezik. Ph.D. thesis, Department of Information Sciences, Faculty of Humanities and Social Sciences, University of Zagreb, Croatia (2009)
7. Grác, M., Jakubíček, M., Kovář, V.: Through low-cost annotation to reliable parsing evaluation. In: Proceedings of the 24th Pacific Asia Conference on Language, Information and Computation, Tokio, Waseda University, pp. 555–562 (2010)

8. Radziszewski, A., Piasecki, M.: A preliminary noun phrase chunker for Polish. In: Proceedings of the Intelligent Information Systems (2010)
9. Maziarz, M., Radziszewski, A., Wieczorek, J.: Chunking of Polish: guidelines, discussion and experiments with Machine Learning. In: Proceedings of the 5th Language & Technology Conference, LTC 2011, Poznań, Poland (2011)
10. Nenadić, G.: Local Grammars and Parsing Coordination of Nouns in Serbo-Croatian. In: Sojka, P., Kopeček, I., Pala, K. (eds.) TSD 2000. LNCS (LNAI), vol. 1902, pp. 57–62. Springer, Heidelberg (2000)
11. Vučković, K., Agić, Ž., Tadić, M.: Improving chunking accuracy on Croatian texts by morphosyntactic tagging. In: Choukri, K., Maegaard, B., Mariani, J., Odijk, J., Piperidis, S., Rosner, M., Tapias, D. (eds.) Proceedings of the Seventh International Conference on Language Resources and Evaluation (LREC 2010), European Language Resources Association (ELRA), Valletta (2010)
12. Przepiórkowski, A., Górski, R.L., Łaziński, M., Pęzik, P.: Recent developments in the National Corpus of Polish. In: Proceedings of the Seventh International Conference on Language Resources and Evaluation, LREC 2010, Valletta, Malta, ELRA (2010)
13. Waszczuk, J., Głowińska, K., Savary, A., Przepiórkowski, A.: Tools and methodologies for annotating syntax and named entities in the National Corpus of Polish. In: Proceedings of the International Multiconference on Computer Science and Information Technology (IMCSIT 2010): Computational Linguistics – Applications, CLA 2010, Wisła, Poland, PTI, pp. 531–539 (2010)
14. Sang, E.F.T.K., Veenstra, J.: Representing text chunks. In: Proceedings of the Ninth Conference on European Chapter of the Association for Computational Linguistics, pp. 173–179. Association for Computational Linguistics, Morristown (1999)
15. Bird, S., Loper, E.: Nltk: The natural language toolkit. In: Proceedings of the ACL Demonstration Session, pp. 214–217. Association for Computational Linguistics, Barcelona (2004)
16. Daelemans, W., Zavrel, J., van der Sloot, K., van den Bosch, A.: TiMBL: Tilburg Memory Based Learner, version 6.3, reference guide. Technical Report 10-01, ILK (2010)
17. Sha, F., Pereira, F.C.N.: Shallow parsing with conditional random fields. In: Proceedings of the 2003 Human Language Technology Conference and North American Chapter of the Association for Computational Linguistics, HLT/NAACL 2003 (2003)
18. Wallach, H.M.: Conditional random fields: An introduction. Technical report, Department of Computer and Information Science, University of Pennsylvania (2004)
19. Radziszewski, A., Śniatowski, T.: A memory-based tagger for Polish. In: Proceedings of the 5th Language & Technology Conference, Poznań, Poland (2011)

Mapping a Lexical Semantic Resource
to a Common Framework of Computational Lexicons

Milena Slavcheva

Institute of Information and Communication Technologies
Bulgarian Academy of Sciences
2, Acad. G. Bonchev St., 1113 Sofia, Bulgaria
http://lml.bas.bg/~milena

Abstract. In recent years the proliferation of language resources has brought up the question of their interoperability, reuse and integration. Currently, it is appropriate not only to produce a language resource, but to connect it to prominent frameworks and global infrastructures. This paper presents the mapping of SemInVeSt – a knowledge base of the semantics of verb-centred structures in Bulgarian, French and Hungarian, to the Lexical Markup Framework (LMF) – an abstract metamodel, providing a common, standardized framework for the representation of computational lexicons. SemInVeSt and LMF share their underlying models, that is, both are based on the four-layer metamodel architecture of the Unified Modeling Language (UML). A two-step mapping of the SemInVeSt and LMF models is considered: the first step provides an LMF conformant schema of SemInVeSt as a multilingual lexical resource with a reference to an external system containing the semantic descriptors of the lexical units; the second step implies an LMF conformant representation of the semantic descriptors themselves, which are a product of the application of the Unified Eventity Representation (UER) – a cognitive theoretical approach to verb semantics and a graphical formalism, based on UML.

Keywords: lexical semantics, multilingual resources, verb-centred structures, eventity frames.

1   Introduction

In recent years the proliferation of language resources has brought up the question of their interoperability, reuse and integration. Currently, it is appropriate not only to produce a language resource, but to connect it to prominent frameworks and global infrastructures. In the last decade a number of frameworks and standardized formats have been developed for producing, maintaining and integrating various linguistic resources. Some of them are standards that provide prescriptive and persistent models and formats for representing linguistic information like, for instance, *Language Resource Management – Linguistic Annotation Framework* (LAF) (ISO:24612), *Language Resource Management - Morphosyntactic Annotation Framework* (MAF) (ISO:24611), *Language Resource Management - Syntactic Annotation Framework* (SynAF) (ISO:24615). There are also standards, which not only provide a format for representing linguistic

P. Sojka et al. (Eds.): TSD 2012, LNCS 7499, pp. 150–157, 2012.

information but also address the mapping and integration of already existing and used formats like, for instance, *Terminological Markup Framework* (TMF) (ISO:16642), and *Lexical Markup Framework* (LMF) (ISO:24613). Since there is a great variety of linguistic resources and tools, there are also frameworks like *Salt* [1] and *LexInfo* [2], based on standards, which, however, provide enriched and extensible high-level models that can incorporate a bigger number of heterogeneous resources and tools.

This paper presents the mapping of *SemInVeSt* (*Sem*antically *In*terpreted *Verb-centred Structures*) – a lexical semantic resource, to the *Lexical Markup Framework* (LMF) – an abstract metamodel that provides a common, standardized framework for the construction and integration of computational lexicons [3,4]. The choice of LMF as a common representational framework is motivated by the fact that *SemInVeSt*, via the application of the *Unified Eventity Representation* (UER) formalism [5], and LMF share their underlying models, that is, both are based on the four-layer metamodel architecture (i.e., user objects, model, metamodel, and meta-metamodel) of the *Unified Modeling Language* (UML) [6]. At the same time, LMF is gaining popularity as a standardized representational framework. It has been used in a number of applications like, for instance, the above mentioned *LexInfo* model, where it allows the users to associate linguistic information to the elements of an ontology; COLDIC – a generic platform for working with computational lexica [7]; an LMF conformant representation of a set of existing bilingual lexicons of the Dutch HLT agency [8]. In the framework of the EU KYOTO project, Wordnet-LMF – an instantiation of LMF for representing wordnets has been developed [9,10]. GermaNet, a lexical semantic network for German, has been converted into a Wordnet-LMF format [11].

The paper is structured as follows. In Section 2 *SemInVeSt* is represented. In Section 3 the basic modeling principles of LMF are provided. Section 4 describes the mapping of *SemInVeSt* to the LMF model. Section 5 contains some concluding remarks.

2   SemInVeSt

SemInVeSt is a knowledge base of the semantics of verb-centred structures in Bulgarian, French and Hungarian. At present a component of *SemInVeSt* has been built, comprising 362 verbs in a reflexive form in Bulgarian (the source language) and their semantic equivalents in French and Hungarian (the target languages). The working copy of the *SemInVeSt* reflexive-verb-component is stored in a relational database. The data are represented in two types of tables: parent and child ones. There are 11 parent tables containing Bulgarian verbs, which correspond to 11 semantic verb classes: *inherent reflexive*, *"whole-part"*, *possessive*, *beneficiary*, *reciprocal*, *motive*, *absolutive*, *deaccusative*, *anticausative*, *passive*, *idiosyncratic* verb units. Each parent table contains a field of a primary key, a field of the non-reflexive counterparts of the reflexive Bulgarian verbs, a field of the reflexive Bulgarian verbs and a field of identifiers of EVENTITY FRAME TEMPLATE diagrams, which provide the semantic description of the lexical objects in the database. The diagrams contain the elements, modeling the verb predicates: static modeling elements representing the characteristics of participants and the relations between them; dynamic modeling elements describing the behaviour of participants and their interactions. Figure 1 shows an example of a descriptor of a subclass of

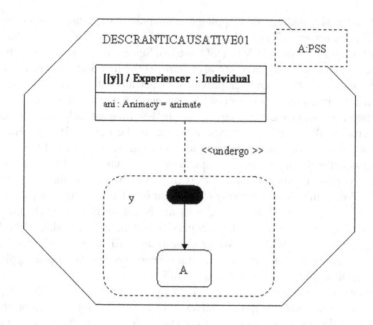

Fig. 1. EVENTITY FRAME TEMPLATE diagram of anticausatives

the anticausative verbs, where the characteristic features and the behaviour of a single participant are represented.

Each parent table is related to child tables, which contain the French and the Hungarian equivalents of the Bulgarian verbs. The relationship among the data is one-to-many, that is, to one Bulgarian verb there are one or more than one equivalents in French or Hungarian. In case of more than one equivalent in French or in Hungarian, the additional equivalents are given in separate tables. Figure 2 provides a schema of the database structure. The left-hand rectangle corresponds to the parent tables, and the right-hand rectangle designates the child tables. The first line in the rectangles indicates the table type, and the rest of the lines denote the fields of the respective table.

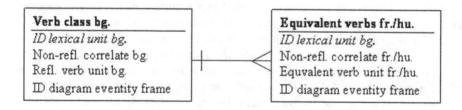

Fig. 2. One-to-many relation between the data tables

3   LMF

The LMF models are represented by UML classes, associations among classes, and data categories that function as UML attribute-value pairs. LMF comprises two types of components: a Core package and Extensions. The Core package is a metamodel that provides a flexible basis for building LMF models, it is the top-level of the hierarchy of lexicon information. The Core package classes that are used in the LMF conformant representation of *SemInVeSt* are: *Global Information* (representing administrative information), *Lexical Resource*, *Lexicon* (containing the lexical entries of a given language within the entire resource), *Lexical Entry* (representing a lexeme in a given language), *Form*, *Sense*. The LMF Extensions are UML packages which are anchored in a subset of the Core package classes. In the work, presented in this paper, the *NLP Semantics* extension and the *NLP Multilingual Notations* extension are used. The LMF models are translated into an XML-based format, suitable for data exchange. The LMF models and exchange formats will be illustrated in the next section, where the LMF-based representation of *SemInVeSt* is discussed.

4   SemInVeSt-LMF

In this section the LMF-based format of *SemInVeSt* is described, which serves as an exchange format of the language resource. It consists of two parts. Using the LMF *Multilingual Notations* model, the first part represents the global structure of *SemInVeSt* as a multilingual lexicon with a reference to the semantic descriptors as an external resource. The second part is the representation of the UER eventity frames in a format, based on the *NLP Semantics* extension of LMF.

4.1   SemInVeSt-LMF Global Structure

Figure 3 depicts the UML class diagram of the global structure of *SemInVeSt* together with the class adornments as attribute-value pairs. The *Lexical Resource* is structured according to the principle of a lexicon per language, that is, it includes three *Lexicon* classes containing the *Lexical Entries* of the three languages represented in *SemInVeSt*. The value of the *id* attribute of an instance of the *Sense* class is a key identifier from the relational database format of *SemInVeSt*. The *Sense Axis* is a class representing the relationship between different closely related senses in different languages. The value of the *id* attribute of an instance of the *Sense Axis* class is an identifier of an eventity frame template diagram (cf. Figure 1). The value of the *label* attribute of the *Sense Axis* is the name of a verb class, for instance, *anticausative*. The LMF authors state that their purpose in not to provide a complex knowledge organization system [4]. Therefore, they provide the possibility to refer to an external representation system such as the UER conformant eventity frame diagrams of *SemInVeSt*. Thus the attributes of the *Interlingual External Ref* class refer to files containing the eventity frame diagrams.

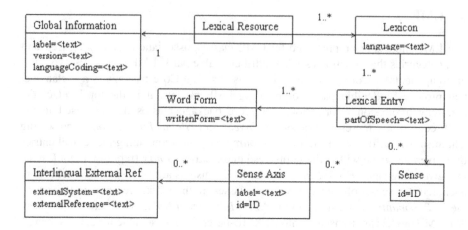

Fig. 3. SemInVeSt-LMF global structure

The LMF instance level representation of the data is in XML format. Below is an XML fragment encoding the information about a Bulgarian verb in a reflexive form. It should be noted that, similar to the KYOTO-LMF [9] and the GermaNet-LMF [11] data representation, the class adornments in *SemInVeSt* are encoded as attributes and values of the XML elements corresponding to the UML classes, and not as attributes and values of a separate feat element as in the XML schema suggested in the LMF specifications [3].

```
<LexicalEntry partOfSpeech="verb">
    <WordForm writtenForm="vdahnovyavam se"/>
    <Sense id="ANTICAUSATIVEBG0002"/>
    <SenseAxis label="anticausative"
               id="DESCRANTICAUSATIVE01"/>
    <InterlingualExternalRef externalSystem="SemInVeSt-Database"
               externalReference="FigDescrAnticausative01.jpg"/>
</LexicalEntry>
```

4.2   SemInVeSt-UER-LMF Semantic Structure

This section is focused on the challenging task of representing the UER eventity frame models in LMF format. For that purpose we rely on the abstract syntax of UML, the common underlying model of UER and LMF, in order to:

1. define the modeling elements that correspond in the UER eventity frame model and the LMF Semantics model;
2. extend the LMF Semantics model with modeling elements, which are necessary for completing the representation of the eventity frames.

The LMF modeling elements which correspond to UER modeling elements are as follows:

- the LMF *Semantic Predicate* corresponds to the UER *Eventity Frame*;
- the LMF *Semantic Argument* corresponds to the UER *Participant Class*.

The elements that extend the LMF model are as follows:

- the *Participant Role* is added to represent the UER idiosyncratic class of *Participant Roles*;
- the *Participant Type* is added to represent the UER reference to ontological types describing the participants;
- the *State Machine* is added to represent the behaviour of the participant it belongs to;
- the *State* is added as part of the *State Machine*;
- the *Transition* is added as part of the *State Machine*.

It should be noted that the UER dynamic structure concepts and notation are based on the UML statechart and activity diagrams (cf. Figure 1). For mapping the UER dynamic modeling elements to the LMF model, the UML abstract syntax is used, which is defined in order to describe, in a notation independent way, the concrete graphic syntax of UML (that is, to describe the UML language with the means of UML itself). The abstract syntax is represented in UML class diagrams showing the metaclasses defining the constructs and their relationships [6, p.18]. Figure 4 depicts the UML class diagram of the UER Eventity Frame model.

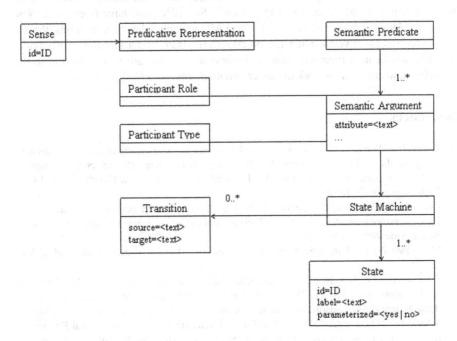

Fig. 4. SemInVeSt-UER-LMF semantic structure

Below is an XML fragment representing the eventity frame diagram in Figure 1.

```xml
<SemanticPredicate>
    <SemanticArgument animacy="animate">
        <ParticipantRole>Experiencer</ParticipantRole>
        <ParticipantType>Individual</ParticipantType>
        <StateMachine>
            <State id="STATE01"
                   label="unspecified state"
                   parameterized=""></State>
            <State id="STATE02"
                   label="passive simple state"
                   parameterized="yes">A</State>
            <Transition source="STATE01"
                        target="STATE02"/>
        </StateMachine>
    </SemanticArgument>
</SemanticPredicate>
```

5   Conclusion

The mapping of *SemInVeSt* as a multilingual semantic resource to the LMF representational model is facilitated by the shared UML metamodel architecture. Where possible, the modeling elements, predefined in the LMF Specifications, have been used. New modeling elements have been added to the LMF metamodel for representing the semantic descriptors of verbs. The UER-based eventity frame diagrams of *SemInVeSt* and the LMF conformant representation of the lexical resource are a contribution to the emerging innovative framework of *object-oriented semantics* [12].

References

1. Zipser, F.S., Romary, L.: A model oriented approach to the mapping of annotation formats using standards. In: Proceedings of a Workshop on Language Resources and Language Technology Standards – State of the Art, Emerging Needs, and Future Developments, LREC 2010, Malta, pp. 7–18 (2010)
2. Cimiano, P., Buitelaar, P., McCrae, J., Sintek, M.: LexInfo: A declarative model for the lexicon-ontology interface. Journal of Web Semantics: Science, Services and Agents on the World Wide Web, 29–51 (2011)
3. LMF: ISO 24613: Language resource management – Lexical markup framework (LMF) (2008)
4. Francopoulo, G., Bel, N., George, M., Calzolari, N., Monachini, M., Pet, M., Soria, C.: Lexical markup framework: ISO standard for semantic information in NLP lexicons. In: (Gesellschaft fuer linguistische Datenverarbeitung) (GLDV), Tuebingen, Germany (2007)
5. Schalley, A.C.: Cognitive Modeling and Verbal Semantics. A Representational Framework Based on UML. In: Trends in Linguistics. Studies and Monographs, vol. 154, Mouton de Gruyter, Berlin (2004)

6. OMG: ISO/IEC 19501 Unified Modeling Language Specification, Version 1.4.2. Object Management Group (OMG) (2005), http://www.omg.org/
7. Bel, N., Espeja, S., Marimon, M., Villegas, M.: COLDIC, a lexicographic platform for LMF compliant lexica. In: Proceedings of LREC 2008, Marrakech, Morocco (2008)
8. Maks, I., Tiberius, C., van Veenendaal, R.: Standardising bilingual lexical resources according to the Lexicon Markup Framework. In: Proceedings of LREC 2008, Marrakech, Morocco (2008)
9. Soria, C., Monachini, M.: Kyoto-LMF: WordNet Representation Format. KYOTO Working Paper: WP02 TR002 V4 Kyoto LMF (2008)
10. Soria, C., Monachini, M., Vossen, P.: Wordnet-LMF: Fleshing out a standardized format for wordnet interoperability. In: IWIC 2009: Proceedings of the 2009 International Workshop on Intercultural Collaboration, pp. 139–146. ACM, New York (2009)
11. Henrich, V., Hinrichs, E.: Standardizing wordnets in the ISO standard LMF: Wordnet-LMF for GermaNet. In: Proceedings of the 23rd International Conference on Computational Linguistics (COLING 2010), Beijing, China, pp. 456–464 (2010)
12. Goddard, C., Schalley, A.C.: Semantic analysis. In: Indurkhya, N., Damerau, F.J. (eds.) Handbook of Natural Language Processing. Chapman and Hall/CRC Machine Learning and Pattern Recognition, pp. 93–120. CRC Press / Taylor and Francis Group, Goshen, Connecticut, USA (2010)

The Rule-Based Approach
to Czech Grammaticalized Alternations*

Václava Kettnerová, Markéta Lopatková, and Zdeňka Urešová

Charles University in Prague
Institute of Formal and Applied Linguistics
{kettnerova,lopatkova,uresova}@ufal.mff.cuni.cz

Abstract. Under the term grammaticalized alternations, we understand changes in valency frames of verbs corresponding to different surface syntactic structures of the same lexical unit of a verb. Czech grammaticalized alternations are expressed either (i) by morphological means (diatheses), or (ii) by syntactic means (reciprocity). These changes are limited to changes in morphemic form(s) of valency complementations; moreover, they are regular enough to be captured by formal syntactic rules.

In this paper a representation of Czech grammaticalized alternations and their possible combination is proposed for the valency lexicon of Czech verbs, VALLEX.

1 Introduction

Prototypically, a single meaning of a verb can be surface syntactically structured in different ways. It follows that changes in valency frame of a verb (usually called alternations, see [1]) must be described either by syntactic rules, or they must be specified in lexicon entries. Here we focus on the changes resulting from the use of specific grammatical means (e.g., passivisation or reciprocity). These changes are referred here to as grammaticalized alternations. Czech grammaticalized alternations are expressed either (i) by morphological means (1a)–(1b) (traditionally referred to as diatheses), or (ii) by syntactic means (2a)–(2b) (reciprocity).

(1) a. *Mobilní operátoři snížili$_{active}$ cenu volání.*
    The mobile network operators reduced$_{active}$ the price of calls.
  b. *Cena volání se snížila$_{deagentive}$.*
    'The price of calls – *refl* – reduced$_{deagentive}$.'
(2) a. *Petr líbá Marii.*
    Peter kisses Mary.
  b. *Petr a Marie se líbají.*
    Peter and Mary kiss (each other).

Whereas the same type of grammatical means cannot be combined together, the combinations of diatheses and reciprocity are allowed within a single surface syntactic structure, see example of the verb *domluvit* 'to arrange' in (3a)–(3b):

---

* This work has been using language resources stored and/or distributed by the LINDAT-Clarin project of MŠMT (project LM2010013). The research reported in this paper has been supported by GA ČR, grant No. GA P406/12/0557 and partially grant No. GA P406/10/0875.

P. Sojka et al. (Eds.): TSD 2012, LNCS 7499, pp. 158–165, 2012.

(3) a. *Petr si domluvil*_{active} *s Janem schůzku.*

Peter arranged_{active} an appointment with John.

b. *Petr a Jan (spolu) mají domluvenu*_{resultative} *schůzku.*

Peter and John have arranged_{resultative} an appointment.

The objective of this paper is to propose a representation of Czech grammaticalized alternations, i.e., diatheses and reciprocity, and their combinations in the valency lexicon of Czech verbs, VALLEX[1]. This lexicon takes the Functional Generative Description (FGD) as its theoretical background. In FGD, valency is related to the layer of linguistically structured meaning, see [2] and [3]. In VALLEX, valency characteristics of a verb are encoded in valency frames which are modeled as sequences of valency slots, each slot standing for a single valency complementation. The slots consist of a functor (coarse-grained semantic role), a list of morphemic form(s) and information on obligatoriness.

In order to satisfy needs of both human users and automated language processing, this lexicon is available in three formats: XML, HTML and PDF version. The information on valency of verbs (including the linguistically adequate and economic description of grammaticalized alternations) can be used for various NLP tasks, as e.g., machine translation, tagging, word sense disambiguation.

We demonstrate that grammaticalized alternations of both types (diatheses and reciprocity) are limited to changes in morphemic form(s) of valency complementations and that these changes are regular enough to be captured by formal syntactic rules. For the representation of combination of diatheses and reciprocity, additional rules are not required. However, explicit rule ordering is necessary to be determined.

From a broader perspective, we address a general question which part of the information needed for a language description is to be captured by general rules (i.e., in the grammar component of the language system) and which part is best recorded in the form of lexicon entries.

2   Grammaticalized Alternations: Theory

2.1   Diatheses

A central type of grammaticalized alternations is characteristic of the relations between surface syntactic structures which differ in the morphological category of voice. In Czech, these relations are referred to as diatheses. The members of diatheses are characterized by a permutation of valency complementations. This permutation affects the prominent surface position of subject from which the valency complementation 'ACTor' is prototypically shifted.

Five types of these relations are determined according to five marked morphological meanings of a verb: *passive, deagentive, dispositional, resultative*, and *recipient passive* diathesis. The syntactic structures characterized by these marked morphological meanings represent the marked members of diatheses whereas the structures with active voice of a verb constitute the unmarked one, see esp. [4].

---

[1] http://ufal.mff.cuni.cz/vallex/2.5/

2.2 Reciprocity

In contrast to diatheses, reciprocity represents rather peripheral type of grammaticalized alternations. It is connected with 'symmetricalization' – a syntactic operation conducted on two (or three) valency complementations, which (if their semantic properties allow for it) are used symmetrically; e.g., (2b) means that *Petr líbá Marii a (zároveň) Marie líbá Petra* 'Peter kisses Mary and (at the same moment) Mary kisses Peter', see esp. [5].

The reciprocal constructions are associated with the shift of the valency complementation expressed in a less prominent surface syntactic position into the more significant syntactic position (subject or direct object) of the other symmetrically used valency complementation. Whereas the prominent position is 'multiplied' (either by coordination, or plural), the less significant position is either deleted from the resulted surface syntactic structure, or it is realized with reflexive pronoun. Reflexive verb forms must be used if 'ACTor' is involved in the reciprocity relation.

Different types of reciprocity can be determined according to valency complementations which enter into the symmetric relation.

The reciprocal constructions can be considered as marked members of this type of alternation whereas the unreciprocal constructions as unmarked ones.

2.3 Combination of Diatheses and Reciprocity

Whereas the same type of grammaticalized alternations cannot be combined together, the combinations of diatheses and reciprocity are allowed within a single surface syntactic structure. Such combinations are applicable on condition that a certain morphological meaning (Section 2.1) can be applied to a verb and at the same time some of its valency complementations can be used symmetrically (Section 2.2). See the following example of the verb *domluvit* 'to arrange' allowing both reciprocal use of its valency complementations 'ACTor' and 'ADDRessee' (4b) and resultative morphological meaning (4c). Then these linguistic means can be combined within a single surface structure (4d):

(4) a. *Petr$_{ACT}$ domluvil$_{active}$ s Janem$_{ADDR}$ schůzku.*
   Peter arranged$_{active}$ an appointment with John.
   b. *(Petr a Jan)$_{rcp:ACT-ADDR}$ si (spolu) domluvili schůzku.*
   (Peter and John)$_{rcp:ACT-ADDR}$ arranged an appointment (together).
   c. *Petr má s Janem domluvenu$_{resultative}$ schůzku.*
   'Peter – has – with John – arranged$_{resultative}$ – an appointment.'
   d. *(Petr a Jan)$_{rcp:ACT-ADDR}$ (spolu) mají domluvenu$_{resultative}$ schůzku.*
   '(Peter and John)$_{rcp:ACT-ADDR}$ – (together) – have – arranged$_{resultative}$ –
   – an appointment.'

However, in some cases, the combination of a certain type of diathesis and reciprocity leads to an ungrammatical structure despite being separately available for a verb, see e.g. the verb *vyhubovat* 'to tell off'. Although this verb allows both for reciprocal use of 'ACTor' and 'PATient' (5b) and for recipient passive morphological meaning (5c), the combination of these linguistic means within a single surface syntactic structure results in an ungrammatical construction (5d).

(5) a. *Jan$_{ACT:Subj}$ vyhuboval$_{active}$ Marii$_{PAT:InObj}$.*
   John$_{ACT:Subj}$ told$_{active}$ Mary$_{PAT:InObj}$ off.
   b. *(Jan a Marie)$_{rcp:ACT-PAT:Subj}$ si (vzájemně) vyhubovali.*
   (John and Mary)$_{rcp:ACT-PAT:Subj}$ told off (each other).
   c. *Marie$_{PAT:Subj}$ dostala (od Jana)$_{ACT:Adv}$ vyhubováno$_{recipient}$.*
   'Mary$_{PAT:Subj}$ – got – (from John)$_{ACT:Adv}$ – told$_{recipient}$ off.'
   d. *dostali (vzájemně) vyhubováno$_{recipient}$ (od Jana a Marie)$_{rcp:ACT-PAT:Adv}$
   'got – (each other) – told$_{recipient}$ off (by John and Mary)$_{rcp:ACT-PAT:Adv}$'

The reason of the ungrammaticality of the combination of recipient passive diathesis and reciprocity in (5d) lies in the fact that the surface syntactic shifts associated with these alternations are contradictory; formally, they result in the surface syntactic structure without subject while the verb form being inadequate. First, reciprocity (putting 'ACTor' and 'PATient' in symmetric relation) leads to the shift of 'PATient' (*Marii* in (5a)) to the subject position (which is multiplied by coordination in (5b)). Second, recipient passive diathesis prototypically consists in the following changes: (i) shifting the valency complementation occupying the subject to an adverbial position while (ii) the vacated subject being filled with the valency complementation corresponding to the cognitive 'Recipient' (expressed originally in dative); (iii) the dative surface position is deleted. However, by applying recipient passive diathesis on reciprocal construction in (5b), both (coordinated) 'ACTor' and 'PATient' (*Jan a Marie*) should be shifted from the subject which remains vacant as no dative 'Recipient' is present; as a result, the verb form is inappropriate for subject-less construction, see (5d).

Let us generalize this observation. Considering the basic postulates – (i) diatheses are characterized by 'ACTor' shifted from the subject syntactic position into a less prominent surface position (Section 2.1) whereas (ii) reciprocity is connected with the shift of a valency complementation occupying the less prominent surface syntactic position into the position of subject or direct object (Section 2.2), see [6], – we can formulate the following hypotheses:

A. If different types of grammaticalized alternations are combined, the order of their application can be prescribed; namely, in certain cases reciprocity should precede diathesis as diathesis may result in a surface construction that do not allow for certain types of reciprocity.

B. Moreover, the combinations of (a certain type of a) diathesis and reciprocity are allowed within a single surface syntactic structure under the condition that the application of reciprocity preserves formal conditions on the application of the particular diathesis.

3   Representation of Grammaticalized Alternations

For the purpose of the representation of grammaticalized alternations, we divide the lexicon into a data component and a rule component. In case of diatheses and reciprocity, the changes in the valency structure of verbs are limited to changes in morphemic form(s) of the valency complementations affected by the shifts in surface syntactic positions; these changes are regular enough to be captured by formal syntactic rules. In Section 3.1, we demonstrate two examples of rules representing diathesis and reciprocity, respectively. For the description of these phenomena, we adopt syntactic rules formulated in the lexicon PDT-VALLEX, see esp. [7].[2] Section 3.2 is focused on the representation of the combination of diatheses and reciprocity.

3.1   Representation of Diatheses and Reciprocity

The **data component** of the lexicon contains lexical entries for individual lexical units. Only the unmarked valency frame, i.e., the valency frame representing use in the active unreciprocal structure, is stored for each lexical unit. A lexical unit is ascribed with the special attributes -diat and -rcp where the applicability of individual diatheses and reciprocity is specified, respectively. Then the **rule component** of the lexicon stores rules describing changes in valency frames associated with individual diatheses and reciprocity; in VALLEX, we make use of the rules designed for PDT-VALLEX [7].

Let us firstly demonstrate our approach on the example of passive diathesis of the verb *seznámit* 'to introduce'. The lexical entry of this verb stored in the data component is structured as follows (the lexical entry is simplified and translated for better understanding):

- **lemma: seznámit**^{pf} 'to introduce'
- **gloss:** *představit* 'to bring together'
- **frame:** ACT_1^{obl} ADDR_4^{obl} PAT_{s+7}^{obl}
- **example:** *seznámit přítele s příbuzným* 'to introduce the friend to the relative'
- **diat:** pass, disp, res1, res2
- **rcp:** ADDR-PAT, ACT-ADDR-PAT

In the rule component, the rule Pass.r given in Table 1 represents changes in valency structure of the verb associated with passive diathesis. (The rule is simplified here for better understanding; the thorough formal rule is split in order to cover all variants of relevant valency frames in the lexicon.)

Table 1. Pass.r rule for passive diathesis

Type	passive	
Action	verbform	replace(active vf → passive vf)
	ACT	replace(nom → instr,od+gen)
	ADDR	replace(acc → nom)

The Pass.r rule allows to derive the marked valency frame (6b) representing the passive surface structure (illustrated by example (7b)) from the valency frame (6a) corresponding to the unmarked active use of the verb (example (7a)).

---

[2] The rules (rule instances) in the cited work are a generalization of rules used in quality checking of the Prague Dependency Treebank 2.0 (PDT).

(6) a. verbform$_{active}$ ACT$_{nom}^{obl}$ ADDR$_{acc}^{obl}$ PAT$_{s+instr}^{obl}$ $\Rightarrow$

    b. verbform$_{passive}$ ACT$_{instr,od+gen}^{obl}$ ADDR$_{nom}^{obl}$ PAT$_{s+instr}^{obl}$

(7) a. *Sára*$_{ACT:Subj}$ *seznámila*$_{active}$ *přítele*$_{ADDR:Obj}$ *se svou matkou*$_{PAT:InObj}$.
    Sara$_{ACT:Subj}$ introduced$_{active}$ her friend$_{ADDR:Obj}$ to her mother$_{PAT:InObj}$.

    b. *Přítel*$_{ADDR:Subj}$ *byl* (*Sárou*$_{ACT:Adv}$) *seznámen*$_{passive}$ *s její matkou*$_{PAT:InObj}$.
    Her friend$_{ADDR:Subj}$ was introduced$_{passive}$ to her mother$_{PAT:InObj}$
    (by Sara$_{ACT:Adv}$).

The same principles can be applied on reciprocities. Only valency frames corresponding to unreciprocal structures are contained in the **data component**. By each relevant
valency frame, valency complementations allowing for symmetrical use are listed in
the special attribute -rcp. For example, the verb *seznámit* 'to introduce' allows for
symmetrical use of 'ADDRessee' and 'PATient' (7c) or even of all three complementations 'ACTor', 'ADDRessee' and 'PATient' (7d). For the sake of simplicity, we limit
our explanation to cases when only two valency complementations are affected.

(7) c. *Sára*$_{ACT:Subj}$ *seznámila* (*přítele a svou matku*)$_{rcp:ADDR-PAT:Obj}$.
    Sara$_{ACT:Subj}$ brought together (her friend and her mother)$_{rcp:ADDR-PAT:Obj}$.

    d. (*Sára, její přítel a matka*)$_{rcp:ACT-ADDR-PAT:Subj}$ *se* (*navázjem*) *seznámili*.
    (Sara, her friend and her mother)$_{rcp:ACT-ADDR-PAT:Subj}$ brought
    together (each other).

The **rule component** of the lexicon stores rules describing changes in valency frames
associated with this type of reciprocity, see Table 2 (again, the rule is simplified here).

Table 2. Rcp.r.ADDR-PAT rule for reciprocity of 'ADDRessee' and 'PATient'

Type	rcp-ADDR-PAT		comment
Action	ADDR	replace(acc $\rightarrow$ h-acc)	(1)
	PAT	replace(s+instr $\rightarrow$!)	(2)
		add (*spolu, navzájem, mezi sebou, ...*; opt)	(3)

Commentary on the Rcp.r.ADDR-PAT rule:
(1) 'ADDRessee' stays in accusative and it must be realized by coordinated nouns, by plural noun
or by an expression of semantic plurality (denoted by the symbol 'h').
(2) 'PATient' is merged with 'ADDRessee' and thus is not expressed in a separate surface
syntactic position (denoted by ' ! ').
(3) In addition, reciprocity may be lexically signalized by expressions such as *spolu* 'together',
navzájem 'mutually', *mezi sebou* 'each other', etc.

The valency frame (8b) representing the reciprocal structure in (7c) can be derived from
the valency frame (8a) corresponding to the unreciprocal structure (7a) on the basis of
the rule given in Table 2:

(8) a. ACT$_{nom}^{obl}$ ADDR$_{acc}^{obl}$ PAT$_{s+instr}^{obl}$ $\Rightarrow$ b. ACT$_{nom}^{obl}$ ADDR-PAT$_{h-acc}^{obl}$

3.2   Representation of Combination of Diatheses and Reciprocity

We can observe that no additional rules are necessary for the representation of combinations of diatheses and reciprocity. However, in many cases explicit ordering of the rules describing the changes in valency frames must be determined. Esp. in cases where reciprocity involves 'ACTor' the rule representing this type of alternation must precede the rule(s) describing certain types of diatheses.

For instance, the passive diathesis and reciprocity of 'ADDRessee' and 'PATient' of the verb *seznámit* 'to bring together' can be combined within a single surface syntactic structure, see (7e):

(7) e. *(Její přítel a matka)*$_{rcp:ADDR-PAT:Subj}$ *byly (navzájem) představeni*$_{passive}$
    (Sárou)$_{ACT:Adv}$.
    (Her friend and her mother)$_{rcp:ADDR-PAT:Subj}$ were brought together$_{passive}$
    (by Sara)$_{ACT:Adv}$.

The valency frame in (9c) describing combination of reciprocity 'PATient' and 'AD-DResee' and passive diathesis (as illustrated in (7e)) can be derived from the valency frame corresponding to active and unreciprocal structure (9a) by the consecutive application of the rule `Rcp.r.ADDR-PAT` (Table 2) and of a rule similar to the `Pass.r` (Table 1):

(9) a. verbform$_{active}$ ACT$_{nom}^{obl}$ ADDR$_{acc}^{obl}$ PAT$_{s+instr}^{obl}$ $\Rightarrow$

   b. verbform$_{active}$ ACT$_{nom}^{obl}$ ADDR-PAT$_{h-acc}^{obl}$ $\Rightarrow$

   c. verbform$_{passive}$ ACT$_{instr,od+gen}^{obl}$ ADDR-PAT$_{h-nom}^{obl}$

The explicit ordering of the rules representing diatheses and reciprocities is enforced by strict conditions imposed on their applications. In case that a lexical unit of a verb allows for applying a certain type of diathesis and at the same time some of its valency complementations can be used reciprocally, the respective rules for this type of diathesis and reciprocity are applied consecutively only in case that they meet the conditions of their applications; in other case this operation is blocked (see also Section 2.3).

4   Conclusion and Future Work

We have proposed the representation of Czech grammaticalized alternations associated with diatheses and reciprocities, with the focus on their combinations. We have demonstrated that these alternations can be described by formal syntactic rules. These rules are stored in the rule component of the lexicon. In the data component, only valency frames corresponding to the unmarked use (i.e., active unreciprocal use) of lexical units are captured; marked (morpho)syntactic uses of a single lexical unit are obtained by applying syntactic rules from the rule component. In cases when diatheses and reciprocity are combined, explicit rule ordering is necessary to be determined.

The proposed representation of Czech grammaticalized alternations has been already partially applied in the valency lexicon of Czech verbs VALLEX (namely deagentive diathesis and reciprocity). Further enrichment of the lexicon is planned for future in two directions: (i) Full adaptation of syntactic rules defined for PDT-VALLEX is under preparation.[3] (ii) Inspired by [8], (semi)automatic identification of the lexical items in the data component that allow for individual types of diatheses is in development. From the theoretical point of view, reciprocity affecting three valency complementations deserves further attention. Moreover, special attention will be paid to the possible combinations of the grammaticalized alternations.

Only the implementation of the tasks mentioned above – the adaptation of formal syntactic rules and identification of relevant lexical items for individual diatheses – allows for thorough evaluation of the outputs of the proposed rules (in a form of valency frames corresponding to marked (morho)syntacic uses of lexical units of Czech verbs). This will be of a great interest since the original rules designed for PDT consistency checking 'over-generated' (in a sense they allowed also wrong surface configurations, relying on the fact that the underlying text analyzed in the corpus was grammatically correct Czech). We suppose that this shortcoming will be eliminated by imposing strict conditions on the application of the rules.

References

1. Levin, B.C.: English Verb Classes and Alternations: A Preliminary Investigation. The University of Chicago Press, Chicago (1993)
2. Sgall, P., Hajičová, E., Panevová, J.: The Meaning of the Sentence in Its Semantic and Pragmatic Aspects. Reidel, Dordrecht (1986)
3. Panevová, J.: Valency Frames and the Meaning of the Sentence. In: Luelsdorff, P.A. (ed.) The Prague School of Structural and Functional Linguistics, pp. 223–243. John Benjamins Publishing Company, Amsterdam (1994)
4. Panevová, J., et al.: Syntax současné češtiny (na základě anotovaného korpusu). Nakladatelství Karolinum, Praha (manuscript)
5. Panevová, J., Mikulová, M.: On Reciprocity. The Prague Bulletin of Mathematical Linguistics 87, 27–40 (2007)
6. Kettnerová, V., Lopatková, M.: Changes in valency structure of verbs: Grammar vs. lexicon. In: Levická, J., Garabík, R. (eds.) Proceedings of Slovko 2009, NLP, Corpus Linguistics, Corpus Based Grammar Research, Bratislava, SAV, pp. 198–210 (2009)
7. Urešová, Z.: Valence sloves v Pražském závislostním korpusu. In: Studies in Computational and Theoretical Linguistics, Institute of Formal and Applied Linguistics, Prague (2011)
8. Skoumalová, H.: Czech Syntactic Lexicon. Ph.D. thesis, Charles University in Prague (2001)

---

[3] The total number of rules described in [7] is 44; since some of the rules serve as templates over varying functors, the number of rule instances is over 100.

Semi-supervised Acquisition
of Croatian Sentiment Lexicon

Goran Glavaš, Jan Šnajder, and Bojana Dalbelo Bašić

University of Zagreb, Faculty of Electrical Engineering and Computing
Unska 3, 10000 Zagreb, Croatia
{goran.glavas,jan.snajder,bojana.dalbelo}@fer.hr

Abstract. Sentiment analysis aims to recognize subjectivity expressed in natural language texts. Subjectivity analysis tries to answer if the text unit is subjective or objective, while polarity analysis determines whether a subjective text is positive or negative. Sentiment of sentences and documents is often determined using some sort of a sentiment lexicon. In this paper we present three different semi-supervised methods for automated acquisition of a sentiment lexicon that do not depend on pre-existing language resources: latent semantic analysis, graph-based propagation, and topic modelling. Methods are language independent and corpus-based, hence especially suitable for languages for which resources are very scarce. We use the presented methods to acquire sentiment lexicon for Croatian language. The performance of the methods was evaluated on the task of determining both subjectivity and polarity at (*subjectivity* + *polarity* task) and the task of determining polarity of subjective words (*polarity only* task). The results indicate that the methods are especially suitable for the *polarity only* task.

1 Introduction

Knowing how other people perceive certain events, entities, and phenomena can be important in various areas of human activity. Sentiment analysis is a subdiscipline of computational linguistics that aims to recognize the subjectivity and attitude expressed in natural language texts. Applications of sentiment analysis are numerous, including sentiment-based document classification (as opposed to traditional topic-based document classification) [1,2], opinion-oriented information extraction [3], and question answering [4].

Sentiment of textual units is usually analysed from two perspectives: *subjectivity* and *polarity*. Subjectivity analysis tries to answer whether the text unit is subjective or objective (neutral) and often precedes polarity analysis, which determines whether a subjective unit expresses positive or negative sentiment. While majority of research approaches [5,6,7] see subjectivity and polarity as categorical terms (i.e., classification problems), there has been some research effort in assessing subjectivity and polarity as graded values [8,9]. The majority of research in the field relies on the existence of sentiment-annotated lexicon containing subjectivity and polarity information of individual words [10,11].

In this paper we present semi-supervised methods for automated acquisition of sentiment lexicon that do not depend on any pre-existing language resource (e.g.,

P. Sojka et al. (Eds.): TSD 2012, LNCS 7499, pp. 166–173, 2012.

WordNet [12]). Our methods are corpus-based and therefore especially suitable for languages such as Croatian, for which language resources are scarce. We compare three different unsupervised methods for automated sentiment lexicon acquisition: latent semantic analysis, graph-based method, and topic modelling. All three methods use the same hand-labeled seed sets as input. All of the employed methods are evaluated on two different tasks. The first task is to determine both subjectivity and polarity of content words appearing in corpus. Put differently, the goal is to mark each corpus word as either *neutral, positive*, or *negative (subjectivity + polarity* task). The second task is to decide only on the polarity of subjective words *polarity only* task. One of the goals was to compare the difficulty of the two tasks. Evaluation is conducted against a human-annotated gold set. Extensive experiments show that the results obtained (especially on the *polarity only* task) are reasonably good considering the lack of explicit semantic links between words in the corpus which exist in semantic networks such as WordNet.

The rest of the paper is structured as follows. Section 2 presents the related work. Section 3 discusses the semi-supervised methods for sentiment lexicon acquisition. In Section 4 we describe for each method the experimental setup and the results obtained. Finally, we conclude in Section 5.

2 Related Work

The task of building sentiment lexicon is also referred to as determining the *prior polarity* of words. In their pioneering work, Hatzivassiloglou and McKeown [5] determine polarity of adjectives based on their co-occurrences in conjunctions such as *and* (same polarity) or *but* (opposite polarity). Turney and Littman [6] use pointwise mutual information (PMI) and latent semantic analysis (LSA) [13] to determine the similarity of the word with positive and negative seed sets. We follow the idea of seed sets and LSA-based comparison, but we also consider objective (i.e., neutral by sentiment) words.

In [14] authors build a graph of adjectives based on synonymy relations gathered from WordNet. Decision on the polarity of the adjective is made by comparing the shortest path distances from positive and negative seed adjectives *good* and *bad*. Esuli and Sebastiani [15] build a graph based on gloss relations from WordNet on which they perform PageRank algorithm [16] in two runs. Word's polarity is decided based on the difference between its PageRank values of the two runs. Our second method follows the idea of building a graph based on seed sets. In the absence of lexical resource such as WordNet for Croatian, we build graph edges based on word co-occurrences in the corpus. Unlike Esuli and Sebastiani, who build a single graph and perform two PageRank runs, we build two different graphs, starting from positive and negative seed sets.

Our third method is based on the hypothesis that words that express similar sentiment are similarly distributed among the topics of the corpus. Latent Dirichlet allocation (LDA) [17,18] is used as a topic model to obtain topic distributions over words. The distribution vector of each word is compared with the vectors from the positive and negative seed sets. To the best of our knowledge, LDA has not been previously used on the task of building prior polarity lexicon.

3   Sentiment Lexicon Acquisition

Our main goal was to create sentiment lexicon for Croatian language because such a resource would be valuable for many applications. We consider only methods that do not depend on external language resources. We used the following seed sets, each containing 10 words for all three semi-supervised methods employed:

positiveSeeds = {dobar (good), lijep (beautiful), pozitivan (positive), pravi (right), sreća (happiness), voljeti (to love), odličan (excellent), uspješan (successful), mir (peace), nov (new)}

negativeSeeds = {loš (bad), nemati (not to have), nesreća (misfortune), negativan (negative), pogreška (mistake), problem (problem), rat (war), napad (attack), crn (black), mrtav (dead)}

Latent Semantic Analysis

Latent semantic analysis is a well-known technique for identifying semantically related concepts and reducing dimensionality [13]. First we create word-document matrix, a sparse matrix whose rows correspond to words and columns correspond to documents. Word-document matrix is decomposed using singular value decomposition (SVD), a well-known linear algebra procedure. Dimensionality reduction is performed by approximating the original matrix using only the top k largest singular values. We build two different word-document matrices using different weighting schemes. The elements of the first matrix were calculated using the *tf-idf* weighting scheme, while for the second matrix the *log-entropy* weighting scheme was used. In the *log-entropy* scheme, each matrix element, $m_{w,d}$, is calculated using logarithmic value of word-document frequency and the global word entropy, as follows:

$$m_{w,d} = \log\left(tf_{w,d} + 1\right) \cdot g_e(w)$$

with

$$g_e(w) = 1 + \frac{1}{\log n} \sum_{d' \in D} \frac{tf_{w,d'}}{gf_w} \log \frac{tf_{w,d'}}{gf_w}$$

where $tf_{w,d}$ represents occurrence frequency of word w in document d, parameter gf_w represents global frequency of word w in corpus D, and n is the number of documents in corpus D.

We then decompose each of the two matrices using SVD and get representation of each word in the vector space of reduced dimensionality k ($k \ll n$). LSA vectors tend to express semantic properties of words, hence the similarity between the vectors may be used as a measure of semantic similarity between the corresponding words. We experimented with different vector similarity measures (cosine similarity, Dice coefficient, Jaccard coefficient). Finally, sentiment prediction for each word was calculated by comparing the similarity between the word vector and the vectors in both seed sets.

PageRank Method

We iteratively build two different undirected graphs (positive and negative), starting from different seed sets. Seed set elements form the initial vertex set of the graph. In each iteration words are introduced as new vertices if they co-occur inside the window of a given size with the words whose vertices already exist in the graph. Each graph edge is assigned a weight denoting the average co-occurrence distance between vertices it connects. Initial value of 1 is assigned to the seed set vertices and 0 to all other vertices. Next, weighted adjacency matrix $\mathbf{W}$ of the graph is row-normalized and the PageRank algorithm is run on the graph. The PageRank algorithm iteratively computes the vector of vertex scores $\mathbf{a}$ as follows:

$$\mathbf{a}^{(k)} = \alpha\mathbf{a}^{(k-1)}\mathbf{W} + (1-\alpha)\mathbf{e}$$

where α is the PageRank damping factor. Vector $\mathbf{e}$ models the normalized internal source of score for all vertices and its elements sum up to 1. We assign the value of e_i to be $\frac{1}{|SeedSet|}$ for the vertices whose corresponding words belong to the seed set and $e_i = 0$ for all other vertices. Sentiment of the word is calculated based on its rank (or value) in both graphs. If some word is not present in one of the graphs, it is assigned value 0 and rank ∞ for that graph.

Topic Modelling

Topic modelling approach is based on the intuition that words with the same sentiment have similar topic distributions (we observe many word pairs that exhibit this property, e.g., *poor – misery*). LDA is a generative topic model that assumes the existence of latent topics according to which documents are created [17]. We use LDA to obtain topically conditioned word probability distributions.

We created the same two word-document matrices as for LSA (*tf-idf* and *log-entropy*). The algorithm for online LDA learning proposed in [18] was run and conditional probabilities of words given topics were obtained. Sentiment of each word was then judged based on its vector similarity with the words from both seed sets. We experimented with varying number of topics.

Recognizing Neutrality

A similar strategy is used for determining neutrality for all semi-supervised methods. As regards LSA, neutral words have similar levels of semantic similarity with words from both positive and negative seed sets. We define word sentiment to be neutral if the absolute difference between its positive and negative similarity is below a given threshold. For PageRank, neutral words are those whose rank and score in the positive graph are similar to those in the negative graph. Word is considered to be of neutral sentiment if the absolute difference between the positive and the negative rank (or score) is below a given threshold. As concerns LDA, we identify neutral words as words whose distribution over topics resembles both distributions of positive and negative seed words in a similar degree. Neutral words are those for which the absolute difference of positive and negative similarity is below a given threshold. For all three methods we experimented with different threshold levels.

Table 1. Inter-annotator agreement

Gold set	Cohen Kappa (%)	Fleiss Kappa (%)	Pairwise accuracy (%)
low	47.72	48.12	77.86
medium	54.70	54.72	82.49
high	68.54	67.92	90.15

4   Evaluation and Results

For our experiments we used Croatian newswire corpus *Vjesnik* consisting of 229,078 documents. The corpus was preprocessed in order to eliminate non-content words and words occurring less than 10 times. Morphological variations were conflated using automatically acquired inflectional lexicon [19].

Human Sentiment Judgements

We evaluate the performance against human-annotated gold set. The list of 2,500 most frequent content lemmas was compiled from the corpus and annotated independently by 12 annotators. Admittedly, selecting most frequent corpus words introduces a bias, as it may be easier to build sentiment models for frequent than for infrequent words. With frequent words, however, we get more reliable statistical estimates (a richer set of contexts in which each word appears) and more conclusive results. It can also be argued that methods that perform bad on frequent words are not likely to perform well on the less frequent words.

Annotators were instructed to annotate the sentiment of the dominant sense of the word with respect to their personal experience. We argue that for most words this sense corresponds to the dominant sense in the *Vjesnik* corpus. They were also instructed to label as neutral the words they considered equally positive and negative. We performed evaluation on three different gold sets depending on the inter-annotator agreement: *low agreement set* (majority label), *medium agreement set* (agreement of at least 8 annotators), and *high agreement set* (agreement of at least 10 annotators). We expected the semi-supervised algorithms to perform better on sets with higher human agreement. In Table 1 we show the inter-annotator agreement measured using pairwise averaged Cohen Kappa, Fleiss Kappa, and pairwise averaged agreement accuracy (which does not account for agreement occurring by chance). Each of the gold standard sets (*low*, *medium*, and *high*) was evenly split into two sets: one for validation and parameter fitting and other for testing. For each of the three gold standard sets we created accompanying polar gold standard set (containing only the polar words from the corresponding gold standard set) in order to test the methods on the *polarity-only* task. All the results reported were obtained on the test sets.

Results

The baseline used for these experiments is a simple majority class prediction model. Although very simple, this baseline achieves high micro-F1 value on the *subjectivity +*

Table 2. Results on the *subjectivity + polarity* task

Gold set	Baseline (%)	LSA (%)	LDA (%)	PageRank (%)	Voting (%)
low	28.5	48.9	44.8	48.7	**50.8**
medium	29.6	51.3	45.2	49.4	**52.5**
high	30.5	**54.2**	50.4	47.6	50.7

Table 3. Results on the *polarity-only* task

Gold set	Baseline (%)	LSA (%)	LDA (%)	PageRank (%)	Voting (%)
low	59.9	78.7	76.0	80.9	**82.9**
medium	53.0	83.4	78.5	**87.5**	**87.5**
high	51.2	82.7	79.1	**89.1**	88.3

polarity task because of the high rate of neutral words (75–85%, depending on the gold set used). Considering the imbalance of both validation and test sets, we performed evaluation using F1 measure averaged over classes (macro-F1). Our unsupervised methods outperform the baseline significantly on both *subjectivity + polarity* task and *polarity-only* task. Results imply that these unsupervised methods are especially suitable for determining the polarity of subjective words.

Subjectivity + Polarity Task. In Table 2 we show the best macro-F1 scores for each of the three semi-supervised methods along with the voting method that combines all three approaches (LSA, LDA, and PageRank each "vote" for its own prediction and the majority vote is predicted as a result). In case when all three methods predict different classes (e.g., LSA predicts *neutral*, LDA predicts *positive*, and PageRank predicts *negative*), LSA prediction is taken because LSA is the best-performing single method. Results are shown for all three gold sets. All three methods outperform the baseline by a wide margin. LSA outperforms both PageRank and LDA on all three test sets, with the margin on the *high* agreement set being wider than on other two sets. This is also the reason why *voting* performs slightly better than LSA on *low* and *medium* agreement sets, and worse on *high* test set (i.e., LDA and PageRank degrade the performance of LSA when the individual performance differences are significant).

Polarity-Only Task. On the *polarity-only* task, all algorithms displayed high performance outperforming the baseline by a wide margin. In Table 3 we show the best results on the *polarity-only* task for each of the three semi-supervised methods along with the voting method that combines all three approaches. The results are shown for all three test sets. On the *polarity-only* task graph-based method significantly outperforms LSA and topic modelling. These results indicate that corpus co-occurrence information is relevant for sentiment propagation. Voting method performs better than or equally well to PageRank on *low* and *medium* test sets, while slightly worse on the *high* test set on which LSA and LDA perform significantly worse than PageRank, degrading the overall voting result.

5   Conclusions and Future Work

We have presented three unsupervised corpus-based methods for automated sentiment lexicon acquisition, requiring no pre-compiled lexical resources. We tested these methods on the tasks of determining subjectivity and polarity of individual words. The evaluation was performed against human-annotated gold sets and our initial results indicate that the algorithms are more suitable for determining the polarity of subjective words than for predicting the subjectivity and polarity simultaneously. Expectedly, unsupervised methods generally performed best on the test set with the highest inter-annotator agreement. Results achieved using our graph-based method on the *polarity-only* task look very promising. On the other hand, none of the methods seems to be fully apt to handle subjectivity and polarity together. In the future we intend to combine supervised methods for subjective-neutral classification and the presented graph-based approach to decide on the polarity of words classified as subjective.

Acknowledgments. We thank the anonymous reviewers for their useful comments. This work has been supported by the Ministry of Science, Education and Sports, Republic of Croatia under the Grant 036-1300646-1986.

References

1. Pang, B., Lee, L., Vaithyanathan, S.: Thumbs up?: Sentiment classification using machine learning techniques. In: Proceedings of the ACL 2002 Conference on Empirical Methods in Natural Language Processing, vol. 10, pp. 79–86. Association for Computational Linguistics (2002)
2. Riloff, E., Patwardhan, S., Wiebe, J.: Feature subsumption for opinion analysis. In: Proceedings of the 2006 Conference on Empirical Methods in Natural Language Processing, pp. 440–448. Association for Computational Linguistics (2006)
3. Hu, M., Liu, B.: Mining opinion features in customer reviews. In: Proceedings of the National Conference on Artificial Intelligence, pp. 755–760 (2004)
4. Somasundaran, S., Wilson, T., Wiebe, J., Stoyanov, V.: QA with attitude: Exploiting opinion type analysis for improving question answering in on-line discussions and the news. In: Proceedings of the International Conference on Weblogs and Social Media (ICWSM), Citeseer (2007)
5. Hatzivassiloglou, V., McKeown, K.: Predicting the semantic orientation of adjectives. In: Proceedings of the Eighth Conference on European Chapter of the Association for Computational Linguistics, pp. 174–181. Association for Computational Linguistics (1997)
6. Turney, P., Littman, M.: Measuring praise and criticism: Inference of semantic orientation from association. ACM Transactions on Information Systems (TOIS) (2003)
7. Wilson, T., Wiebe, J., Hoffmann, P.: Recognizing contextual polarity: An exploration of features for phrase-level sentiment analysis. Computational Linguistics 35, 399–433 (2009)
8. Andreevskaia, A., Bergler, S.: Mining WordNet for fuzzy sentiment: Sentiment tag extraction from WordNet glosses. In: Proceedings of EACL, vol. 6, pp. 209–216 (2006)
9. Thelwall, M., Buckley, K., Paltoglou, G.: Sentiment strength detection for the social web. Journal of the American Society for Information Science and Technology (2011) (in press)
10. Wilson, T., Wiebe, J., Hoffmann, P.: Recognizing contextual polarity in phrase-level sentiment analysis. In: Proceedings of the Conference on Human Language Technology and Empirical Methods in Natural Language Processing, pp. 347–354. Association for Computational Linguistics (2005)

11. Taboada, M., Brooke, J., Tofiloski, M., Voll, K., Stede, M.: Lexicon-based methods for sentiment analysis. Computational Linguistics, 1–41 (2011)
12. Fellbaum, C.: WordNet. In: Theory and Applications of Ontology: Computer Applications, pp. 231–243 (2010)
13. Dumais, S.: Latent semantic analysis. Annual Review of Information Science and Technology 38, 188–230 (2004)
14. Kamps, J., Marx, M., Mokken, R., De Rijke, M.: Using WordNet to measure semantic orientations of adjectives (2004)
15. Esuli, A., Sebastiani, F.: PageRanking WordNet synsets: An application to opinion mining. In: Annual Meeting-Association for Computational Linguistics, vol. 45, p. 424 (2007)
16. Page, L., Brin, S., Motwani, R., Winograd, T.: The PageRank citation ranking: Bringing order to the web (1999)
17. Blei, D., Ng, A., Jordan, M.: Latent Dirichlet allocation. The Journal of Machine Learning Research 3, 993–1022 (2003)
18. Hoffman, M., Blei, D., Bach, F.: Online learning for latent Dirichlet allocation. In: Advances in Neural Information Processing Systems, vol. 23, pp. 856–864 (2010)
19. Šnajder, J., Dalbelo Bašić, B., Tadić, M.: Automatic acquisition of inflectional lexica for morphological normalisation. Information Processing & Management 44, 1720–1731 (2008)

Towards a Constraint Grammar Based Morphological Tagger for Croatian

Hrvoje Peradin[1] and Jan Šnajder[2]

[1] University of Zagreb, Faculty of Science, Department of Mathematics
Bijenička cesta 30, 10000 Zagreb, Croatia
hperadin@student.math.hr
[2] University of Zagreb, Faculty Electrical Engineering and Computing
Unska 3, 10000 Zagreb, Croatia
jan.snajder@fer.hr

Abstract. A Constraint Grammar (CG) uses context-dependent hand-crafted rules to disambiguate the possible grammatical readings of words in running text. In this paper we describe the development of a CG-based morphological tagger for Croatian language. Our CG tagger uses a morphological analyzer based on an automatically acquired inflectional lexicon and an elaborate tagset based on MULTEXT-East and the Croatian Verb Valence Lexicon. Currently our grammar has 290 rules, organized into cleanup and mapping rules, disambiguation rules, and heuristic rules. The grammar is implemented in the CG3 formalism and compiled with the vislcg3 open-source compiler. The preliminary tagging performance is P: 96.1%, R: 99.8% for POS tagging and P: 88.2%, R: 98.1% for complete morphosyntactic tagging.

1 Introduction

A Constraint Grammar (CG) is a methodological paradigm for natural language processing that directly addresses the notorious problem of linguistic ambiguity [1]. CG uses context-dependent hand-crafted rules to constrain and therefore disambiguate the possible grammatical readings of words in running text. The readings can be morphological, syntactic, or semantic, and may correspond to different levels of linguistic processing, ranging from morphological tagging and parsing to semantic labeling, anaphora resolution, information extraction, and machine translation. The CG formalism has been successfully applied to a number of languages, including English [1,2], Basque [3], Portuguese [4], French [5], Danish [6], Norwegian [7], and Spanish [8]. Recently, CG-based tools for other (resource-scarce) languages have been developed for the Apertium MT platform [9]. The main advantages of CG-based taggers over their statistic-based counterparts are robustness and accuracy. Because the last remaining reading cannot be removed, a CG tagger will assign reading even to unconventional language input. Performance of mature CG-based taggers are exceptionally high, with F-scores typically around 99% for POS tagging and 95% for syntactic function labeling. The CG is essentially a glass box approach, facilitating rule interpretability, grammar modularity, and grammatical adaptability to different domains.

P. Sojka et al. (Eds.): TSD 2012, LNCS 7499, pp. 174–182, 2012.

This paper describes the development of a CG-based morphological tagger for Croatian language. Previous work on morphological tagging of Croatian focused on the use of stochastic taggers [10,11]. With our CG-based tagger we seek to provide a robust and high-performance alternative. Our inspiration comes from the work on the Serbo-Croatian CG within the Apertium MT framework [9], developed as part of the open-source Serbo-Croatian-to-Macedonian language pair implementation (apertium-sh-mk).[1] Based on the experience gained with Apertium, we developed a new CG grammar for morphological disambiguation from scratch. The new grammar is more elaborate and more structured, uses a more elaborate tagset and a different morphological analyzer as input, and focuses specifically on Croatian language. In this paper we describe the currently developed tagger, focusing on the CG disambiguation rules and how these address the ambiguities arising in Croatian language, and give a preliminary experimental evaluation of tagger's performance. We see our CG-based tagger as a first step in the development of a robust and modular CG-based NLP pipeline for Croatian.

The rest of this paper is structured as follows. In the next section we describe our tagset. In Section 3 we give an overview of the CG-based tagger, while in Section 4 we describe the CG disambiguation rules. In Section 5 we present the evaluation results. Section 6 concludes the paper and outlines future work.

2 The Tagset

Previous work on tagging of Croatian [10,11] uses a tagset based on MULTEXT-East morphosyntactic descriptors (MSDs) [12]. The MULTEXT-East provides an elaborate and harmonized lexical specification for nine Eastern-European languages, including Croatian, thereby greatly facilitating interoperability and standardization. Despite the wide adoption of MULTEXT-East, we decided not to use it for our tagger. The main reason is that MULTEXT-East uses positional encoding, which is impractical for use with CG. We therefore developed a new tagset that uses non-positional encoding (Table 1). The tagset is based on the one used in apertium-sh-mk, but extended to cover most of the MSDs for Croatian. In some aspects, our tagset is more elaborate than MULTEXT-East. E.g., we include tags for transitivity and aspect of verbs, distinct tags for future I and future II, as well as punctuation tags. We also include a number of tags for the semantic classes of nouns, which might be useful for later processing at semantic levels. As the core of our tagset is essentially a non-positional variant of MULTEXT-East, it can easily be converted back to positional tags.

3 System Overview

A CG-based morphological tagger typically is a two step process. In the first step, a morphological analyzer is applied to the input token and results in a set of ambiguous morphological readings (lemmas and tags), called *cohorts*. In the second step, the CG rules are used to disambiguate the cohorts yielding, ideally, a single and correct reading for each cohort. The CG rules are used to vote out, select correct analyzes, or assign new tags to tokens based on the their context.

---

[1] http://wiki.apertium.org/wiki/Serbo-Croatian_and_Macedonian

Table 1. Morphological tagset for Croatian language

POS tag	Value tags
abbr (abbreviation)	**n** (nominal), **adv** (adverbial), **adj** (adjectival) **spl** (simple), **cpx** (compound), **ant** (antroponym), **cog** (cogname), **top** (toponym), **alt** (alternative), **sg** (singular), **pl** (plural) + **CASE** + **GEND**
adv (adverb)	**pst** (positive), **cmp** (comparative), **sup** (superlative)
adj (adjective)	**qlf** (qualificative), **pos** (possessive), **pst** (positive), **cmp** (comparative), **sup** (superlative), **sg, pl, ind** (indefinite), **def** (definite) + **CASE** + **GEND**
cnj (conjunction)	**coo** (coordinating), **sub** (subord.), **spl** (simple), **cpx** (compound)
ij (interjection)	**spl** (simple), **cpx** (compound)
n (noun)	**com** (common), **prp** (proper), **ant** (antroponym), **cog** (cogname), **top** (toponim), **alt** (other), **sg, pl** + **CASE** + **GEND**
num (numeral)	**crd** (cardinal), **ord** (ordinal), **mlt** (multiple), **dgt** (digit), **rom** (roman), **ltr** (letter) + **CASE** + **GEND**
part (particle)	**neg** (negative), **int** (interrog.), **mod** (modal), **aff** (affirmative)
pr (preposition)	**spl** (simple), **cpx** (compound) + **CASE**
prn (pronoun)	**prs** (personal), **dem** (demonstrative), **ind** (indefinite), **pos** (possessive), **int** (interrogative), **rel** (relative), **rfx** (reflexive) **p1** (first), **p2** (second), **p3** (third), **sg, pl, clt** (clitic) + **CASE** + **GEND**
vb (verb)	**lex** (lexical), **aux** (auxiliary), **mod** (modal), **cop** (copula) **ind** (indicative), **imp** (imperative), **cnd** (conditional), **inf** (infinitive), **pp** (participle), **prs** (present), **ipf** (imperfect), **aor** (aorist), **futI** (future I), **futII** (future II), **pct** (perfect), **pqp** (pluperfect) **p1** (first), **p2** (second), **p3** (third) **neg** (negative) **perf** (perfective) **imperf** (imperfective) **tv** (transitive) **iv** (intransitive) **act** (active), **psv** (passive)

CASE = nom, gen, dat, acc, voc, loc, ins
GEND = **m** (masculine), **mi** (masc. inanimate), **ma** (masc. animate), **nt** (neuter), **f** (feminine)

3.1 Morphological Analysis

The morphological analysis is performed using the inflectional lexicon from [13]. This lexicon has been acquired semi-automatically from corpora and currently covers 66,500 lemmas and 3.5 million word-forms, with a total of 318 unique tags. The number of unique tags appearing in our gold set is 487, while the theoretical upper bound of our tagset is 6200. For each word-form, the lexicon outputs a cohort of lemmas and MULTEXT-East MSDs. Before the cohorts can be disambiguated using CG rules, the MSD tagset needs to be mapped to our tagset. This mapping is non-injective as in some cases two MSD tags are merged into a single tag. (e.g., masculine animate ma and masculine inanimate mi tags), whereas in some cases we omit a tag considered as default (e.g., indicative or active tag for verbs). The MSD tagset does not contain tags for reflexivity, transitivity, and perfectivity. Because in many cases these tags are required for full morphological disambiguation of Croatian (cf. Section 4.4), we use additional resources to supplement the MSD tags where possible. To this end we use the Croatian Verb Valence Lexicon [14] and the morphological analyzer from apertium-sh-mk. Together they cover roughly 2,340 different verb lemmas.

3.2 CG Rules

Currently our system has 290 disambiguation rules, most of which are elaborated and restructured rules from apertium-sh-mk. This is more than the number of rules in apertium-sh-mk (currently 170), but still much less than in mature CG systems, such as [2] (almost 1,500 rules). The rules are implemented in the CG3 formalism and compiled with the vislcg3 open-source compiler.[2] As a development set, we

---

[2] http://beta.visl.sdu.dk/cg3.html

have used a sentence-segmented corpus derived from newspaper articles of the Croatian newspaper *Vjesnik* totaling 1M tokens.

The CG is structured in three main sections. First, the output of the analyzer is cleaned up, and modified to be more convenient for further use. Some readings are contracted, while some are given preliminary syntactic tagging which are later used for quicker removal of incorrect readings. Also, some heuristics are applied here, like inferring proper noun readings in adjectives and nouns beginning with a capital letter. The next two rule sections mostly consist of SELECT and REMOVE rules. The batches of rules are grouped by POS tags, following the principle that the rules for removing morphologically implausibilities appear first, followed by heuristic rules that reflect linguistic intuition.

4   Examples of CG Rules

In what follows we give some examples of rules for resolving common cases of morphological ambiguity in Croatian. In general, the CG rules consist of a sequence of tests on tokens surrounding the currently processed token, with numbers specifying the relative positions of the context tokens with respect to the current token. If all tests specified by a rule are satisfied, then a selection, removal, or a substitution operation is performed, depending on the rule type. For a more detailed explanation of the CG rule syntax, the reader is referred to the vislcg3 documentation.[2]

4.1   Basic Proper Noun Detection

One of the simplest rules allows for detection of proper nouns:

```
SUBSTITUTE $$Nominal $$Nominal + (np) TARGET ("<[A-Z].*>"r) + $$Nominal
    IF (0C Noun OR Adjective) (NOT -1 BOS)
```

The $$ prefix denotes the iteration over the set of all nominal words. This heuristic adds a proper noun attribute to an adjective or a noun beginning with a capital letter inside a sentence, while preserving the original reading.

This opens up possibilities for applying more heuristics to proper noun candidates. E.g., in (1), by knowing that *Lusaka* is a proper noun, we can infer that it receives case from the preceding noun.

(1) ... *u glavnom gradu Lusaki* ... *(In the capital city of/to Lusaka)*

4.2   Adverb or Adjective

Some adverbs in Croatian are identical to neuter singular adjectives. E.g., in:

(2) *Dobro dijete. (A good child.)*
(3) *Dobro izgledaš. (You look good.)*

In (2), *dobro* is a neuter adjective in nominative, whereas in (3) it is an adverb. The correct reading for *dijete* is a singular neuter noun in nominative, and the word *izgledaš* is a 2nd person singular verb in the present tense. The adjective reading should be selected if the word is followed by a neuter noun, and the adverb reading if it is followed by a verb. This is accomplished by the following rules:

```
SELECT Adjective IF (0 Adverb OR Adjective)(1 Noun + Neuter)
SELECT Adverb IF (0 Adverb OR Adjective)(1 Verb)
```

4.3 Disambiguating Relative Pronouns

A relative pronoun agrees in number and gender with the noun it relates to. E.g., in (4) pronoun *kojima* refers to noun *hitovima*.

> (4) ... *hitovima s prošla dva albuma, medju kojima pjesmama* ... *(... with hits from the last two albums, among which with songs ...)*

The word *kojima* is ambiguous for feminine, masculine and neuter plural, while *hitovima* is a plural masculine noun. The pronoun refers to *hitovima*, so a heuristic rule may be used to remove the feminine and neuter reading.

```
SELECT $$Gender IF (0 ("<kojima>"i) )
        (-1* Noun + Plural + $$Gender BARRIER ("koji"i))
SELECT $$Gender IF (0 ("<kojima>"i) )
        (-1* Pronoun + Plural + $$Gender BARRIER ("koji"i))
```

The asterisk and the BARRIER keyword mean that we are searching the token stream to the left until we encounter the argument of the BARRIER, which in this case is the token *koji*.

4.4 Accusative and Transitivity

A common ambiguity in Croatian is between nominative/accusative and genitive/accusative masculine inanimate and masculine animate singular. The former is illustrated by (5), in which the verb *pada* is intransitive, and *snijeg* is ambiguous between accusative and nominative.

> (5) *Vani pada snijeg. (It's snowing outside.)*

The following rule removes the accusative reading, as there is no transitive active verb in the sentence:

```
REMOVE Accusative IF (NOT 0* Verb + Transitive BARRIER BOS OR EOS)
```

The keywords BOS and EOS denote the beginning and the end of a sentence, respectively. A similar ambiguity arises in (6). Because the verb *dozivam* is transitive, the noun phrase *dobroga prijatelja* is assigned an accusative reading.

> (6) *Dozivam dobroga prijatelja. (I'm calling the good friend of mine.)*

The rule to disambiguate this is as follows:

```
SELECT Accusative IF (0C Genitive OR Accusative) (-1C Verb + Transitive)
```

A full disambiguation is obtained using rules for noun phrases described below.

4.5 Noun Phrases

A great deal of ambiguities in noun phrases can be easily resolved by looking at gender/number/case agreements of their constituents. In (7), which contains a locative noun phrase, the preposition *na* has two case readings (locative/accusative), the adjectives *lijepom* and *plavom* have several gender and three case readings (dative/locative/instrumental), and *Danube* is ambiguous between dative and locative.

(7) *Na lijepom plavom Dunavu. (On the beautiful blue Danube.)*

Disambiguation is accomplished by applying a chain of rules. First the preposition is disambiguated for locative, and afterwards the instrumental and dative reading are removed from the noun and the adjectives. Finally the genders not agreeing with the gender of *Danube* are removed from the adjectives.

Because dative and locative are orthographically identical, the exemplified ambiguity is always present. Locative always follows a preposition, so in cases where the phrase is not prepositional, the locative reading can be removed:

```
REMOVE Locative IF (0 Nominal + Locative)
               (-1* Preposition + Locative BARRIER Word - Modifier)
```

4.6 Numeral Phrases

Numeral phrases further complicate readings, because the numeral is the head of the phrase, and the rest of the phrase is assigned a different ending, which is either the remnant of the old dual declension or is identical to genitive (singular for numbers ending in 2, 3, 4, and plural for the rest). The former case is morphologically more complex, but the disambiguation is easier. The remnant of the dual declension consists of three distinct endings, the ones for nominative, accusative, and vocative are identical to morphological genitive, the dative/locative/instrumental endings are identical to dative/locative/instrumental plural, and the genitive receives a special ending *-ju*.

(8) *Govorio je o dvjema ženama. (He spoke of two women.)*

The prepositional phrase rules described above suffice to resolve this ambiguity (preposition *o* governs locative, thus the phrase receives a locative reading).

The other case is exemplified by (9). The number is in a morphological nominative and the rest of the phrase is in genitive. This style is more characteristic of colloquial speech, however it also frequently appears in newspaper articles.

(9) *Govorio je o dvije žene. (He spoke of two women.)*

In this case, numeral phrase constituents (except the numeral) are assigned genitive plural or singular, depending on the number, and the numeral and the preposition are disambiguated independently.

4.7 General Remarks

In general, a great number of grammatical case ambiguities can be reduced by looking at prepositions. However, in some cases (especially in numeral phrases or in classes of words that do not have a rich inflection), it is difficult to properly disambiguate. A case in point is (10), in which the prepositional phrase *sa svjetlosti* gives rise to three interpretations.

(10) *Došao je sa svjetlosti. (He came with light / off of the light / off of the lights.)*

This is also the case when there are no prepositions, i.e., for the common plural dative/locative/instrumental ambiguity. Another frequent ambiguity that can only be resolved at the semantic level is the genitive singular/plural ambiguity:

(11) *... prodajom udjela državnih poduzeća ... (by selling a share / shares of state firms.)*

5 Evaluation

We have evaluated the current CG implementation on a manually annotated gold set consisting of 5,000 tokens. Ambiguities that can only be resolved at the semantic level (like the ones discussed above) are left unresolved in the gold set.

The evaluation is based on readings for each cohort; a reading from the test set is considered as true positive if it exactly matches a reading from the gold set. Here we consider two tasks: POS tagging (13 tags) and full morphosyntactic tagging (the complete tagset from Table 1). The performance is evaluated in terms of precision, recall, F-score, and percent of readings disambiguated, as computed on a per cohort basis. The results are given in Table 2. Because some readings are intentionally left ambiguous in the gold set, they account for higher precision in the cases when the words are not fully disambiguated by the tagger. The `vislcg3` tagging speed, measured on a desktop computer with Pentium 2.30GHz CPU and 3GB memory, was approx. 1,500 tokens per second.

Table 2. Tagging performance (%)

	Precision	Recall	F-score	Disambiguated
POS tagging	96.08	99.76	97.88	95.30
Morphosyntactic tagging	88.16	98.13	92.88	86.36

A comparison to existing work for POS/MSD tagging of Croatian is not straightforward due to differences in the test set and tagset used; we leave a thorough performance comparison for future work. However, a comparison to CG results obtained for other languages suggests that there is room for improvement. The main problem is that many readings are still not fully disambiguated, yielding a relatively low precision. This can be attributed to three causes.

Firstly, our rule set is still under development. Some rules are not thoroughly tested and certain grammatical constructions are not fully covered. A case in point is the nominative/accusative ambiguity, which is only partially covered. Analyses are removed if there is no transitive verb in the sentence, and verb/object relationship is considered only if the verb is immediately to the left of the noun. However, there currently exist no long distance rules for subject/object relation of nominative/accusative.

Another problem are the morphological analyses. The morphological lexicon was obtained automatically from corpus, and while it has a very wide coverage, some words have wrong analyses, while some analyses are incomplete. A wrong or incomplete analysis of a word prevents the disambiguation of a word in question, but it also often obscures the disambiguation of other neighboring words.

Finally, a considerable number of the unresolved ambiguities are due to out-of-vocabulary words, as our morphological analyzer currently does not support the analysis the unknown words.

6   Conclusion

We have described a prototype CG-based morphological tagger for Croatian. The tagger uses a morphological analyzer based on an inflectional lexicon and a rather elaborate tagset. The current grammar has 290 disambiguation rules, which is far less than mature CG systems. Preliminary evaluation reveals that more rules and a better morphological analysis will be needed to obtain satisfactory tagging performance. For future work, we intend to address all these deficiencies and perform a more thorough evaluation of tagging performance. Further improvements may be obtained using hybrid approaches that combine CG-based and stochastic tagging.

Acknowledgments. We thank the anonymous reviewers for their useful comments. This work has been supported by the Ministry of Science, Education and Sports, Republic of Croatia under the Grant 036-1300646-1986.

References

1. Karlsson, F.: Constraint Grammar: A language-independent system for parsing unrestricted text, vol. 4. Walter de Gruyter (1995)
2. Bick, E.: Degrees of orality in speech-like corpora: Comparative annotation of chat and e-mail corpora. In: Proc. of the 24th Pacific Asia Conference on Language, Information and Computation, pp. 721–729. Waseda University, Sendai (2010)
3. Aduriz, I., Arriola, J.M., Artola, X., de Ilarraza, A.D.: et al.: Morphosyntactic disambiguation for Basque based on the constraint grammar formalism. In: Proceedings of RANLP 1997, pp. 282–288 (1997)
4. Bick, E.: The Parsing System "Palavras": Automatic Grammatical Analysis of Portuguese in a Constraint Grammar Framework. Aarhus Univ. Press (2000)
5. Bick, E.: Parsing and evaluating the French Europarl corpus. Méthodes et outils pour lévaluation des analyseurs syntaxiques Journée ATALA, 4–9 (2004)
6. Bick, E.: A CG & PSG hybrid approach to automatic corpus annotation. In: Proceedings of SProLaC 2003, pp. 1–12 (2003)
7. Johannessen, J., Hagen, K., Nøklestad, A.: A constraint-based tagger for Norwegian. In: 17th Scandinavian Conference of Linguistics, Odense Working Papers in Language and Communication, vol. 19, pp. 31–47 (2000)

8. Bick, E.: A constraint grammar parser for Spanish. In: Proceedings of TIL (2006)
9. Forcada, M., Tyers, F., Ramírez-Sánchez, G.: The Apertium machine translation platform: five years on. In: Proceedings of the First International Workshop on Free/Open-Source Rule-Based Machine Translation, pp. 3–10 (2009)
10. Agić, Ž., Tadić, M.: Evaluating morphosyntactic tagging of Croatian texts. In: Proc. of the 5th Int. Conference on Language Resources and Evaluation (2006)
11. Agić, Ž., Tadić, M., Dovedan, Z.: Improving part-of-speech tagging accuracy for Croatian by morphological analysis. Informatica 32, 445–451 (2008)
12. Erjavec, T.: MULTEXT-East version 3: Multilingual morphosyntactic specifications, lexicons and corpora. In: Fourth Int. Conference on Language Resources and Evaluation, LREC, vol. 4, pp. 1535–1538 (2004)
13. Šnajder, J., Dalbelo Bašić, B., Tadić, M.: Automatic acquisition of inflectional lexica for morphological normalisation. Information Processing & Management 44, 1720–1731 (2008)
14. Preradović, N., Boras, D., Kišiček, S.: CROVALLEX: Croatian verb valence lexicon. In: Proceedings of the ITI 2009 31st International Conference on Information Technology Interfaces, ITI 2009, pp. 533–538 (2009)

A Type-Theoretical Wide-Coverage Computational Grammar for Swedish

Malin Ahlberg and Ramona Enache

Department of Computer Science & Engineering
Gothenburg University
Box 8718, SE-402 75 Gothenburg, Sweden
ahlberg.malin@gmail.com, enache@chalmers.se

Abstract. The work describes a wide-coverage computational grammar for Swedish. It is developed using GF (Grammatical Framework), a functional language specialized for grammar programming. We trained and evaluated the grammar by using Talbanken, one of the largest treebanks for Swedish. As a result 65% of the Talbanken trees were translated into the GF format in the training stage and 76% of the noun phrases were parsed during the evaluation. Moreover, we obtained a language model for Swedish which we use for disambiguation.

Keywords: grammar-formalism, computational grammar, treebank, Swedish.

1 Introduction

Two main approaches have divided the research in computational linguistics. The first and most prominent one is the wide-coverage method, mostly based on statistical methods, which compensates its usually shallow analysis with its coverage. The second one is the opposite – provides a deep analysis, but has a limited coverage. Our work on a computational grammar from Swedish falls into the second category, but aims at getting the best of both worlds by enlarging the coverage in order to fit free text, without affecting the quality of the model.

2 Background

The computational grammar was developed within **GF** (Grammatical Framework) [1]. GF is a dependently-typed grammar formalism, based on Martin-Löf type theory, which is mainly used for multilingual natural language applications.

GF grammars are composed of an abstract syntax, the interlingua of the grammar which describes the semantics on a language-independent level and a number of concrete syntaxes, usually corresponding to natural languages. These describe how the concepts from the abstract syntax are described in the language. By defining such a grammar, one also obtains a parser for the language fragment that is described by a concrete grammar and a natural language generation tool for constructions from the same language fragment. In this way, due to the division of the GF grammars into

P. Sojka et al. (Eds.): TSD 2012, LNCS 7499, pp. 183–190, 2012.

abstract and concrete, it is possible to achieve semantics-preserving translation between any language pair.

However, because the grammar is strict, we can only deal with constructions that the grammar defines and nothing more. One could classify the limitations of the natural language grammar into two categories — missing lexical items and missing syntactic structures. For this reason, GF was mainly used so far for controlled languages [2], dialogues systems [3] and interactive systems where the user is guided to stay within the bounds of the grammar by a predictive parser, such as a multilingual tourist phrasebook [4]. However, there is recent work on parsing free text — a grammar for patent claims from the biomedical domain [5]. This work highlighted a number of caveats of real-world text, but in the same time gave valuable insights for future work on wide-coverage GF grammars.

The current work aims at overcoming the second limitation of GF grammars — adding support for the most common syntactic constructions that could appear in Swedish texts. For some of these we use dependent types to encode the syntactic phenomena in an elegant way. The lexicon limitation is not a problem, as we use an existing large Swedish lexicon [6], which is imported to GF format from an electronic dictionary and which contains more than 100,000 entries.

Swedish is a North-Germanic language, sharing most of its grammatical structure with Norwegian and Danish. It spoken by approximately nine million people, in Sweden and parts of Finland. Although Swedish syntax is often similar to English, the morphology is richer and the word order more intricate. It is a verb-second language: the second constituent of a declarative main clause must consist of a verb. The normal word order is subject-verb-object, but fronting other constituents (topicalisation) is very common, especially for temporal and locative adverbial phrases. Fronting the finite verb marks questions. Special reflexive pronouns and reflexive possessive pronouns for the third person exist, distinct from the normal third person forms.

For testing and evaluation of the grammar and lexicon, we needed a reliable source for comparison. We have used **Talbanken** [7], a freely available, manually annotated, large-scale treebank. The section used for our work contains more than 6,000 sentences of professionally written Swedish gathered from newspapers, brochures and textbooks. It is also redistributed in an updated version, Talbanken05 [8].

3   Related Work

There exist a number of parsers for Swedish already, such as the data-driven Malt parser [9], also trained on Talbanken, the cascaded finite state parser CassSwe [10], The Swedish Constraint Grammar [11] and the Swedish FDG, which uses the Functional Dependency Grammar [12]. Among computational grammars for Swedish, we mention the Swedish version of the Core Language Engine [13], which provides a comprehensive description of syntax and semantics, as well as a translation to English. Unfortunately, the resource is not available for comparison. Other grammars are BiTSE [14], a Swedish grammar that uses the HPSG format [15] and developed within the LinGO Matrix library [16].

Besides the usages mentioned before, GF is currently the leading technology in the European project MOLTO[1], which aims at developing tools for translating between 15 languages in real-time and with high quality. Previous examples of larger type-theoretical GF grammars that use dependent-types are SUMO-GF [17], a GF representation of SUMO[2], the largest open-source ontology and a natural language generation grammar via Montague semantics [18].

4   Grammar

The work focuses on the syntactical dimension and the following section illustrates the work on describing grammatical constructions in the GF formalism.

Resource Grammar. The GF package provides an useful resource library [19], covering the fundamental morphology and syntax of more than 20 languages. There is also a small test lexicon included, containing a few hundred common words. The grammars describe how to construct phrases and sentences and how to decline words. They cover the word order, agreement, tense, basic conjunction, etc. Due to the syntactic similarities between the Scandinavian languages, much of the implementation for Swedish is shared with Norwegian and Danish. The modules that concern the lexical aspects are separate, while 85% of the syntax description is shared. There are about 80 functions, which describe the rules for building phrases and clauses, such as functions for predication and complementation:

```
PredVP      : NP -> VP -> Cl ;       - Predication
ComplSlash : VPSlash -> NP -> VP ;   - Complementation
```

This is the function types – PredVP returns a clause when given its input arguments: a noun phrase and a verb phrase. In addition to the core resource grammars, which is shared with all other languages implemented in the library, there is also an extra module where language specific constructions may be added.

Lexicon. As mentioned before, our main lexical resource is SALDO[3] from which a GF lexicon has been extracted [6]. Valency information, which is a key feature to good parsing using GF, is extracted from the lexicon Lexin[4]. This gives us a dictionary with more than 100,000 entries, covering all but 500 words from Talbanken (excluding names, compounds and numerical expressions).

4.1   New Features

Earlier it has been hard to identify missing constructions of the Swedish implementation since there was no large resource available to evaluate it on. Our evaluations are based

---

[1] http://www.molto-project.eu/
[2] http://www.ontologyportal.org/
[3] http://spraakbanken.gu.se/resurs/saldo
[4] http://spraakbanken.gu.se/lexin/

on Talbanken and when first conducting tests we found much room for improvement. The items listed below are examples of constructions implemented during this work.

The S-Passive. Swedish has two ways of forming passive verb phrases: the periphrastic passive, formed by using the modal auxiliary verb *bli* ("become") and the s-passive which is formed by adding an *s* to the verb. Passive voice is often used, especially the s-passive. It is however not so common in the other Scandinavian languages, where not all words have passive forms for all tenses. The resource grammar for Scandinavian therefore only implemented the periphrastic passive. During this project, an implementation of the s-passive was added. A ditransitive verb – *ge* ("give") (1a) – gives rise to two passives, (1b) and (1c), both covered by our grammar.

(1) a. **Active use of two-place verb**
        Vi erbjöd henne jobbet   (We offered her the job)
        we offered her   the job
     b. **Passive use, first place** c. **Passive use, second place**
        Hon erbjöds   jobbet   Jobbet erbjöds   henne
        she   offered+s the job   the job offered+s her
        (She was offered the job)   (The job was offered to her)

Impersonal Constructions. Formal subjects are often used in Swedish.

(2) Det sitter en fågel på taket   (There is a bird sitting on the roof)
    it   sits   a   bird on the roof

Restrictions on both the verb and the noun phrase are covered by the grammar: the determiner of the noun phrase must be of such type that it requires both the noun and its modifiers to appear in indefinite form. The verb phrase must consist of an intransitive verb or be in passive form.

Formalizing the Rules for Reflexive Pronouns by Using Dependent Types. An important area in a Swedish grammar is the treatment of the reflexive pronouns and the reflexive possessive pronouns. The reflexives require an antecedent with which they agree in number and person. They must not be used in subject noun phrases of finite sentences, as shown by the ungrammatical examples (3b) and (4b). Still our grammar should accept the sentences (3a) and (4a) :

(3) a. Sina vantar hade han tappat. (SELF'S gloves, he had lost.)
     b. *Sina vantar var kvar på tåget. (SELF'S gloves was left on the train.)

(4) a. Han är längre än sin kompis. (He is taller than SELF'S friend.)
     b. *Han och sin kompis leker. (He and SELF'S friend are playing.)

In the standard GF analysis, which is performed bottom-up starting from the POS-tags, information about syntactic roles are given by the functions, not by the categories. For example, the first argument of the function `PredVP` always acts as the subject, but the noun phrase itself does not carry information about its syntactic role. As noun phrases containing reflexive pronouns may be used as ordinary noun phrases – apart from the restrictions stated above – we do not want to differentiate them from other NPs on the type level since this would require code duplication. Still, the type system should prevent noun phrases containing reflexive pronouns from being used as subjects. This does not only concern noun phrases, but also adverbial and adjectival phrases (4a).

Our solution introduces the use of dependent types. We make a difference between subjects and objects by letting the type NP depend on an argument, which may either be Subject or Object.

```
PredVP : NP Subject -> VP -> Cl ;       - predication
ComplSlash : VPSlash -> NP Object -> VP ; - complementation
PrepNP : (a: NPType) -> Prep -> NP a -> Adv a ; - adverbial phrase
```

We hence combine the-part-of-speech driven analysis normally performed by GF with a part-of-sentence analysis, where the dependent types give the information we were missing. This is the first example in GF making use of the dependent types for describing the syntax of a natural language. The approach could be extended to describing similar relations in other languages.

Overgeneration. Another aspect of the grammar implementation was the avoidance of overgeneration. As the Swedish resource grammar had not been used for larger projects, many examples of overgeneration were present that did not cause problems when working with small lexicons and controlled language. However, when inspecting the test output from Talbanken, unexpected compositions of functions were identified and there after fixed.

5   A GF Treebank and Model-Based Disambiguation

Talbanken contains very valuable information about phrase structures and word usage. One part of this project has focused on translating trees from Talbanken to GF by constructing a mapping, which automatically transforms trees in the Talbanken format to GF abstract syntax trees. We hence get a comparison between the two annotations and at the same time we extract a GF treebank. Nodes that fail to be translated are represented by a meta variable, annotated ?. In Figure 1, the word *lätt* ("easy") was unknown and therefore represented by a meta variable in the translated tree. Meta variables are also used for connecting sister nodes that can be translated on their own, but not joined into one tree.

Talbanken05 uses three kinds of tags: categories, edge labels and POS-tags. While the POS-tags are reserved for words, the categories give grammatical information: S, NP and VP. The edge labels give the part of sentence: SS for subject, OO for object, etc.

The mapping could in some cases be performed tag-by-tag, but annotational differences complicated the translation. One example is valency information, which is given implicitly by the complements of a word in Talbanken. If a verb is followed by an object, OO containing an S, we can conclude that this is a sentence-complement verb. In GF, the valency information is given explicitly for each entry in the lexicon. A sentence-complement verb has the type VS and the function ComplVS must be used for combining the verb and a sentence into a complete verb phrase.

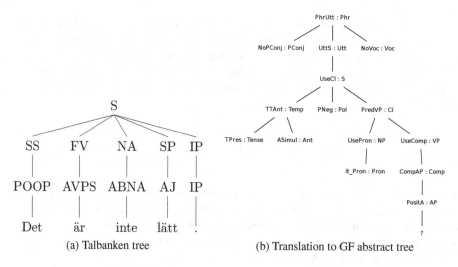

(a) Talbanken tree   (b) Translation to GF abstract tree

Fig. 1. The sentence *Det är inte lätt* ("It is not easy")

The translation of Talbanken to GF gives us a GF treebank consisting of more than 6,000 sentences. Moreover, the treebank could be mapped to another language from the resource library provided that one defines transfer rules for lexical items and implements the extra syntactic constructions.

The trees give valuable information about which lexical entries and functions are needed to parse the sentence. GF has built-in support for probabilities, and when feeding the output from the mapping to our grammar, we get a model of the language. After the parsing is completed, all trees are ranked, measuring the probability of their constituents. As the data is extracted from manually annotated, real-world sentences, it constitutes a reliable source for disambiguation.

6   Chunk Parsing

The grammar, which is hand written, cannot be expected to cover all possible expressions of the language. Hence, we combine it with statistical means to get an even better coverage. The parser should give as much output as possible even when encountering unknown words and grammatical constructions, idioms and ellipses. Our approach uses of the rich annotation in Talbanken – we perform chunk parsing relying on the tags given in the treebank. Consequentially, whenever a sentence is not recognized by the grammar, it is parsed chunk by chunk. Each chunk is analysed by the grammar-driven parser, and resulting trees are returned. Noun, adverbial and adjectival phrases are considered. Due to the differences in the annotation between Talbanken and GF mentioned is Section 5, verbs are not considered as parsable chunks. Therefore, our focus is on whole sentences and noun phrases. At the first stage, we try to parse the basic sentence structure, and therefore allow the parser to treat complicated chunks as dummy words. This way, we don't lose the high-level sentence analysis given by GF. Subsequently, we try to parse the chunks separately as far as possible.

7   Evaluation

For the extraction of a GF treebank from Talbanken (Section 5), the evaluation measures the numbers of meta variables in each translated tree. Overall we get a number of 65%, and if we limit our input to simple sentences with no more than 10 words, all known to our lexicon, we reach 85%.

The evaluation of the parser is performed by testing it on Talbanken sentences and is still being carried out. So far, our evaluation results cover 600 sentences – about 10% of the treebank. For noun phrases, we cover 76%. By using the chunk parsing described in Section 6, we identify the structure on 65% of the sentences. That is, the parser can identify the verbs and how they relate to the noun phrases, prepositions and particles. However, if a verb is unknown to the grammar, or if it is used with another valency, other prepositions or particles than the lexicon has assigned it, the sentence structure cannot be identified. In that case the parser returns chunks, showing the parse trees of identifiable subphrases.

8   Future Work

Evaluation. A more thorough evaluation is still to be done. This should be carried out by measuring the agreement between all parsed chunks and their respective translation from Talbanken. Evaluation should also be done manually. For this, we rely on an expert in Swedish.

Grammar. The grammar should cover the most prominent Swedish features. While statistical methods compensate for constructions not covered by the grammar, we still aim for an even broader grammar coverage. As examples of constructions to be added, we mention pronominal object shift, bare indefinites and distinguish between object and subject control verbs.

9   Conclusion

Our work has resulted in a wide-coverage grammar for Swedish which – combined with statistical resources – can be used for parsing open-domain text. The grammar covers a large number of syntactic phenomena and uses an extensive morphological lexicon. To our knowledge, it is the most comprehensive computational grammar for Swedish and the most complex GF grammar to date. Besides parsing, the grammar may well be used for language generation. All parts of the project are open-source and the grammar and other parts are modular and could be reused and further developed.

References

1. Ranta, A.: Grammatical Framework: Programming with Multilingual Grammars. CSLI Publications, Stanford (2011)
2. Angelov, K., Ranta, A.: Implementing Controlled Languages in GF. In: Fuchs, N.E. (ed.) CNL 2009. LNCS, vol. 5972, pp. 82–101. Springer, Heidelberg (2010)

3. Ljunglöf, P., Bringert, B., Cooper, R., Forslund, A.C., Hjelm, D., Jonson, R., Ranta, A.: The talk grammar library: an integration of gf with trindikit (2005)
4. Ranta, A., Enache, R., Détrez, G.: Controlled Language for Everyday Use: The MOLTO Phrasebook. In: Rosner, M., Fuchs, N.E. (eds.) CNL 2010. LNCS, vol. 7175, pp. 115–136. Springer, Heidelberg (2012)
5. España-Bonet, C., Enache, R., Slaski, A., Ranta, A., Marquez, L., Gonzalez, M.: Patent translation within the MOLTO project. In: Proceedings of the 4th Workshop on Patent Translation, MT Summit XIII, pp. 70–78 (2011)
6. Ahlberg, M., Enache, R.: Combining Language Resources Into A Grammar-Driven Swedish Parser. In: Proceedings of LREC (2012)
7. Einarsson, J.: Talbankens skriftspråkskonkordans. Lund University: Department of Scandinavian Languages (1976)
8. Nivre, J., Nilsson, J., Hall, J.: Talbanken05: A Swedish treebank with phrase structure and dependency annotation. In: Proceedings of the Fifth International Conference on Language Resources and Evaluation, LREC 2006, pp. 24–26 (2006)
9. Hall, J., Nivre, J., Nilsson, J.: A Hybrid Constituency-Dependency Parser for Swedish. In: Proceedings of NODALIDA 2007, pp. 284–287 (2007)
10. Kokkinakis, D., Kokkinakis, S.J.: A Cascaded Finite-State Parser for Syntactic Analysis of Swedish. In: Proceedings of the 9th EACL, pp. 245–248 (1999)
11. Birn, J.: Swedish Constraint Grammar. Technical report, Lingsoft Inc. (1998)
12. Tapanainen, P., Järvinen, T.: A non-projective dependency parser. In: Proceedings of the 5th Conference on Applied Natural Language Processing (1997)
13. Gambäck, B.: Processing Swedish Sentences: A Unification-Based Grammar and Some Applications. Ph.D. thesis, The Royal Institute of Technology and Stockholm University, Dept. of Computer and Systems Sciences (1997)
14. Stymne, S.: Swedish-English Verb Frame Divergences in a Bilingual Head-driven Phrase Structure Grammar for Machine Translation. Master's thesis, Linköping University (2006)
15. Pollard, C., Sag, I.: Head-Driven Phrase Structure Grammar. University of Chicago Press (1994)
16. Bender, E.M., Flickinger, D., Oepen, S.: The Grammar Matrix: An Open-Source Starter-Kit for the Rapid Development of Cross-Linguistically Consistent Broad-Coverage Precision Grammars. In: Proceedings of the Workshop on Grammar Engineering and Evaluation at the 19th International Conference on Computational Linguistics, pp. 8–14 (2002)
17. Angelov, K., Enache, R.: Typeful Ontologies with Direct Multilingual Verbalization. In: Rosner, M., Fuchs, N.E. (eds.) CNL 2010. LNCS, vol. 7175, pp. 1–20. Springer, Heidelberg (2012)
18. Ranta, A.: Computational Semantics in Type Theory. Mathematics and Social Sciences 165, 31–57 (2004)
19. Ranta, A.: The GF Resource Grammar Library. Linguistic Issues in Language Technology (2009)

Application of Lemmatization and Summarization Methods in Topic Identification Module for Large Scale Language Modeling Data Filtering

Lucie Skorkovská

University of West Bohemia, Faculty of Applied Sciences, Dept. of Cybernetics
Univerzitní 8, 306 14 Plzeň, Czech Republic
`lskorkov@kky.zcu.cz`

Abstract. The paper presents experiments with the topic identification module which is a part of a complex system for acquisition and storing large volumes of text data. The topic identification module processes each acquired data item and assigns it topics from a defined topic hierarchy. The topic hierarchy is quite extensive – it contains about 450 topics and topic categories. It can easily happen that for some narrowly focused topic there is not enough data for the topic identification training. Lemmatization is shown to improve the results when dealing with sparse data in the area of information retrieval, therefore the effects of lemmatization on topic identification results is studied in the paper. On the other hand, since the system is used for processing large amounts of data, a summarization method was implemented and the effect of using only the summary of an article on the topic identification accuracy is studied.

Keywords: topic identification, lemmatization, summarization, language modeling.

1 Introduction

In order to robustly estimate the parameters of statistical language models for natural language processing (automatic speech recognition, machine translation, etc.) the extensive amount of training data is required. It has been shown that not only the size of the training data is important, but also the right scope of the language models training texts is needed [9].

A complex system for acquisition and storing large volumes of text data was implemented [12] and one of its parts is a topic identification module used for large scale language modeling data filtering [11]. The module uses a language modeling based approach for the implementation of the topic identification.

Lemmatization has been shown to improve the results when dealing with sparse data in the area of information retrieval [4] and spoken term detection [10] in highly inflected languages, therefore the effects of lemmatization on topic identification accuracy is studied in the paper. On the other hand, since the system is used for processing large amounts of data, a summarization method was implemented and the effect of using only the summary of an article on the topic identification accuracy is studied.

P. Sojka et al. (Eds.): TSD 2012, LNCS 7499, pp. 191–198, 2012.

2 System for Acquisition and Storing Data

The topic identification module is a part of a system designed for collecting a large text corpus from Internet news servers described in [12]. The system consists of a SQL database and a set of text processing algorithms which use the database as a data storage for the whole system. One of the important features of the system is its modularity – new algorithms can be easily added as modules.

For the topic identification experiments the most important parts of the system are the text preprocessing modules. Each new article is obtained as a HTML page, then the *cleaning* algorithm is applied – it extracts the text and the metadata of the article. Then the *tokenization* and *text normalization* algorithms are applied – text is divided into a sequence of tokens and the non-orthographical symbols (mainly numbers) are substituted with a corresponding full-length form. The tokens of a normalized text are processed with a *vocabulary-based substitution* algorithm. Large vocabularies prepared by experts are used to fix the common typos, replace sequences of tokens with a multiword or to unify the written form of common terms. *Decapitalization* is also performed - substitutes the capitalized words at the beginning of sentences with the corresponding lower-case variants. The result of each of the preprocessing algorithm is stored as a text record in the database.

For the experiments described in this paper two new modules was added – automatic lemmatization module and automatic text summarization module.

2.1 Lemmatization Module

The task of the automatic lemmatization is to find a basic word form or a "lexical headword" of a given word. The use of some lemmatization preprocessing has been shown to be especially important for the highly inflected languages in the various tasks of natural language processing, such as keyword spotting or information retrieval.

The "lexical headword" of a word can be any of its forms, usually for example for a noun its singular nominativ is used or for a verb its infinitive. The particular base form of a word is given by creating the assignment between the word and its base form. The use of lemmatization leads to the reduction of the number of processed words – different forms of a word can be treated as one (its base form) – so it is especially advantageous in highly inflected languages, such as Czech, where for example a verb can have as much as fifty different forms.

The lemmatization module uses a lemmatizer described in the work [5]. This lemmatizer is automatically created from the data containing the pairs full word form – base word form. Based on this data a set of lemmatization rules and a vocabulary of base word forms is created. Also, a set of lemmatization examples is extracted, which is used for the lemmatization of the words that are not contained in the base word forms vocabulary. A lemmatizer created in this way has been shown to be fully sufficient in the task of information retrieval [6].

2.2 Summarization Module

Automatic text summarization is a process of creating a compression of a given text (or texts) with the preservation of as much information as is possible. The content of the information can be either generic or topic related. Summarization can be done in two ways – *summary by extraction*, where the text of the summary is extracted from the original text, or *summary by abstraction*, where the text of the summary is automatically generated to rephrase the contained information.

The task of the extractive summarization is to choose a subset of sentences with the maximal information content. Statistic algorithms for extractive text summarization often requires large training data – for example the work [7] uses a Bayesian classifier trained on a large corpus of professionally created abstracts to chose the sentences to include in the summary. On the other hand, when such training data is not accessible, the selection of the sentences is based on some heuristic features such as word frequency [8], position of sentences [2] in the text or the relation between sentences.

For the automatic summarization module an extractive generic summarization was chosen, as we want our summaries to preserve all the information contained in the original text, so the topic identification module can assign the correct topics. The implemented summarization algorithm selects the most important sentences in a text, where an importance of a sentence is measured by the importance of its words. One of the most commonly used measure for assessing the word importance in information retrieval area is the TF-IDF[1] measure, so we have decided to use it as well. The summary is created in a following way:

1. Split text to sentences and sentences to words.
2. For each term t in the document compute an *idf* weight:

$$idf_t = \log \frac{N}{N_t} \qquad (1)$$

where N is the total number of sentences in the document and N_t is the number of sentences containing the term t.
3. For each sentence s compute a term frequency $tf_{t,s}$ for each term. We have used the normalization of the term frequency by the maximum term frequency in the sentence.
4. The importance score S of each sentence in the document is computed as:

$$S_s = \sum_{t \in s} tf_{t,s} \cdot idf_t \qquad (2)$$

5. The five sentences with the highest score S are included in the summary.

2.3 Topic Identification Module

The main purpose of the topic identification module is to filter the huge amount of data according to their topics for the future use as the language modeling training data. Currently, the topic identification module uses a language modeling based classification algorithm and assigns 3 topics to each article. Topics are chosen from a hierarchical system – a "topic tree". Further information about the topic identification module can be found in [11].

---

[1] Term Frequency – Inverse Document Frequency.

Topic Tree. The topic hierarchy was built in a form of a topic tree, it is based on our expert findings in topic distribution in the articles on the Czech favorite news servers like *ČeskéNoviny.cz* or *iDnes.cz*. At present the topic tree has 32 generic topic categories like politics, schools or sports, each of this main category has its subcategories with the "smallest" topics represented as leaves of this tree. The deepest path in the tree has a length of four nodes(industry - energetics - energy - solar), an example can be seen on Figure 1.

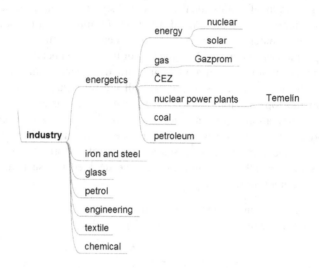

Fig. 1. Branch of the topic tree representing the industry topics

For the experiments the same topic tree as in [11] was used – it contains about 450 topics and topic categories, which correspond to the keywords assigned to the articles on the mentioned news servers. The articles with these "originally" assigned topics were used as training texts for identification algorithms.

Identification Algorithm. Current version of the topic identification module uses a language modeling based approach chosen due to the results of experiments published in [11]. This approach is similar to the Naive Bayes classifier, used for example in the work [1].

The probability $P(T|A)$ of an article A belonging to a topic T is computed as

$$P(T|A) \propto P(T) \prod_{t \in A} P(t|T) \tag{3}$$

where $P(T)$ is the prior probability of a topic T and $P(t|T)$ is a conditional probability of a term t given the topic T. The probability is estimated by the maximum likelihood estimate as the relative frequency of the term t in the training articles belonging to the topic T:

$$\hat{P}(t|T) = \frac{tf_{t,T}}{N_T} \tag{4}$$

where $tf_{t,T}$ is the frequency of the term t in T and N_T is the total number of tokens in articles of the topic T.

The goal of this language modeling based approach is to find the most likely topics T of an article A - for each article the three topics with the highest probability $P(T|A)$ are chosen. The prior probability of the topic $\hat{P}(T)$ is not used.

3   Evaluation

For the experiments on the effect of the automatic lemmatization and summarization algorithms three smaller collections containing the articles from the news server *ČeskéNoviny.cz* was separated from the whole corpus. This collections contain 5,000, 10,000 and 31,419 articles, details about the exact number of articles and the separation to training / testing data is shown in Table 1. The articles from *ČeskéNoviny.cz* have included the originally assigned keywords from their authors, which were used as the training and reference topics.

Table 1. Number of articles in the test collections and the training / testing data parts

collection name	articles	training	test
5k	5,000	4,000	1,000
10k	10,000	8,000	2,000
30k	31,419	27,000	4,419

Evaluation from the point of view of information retrieval (IR) was performed on the collections, where each newly downloaded article is considered as a query in IR and precision (P) and recall (R) is computed for the answer topic set:

$$P = \frac{T_C}{T_A}, \qquad R = \frac{T_C}{T_R} \tag{5}$$

where T_A is the number of topics assigned to the article, T_C is the number of correctly assigned topics and T_R is the number of relevant reference topics. An average of these measures is then computed across a set of testing articles. The F_1-measure is then computed from the (P) and (R) measures:

$$F_1 = 2\frac{P \cdot R}{P + R} \tag{6}$$

The training of the topic identification is done by counting the statistics containing the number of occurrences of each word in the whole collection, number of occurrences of each word in each document and the number of occurrences of each word in the documents belonging to a topic. These statistics can be trained from each kind of a text record in the database (result of a preprocessing step). For our experiments, the statistics for each of the collections were trained from the text preprocessed by following modules:

- *Replaced* - tokenization, text normalization, decapitalization and vocabulary-based substitution modules

– *Lemma* - tokenization, text normalization and lemmatization modules
– *Summary* - tokenization, text normalization, decapitalization, vocabulary-based substitution and summarization modules

The results of the topic identification on the different sized and preprocessed collections can be seen in Table 2. Testing articles were preprocessed in the same way as the collections – *replaced, lemma* and *summary* (first 3 columns of the table). For the summarization testing, there was done also the summary from the lemmatized testing articles (for the combination with *lemma* preprocessed collection – column 4 of the table) and from the *replaced* testing articles (combination with *replaced* preprocessed collection – column 5 of the table).

Table 2. Results of topic identification on different collections

coll./art.		replaced	lemma	summary	lemma/summary	replaced/summary
5k	P	0.5366	**0.5547**	0.5028	*0.5457*	0.5374
	R	0.5544	**0.5754**	0.5155	*0.5686*	0.5546
	F_1	0.5454	**0.5649**	0.5091	*0.5569*	0.5459
10k	P	0.5481	**0.5536**	0.5024	*0.5378*	0.5293
	R	0.5472	**0.5555**	0.4979	*0.5421*	0.53
	F_1	0.5476	**0.5546**	0.5002	*0.54*	0.5296
30k	P	0.5864	**0.5859**	0.5387	*0.5588*	0.5598
	R	0.6125	**0.6155**	0.5616	*0.5921*	0.5884
	F_1	0.5992	**0.6003**	0.5499	*0.575*	0.5737

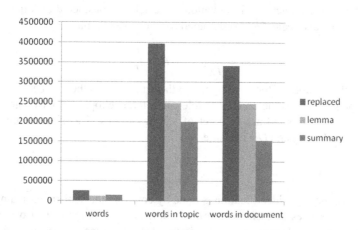

Fig. 2. Comparison of the number of lines of the database tables used to store the topic identification statistics for the 3 types of the training articles preprocessing

From the table we can draw following conclusions:

– The summarized text is not suitable for training topic identification statistics, results in column summary are the worst for all sizes of collections. This is not surprising,

as much less text is used for counting the statistics so the topic important words may be missing.
- The use of lemmatization seems to improve the topic identification results, especially on the smaller collections 5k and 10k. Lemmatization also reduces the size of the database tables used to store the topic identification statistics (for comparison see Figure 2).
- The most interesting finding of our experiments can be seen in column lemma/summary. When needed, a faster computation of topic identification using summarized and lemmatized texts can be used with a minimum loss on the topic identification accuracy.

4   Conclusions and Future Work

The evaluation of topic identification accuracy suggests that the lemmatization algorithm should be used for text preprocessing, as the accuracy of topic identification are slightly better than without the lemmatization. On the top of that, is was shown that the use of lemmatization reduces the size of stored database word statistics tables almost about a half. The lemmatized database with the combination of summarized test articles has only slightly worse topic identification accuracy, but the time needed for the topic identification of an article is reduced as the computation of the probability $P(T|A)$ of an article belonging to a topic is done over a reduced set of words.

For the future work, we would like to implement more complex summarization algorithm, currently the Graph-based LexRank method [3] for text summarization is being implemented. It may have an effect on the topic identification accuracy if the sentences in the summary will be chosen better. Also, we would like to use the summarization module for example for the multi-document topic oriented summaries or for the actuality summaries – summary of what happened in the last week for example.

Acknowledgments. This research was supported by the Ministry of Education, Youth and Sports of the Czech Republic project No. LM2010013 and by the grant of the University of West Bohemia, project No. SGS-2010-054.

References

1. Asy'arie, A.D., Pribadi, A.W.: Automatic news articles classification in indonesian language by using naive bayes classifier method. In: Proceedings of the 11th International Conference on Information Integration and Web-based Applications & Services, iiWAS 2009, pp. 658–662. ACM, New York (2009)
2. Edmundson, H.P.: New methods in automatic extracting. J. ACM 16(2), 264–285 (1969)
3. Erkan, G., Radev, D.R.: Lexrank: graph-based lexical centrality as salience in text summarization. J. Artif. Int. Res. 22(1), 457–479 (2004)
4. Ircing, P., Müller, L.: Benefit of Proper Language Processing for Czech Speech Retrieval in the CL-SR Task at CLEF 2006. In: Peters, C., Clough, P., Gey, F.C., Karlgren, J., Magnini, B., Oard, D.W., de Rijke, M., Stempfhuber, M. (eds.) CLEF 2006. LNCS, vol. 4730, pp. 759–765. Springer, Heidelberg (2007)

5. Kanis, J., Müller, L.: Automatic Lemmatizer Construction with Focus on OOV Words Lemmatization. In: Matoušek, V., Mautner, P., Pavelka, T. (eds.) TSD 2005. LNCS (LNAI), vol. 3658, pp. 132–139. Springer, Heidelberg (2005)
6. Kanis, J., Skorkovská, L.: Comparison of Different Lemmatization Approaches through the Means of Information Retrieval Performance. In: Sojka, P., Horák, A., Kopeček, I., Pala, K. (eds.) TSD 2010. LNCS, vol. 6231, pp. 93–100. Springer, Heidelberg (2010)
7. Kupiec, J., Pedersen, J., Chen, F.: A trainable document summarizer. In: Proceedings of the 18th Annual International ACM SIGIR Conference on Research and Development in Information Retrieval, SIGIR 1995, pp. 68–73. ACM, New York (1995)
8. Luhn, H.P.: The automatic creation of literature abstracts. IBM J. Res. Dev. 2(2), 159–165 (1958)
9. Psutka, J., Ircing, P., Psutka, J.V., Radová, V., Byrne, W., Hajič, J., Mírovský, J., Gustman, S.: Large vocabulary ASR for spontaneous Czech in the MALACH project. In: Proceedings of Eurospeech 2003, Geneva, pp. 1821–1824 (2003)
10. Psutka, J., Švec, J., Psutka, J.V., Vaněk, J., Pražák, A., Šmídl, L., Ircing, P.: System for fast lexical and phonetic spoken term detection in a czech cultural heritage archive. EURASIP J. Audio, Speech and Music Processing 2011 (2011)
11. Skorkovská, L., Ircing, P., Pražák, A., Lehečka, J.: Automatic Topic Identification for Large Scale Language Modeling Data Filtering. In: Habernal, I., Matoušek, V. (eds.) TSD 2011. LNCS, vol. 6836, pp. 64–71. Springer, Heidelberg (2011)
12. Švec, J., Hoidekr, J., Soutner, D., Vavruška, J.: Web Text Data Mining for Building Large Scale Language Modelling Corpus. In: Habernal, I., Matoušek, V. (eds.) TSD 2011. LNCS, vol. 6836, pp. 356–363. Springer, Heidelberg (2011)

Experiments and Results with Diacritics Restoration in Romanian

Cristian Grozea

Fraunhofer Institute FIRST, Kekulestrasse 7, 12489 Berlin, Germany

Abstract. The purpose of this paper is (1) to make an extensive overview of the field of diacritics restoration in Romanian texts, (2) to present our own experiments and results and to promote the use of the word-based Viterbi algorithm as a better accuracy solution used already in a free web-based TTS implementation, (3) to announce the production of a new, high-quality, high-volume corpus of Romanian texts, twice the size of the Romanian language subset of the JRC-Acquis.

1 Introduction

In many European languages electronically stored texts have been written by replacing the diacritics-decorated letters with place-holders composed of one or more standard Latin alphabet letters, a process named *latinization*. Historically, the latinization was used as a way to achieve plain text portability across computer platforms.

In some languages, such as Romanian, the practice of writing without diacritics has not ceased yet. It is still used in informal communication such as emails, text chat, SMS and others.

Romanian is also one of the languages where he latinization leads to numerous ambiguities and the process of automatically restoring the diacritics is not trivial (as it would be in German). It is interesting to note that the Romanian teenagers tend to write on text messages media (like SMS or text chat) with an enhanced but not universally accepted substitutions set for the diacritics, which is closer to a phonetic notation, facilitating therefore the understanding/diacritics restoration process (Table 1).

Only to increase the complexity of the issue, the return in 1993 to a pre-1953 orthography, decided by the Romanian Academy, led to the existence of two types of electronic texts without diacritics, each corresponding to a different orthography, the pre-1993 orthography and the post-1993 orthography. Two of the changes in the 1993

Table 1. Latinization of the diacritics in Romanian

Source letter	Replacement	Alternative unofficial replacement
Â/â	A/a	A/a
Ă/ă	A/a	A/a
Î/î	I/i	I/i
Ş/ş	S/s	Sh/sh
Ţ/ţ	T/t	Tz/tz

P. Sojka et al. (Eds.): TSD 2012, LNCS 7499, pp. 199–206, 2012.

orthography reform have effect on the diacritics restoration: in the case of a verb of Latin origin, "î" has been replaced by "u" (e.g. "sîntem" changed to "suntem"), which could have made the restoration of the diacritics slightly easier. Unfortunately the other change has a much higher statistical impact, allmost all other occurrences of "î" inside words have been changed to "â" (e.g. "pîine" changed to "pâine").

A need for good solutions to this problem comes from the field of technologies for accessibility. Without diacritics restoration, the texts are unintelligible when passed through text to speech systems. Without automatic restoration of the diacritics, the screen readers, text readers and other assistive appliances fail to perform their job properly, rendering the information once again inaccessible (or hardly accessible) to the blind persons. Those disabled persons have little to no control on the texts they are faced with, especially in the context of e-mail and internet. E.g., Ejobs.ro[1] – the biggest website for finding/posting jobs in Romania, `Ziare.com`[2] and `Hotnews.ro`[3] – the main online news websites in Romanian, are all written consistently without diacritics.

1.1   An Overview of the Published Methods

The existing methods for diacritics restoration in Romanian can be grouped into two categories: word-based and letter-based. The evaluation of any method can be done on words (counting the number of words for which the diacritics have been restored wrongly) or on letters (counting the percentage of the ambiguous letters for which the diacritics have been restored wrongly).

A summary of the existing literature is given in Table 2.

The word-based methods for Romanian have appeared first in [1,2], where a simple dictionary lookup is enhanced by using probabilistic POS tagging. The main disadvantage of this method (as of all dictionary-based ones) is that it cannot handle new words, not present in the dictionary used.

In [3], a corpus-based method for learning the diacritics restoration at letter level was first proposed for Eastern-European languages, as a way to compensate for the lack of good dictionaries with diacritics. In [4] the same method is evaluated in slightly more detail, but the performance reported is lower - it equals the level reported in [3] when a certain ambiguity is ignored (between "a" and "ă" at the end of the words).

In [5] the same methodology as in [4] is evaluated once more on the same corpus for Romanian (tests where done also with other languages), which provides the previously unreported performance on words of this letter-based method and brings an independent verification of its accuracy on letters. The accuracy reported is lower than both [3] and [4], and the authors – who use 10-fold cross-validation to evaluate the performance – comment this difference thus: *"Interestingly, the grapheme accuracy scores for Czech and Romanian are well below those reported in [3,4]. Since we use the same machine learning algorithm and same data, we hypothesize that the difference is due to evaluating the task on unseen words, rather than evaluating it on graphemes, extracted from a combination of known and unknown words".*

---

[1] http://www.ejobs.ro

[2] http://www.ziare.com

[3] http://www.hotnews.ro

Table 2. Previous Methods

Paper	Method	Accuracy on letters	Accuracy on words	Comments
Tufiş and Chiţu 1999	dictionary lookups and probabilistic tagging	not reported	97.08% ...98.47%	works only for words in dictionary
Mihalcea and Năstase 2002	TIMBL or C4.5	98.3%	not reported	letter-based
Mihalcea 2002	TIMBL or C4.5	99%	not reported	letter-based
De Pauw et al. 2007	same as Mihalcea and Nastase 2002	97.3%	83.2%	same as Mihalcea and Năstase 2002, on the same corpus, independent evaluation
Bobicev 2007	prediction by partial matching (PPM)	99.3%	not reported	
Tufiş and Ceauşu 2007	dictionary lookups and probabilistic tagging	not reported	97.79%	works only for words in dictionary
Tufiş and Ceauşu 2008	dictionary lookups and probabilistic tagging, fallback to letter-based for unknown words	99.4%	97.75%	it includes letters without ambiguity in the computation of the letter-level error
our contribution	Viterbi on word-based N-grams	99.6%	99.3%	

There is also another follow-up to the word-based methods series initiated by [1], the paper [6], where a letter-based method is added as a fall-back for the handling of the unknown words (not in the dictionary).

Letter frequency models (e.g. is "a", "ă" or "â" more frequent?), word frequency models, and symmetric N-grams centered on the target letter (similar to [3], but using the statistics of the N-grams extracted from the training corpus for deciding on the higher probability alternative) are tested in [7], where the author concludes that the later method is the best of all three.

1.2 Objectives

Given these fairly strong experimental results, is the problem solved? To answer this, one should consider to what extent a word-error rate of 2.21% (the lowest error rate reported) would affect the texts being restored. It is written in [1] that "In Romanian, every second word might contain at least one diacritical character"; in [6] it is estimated that 30% of the words in the written texts in certain fields contain diacritics. Our own measurements on three corpora (Table 3) show that between 25% and 40% of all words instances in texts with the diacritics removed could have had diacritics in the original text. If more than 2% in average of all words will still contain errors after the restoration of the diacritics and given that in general a (printed) page contains an average of 300–400 words, this means that one should expect in average 6–8 erroneous words on every page after restoring the diacritics. These numbers render the problem as far from completely solved.

Table 3. Corpora Statistics – Words and Bigrams

1.Corpus words	2.Count	3.Distinct words	4.Distinct after latinization	5.Ambiguous occurrences (% out of Column 2)
JRC-Acquis (JRC)	17,966,937	216,788	211,984	32.01%
Ec. Journal (ECON)	1,242,830	50,754	48,699	24.98%
Parliament (PAR)	38,350,406	164,353	155,403	39.99%
1.Corpus bigrams	2.Count	3.Distinct bigrams	4.Distinct after latinization	5.Ambiguous occurrences (% out of Column 2)
JRC-Acquis (JRC)	16,906,290	1,747,196	1,719,639	20.55%
Ec. Journal (ECON)	1,271,556	395,177	389,470	12.09%
Parliament (PAR)	38,364,261	2,981,063	2,912,247	28.38%

2 Methods

2.1 Corpora

We have used three corpora – statistics are given in Table 3; the fifth column contains the number of ambiguous word occurrences after the removal of the diacritics (instances of words that have more ways of restoring the diacritics). One can comment from these numbers that there aren't that many distinct words with potential ambiguity, but they are used quite frequently.

The first corpus (JRC) is the Romanian language subset of the JRC-Acquis version 3.0 [8]. We had expected the quality of the orthography to be high, but it is not entirely so. There are texts having most of the diacritics missing. The paper [2] also mentions various errors within this corpus. Another problem is that many words are simply not in Romanian. There are even entire Greek paragraphs in one of the texts there. The topics are various, covering an extremely wide range of subjects, such as regulations concerning the car's lights, regulations concerning the measurement of the level of dioxin in the food the living stock is fed with, regulations concerning the commerce with other countries and so on. The language style is that of the laws, very formal, with many juridical and technical terms. Therefore its stylistic spectrum is severely limited, despite its big size.

The second corpus (ECON) is 2 years worth of a Romanian economic journal[4]. While the journal itself is very good, the quality of the orthography on the web archive is low, as many of the articles are entirely without diacritics, others have paragraphs with, and paragraphs without diacritics (obvious traces of copy/paste use). This shows again to what extent the problem of diacritics is disregarded by the native speakers, who are in general perfectly able to read – and find much easier to type – texts without diacritics. We have eliminated the articles entirely without diacritics, and kept all the others. Therefore it is still of mixed quality with respect to the diacritics and only

---

[4] http://www.capital.ro

suitable for human inspection of the restoration results. It contains mostly economic news written in third-person, but also some interviews containing dialogs.

The third corpus (PAR) is the transcription of the parliamentary debates in the Romanian Parliament[5], from 1996 to 2007. The quality of the orthography is very high. It contains a balanced mix of dialog and writing in the third-person, either of the speakers (about external circumstances), or of the transcribers, describing the scene and the circumstances (such as "the right wing applauds the speaker"). The subjects are sometimes political, sometimes economical or social. Each and every important issue in Romania in these more than 10 years is reflected there and discussed thoroughly, by several speakers. It is more diverse and more surprising that one could even imagine. It even contains a full day praising of a 15th century Moldavian king (Stephen the Great[6]), poetry (one former senator is a poet), history references, EU references, and other topics.

2.2 Algorithms

After reimplementing and testing some of the methods presented in the literature overview, we decided to attempt a classical statistical approach, the Viterbi algorithm [9], that we use to rebuild the most probable sequence of word alternatives for each phrase given the statistics of the language inferred from the training set - the in-phrase word bigrams.

Formally, assuming that a phrase is composed of n words, and that for the word on position i there are k_i alternatives $a_{i,j}$, for $j = 1 \ldots k_i$ of restoring the diacritics, the method computes recursively (using the technique named dynamic programming) the probability $P_{i,j}$ for each $i = 1 \ldots n$ and each $j = 1 \ldots k_i$ that the prefix of the phrase ending at the position i ends with the alternative $a_{i,j}$ using $P_{1,j} = P(a_{1,j})$ and the Equation 1 for $i > 1$.

$$P_{i,j} = \sum_{q=1\ldots k_{i-1,j}} P_{i-1,q} \cdot P(a_{i,j}|a_{i-1,q}) \tag{1}$$

For the previously unseen words w we use $P(w) = 1/N$ where N is the length in words of the training corpus; for the previously unseen word transitions xy, we use $P(y|x) = P(y)/2N$. The largest resulting $P_{n,\cdot}$ is retained and from it the optimal combination of alternatives is inferred.

Our experimental procedure is the following: we build a model on a corpus and we test it on another (or the same) corpus. Whenever we train and test on the same corpus, we permute randomly its files, use half of the files for training, and the other half of the files for testing. Therefore, even when the same corpus is used for both training and testing, no single file is both in the training set and in the test set – it is a random-split two-fold cross validation.

3   Results and Analysis

The results are summarized in Tables 4 and 5. The rows contain the corpus used for training, and the columns contain the corpus used for testing.

---

[5] http://www.cdep.ro, http://www.senat.ro
[6] http://en.wikipedia.org/wiki/Stephen_III_of_Moldavia

Table 4. Results - Letter-Level Error (%)

Trained:Tested	JRC	ECON	PAR
JRC	**0.40**	1.72	3.06
ECON	1.73	1.33	2.15
PAR	**0.90**	0.69	**0.39**

Table 5. Results - Word-Level Error (%)

Trained:Tested	JRC	ECON	PAR
JRC	**0.78**	2.95	5.20
ECON	3.22	2.26	3.61
PAR	**1.75**	1.25	**0.70**

Precise details on the definition of the error rates are the following: The letter-level error is measured by applying the diacritics on a copy of the original text, then comparing letter by letter and counting the differences. The count of the differences is divided to the number of occurrences of the four letters that could have had diacritics (a,i,s and t). The result is the observed error rate reported.

The word-level error is measured by comparing the list of words after tokenization. The tokenization is oriented towards Romanian words and ignores words in Greek, for example. Numbers are not counted as words. This is why on the Romanian subset of the JRC-Acquis we count fewer words than mentioned on the JRC-Acquis webpage[7]. Our way to proceed is cautious, it can only underestimate but not overestimate the accuracy we report.

The results are very good for 2-fold cross-validation on a single corpus (the diagonal elements in Tables 4 and 5), provided the corpus has consistently good quality orthography. ECON is very inconsistent and of somewhat lower quality, as mentioned in a previous section.

It is always interesting to see to what extent the model learned on a corpus is portable to other corpora. The distributions of the words and of the word senses are expected to differ, changing the implicit prior probability distribution of the algorithm. Training on PAR and testing on JRC gives fairly good results, while reverting their roles leads to higher errors.

3.1   Letters and Words Contributing Most to the Error

The two words **ca** (translation: *as* – comparison) and **că** (translation: *that* – as in "He knew *that* she was there") contribute most to the error, in the case of most corpora pairs. The contribution usually sums up to about 1/1000 of the number of words. If we consider also the word-level errors for the pairs PAR/PAR and JRC/JRC we can see that by solving the ambiguity of this single word pair we could decrease the error by at least 10% of it.

The very high frequency of this missclassified pair seems to invalidate the assumption (in [3]) that the word-ending "a"-s can be resolved by simply distinguishing between articulated and non-articulated words. If this assumption is not valid – and this troublesome pair **ca** and **că** suggests it is not, as they are not in this relation – the error rate of the letter based method [3] would increase from less than 1% to very close to 2%, corresponding to an increase of error rate on the symbol "a" from 0.86% to 4.44% and would thus match the other paper [4], and the independent verification in [5].

---

[7] http://langtech.jrc.it/JRC-Acquis.html

Table 6. Detailed Results – Letter-Level Errors (all are under 1%)

Corpus	JRC	PAR		JRC	PAR
error on letters	0.40%	0.39%	error on the letter i	0.06%	0.07%
error on words	0.78%	0.70%	error on the letter s	0.15%	0.16%
error on the letter a	**0.97%**	**0.90%**	error on the letter t	0.19%	0.15%

We were able to achieve an error rate of less than 1% on the symbol "a" (Table 6), without any simplifying assumption, and despite the potentially higher ambiguity (the currently valid orthography specifies that most î inside the words should be replaced by â, thus potentially clashing with ă and a). All corpora used here followed the current orthography, reintroduced by the Romanian Academy in 1993, as explained in the introduction.

4 Discussions and Conclusion

4.1 The Best Corpus for Training

As it can be observed from the numbers in the Tables 3 and 4, the best corpus for training is almost always the corpus that is being tested on. There is a notable exception, training on PAR and testing on ECON gives better results than when training on ECON itself. We think PAR is diverse enough to cover the economical topics and the writing styles used in ECON.

What we have found is that training on a bigger corpus is not necessarily better, as even training on the partly faulty and smallish ECON is better than training on the 15 times bigger JRC – except for testing on the JRC corpus itself. We explain this through the severely narrow stylistic diversity (writing styles) of the JRC corpus.

Although none of the corpora considered here can be considered an universal training set, but PAR[8] is a very good candidate that covers well the daily news, the economical, social and political issues (although very specialized fields like medicine won't be covered enough).

4.2 Conclusion

The system exhibited for the biggest corpus smaller error rates than the previously published results, both in terms of word and letter rates. Especially our word-level error on PAR is interesting, as it is at least three times lower than any word-level error on any dataset reported in the literature relevant for the diacritics restauration for Romanian. For cross-corpus it can still give very good results, comparable to the previously published ones, if the corpus to train on is wisely chosen. The "PAR" corpus proved to be the best in this respect. The take-home message is that the corpus should not be too small or too specialized (in topics, writing styles).

---

[8] We can provide on request the corpus "PAR" to the interested researchers, as we see it as a valuable resource for the computational linguists interested in Romanian.

We have embedded an early implementation in a TTS system for Romanian, freely usable on the web with no restrictions on the size of the texts[9].

We intend to enhance the system with explicit high-performance word sense disambiguation [10], automatic rough categorization of the text (e.g: "juridical", "economical", "medicine" , "technical" or "culture"), and to apply it to other Eastern European languages.

References

1. Tufiş, D., Chiţu, A.: Automatic diacritics insertion in Romanian texts. In: Proceedings of COMPLEX 1999 International Conference on Computational Lexicography (1999)
2. Tufiş, D., Ceauşu, A.: Diacritics Restoration in Romanian Texts. In: Pascaleva, E., Slavcheva, M. (eds.) A Common Natural Language Processing Paradigm for Balkan Languages, pp. 49–55 (2007)
3. Mihalcea, R.F.: Diacritics Restoration: Learning from Letters versus Learning from Words. In: Gelbukh, A. (ed.) CICLing 2002. LNCS, vol. 2276, pp. 339–348. Springer, Heidelberg (2002)
4. Mihalcea, R., Năstase, V.: Letter level learning for language independent diacritics restoration. In: International Conference on Computational Linguistics, pp. 1–7. Association for Computational Linguistics, Morristown (2002)
5. De Pauw, G., Wagacha, P.W., de Schryver, G.-M.: Automatic Diacritic Restoration for Resource-Scarce Languages. In: Matoušek, V., Mautner, P. (eds.) TSD 2007. LNCS (LNAI), vol. 4629, pp. 170–179. Springer, Heidelberg (2007)
6. Tufis, D., Ceausu, A.: Diac+: a professional diacritics recovering system. In: LREC, European Language Resources Association (2008)
7. Bobicev, V.: Statistical Methods and Algorithms of Text Processing (based on Romanian texts). Ph.D. thesis, Technical University of Moldova (2007)
8. Steinberger, R., Pouliquen, B., Widiger, A., Ignat, C., Erjavec, T., Tufis, D., Varga, D.: The JRC-acquis: A multilingual aligned parallel corpus with 20+ languages. In: Proceedings of the 5th International Conference on Language Resources and Evaluation, LREC 2006, Genoa, Italy (2006)
9. Viterbi, A.: Error bounds for convolutional codes and an asymptotically optimum decoding algorithm. Information Theory 13, 260–269 (1967)
10. Grozea, C.: Finding optimal parameter settings for high performance word sense disambiguation. In: Proceedings of Senseval-3: The Third International Workshop on the Evaluation of Systems for the Semantic Analysis of Text (2004)

---

[9] http://www.phobos.ro/demos/tts

Supervised Distributional Semantic Relatedness

Alistair Kennedy[1] and Stan Szpakowicz[1,2]

[1] School of Electrical Engineering and Computer Science
University of Ottawa, Ottawa, Ontario, Canada
{akennedy,szpak}@eecs.uottawa.ca
[2] Institute of Computer Science, Polish Academy of Sciences, Warsaw, Poland

Abstract. Distributional measures of semantic relatedness determine word similarity based on how frequently a pair of words appear in the same contexts. A typical method is to construct a *word-context* matrix, then re-weight it using some measure of association, and finally take the vector distance as a measure of similarity. This has largely been an unsupervised process, but in recent years more work has been done devising methods of using known sets of synonyms to enhance relatedness measures. This paper examines and expands on one such measure, which learns a weighting of a *word-context* matrix by measuring associations between words appearing in a given context and sets of known synonyms. In doing so we propose a general method of learning weights for *word-context* matrices, and evaluate it on a word similarity task. This method works with a variety of measures of association and can be trained with synonyms from any resource.

1   Introduction

Measures of Semantic Relatedness (MSRs) are central to a variety of NLP tasks. In general, there are three methods of measuring semantic relatedness: resource-based methods, such as those using *WordNet* or *Roget's Thesaurus*; distributional methods, using large corpora; and hybrid methods, combining the two. Distributional MSRs rely on the hypothesis that the interchangeability of words is a strong indication of their relatedness (see [1] for an overview). If two words tend to appear in the same contexts regularly, they are more likely to be synonyms than those that do not. Usually some measure of association is used to determine the dependency between a word and the context in which it appears. This is an essentially unsupervised process. Recent work on MSRs that mix distributional and task-specific information includes [2,3,4,5]. We consider many of these methods partially supervised because they employ known sets of related words to train their system.

We describe an expansion of our supervised MSR first proposed in [6]. Our MSR reweighed a *word-context* matrix using Pointwise Mutual Information (PMI) to increase weight of contexts that tend to contain synonyms, while decreasing the weight of other contexts. We found the best results when combining supervised and unsupervised MSRs. Cosine similarity was used to measure vector distance. Our MSR resembles work where a function was learned to re-weight a matrix for measuring document

P. Sojka et al. (Eds.): TSD 2012, LNCS 7499, pp. 207–214, 2012.

$$x \in X \begin{array}{c} y \in Y \quad y \notin Y \\ \left[\begin{array}{cc} O_{0,0} & O_{0,1} \\ O_{1,0} & O_{1,1} \end{array} \right] \end{array} \qquad\qquad x \in X \begin{array}{c} y \in Y \quad y \notin Y \\ \left[\begin{array}{cc} E_{0,0} & E_{0,1} \\ E_{1,0} & E_{1,1} \end{array} \right] \end{array}$$

$x \notin X$ row for observed on left, $x \notin X$ row for expected on right.

Fig. 1. Confusion matrix of observed values   **Fig. 2.** Confusion matrix of expected values

similarity [7,8]. In [6] we used a tool called *SuperMatrix* [9], while now we use of our own implementation.[1] The following contributions add to our methodology from [6]:

- Evaluate several measures of association for *word-context* matrix re-weighting.
- Propose and evaluate an expansion to our supervised MSR.
- Evaluate the supervised MSR on verbs and adjectives, in addition to nouns.
- Explore training data from *WordNet* as well as *Roget's Thesaurus*.

Section 2 describes how we measure association, while Section 3 describes how these measures are applied to learning MSRs. Section 4 describes our experiments determining the best parameters for the MSRs and Section 5 concludes this work.

2   Measuring Association

A measure of association measures the dependency between two random variables, X and Y. Counts of co-occurring events $x \in X$, $y \in Y$, $x \notin X$ and $y \notin Y$ are recorded in a matrix of observed values, illustrated in Figure 1.[2] Using the observed counts the expected counts (Figure 2) are calculated with Equation 1.

$$E_{i,j} = \frac{\sum_y O_{i,y} \ \sum_x O_{x,j}}{\sum_{x,y} O_{x,y}} \tag{1}$$

From the observed and expected counts, we calculate the dependency between X and Y using six measures of association: Pointwise Mutual Information (PMI) (Equation 3); Z-score (Equation 4); T-score (Equation 5); χ^2 (Equation 6); Log Likelihood (LL) (Equation 7); and Dice (Equation 2).

$$Dice = \frac{2 * O_{0,0}}{\sum_j O_{0,j} + \sum_i O_{i,0}} \tag{2} \qquad\qquad PMI = \log \frac{O_{0,0}}{E_{0,0}} \tag{3}$$

$$Z\text{-}score = \frac{O_{0,0} - E_{0,0}}{\sqrt{E_{0,0}}} \tag{4} \qquad\qquad T\text{-}score = \frac{O_{0,0} - E_{0,0}}{\sqrt{O_{0,0}}} \tag{5}$$

$$\chi^2 = \sum_{i,j} \frac{(O_{i,j} - E_{i,j})^2}{E_{i,j}} \tag{6} \qquad\qquad LL = 2 \sum_{i,j} O_{i,j} \log \frac{O_{i,j}}{E_{i,j}} \tag{7}$$

[1] The code used in these experiments is available as a Java package called *Generalized Term Semantics* (GenTS)
http://eecs.uottawa.ca/~akennedy/Site/Resources.html
[2] The notation we use to describe this process is derived from that in [10].

Table 1. Counts of unique words, contexts and non-zero entries in the *word-context* matrices

POS	Terms	Contexts	Non-zero Entries
Nouns	43 834	1 050 178	28 296 890
Verbs	7 141	1 423 665	25 239 485
Adjectives	17 160	360 436	8 379 637

These measures can be divided into three groups. LL and χ^2 use all observed and expected values from Figures 1 and 2. PMI, T-score and Z-score use only $O_{0,0}$ and $E_{0,0}$. Dice measures vector overlap.

3 Measures of Semantic Relatedness

This section describes how the Measures of Semantic Relatedness (MSRs) are implemented using the measures of association from Section 2. In all cases, we use cosine similarity to measure distance; the difference is how the *word-context* matrix is re-weighted. Before we can evaluate the MSRs, we must first build a *word-context* matrix.

We build a *word-context* matrix using a common procedure [11]. We use a Wikipedia dump as a corpus,[3] and parse it with *Minipar* [11] to create a set of dependency triples. An example of a triple is $\langle settle, obj, question \rangle$: the noun "question" appears as the object of the verb "settle". A dependency triple $\langle w_1, r, w_2 \rangle$ generates *word-context* pairs $(w_1, \langle r, w_2 \rangle)$ and $(w_2, \langle w_1, r \rangle)$. When the words w_1 and w_2 are used as part of a context, they can be of any part-of-speech, and all relations r are allowed. When w_1 and w_2 are the words, they must be single words with no upper case letters, digits or symbols. From these triples, we built three matrices for nouns, verbs and adjectives/adverbs.[4] One problem is that some words and contexts appear very infrequently. To remedy this, we only use nouns and adjectives that appear 35 times or more, and verbs 10 times or more. Likewise a context had to be used twice to be included.[5] We report the sizes of our matrices in Table 1.

3.1 Unsupervised Learning of Context Weights

When measuring semantic relatedness in an unsupervised fashion, we take $x \in X$ to be the appearance of a word, while $y \in Y$ is the appearance of a context. We count the following observed values:

- $O_{0,0}$ $[x \in X \land y \in Y]$: w_i is found in context c_j;
- $O_{0,1}$ $[x \in X \land y \notin Y]$: w_i is found in a context other than c_j;
- $O_{1,0}$ $[x \notin X \land y \in Y]$: a word other than w_i is found in context c_j;
- $O_{1,1}$ $[x \notin X \land y \notin Y]$: a word other than w_i is found in a context other than c_j.

The unsupervised MSR uses these counts to create a unique score for every *word-context* pair. The matrix is then re-weighted with these scores.

---

[3] Downloaded in August 2010.

[4] *Minipar* uses the symbol "A" for adjectives and adverbs, so we placed them in the same matrix.

[5] We found numbers by experimenting (selecting random words and generating lists of synonyms) and found that they make matrices fairly reliable.

3.2 Supervised Learning of Context Weights

A supervised MSR would use measures of association not just between words and contexts, but between pairs of words co-occurring in a context and pairs of words from our training data known to be synonyms. We calculate an association score for every context c_k. In this case, $x \in X$ represents a word pair's co-occurrence in context, $y \in Y$ – a pair of synonymous words. We explore three sources of training data coming from the 1911 and 1987 editions of *Roget's Thesaurus* and *WordNet* 3.0. We identify synonyms by selecting words from the same synset in *WordNet* or from the same Semicolon Group in *Roget's*.[6] A few examples of synonyms from the 1911 *Roget's Thesaurus*: $\langle calculator, algebraist, mathematician \rangle$ and $\langle boating, yachting \rangle$.

To calculate the association, we count pairs of words $\langle w_i, w_j \rangle$ for each context c_k:

- $O_{0,0} [x \in X \land y \in Y]$: $\langle w_i, w_j \rangle$ are synonyms and both appear in c_k;
- $O_{0,1} [x \in X \land y \notin Y]$: $\langle w_i, w_j \rangle$ are synonyms and only one appears in c_k;
- $O_{1,0} [x \notin X \land y \in Y]$: $\langle w_i, w_j \rangle$ are not synonyms and both appear in c_k;
- $O_{1,1} [x \notin X \land y \notin Y]$: $\langle w_i, w_j \rangle$ are not synonyms and only one appears in c_k.

When taking these counts, a pair can be counted multiple times if both its words appear more than once in a given context. This takes care of situations when context c_k contains a large set of unrelated words with low counts, and a small set of related words but with high counts. Now $score(c_k)$ can be calculated for every context c_k using one of the measures of association. (Negative scores are rounded up to 0.) The scores are normalized so that their average is 1.0. We then multiply the count of each word in c_k by $score(c_k)$. Some contexts contain no words from the training data, so a weight cannot be calculated. We give such contexts a score of 1.0.

We propose a second version of this training methodology. Our version finds a unique weight for every relationship r and then applies that weight to all contexts $\langle r, w_i \rangle$ .. $\langle r, w_j \rangle$. We use the same method as described above, but we combine the counts for contexts that share a common relation r. In this experiment, rather than learning contexts most appropriate for measuring semantic relatedness, we are learning which syntactic relationships best indicate semantic relatedness. The hypothesis behind this method is that the syntactic relationship is more important than the word in any given context. These two training methodologies will be distinguish by referring to them as learning at the "context" level and the "relation" level.

3.3 Combined Learning of Context Weights

In [6] we found that the best results came when mixing supervised and unsupervised learning. Supervised weighing is first performed on the matrix and then unsupervised weighting is run to reweight the matrix a second time. One problem with this methodology is identifying optimal parameters – measures of association and training type – before building a combined method. We run experiments first to identify those parameters and then construct and test this combined method.

---

[6] A Semicolon Group in *Roget's* contains near-synonyms, just like synset members in *WordNet*.

4   Evaluating the Measures

We first evaluate the individual supervised and unsupervised systems on a tuning set and then use them to build the combined method to be evaluated on a test set. Our evaluation task is to determine whether two words appears in the same Head in *Roget's Thesaurus*. *Roget's Thesaurus* divides the English lexicon into approximately 1000 broad categories named Heads, which represent such broad concepts as *Existence, Nonexistence, Materiality, Immateriality, Advice, Council, Reward* and *Punishment*. Each Head can contain nouns, verbs, adjectives and adverbs. To create the tuning set and the test set, we randomly create two sets of 1000 nouns, two sets of 600 verbs and two sets of 600 adjectives that appear in the 1987 *Roget's Thesaurus*. All words from the tuning set or test set are removed from the training data selected from *WordNet*, or *Roget's Thesaurus* for the supervised MSRs. We generate a long lists of all nearest neighbours for each of these words, with each measure. For example the four most related words to *psychology*, with their scores are: *sociology (0.720), anthropology (0.707), linguistics (0.582), economics (0.572)*. We evaluate these measures at a variety of recall points: the top 1, 5, 10, 20, 50 and 100 nearest neighbours. We also include an unweighted matrix as a baseline to these experiments.

4.1   Tuning Our Measure of Semantic Relatedness

We perform experiments with six different measures of association applied to three parts of speech. We use unsupervised and two kinds of supervised training, with three different training sets. In effect, there are far too many experiments to report the results in a single paper. Instead we describe and summarize the results using graphs.

The first experiment is to identify which MSR performed best in an unsupervised setting; this is summed up in Figure 3. We only present results for nouns. The results for verbs and adjectives are quite similar. Our basic findings are that most measures show a noticeable improvement over the baseline, with the exception of LL, where there is no improvement. χ^2 also perform poorly. Without exception, PMI is the superior measure, performing best at all recall points.

The second experiment is to identify the best measure of association for a supervised MSR. Once again our findings are the same for all POSs, all training data and with supervision at both the context and relation level; see Figure 4. PMI is clearly superior, though – unlike the supervised case – most other measures of association are worse than the unweighted baseline. Dice is frequently very close to the baseline, while χ^2 and LL are almost never superior at any recall point.

Having established the best measure of association, we now look to which kind of training – context or relation level – actually yields the best results. For nouns and verbs we find consistently that training at the context level is superior; see Figure 5. The Figure shows that, at most recall points, training at the context level outperforms training at the relation level. This is consistently true across all three sources of training data. For adjectives we find quite different results; see Figure 6. In this case there is a very small difference between training at the relation and context level, but more often than not training at the relation level is superior. One possible reasons for this is that the adjective matrix is smaller than the noun and verb matrices, making it more difficult to find large groups of related or unrelated words in a given context.

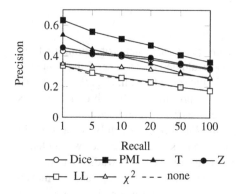

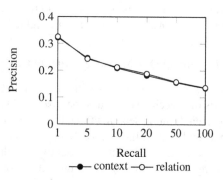

Fig. 3. Scores for nouns, unsupervised

Fig. 4. Scores for nouns, supervised by context with *Roget's* 1911

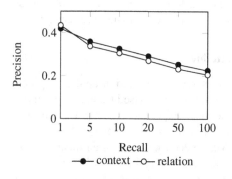

Fig. 5. Context and relation scores for nouns, trained with *Roget's* 1911

Fig. 6. Context and relation scores for adjectives, trained with *Roget's* 1911

4.2 Testing Our Measure of Semantic Relatedness

We have now identified the parameters for our supervised and unsupervised systems. Both the supervised and unsupervised MSRs use PMI weighting for all three POSs, while the supervised MSRs use learning at the context level for nouns and verbs and learn at the relation level for adjectives. In the tuning phase we did not attempt to identify which source of training data worked best, nor did we experiment with the combined method. This section examines both of these. The unweighted matrix makes up a lower baseline, while the unsupervised PMI re-weighted matrix makes up the high baseline. We compare the three supervised systems and three combined systems against these baselines on all three POSs; see Table 2.

The findings in Table 2 show that all supervised methods, while consistently outperforming the unweighted baselines, do not outperform the higher baseline of unsupervised PMI. The combined systems fare much better. By harnessing elements of both supervised and unsupervised matrix re-weighting, often we can find a statistically

Table 2. Evaluation of the various MSRs with statistically significant improvements over the Unsupervised-PMI baseline in bold

POS	Measure	Top 1	Top 5	Top 10	Top 20	Top 50	Top 100
Nouns	Unweighted	0.376	0.296	0.262	0.239	0.207	0.186
	Unsupervised-PMI	0.645	0.579	0.537	0.490	0.423	0.374
	context-1911	0.440	0.363	0.330	0.303	0.262	0.233
	context-1987	0.456	0.376	0.334	0.296	0.252	0.223
	context-WN	0.466	0.370	0.333	0.291	0.252	0.224
	Combined-1911	0.659	**0.588**	**0.548**	**0.501**	**0.431**	**0.382**
	Combined-1987	0.651	0.584	**0.549**	**0.501**	**0.430**	**0.381**
	Combined-WN	0.654	0.586	0.541	**0.495**	**0.430**	**0.380**
Verbs	Unweighted	0.398	0.331	0.318	0.299	0.276	0.256
	Unsupervised-PMI	0.582	0.526	0.487	0.444	0.396	0.357
	context-1911	0.468	0.394	0.368	0.334	0.303	0.283
	context-1987	0.480	0.418	0.382	0.356	0.318	0.299
	context-WN	0.482	0.426	0.393	0.365	0.324	0.303
	Combined-1911	0.605	0.533	**0.500**	**0.455**	**0.401**	**0.362**
	Combined-1987	0.588	**0.537**	**0.499**	**0.453**	0.399	**0.360**
	Combined-WN	0.587	0.531	**0.495**	**0.451**	0.395	0.356
Adjectives	Unweighted	0.317	0.259	0.224	0.205	0.163	0.139
	Unsupervised-PMI	0.600	0.480	0.431	0.368	0.295	0.247
	relation-1911	0.358	0.273	0.243	0.212	0.175	0.148
	relation-1987	0.357	0.277	0.250	0.217	0.179	0.153
	relation-WN	0.353	0.278	0.242	0.213	0.175	0.148
	Combined-1911	0.602	0.484	0.431	0.368	0.296	0.247
	Combined-1987	0.603	0.483	0.431	0.367	0.296	0.247
	Combined-WN	0.595	0.483	0.430	0.368	0.296	0.247

significant improvement (the bold results in Table 2) over the high baseline of unsupervised PMI for both nouns and verbs. For adjectives, we we do not find a significant improvement using the combined MSRs. We hypothesize that this is due to the smaller matrix size. Perhaps a larger amount of data is needed before supervision can offer a meaningful benefit.

In terms of sources of training data, it would appear that the 1911 and 1987 versions of *Roget's* performed comparably. The combined system trained with the 1911 *Roget's Thesaurus* shows a significant improvement on 9 out of 12 recall points for nouns and verbs. The 1987 version significantly improves on 8 out of 12 recall points. *WordNet* 3.0 still can improve the MSRs at a statistically significant level, although only 5 times. This may not seem surprising, because we evaluate our MSRs on *Roget's Thesaurus*, so those trained using data from *Roget's Thesaurus* could have an edge. That said, it is not completely clear that identifying words in the same *Roget's* Head should benefit more from training with *Roget's* Semicolon Groups than training with *WordNet* synsets.

5   Conclusion and Discussion

We have expanded on the methods in [6] to show how our MSRs can be implemented with a variety of measures of association and applied to different parts-of-speech. We have also noted that the supervised matrix weighting can be applied in two ways: learning at the relation level and learning at the context level. Finally we explore the use of *WordNet* as training data for identifying words in the same *Roget's* Head. We have found PMI to be the strongest measure of association for all of our measures. Learning at the context level worked best for nouns and verbs, but learning at the relation level was best for adjectives. Training data from *Roget's Thesaurus* proved superior to *WordNet*'s data on our task, though *WordNet* still improved over our high baseline. Ultimately, our combined MSR has a statistically significant improvement for nouns and verbs, though for adjectives the differences are too small to determine significance.

Acknowledgments. Partially funded by the Natural Sciences and Engineering Research Council of Canada.

References

1. Turney, P.D., Pantel, P.: From frequency to meaning: Vector space models of semantics. Journal of Artificial Intelligence Research 37, 141–188 (2010)
2. Patwardhan, S.: Incorporating dictionary and corpus information into a vector measure of semantic relatedness. Master's thesis, University of Minnesota, Duluth (2003)
3. Weeds, J., Weir, D.: Co-occurrence retrieval: A flexible framework for lexical distributional similarity. Comput. Linguist. 31, 439–475 (2005)
4. Mohammad, S., Hirst, G.: Distributional measures of concept-distance: A task-oriented evaluation. In: Jurafsky, D., Gaussier, É. (eds.) EMNLP, pp. 35–43. ACL (2006)
5. Hagiwara, M., Ogawa, Y., Toyama, K.: Supervised synonym acquisition using distributional features and syntactic patterns. Journal of Natural Language Processing 16, 59–83 (2005)
6. Kennedy, A., Szpakowicz, S.: A Supervised Method of Feature Weighting for Measuring Semantic Relatedness. In: Butz, C., Lingras, P. (eds.) Canadian AI 2011. LNCS, vol. 6657, pp. 222–233. Springer, Heidelberg (2011)
7. Yih, W.T.: Learning term-weighting functions for similarity measures. In: Proceedings of the 2009 Conference on Empirical Methods in Natural Language Processing, EMNLP 2009, vol. 2, pp. 793–802. Association for Computational Linguistics, Morristown (2009)
8. Hajishirzi, H., Yih, W.T., Kolcz, A.: Adaptive near-duplicate detection via similarity learning. In: Proceeding of the 33rd International ACM SIGIR Conference on Research and Development in Information Retrieval, SIGIR 2010, pp. 419–426. ACM, New York (2010)
9. Broda, B., Piasecki, M.: Supermatrix: a general took for lexical semantic knowledge acquisition. Technical report, Institute of Applied Informatics, Wroclaw University of Technology, Poland (2008)
10. Evert, S.: The statistics of word cooccurrences: word pairs and collocations. Ph.D. thesis, Institut für maschinelle Sprachverarbeitung, Universität Stuttgart (2004)
11. Lin, D.: Automatic retrieval and clustering of similar words. In: Proceedings of the 17th International Conference on Computational Linguistics, pp. 768–774. Association for Computational Linguistics, Morristown (1998)

PSI-Toolkit

How to Turn a Linguist into a Computational Linguist

Krzysztof Jassem

Faculty of Mathematics and Computer Science, Adam Mickiewicz University, Poznań, Poland

Abstract. The paper presents PSI-Toolkit, a set of text processing tools, being developed within a project funded by the Polish Ministry of Science and Higher Education. The toolkit serves two objectives: to deliver a set of advanced text processing tools (with the focus set on the Polish language) for experienced language engineers and to help linguists without any technological background learn using linguistics toolkits. The paper describes how the second objective can be achieved: First, a linguist, thanks to PSI-Toolkit, becomes a conscious user of NLP tools. Next, he designs his own NLP applications.

Keywords: Tagging, Classification and Parsing of Text, NLP Toolkits.

1 Introduction

Our experience shows that most computational linguists are computer scientists, who have learned linguistics as their second major field of study, rather than the other way round. We are of the opinion that so few *pure* linguists become computational linguists because NLP tools are presented to them in an unappealing way. Let us present a few examples that may support this thesis.

1.1 The Stanford Natural Language Processing Group Toolkit

Stanford CoreNLP [3] is a suite of NLP tools designed for the analysis of raw English texts. "The goal of the project is to enable people to quickly and painlessly get complete linguistic annotations of natural language texts." However, the first sentence explaining the usage of the toolkit reads: "Before using Stanford CoreNLP, it is usual to create a configuration file (a Java Properties file)." If a linguist is not discouraged by the *Java Properties file*, here what comes next: In particular, to process the included sample file input.txt you can use this command in the distribution directory: *Stanford CorNLP Command*

```
java -cp stanford-corenlp-2012-01-08.jar:
stanford-corenlp-2011-12-27-models.jar:
xom.jar:joda-time.jar-Xmx3g edu.stanford.nlp.pipeline.StanfordCoreNLP
-annotators tokenize,ssplit,pos,lemma,ner,parse,dcoref -file input.txt
```

The output of the Stanford processor is an XML file – a format more suitable for machine analysis rather than for human reading. A drawback of Stanford CoreNLP from the point of view of language engineers is that the toolkit is licensed under the General Public License, which does not allow for using the code in the proprietary software.

P. Sojka et al. (Eds.): TSD 2012, LNCS 7499, pp. 215–222, 2012.

1.2   UIMA Project

"Unstructured Information Management applications are software systems that analyze large volumes of unstructured information in order to discover knowledge that is relevant to an end user" ([4]). The tools (annotators) in the UIMA Project are available only as Java source codes and need compilation under a Java development environment. A potential user needs at least a preliminary course in Java programming in order to benefit from the UIMA project.

1.3   Teaching Basic Programming to Linguists

An attempt to use Stanford CoreNLP, UIMA or similar toolkits may tempt a linguist to learn basic programming skills. One of the manuals intended for linguists is Martin Wieser's [2]. The introduction sounds encouraging: "This book is mainly intended as an introduction to programming for linguists without any prior programming experience". The author of this paper tried to follow this approach with a group of bright students at interdisciplinary Ph.D. studies: Language, Society, Technology and Cognition funded by the European Social Fund ([5]). At the end of the 16-hour course students had the skills to write Perl programs that could lemmatize, form frequency lists or concordance lists. When the course and the exam were over, the students asked why they should write programs like those instead of using existing tools. The argument that existing tools may not always satisfy their specific needs, did not seem convincing to them.

1.4   Natural Language Toolkit (NLTK)

Natural Language Toolkit [6] is geared towards less experienced users. The user downloads three files and soon is offered valuable results of text processing. There are two small "buts": 1) NLTK basic tools work on texts delivered by the authors or textual files, which first have to be converted to the NLTK format. 2) The Python GUI is in fact a form of the command line (not a graphical interface a linguist is used to work with).

1.5   General Architecture for Language Engineering (GATE)

GATE ([7]) is one of the most mature NLP toolkits, dating from 1995. The system is intended for both language engineers who can develop their programs including GATE modules and for linguists who can write their own grammars for GATE tools. A linguist can process his own (set of) documents (of various formats). However, in order to obtain the first annotation, the user must overcome a few difficulties. First, he has to load a CREOLE plugin. Then, it is required to create processing resources. Next, an application has to be set up from the processing resources. The application is not likely to work, as two hardly obvious conditions must be fulfilled: 1) Each process forming the application must be assigned to a specific document, 2) The pipelines of processes must be "logical". Contrary to reasonable expectation the user will not see the immediate result of processing displayed on the screen, but has to go to *Language resources* to be able to view them.

1.6 Apertium

Apertium ([8]) is a free open-source machine translation platform that provides an engine for using existing machine translation systems as well as building new ones. Two features make the toolkit attractive for linguists: 1) the translation window, thanks to which a user may easily and immediately obtain the result of translation, 2) user-friendly means of creating self-developed systems (it suffices to edit a few dictionary files and a rule file to design an MT system).

We would like to take the Apertium approach in PSI-Toolkit. However, we would like to expand the domain of the toolkit so that it could assist in various linguistic tasks, not only Machine Translation.

2 PSI-Toolkit

2.1 PSI-Toolkit Outline

PSI-Toolkit is a set of tools designed for text processing, released under Lesser General Public License. This is a free license, which, in addition to GPL, could be linked with proprietary software and further distributed under any terms. The toolkit is intended for both language engineers and pure linguists (without computer education). The former group of users should be satisfied with Java, Perl and Python libraries, whereas the latter group is encouraged to use PSI-Toolkit by means of a user-friendly web portal.

2.2 PSI-Toolkit Processors

The programs included in the PSI-Toolkit are called PSI-processors. There are three types of PSI-processors: readers, annotators and writers. Readers process the text (input either from keyboard or from a file) in order to initialize the main data structure, a so-called PSI-lattice (see [1] for the description of the PSI-lattice). Annotators (e.g. a tokenizer, a lemmatizer, a parser) add new edges to the PSI-lattice. Writers convert the PSI-lattice to a graphical or textual format and re-direct the result to the output device.

PSI-Toolkit Readers. PSI-Toolkit readers read all formats of files listed in [9] and additionally .pdf files. A reader initializes the PSI-Lattice by converting individual characters of the textual content into vertices of the lattice. For instance, in the web version of the toolkit, the text typed into the edit window is processed by *txt-reader*.

PSI-Toolkit Annotators. PSI-Toolkit annotators are core processes in the system. The current version supports, among others, a tokenizer, a sentence-splitter, a lemmatizer and a shallow parser. A unique annotator is a bilexicon processor, which returns all the equivalents of all possible translations of input lemmas.

PSI-Toolkit Writers. PSI-Toolkit writers are a distinguishing feature of the project. There are two types of PSI writers: textual and graphical. The former return a textual representation of the PSI-lattice, whereas the latter display its visual representation. Both types can be customized by a user, who can filter the labels of edges to be returned. Also, the output of the textual writer may be simplified by filtering the information attached to edges.

2.3 PSI-Toolkit Web Portal

The PSI-Toolkit portal is accessible via a web browser at [10]. Fig. 1 shows an example of using the toolkit in a web window.

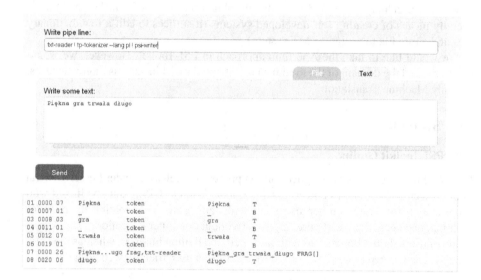

Fig. 1. PSI-Toolkit web portal

The user types a text into the edit window, e.g. *Piękna gra trwała długo. (Eng. A beautiful game lasted long.)* and specifies a command: a sequence of processors to execute. (This is facilitated by a list of prompts: most often used commands.) The processors are run in the order specified in the command. The exemplary sentence consists of ambiguous words (word ambiguity in the Polish language is an irregular phenomenon). *Piękna* is the feminine form of the adjective *piękny* or the genitive form of the noun *piękno (Eng. beauty). Gra* is the base form of the noun *gra (Eng. game)* or a form of the verb *grać (Eng. to play). Trwała* is the past form of the verb *trwać (Eng. to last)* or the feminine form of the adjective *trwały (Eng. long-lasting)*, or the base form of the noun *trwała (Eng. durable haircut)*. The only unambiguous word, the adverb *długo (Eng. long)* determines the syntactic interpretation of the whole expression. The PSI output lists each edge of the PSI-lattice in a separate line. In Fig. 1: Line 01 corresponds to the edge spanning over the first 6 characters of the input (start position is equal to 0000, offset counted in bytes is equal to 0007, as *ę* is represented by two bytes). Line 02 describes the space between the first and the second token of the input. Line 07 corresponds to the edge spanning over the whole input (the edge has been constructed by *txt-reader*).

2.4   Linux Distribution

PSI-Toolkit is also distributed in the form of two Linux binaries: PSI-Pipe and PSI-Server. PSI-Pipe may be installed and used on personal computers, whereas PSI-server allows for the creation of other PSI-Toolkit web pages.

2.5   PSI Pipeline

The PSI-toolkit command is specified as a pipeline of processors. If PSI-Toolkit is used under Linux on a personal computer, the processors should be invoked in a bash-like manner. For example, in order to process the string *Piękna gra trwała długo* in the way equivalent to that shown in Figure 1, the following pipeline should be formed:

```
>echo 'Piękna gra trwała długo' |
psi_pipe ! txt_reader ! tp-tokenizer --lang pl ! psi-writer
```

Interestingly this approach allows for using standard Linux commands e.g. *sort*, *grep* besides PSI pipelines. Moreover, PSI processors may be replaced or supplemented by external tools in the PSI pipeline. The PSI engine will add annotations provided by external tools to the PSI-lattice. It is admissible to use two different processors of the same type in the same pipeline. For example, running two different sentence splitters in the same process (in any order) may result in two different sentence splits. The ambiguity is stored in the PSI-lattice (it may or may not be resolved in further processing).

2.6   Switches

The use of the PSI processor may be customized be means of switches (options). Table 1 shows the list of admissible switches for *simple-writer*, the processor for printing the PSI-lattice in a simple, human-readable way.

Table 1. Use of switches for *simple-writer*

–sep arg	set separator between basic edges
–alt-sep arg	set separator between alternative edges
–linear	skip cross-edges
–no-alts	skip alternative edges
–with-blank	do not skip edges for 'blank' characters
–tag arg	filter edge tags

2.7   Examples of Usage

A combination of processors may result in an output format customized to the user's needs. Table 2 shows the results printed by *simple-writer* customized to show respectively: a) only tokens, b) only lemmas or c) only lexemes.

Table 2. Output of *simple-writer* depends on tag filtering

Piękna gra trwała długo	pięknolpiękny gralgrać trwaćltrwałaltrwały długo	piękno+substlpiękny+adj gra+subslgrać+verb trwać+verbltrwała+substltrwały+adj długo+adv
a) tokens	b) lemmas	c) lexemes

Table 3 shows the output if the pipeline contains the *bilexicon* command.

Table 3. Output of *simple-writer* depends on tag filtering

beauty+substlbeautiful+adjlexcellent+adj game+substlplay+verb last+verblstay+verblabide+verblpermanent+adjldurable+adjllasting+adj long+adv

By the end of the project a combination of PSI-processors will deliver on-line translation.

The visual writer displays either the whole lattice built by the processors or a part of it, if tag-filtered.

Fig. 2 shows the whole lattice after the lemmatization .

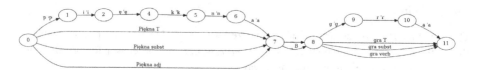

Fig. 2. Output of the PSI graphical writer: PSI-lattice after lemmatization

Fig. 3 displays the lattice without character vertices.

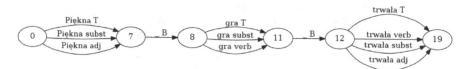

Fig. 3. Output of the PSI graphical writer: PSI-lattice without character vertices

2.8   Java, Perl, Python Libraries

PSI tools are also accessible as libraries of selected programming languages. The user may include the PSI tools as Java, Perl or Python modules.

2.9   Natural Languages Processed by PSI-Toolkit

There are no restrictions on languages analyzed by PSI processors (UTF-8 is used). One PSI pipeline may consist of processors defined for various languages. However, the tools delivered by the authors are oriented mainly towards the Polish language (some processors are also defined for English). The authors hope that PSI-Toolkit will bring together the Polish language processing tools, which are dispersed (see [11] for an exhaustive list of NLP tools and resources for Polish). The proof of this concept has been implemented on the morphosyntactical level: the external morphological lemmatizer *Morfologik* ([12]) has been incorporated into the PSI toolkit and can be run in the PSI pipeline instead of or besides (!) the PSI-dedicated lemmatizer.

3   PSI-Toolkit as a Didactic Tool for Linguists

This section describes one of the objectives for the creating PSI-Toolkit: *turning a linguist into a computational linguist*. The training path involves building computer skills of a *pure* linguist, by means of PSI-Toolkit:

1. With the aid of PSI-Toolkit perform some basic linguistic tasks
(e.g. tokenize, sentence split, lemmatize, parse)
    Show how syntactical parsing may solve word disambiguities
    Introduce PSI-writers in the appropriate order:
        Use the graphical writer (*gv-writer*) first
        Then proceed with *simple-writer*
        Finally use the full *psi-writer*
2. Introduce the 'bash' shell
    Teach the basics of the 'bash' shell
    Define simple exercises that require sorting or greping
    Teach the usage of the 'grep' and 'sort' commands
3. Use PSI-Toolkit in the 'bash' shell
    Download the PSI-Tools package from the PSI portal
    Install the package
    Define PSI pipelines including 'bash' commands
4. Introduce a "linguistic" programming language e.g. Perl or Python
    Define simple tasks that cannot be handled with the knowledge acquired so far
(e.g. generating a text index or a frequency list)
    Teach the basics of the language until students are able solve the above problems
    Suggest using PSI-Toolkit libraries in the student's applications
5. Design your own toolkit portal
    Download the PSI–Server package
    Publish your own PSI-Toolkit web-page
6. Introduce other NLP toolkits such as: NLTK, UIMA.

4   Conclusions

The paper presents the main ideas and the architecture of PSI-Toolkit, an open source NLP toolkit. The PSI tools may be accessed via a web browser or run locally. A PSI processing command should be formed as the pipeline, which may also include external tools. The PSI wrappers make it possible to use the PSI tools in Java, Perl or Python applications.

The toolkit is being developed within a project funded by the Polish Ministry of Science and Higher Education, which runs from April 2011 to April 2013. The progress of the project may be traced systematically at the Psi-Toolkit portal. At the moment this paper was written the portal provided 17 NLP processors. Providing tools for deep parsing and Machine Translation is scheduled for the months to come.

One of the main objectives underlying the creation of the toolkit was to encourage pure linguists to use and create text processing tools. The paper presents a path from a linguist to a computational linguist or even a language engineer. The incentive is a user-friendly web application, which returns comprehensive results for users without any programming background. Further steps to take depend on the increasing needs of a linguist already convinced that computer education could be easy and useful.

References

1. Jassem, K., Gralinski, F., Junczys-Dowmunt, M.: PSI-toolkit: A Natural Language Processing Pipeline. Computational Linguistics - Application. Springer, Heidelberg (to appear)
2. Wiesser, M.: Essential progamming for lingusts. Edinburgh University Press Ltd, Edinburgh (2009)
3. The Stanford Natural Language Processing Group,
   http://nlp.stanford.edu/software/corenlp.shtml
4. Apache UIMA, http://uima.apache.org/
5. LSTC, http://unikat.amu.edu.pl/studia/lstc/eng
6. NLTK, http://www.nltk.org/
7. GATE, http://gate.ac.uk/
8. Apertium, http://www.apertium.org
9. Apertium format handling, http://wiki.apertium.org/wiki/Format_handling
10. PSI-Toolkit portal, http://psi-toolkit.wmi.amu.edu.pl
11. CLIP, http://clip.ipipan.waw.pl/
12. Morfologik, http://morfologik.blogspot.com/

Heterogeneous Named Entity Similarity Function

Jan Kocoń and Maciej Piasecki

Institute of Informatics, Wrocław University of Technology
Wybrzeże Wyspiańskiego 27, Wrocław, Poland
maciej.piasecki@pwr.wroc.pl

Abstract. Many text processing tasks require to recognize and classify Named
Entities. Currently available morphological analysers for Polish cannot handle
unknown words (not included in analyser's lexicon). Polish is a language with
rich inflection, so comparing two words (even having the same lemma) is a non-
trivial task. The aim of the similarity function is to match unknown word form
with its word form in named-entity dictionary. In this article a complex similarity
function is presented. It is based on a decision function implemented as a Logistic
Regression classifier. The final similarity function is a combination of several
simple metrics combined with the help of the classifier. The proposed function is
very effective in word forms matching task.

Keywords: similarity of Proper Names, word similarity metric, logistic regression, named entities recognition, information extraction.

1 Introduction

Gazetteers[1] mostly store only basic morphological forms (henceforth lemmas) of
Proper Names (PNs). This can much reduce applicability of a gazetteer in Named Entity
Recognition (NER) for highly inflected languages. Most inflected forms of PNs found
in texts cannot be straightforwardly matched in the gazetteer. This is a specific case
of a more general problem of the unknown word recognition, i.e. words not covered
by the existing morphological analyser [1,2]. They can be very rare, domain-specific
or specific PNs. Another class comprises misspelled words that can be matched to a
known word.

Our goal is to construct the similarity function for Named Entities (called *NamEn-*
Sim) to be used in effective recognition and classification of unknown word forms in
Polish texts, in general, but especially in recognition of the unknown inflected forms of
the PNs included in a large gazetteer. *NamEnSim* can be also used in more sophisticated
NER task [3]. Gazetteers include a limited number of PN morphological forms, mostly
only lemmas and are relatively large that increases complexity of searching and matching. We assume that unknown word recognition is done out of the context. Polish is a
language of rich inflection and there are many morphological forms. Similarity levels
that are meaningful for applications in NER, i.e. in order to identify proper matches in
the gazetteer, can be difficult to be established.

---

[1] Gazetteers are lexicons of PNs, often classified into a limited number of classes.

P. Sojka et al. (Eds.): TSD 2012, LNCS 7499, pp. 223–231, 2012.
© Springer-Verlag Berlin Heidelberg 2012

2   Related Works

Several word similarity metrics were proposed in literature: they take two strings on the input and produce a real value from the range *[0,1]*. Their evaluation is not an easy task: in general, many words are similar to each other to some extent. The interpretation of similarity values is an important problem. A possible solution is to evaluate the performance of a tool utilising the metric in some text processing task. In [4,5] several known and unique metrics (e.g. *Common Prefix δ, $CP_δ$*) for Polish were evaluated in a task of assigning named entities (NEs) included in a gazetteer to text words. The best results in one-word NE matching were obtained with *Common Prefix δ*, mainly due to favouring certain suffix pairs by the δ parameter [4]. Consider the following matching examples, containing items defined as: NE lemma $L_{case,gender}$ − NE word form $W_{case,gender,lemma}$; value of $CP_δ(L, W)$ (rounded to 2 decimal places) ; δ value:

zdzisława$_{nom,f}$ − zdzisławą$_{inst,f,zdzisława}$; 1.00 ; 1 (correct matching)
zdzisław$_{nom,m}$ − zdzisławem$_{inst,m,zdzisław}$; 0.80 ; 0 (correct)
zdzisław$_{nom,m}$ − zdzisławą$_{inst,f,zdzisława}$; 0.89 ; 0 (incorrect matching)
zdzisława$_{nom,f}$ − zdzisławem$_{inst,m,zdzisław}$; 0.71 ; 0 (incorrect matching)

These examples explain the δ parameter usage. It favours correct $L_{nom,f}$ − $W_{...,f,L}$ pairs for most cases and improves the matching accuracy. In $CP_δ$, the δ value is set empirically. *Common Prefix δ* is calculated as:

$$CP_δ(s, t) = \frac{(|lcp(s, t)| + δ)^2}{|s| \cdot |t|} \tag{1}$$

where *lcp(s, t)* – the longest common prefix of: *s & t*, δ – a parameter: *1* if one of given strings ends with *a*, and the second ends with *o, y, ą* or *ę*, else *0*.

Several metrics for PN matching task were analysed in [6], like *Overlap coefficient*, *Soundex* or *Levensthein*. The experimental results in [6] on different real data sets showed that there is no single best technique available.

We assumed that a combination of the different metrics can improve the overall result. First, we considered combining selected metrics from *SimMetrics*[2] library, namely: *Jaro, Jaro-Winkler, Soundex, Matching Coefficient, Overlap Coefficient, Euclidean Distance, Cosine Similarity, Q-grams Distance*. The first two were also analysed in [4,5], in which a very efficient *Common Prefix δ (CP_δ)* metric was proposed and added to our set. The last one considered is our *Modified Longest Substring (MLS)* defined below:

Modified Longest Substring is a metric based on the longest common subsequence of two strings. It does not contain any additional rules.

$$MLS(s, t) = \frac{2 \cdot max(|lcs(s, t)| − α, 0)}{|s| + |t|}$$

where *lcs(s, t)* – the longest common subsequence of given strings: *s & t*, α – greater position of *lcs(s, t)* first letter occurrence in either *s* or *t*.

---

[2] http://sourceforge.net/projects/simmetrics

We aimed at combining the individual metrics into a complex one. As expected, the dependency of the overall result on the constituents can be very complicated, so we decided to use a classifier-based approach: a vector of individual single metric values is classified into two classes: *similar* and *non-similar*. A decision function value is produced as an additional description. Two classification algorithms were considered: *Logistic Regression (LR)* and *Support Vector Machine (SVM)*. Aggregation of different string metrics using SVM in named-entity similarity task (for multi-word entities) was proposed in [7,8]. However as the results obtained by SVM and LR were nearly identical (statistically insignificant differences), we chose LR as more efficient.

LR (and SVM too) provides binary classification (similar/not similar), but it can also produce decision function values [9], which can be used to describe word pair similarity strength. Initial test showed also that words sharing the same lemma are mostly assigned high values of the function. However, as this correlation is not straightforward, and the function values must be interpreted in the context of a particular task. Manual analysis of test cases showed that the higher decision function value, the higher chance that the compared words share the same lemma. It is not the rule, so sometimes it is necessary to perform an additional interpretation of this value, when using *NamEnSim* in different tasks.

3   Complex Word Similarity Function

A simple morphological filtering is applied to the compared words to reduce computation complexity. Only words longer than $k = 2$ (defined empirically) are processed. If the words are not identical, then it is checked with a morphological analyser whether one is not the lemma of the other. Finally, we check whether the words share a lemma. Positive training examples are pairs of similar words confirm that to the above rule of simple morphological filtering, e.g. *Aleksander* 'a first name'$_{case=nom}$ − *Aleksandra*$_{case=(gen \vee acc)}$, and (values of *Soundex, Overlap coefficient, Euclidean distance, CP_δ, MLS*, rounded, follow the pair):

1;bzowy;bzowa;0.922;0.667;0.753;1.000;0.800;
1;dyplomatów;dyplomatą;1.000;0.800;0.837;0.711;0.842;

The construction of the negative set is a much harder task. First of all, word pairs not fulfilling the simple filtering rule are candidates to be included in it. However, such set would be huge and the vast majority of examples would be trivial. Thus, we aimed at selecting only the most 'similar' negative examples. We used Levensthein metric for this purpose, as a kind of baseline metric, defined to be the smallest number of edit operations (insertions, deletions and substitutions) required to change one string into another [6,10].

For word pairs selected randomly (subsampling of the large space), such that they do not match simple morphological filtering, if Levensthein metric for the pair is less than a maximum value across all examples in positive set, then the given example is added to the negative set, e.g.

0;mariaszowy;marszowymi;0.944;0.636;0.796;0.090;0.000;
0;niestandardowy;niestacjonarne;0.933;0.467;0.802;0.184;0.429;
0;surowców;surowiczymi;0.867;0.556;0.780;0.284;0.526;

Finally, we generated a training set containig 983,148 positive and 993,411 negative examples, that were acquired mainly from *NELexicon* gazetteer including 1,556,825 PNs: lemmas and their inflected forms – 514,195 one-word unique items. *NELexicon* is based on resources described in [11] and complemented by a large number of PNs collected from Polish Internet. The training set contains mainly examples collected from Polish inflected forms included in *Słownik Języka Polskiego* [3] – a Polish electronic morphological dictionary. About 10% of examples in training set is also in *NELexicon*, training and test sets are disjoint.

As *NELexicon* contains about 500,000 one-word PNs, a kind of relaxed similarity metric is required to select a subset of PNs for which the exact function can be calculated. The most effective solution was to group PNs by *letter trigrams* – it increases memory complexity, but also limits search space size and increases performance. Two simple word similarity rules referring to trigrams were introduced. Let w_I (the input one) and w_L (word stored in *NELexicon*) are words to compare: an *infrequent trigram* is a trigram, which occurs in less than $max = 15,000$ unique PNs in *NELexicon* (max value was set experimentally).

Relaxed similarity rules w_I and w_L are similar if:

1. The length of $w_I < 7$ and there is at least one infrequent trigram t such that t occurs in both: w_I and w_L.
2. The length of $w_I > 6$ and there are at least two infrequent trigrams: t_1, t_2 that occur in both w_I and w_L.

Experiments showed that more than 99% of pairs of similar words were covered by the rules (i.e. the error of omission is below 1%). In addition, the pre-selection caused reduction of the search space to 0.4% of initial size.

As we considered combining selected metrics from *SimMetrics* with classifier, we also performed feature selection to evaluate the similarity function with reduced metric set. In tests two similarity functions are used:

1. *NamEnSim10* – based on full set, containing 10 metrics (mentioned in 2)
2. *NamEnSim5* – based on reduced set, containing 5 metrics: *Soundex, Overlap Coefficient, Euclidean Distance, CP_δ* and *MLS*

The full set reduction was done with the proposed cyclic top-down feature selection method based on the similarity function evaluation (see 4). The process (one cycle) of removing one attribute from the set S of n metrics, with the initial $best_a$, $cyclebest_a$, $worst_a$ as the value of *mACC* is presented below:

1. Create the set S' by removing j-th attribute from the set S.
2. Train the classifier using set S'.
3. Evaluate the classifier using the chosen test set (see 4), get a as the value of $mACC$ evaluation result.
4. If $a > best_a$, set $best_a = a$, $cyclebest_a = a$
5. If $a > cyclebest_a$, set $cyclebest_a = a$
6. If $a < worst_a$, set $worst_a = a$, $worst_{id} = j$

---

[3] http://www.sjp.pl/slownik/odmiany/

7. $j = j + 1$

8. If $j > n$ stop cycle

After the cycle the attribute with index $worst_{id}$ can be removed from S if $cyclebest_a - best_a < eps$ (whole cyclic process stop condition), where eps is the given by the user. For the next cycle only $cyclebest_a$, $worst_a$ values are set as $mACC$ evaluation result of partially reduced set. Selection was performed on $person\_first\_nam1$ test set (see: 4) with $eps = 0.005$.

4   Evaluation

The basic application of the complex similarity function is to find for an input word w_I its lemma or an morphological word form in *NELexicon* if it exists. In practice, similar word set P returned by a similarity function for w_I should contain its proper lemma w_L and the decision function value for all other pairs: $\langle w_I, w_O \rangle$ where $w_O \in P$ and $w_O \neq w_L$ should be lower than for $< w_I, w_L >$. This task is different than morphological guessing, e.g. [1] which is aimed at the generation of lemmas for unknown words on the basis of *a tergo* index.

For the sake of comparison with [4], we reproduced the test sets from [4], i.e. analogical test sets were prepared on the basis of the description in [4]. During experiments with single similarity metrics (baseline tests) we obtained the same results as presented by the authors. So the reproduced test sets seem to be a good approximation of the original ones and can be used in comparing our own solutions with the methods of [4]. They concentrated on selecting a lemma (from search space) for a PN inflected form on the input. On the basis of *NELexicon* the following test sets of pairs: lemma – inflected form were generated:

1. *person_first_nam* – a set of Polish person first names,
2. *country_nam* – Polish country names,
3. *city_nam* – Polish city names – not used in [4],
4. *person_full_nam* – Polish person full names (multi-word).

Because our complex similarity function is limited to only one-word PNs, the last test set was not used. Following [4,5] all experiments were performed in two variants, each of different search space size: a *small search space (0 mode)* – only base forms of the test examples, a *large search space (1 mode)* – all base forms from a named entity category. Table 1 shows size of test sets and search spaces for different experiment modes and categories.

- Let a be the number of all test examples,
- s – the number of tests, in which single result was returned,
- m – the number of tests with more than one result returned,
- sc (*single correct*) – the number of tests, in which a single result was returned and it was correct,
- mc (*multiple and correct*) – the number of tests with more than one result, but among them the correct one,

Table 1. Test sets and search spaces for different experiment modes and categories

Category	Size		
	Small search space (0 mode)	Large search space (1 mode)	Tests
person_first_nam	480	15208	1720
country_nam	157	332	621
city_nam	8144	38256	30323

Table 2. Baseline test for person_first_nam

Similarity metric	Small search space			Large search space		
	AA	SR	RAA	AA	SR	RAA
ChapmanLengthDeviation	0.32260	0.57387	0.53503	0.06	0.30721	0.40293
Jaro	0.83164	0.86895	0.87062	0.30501	0.64447	0.63590
JaroWinkler	0.84859	0.87275	0.87514	0.55599	0.64407	0.66517
MatchingCoefficient	0.74011	0.96608	0.97119	0.32133	0.93148	0.94260
Soundex	0.66158	0.97502	0.97401	0.63084	0.68395	0.69893
OverlapCoefficient	0.76780	0.82815	0.83164	0.61171	0.67016	0.68149
QGramsDistance	0.85198	0.86717	0.86893	0.61902	0.67568	0.68542

– mc2 (*multiple with best one correct*) – the number of tests with more than one result and with the correct result as the top one.

Evaluation measures for *NamEnSim* are modified, because of its ability to score the result. We used three measures from [4]: *All answer accuracy:* $AA = \frac{sc}{s+m}$, *Single result accuracy:* $SR = \frac{sc}{s}$, *Relaxed all AA:* $RAA = \frac{sc+mc}{s+m}$.

AA measure was modified by adding mc2 parameter in order to better analyse the cases in which more than one result was returned. Because complex similarity function always returns decision function value as a similarity value, not only binary decision, we proposed additional measure $mACC$ aimed at comparison of different similarity functions in the domain of the values produced: *Modified all answer accuracy:* $mAA = \frac{sc+mc2}{s+m}$, *Modified global accuracy:* $mACC = \frac{sc+mc2}{a}$.

As a baseline we used the similarity metrics included in *SimMetrics* package. The experiment was performed on the Person First Name Test Set in two variants: with small and large search space. The results are presented in Tab. 2. Single metrics presented good accuracy in tests with small search spaces, but not in tests with large search spaces. Values for the modified evaluation measures are not presented in Tab. 2. Table 3 presents returned sets size. Evaluation measures are based on these results. Resources with suffix *0* are variants with small search space, and with suffix *1* are variants with large search space.

Results presented in Tab. 4 for similarity functions are comparable with single similarity metric CP_δ in [4]. A complex similarity function trained on the reduced set of single metrics (*NamEnSim5*) achieves better results in most categories and cases, specially for two measures: RAA and mAA. CP_δ has better global accuracy only for test cases with a small search space (statistically insignificant), but in real situations (where search space is large) results of CP_δ are not satisfactory. In most cases

Table 3. Sets size returned in experiments (description of parameteres in 4)

similarity metric	resource	a	s	m	sc	mc	mc2
$NamEnSim10$	person_first_nam0	1720	1308	362	1301	361	344
	person_first_nam1	1720	792	892	786	876	689
	country_nam0	621	577	23	577	23	21
	country_nam1	621	510	91	509	91	84
	city_nam0	29492	20806	7689	20584	7608	6185
	city_nam1	29492	12539	16418	12210	15982	11653
$NamEnSim5$	person_first_nam0	1720	1296	372	1289	371	356
	person_first_nam1	1720	758	923	751	909	746
	country_nam0	621	572	27	572	27	25
	country_nam1	621	501	99	500	99	93
	city_nam0	29492	20495	8121	20258	8022	6674
	city_nam1	29492	11858	17177	11579	16701	12404
CP_δ	person_first_nam0	1720	1682	22	1650	22	
	person_first_nam1	1720	1569	143	1355	128	
	country_nam0	621	604	4	601	4	
	country_nam1	621	598	12	590	12	
	city_nam0	29492	29005	465	27039	385	
	city_nam1	29492	27922	1565	22806	1285	

Table 4. Results for complex similarity functions: *NamEnSim10* and *NamEnSim5*

similarity metric	resource	mAA	SR	RAA	$mACC$
$NamEnSim10$	person_first_nam0	0.9850	**0.9946**	**0.9952**	0.9564
	person_first_nam1	0.8759	**0.9924**	0.9869	0.8576
	country_nam0	**0.9967**	**1.0000**	**1.0000**	0.9630
	country_nam1	0.9867	**0.9980**	**0.9983**	**0.9549**
	city_nam0	0.9394	**0.9893**	**0.9894**	0.9077
	city_nam1	0.8241	0.9738	0.9736	0.8091
$NamEnSim5$	person_first_nam0	**0.9862**	**0.9946**	**0.9952**	0.9564
	person_first_nam1	**0.8905**	0.9908	**0.9875**	**0.8703**
	country_nam0	**0.9967**	**1.0000**	**1.0000**	0.9614
	country_nam1	**0.9883**	**0.9980**	**0.9983**	**0.9549**
	city_nam0	**0.9412**	0.9884	0.9883	0.9132
	city_nam1	**0.8260**	**0.9765**	**0.9740**	**0.8132**
CP_δ	person_first_nam0	0.9683	0.9810	0.9812	**0.9593**
	person_first_nam1	0.7915	0.8636	0.8662	0.7878
	country_nam0	0.9885	0.9950	0.9951	**0.9678**
	country_nam1	0.9672	0.9866	0.9869	0.9501
	city_nam0	0.9175	0.9322	0.9306	**0.9168**
	city_nam1	0.7734	0.8168	0.8170	0.7733

results achieved by *NamEnSim10* and *NamEnSim5* are much better than similar results achieved on lemmatization of Slovene words, where the proposed method (statistics-based trigram tagger) achieves an accuracy of 81% [12]. Results achieved by similarity function are also better than one-word similarity methods described in [4,5], mainly comparing to CP_δ.

5   Summary

The experiments showed, that it is possible to create a suitable function for the similarity estimation between two words, which does not use a complete dictionary of inflected forms. The quality of finding corresponding base forms achieved by proposed similarity function is very high. The best results were achieved with *NamEnSim5* complex similarity function, based on logistic regression applied to a reduced number of single similarity metrics. The proposed method is language independent and should achieve similar results also for other languages with rich inflection. We used realistic data set (containing also mispelled words), that might cause slightly worser results than expected, but still the results achieved by similarity function are better than methods proposed in [4,5,12].

Acknowledgments. Partially financed by the Polish National Centre for Research and Development project SyNaT.

References

1. Piasecki, M., Radziszewski, A.: Polish morphological guesser based on a statistical a tergo index. In: Proc. of IMCSIT — 2nd International Symposium Advances in Artificial Intelligence and Applications, AAIA 2007, pp. 247–256 (2007)
2. Woliński, M.: Morfeusz —a Practical Tool for the Morphological Analysis of Polish. In: Kłopotek, M.A., Wierzchoń, S.T., Trojanowski, K. (eds.) Morfeusz a Practical Tool for the Morphological Analysis of Polish. Advances in Soft Computing, vol. 5, pp. 511–520. Springer, Heidelberg (2006)
3. Piskorski, J.: Named-Entity Recognition for Polish with SProUT. In: Bolc, L., Michalewicz, Z., Nishida, T. (eds.) IMTCI 2004. LNCS (LNAI), vol. 3490, pp. 122–133. Springer, Heidelberg (2005)
4. Piskorski, J., Sydow, M., Wieloch, K.: Comparison of String Distance Metrics for Lemmatisation of Named Entities in Polish. In: Vetulani, Z., Uszkoreit, H. (eds.) LTC 2007. LNCS, vol. 5603, pp. 413–427. Springer, Heidelberg (2009)
5. Piskorski, J., Sydow, M., Kupść, A.: Lemmatization of polish person names. In: Proc. of the Workshop on Balto-Slavonic Natural Language Processing: Information Extraction and Enabling Technologies, ACL 2007, pp. 27–34. ACL, USA (2007)
6. Christen, P.: A comparison of personal name matching: Techniques and practical issues. In: International Conference on Data Mining Workshops, pp. 290–294 (2006)
7. Cohen, W.W., Ravikumar, P., Fienberg, S.E.: A Comparison of String Distance Metrics for Name-Matching Tasks. In: Proceedings of IJCAI 2003 Workshop on Information Integration, pp. 73–78 (2003)

8. Cohen, W.W., Ravikumar, P., Fienberg, S.E.: A comparison of string metrics for matching names and records. In: Proceedings of the KDD 2003 Workshop on Data, Washington, DC, pp. 13–18 (2003)

9. Lubenko, I., Ker, A.D.: Steganalysis using logistic regression. In: Proc. SPIE 7880, 78800K (2011)

10. Levenshtein, V.I.: Binary codes capable of correcting deletions, insertions, and reversals. Technical Report 8 (1966)

11. Savary, A., Piskorski, J.: Lexicons and grammars for named entity annotation in the National corpus of Polish. In: Proceedings of the 18th International Conference Intelligent Information Systems, IIS 2010, Siedlce, Poland (2010)

12. Džeroski, S., Erjavec, T.: Learning to Lemmatise Slovene Words. In: Cussens, J., Džeroski, S. (eds.) LLL 1999. LNCS (LNAI), vol. 1925, pp. 69–88. Springer, Heidelberg (2000)

Joint Part-of-Speech Tagging and Named Entity Recognition Using Factor Graphs[*]

György Móra[1] and Veronika Vincze[2,3]

[1] University of Szeged, Department of Informatics,
6720 Szeged, Árpád tér 2., Hungary
gymora@inf.u-szeged.hu
[2] MTA-SZTE Research Group on Artificial Intelligence,
6720 Szeged, Tisza Lajos krt. 103., Hungary
[3] Universität Trier, Linguistische Datenverarbeitung,
54286 Trier, Universitätsring, Germany
vinczev@inf.u-szeged.hu

Abstract. We present a machine learning-based method for jointly labeling POS tags and named entities. This joint labeling is performed by utilizing factor graphs. The variables of part of speech and named entity labels are connected by factors so the tagger jointly determines the best labeling for the two labeling tasks. Using the feature sets of SZTENER and the POS-tagger magyarlanc, we built a model that is able to outperform both of the original taggers.

Keywords: POS tagging, named entity recognition, joint labeling, factor graphs.

1 Introduction

In syntax, proper nouns behave in a similar way to a common noun (e.g. in the sentence *Have you seen "Interview with the vampire"?*, the title of the movie fulfils the function of the object and could be substituted by the common noun *movie*) and hence they are considered a subclass of nouns. Some morphological coding systems give a distinct code to proper nouns (such Np-s* in the Hungarian MSD coding system or NNP in the Penn Treebank system), but the members of multiword proper nouns may belong to other parts of speech on their own (in the above example, we may have a preposition (*with*) and an article (*the*) as well). In such cases, a possible solution is to duplicate the part of speech codes (i.e. to add a proper noun code to every word) because in fact every word with any part of speech code can be part of a proper noun. Thus, in this example all the four words within the title would have the part of speech (POS) code of a proper noun. However, this duplication would make the POS tagging more expensive (each word would have at least two possible codes, from which the POS tagger should select the correct one) and the POS tagger should be able to recognize proper nouns, which is normally the task of a named entity recognition (NER) system.

Here, we propose a solution to solve both problems – POS tagging and named entity recognition – in a parallel way. Our approach separates the two subtasks by assigning

[*] This work was supported in part by the National Innovation Office of the Hungarian government within the framework of the projects BELAMI and MASZEKER.

P. Sojka et al. (Eds.): TSD 2012, LNCS 7499, pp. 232–239, 2012.

ordinary POS codes to the parts of the proper nouns, which are tagged separately. The two tagging processes work in a parallel way but the label sequences are determined depending on each other. This joint labeling is carried out by utilizing factor graphs. The variables of part of speech and named entity labels are connected by factors, hence the tagger jointly determines the best labeling for both labeling tasks. Sequential models where different subtasks are performed subsequentially usually aggregate tagging errors and one tagger in the processing pipeline may utilize labels created by the previous taggers. Performing the processes in parallel, both taggers can use the other's labels as features. In this paper, we carry out experiments on English and Hungarian texts and we find that parallel labeling can improve the performance and quality of both labeling processes.

2 Morphology and Proper Nouns

Proper nouns are usually considered to be rigid designators, which constantly refer to the same entity [1]. Rigidity here means that the relationship between the designator and the designated is constant but we argue that rigidity can be applied to morphology as well. In agglutinative languages, proper nouns can be inflected or some derivational suffixes could be added, but their base form does not change. It is most salient when a noun with a morphologically irregular behaviour acts as a proper noun as in the following Hungarian examples: *Fodor* 'Fodor as a proper noun' – *Fodort* 'Fodor-ACC' vs. *fodor* 'frill' – *fodrot* 'frill-ACC'.

The common noun *fodor* has a vowel-deleting stem, which means that the last vowel of the stem is deleted before certain suffixes. However, when it functions as a proper noun, this rule is no longer valid (i.e. the last vowel is preserved), which may be exploited in named entity recognition. As an accusative form of *fodor*, the morphological analyzer would expect to get *fodrot*. If it gets *Fodort* as input, it can only analyze this word form with the help of guessing, separating it into the morphs *fodor* 'frill' and *t* 'accusative suffix'. If this lemma is listed in the morphological database, but with a different analysis (fodr+ot vs. fodor+t), then it is highly probable that it is an instance of a proper noun.

Some proper nouns contain an inflectional (or derivational) suffix even within their lemmas. One such case is *McDonald's* in English, where we can see a possessive marker as part of the original name. However, when it comes to speaking about things owned by McDonald's, we get the form *McDonald's'*. If it is supposed that the morphological analyzer does not include a list of companies, this latter form is analyzed by the guesser, whereby the morphs *McDonald's* and ' are produced. Since the lemma already contains a possessive suffix, it is again suggested that it is a proper noun.

From a morphological analysis view point, named entity recognition that is carried out in parallel can help in accelerating the process. If an element is recognized as a named entity, the morphological analyzer can immediately call the guesser instead of analyzing the element in the traditional way.

3 Joint Labeling Approaches

Different labeling tasks (such as POS tagging, chunking, NER) are usually performed in sequential steps and are defined as separate machine learning problems. Using sequential processing pipelines, a labeler can only use labels as features produced by a previously performed labeling step. Another drawback of methods which use sequential analysis is that the errors made by separate steps may aggregate over the pipeline. To overcome these problems, multiple labeling tasks should be performed in a single step.

Combining the label-spaces of multiple labeling tasks produces a single label-space and the separate machine learning problems become a single machine learning task. If the separated label-spaces were large, the size of the combined space might be intractable. The combined label-space may contain labels which are rare or they do not even occur in the training data and the detection of these combinations cannot be learned properly. A single feature space may not be ideal for all of the labeling subtasks.

In our experiments we utilized an approach based on probabilistic graphical models to perform joint labeling. The MALLET GRMM [2] and FactorIE [3] software packages enable us to define arbitrary conditional dependencies between labels and feature sets instead of using classical linear chain conditional random fields. A token may hold multiple types of labels and feature sets. The conditional dependencies between the labels and the features can be described by factors. Factors between POS and NER labels permit the interaction between labels during the learning and tagging process, but we can still have separate feature sets for each labeling task. This method can be adapted to other tasks such as chunking and it can also handle three or more label types.

Experiments carried out on English language texts reveal that the joint learning of POS and chunk tags gives better results than just performing these task sequentially. Our experiments showed that the accuracy of POS tagging rose from 62.42% to 72.87% and the accuracy of chunking rose from 83.95% to 85.76% by performing joint labeling using the same feature sets as that in the separated cases. The labels act as dynamic features during parallel training, improving the accuracy scores of both labeling tasks. The experiments were performed on a subset of the CoNLL-2000 Shared Task data.

In experiments carried out on the Spanish language dataset of the language independent NER task of the CoNLL2003 Shared Task, we found that both the POS and NER labeling can be improved using joint labeling. With a basic feature set and separate labeling we achieved an accuracy of 88.6% in POS labeling and an F-measure of 39.5 in NER. These scores rose to 88.7% and 42.2, respectively, with our joint labeling approach.

4 Named Entity Recognition

Named entity recognition is a key part of all information extraction systems as named entities are the main building bricks of relations and events. The classification of the entities is a more challenging problem than the simple recognition and it often needs information based on the environment of the token.

The recognition of named entities may be token or sequence-based. The token based approach assigns a label to each individual token independently of the labels of the

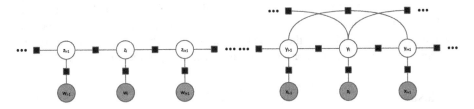

Fig. 1. First-order model used for named entity recognition and second-order model used for POS tagging

other tokens. Support Vector Machines [4] or Maximum Entropy methods [5] are most widely used as machine learning models. The different methods can be combined in a multilayer classification scheme [6].

Sequence labeling approaches like Hidden Markov Models [7] and Conditional Random Fields (CRF) [8] label a sequence of tokens and the most likely sequence is determined instead of separate token classification. The results of the CoNLL-2002 and CoNLL-2003 Shared Tasks on NER revealed that the sequence labeling approaches usually outperform token labeling methods [9,10].

Our NER system is based on the feature set of the SZTENER language independent Named Entity Tagger [11]. It uses a first order CRF machine learning model implemented in the MALLET [2] machine learning tool. The tagger utilizes orthographic, frequency-based and dictionary-based features. In our machine learning settings, we applied the feature-vectors extracted from SZTENER.

In order to compare the two approaches in the same machine learning framework, we implemented a similar first order chain (see Figure 1) in the FactorIE probabilistic programming framework. A modified version of the Gibbs Sampler was employed to train our models.

5   Part of Speech Tagging

POS tagging is a key step in syntactical analysis and many systems use POS codes as features. POS tagging is a token classification task where a label is assigned to each token from a coding system. Here, we used the simplified Hungarian MSD coding system, which is more suitable for machine learning.

Several POS taggers are available for Hungarian (see [12]). Our POS tagger is based on *magyarlanc*, which is a modified version of the Stanford POS Tagger [13]. It utilizes a Cyclic Dependency Network with Maximum Entropy classifier. The feature set adapted to Hungarian consists of character prefixes and suffixes, the word forms and the token patterns of the words.

The Cyclic Dependency network used by the original POS tagger was not directly implementable in FactorIE, but the main structure of dependencies was kept in a factor graph (Figure 1). The resulting model is similar to the NER model, but it has a second order chain and various factors emulating the label and token combination features of the original system.

The set of possible POS codes were added to the model as a second feature vector which allows the learner to incorporate the output of the morphological analyzer, but the possible tags are not limited to these labels. In some cases the tagger chose the correct label despite the fact that the morphological analyzer failed to recommend it.

6   Results

By connecting the graphical models of the NER and POS tagging tasks with factors between the two label sequences, a joint model was created (Figure 2). The NER and POS label variables of the same token and label variables for neighbouring tokens were connected by factors.

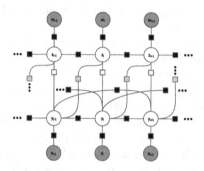

Fig. 2. Unification of the two independent models

We evaluated the original systems and the independent and joint models of our approach on the subcorpus containing business newswire texts of the *Szeged Corpus*, where the gold standard Named Entities are annotated [14,15]. It is a Hungarian language corpus that contains over 220,000 tokens in 9400 sentences. We split the corpus into training and test sets in a 70/30 ratio.

In the original MSD annotation, proper nouns had the code Np-* and multiword NEs were contracted. Before our evaluation, multiword NEs were split into parts and their members were reannotated; moreover, proper and common nouns were not distinguished, both having the POS code 'noun'. Thus, the multiword NE *Magyar Nemzeti Bank* "National Bank of Hungary" was retagged with the A A N POS sequence.

6.1   The Evaluation of Named Entity Recognition

Here, we applied a phrase-based evaluation of named entity recognition. This means that the labeling of multiword NEs was only accepted if all of its members were labeled correctly and no other neighbouring words were marked. For the sake of comparison, all models were trained and evaluated on the same training and test sets, and the same metrics were applied. The phrase-based F-score was used in the evaluation process of the CoNLL-2003 shared task on NER, which we also applied.

Table 1 lists the results obtained by the base model and our joint and sequential NER models running the learning algorithm for 2 and 5 iterations.

Our results confirm that using the same feature space, joint tagging improves the quality of NER compared with the independent model. The independent model improves its performance from 83.86 (which was worse than the base model) to 88.93. The inefficiency of the feature space is indicated by the fact that the results, which are worse than those of the original model, are actually made worse by increasing the number of iterations, most probably due to overfitting. This lack of information may be compensated by the presence of POS codes in the joint learning process.

6.2   The Evaluation of POS-Tagging

POS-tagging was trained and evaluated on a reduced set of MSD codes [16], only those codes being distinguished where the word form does not unambiguously determine the POS-code (e.g. *tőrnek* can mean both "of (the) dagger" and "for (the) dagger"). The reduction of the original set with several hundred codes was necessary because it would have been unfeasible for the machine learning algorithm to treat them properly. Since the original codes can be recovered from the reduced ones, this reduction does not have any substantial effect on the results.

We also evaluated the results concerning just the first character (i.e. the one denoting the main part of speech) of the codes. Hence, it could be seen what the differences were between the two POS-tagging methods that achieved almost the same results on the reduced set of MSD-codes.

In contrast with NER, we used accuracy to measure the performance of the systems, but macro F-scores were also provided for each POS class. Accuracy reflects the average performance of the system, while macro F-score is the average of the F-scores of the classes. If it is only the frequent POS tags that the system identifies correctly, the average of F-scores per class will be low due to the high number of mistagged POS classes with only a few members.

Table 1. Results obtained for NER and POS tagging

				Part of Speech			
				Reduced MSD		Main part of speech	
It.	Model	Precision	Recall $F_{\beta=1}$	Accuracy	$F_{\beta=1macro}$	Accuracy	$F_{\beta=1macro}$
			Named Entity				
	SZTENER	86.81	88.71 87.75				
	magyarlanc			97.11	67.81	97.98	85.18
2	Independent	86.81	81.11 83.86	97.75	71.03	98.60	84.12
	Parallel	88.57	89.27 88.93	97.78	72.48	98.68	86.32
5	Independent	84.73	81.60 83.13	**98.00**	71.33	98.78	86.44
	Parallel	**89.71**	**90.04 89.87**	97.99	**73.32**	**98.81**	**88.77**

The improvement in the POS-tagging results can be primarily attributed to the proper analysis of words that begin with a capital letter. In Hungarian, it is mostly sentence-initial words and named entities that start with a capital letter. With the parallel POS tagging and NER, sentence-initial named entities were easier to find, so it was easier to assign the proper MSD code to the rest of sentence-initial elements. For instance, the sentence-initial word *Szerinte* was tagged as a noun in the sequential tagging, while it was assigned the proper code 'adverb' in the parallel tagging. The POS tagging of abbreviations rose by 17.68%, which can be attributed to the correct identification of *Dr.* and *Jr.*, which are parts of named entities. The tagging of some NEs ending in pseudo-interjections like *Palotainé* "Mrs. Palotai" was also improved using the parallel NER approach.

Overall, we may conclude that the biggest differences between the systems could be observed in the case of the rare POS classes, while there were no great differences in the case of frequent POS classes. However, the accuracy on the latter class was high (above 97%) when tagging with the sequential model hence the addition of NER did not significantly affect the results.

Although the absolute difference between the accuracies may seem small, in the case of parallel tagging the quality of POS tagging was improved. Macro averages in Table 1 show that the parallel system performs better with POS classes having only a few members, hence it is more balanced. When taking just the main POS into account, it is seen that the parallel system identifies the main POS code slightly better than the sequential system; that is, the errors made by the former are less serious than those of the latter.

7 Conclusions

Here, we presented our system for the joint labeling of part of speech tags and named entities. Our results show that the performance on both tasks can be slightly improved, compared with the traditional sequential models. Although the improvement is less substantial in the case of POS-tagging, our method was still able to raise the overall quality. In our experiments we found that joint labeling is able to exploit labels of one task as features in the other task. These features are not independent of each other from a linguistic point of view, but this joint model is linguistically more feasible than single model approaches. The creation of joint models like this seems to be a promising direction for further research.

References

1. Kripke, S.: Naming and Necessity. Basil Blackwell, Oxford (1980)
2. McCallum, A.K.: Mallet: A machine learning for language toolkit (2002), http://mallet.cs.umass.edu
3. McCallum, A., Schultz, K., Singh, S.: FACTORIE: Probabilistic Programming via Imperatively Defined Factor Graphs. In: Advances in Neural Information Processing Systems, vol. 22, pp. 1–9 (2009)
4. Mayfield, J., Mcnamee, P., Piatko, C.: Named entity recognition using hundreds of thousands of features. In: Proceedings of CoNLL 2003, pp. 184–187 (2003)
5. Chieu, H.L., Ng, H.T.: Named entity recognition with a maximum entropy approach. In: Daelemans, W., Osborne, M. (eds.) Proceedings of CoNLL 2003, Edmonton, Canada, pp. 160–163 (2003)

6. Florian, R., Ittycheriah, A., Jing, H., Zhang, T.: Named entity recognition through classifier combination. In: Daelemans, W., Osborne, M. (eds.) Proceedings of CoNLL 2003, Edmonton, Canada, pp. 168–171 (2003)
7. Miller, S., Crystal, M., Fox, H., Ramshaw, L., Schwartz, R., Stone, R., Weischedel, R., Group, T.A.: Algorithms That Learn To Extract Information BBN: Description of The Sift System as Used For MUC-7. In: Proceedings of MUC-7 (1998)
8. Lafferty, J.D., McCallum, A., Pereira, F.C.N.: Conditional random fields: Probabilistic models for segmenting and labeling sequence data. In: Proceedings of the Eighteenth International Conference on Machine Learning, ICML 2001, pp. 282–289. Morgan Kaufmann Publishers Inc., San Francisco (2001)
9. Tjong Kim Sang, E.F.: Introduction to the CoNLL-2002 Shared Task: Language-Independent Named Entity Recognition. In: Proceedings of CoNLL 2002, Taipei, Taiwan, pp. 155–158 (2002)
10. Tjong Kim Sang, E.F., De Meulder, F.: Introduction to the CoNLL-2003 Shared Task: Language-Independent Named Entity Recognition. In: Daelemans, W., Osborne, M. (eds.) Proceedings of CoNLL 2003, Edmonton, Canada, pp. 142–147 (2003)
11. Szarvas, Gy., Farkas, R., Kocsor, A.: A Multilingual Named Entity Recognition System Using Boosting and C4.5 Decision Tree Learning Algorithms. In: Todorovski, L., Lavrač, N., Jantke, K.P. (eds.) DS 2006. LNCS (LNAI), vol. 4265, pp. 267–278. Springer, Heidelberg (2006)
12. Halácsy, P., Kornai, A., Oravecz, C.: HunPos: an open source trigram tagger. In: Proceedings of the ACL 2007 Demo and Poster Sessions, pp. 209–212. Association for Computational Linguistics, Stroudsburg (2007)
13. Toutanova, K., Klein, D., Manning, C.D., Singer, Y.: Feature-rich part-of-speech tagging with a cyclic dependency network. In: Proceedings of NAACL 2003, pp. 173–180. Association for Computational Linguistics, Stroudsburg (2003)
14. Csendes, D., Csirik, J., Gyimóthy, T.: The Szeged Corpus. A POS Tagged and Syntactically Annotated Hungarian Natural Language Corpus. In: Hansen-Schirra, S., Oepen, S., Uszkoreit, H. (eds.) COLING 2004 5th International Workshop on Linguistically Interpreted Corpora, Geneva, Switzerland, pp. 19–22 (2004)
15. Szarvas, Gy., Farkas, R., Felföldi, L., Kocsor, A., Csirik, J.: A highly accurate Named Entity corpus for Hungarian. In: Proceedings of LREC 2006 (2006)
16. Zsibrita, J., Vincze, V., Farkas, R.: Ismeretlen kifejezések és a szófaji egyértelműsítés. In: Tanács, A., Vincze, V. (eds.) MSzNy 2010 – VII. Magyar Számítógépes Nyelvészeti Konferencia, Szeged, Hungary, University of Szeged, pp. 275–283 (2010)

Assigning Deep Lexical Types

João Silva and António Branco

University of Lisbon
Departamento de Informática, Edifício C6
Faculdade de Ciências, Universidade de Lisboa
Campo Grande, 1749-016 Lisboa, Portugal
{jsilva,antonio.branco}@di.fc.ul.pt

Abstract. Deep linguistic grammars provide complex grammatical representations of sentences, capturing, for instance, long-distance dependencies and returning semantic representations, making them suitable for advanced natural language processing. However, they lack robustness in that they do not gracefully handle words missing from the lexicon of the grammar. Several approaches have been taken to handle this problem, one of which consists in pre-annotating the input to the grammar with shallow processing machine-learning tools. This is usually done to speed-up parsing (supertagging) but it can also be used as a way of handling unknown words in the input. These pre-processing tools, however, must be able to cope with the vast tagset required by a deep grammar. We investigate the training and evaluation of several supertaggers for a deep linguistic processing grammar and report on it in this paper.

Keywords: supertagging, deep grammar.

1 Introduction

Parsing is one of the fundamental tasks in Natural Language Processing and a critical for many applications. Many of the most commonly used parsers rely on probabilistic approaches and are obtained by inferring a language model over a dataset of annotated sentences. Though these parsers always produce some analysis of their input sentences, they do not go into deep linguistic analysis.

Deep grammars seek to make explicit highly detailed linguistic phenomena and produce complex grammatical representations for their input sentences. In particular, they are able to capture long-distance dependencies and produce the semantic representation of a sentence. Although there is a great variety of parsing algorithms (see [1] for an overview), they all require a lexical look-up initialization step that, for each word in the input, returns all its possible syntactic categories. From this it follows that if any of the words in a sentence is not present in the lexicon—an *out-of-vocabulary* (OOV) word—a full parse of that sentence is impossible to obtain. Given that novelty is one of the defining characteristics of natural languages, unknown words will eventually occur. Hence, being able to handle OOV words is of paramount importance if one wishes to use a parser to analyze unrestricted texts with an acceptable coverage. Another important issue is that of words that may bear more than one syntactic category. The combinatorial explosion of lexical and syntactic ambiguity may also hinder parsing due to the

P. Sojka et al. (Eds.): TSD 2012, LNCS 7499, pp. 240–247, 2012.

increased requirements in terms of parsing time and memory usage. Thus, even if there are no OOV words in the input, assigning syntactic categories to words prior to parsing may be desirable for efficiency reasons.

For shallower approaches, like constituency parsing, it suffices determining the part-of-speech (POS), so pre-processing the input with a POS tagger is a common and effective way to tackle either of these problems. However, the information contained in the lexicon of a deep grammar is much more fine-grained, including, in particular, the subcategorization frame of the word, which further constraints what can be taken as a well-formed sentence by imposing several restrictions on co-occurring words. Thus, what for a plain POS tagger corresponds to a single tag is often expanded into hundreds of different distinctions, and hence tags, when at the level of detail required by a deep grammar. For instance, the particular grammar we use for the study reported here has a lexicon with roughly 160 types of verb and 200 types of common noun.[1] While the grammar may proceed with the analysis knowing only the POS category of a word, it does so at the cost of vastly increased ambiguity which may even lead the grammar to accept ungrammatical sentences as valid. This has lead to research that specifically targets annotation with a tagset suitable for deep grammars.

In this paper we investigate several machine-learning approaches to supertagging and compare them with state-of-the-art results. In particular, we experiment with a support vector machine algorithm, a novel approach to supertagging.

2   Related Work

The two main approaches to assigning lexical types for a deep grammar can be divided in terms of how they resolve lexical ambiguity. In lexical acquisition are approaches that try to discover all the types a given unknown word may occur with, effectively creating a new lexical entry. However, at run-time, it is still up to the grammar using the newly acquired entry to choose which of its possible types is the correct one for each particular occurrence of that word. In supertagging are approaches that assign, typically on-the-fly at run-time, a single lexical type to a particular occurrence of a word. Their rationale is not to acquire a new entry to record in the lexicon, but to allow the grammar to keep parsing despite the occurrence of OOV words, or to ease parsing by reducing ambiguity. The approach reported in this paper falls under the umbrella of supertagging.

2.1   Supertagging

POS tagging relies on a small window of context to achieve a limited form of syntactic disambiguation [2]. As such, it is commonly used prior to parsing as a way of reducing ambiguity by restricting words to a certain category, leading to a greatly reduced search space, faster parsing and less demanding memory requirements. Supertagging can be seen as an extension of this idea to a richer tagset, in particular to one that includes information on subcategorization frames.

---

[1] For instance, in the deep grammar we are using for the study reported here, the lexical type `noun-common-2comps_de-com` is assigned to common nouns that have two complements, the first introduced by "de" and the second by "com".

In [3], a supertagger using a trigram model was applied to a Lexicalized Tree Adjoining Grammar (LTAG), where each lexical item is associated with one or more trees that localize information on dependencies, even long-range ones, by requiring that all and only the dependents be present in the structure [4]. Training data was obtained by taking sentences from the Wall Street Journal and parsing them with XTAG, a wide-coverage LTAG. In addition, in order to reduce data-sparseness, POS tags were used in training instead of words. Evaluation was performed over 100 held-out sentences from the Wall Street Journal. For a tagset of 365 items, this supertagger achieved 68% accuracy. In a later experiment, this is improved to 92% by smoothing model parameters and additional data [5]. The supertagger can also assign the n-best tags, which increases the chances of it assigning the correct supertag at the cost of leaving more unresolved ambiguity. With 3-best tagging, it achieved 97% accuracy.

In [6,7,8], a supertagger is used with a Combinatory Categorial Grammar (CCG), which uses a set of logical combinators to manipulate linguistic constructions. For our purposes here, it matters only that lexical items receive complex tags that describe the constituents they require to create a well-formed construction. The set of 409 categories to be assigned was selected by taking those that occur at least 10 times in sections 02–21 of a CCG automatic annotation of Penn Treebank. Evaluation was performed over section 00, and achieved 92% accuracy. As with [5], assigning multiple tags increases accuracy. However, instead of using a fixed n-best number of tags—which might be to low, or too high, depending on case at hand—the CCG supertagger assigns all tags with a likelihood within a factor β of the best tag. A value for β as small as 0.1 is enough to boost accuracy up to 97% with an average of only 1.4 tags per word.

2.2 Supertagging for HPSG

[9] present an supertagger for the Alpino Dutch grammar based on hidden Markov models. An interesting feature of their approach is that the training data (2 million sentences of newspaper text) is the output of the parser, thus avoiding the need for a hand-annotated dataset. A gold standard was created by having Alpino choose the best parse for a set of 600 sentences. When assigning a single tag (from a tagset with 2,392 tags), the supertagger achieves an accuracy close to 95%. It is unclear to what extent this can be affected by some bias in the disambiguation module, given that the lexical types in the training dataset and in the gold standard are both automatically picked by Alpino.

[10] use a supertagger with the Enju grammar for English. The novelty in their work comes from filtering invalid tag sequences produced by the supertagger before running the parser. In this approach, a CFG approximation of the HPSG is created with the key property that the language it recognizes is a superset of the parsable supertag sequences. Hence, if the CFG is unable to parse a sequence, that sequence can be safely discarded. This approach achieved a labeled precision and recall for predicate-argument relations of 90% and 86%, respectively, over 2,300 sentences with up to 100 words in section 23 of the Penn Treebank.

[11] uses a supertagger with ERG, another grammar for English. Evaluated over 1,798 sentences, it achieves an accuracy of 91% with a tagset of 615 lexical types. Also for ERG, [12] trains two supertaggers over a 158,000 token dataset, one using the

Table 1. Sentence, token and type counts for the various dataset sizes

dataset	sent.	all	tokens verbs	nouns	all	lexical types verbs	nouns
base	5,422	51,483	6,453 [12.5%]	9,208 [17.9%]	581	129	140
+ Público	10,727	139,330	14,540 [10.4%]	25,301 [18.2%]	626	133	157
+ Wiki	15,108	205,585	20,869 [10.2%]	37,066 [18.0%]	646	139	160
+ Folha	21,217	288,875	29,683 [10.3%]	52,337 [18.1%]	668	140	165

TnT POS tagger [13] and another using the C&C supertagger [6,7,8], and experiments with various tag granularities. For instance, assigning only POS—a tagset with only 13 tags—is the easiest task, with 97% accuracy, while highly granular supertags formed by POS concatenated with selectional restrictions increases the number of tags to 803, with accuracy dropping to 91%.

3   The Grammar and the Dataset

The deep grammar used here is LXGram, an HPSG grammar for Portuguese [14,15]. It supported the annotation of a corpus by providing the set of possible analyses for a sentence (the parse forest), which is then disambiguated by manually picking the correct analysis from among those returned. This ensures that the syntactic and semantic annotation layers are consistent. The corpus is composed mostly by text from the Público newspaper, which was previously annotated with manually verified shallow information, such as POS and lemmas. After running LXGram and disambiguating the result, we were left with 5,422 sentences annotated with all the information provided by LXGram, though, for this paper, we only keep the lexical type assigned to each token. Type distribution is highly skewed. For instance, the 2 most frequent common noun types cover 57% of all the common noun tokens. Such distributions are usually a problem for machine-learning approaches since the number of instances of the rare categories is too small to properly estimate the parameters of the model.

3.1   Dataset Extension

Ahead we will be interested in determining learning curves for the various classifiers under study, but the current dataset is still small. We thus opted for extending it with automatically annotated data obtained by taking additional sentences from Público (4,381), the Portuguese Wikipedia (5,305) and the Folha de São Paulo newspaper (6,109), processing them with a POS tagger, and running them through LXGram. The cumulative sizes are shown in Table 1.

This is made possible because LXGram has a stochastic disambiguation module that chooses the most likely analysis in the parse forest, instead of requiring a manual choice by a human annotator [15]. Manual evaluation of a 50 sentence sample indicates that this module picks the correct reading in 40% of the cases. If this ratio holds, 60% of the sentences will have an analysis that is wrong in some way, though it is not clear how this translates into errors in the assigned lexical types. For instance, when faced with a case of PP-attachment ambiguity, the module may pick the wrong attachment, which gives the wrong analysis though the lexical types assigned to the tokens may be correct.

The lexical types in the corpus are a subset of all types known to the grammar. Table 1 shows the token count for verbs and nouns and, in brackets, their relative frequency. In addition, it provides a breakdown of the number of types for each dataset. As expected, in a true Zipfian way, the dataset must have a major increase in size for the more rare types to appear. For instance, though there nearly a 6-fold difference in tokens between the base and the largest datasets, the latter only contains 11 verb and 25 noun types more. The reader may recall, from Section 1, that the lexicon of LXGram has 160 types of verb and 200 types of common noun. The largest corpus comes close to covering all of these, but there are still a few that, by not occurring in the dataset, cannot be learned. It is worth noting that, for the experiments reported in this paper, and in order to mitigate data-sparseness issues, words were replaced by their lemmas.

4   The Supertaggers

Assigning HPSG lexical types can be envisaged as POS tagging with an unusually detailed tagset. Hence, the most direct way to quickly create a supertagger is to train an off-the-shelf POS tagger over a corpus where tokens are annotated with the HPSG lexical type assigned by the grammar. Though there is a great variety of approaches to POS tagging, they all now achieve roughly the same scores, indicating that a performance ceiling has likely been reached by machine learning approaches to this task.[2] For the current study we opted for the following tools:

TnT is well known for its efficiency and for achieving state-of-the-art results despite having a simple underlying model. It is based on a second-order hidden Markov model extended with linear smoothing of parameters to address data-sparseness issues and suffix analysis for handling unknown words [13].

SVMTool is a tagger based on support-vector machines (SVM). It is extremely flexible in allowing to define which features should be used in the model (e.g. size of word window, number of POS bigrams, etc.) and the tagging strategy (left to right, bidirectional, number of passes, etc). Due to this flexibility, it is described as being a tagger generator [18].

C&C is a package that includes a CCG parser, a computational semantics tool and a supertagger. It is the latter component, the supertagger, that will be used in our study. It is based on the maximum entropy model described in [6]. From the three tools, it is the only one actually described as being a supertagger. It expects, for instance, that the input tokens be already annotated with base POS categories.

---

[2] This might not be exactly true. Studies have found that, albeit for a simpler task, learning curves still grow as corpora increases exponentially in size into the billion of tokens [16,17]. However, it is doubtful that a manually revised POS tagged corpora of such size could be created to test whether this effect holds for POS tagging.

Table 2. Accuracy scores (%) for various dataset sizes

dataset	all types			verb types			noun types		
	TnT	C&C	SVM	TnT	C&C	SVM	TnT	C&C	SVM
base	88.99	78.65	88.83	90.01	69.72	90.83	87.11	75.46	88.55
+ Público	90.02	80.93	90.30	89.48	74.06	91.08	92.65	85.31	93.20
+ Wiki	90.80	82.45	91.20	89.58	76.18	91.52	93.84	87.98	94.20
+ Folha	91.52	83.60	92.06	90.94	78.03	92.63	94.95	90.48	95.18

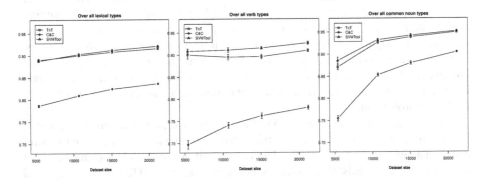

Fig. 1. Learning curves for all types, verb types and noun types

All were run using default parameters. SVMTool used the simplest setting, "M0 LR".[3] The results discussed in the following Section can thus be seen as a baseline for these tools over this task.

5 Evaluation

Evaluation was performed following a 10-fold cross-validation approach over a random shuffle of the sentences in the corpus. Table 2 shows the results obtained over the base dataset of 5,422 sentences. TnT and SVMTool have very similar performance scores, with non-significant differences in accuracy (cf. Fig. 1 for confidence intervals). C&C shows worse performance even though it relies on a more complex model and makes use of the POS tags that are already in the text.[4] Though this might at first seem surprising, it is in line with [12], where C&C also performed worse than TnT. The best results are also close to those of [11] and [12], who got similar overall accuracy scores in their experiments.

To assess the impact of extra training data, we turn to the automatically extended datasets described in Section 3.1. The accuracy scores are summarized in Table 2 while Fig. 1 shows the corresponding plots with the addition of errors bars that represent a 95% confidence interval.

---

[3] Model 0, left-to-right tagging direction. See [19] for an explanation of these settings.

[4] Recall that the datasets were pre-annotated with POS tags.

There is an improvement in the performance of all taggers as the dataset increases in size. The C&C supertagger was seen to have worse performance over the base dataset. This might be due to it using a more complex model, which needs additional data to properly estimate the parameters, and extra training data would allow it to close the gap to the other tools. The learning curves, however, indicate that this is not the case since the performance of C&C does not seem to be increasing at a fast enough rate for that to happen. TnT and SVMTool show similar scores as the corpus increases in size. Upon reaching the largest dataset, SVMTool pushes ahead with an advantage that, though only of 0.5% points, is likely significant for such a dataset size. The conclusion that can be taken from the remaining tests is similar to what was seen when evaluating over all lexical types—C&C is the worst supertagger, while TnT and SVMTool have similar performance scores. The curves over verb types are quite flat, C&C being the only one that shows any significant improvement, though it might be that, due to its low scores, it still has much room for increasing its accuracy. Since accuracy over verbs does not improve much, the raising curve that is seen over all lexical types is due to an increase in accuracy over other categories, such as common nouns, which show a much more marked improvement.

Each tool has different strengths. SVMTool is better than TnT at annotating verbs, having higher accuracy when tagging this category from early on. This is not so with common nouns, where TnT actually closes up the initial gap to SVMTool and both end up having an indistinguishable score. The best supertagger, SVMTool, shows very good results. Despite having ran using the simplest model, it has better performance than the other supertaggers.

6 Final Remarks and Future Work

In this paper we report on experiments where three supertaggers for a Portuguese HPSG grammar were induced over differently sized datasets. Over the larger dataset, the best supertaggers showed state-of-the-art accuracy of 91–93%, similar to that obtained in related work for English, in particular [11] and [12]. SVM technology had yet to be applied to supertagging. The experiments reported here show that, like in other NLP areas where it was used, it improves over existing techniques. Given how SVM easily incorporates many features, and the flexibility of SVMTool, future work will test how features and tagging strategy can be adjusted to improve performance. For instance, despite having better performance, the configuration used seems poorly suited for an effective annotation of verbs. This makes developing features specifically suited for discriminating the various verb types a promising avenue of research.

References

1. Mitkov, R. (ed.): The Oxford Handbook of Computational Linguistics. Oxford University Press (2004)
2. Manning, C., Schütze, H.: Foundations of Statistical Natural Language Processing, 1st edn. MIT Press (1999)

3. Bangalore, S., Joshi, A.: Disambiguation of super parts of speech (or supertags): Almost parsing. In: Proceedings of the 15th Conference on Computational Linguistics (COLING), pp. 154–160 (1994)
4. Joshi, A., Schabes, Y.: Handbook of Formal Languages and Automata. In: Tree-Adjoining Grammars. Springer (1996)
5. Bangalore, S., Joshi, A.: Supertagging: An approach to almost parsing. Computational Linguistics 25(2), 237–265 (1999)
6. Clark, S., Curran, J.: Log-linear models for wide-coverage CCG parsing. In: Proceedings of the 8th Conference on Empirical Methods in Natural Language Processing (EMNLP), pp. 97–104 (2003)
7. Clark, S., Curran, J.: The importance of supertagging for wide-coverage CCG parsing. In: Proceedings of the 20th Conference on Computational Linguistics (COLING), pp. 282–288 (2004)
8. Clark, S., Curran, J.: Wide-coverage efficient statistical parsing with CCG and log-linear models. Computational Linguistics 33, 493–552 (2007)
9. Prins, R., van Noord, G.: Reinforcing parser preferences through tagging. Traitement Automatique des Langues 44, 121–139 (2003)
10. Matsuzaki, T., Miyao, Y., Tsujii, J.: Efficient HPSG parsing with supertagging and CFG-filtering. In: Proceedings of the 20th International Joint Conference on Artificial Intelligence (IJCAI), pp. 1671–1676 (2007)
11. Blunsom, P.: Structured Classification for Multilingual Natural Language Processing. Ph.D. thesis, University of Melbourne (2007)
12. Dridan, R.: Using Lexical Statistics to Improve HPSG Parsing. Ph.d. thesis, University of Saarland (2009)
13. Brants, T.: TnT — a statistical part-of-speech tagger. In: Proceedings of the 6th Applied Natural Language Processing Conference and the 1st North American Chapter of the Association for Computational Linguistics, pp. 224–231 (2000)
14. Branco, A., Costa, F.: A computational grammar for deep linguistic processing of Portuguese: LX-Gram, version A.4.1. Technical Report DI-FCUL-TR-08-17, University of Lisbon (2008)
15. Costa, F., Branco, A.: LXGram: A Deep Linguistic Processing Grammar for Portuguese. In: Pardo, T.A.S., Branco, A., Klautau, A., Vieira, R., de Lima, V.L.S. (eds.) PROPOR 2010. LNCS (LNAI), vol. 6001, pp. 86–89. Springer, Heidelberg (2010)
16. Banko, M., Brill, E.: Mitigating the paucity of data problem: Exploring the effect of training corpus size on classifier performance for NLP. In: Proceedings of the 1st Human Language Technology (HLT) Conference (2001)
17. Banko, M., Brill, E.: Scaling to very very large corpora for natural language disambiguation. In: Proceedings of the 39th Annual Meeting of the Association for Computational Linguistics and 10th Conference of the European Chapter of the Association for Computational Linguistics, pp. 26–33 (2001)
18. Giménez, J., Màrquez, L.: SVMTool: A general POS tagger generator based on support vector machines. In: Proceedings of the 4th Language Resources and Evaluation Conference (LREC) (2004)
19. Giménez, J., Màrquez, L.: SVMTool: Technical Manual v1.3. TALP Research Center, LSI Department, Universitat Politecnica de Catalunya (2006)

SBFC: An Efficient Feature Frequency-Based Approach to Tackle Cross-Lingual Word Sense Disambiguation

Dieter Mourisse[1], Els Lefever[1,2], Nele Verbiest[1], Yvan Saeys[3,4],
Martine De Cock[1], and Chris Cornelis[1,5]

[1] Department of Applied Mathematics and Computer Science, Ghent University, Gent, Belgium
[2] LT3, University College Ghent, Gent, Belgium
[3] Department of Plant Systems Biology, VIB, Gent, Belgium
[4] Department of Molecular Genetics, Ghent University, Gent, Belgium
[5] Granada University, Granada, Spain
{Dieter.Mourisse,Nele.Verbiest,
Yvan.Saeys,Martine.DeCock}@UGent.be, Els.Lefever@HoGent.be,
chriscornelis@ugr.es

Abstract. The Cross-Lingual Word Sense Disambiguation (CLWSD) problem is a challenging Natural Language Processing (NLP) task that consists of selecting the correct translation of an ambiguous word in a given context. Different approaches have been proposed to tackle this problem, but they are often complex and need tuning and parameter optimization.

In this paper, we propose a new classifier, Selected Binary Feature Combination (SBFC), for the CLWSD problem. The underlying hypothesis of SBFC is that a translation is a good classification label for new instances if the features that occur frequently in the new instance also occur frequently in the training feature vectors associated with the same translation label.

The advantage of SBFC over existing approaches is that it is intuitive and therefore easy to implement. The algorithm is fast, which allows processing of large text mining data sets. Moreover, no tuning is needed and experimental results show that SBFC outperforms state-of-the-art models for the CLWSD problem w.r.t. accuracy.

1 Introduction

Word Sense Disambiguation (WSD) is the Natural Language Processing (NLP) task that consists of assigning the correct sense of an ambiguous word in a given context. Traditionally, the sense label is chosen from a predefined monolingual sense inventory such as WordNet [1]. The computational WSD task can be defined as a classification task where the possible word senses are the classes and each new occurrence of an ambiguous word is assigned to the correct sense class based on the surrounding context information of the ambiguous word.

The information that is traditionally used for WSD consists of a selection of very local context features (preceding and following words and grammatical information) and a bag-of-words feature set that reflects the presence or absence of a large set of possible content words in the wider context of the ambiguous word. As only a few of

P. Sojka et al. (Eds.): TSD 2012, LNCS 7499, pp. 248–255, 2012.

these bag-of-words features are actually present for a given occurrence of an ambiguous word, this results in very large and sparse feature vectors.

A wide range of supervised and unsupervised approaches have been proposed to tackle the WSD problem. For a detailed overview of these approaches we refer to [2]. Amongst these approaches we find all major machine learning techniques that are deployed for NLP tasks, such as memory-based learning algorithms, probabilistic models, linear classifiers, kernel-based approaches, etc. The main disadvantages of the existing classification methods are their complexity and need for tuning and parameter optimization during the training phase. As we typically work with these very large and sparse feature vectors for WSD, this leads to very complex training and classification cycles.

Cross-Lingual Word Sense Disambiguation (CLWSD) is the multilingual variant of WSD that consists of selecting the correct translation (instead of a monolingual sense label as is the case for WSD) of an ambiguous word in a given context.

This paper describes a classification algorithm that is specifically designed for the CLWSD problem, named Selected Binary Feature Combination (SBFC). We consider the CLWSD problem as a classification problem; in order to predict a correct translation of an ambiguous noun in one target language, English local context features and translation features from four other languages are incorporated in the feature vector. The main idea behind the SBFC method is that the features that occur frequently in the new instance should also occur frequently in the training instances with the predicted translation label for the new instance. The SBFC algorithm is easy to understand and fast, and can hence be used to process large-scale text mining data sets. Its advantage over other CLWSD algorithms is that it does not need tuning and is parameter independent.

The structure of this paper is as follows. Section 2 describes the data set and extracted feature set we used for our Cross-Lingual WSD experiments. Section 3 introduces our novel classification algorithm, while Section 4 provides a detailed overview of the experimental setup and results. Finally, Section 5 concludes this paper and gives some directions for future work.

2   Data

To construct the training feature set, we used the six-lingual sentence-aligned Europarl corpus that was also used in the SemEval-2010 "Cross-Lingual Word Sense Disambiguation" (CLWSD) task [3]. This task is a lexical sample task for English ambiguous nouns that consists in assigning a correct translation in the five supported target languages (viz. French, Italian, Spanish, German and Dutch) for an ambiguous focus word in a given context. In order to detect the relevant translations for each of the ambiguous focus words, we ran automatic word alignment [4] and considered the word alignment output for the ambiguous focus word to be the label for the training instances for the corresponding classifier (e.g. the Dutch translation is the label that is used to train the Dutch classifier).

For our feature vector creation, we combined a set of English local context features and a set of binary bag-of-words features that were extracted from the aligned translations. All English sentences were preprocessed by means of a memory-based shallow parser (MBSP) [5] that performs tokenization, Part-of-Speech (PoS) tagging and text chunking. The preprocessed sentences were used as input to build a set of commonly used WSD features related to the English input sentence:

- features related to the **focus word itself** being the word form of the focus word, the lemma, Part-of-Speech and chunk information
- **local context features** related to a window of three words preceding and following the focus word containing for each of these words their full form, lemma, Part-of-Speech and chunk information

In addition to these monolingual features, we extracted a set of binary bag-of-words features from the aligned translations that are not the target language of the classifier (e.g. for the Dutch classifier, we extract bag-of-words features from the Italian, Spanish, French and German aligned translations). Per ambiguous focus word, a list of content words (verbs, nouns, adjectives and adverbs) was extracted that occurred in the aligned translations of the English sentences containing the focus word. One binary feature per selected content word was then created per ambiguous word: '0' in case the word does not occur in the aligned translation of this instance, and '1' in case the word does occur. For the creation of the feature vectors for the test instances, we follow a similar strategy as the one we used for the creation of the training instances. For the construction of the bag-of-words features however, we need to adopt a different approach as we only have the English test instances at our disposal, and no aligned translations for these English sentences. Therefore we decided to deploy the Google Translate API[1] to automatically generate a translation for all English test instances in the five supported languages.

3   Method

As described in Section 2, the CLWSD feature vectors consist of two parts. The first part, that covers the local context features, contains non-binary but discrete data. Before applying the SBFC algorithm, we used a straightforward procedure to make this data binary. For each value v of a non-binary feature f in our training set, a new binary feature is generated that is 1 if the instance has value v for f and 0 otherwise.

As a result, we have training data that consists of n ambiguous words $w_1, \ldots, w_n$, described by m binary features $f_1, \ldots, f_m$. The translation of a word w is denoted by $T(w)$, the value of a word w for a feature f_i is denoted by $f_i(w)$ and is a value in $\{0, 1\}$. We say that a feature f_i *occurs* in a word w if $f_i(w) = 1$. The task is now to predict a translation $T(t)$ for a test word t described by the binary vector $(t_1, \ldots, t_m)$.

Before the actual method is carried out, we apply a preprocessing step in order to remove the training instances with a unique translation because this very often occurs in the presence of noise. An example is given in Table 1, that lists part of the Italian training data for the ambiguous word *mood*. The word w_4 is removed, as its class label *umore* only occurs once in the training data.

The SBFC method that we designed for the CLWSD problem reflects the following ideas: a translation is a good classification label for new instances if the features that occur in the new instance occur.

---

[1] http://code.google.com/apis/language/

Table 1. Training data for translating the ambiguous word *mood* before (left-hand-side) and after (right-hand-side) preprocessing

	f_1	f_2	f_3	f_4	T
w_1	1	0	1	0	clima
w_2	0	0	1	1	clima
w_3	1	1	0	1	atmosfera
w_4	1	0	1	0	umore
w_5	1	0	0	1	atmosfera
w_6	0	0	1	0	atmosfera

	f_1	f_2	f_3	f_4	T
w_1	1	0	1	0	clima
w_2	0	0	1	1	clima
w_3	1	1	0	1	atmosfera
w_5	1	0	0	1	atmosfera
w_6	0	0	1	0	atmosfera

- I1: **at least once** in the training feature vectors associated with the same translation label
- I2: **frequently** in the training feature vectors associated with the same translation label

Note that I1 is contained in I2: if a feature occurs frequently in a feature vector, it will of course appear at least once in the feature vector. The reason why we handle these cases separately is that we want to penalize classification labels for which features that occur in the test vector never occur in training instances with this classification label. Moreover, we show in the experimental section that combining both ideas results in the best accuracies.

If we apply these hypotheses on the example in Table 1, we come to the following predictions for a given test vector $(1, 0, 0, 1)$. *Clima* might be a good translation for this test vector, because feature f_1 and f_4 occur at least once in a training feature vector with translation *clima*. On the other hand, these features do not occur so often. By consequence, *atmosfera* is a better translation candidate, because f_1 and f_4 occur at least once in a word with translation *atmosfera*, and these features occur each twice in a word with translation *atmosfera*.

To formalize this idea, we first translate the training data into the so-called *model matrix M*. This matrix has dimensions $m \times c$ with m being the number of features and c the number of different translations appearing in the data set. The entry M_{ij} is the number of times that the ith feature occurs in a word with the jth translation in the training data. The model matrix of the example is given on the left-hand-side of Table 2.

The columns of translations that occur often will generally contain higher values and will therefore be favored in the final algorithm. To prevent this, we scale the matrix by dividing the values in the column of a translation by the times this translation occurs. This is shown in the right-hand-side of Table 2.

The scaled model matrix can now be used to predict the translation label of new test instances with vector $(t_1, \ldots, t_m)$. We first assign a score to each class label (translation) in the training data as follows: suppose the translation considered is C, then we look up the column in the scaled model matrix corresponding to this translation and count how many features occurring in the test vector have a value different from zero in this column. This measure reflects the idea in I1. To express the idea in I2, we sum the values in the column corresponding to features that occur in the test vector. The resulting score of translation C is then the product of these two measures. Finally, the translation label

Table 2. Model matrix of the example in Table 1 before (left-hand-side) and after (right-hand-side) scaling

	clima	atmosfera
f_1	1	2
f_2	0	1
f_3	2	1
f_4	1	2

	clima	atmosfera
f_1	0.5	0.67
f_2	0	0.33
f_3	1	0.33
f_4	0.5	0.67

that will be predicted by the algorithm for the given test vector will be the translation with the highest overall score.

Formally, the first measure for the jth class is given by:

$$\text{score}_1(j) = \sum_{i=1}^{m} h(M_{ij}) \cdot t_i,$$

where $h(M_{ij})$ is one if M_{ij} is different from zero and zero otherwise, and by

$$\text{score}_2(j) = \sum_{i=1}^{m} M_{ij} \cdot t_i$$

for the second measure. The final score of the jth class is then:

$$\text{score}(j) = \text{score}_1(j) \cdot \text{score}_2(j).$$

This algorithm works fast and only needs limited storage: constructing the model matrix M needs $\mathcal{O}(nm)$ operations, and the model matrix itself has dimensions $m \times c$. The original training data does not need to be stored anymore during the test phase. To classify a new test vector, $\mathcal{O}(m)$ operations are required.

Suppose the test vector in the running example is $(0, 1, 1, 0)$. To calculate the score of the translation *clima*, we look at the first column in the scaled model matrix. Features f_2 and f_3 occur in the test vector, but only one of them has a non-zero value in the model matrix, so the first measure for *clima* is score$_1$(clima)= 1. Next, we sum the values for f_2 and f_3, which results in score$_2$(clima)= 1. The final score for the translation *clima* is score(clima)= $1 \cdot 1 = 1$. To determine the score of *atmosfera*, we look at the second column. For both f_2 and f_3, the values are different from zero, so the first measure is score$_1$(atmosfera)= 2. Next, we sum the values for f_2 and f_3, resulting in score$_2$(atmosfera)= 0.66. The final value for translation is score(clima)= $2 \cdot 0.66 = 1.32$, which is higher than the score for *clima*. By consequence, the translation *atmosfera* is returned by the algorithm for this given test instance.

4   Experimental Evaluation

To evaluate our classification algorithm for the five target languages, we used the sense inventory and test set of the SemEval "Cross-Lingual Word Sense Disambiguation" task. A more detailed description of the construction of the data set can be found in [7].

4.1 Experimental Set-Up

We consider three versions of the SBFC method: $SBFC_1$ (resp. $SBFC_2$) only uses $score_1$ (resp. $score_2$) to measure the quality of the class labels, while $SBFC$ uses the product of the two scores. We make this distinction to show that both ideas I1 and I2 as described in Section 3 have to be taken into account. We apply the SBFC method to the CLWSD data sets for the ambiguous words *coach, education, execution, figure, letter, match, mission, mood, paper, post, pot, range, rest, ring, scene, side, soil, strain* and *test* and compare it to a baseline, three state-of-the-art CLWSD systems and Naive Bayes:

- As a **baseline**, we select the most frequent lemmatized translation that resulted from the automated word alignment.
- The **ParaSense** system [8] uses the same set of local context and translation features as described in Section 2 and a memory-based learning algorithm implemented in TIMBL [5].
- The **UvT-WSD** system [9] uses a k-nearest neighbor classifier and a variety of local and global context features and obtained the best scores for Spanish and Dutch in the SemEval CLWSD competition.
- The **T3-COLEUR** system [10] participated for all five languages and outperformed the other systems in the SemEval competition for French, Italian and German. This system adopts a different approach: during the training phase a monolingual WSD system processes the English input sentence and a word alignment module is used to extract the aligned translation. The English senses together with their aligned translations (and probability scores) are then stored in a word sense translation table, in which look-ups are performed during the testing phase.
- The **Naive Bayes (NB)** [11] classifier is a probabilistic classifier that assumes that the features are independent. We compare with this classifier because it has similarities with our new approach, that is, it is also based on the frequencies of the features, but it does not take into account the sparse nature of the data.

As evaluation metric, we use a straightforward accuracy measure that divides the number of correct answers by the total amount of test instances.

4.2 Results

Table 3 lists the average results over the different test words per language.

Table 3. Accuracy values averaged over all nineteen test words

	SBFC	SBFC$_1$	SBFC$_2$	baseline	ParaSense	UvT-WSD	T3-COLEUR	NB
Dutch	**0.70**	0.64	0.17	0.61	0.68	0.64	0.42	0.07
French	**0.75**	0.71	0.17	0.65	**0.75**	-	0.67	0.11
Italian	**0.66**	0.61	0.20	0.54	0.63	-	0.56	0.07
Spanish	**0.73**	0.67	0.23	0.59	0.68	0.70	0.58	0.07
German	**0.69**	0.67	0.22	0.54	0.67	-	0.57	0.11

As can be seen in Table 3, SBFC$_2$ does not score well. SBFC$_1$ scores better, but is outperformed by the ParaSens and UvT-WSD system. However, the SBFC method, that combines the scores used in SBFC$_1$ and SBFC$_2$, outperforms all other methods for all considered languages, and although there are some similarities with the Naive Bayes classifier, SBFC does a far better job in selecting the correct translation for a word. It is furthermore the case that, since we only use features that occur in a particular test word (i.e. features whose value is one), this algorithm works very fast on sparse binary data.

5 Conclusion

We presented the new classifier SBFC (Selected Binary Feature Combination) to tackle the Cross-lingual Word Sense Disambiguation task. The algorithm merely relies on feature frequencies and is therefore very efficient. In addition, the method is very intuitive, and hence easy to implement and comprehend; it is by consequence easy to adapt to make it more suitable for different classification data sets. Experimental results show that the SBFC algorithm outperforms state-of-the-art CLWSD systems for all five considered languages.

In future work, we will apply the SBFC method to other classification data sets and investigate alternative methods to combine the two different scores (i.e. making one score more important than the other). In addition, we will also examine other techniques to scale the model matrix.

References

1. Fellbaum, C.: WordNet: An Electronic Lexical Database. MIT Press (1998)
2. Agirre, E., Edmonds, P.: Word Sense Disambiguation. In: Algorithms and Applications. Text, Speech and Language Technology series. Springer (2006)
3. Lefever, E., Hoste, V.: SemEval-2010 Task 3: Cross-Lingual Word Sense Disambiguation. In: Proceedings of the 5th International Workshop on Semantic Evaluation, ACL 2010, Uppsala, Sweden, pp. 15–20 (2010)
4. Och, F., Ney, H.: A systematic comparison of various statistical alignment models. Computational Linguistics 29(1), 19–51 (2003)
5. Daelemans, W., van den Bosch, A.: Memory-based Language Processing. Cambridge University Press (2005)
6. Schmid, H.: Probabilistic part-of-speech tagging using decision trees. In: Proceedings of the International Conference on New Methods in Language Processing, Manchester, UK (1994)
7. Lefever, E., Hoste, V.: Construction of a Benchmark Data Set for Cross-Lingual Word Sense Disambiguation. In: Proceedings of the Seventh International Conference on Language Resources and Evaluation (LREC 2010). European Language Resources Association (ELRA), Valletta, Malta (2010)
8. Lefever, E., Hoste, V., De Cock, M.: ParaSense or How to Use Parallel Corpora for Word Sense Disambiguation. In: Proceedings of the 49th Annual Meeting of the Association for Computational Linguistics: Human Language Technologies, pp. 317–322. Association for Computational Linguistics, Portland (2011)

9. van Gompel, M.: UvT-WSD1: A Cross-Lingual Word Sense Disambiguation System. In: Proceedings of the 5th International Workshop on Semantic Evaluation, ACL 2010, pp. 238–224. Association for Computational Linguistics, Uppsala (2010)
10. Guo, W., Diab, M.: COLEPL and COLSLM: An Unsupervised WSD Approach to Multilingual Lexical Substitution, Tasks 2 and 3 SemEval 2010. In: Proceedings of the 5th International Workshop on Semantic Evaluation, ACL 2010, pp. 129–133. Association for Computational Linguistics, Uppsala (2010)
11. Maron, M.E.: Automatic Indexing: An Experimental Inquiry. Journal of the ACM (JACM) 8(3), 404–417 (1961)

Dependency Relations Labeller for Czech

Rudolf Rosa and David Mareček

Charles University in Prague, Faculty of Mathematics and Physics
Institute of Formal and Applied Linguistics, Prague, Czech Republic
{rosa,marecek}@ufal.mff.cuni.cz
http://ufal.mff.cuni.cz

Abstract. We present a MIRA-based labeller designed to assign dependency relation labels to edges in a dependency parse tree, tuned for Czech language. The labeller was created to be used as a second stage to unlabelled dependency parsers but can also improve output from labelled dependency parsers. We evaluate two existing techniques which can be used for labelling and experiment with combining them together. We describe the feature set used. Our final setup significantly outperforms the best results from the CoNLL 2009 shared task.

Keywords: natural language processing, dependency parsing, sequence labelling.

1 Introduction

Dependency trees have for a long time been the primary structure to represent Czech syntax, as well as in other, especially highly non-projective, languages. In recent years, dependency trees have proven to be a valuable alternative to phrase-structure trees even for relatively projective languages, and much effort has been invested into creating various dependency parsers, such as the MST Parser [1] or the MALT Parser [2].

As opposed to phrase-structure trees, assigning a correct label to each dependency relation is often an important task as well. Some parsers perform joint parsing and labelling, producing a labelled dependency tree as their output. Others produce only an unlabelled tree, requiring a standalone labeller to assign the labels in a second step. We present such a labeller in our paper.

In Section 2, we review two existing techniques which can be used to assign labels to dependency relations and try to combine them together. Section 3 contains a description of our feature set. We evaluate our results in Section 4.

2 Labelling Methods

We make use of several algorithms that can be used and combined in various ways in the task of dependency relations labelling, building upon [3] and [4].

A *labelling* $\mathcal{L}$ of a dependency tree is a mapping $\mathcal{L}\ E \to L$, which assigns a label $l \in L$ to each edge $e \in E$ of the dependency tree. The general approach that we follow to find the best labelling is using edge-based factorization, i.e. to assign a score to each possible pair e, l, and to try to find a labelling with a maximum overall score. Based on the scoring method used, the score can be a probability or an additive score.

P. Sojka et al. (Eds.): TSD 2012, LNCS 7499, pp. 256–263, 2012.
© Springer-Verlag Berlin Heidelberg 2012

Following [3], we treat the dependency relations labelling as a *sequence labelling problem*. Starting at the root node of the tree, we label all edges going from the current node to its children from left to right (these adjacent edges form a sequence), and then continue in the same way on lower levels of the tree. This implies that at each step we have already processed all ancestor edges and left sibling edges. We utilize this fact in designing the feature set.

To label the sequence, we use the well-known Viterbi algorithm, as described in [3].

Let $\vec{e} = (e_1, e_2, ..., e_n)$ be a sequence of edges and $\vec{l} = (l_1, l_2, ..., l_n)$ one of its possible labellings. The score $S_{\vec{e},\vec{l}}$ of $\vec{e}$ being labelled by $\vec{l}$, which Viterbi tries to maximize, is defined differently for additive scores and for probabilities. For additive scores, it is computed as a sum:

$$S_{\vec{e},\vec{l}} = \sum_{i=1}^{n} score(e_i, l_i) \tag{1}$$

whereas for probabilities, it is computed as a product:

$$S_{\vec{e},\vec{l}} = \prod_{i=1}^{n} p(e_i, l_i) \tag{2}$$

We use n-best Viterbi, which always keeps n best partial labellings at each step, selecting the best scoring labelling at the end.

2.1 Maximum Likelihood Estimate

We use Maximum Likelihood Estimate (MLE) as a simple baseline method, trying to estimate the probability distribution of labels that can be assigned to an edge, based on its features and the previously assigned labels.

Suppose that we want to find the probability of an edge e with binary features F to be labelled with the label l from the set of all labels L.

We use MLE to estimate the probability distribution of various labels that can be assigned to an edge, based on its features.

We estimate the *emission probability* as an average probability of an edge being labelled by label l given the set of the features F of the edge:

$$p_{em}(l|F) = \frac{1}{|F|} \sum_{f \in F} p_{em}(l|f) \text{ where } p_{em}(l|f) = \frac{count(f \wedge l)}{count(f)} \tag{3}$$

where $count()$ refers to number of occurrences of the feature f or pair of the feature f and the label l in the training data. The averaging can be regarded as lambda smoothing with all lambdas equal to $1/|F|$.

We estimate the *transition probability* as a conditional probability of a label l given the previous label l_{prev}, smoothed together with marginal probability of the label l and a constant parameter:

$$p_{tr}(l|l_{prev}) = \lambda_2 \cdot p_{bi}(l|l_{prev}) + \lambda_1 \cdot p_{uni}(l) + \lambda_0 \tag{4}$$

where

$$p_{bi}(l|l_{prev}) = \frac{count(l \wedge l_{prev})}{count(l_{prev})} \text{ and } p_{uni}(l) = \frac{count(l)}{|L|} \quad (5)$$

The smoothing parameters λ_2, λ_1 and λ_0 are estimated using the Expectation Maximization (EM) algorithm on held-out data.

MLE is very fast and also very easy to implement. However, it did not perform well enough.

2.2 Margin Infused Relaxed Algorithm

Following [4] and [3], we decided to use the Margin Infused Relaxed Algorithm (MIRA), described in [5]. MIRA is an online learning algorithm for large-margin multiclass classification, successfully used in [1] for dependency parsing and suggested to be used for dependency relations labelling in a second stage labeller.

MIRA assigns a score for each label that can be assigned to a dependency relation. Feature-based factorization is used, thus the score of a label l to be assigned to an edge e which has the features F is computed as follows:

To find the probability of an edge e with features F to be labelled with the label l from the set of all labels L, we compute a *score*:

$$score(l, e) = \sum_{f \in F} score(l, f) \quad (6)$$

where $score(l, f)$, also called the *weight* of the feature (l, f), is computed by MIRA, trying to minimize the classification error on the training data, iterating over the whole dataset several times. The final scores are then averaged to avoid overtraining.

We used the *single-best MIRA* variant in our experiments.

2.3 MLE and MIRA

Having already implemented both MLE and MIRA, we also tried to combine them together. We have observed that the emission probabilities are crucial for assigning the correct label, whereas the transition probabilities have a rather small impact. We therefore decided to use MLE to estimate transition probabilities the same way as in MLE approach, but to use MIRA to estimate emission scores instead of MLE emission probabilities (this time not using features based on left sibling's label). However, a major issue with this approach is to combine the transition probabilities and emission scores correctly.

Product. Our first attempt was to simply use

$$score(l, e) = p_{tr}(l|l_{prev}) \cdot score_{MIRA}(l, e) \quad (7)$$

as in MLE approach, and to sum these together in Viterbi as in MIRA approach. Although being severely mathematically incorrect (there are grave issues with ordering of the scores and even greater issues with negative scores), this approach did lead to competitive results.

Sigmoid. We then tried to use a mathematically sound way by recomputing the emission score into a probability-like number using the sigmoid function:

$$score(l, e) = p_{tr}(l|l_{prev}) \cdot \frac{1}{1 + e^{-score_{MIRA}(l,e)}} \tag{8}$$

Interestingly, the results of this approach were significantly worse than these of the Product approach. However, as the MIRA-only method proved to outperform both of these combination methods, we did not investigate this further.

3 Feature Set

We construct the feature set by combining features suggested in [1,3,6] and [7] and tuning it to maximize performance on Czech data.

Our input for labelling is the parsed sentence, together with its morphological analysis – i.e., for each word, we know its form, lemma and morphological tag. We use coarse Czech tags as described in [8].

We join these fields (and feature functions defined in following sections) in various ways to form feature templates for the labeller. For example, feature template that consists of the parent and child word forms and coarse tag of the child is defined as "FORM|form|coarse_tag". We denote a field of the parent node of the edge by using uppercase; the child node is indicated by using lowercase.

Each of the feature values is then joined with its feature template and the label of the edge.

For example, for an edge "Rudolf/N1 ← relaxuje/VB" (Rudolph ← relaxes) with a label "Sb" (subject), the feature template "FORM|form|coarse_tag" would return the value "FORM|form|coarse_tag:*relaxuje*|*Rudolf*|*N1*:Sb".

Most of the feature templates are created by joining one or more fields or feature functions with the COARSE_TAG and coarse_tag fields, as this has been observed to lead to the best results. For instance, feature templates incorporating *attachment direction* feature function attrdir() have the following shapes:

- attdir()|coarse_tag,
- attdir()|COARSE_TAG,
- attdir()|coarse_tag|COARSE_TAG

Also the feature functions that can take a field as a parameter are usually declared with coarse_tag being the parameter. The complete feature set is available in the TectoMT Share repository.[1]

Thanks to the fact that we get the whole dependency tree on input, we do not have to limit ourselves to local features computed over the edge; we can easily include information about any nodes in the tree.

---

[1] http://ufallab.ms.mff.cuni.cz/tectomt/share/data/models/
labeller_mira/cs/pdt_labeller_best.config

3.1 First-Order Features

Our basic feature set is based on the first order features described in [1] (first-order features are defined as features on single edges), which were designed for unlabelled parsing but proved to be useful for labelling as well. However, we do not conjoin our feature templates with distance of words or attachment direction, as these are more relevant for parsing than for labelling. Also we use morphological lemmas instead of 5-gram prefixes. The combinations of features were modified a little to maximize performance on development data.

The first-order feature templates consist of the basic fields available on input, and of several context features, providing information about nearby words.

- COARSE_TAG / coarse_tag
- FORM / form
- LEMMA / lemma
- PRECEDING(*field*) / preceding(*field*) – a field of the word immediately preceding the parent/child node, e.g. PRECEDING(coarse_tag) – the coarse tag of the word which precedes the parent node of the edge
- FOLLOWING(*field*) / following(*field*) – a field of the word immediately following the parent/child node
- between(*field*) – a field of each of the words between the parent node and the child node; this function can return multiple values, creating several features from one feature template
- attdir() – the direction of attachment of the child to the parent

3.2 Non-local Features

Based on non-local features described in [3], we extend our feature set with labelling-specific features, which make use of knowledge of potentially the whole parse tree:

- CHILDNO() / childno() – number of child nodes of the parent/child node
- isfirstchild() / islastchild() – whether the child node is the first/last child of the parent node

3.3 Higher-Order Features

Higher order features are based on multiple edges. These can be sibling features as described in [6], parent-child-grandchild features as described in [7], or other variations and conjunctions of these concepts.

The possibility to use some of the features depends on the order of assigning labels to edges in the dependency tree. Following [3], we express the task as a sequence labelling problem, with a sequence defined as a sequence of adjacent edges, i.e. edges sharing the same parent node. This enables us to use the information about the label assigned to sibling edges which are on the left from the current edge.

Furthermore, we decided to go through the tree in a top-down direction, i.e. first labelling the sequence of edges whose parent is the root node, continuing to their children. Thus we can also make use of the knowledge of the label assigned to the edge between the parent and the grandparent of the child node, and we even make use of the label assigned to the edge between the grandparent and the great-grandparent.

The grandparent features belong to features with the biggest influence on accuracy. Their inclusion in the feature set immediately led to an improvement of accuracy by 2%, whereas most of the other features contribute with less than 0.5%. They are especially useful for e.g. prepositional and coordination structures.

- LABEL() – label assigned to the edge between the parent and the grandparent of the child node
- l.label() – label assigned to the left sibling edge, i.e. the edge between the parent and the left sibling of the child node
- r.coarse_tag – coarse tag of the right sibling of the child node
- G.coarse_tag – coarse tag of the grandparent of the child node
- G.label() – label assigned to the edge between the grandparent and the great-grandparent of the child node
- G.attdir() – whether the grandparent node precedes or follows the child node in the sentence

4   Experiments and Results

We use the Prague Dependency Treebank 2.0 [9] for training and testing. We use 68500 sentences as training data, 4500 sentences as development data and 4500 sentences as test data.

In Table 1, we present a comparison of the methods that we tried to use for labelling. We measure labelling accuracy on golden trees. MIRA clearly outperforms all the other methods.

Table 1. Comparison of several labelling methods. The results obtained on the test-data golden trees using the feature set tuned on the developement data.

labelling method	labelling accuracy [%]
MLE	65.7
MLE + MIRA (sigmoid)	89.0
MLE + MIRA (product)	91.9
MIRA	94.1

To reliably compare our results with results of other existing labelled parsers, we used the datasets from the CoNLL 2009 Shared Task [10]. On the website[2] of the task, the training and test data are available, together with outputs of the participating systems.

We decided to compare our results with the four best-performing systems. We trained our labeller on the official training data and then used it to relabel the outputs of the

---

[2] http://ufal.mff.cuni.cz/conll2009-st/

four systems. Labelled Attachment Score (LAS) was then measured by comparing the results with the golden test data.

The results in Table 2 show that, although we did not improve LAS of each of the systems, we managed to improve LAS of the *che* system [11] by 1.2%, leading to a LAS of 81.2%, which significantly outperforms all of the four original systems, leading to a state-of-the-art labelling – that is, at least on this task.

Manual inspection of the original and the relabelled outputs of the *che* system showed that most of the differences are in differentiating between the Attribute, Adverbial and Object, and between the Object and Subject.

Table 2. Labelled Attachment Scores on CoNLL2009 shared task best systems outputs

system	LAS original [%]	LAS relabelled [%]
merlo	**80.4**	79.9
bohnet	80.1	81.0
che	80.0	**81.2**
chen	79.7	79.5

5 Conclusion

We created a labeller which provides to our knowledge state-of-the-art dependency relations labelling for Czech language. It is a standalone labeller and therefore can be used as a second-stage to any unlabelled dependency parser. Moreover, it might even improve the accuracy of a labelled parser by relabelling its output. It uses the Margin Infused Relaxed Algorithm, which makes it possible to use a large number of features, both first- and higher-order.

In future we are planning to experiment with using labels assigned by another labelled parser as input features for the labeller, which might lead to even better results.

It would be favourable to reduce the feature set and prune the model if possible to achieve lower time and memory consumption with similar accuracy. We leave this for our future work as well.

The labeller is published on CPAN[3] as a part of the MSTperl package. Various trained models for the labeller can be obtained in the TectoMT Share repository.[4]

Acknowledgements. This research has been supported by the EU Seventh Framework Programme under grant agreement n° 247762 (Faust).

References

1. McDonald, R., Crammer, K., Pereira, F.: Online large-margin training of dependency parsers. In: Proceedings of the 43rd Annual Meeting on Association for Computational Linguistics, pp. 91–98. Association for Computational Linguistics (2005)

---

[3] http://www.cpan.org/
[4] http://ufallab.ms.mff.cuni.cz/tectomt/share/
data/models/labeller_mira/cs/

2. Nivre, J., Hall, J., Nilsson, J.: Maltparser: A data-driven parser-generator for dependency parsing. In: Proceedings of LREC, vol. 6, pp. 2216–2219 (2006)
3. McDonald, R., Lerman, K., Pereira, F.: Multilingual dependency analysis with a two-stage discriminative parser. In: Proceedings of the Tenth Conference on Computational Natural Language Learning, CoNLL-X 2006, pp. 216–220. Association for Computational Linguistics, Stroudsburg (2006)
4. Rosenfeld, B., Feldman, R., Fresko, M.: A systematic cross-comparison of sequence classifiers. In: SDM, pp. 563–567 (2006)
5. Crammer, K., Singer, Y.: Ultraconservative online algorithms for multiclass problems. The Journal of Machine Learning Research 3, 951–991 (2003)
6. McDonald, R., Pereira, F.: Online learning of approximate dependency parsing algorithms. In: Proceedings of EACL, vol. 6, pp. 81–88 (2006)
7. Carreras, X.: Experiments with a higher-order projective dependency parser. In: Proceedings of the CoNLL Shared Task Session of EMNLP-CoNLL, vol. 7, pp. 957–961 (2007)
8. Collins, M., Ramshaw, L., Hajič, J., Tillmann, C.: A statistical parser for Czech. In: Proceedings of the 37th Annual Meeting of the Association for Computational Linguistics on Computational Linguistics, ACL 1999, pp. 505–512. Association for Computational Linguistics, Stroudsburg (1999)
9. Hajič, J., Hajičová, E., Panevová, J., Sgall, P., Pajas, P., Štěpánek, J., Havelka, J., Mikulová, M.: Prague Dependency Treebank 2.0. CD-ROM, Linguistic Data Consortium, LDC Catalog No.: LDC2006T0 1, Philadelphia (2006)
10. Hajič, J., Ciaramita, M., Johansson, R., Kawahara, D., Martí, M., Màrquez, L., Meyers, A., Nivre, J., Padó, S., Štěpánek, J., et al.: The CoNLL 2009 shared task: Syntactic and semantic dependencies in multiple languages. In: Proceedings of the Thirteenth Conference on Computational Natural Language Learning: Shared Task, pp. 1–18. Association for Computational Linguistics (2009)
11. Chen, W., Zhang, Y., Isahara, H.: A two-stage parser for multilingual dependency parsing. In: Proceedings of the CoNLL Shared Task Session of EMNLP-CoNLL, pp. 1129–1133 (2007)

Preliminary Study on Automatic Induction of Rules for Recognition of Semantic Relations between Proper Names in Polish Texts

Michał Marcińczuk and Marcin Ptak

Wrocław University of Technology, Wrocław, Poland
michal.marcinczuk@pwr.wroc.pl, mp.marcin.ptak@gmail.com

Abstract. In the paper we present a preliminary work on automatic construction of rules for recognition of semantic relations between pairs of proper names in Polish texts. Our goal was to check the feasibility of automatic rule construction using existing inductive logic programming (ILP) system as an alternative or supporting method for manual rule creation. We present a set of predicates in first-order logic that is used to represent the semantic relation recognition task. The background knowledge encode the morphological, orthographic and named entity-based features. We applied an ILP on the proposed representation to generate rules for relation extraction. We have utilized an existing ILP system called Aleph [1]. The performance of automatically generated rules was compared with a set of hand-crafted rules developed on the basis of training set for 8 categories of relations (affiliation, alias, creator, composition, location, nationality, neighbourhood, origin). Finally, we proposed several ways how to improve to preliminary results in the future work.

Keywords: Semantic Relations, Named Entities, Proper Names, Rule Induction, ILP, Polish.

1 Introduction

Recognition of semantic relations between named entities is one of the information extraction tasks, which goal is to identify pairs of named entities linked with a pre-defined categories of semantic relations, for example physical location, social relation, etc. The list of relations is open and depend on several factors, like desired application, text domain, categories of proper names being recognized in the texts. For example ACE [2] defines a set of 8 general relations (i.e., physical, part-whole, personal-social, ORG-affiliation, agent-artifact and gen-affiliation), while in bioinformatic domain BioNLP [3] defines two categories of relations between genes, proteins and associated entities — PROTEIN-COMPONENT and SUBUNIT-COMPLEX.

There are two major approaches to relation recognition — construction of human-readable rules and construction of statistical models. In the rule-based approach the rules can be created manually (information extraction from diabetic patients' hospital documentation [4]), semi-automatically [5] or automatically (RAPIER [6], SRV [7]). The advantage of rule-based approach over statistical models is the traceability of

P. Sojka et al. (Eds.): TSD 2012, LNCS 7499, pp. 264–271, 2012.

decisions taken by the system. On one hand rule-based system can be easily manually tuned or corrected but on the other hand creation and maintenance of large set of rules might be challenging.

According to our best knowledge this is the first work on recognition semantic relations for Polish. For other languages, mainly English, it is common to use sentence dependency parser in this task but for Polish the existing parser is not effective due to two reasons: (1) it generates hundreds tree-dependencies for single sentence without ranking and (2) the sentence must be grammatically correct and cannot contains words out of dictionary. In other way it will not generate any tree structure. The works on tree-parsers for Polish are under intensive development [8].

2   Task Definition

We intend to recognise 8 categories of semantic relations between proper names. The relations are: *affiliation* (person is a member of organization or band, etc.), *alias* (alternative or old name), *nationality* (person has nationality of a country), *origin* (person or band comes from a city or a country), *location* (physical location of one entity within another entity), *creator* (creator or author of an entity), *neighbourhood* (physical location, one entity is located next to another entity) and *composition* (one entity is an integral part of another entity). The task is also limited to relations between named entities present in the same sentence. The other assumption is that the relations must be supported by some premises stated in the sentence — we do not intend to recognise relations supported only by external knowledge. We also want to recognize every single instance of a relation in the document (no information aggregation is performed). We assume that the named entities will be recognized beforehand with the NER tool presented in [9] extended to 56 types of proper names.

3   Corpora

The supervised learning and evaluation was performed on the KPWr corpus (Wrocław University of Technology Corpus) [10]. To our best knowledge this is the first corpus annotated with semantic relations between named entities for Polish. The corpus is diversified as it contains texts from various domains, like blogs, science, stenographic recordings, dialogue, contemporary prose and 9 other domains. We have taken all documents annotated with semantic relations (761 out of 1,300 documents). The selected documents were divided into three sets: training, held-out and testing set. The training and held-out sets were used for training and parameters tuning. The testing set was used to compare the results for final configurations. The detailed statistics of the KPWr corpus divided into three sets are presented in Table 1.

4   Base Line

As a baseline for the automatic rule induction we manually created a set of rules. For every category of relations except *origin* a linguist spent 2 hours on developing the best

Table 1. Statistics of KPWr corpus and training, held-out and testing sets

	KPWr	**Training**	**Held-out**	**Testing**
	General statistics			
documents	761	418	135	208
tokens	204,354	105,613	43,184	55,557
relations	3,641	2,317	516	808
	Detailed relation statistics			
affiliation	446	250	61	135
alias	187	87	33	67
composition	385	277	50	58
creator	187	94	31	62
location	1141	816	156	169
nationality	27	14	6	7
neigbourhood	159	98	28	33
origin	106	70	12	26

set of rules. For *origin* we spent more than 2 days on developing the set of rules. In total the linguist spent 4 days on developing the rules on the basis of training set. Figure 1 presents a sample WCCL rule (the WCCL language was extended by an additional operator *link*) which recognize the *origin* relation between a person name and a city name.

```
apply(
  match(
    is('person_nam'),          // match annotation of type person_nam
    equal( base[0], 'urodzić'), // match word with base form 'born'
    equal( base[0], 'się'),
    equal( base[0], 'w'),       // match word with base form 'in'
    is('city_nam'),             // match annotation of type city_nam
  ),
  actions(
    link(1, 'person_nam', 5, 'city_nam', 'origin')
  )
)
```

Fig. 1. A sample WCCL rule created by linguist to recognize *origin* relation between a person name and a city name (WCCL with additional operator *link*)

In the limited time the linguist created 117 rules (25 for *alias* and *affiliation* each, 22 for *location*, 15 for *creator*, 14 for *composition* and *origin* each, 9 for *neighbourhood* and 7 for *nationality*). The rules were evaluated on the training, held-out and testing sets and the results are presented in Table 2. As expected, the best results were obtained for *origin* (as more time was spent to develop the rules) on each set, from 75.97% of F-measure on training set to 55% on testing set. Analysing results for the other relations we observed, that the best results were obtained for relations with the smallest number of examples — 40% of F-measure for *nationality* containing only 14 instances, 23.02% for *alias* with 87 instances. The low recall shows that the information about relations can be encoded in many different ways and small set of rules is insufficient. Also the high precision and very low recall on held-out and testing sets show that 2 hours per relation category is insufficient to develop general rules.

Table 2. Performance of hand-crafted rules

| Set | Training | | | Held-out | | | Testing | | |
Relation	P [%]	R [%]	F [%]	P [%]	R [%]	F [%]	P [%]	R [%]	F [%]
affiliation	72.22	9.56	**16.88**	33.33	2.86	**5.26**	80.95	11.49	**20.12**
alias	100.00	13.01	**23.02**	80.00	23.53	**36.36**	100.00	7.69	**14.29**
composition	79.17	5.56	**10.38**	50.00	2.08	**4.00**	100.00	18.87	**31.75**
creator	77.78	6.25	**11.57**	0.00	0.00	**0.00**	100.00	1.49	**2.94**
location	88.79	11.16	**19.83**	100.00	11.39	**20.45**	100.00	9.36	**17.11**
nationality	66.67	28.57	**40.00**	0.00	0.00	**0.00**	66.67	28.57	**40.00**
neigbourhood	85.71	5.36	**10.08**	0.00	0.00	**0.00**	100.00	2.78	**5.41**
origin	87.50	67.12	**75.97**	85.71	46.15	**60.00**	84.62	40.74	**55.00**

5 Rule Induction

5.1 Background Knowledge Definition

Data Types which are used in the predicates:

- token — every token has unique identifier among all documents,
- tag — every morphological tag has unique identifier among all documents,
- word — enumeration of all orthographic forms and base forms.
- annotation — every annotation has unique identifier among all documents,
- annotation_type — enumeration of all annotation types (56 labels),
- relation_type — enumeration of all relation types (8 labels),

Basic Predicates. are used to describe the "raw" structure of the sentence and the predicates are explicitly stated in the background knowledge file. Sentence is represented as a sequence of tokens and a set of annotations spanning over tokens. Relation is a triple of source annotation, target annotation and relation type. Every token is described with an orthographic form, a set of morphological tags and additional features. A sample sentence encoded as a set of predicates is presented in Figure 2.

- token_after_token(token, token) — describes the order of tokens,
- token_orth(token, word) — word is an orthographic form of token,

- token_tag(token, tag) — tag is a morphological tag assigned to token,
- token_pattern(token, pattern) — token orthographic form is pattern, where *pattern* is one of *upper case*, *lower case*, etc. (see [11]),
- tag_base(tag, word) — word is a base form for tag,
- tag_morph(tag, morph) — morph is a morphological feature for tag,
- annotation_type(annotation, annotation_type) — annotation is type of annotation_type,
- annotation_range(annotation, token, token) — annotation starts from first token and ends on the second token.

```
token(d1_s1_t1). token_orth(d1_s1_t1, 'word_jan').
   token_pattern(d1_s1_t1, 'PATTERN_UPPERFIRST'). tag(d1_s1_t1_1).
   token_tag(d1_s1_t1, d1_s1_t1_1). tag_base(d1_s1_t1_1. 'word_jan').
token(d1_s1_t2). token_orth(d1_s1_t2, 'word_from').
   token_pattern(d1_s1_t2, 'PATTERN_LOWERCASE'). tag(d1_s1_t2_1).
   token_tag(d1_s1_t2, d1_s1_t2_1). tag_base(d1_s1_t2_1. 'word_2').
token(d1_s1_t3). token_orth(d1_s1_t3, 'word_warsaw').
   token_pattern(d1_s1_t3, 'PATTERN_UPPERFIRST'). tag(d1_s1_t3_1).
   token_tag(d1_s1_t3, d1_s1_t3_1). tag_base(d1_s1_t1_1. 'word_warszawa').

token_after_token(d1_s1_t1, d1_s1_t2). token_after_token(d1_s1_t2, d1_s1_t3).

annotation(d1_a1). annotation_type(d1_a1, 'person_nam').
   annotation_range(d1_a1, d1_s1_t1, d1_s1_t1).
annotation(d1_a2). annotation_type(d1_a2, 'city_nam').
   annotation_range(d1_a2, d1_s1_t3, d1_s1_t3).

relation(d1_a1, d1_a2, 'origin').
```

Fig. 2. A sample sentence "Jan z Warszawy" (Eng. *John from Warsaw*) represented as a set of predicates in first-order logic

Extended Predicates have the form of rules utilizing basic predicates. The role of these predicates is to simplify the search space by shortening the chain of predicates. For instance predicate next_orth(A, B, C) is a shortcut for token_after_token(A, B) and token_orth(B, C).

- next_orth(token-1, token-2, word) — token-2 has orthographic form word and appears directly after token-1,

      next_orth(A, B, C) :-
      token_after_token(A, B),
      token_orth(B, C).

- next_pattern(token-1, token-2, pattern) — token-2 matches pattern and appears directly after token-1,
- prev_orth(token-1, token-2, word) — token-2 has orthographic form word and appears directly before token-1,
- prev_pattern(token-1, token-2, pattern) — token-1 matches pattern and appears directly before token-2.

5.2 ILP System

There are several available ILP systems, like Aleph [1], Foil [12], Golem [13], Progol [14], etc. We have decided to use Aleph because of its comprehensive documentation and flexible configuration options.

In the preliminary experiment we set the configuration to `i=8` (upper bound on layers of new variables), `clauselength=8` (upper bound on number of literals in clause), `nodes=320000` (upper bound on the nodes to be explored), `minpos=2` (lower bound on the number of positive examples to be covered by a clause), `noise=10` (upper bound on number of negative examples covered by clause). This configuration allowed to generate rules describing a continuous sequence of up to 7 tokens or several shorter sequences around annotations. As we do not apply any advanced control of space exploration the limitations were required to make the search feasible. `minpos` was used to eliminate rules covering only single examples and `noise` was used to give tolerance for errors in the data.

5.3 Evaluation

After applying inductive logic programming we generated 276 rules (110 for *location*, 52 for *affiliation*, 38 for *composition*, 27 for *neighbourhood*, 21 for *creator*, 12 for *origin* and 4 for *nationality*) — see sample rule for *origin* on Figure 3. The automatically generated rules outperformed manually created rules in almost all test cases. Only handcrafted rules for *origin* were better by only 5% of F-measure on testing set. Rules for *nationality* were also better but due to low number of relations it is a matter of only 2 correctly recognized instances.

```
relation(A,B,origin) :-
   annotation_range(B,C,D),
   prev_orth(E,C,word_w),                // word_w means ''in''
   annotation_range(A,F,G),
   next_orth(G,H,meta_BRACKET_LEFT),
   next_orth(H,I,word_ur),               // word_ur stands for ''born''
   next_orth(I,J,meta_DOT),
   next_pattern(J,K,'PATTERN_NUM'),
   next_pattern(K,L,'PATTERN_LOWERCASE').
```

Fig. 3. A sample rule in first-order logic which recognize *origin* relation between two proper names and covers 32 positives and 8 negative examples from the training set

Evaluation on the held-out set showed two common errors: problem of rules generality and large number of false positives.

The rule generality can be improved by introducing hyperonyms and Kleene operator. Both modification will cause explosion of rule combination so it will require a better search space control. For example, if we allow long jumps over sequence of token we should constrain the jumps to one direction to avoid "jumping back and forth". The other way to limit the search space is to add constrain that every token

Table 3. Results for automatic rule induction

Set Relation	Training			Held-out			Testing		
	P [%]	R [%]	F [%]	P [%]	R [%]	F [%]	P [%]	R [%]	F [%]
affiliation	91.97	93.09	**92.53**	53.33	39.34	**45.28**	43.79	49.63	**46.53**
alias	84.52	81.61	**83.04**	93.75	45.45	**61.22**	62.86	32.84	**43.14**
composition	92.91	93.91	**93.40**	57.45	54.00	**55.67**	40.63	44.83	**42.62**
creator	85.11	85.11	**85.11**	37.04	32.26	**34.48**	30.77	12.90	**18.18**
location	84.87	92.77	**88.64**	50.00	51.28	**50.63**	31.30	42.60	**36.09**
nationality	100.00	64.29	**78.26**	0.00	0.00	**0.00**	0.00	0.00	**0.00**
neigbourhood	84.04	80.61	**82.29**	20.00	10.71	**13.95**	12.00	9.38	**10.53**
origin	91.18	88.57	**89.86**	87.50	58.33	**70.00**	48.28	53.85	**50.91**

referred in the rule can be constrained with only one predicate, i.e. for given tokens A and B rule can contain only one of the following predicates next_orth(A, B, ?), next_base(A, B, ?), next_pattern(A, B, ?) or next_hyperonym(A, B, ?). Also the length of token sequences can be shortened by discarding infrequent words.

The deeper analysis of false positives showed, that only a small part of them are missing annotations (less than 10%). The majority number of false positives is caused by the fact that some of the generated rules have a form of two independent patterns for contexts surrounding source and target annotation — there is no path or dependency in the rules between them. In sentences with more than two annotations such rules recognize relations between every combination of the annotations. To solve this problem some constrains on the distance between annotations or on the context between annotations will be required.

6 Summary

In the paper we presented a set of predicates which can be used to encode the task of semantic relation recognition between proper names in first-order logic. Then we applied ILP method and compared the automatically generated rules with hand-crafted rules. The rules were evaluated on the KWPr corpus.

Despite the background knowledge utilize very simple patterns (continuous sequences of orthographic forms, base forms and orthographic patterns) automatically induced rules outperformed the manually created rules (developed with time limitation). However, the performance of generated rules can be improved in several ways, what will done in future work.

References

1. Srinivasan, A.: The Aleph Manual (2006),
   http://www.cs.ox.ac.uk/activities/machlearn/Aleph/aleph.html
2. Linguistic Data Consortium (LDC). ACE (Automatic Content Extraction) English Annotation Guidelines for Relations (2008)

3. Pyysalo, S., Ohta, T., Tsujii†, J.: Overview of the Entity Relations (REL) supporting task of BioNLP Shared Task 2011. In: Proceedings of BioNLP Shared Task 2011 Workshop, June 24, pp. 83–88. Association for Computational Linguistics, Portland (2011)
4. Marciniak, M., Mykowiecka, A.: Automatic processing of diabetic patients' hospital documentation. In: Annual Meeting of the ACL (2007)
5. Patwardhan, S., Riloff, E.: Learning Domain-Specific Information Extraction Patterns from the Web. In: ACL 2006 Workshop on Information Extraction Beyond the Document (2006)
6. Califf, M.E.: Relational learning techniques for natural language information extraction. Doctor of philosophy, The University of Texas at Austin (1998)
7. Freitag, D.: Machine learning for information extraction in informal domains. Doctor of philosophy. Carnegie Mellon University (1998)
8. Wróblewska, A., Woliński, M.: Preliminary Experiments in Polish Dependency Parsing. In: Bouvry, P., Kłopotek, M.A., Leprévost, F., Marciniak, M., Mykowiecka, A., Rybiński, H. (eds.) SIIS 2011. LNCS, vol. 7053, pp. 279–292. Springer, Heidelberg (2012)
9. Marcińczuk, M., Janicki, M.: Optimizing CRF-Based Model for Proper Name Recognition in Polish Texts. In: Gelbukh, A. (ed.) CICLing 2012, Part I. LNCS, vol. 7181, pp. 258–269. Springer, Heidelberg (2012)
10. Broda, B., Marcińczuk, M., Maziarz, M., Radziszewski, A., Wardyński, A.: KPWr: Towards a Free Corpus of Polish. In: Proceedings of the 8th ELRA Conference on Language Resources and Evaluation LREC 2012, Istanbul, Turkey (2012)
11. Marcińczuk, M., Stanek, M., Piasecki, M., Musiał, A.: Rich Set of Features for Proper Name Recognition in Polish Texts. In: Bouvry, P., Kłopotek, M.A., Leprévost, F., Marciniak, M., Mykowiecka, A., Rybiński, H. (eds.) SIIS 2011. LNCS, vol. 7053, pp. 332–344. Springer, Heidelberg (2012)
12. Quinlan, J.R., Cameron-jones, R.M.: FOIL: A Midterm Report. In: Brazdil, P.B. (ed.) ECML 1993. LNCS, vol. 667, pp. 3–20. Springer, Heidelberg (1993)
13. Muggleton, S., Feng, C.: Efficient induction in logic programs. In: Muggleton, S. (ed.) Inductive Logic Programming, pp. 281–298. Academic Press (1992)
14. Muggleton, S.: Inverse Entailment and Progol. New Generation Computing Journal 13, 245–286 (1995), http://www.doc.ic.ac.uk/~shm/progol.html

Actionable Clause Detection
from Non-imperative Sentences in Howto Instructions:
A Step for Actionable Information Extraction

Jihee Ryu[1], Yuchul Jung[2], and Sung-Hyon Myaeng[1]

[1] Division of Web Science and Technology, KAIST, Daejeon, Korea
[2] Future Internet Service Research Team, ETRI, Daejeon, Korea
{jiheeryu,myaeng}@kaist.ac.kr, enthusia77@gmail.com

Abstract. Constructing a sophisticated experiential knowledge base for solving daily problems is essential for many intelligent human centric applications. A key issue is to convert natural language instructions into a form that can be searched semantically or processed by computer programs. This paper presents a methodology for automatically detecting actionable clauses in how-to instructions. In particular, this paper focuses on processing non-imperative clauses to elicit implicit instructions or commands. Based on some dominant linguistic styles in how-to instructions, we formulate the problem of detecting actionable clauses using linguistic features including syntactic and modal characteristics. The experimental results show that the features we have extracted are very promising in detecting actionable non-imperative clauses. This algorithm makes it possible to extract complete action sequences to a structural format for problem solving tasks.

Keywords: Problem solving tasks, action sequence extraction, actionable expression detection, how-to instructions.

1 Introduction

People's experiential problem solving knowledge is useful in various daily situations especially if it becomes available immediately on the spot. When having to go to a decent local food restaurant in a strange city, for example, a number of questions may arise: how should I choose a restaurant that goes well with my appetite, what's the best transportation to get there on time, which segment of the bus line is optimal for avoiding traffic jam during the rush hour, how do I buy a bus ticket without cash, etc.

Such knowledge has been made available online in digital form and has been increasing with the help of personal blogs and Web 2.0 sites, e.g., QA and wiki pages. Due to the sheer amount of such knowledge, which keeps increasing by the second, and the need to get such knowledge quickly on the spot, it becomes more and more important to make how-to instructions in natural language into the form that can be processed by a computer program or searched accurately and understood clearly by humans. For instance, an instructional sentence *"Clean the bowl completely with mineral spirits on a rag"* can be converted into the following manner: {ACTION: clean, TARGET: bowl, INSTRUMENT: mineral spirits on a rag}.

P. Sojka et al. (Eds.): TSD 2012, LNCS 7499, pp. 272–281, 2012.

Converting human-generated problem-solving instructions written in English into a structured format is very challenging, since the instructions are written in different styles at different levels of abstraction, and also by various people with different educational and cultural backgrounds. It is more so when the domain for problem solving is not limited. One important characteristic of this task, however, is that how-to instructions are comprised of instructional steps, which form a sub-language. For this reason, the task is more amenable to automation and practical than the general-purpose, domain-independent information extraction problem.

The steps in how-to instructions are mainly written in an imperative form, making the task of identifying key components like ACTION, TARGET, INSTRUMENT, TIME and PLACE somewhat easier to deal with than general sentences. Lau et al. [1] tackled the problem of interpreting imperative sentences in instructional steps for Web commands and labeling ACTION, VALUE, and TARGET TYPE. Jung et al. [2,3] attempted to identify ACTION, OBJECT, TIME, and PLACE from imperative sentences in instructional steps.

Performable actions are usually assumed to be expressed explicitly in the form of "do this" in an imperative sentence. However, they are also implicitly expressed in a non-imperative style, as shown in the following examples.

(1) So you should *change your cover letter* to reflect that.
(2) I encourage you to *try out for a sport*.
(3) You need to *practice twice a week*.
(4) You can *find articles about Silly Bandz* in magazines and newspaper.

Each sentence above contains a performable action with a different degree of importance to achieving a goal. When a sentence or expression contains a performable action, we call it actionable.

This paper focuses on the problem of automatically detecting actionable expressions in instructions. Sentences in instructional steps are segmented into expressions so that the decision is made on the smallest possible units. Because of the variations in instructional steps, it is important to ensure that all performable actions are captured. We analyzed a variety of instructions and identified syntactic and modal features that characterize actionable expressions, which are then used for classification of expressions into actionable and non-actionable classes.

Our contribution of this paper lies in the following: 1) formulation of the concept of actionable clause detection for how-to instructions, which helps finding required actions for completing a problem solving task; 2) an introduction to features and their generation methods for actionable clause detection, which are shown to be highly effective in identifying actionable expressions; 3) experiments that show the values of feature types in actionable expression detection as well as the very high overall performance we obtained, which shows feasibility and provides a key step for building a full function of action information extraction from how-to instructions across all domains.

The rest of the paper is organized as follows. We begin with a brief overview of related work in Section 2. Section 3 describes various features for actionable expression detection. We report on the experiments and present the results in Section 4. We then close with a conclusion and future work.

2   Related Work

Actionable information extraction in how-to instruction is a relatively new task where natural language processing and text mining techniques play a significant role. Although there is little prior work for the specific problem addressed in this paper, some related work exists that attempts to understand and interpret action information in email, dialogue, weblogs, and how-to instructions. Interestingly, most approaches to process how-to instructions undergo difficulties in tackling non-imperative sentences and their interpretations for inferring actionable sentences.

Detecting action item in email [4,5] and dialogue structure [6,7] was attempted in the past. To understand email through detection of action items, Cohen et al. [4] described an ontology of 'speech acts' that subsumes action items. The ontology consist of activity verbs (e.g., propose, request, commit, and refuse) and their associated nouns (including meetings, and short-term/long-term activities). It treats each email as a learning instance with the speech acts class-label. Meanwhile, Bennett and Carbonell [5] considered whether accuracy can be increased by marking the specific sentences and phrases of interest rather than in the unit of documents. Finer-grained sentence level detection outperformed coarse-grained document level detection.

For action item detection in dialogues, Purver et al. [6] employed individual classifiers to detect a set of distinct action item dialogue act (AIDA) classes (i.e. task description, timeframe, ownership and agreement) in a small corpus of simulated meetings. They extended the problem into open-domain conversational speech [7]. While the detection problem is difficult, they improve accuracy by considering local dialog structure. Especially, multiple independent sub-classifiers are used to build a hierarchical classifier which exploits the local dialog structure and utterance features, such as n-gram, utterance (durational and locational features), prosodic, time expression tags, general dialogue act class from the ICSI-MRDA annotation [8], and context (e.g. preceding 5 utterances).

Park et al. [9] proposed a method of detecting experience-revealing sentences from weblogs. They observed that experience-revealing sentences have a certain linguistic style and employed various linguistic features (including tense, mood, aspect, modality, experiencer, and verb classes) for the detection task. As a key enabler for that, they automatically constructed activity/event verb lexicon based on Vendler's theory [10] and statistics utilizing a web search engine.

As a recent approach to understand and interpret human-written instructions, Lau et al. [1] compared three different methods: keyword-based, grammar-based, and machine learning based methods. Their interpreted results were represented as a command, C = (A: the type of action to perform, V: the textual parameter to an action, T: the target on which to act). They considered only imperative sentences which cover more than 50% of all data, and their interpretation modules were limited to sentences that can be converted into a tuple (A, V, T) format. It was reported that their interpreters fail to correctly interpret many of the instructions due to noisiness and incompleteness of human-written instructions. Moreover, their task was limited to the domain of web tools consisting of only 43 tasks.

Jung et al. [2,3] successfully extracted activity knowledge from how-to instructions. Tenorth et al. [11] were able to extract complex task descriptions from the web and translate them into executable robot plans. However, their modules' application was limited to imperative statements. In many cases, this is not enough for a successful execution, and the missing pieces of information need to be processed to complete the plans.

Chen and Mooney's model [12] learns how to interpret and follow free-form navigation instructions without prior linguistic knowledge by observing how humans follow instructions. Goldwasser and Roth's model [13] and Branavan et al.'s model [14] learn ways to win by reading game manuals based on machine learning methods. One common motivation is to rid of manual feedback in gaming environments.

This paper focuses on detecting actionable expressions in how-to instructions for the purpose of building full action sequences from the instructions. Unlike previous work, our work deals with non-imperative expressions without any domain or coverage limitations.

3   Features for Actionable Expression Detection

A how-to instruction corresponds to a problem solving task and consists of steps to solve the problem. Steps in each how-to instruction have one or more sentences, and each sentence can be segmented into expression units. An expression unit is defined as a clause that contains at least one verb phrase which also corresponds to a performable action. As such, a how-to instruction is converted into a sequence of steps, each of which has one or more performable action units. In this paper, however, we concentrate on detecting individual expression units in non-imperative sentences, whose results will be essential for the conversion process.

In order to devise a way to detect actionable expressions, we first performed a qualitative analysis on how-to instructions (i.e. wikiHow articles). We found that actionable expressions can be found based on two different types of computable features: syntax and modality. Because how-to instructions have some dominant linguistic characteristics, we composed feature vectors representing the characteristics.

3.1   Syntactic Features

Based on the linguistic characteristics we have obtained by using Stanford PCFG Parser for English [15] and Stanford Dependency Parser [16], we automatically captured the following syntactic features and used them as actionable expression detection classifiers. These primitive features are sometimes used jointly to detect higher level features as can be seen in Section 3.2.

Clause Type. The clause type of an expression unit has a discriminating ability, since a usual subordinate clause represents suppositional information or mentions just a fact, and this indicates that the expression unit is not actionable.

Person. The person of an expression unit indicates whether or not the main subject of an action is the actual reader. If it is not the second person, the expression unit is usually non-actionable.

Auxiliary Verb. An auxiliary verb usually conveys the modality of a sentence such as ability, possibility, and obligation, which in turn help detecting actionable expressions. We identified 12 auxiliary verbs that are linked to the main predicate using the dependency relations in parsing results.

Voice. The voice of an expression unit shows whether the main predicate is expressed actively or passively. In most cases, active voice of the main predicate makes the expression actionable.

Tense. Tense of an expression unit also has a discriminating ability since actionable expressions usually occur with a present tense. Past tense usually indicates events or facts that already occurred whereas future tense indicates events or guesses the occurrence at a later time.

Polarity. Polarity of an expression unit has a discriminating ability since actionable expressions usually occur with affirmative polarity whereas negative polarity usually means a prohibited or irrelative action for the problem solving task.

3.2   Modality Features

Our qualitative analysis of the data set revealed that instructional clauses can be classified into four categories bearing on linguistic modality. We have identified two usual deontic modality types, *obligation, permission*, which are commonly found in non-imperative clauses. In addition, we have identified two epistemic modality types that are also common in instructional clauses, *explanation* and *supposition*.

Obligation. An instructional clause belonging to this category forces the readers into or forbids the readers from following the instruction. This kind of instructional clauses usually contains a deontic auxiliary verb in their expression.

> (5) You *have to* ask about the car.   [actionable]
> (6) You *must* not communicate this.   [non-actionable]

Permission. An instruction belonging to this category gives the reader an option to take a certain action. This kind of instructional clause is usually expressed with an auxiliary verb such as 'can' or 'may'.

> (7) You *can* search for the world weather.   [actionable]
> (8) You *do not have to* buy it.   [non-actionable]

Explanation. An instructional clause belonging to this category provides an interpretation or assertion about a fact or current situation. This kind of clauses is usually expressed in declarative style.

> (9) The cost for delivery is already included.   [non-actionable]
> (10) Soda drinks have been part of our culture.   [non-actionable]

Supposition. An instructional clause belonging to this category adds epistemic modality to a proposition for less certainty. This type of clauses contain auxiliary like 'would' and 'may' but are non-actionable.

(11) You will have access to the world weather.   [non-actionable]
(12) You may not even notice.   [non-actionable]

Once instructional clauses are identified as belonging to one of the four categories, they can be easily classified into either actionable or non-actionable. In other words, those that belong to the *obligation* and *permission* categories are likely to be actionable, while others that belong to the *explanation* and *supposition* categories are not. For higher certainty, additional features such as polarity must be used together.

4 Experiments

In order to evaluate the features introduced in Section 3, we built classifiers for detecting actionable expression units as well as for determining the feature classes. The latter is used to evaluate accuracy of feature generation methods, which may influence accuracy of the actionable expression detection method. Our experiments using machine learning methods are conducted in WEKA [17].

We built a how-to instruction collection by crawling all the article pages in wikiHow. To train and test our classifiers, we sampled 300 articles randomly from all the domains in wikiHow. Our collection contains about 11,000 expression units, 44% of which are non-imperative clauses.[1] To determine actionable units and modality of the expressions, each expression unit was manually tagged by trained annotators for its modality and whether it is actionable or not. The annotators were instructed to consider our purpose of building a problem-solving knowledge base when the decisions were made. The inter-annotator agreements were checked with the Cohen's Kappa [18]. The average agreement rate was 0.902 for modality feature annotation and 0.889 for actionable clause annotation. This indicates that the agreements among annotators were quite high although it was less clear to determine actionable expressions.

4.1 Feature Generation

Our syntactic feature generation methods work on the result of the syntactic parsers we employed. We first ran an experiment for accuracy of generating the features by using 30 how-to articles in our collection to check for errors from the design of our method or the parsers. The result of automatic generation methods was compared to that of manual generation. As in Table 1, precision values are very high regardless of the features with a possible exception of voice. Note that performance of auxiliary verb detection is the weighted sum of the binary classifications for 12 types of auxiliary verbs.

For the generation of modality features, we trained four binary classifiers using the syntactic features generated automatically. We used Naive Bayes (NB), Decision Trees

---

[1] The dataset with human annotations is available for research purpose in
http://jihee.kr/research/non-imperative/

Table 1. Syntactic feature generation

Feature type	Precision	Recall	F1
Clause Type	0.994	0.856	0.920
Person	0.961	0.880	0.917
Auxiliary Verbs	0.956	0.861	0.905
Voice	0.886	0.870	0.876
Tense	0.972	0.903	0.936
Polarity	0.992	0.992	0.992

(DT), and Support Vector Machines (SVM) as learning algorithms. As in Table 2, the modality features were also generated with very high accuracy for all the algorithms. The high performance for the feature generation methods allowed us to give high credibility to the result of the actionable expression detection method, which is our ultimate goal in this paper. The second column in Table 2 shows the number of expression units containing each feature out of a total of 4,898 expression units in our test set.

Table 2. Modality feature generation

Feature type	# Clauses	F1(NB)	F1(DT)	F1(SVM)
Obligation	188	0.932	0.956	0.966
Permission	698	0.886	0.942	0.950
Explanation	1742	0.957	0.994	0.998
Supposition	2350	0.975	0.991	0.992

4.2   Actionable Expression Detection

We aim to validate the usefulness of the proposed features. We compared the performance in terms of F1 measure. Table 3 shows the overall performance of using different feature types. Using syntactic features alone, the performance was already quite high. The performance was improved with the addition of the generated modality features. The overall result indicates that using the combination of syntactic features and the modality features is very effective in detecting actionable expressions in how-to instructions. To the best of our knowledge, it is the first time to define and tackle this problem, and it turned out that within the dataset, the problem is quite manageable if we use the selected features.

Table 3. Actionable meaning detection

Trained features	F1(NB)	F1(DT)	F1(SVM)
Syntactic features	0.933	0.942	0.948
+ Modality features	0.862	0.963	0.966

It was observed that when the modality features were added to the DT classifier, first division in forming a decision tree was done with the modality features and then the next by syntactic features. It indicates that modality features have higher discriminating power. We conjecture that even though modality features were derived from syntactic features, using *macro* features in addition to *micro* features (i.e. syntactic ones here) contributes to increased effectiveness.

4.3   Failure Analysis

We reviewed misclassified expression units and analyzed why they failed to detect actionable clauses. First, one obvious reason was the dynamic expression styles of actionable clauses. For instance, the clause *"Men must wear long pants"* was classified as non-actionable. We assumed that the model had been trained by the evidence that most actionable clauses were expressed with a second person subject, but this actionable clause is expressed with a third person subject. Moreover, similar syntactic characteristics make it difficult to build a distinct decision boundary. For instance, the clause *"You can be misunderstood"* was classified as actionable because 'can' is usually expressed as *permission*. However, human annotators interpreted it as *supposition*, and it should be a non-actionable clause.

Second, the failure of correct syntactic feature generation has resulted in incorrect detection. For instance, a clause starting with an additional prepositional phrase sometimes caused an error in person feature generation. To make matters worse, the clause like *"Your should keep your pace"* contain a critical typographical error. Human readers can comprehend the intended meaning and subject, but the feature generation module could not process it properly. The worst case was a radically shortened expression like *"You can!"*, from which our module could get almost no evidence.

Third, the subtleness of an actionable clause sometimes induces a different opinion and makes the decision subjective. For instance, the clause *"You should also consider the strength of the router's signal"* is tagged as non-actionable by two annotators, but the third annotator and our trained model regarded it as actionable. The two annotators interpreted it as a cognitive process and not containing an action to take physically while it has usual characteristics of an actionable clause. Our trained model failed to do a perfect job in actionable clause detection due to the aforementioned reasons, but we believe some of them are also somewhat difficult for human to make a correct decision.

5   Conclusion

We introduced the actionable clause detection problem as an essential task for building a problem solving knowledge base from how-to instructions. Detecting actionable clause is challenging because many clauses in instructional steps are non-imperative sentences. Viewing the detection task as a classification problem, we focused on generation and examination of various linguistic features based on syntax and modality. For modality, we defined four features that determine the degree of freedom to act. The experimental results show our proposed features are very effective for actionable clause detection. Each feature type plays a mutually complementary role so that the strong classification boundary can be built.

Our effort for identifying key features in detecting actionable non-imperative sentences was found to be very successful with a relatively cleaned data. The problem we tackled, for the first time to the best of knowledge, turned out to be quite manageable with the features. However, there still remain some issues to be resolved such as anaphora resolution and implicit meaning/intent resolution. The next step would be to investigate how to extract key elements from actionable clauses, such as targets and instruments as well as actions, so that it becomes more amenable to provide solutions or recommendable actions for a given problem at hand.

Acknowledgements. This research was supported by WCU (World Class University) program under the National Research Foundation of Korea and funded by the Ministry of Education, Science and Technology of Korea (Project No: R31-30007). We also thank the anonymous reviewers and the KAIST Activity Research Group for their invaluable feedback on our work, and Hwon Ihm for his helpful comments that lead to significantly better presentations.

References

1. Lau, T., Drews, C., Nichols, J.: Interpreting Written How-To Instructions. In: 21st International Joint Conference on Artificial Intelligence, pp. 1433–1438 (2009)
2. Jung, Y., Ryu, J., Kim, K., Myaeng, S.H.: Automatic Construction of a Large-Scale Situation Ontology by Mining How-to Instructions from the Web. Web Semantics: Science, Services and Agents on the World Wide Web 8(2-3), 110–124 (2010)
3. Ryu, J., Jung, Y., Kim, K., Myaeng, S.H.: Automatic Extraction of Human Activity Knowledge from Method-Describing Web Articles. In: 1st Workshop on Automated Knowledge Base Construction, pp. 16–23 (2010)
4. Cohen, W.W., Carvalho, V.R., Mitchell, T.M.: Learning to Classify Email into "Speech Acts". In: 2004 Conference on Empirical Methods in Natural Language Processing, pp. 309–316 (2004)
5. Bennett, P.N., Carbonell, J.: Detecting Action-Items in E-mail. In: 28th Annual International ACM SIGIR Conference on Research and Development in Information Retrieval, pp. 585–586 (2005)
6. Purver, M., Ehlen, P., Niekrasz, J.: Detecting Action Items in Multi-party Meetings: Annotation and Initial Experiments. In: Machine Learning for Multimodal Interaction, pp. 48–59 (2006)
7. Purver, M., Dowding, J., Niekrasz, J., Ehlen, P., Noorbaloochi, S., Peters, S.: Detecting and Summarizing Action Items in Multi-Party Dialogue. In: 8th SIGdial Workshop on Discourse and Dialogue, pp. 18–25 (2007)
8. Shriberg, E., Dhillon, R., Bhagat, S., Ang, J., Carvey., H.: The ICSI Meeting Recorder Dialog Act (MRDA) Corpus. In: 5th SIGdial Workshop on Discourse and Dialogue, pp. 97–100 (2004)
9. Park, K.C., Jeong, Y., Myaeng, S.H.: Detecting Experiences from Weblogs. In: 48th Annual Meeting of the Association for Computational Linguistics, pp. 1464–1472 (2010)
10. Vendler, Z.: Linguistics in Philosophy. Cornell University Press (1967)
11. Tenorth, M., Nyga, D., Beetz, M.: Understanding and Executing Instructions for Everyday Manipulation Tasks from the World Wide Web. In: 2010 IEEE International Conference on Robotics and Automation, pp. 1486–1491 (2010)

12. Chen, D.L., Mooney, R.J.: Learning to Interpret Natural Language Navigation Instructions from Observations. In: 25th AAAI Conference on Artificial Intelligence, pp. 859–865 (2011)
13. Goldwasser, D., Roth, D.: Learning from Natural Instructions. In: 22nd International Joint Conference on Artificial Intelligence, pp. 1794–1800 (2011)
14. Branavan, S., Silver, D., Barzilay, R.: Learning to Win by Reading Manuals in a Monte-Carlo Framework. In: 49th Annual Meeting of the Association for Computational Linguistics, pp. 268–277 (2011)
15. Klein, D., Manning, C.D.: Accurate Unlexicalized Parsing. In: 41st Annual Meeting of the Association for Computational Linguistics, pp. 423–430 (2003)
16. De Marneffe, M.C., MacCartney, B., Manning, C.D.: Generating Typed Dependency Parses from Phrase Structure Parses. In: 5th International Conference on Language Resources and Evaluation, pp. 449–454 (2006)
17. Hall, M., Frank, E., Holmes, G., Pfahringer, B., Reutemann, P., Witten, I.H.: The WEKA Data Mining Software: An Update. SIGKDD Explorations 11(1), 10–18 (2009)
18. Carletta, J.: Assessing Agreement on Classification Tasks: The Kappa Statistic. Computational Linguistics 22(2), 249–254 (1996)

Authorship Attribution: Comparison of Single-Layer and Double-Layer Machine Learning

Jan Rygl and Aleš Horák

Natural Language Processing Centre
Faculty of Informatics, Masaryk University, Brno, Czech Republic
{xrygl,hales}@fi.muni.cz

Abstract. In the traditional authorship attribution task, forensic linguistic specialists analyse and compare documents to determine who was their (real) author. In the current days, the number of anonymous documents is growing ceaselessly because of Internet expansion. That is why the manual part of the authorship attribution process needs to be replaced with automatic methods. Specialized algorithms (SA) like delta-score and word length statistic were developed to quantify the similarity between documents, but currently prevailing techniques build upon the machine learning (ML) approach.

In this paper, two machine learning approaches are compared: Single-layer ML, where the results of SA (similarities of documents) are used as input attributes for the machine learning, and Double-layer ML with the numerical information characterizing the author being extracted from documents and divided into several groups. For each group the machine learning classifier is trained and the outputs of these classifiers are used as input attributes for ML in the second step.

Generating attributes for the machine learning in the first step of double-layer ML, which is based on SA, is described in detail here. Documents from Czech blog servers are utilized for empirical evaluation of both approaches.

Keywords: double layered machine learning, authorship attribution, similarity of documents.

1 Introduction

Since medieval times, authorship attribution is one of the topics of forensic linguistics. The task is to ascertain whether two documents were written by the same author. Despite the fact that authors frequently hide behind pseudonyms, it is usually possible to obtain a collection of documents that can be with certainty ascribed to the analyzed author. A credible method for judging the authorship of two documents is needed to identify anonymous writers [1,2].

Authorship identification studies were initiated by the work of Mosteller and Wallace [3]. Nowadays, machine learning, information retrieval, and natural language processing are utilized to solve the authorship identification problem. One of the most important approaches are similarity-based models (proposed by [4]).

The process of identification consists of two steps: 1) Document properties are quantified and authors' characteristics (similarities according to these properties) are

P. Sojka et al. (Eds.): TSD 2012, LNCS 7499, pp. 282–289, 2012.

calculated. 2) The characteristics are utilized as attributes for the machine learning (ML). A classifier is trained for the decision whether two documents are written by the same author.

In this paper, two machine learning approaches are compared: Single-layer ML, where the results of heuristic algorithms are used as input attributes for the machine learning, and Double-layer ML with the heuristics being replaced by an another layer of ML. Two-layered ML has been used in other fields, such as biology [5], but it has never been applied to authorship identification. The experiments and evaluation of such application are presented further in this text.

2 Machine Learning

Support Vector Machines method (*SVM*) achieves the best results in comparisons of machine learning approaches to authorship verification problem [6]. For the purposes of experiments the *Orange* [7] implementation of *SVM* was selected. *Orange* is a data mining tool and its implementation of *SVM* is based on the *LibSVM* [8] and *LIBLINEAR* [9] libraries.

The model was trained to estimate the probability of the same authorship instead of class labels (SVM predicts only class label without probability information. The LIB-SVM implementation is extending SVM to give probability estimates [10]). The resulting classifier indicates that two documents have the same authorship if the probability of the same authorship is greater or equal to 50 %.

2.1 Single-Layer ML

The single-layer ML process consists of two components: *heuristic algorithms* and *machine learning*.

1. The first component uses five heuristic algorithms (each heuristic represents similarity of two documents based on the selected author's characteristic – these characteristics are described in Section 3). To verify the same authorship of two documents, the heuristics compare differences between documents according to corresponding authors' characteristics and return five values in the interval $\langle 0, 1 \rangle$. The values describe the similarity of two documents (1 means identical documents, 0 is for entirely different authors' styles).
2. The second component, ML, learns a classifier that accepts 5-tuples of numbers. The value returned by the classifier is the probability of the same authorship.

2.2 Double-Layer ML

The double-layer machine learning has three phases: 1) Comparison of attributes of documents, 2) the first layer of ML, and 3) the second layer of ML. The whole process is illustrated by an example in Figure 1.

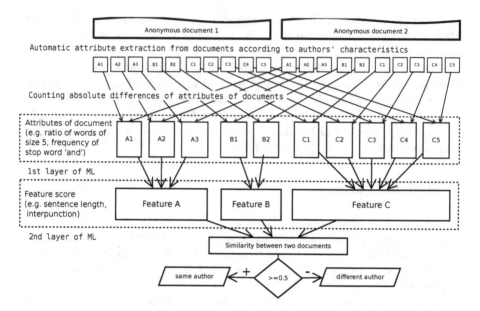

Fig. 1. The process of double-layer ML

1. Five tuples are extracted according to authors' characteristics from both documents. The tuples describe properties of documents and the lengths of the tuples are variable for each authors' characteristics (see Section 3 for details).

   To compare two documents, we are given five tuples for the first document: D_1^1, D_1^2, D_1^3, D_1^4, D_1^5 and five tuples representing the second document: D_2^1, D_2^2, D_2^3, D_2^4, D_2^5. To enumerate the similarity of documents, absolute differences of the corresponding tuples are calculated:

$$D_{1,2}^x = |D_1^x \setminus D_2^x|, x \in \{1, 2, 3, 4, 5\}$$

   where the operation $\setminus$ is defined as: $A \setminus B = \{|a_1 - b_1|, |a_2 - b_2|, \ldots, |a_n - b_n|\}$.

2. Each tuple $D_{1,2}^x$ consists of values representing the similarity of documents. These values are used as attributes for the classifier of the first layer of the machine learning component. The classifiers correspond with the input tuples, therefore there are 5 classifiers (the first phase produces 5 n-tuples of varying length), which return the probability of the same authorship according to 5 authors' characteristics.

3. The second layer of ML is composed of a single classifier that receives five attributes, which are obtained using 5 classifiers from the first layer of ML. Five probabilities of the identical authorship are transformed to the single probability that both documents were written by the same author.

3 Authors' Characteristics

Authors' characteristics are used to quantify the similarity between two documents. Each characteristic associates documents with their resemblance according to one of several criteria. For the ML purposes, the characteristics can be divided into three groups:

1. *Algorithms developed to solve the authorship identification problem:*
   The input consists of two unknown documents and the output is the similarity of their authors. These methods can be utilized as attributes for single-layer machine learning. Since the methods are intended as black box models, replacing the algorithm by machine learning layer is inapplicable. An example of these methods is δ-*score* [4].
2. *Models based on X ($\approx$ 1000 or more) attributes for each document:*
   An absolute differences of two input X-tuples is computed and the resulting tuple is analyzed (e.g. an increasing sum of differences correlates with the decreasing similarity of authors) in the heuristic approach.

   The ML approach requires training data that contain enough examples to cover all attributes. For example, if bigram frequencies are used, the number of new bigrams per document is decreasing with increasing number of documents.
3. *Models with small number of attributes:*
   If an algorithm expects only a few numeric attributes obtained from the document, there are two possible approaches:
   a) the heuristic algorithm can be designed according to the data analysis by an expert in computational linguistics,
   b) a machine learning classifier is trained capturing those connections between attributes which are not necessarily apparent.

In this paper, we focus on the analysis of the third group of authors' characteristics, i.e. tasks processing small number of attributes. The general performances of heuristic approaches (single-layer ML, authors are identical if *sim* $\approx$ 1) and machine learning approaches (double-layer ML) are compared on five frequently used authors' characteristics: Sentence Length, Punctuation, Stoplist words, Word classes, and Basic typography errors.

3.1 Sentence Length

Heuristic algorithm: An average sentence length of the document is computed as a ratio of all words to the number of sentences. The document similarity is expressed as a ratio of the minimum (the lesser) of the two document average sentence lengths and the maximum (the greater) of these sentence lengths.

– Average sentence length $len(doc) = \frac{|doc\ word\ count\,|}{|doc\ sentence\ count\,|}$

Attribute extraction for the first layer of ML: Return 41-tuple, x. item of tuple is counted as a ratio of the number of sentences containing exactly $x - 1$ words to the number of all sentences in the document.

Heuristic similarity: $sim = \frac{min(len(D_1),len(D_2))}{max(len(D_1),len(D_2))}$

2-layer ML atrributes: $\left[\ \frac{|doc\ x\text{-}word\ sentence\ count\,|}{|doc\ sentence\ count\,|}\ \middle|\ x \in \{0,\dots,40\}\ \right]$

3.2 Punctuation

Heuristic algorithm: Sum absolute differences of relative frequencies of punctuation symbols. The frequency is counted as ratio of symbol occurrences in the document to the number of all characters in the document.

- Recognized punctuation symbols $Int = \{\text{'.'}, \text{':'}, \text{','}, \text{';'}, \text{'?'}, \text{'!'}, \text{'-'}, \text{'('}\}$.
- Relative frequency $freq(sym, doc) = \frac{|doc\, sym\, count|}{|doc\, character\, count|}$ $(sym \in Int)$

Attribute extraction for the first layer of ML: Return 8-tuple of relative frequencies of selected punctuation symbols (identical list as list *Int* above).

Heuristic similarity: $\quad sim = 1 - \frac{\sum_{sym \in Int} |freq(sym, D_1) - freq(sym, D_2)|}{2}$

2-layer ML atrributes: $[\, freq(sym, doc) \mid sym \in Int\,]$

3.3 Stoplist Words

Heuristic algorithm: Sum absolute differences of relative frequencies of the stoplist words (the *n*-most frequent words in a language). The frequency is counted as ratio of stoplist occurrences in the document to the number of words in the document.

- Recognized stopwords $S = \{$ 'být', 'v', 'a', 'sebe', 'na', 'ten', 's', 'z', 'že', 'který', 'o', 'mít', 'i', 'do', 'on', 'k', 'pro', 'tento', 'za', 'by', 'moci', 'svuj', 'ale', 'po', 'rok', 'jako', 'však', 'od', 'všechen', 'dva', 'nebo', 'tak', 'při', 'jeden', 'podle', 'Praha', 'jen', 'další', 'jeho', 'aby', 'co', 'český', 'jak', 'veliký', 'nový', 'až', 'už', 'muset', 'než', 'nebýt', 'člověk', 'jenž', 'léto', 'firma', 'první', 'náš', 'také', 'my', 'jejich', 'když', 'před', 'doba', 'chtít', 'jiný', 'mezi', 'ještě', 'já', 'ani', 'cena', 'již', 'jít', 'strana', 'či', 'druhý'$\}$ (the stoplist is from [11]).
- Relative frequency $freq(stop\_w, doc) = \frac{|doc\, stop\_w\, count|}{|doc\, word\, count|}$ $(stop\_w \in S)$

Attribute extraction for the first layer of ML: Return 74-tuple of relative frequencies of stop words (identical list as list *S* above).

Heuristic similarity: $\quad sim = 1 - \frac{\sum_{stop\_w \in S} |freq(stop\_w, D_1) - freq(stop\_w, D_2)|}{2}$

2-layer ML atrributes: $[\, freq(stop\_word, doc) \mid stop\_word \in S\,]$

3.4 Word Classes

Heuristic algorithm: Sum absolute differences of relative frequencies of word classes. The frequency is computed as a ratio of word class occurrences in the document to the number of words in the document.

- Recognized word classes $WC = \{$ 'Noun', 'Adjective', 'Pronoun', 'Cardinal number', 'Verb', 'Adverb', 'Preposition', 'Conjunction', 'Participle', 'Interjection'$\}$
- Relative freq. $freq(word\_class, doc) = \frac{|doc\, word\_class\, count|}{|doc\, word\, count|}$ $(word\_class \in WC)$

Attribute extraction for the first layer of ML: Return 10-tuple of relative frequencies of word classes (identical list as list WC above).

Heuristic similarity: $\quad sim = 1 - \frac{\sum_{w\_class \in WC} |freq(w\_class, D_1) - freq(w\_class, D_2)|}{2}$

2-layer ML atrributes: $[\ freq(w\_class, doc)\ |\ word\_class \in WC\ ]$

3.5   Basic Typography Errors

Heuristic algorithm: Sum absolute differences of relative frequencies of selected typography errors. The frequency is computed as a ratio of error occurrences in the document to the number of characters in the document.

– Recognized typography errors: $Er = \{$'two dots', '4 or more dots', 'two spaces', '3 or more spaces', 'a missing space after punctuation sign', 'incorrect use of hyphen'$\}$.
– Relative frequency $freq(error, doc) = \frac{|doc\ error\ count|}{|doc\ character\ count|}$ $(error \in Er)$

Attribute extraction for the first layer of ML: Return 6-tuple of relative frequencies of selected typography errors (identical list as list Er above).

Heuristic similarity: $\quad 1 - \frac{\sum_{error \in Er} |freq(error, D_1) - freq(error, D_2)|}{2}$

2-layer ML atrributes: $[\ freq(error, doc)\ |\ error \in Er\ ]$

4   Experimental Results

4.1   Test Data

A specialized corpus of Czech documents from the Internet domain was used for the evaluation. The texts are collected from Czech blogs and Internet discussions connected to these blogs. There are 9,565 blog documents and 3,868 discussion documents from servers `blog.ihned.cz`, `aktualne.centrum.cz` and `blog.lupa.cz` (approximately 3,000,000 tokens). The texts are tokenized, split into sentences (a fast algorithm using punctuation, capital letters and rules for numbers is used) and morphologically processed by the Czech tagger *Desamb* [12].

The data were divided into two groups:

– Training data: These documents were exclusively utilized to train the ML classifier. There is 1 to 15 documents per author. 6,000 documents were used to extract 6,000 pairs of documents with the same authorship and 6,000 pairs of documents with different authorship.
– Test data: There are 2,000 test documents and they are different from training data, but the intersection of authors from the training and test data is not empty. The pairs of documents were extracted randomly and the number of positive examples (pairs with the same authorship) and negative examples (pairs with the different authorship) is identical.

4.2 Experiments

The qualities of both (single-layer and double-layer) approaches were measured and compared. The results are displayed in confusion matrices. Two experiments were conducted:

1. Training data: 6,000 positive examples and 6,000 negative examples
   Test data: 12,000 positive examples and 12,000 negative examples
   Results: The 1-layer SVM classifier preferred to classify documents as different instead of the correspondence of authorships (over-learning occurred). This negatively affected the accuracy – only 45 % documents were attributed correctly. The double-layer SVM methods achieved 64 % accuracy and outperformed the first method by 19 %. For details, see Table 1.
2. Training data: 1,000 positive examples and 1,000 negative examples
   Test data: 2,000 positive examples and 2,000 negative examples
   Results: Despite the fact that the data used for the second experiment were six times smaller than the data in the previous test, the results were consistent. The nearly identical results were caused by the SVM algorithm: to train the model only several support vectors are used in computations. For details, see Table 2.

Table 1. 12,000 training pairs of documents

1-layer SVM			
Correct:	10, 768	*Predicted*	
Wrong:	13, 232	*Same*	*Diff*
Actual	*Same*	0.11	0.39
	Diff	0.16	0.34

2-layer SVM			
Correct:	15, 451	*Predicted*	
Wrong:	8, 549	*Same*	*Diff*
Actual	*Same*	0.29	0.21
	Diff	0.15	0.35

Table 2. 2,000 training pairs of documents

1-layer SVM			
Correct:	1, 767	*Predicted*	
Wrong:	2, 233	*Same*	*Diff*
Actual	*Same*	0.1	0.4
	Diff	0.16	0.34

2-layer SVM			
Correct:	2, 596	*Predicted*	
Wrong:	1, 404	*Same*	*Diff*
Actual	*Same*	0.29	0.21
	Diff	0.15	0.35

5   Conclusion and Future Work

The results indicate that minimization of choices that are not based on the data improve performance of machine learning in the authorship identification task. For cases, where sufficient amount of training data are available, we recommend to use machine learning techniques instead of utilization of heuristics. Double-layer ML outperformed single-layer ML by 19 %: 64 % of document pairs were classified correctly by the double-layer ML.

Many authors' characteristics are not suitable for ML methods, therefore we plan to conduct experiments combining single-layer and double-layer approaches. This is possible, since outputs of authors' characteristics are identical in both approaches – all methods quantify similarity of documents as the likelihood that two documents were written by the same author.

Acknowledgements. This work has been partly supported by the Ministry of the Interior of CR within the project VF20102014003.

References

1. Abbasi, A., Chen, H.: Applying authorship analysis to extremist-group web forum messages. IEEE Intelligent Systems 20, 67–75 (2005)
2. Chen, H., Atabakhsh, H., Zeng, D., et al.: COPLINK: visualization and collaboration for law enforcement. In: Proceedings of the 2002 Annual National Conference on Digital Government Research. dg.o 2002, pp. 1–7. Digital Government Society of North America (2002)
3. Mosteller, F., Wallace, D.L.: Inference and Disputed Authorship: The Federalist. Addison-Wesley (1964)
4. Burrows, J.: Delta': a measure of stylistic authorship 1. Literary and Linguistic Computing 17, 267–287 (2002)
5. Kim, E., Song, Y., Lee, C., Kim, K., Lee, G.G., Yi, B.K., Cha, J.: Two-phase learning for biological event extraction and verification 5, 61–73 (2006)
6. Koppel, M., Schler, J., Argamon, S.: Computational methods in authorship attribution. J. Am. Soc. Inf. Sci. Technol. 60, 9–26 (2009)
7. Curk, T., Demšar, J., Xu, Q., Leban, G., Petrovič, U., Bratko, I., Shaulsky, G., Zupan, B.: Microarray data mining with visual programming. Bioinformatics 21, 396–398 (2005)
8. Chang, C., Lin, C.: LIBSVM: a library for support vector machines (2001), http://www.csie.ntu.edu.tw/~cjlin/libsvm
9. Fan, R.E., Chang, K.W., Hsieh, C.J., Wang, X.R., Lin, C.J.: LIBLINEAR: A library for large linear classification. Journal of Machine Learning Research 9, 1871–1874 (2008)
10. Huang, T.K., Weng, R.C., Lin, C.J.: Generalized Bradley-Terry Models and Multi-class Probability Estimates. Journal of Machine Learning Research 7, 85–115 (2006)
11. NLP Centre: (Czech lemma stoplist), http://nlp.fi.muni.cz/cs/Stoplist_zakladnich_tvaru
12. Šmerk, P.: K počítačové morfologické analýze češtiny (in Czech, Towards Computational Morphological Analysis of Czech). Ph.D. thesis, Faculty of Informatics, Masaryk University (2010)

Key Phrase Extraction
of Lightly Filtered Broadcast News

Luís Marujo[1,2,4], Ricardo Ribeiro[2,3], David Martins de Matos[2,4], João P. Neto[2,4],
Anatole Gershman[1], and Jaime Carbonell[1]

[1] LTI/CMU, USA
[2] L2F - INESC ID Lisboa, Portugal
[3] Instituto Universitário de Lisboa (ISCTE-IUL), Portugal
[4] Instituto Superior Técnico - Universidade Técnica de Lisboa, Portugal
{luis.marujo,ricardo.ribeiro,david.matos,joao.neto}@inesc-id.pt
{anatoleg,jgc}@cs.cmu.edu

Abstract. This paper explores the impact of light filtering on automatic key
phrase extraction (AKE) applied to Broadcast News (BN). Key phrases are words
and expressions that best characterize the content of a document. Key phrases
are often used to index the document or as features in further processing. This
makes improvements in AKE accuracy particularly important. We hypothesized
that filtering out marginally relevant sentences from a document would improve
AKE accuracy. Our experiments confirmed this hypothesis. Elimination of as
little as 10% of the document sentences lead to a 2% improvement in AKE
precision and recall. AKE is built over MAUI toolkit that follows a supervised
learning approach. We trained and tested our AKE method on a gold standard
made of 8 BN programs containing 110 manually annotated news stories. The
experiments were conducted within a Multimedia Monitoring Solution (MMS)
system for TV and radio news/programs, running daily, and monitoring 12 TV
and 4 radio channels.

Keywords: Keyphrase extraction, Speech summarization, Speech browsing,
Broadcast News speech recognition.

1 Introduction

With the overwhelming amount of News video and audio information broadcasted daily
on TV and radio channels, users are constantly struggling to understand the big picture.
Indexing and summarization provide help, but they are hard for multimedia documents,
such as broadcast news, because they combine several sources of information, e.g.
audio, video, and footnotes. We use light filtering to improve the indexing, where AKE
is a key element.

AKE is a natural language procedure that selects the most relevant phrases (key
phrases) from a text. The key phrases are phrases consisting of one or more significant
words (keywords). They typically appear verbatim in the text. Light filtering removes
irrelevant sentences, providing a more adequate search space for AKE. AKE is
supposed to represent the main concepts from the text. But even for a human being, the

P. Sojka et al. (Eds.): TSD 2012, LNCS 7499, pp. 290–297, 2012.

manual selection of key phrases from a document is context-dependent and needs to rely more on higher-level concepts than low-level features. That is why filtering improves AKE.

In general, AKE consists of two steps [9,18,19]: candidate generation and filtering of the phrases selected in the candidate generation phrase. Several AKE methods have been proposed. Most approaches only use standard information retrieval techniques, such as N-gram models [3], word frequency, TFxIDF (term frequency × inverse document frequency) [17], word co-occurrences [12], PAT tree or suffix-based for Chinese and other oriental languages [2]. In addition, some linguistic methods, based on lexical analysis [4] and syntactic analysis [7], are used. These methodologies are classified as unsupervised methods [8], because they do not require training data. On the other hand, supervised methods view this problem as a binary classification task, where a model is trained on annotated data to determine whether a given phrase is a key phrase or not. Because supervised methods perform better, we use them in our work. In general, the supervised approach uses machine-learning classifiers in the filtering step (e.g.: C4.5 decision trees [13], neural networks [18]).

All of the above methods suffer from the presence of irrelevant or marginally relevant content, which leads to irrelevant key phrases. In this paper, we propose an approach that addresses this problem through the use of light filtering based on summarization techniques.

A summary is a shorter version of one or more documents that preserves their essential content. Compression Ratio (CR) is the ratio of the length of the removed content (in sentences) to the original length. Light filtering typically involves a CR near 10%. Light filtering is a relaxation to the summarization problem because we just remove the most irrelevant or marginally relevant content. This relaxation is very important because the summarization problem is especially difficult when processing spoken documents: problems like speech recognition errors, disfluencies, and boundaries identification (both sentence and document) increase the difficulty in determining the most important information. This problem has been approached using shallow text summarization techniques such as Latent Semantic Analysis (LSA) [6] and Maximal Marginal Relevance (MMR) [1], which seem to achieve comparable performance to methods using specific speech-related features [15], such as acoustic/prosodic features.

This work here addresses the use of light filtering to improve AKE. The experiments were conducted within a Media Monitoring Solution (MMS) system.

This paper is organized as follows: Section 2 presents the overall architecture; the description of the summarization module included in the MMS system is the core of Section 3, results are described in Section 4, and Section 5 draws conclusions and suggests future work.

2   Overall Architecture

The main workflow of the complete MMS system [11,14], depicted in Figure 1, is the following: a Media Receiver captures and records BN programs from TV and radio.

Then, the transcription is generated and enriched with punctuation and capitalization. Subsequently, each BN program is automatically segmented into several stories. News stories are lightly filtered (90% of the original size or remains unchanged if the number of sentences in the summarized version is less than or equal to 3). The remaining text is passed to the key phrase extraction process. Each news story is topic-indexed or topic-classified. Finally, each news story is stored in a metadata database (DB) with the respective transcription, key phrases, and index, besides program/channel and timing information. A Key phrase Cloud Generator creates/updates 3D key phrase cloud based on the interaction with the Metadata DB and links with the videos that are shown when a user accesses the system. A 3D key phrase cloud is a tag/word cloud, which is a visual representation of the most frequently used words in text data. The most frequent tags are usually displayed in larger fonts in 2D clouds or at the front in 3D clouds (rotating the 3D cloud allow access to the less frequent/relevant tags). Typically, tags are keywords or single words; key phrases extend this concept to several words.

The gray blocks are the focus of our work. A summarization module [16], responsible for the light filtering step, was included in the workflow and its impact on the key phrase extraction module is analyzed.

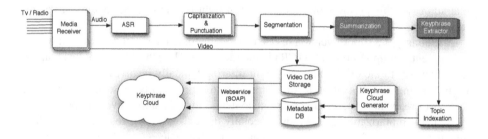

Fig. 1. Component view of the system architecture

3   Key Phrase-Cloud Generation Based on Light Filtering

3.1   Filtering

The automatic filtering step applied in this work is performed by a summarization module that follows a centrality-as-relevance approach. Centrality-as-relevance methods base the detection of the most important content on the determination of the most central passages of the input source(s), considering an adequate input source representation (e.g.: graph, spatial). Although pioneered in the context of text summarization, this kind of approaches has drawn some attention in the context of speech summarization, either by trying to improve them [5] or using them as baseline [10]. Even in text summarization, the number of up-to-date examples is significant.

The summarization model we use [16] does not need training data or additional information. The method consists in creating, for each passage of the input source, a set containing only the most semantically related passages, designated support set. Then, the determination of the most relevant content is achieved by selecting the passages that occur in the largest number of support sets. Geometric proximity (Manhattan, Euclidean, Chebyshev are some of the explored distances) is used to compute semantic relatedness. Centrality (relevance) is determined by considering the whole input source (and not only local information), and by taking into account the presence of noisy content in the information sources to be summarized. This type of representation diminishes the influence of the noisy content, improving the effectiveness of the centrality determination method.

3.2   Automatic Key Phrase Extraction

AKE extracts key concepts. Our AKE process was designed to take into account the extraction of few key phrases (e.g.: 10 used in 3D Key phrase Cloud) and large number of key phrases (e.g.: 30 used for indexing). We privileged precision over recall when extracting fewer key phrases because we want to mitigate visible mistakes in the 3D Key phrases Cloud. On the other hand, recall gains importance when we extract many key phrases because we want to have the best coverage possible. During our experiments, we observed that the most general and at the same time relevant concepts can be directly linked with an index topic (examples: soccer/football $\rightarrow$ sports, PlayStation $\rightarrow$ technology). However, they are frequently captured by the previous methods with low confidence (<50%). Since filtering reduces irrelevant content, it increases the confidence of capturing the best key phrases. The AKE system we use [11], developed for European Portuguese BN, is an extended version of Maui-indexer toolkit [13] (a state-of-art supervised key phrase extraction toolkit), which is in turn an improved version of KEA [19]. Training data is used to train a machine learning classifier (bagging over C4.5 decision tree). The output is a model that uses extracted features to classify whether a word or phrase is a key phrase. The same CR (filtering) is used to train the models and evaluate them at the test sets. This allows the models to be more robust. The Maui-indexer feature extraction process was enriched with the following 5 features: number of characters; the number of named entities using the MorphoAdorner name recognizer; number of capital letters; count of POS tags; and probability of the key phrase in a 4-gram domain model (about 58K unigrams, 700M bigrams, 1.500M trigrams, and 10.000M 4-grams). We have previously demonstrated that these features improved AKE [11].

4   Evaluation

We used a BN gold standard corpus annotated with the corresponding key phrases, created in previous work. The gold standard consists of 8 BN programs transcribed from the European Portuguese ALERT BN database. The news transcriptions were produced by AUDIMUS, an ASR for Portuguese, with low WER (14,56% on average); and punctuated and capitalized automatically using in-house tools. Those news programs were automatically split into a total of 110 news stories. Later, each news story was manually examined to fix segmentation errors. Afterward, one annotator was asked to extract all key phrases that represent a relevant concept in each news story. The gold

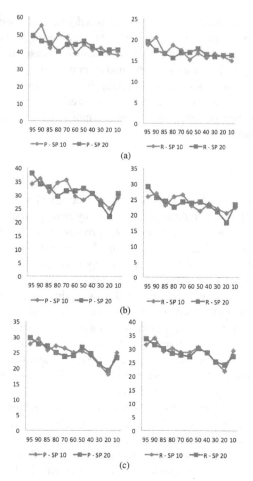

Fig. 2. The percentage of the original text in X-axis vs. AKE metrics in Y-axis. The evaluation performed in the test set used the Manhattan metric, 10% and 20% SSC obtained The Precision and Recall extracting: (a) 10, (b) 20 and (c) 30 key phrases.

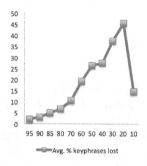

Fig. 3. Avg. key phrase percentage lost in summarization. The results were obtained when extracting 10 keyphrases in the test set using 10% SSC and Manhattan distance.

Table 1. AKE results obtained in the test set using light filtering (p-value ≈ 0.1)

#Key. Extr.	%orig. text	SSC	Dist.Metric	#Key.Ident.	P	R	F1
10	100%	-	-	5.3	53	20.63	29.7
10	90%	20%	chebyshev	4.7	47.00	18.45	26.50
10	90%	10	chebyshev	5.3	53.00	19.57	28.59
10	**90%**	**10%**	**manhattan**	**5.5**	**55**	**20.45**	**29.81**
10	90%	5	manhattan	5.0	45.00	17.88	26.05
10	90%	20	manhattan	5.3	53.00	20.67	29.71
10	90%	10%	minkowski	4.8	48.00	18.27	26.46
10	90%	8	minkowski	4.6	46.00	17.45	25.30
10	90%	20	minkowski	5.1	51.00	18.84	27.52
10	90%	20%	cosine	5.1	51.00	19.34	28.04
10	90%	20%	euclidean	4.8	48	18.67	26.88
10	90%	5	euclidean	5.0	50.00	19.37	27.93
20	100%	-	-	7.4	37	28.21	32.01
20	90%	20%	manhattan	6.8	34	25.45	29.11
20	90%	20	manhattan	5.2	52.00	19.63	28.51
20	90%	10%	minkowski	7.1	35.5	27.74	31.14
20	**90%**	**8**	**minkowski**	**7.6**	**38.00**	**29.08**	**32.95**
20	90%	20%	euclidean	7.5	37.5	28.43	32.34
20	90%	21	euclidean	7.6	37.50	28.13	32.33
30	100%	-	-	9.2	30.67	35.12	32.74
30	90%	10%	manhattan	8.8	29.33	33.75	31.39
30	90%	5	manhattan	8.9	29.67	33.21	31.34
30	90%	20%	minkowski	8.6	28.67	34.48	31.31
30	**90%**	**8**	**minkowski**	**9.5**	**31.67**	**35.99**	**33.69**
30	90%	20%	euclidean	8.8	29.33	32.81	30.97
30	90%	25	euclidean	9.2	30.67	34.57	32.50
40	100%	-	-	10.3	25.75	38.87	30.98
40	90%	10%	manhattan	10.1	25.25	38.44	30.48
40	90%	20%	manhattan	9.3	23.25	35.50	28.10
40	**90%**	**8**	**minkowski**	**10.6**	**26.50**	**40.82**	**32.14**
40	90%	20%	minkowski	9.6	24.00	38.00	29.42
40	90%	10%	euclidean	9.3	23.25	35.64	28.14
40	90%	20%	euclidean	10.3	25.75	38.99	31.02

standard was divided in training (100 news stories containing on average 24 key phrases and 19 sentences) and test set (10 news stories containing on average 29 key phrases and 17 sentences). In our experiments, light filtering improved AKE precision and recall by 2%. We have also tested higher CR (Figure 2) and restricting the summary length to 4 sentences (roughly the average size of a paragraph). However, we did not observed improvements in the results. The average percentage of key phrases lost by the filtering process was less than 5% (Figure 3).

5 Conclusions and Future Work

This paper explores a novel method to improve key phrase extraction from BN by using light filtering. The key phrases are extracted to create a hierarchical 3-layer representation of news. The key phrases of top news are visualized in tag cloud to allow users to skim their content and jump to the most relevant news faster.

Based on the results, we show that light filtering improves automatic key phrase extraction. We included light filtering, constrained to have at least 4 sentences in the summary in the MMS system. This step is done before extracting 10 key phrases of each news story. In addition, we show that even changing the number of key phrases extracted the light filtering still improves the AKE process. We also show that filtering up to 50% of the original size corresponds to about 26% in key phrases loss. That corresponds to less than 5% in terms of AKE performance metrics degradation. This is an important result because we take advantage of the summary shown in the MMS interface to reduce AKE computational resources, such as processing time, while the AKE performance degradation is very low. Nevertheless, we create news summaries at both 10% and 50% CR to use before the AKE and to shown in the MMS interface. At the present time, the MMS interface uses AKE process to identify the 10 top ranked key phrases from top news from 12 TV and 4 Radio channels and generate the 3D key phrase cloud. Although 50% CR seem enough to us, we would like to analyze in future research what percentages of CR users prefer. Alternatively, they could prefer to customize this value based on the amount of time available to interact with the system.

In the future, we plan to augment the centrality-based summarization with AKE.

Acknowledgments. This work was supported by Carnegie Mellon Portugal Program and under FCT grant SFRH/BD /33769/2009.

References

1. Carbonell, J., Goldstein, J.: The Use of MMR, Diversity-Based Reranking for Reordering Documents and Producing Summaries. In: ACM SIGIR 1998, pp. 335–336 (1998)
2. Chien, L.: Pat-tree-based keyword extraction for Chinese information retrieval. In: ACM SIGIR 1997, pp. 50–58. ACM, New York (1997)
3. Cohen, J.D.: Highlights: Language- and Domain-Independent Indexing Terms for Abstracting Automatic. English 46(3), 162–174 (1995)
4. Ercan, G., Cicekli, I.: Using lexical chains for keyword extraction. Information Processing & Management 43(6), 1705–1714 (2007); text Summarization
5. Garg, N., Favre, B., Reidhammer, K., Hakkani-Tür, D.: ClusterRank: A Graph Based Method for Meeting Summarization. In: Interspeech 2009, pp. 1499–1502. ISCA (2009)
6. Gong, Y., Liu, X.: Generic Text Summarization Using Relevance Measure and Latent Semantic Analysis. In: ACM SIGIR 2001, pp. 19–25. ACM (2001)
7. Harabagiu, S., Lacatusu, F.: Topic Themes for Multi-Document Summarization. In: ACM SIGIR 2005, pp. 202–209. ACM (2005)
8. Hasan, K., Ng, V.: Conundrums in unsupervised keyphrase extraction: making sense of the state-of-the-art. In: ACL 2010, pp. 365–373. ACL (2010)
9. Hulth, A., Karlgren, J., Jonsson, A., Boström, H., Asker, L.: Automatic Keyword Extraction Using Domain Knowledge. In: Gelbukh, A. (ed.) CICLing 2001. LNCS, vol. 2004, p. 472. Springer, Heidelberg (2001)

10. Lin, S.H., Yeh, Y.M., Chen, B.: Extractive Speech Summarization – From the View of Decision Theory. In: Proceedings of Interspeech 2010. ISCA (2010)
11. Marujo, L., Coheur, L., Trancoso, I.: Keyphrase Cloud Generation of Broadcast News. In: Interspeech 2011. ISCA (September 2011)
12. Matsuo, Y., Ishizuka, M.: Keyword extraction from a single document using word co-ocurrence statistical information. Inter. Journal on A.I. Tools 13, 157–170 (2004)
13. Medelyan, O., Perrone, V., Witten, I.H.: Subject metadata support powered by Maui. In: Proceedings of JCDL 2010, p. 407. ACM, New York (2010)
14. Neto, J., Meinedo, H., Viveiros, M.: A media monitoring solution. In: Proceedings of ICASSP 2011, Prague, Czech Republic (2011)
15. Penn, G., Zhu, X.: A Critical Reassessment of Evaluation Baselines for Speech Summarization. In: Proceeding of ACL 2008: HLT, pp. 470–478. ACL (2008)
16. Ribeiro, R., de Matos, D.M.: Revisiting Centrality-as-Relevance: Support Sets and Similarity as Geometric Proximity. Journal of A.I. Research 42, 275–308 (2011)
17. Salton, G., Yang, C.S., Yu, C.T.: A theory of term importance in automatic text analysis. Tech. rep., Ithaca, NY, USA (1974)
18. Sarkar, K., Nasipuri, M., Ghose, S.: A new approach to keyphrase extraction using neural networks. Inter. Journal of Computer Science Issues 7(2,3), 16–25 (2010)
19. Witten, I., Paynter, G., Frank, E., Gutwin, C., Nevill-Manning, C.: KEA: Practical automatic keyphrase extraction. In: Proceedings of the Fourth ACM Conference on Digital Libraries, pp. 254–255. ACM (1999)

Using a Double Clustering Approach to Build Extractive Multi-document Summaries

Sara Botelho Silveira and António Branco

University of Lisbon, Portugal
Edifício C6, Departamento de Informática
Faculdade de Ciências, Universidade de Lisboa
Campo Grande, 1749-016 Lisboa, Portugal
{sara.silveira,antonio.branco}@di.fc.ul.pt

Abstract. This paper presents a method for extractive multi-document summarization that explores a two-phase clustering approach. First, sentences are clustered by similarity, and one sentence per cluster is selected, to reduce redundancy. Then, in order to group them according to topics, those sentences are clustered considering the collection of keywords that represent the topics in the set of texts. Evaluation reveals that the approach pursued produces highly informative summaries, containing many relevant data and no repeated information.

1 Introduction

Automatic text summarization is the process of creating a summary from one or more input text(s) through a computer program. It seeks to combine several goals: (1) the preservation of the idea of the texts; (2) the selection of the most relevant content of the texts; (3) the reduction of eventual redundancy; and (4) the organization of the final summary. While meeting these demands, it must be ensured that the final summary complies with the desired compression rate.

This paper presents a multi-document summarization system, SIMBA, that receives a collection of texts and returns an extract summary. The main goals of multi-document summarization are tackled through a double clustering approach, which includes a similarity clustering phase and a keyword clustering phase. Redundancy is addressed by grouping the sentences based on a similarity measure. Afterwards, the sentences are assembled by topics, using the keywords retrieved from the collection of texts. Furthermore, to support the compression process, this system includes a sentence simplification module, which aims to produce simpler and more incisive sentences, allowing more relevant content to enter the summary. Finally, an automatic evaluation of SIMBA is presented.

Previous works addressed multi-document summarization in different ways. MEAD [1] is a multi-document system that summarizes clusters of news articles automatically grouped by a topic detection system. It uses information from the centroids of the clusters to select the sentences that are most likely to be relevant to the cluster topic. By identifying similarities and important differences across sets of documents, Newsblaster [2] builds summaries from on-line news sources. It performs summarization through different modules that use different strategies depending on the type of

P. Sojka et al. (Eds.): TSD 2012, LNCS 7499, pp. 298–305, 2012.

the documents in the input set. In order to produce extracts from a set of documents, NeATS [3] selects relevant portions about a topic and presents them in coherent order, using several metrics: term frequency, sentence position, stigma words and maximum marginal relevance. Concerning the Portuguese language, GistSumm [4] was the first single-document summarizer built. It is based on the notion of gist, which is the most important passage of the text, conveyed by just one sentence that best expresses the text's main topic. The system algorithm relies on this sentence to produce extracts. GistSumm is available on-line and though it has been built to produce summaries from a single-document, it also performs multi-document summarization by means of an option in its interface. In their work, [5] treat multi-document summarization as a classification problem, by combining features, as sentence position and sentence size, with sophisticated linguistic features, given by the CST model, such as semantic relations between sentences from different texts.

Henceforth, this paper is organized as follows: Section 2 describes our system, Section 3 reports system evaluation, and in Section 4 some conclusions are drawn.

2   The SIMBA System

SIMBA is an extractive multi-document summarizer for the Portuguese language, that receives a collection of texts, from any domain, and produces informative summaries, for a generic audience. Summarization is performed by means of two main phases executed in sequence: clustering by similarity and clustering by keywords. The length of the summaries is determined by a compression rate value that is submitted by the user.

2.1   Methodology

The procedure starts by processing automatically the documents submitted to be summarized. A set of shallow processing tools for Portuguese, LX-Suite [6], is used to annotate the texts. Sentence and paragraph boundaries are identified and words are tagged, with its corresponding POS and lemmata. Also, a parse tree representing each sentence syntactic structure is built, using LX-Parser [7]. Henceforth, the collection of texts is handled as a set of sentences.

Afterwards, sentence scores are computed. The scoring procedure includes two scores, the main score and the extra score, that are combined to define the sentence final score – named complete score. The main score reflects the sentence relevance in the overall collection of sentences, and is the sum of the `tf-idf` score (computed considering the word lemma) of each word of the sentence, smoothed by the number of words in the sentence. The extra score is used during the clustering phases to reward or penalize the sentences by adding or removing predefined score values[1]. The complete score is the sum of these two scores.

The next stages of processing aim to identify relevant information in the collection of texts in two steps: similarity clustering and keyword clustering.

---

[1] The predefined extra score value is set to 0.1, both for the reward and the penalty values. This value has been determined empirically, through a set of experiments.

Similarity clustering. In order to identify sentences conveying the same information, they are clustered considering their degree of similarity.

The similarity between two sentences (Equation 3) comprises two dimensions, computed considering the word lemmas: the sentences subsequences (Equation 1) and the word overlap (Equation 2).

$$subsequences(s_1, s_2) = \frac{\sum_i \left(\frac{subsequence_i}{totalWords_{s_1}} + \frac{subsequence_i}{totalWords_{s_2}} \right)}{total Subsequences} \tag{1}$$

$$overlap(s_1, s_2) = \frac{\sum commonWords(s_1, s_2)}{totalWords_{s_1} + totalWords_{s_2} - \sum commonWords(s_1, s_2)} \tag{2}$$

$$similarity(s_1, s_2) = \frac{subsequences(s_1, s_2) + overlap(s_1, s_2)}{2} \tag{3}$$

$overlap(s_1, s_2)$ — number of overlapping words between the two sentences.

$subsequence(s_1, s_2)$ — number of overlapping words in the subsequences between the two sentences.

$commonWords(s_1, s_2)$ — common words between the two sentences.

$totalWords_{s_i}$ — total words in the sentence i.

$subsequence_i$ — number of words of the subsequence i.

$totalSubsequences$ — number of subsequences between the two sentences.

The subsequence value is inspired in ROUGE-L and consists of the sum of the number of words in all the subsequences common to each sentence, smoothed by the total number of words of each sentence being considered, and divided by the total number of subsequences found between the two sentences. The overlap value is computed using the Jaccard index [8].

The similarity value is the average of both these values: the overlap and the subsequences value. It is then confronted with a predefined threshold – similarity threshold[2] –, set to 0.75, meaning that sentences must have at least 75% of common words or subsequences to be considered as conveying the same information.

Two examples are discussed below. Taking into account this threshold, the sentences in the following example are considered to be similar.

Sentence#1:

A casa que os Maias vieram habitar em Lisboa, no outono de 1875, era conhecida pela casa do Ramalhete.

The house in Lisbon to which the Maias moved in the autumn of 1875, was known as the Casa do Ramalhete.

Sentence#2:

A casa que os Maias vieram habitar, no outono de 1875, era conhecida pela casa do Ramalhete.

The house to which the Maias moved in the autumn of 1875, was known as the Casa do Ramalhete.

Overlap	Subsequences	Similarity Value
0.89	0.95	0.92

These sentences share most of the words, but there is a leap (*"em Lisboa"*) between Sentence#1 and Sentence#2. Both the overlap and the subsequences values are high,

---

[2] This threshold was determined empirically, using a set of experiments.

so the similarity value is also high (0.92), and thus the sentences are considered to be similar.

The two sentences in the following example are not similar, despite having many words in common.

Sentence#1:

A casa que os Maias vieram habitar em Lisboa, no outono de 1875, era conhecida pela casa do Ramalhete.

The house in Lisbon to which the Maias moved in the autumn of 1875, was known as the Casa do Ramalhete.

Sentence#2:

A casa que os Maias vieram habitar em Lisboa, no outono de 1875, era conhecida na vizinhança da Rua de S. Francisco de Paula, pela casa do Ramalhete ou simplesmente o Ramalhete.

The house in Lisbon to which the Maias moved in the autumn of 1875, was known in Rua S. Francisco de Paula, as the Casa do Ramalhete or, more simply, as Ramalhete.

Overlap	Subsequences	Similarity Value
0.59	0.79	0.69

Despite Sentence#1 is being contained in Sentence#2, both sentences are considered not to be similar, since their similarity value is below the threshold.

Afterwards, sentences are clustered considering their similarity value. As the primary goal of this phase is to determine the sentences that represent each cluster, a simple algorithm, that seeks to optimize sytem execution, is used.

A cluster contains a collection of sentences, a similarity value, and a centroid (the highest scored sentence of the collection). The algorithm starts with an empty set of clusters. All sentences in the collection of texts are considered. The first sentence of the collection creates the first cluster. Then, each sentence in the collection of sentences is compared with the sentences already clustered. For each cluster, the similarity value is computed between the current sentence being compared and all the sentences in the collection of sentences of the cluster. The similarity value considered is the highest between the current sentence and all the sentences in the collection of sentences of the current cluster. If the similarity value is higher than the similarity threshold, the sentence is added to that cluster.

When a sentence is added to a cluster, its centroid is updated. If the score of this sentence is higher than the centroid one, the newly added sentence becomes the centroid of the cluster, and is rewarded with an extra score. Likewise, an extra score value is subtracted from the sentences which are replaced as centroids.

Finally, if all the clusters have been considered, and the sentence was not added to any cluster, a new cluster with this sentence is created, meaning that this sentence does not repeat information previously considered.

Once the procedure is finished, sentences with redundant information are grouped in the same cluster and the one with the highest score (the centroid) represents them. So, this phase returns a collection of sentences built by selecting only the centroid of each similarity cluster. The sentences in the collection of sentences, by being redundant, are discarded.

Keyword clustering. Our system produces a generic summary, so it is not focused on a specific matter. Thus, the keywords that represent the global topic within the collection of texts are identified. A list with the candidate keywords is constructed containing words that are common and proper names, since these words identify ideas or themes. Words are compared considering their lemmas, to ensure that the words in the collection are unique. Thereafter, the list is ordered considering the score of each word. We define k, the number of keywords, as $k = \sqrt{\frac{N}{2}}$, where N is the total number of words in the collection of documents. The final list of keywords contains the first k words of the list of candidate keywords. Sentences are clustered based on that final list of keywords.

In this phase, a cluster is identified by a keyword, and contains a centroid (a sentence), and a collection of sentences (related to the keyword). The algorithm that clusters sentences by keywords is an adapted version of the K-means algorithm [9], and follows the steps described below:

1. Choose the number of clusters, k, defined by the number of keywords;
2. Create the initial clusters, represented by each keyword obtained;
3. Consider each sentence:
   (a) Compute the occurrences of each keyword in the sentence;
   (b) Assign the sentence to the cluster whose keyword occurs more often;
4. Recompute the cluster centroid. If the current sentence has more occurrences of the keyword than the previous centroid sentence had, the newly added sentence becomes the cluster centroid;
5. If the sentence does not contain any keywords, it is added to a specific set of sentences which do not have any keyword ("no-keyword" set);
6. Recompute the set of keywords if:
   (a) All the sentences have been considered;
   (b) The "no-keyword" set contains new sentences.
7. Repeat previous steps (2 – 6) while the "no-keyword" set of sentences remains different in consecutive iterations.

As in the similarity algorithm, when the centroid is changed, the extra scores of the current centroid and of the previous centroid are updated.

In addition, an extra score is also assigned to the sentences in the clusters that represent the initial set of keywords. These sentences are considered more significant than the others, since they address the main topics conveyed by the collection of texts. Still, sentences in the "no-keyword" set are ignored, since they do not convey relevant information concerning the overall collection of texts.

The next step of the summarization procedure orders sentences based on their complete score, defining the order of the sentences to be included in the summary. Afterwards, this set of sentences is compressed in order to select the ones composing the summary. Compression is applied in two ways. First, the compression rate given by the user is applied to the collection of sentences. When the total number of words of the sentences already added to the summary reaches or surpasses the maximum compression, no more sentences are selected. Afterwards, a sentence simplification procedure [10] removes, from a sentence, syntactic structures whose removal is less detrimental to the comprehension of the text. Simplification is performed by identifying and removing appositions, parentheticals, and relative clauses from the sentence parse tree. This process seeks to make room for more relevant data to be included in the summary, aiming to produce a more informative text. As the simplification process removes words from the already selected set of sentences, more sentences are added to this set in order to achieve the desired number of words again. These two steps, compression and simplification, are repeated until no more new sentences are added to the set of sentences that defines the summary. Finally, the summary is delivered to the user in the form of a text file.

3   Evaluation

In order to perform evaluation, *CSTNews* [11], an annotated corpus of texts in Portuguese, was used. It contains 50 sets of news texts from several domains, for a total of 140 documents, 2,247 sentences, and 47,350 words. Each set contains, on average, 3 documents which address the same subject. The texts were retrieved from five Brazilian newspapers. Also, each set of texts contains a manually built summary – the ideal summary. There are 50 ideal summaries, containing an average of 137 words, resulting in an average compression rate of 85%.

In order to understand the impact of this approach, Table 1 details the sentences considered before and after the clustering phases have been executed.

Table 1. Sentences involved in the clustering phases

Clustering by similarity:			Clustering by keywords:		
Before	After	Difference	Before	After	Difference
2,247	2,115	132	2,115	1,599	516

The similarity clustering is the first step of the summarization process, so that it takes all the sentences in the corpus (2,247). After executing this step, 132 sentences have been considered redundant. This corresponds to 5% of the sentences in the corpora. Thus, these sentences are not considered in the next steps of the summarization procedure. If these sentences have been selected to be part of the final summaries, those would contain many superfluous data that would impact negatively on their informativity.

The double clustering approach executes both phases in sequence. Thus, after the set of sentences has been filtered in the similarity clustering, the remainder of the sentences (2,115) are clustered by keywords. Considering all the document sets, there were 516 sentences that do not mention the main topics of the texts. This corresponds to 24% of the sentences considered in this phase. Therefore, only 76% of the sentences considered in the keywords clustering phase are indeed relevant to the topic. This way, after executing the double clustering procedure, a total of 648 sentences were discarded, either by being redundant or irrelevant.

Concerning the evaluation itself, we compared the summaries generated automatically by SIMBA with summaries produced by GISTSUMM. The compression rate used was the one of the ideal summaries (85%), meaning that the summary contains 15% of the words contained in the set of texts.

Afterwards, the summaries were compared with the ideal summaries using ROUGE [12]. In fact, a more precise metric of ROUGE was used, ROUGE-L (longest common subsequence), since it identifies the common subsequences between two sequences. As the simplification process introduces gaps in the extracted sentences, this is considered a fairer metric.

Table 2 details precision, recall and F-measure metrics for both summarizers.

Table 2. Multi-document evaluation metrics

	GISTSUMM	SIMBA
Precision	0.43616	0.48534
Recall	0.38469	0.54014
F-measure	0.40398	0.50752

The SIMBA process has an overall better performance than the baseline.

The recall values obtained by SIMBA are very encouraging. These values indicate that there is a high density of words that are both in SIMBA summaries and in the ideal summaries. Retrieving the most relevant information in a sentence by discarding the less relevant data ensures that the summary indeed contains the most important information conveyed.

The precision values of the two systems are closer than the ones concerning recall. Intuitively, the precision values should be similar or even decrease, since, in comparison to the ideal summary, less in-sequence matches are likely to be found in SIMBA summaries due to the simplification process. Still, SIMBA has a higher precision value than GISTSUMM, meaning that its summaries cover more significant topics than the ones produced by GISTSUMM. Both the precision and recall values attained by SIMBA are a direct result of the combination of both clustering phases, along with the simplification process.

Consequently, when computing the F-measure value, by combining both precision and recall, the claim that SIMBA produces better summaries than GISTSUMM can be confirmed.

4   Concluding Remarks

The results reported in this paper show that the quality of an automatic summary can be improved by (1) performing specific multi-document tasks – such as removing redundant information, or considering all the texts in each set as a single information source – and (2) executing an algorithm that seeks to optimize the content selection and allows the addition of more relevant information.

The multi-document summarizer presented relies on statistical features to perform summarization of a collection of texts in Portuguese. Despite the core algorithm being language-independent, this system uses language-specific tools that aim to improve not only the content selection, but also the general quality of a summary produced from a collection of texts written in Portuguese.

The final evaluation demonstrates promising results. Both F-measure and recall values are very encouraging, since they reflect the high relevance of the sentences present in the summaries produced by SIMBA.

This approach impacts on the content of the final summary in two ways. On the one hand, similarity clustering ensures that the content is not repetitive. On the other hand, keyword clustering insures the selection of relevant content, and the preservation of the idea of the input texts. Thus, the combination of these two clustering phases allows the creation of highly informative summaries.

References

1. Radev, D.R., Jing, H., Budzikowska, M.: Centroid-based summarization of multiple documents: sentence extraction, utility-based evaluation, and user studies. In: Proceedings of the 2000 NAACL-ANLP Workshop on Automatic Summarization, NAACL-ANLP-AutoSum 2000, pp. 21–30. ACL (2000)
2. McKeown, K.R., Hatzivassiloglou, V., Barzilay, R., Schiffman, B., Evans, D., Teufel, S.: Columbia multi-document summarization: Approach and evaluation. In: Proceedings of the Document Understanding Conference, DUC 2001 (2001)
3. Lin, C.Y., Hovy, E.: From single to multi-document summarization: A prototype system and its evaluation. In: Proceedings of the ACL, pp. 457–464. MIT Press (2002)
4. Pardo, T.A.S., Rino, L.H.M., das Graças Volpe Nunes, M.: GistSumm: A Summarization Tool Based on a New Extractive Method. In: Mamede, N.J., Baptista, J., Trancoso, I., Nunes, M.d.G.V. (eds.) PROPOR 2003. LNCS, vol. 2721, pp. 210–218. Springer, Heidelberg (2003)
5. Jorge, M.L.C., Agostini, V., Pardo, T.A.S.: Multi-document summarization using complex and rich features, pp. 1–12 (July 2011)
6. Branco, A., Silva, J.: A suite of shallow processing tools for portuguese: Lx-suite. In: Proceedings of the 11th Conference of the European Chapter of the Association for Computational Linguistics, EACL 2006 (2006)
7. Silva, J., Branco, A., Castro, S., Reis, R.: Out-of-the-box robust parsing of Portuguese. In: PROPOR 2010: Proceedings of the 9th Encontro para o Processamento Computacional da Língua Portuguesa Escrita e Falada, pp. 75–85 (2010)
8. Jaccard, P.: Nouvelles recherches sur la distribution florale. Bulletin de la Sociète Vaudense des Sciences Naturelles 44, 223–270 (1908)
9. MacQueen, J.B.: Some methods for classification and analysis of multivariate observations. In: Proceedings of the 5th Berkeley Symposium on Mathematical Statistics and Probability, pp. 281–297. University of California Press (1967)
10. Silveira, S.B., Branco, A.: Enhancing multi-document summaries with sentence simplification. In: ICAI 2012: International Conference on Artificial Intelligence (2012)
11. Aleixo, P., Pardo, T.A.S.: CSTNews: Um córpus de textos jornalísticos anotados segundo a teoria discursiva multidocumento CST (cross-document structure theory). Technical report, Universidade de São Paulo (2008)
12. Lin, C.Y.: Rouge: A package for automatic evaluation of summaries. In: Text Summarization Branches Out: Proceedings of the ACL 2004 Workshop, Barcelona, Spain, pp. 74–81 (July 2004)

A Comparative Study
of the Impact of Statistical and Semantic Features
in the Framework of Extractive Text Summarization

Tatiana Vodolazova, Elena Lloret, Rafael Muñoz, and Manuel Palomar

Dept. Lenguajes y Sistemas Informáticos, Universidad de Alicante
Apdo de correos, 99, E-03080, Alicante, Spain
{tvodolazova,elloret,rafael,mpalomar}@dlsi.ua.es

Abstract. This paper evaluates the impact of a set of statistical and semantic features as applied to the task of extractive summary generation for English. This set includes word frequency, inverse sentence frequency, inverse term frequency, corpus-tailored stopwords, word senses, resolved anaphora and textual entailment. The obtained results show that not all of the selected features equally benefit the performance. The term frequency combined with stopwords filtering is a highly competitive baseline that nevertheless can be topped when semantic information is included. However, in the selected experiment environment the recall values improved less than expected and we are interested in further investigating the reasons.

1   Introduction

The research in Text Summarization (TS) is still mainly focused on extractive approaches, that attempt to determine the most relevant segments of the input document by computing their weights based on different techniques and features. These features commonly include the segment position within the original text [12], presence of the cue phrases [3], term frequency of the topic terms [5] and length of the segment [2] among others. In recent years the research in this area has been directed towards incorporating more semantic knowledge in the set of analysis features. A graph-based method combined with the word sense disambiguation (WSD) was proposed in [11]. The LeLSA+AR is a Latent Semantic Analysis-based system for TS supplied with anaphoric information [13]. Textual Entailment (TE) as well was reported to benefit the TS task [8]. However, to the best of our knowledge, there has been no research exploring in detail the impact of each of the aforementioned features and their combinations in the identical evaluation environment.

The present paper reports on the initial results of the ongoing work investigating the impact of different semantic and statistical features on the performance of extractive summary generation system. Those are represented by term frequency, inverse term and sentence frequencies, word senses, resolved anaphora, textual entailment and corpus-tailored stopword list. Although the system performance improved less than expected, the obtained results clearly show the relative importance of each of the selected features.

This paper is organized as follows. Section 2 introduces the features selected for evaluation with the brief explanations and related work reporting their usage. Section 3

P. Sojka et al. (Eds.): TSD 2012, LNCS 7499, pp. 306–313, 2012.

describes the evaluation environment. The results are reported in Section 4. Finally, the conclusions together with the future work can be found in Section 5.

2 Selected Features and Related Work

2.1 Term Frequency

Term frequency (TF) is one of the very first features used for automatic TS [9]. The impact of TF isolated from all the other features was analyzed in [10] . It was shown that the likelihood of a word appearing in a human summary depends on the word's frequency in the original text. Thus it can be hypothesized that the sentences with the highest number of most frequent words should be selected for the final summary.

For a sentence $S_j = \{t_1, t_2, ..., t_m\}$ with m tokens, its score based on the TF will be calculated using the following formula:

$$Sc_{tf(S_j)} = \frac{\sum_{i=1}^{m} tf_i}{n} \tag{2.1}$$

where tf_i is the frequency of the i_{th} token (or its stem/lemma) in the text j; n is the number of sentence tokens (stopwords removed).

2.2 Inverse Term and Sentence Frequencies

TFIDF is a common keyword identification method successfully used in the information retrieval. It involves calculating term frequencies, that as it has been hypothesized in the previous section represent a reliable measure of the sentences to be selected for the final summary. In the context of single document summarization the keywords of each single document do not depend on the other documents in the collection. The three language models were compared in [1]: Inverse Document Frequency(IDF), Inverse Sentence Frequency(ISF) and Inverse Term Frequency(ITF). It was speculated on using ISF for the systems identifying sentences and ITF for the systems identifying terms as their smallest compositional unit. Both ISF and ITF can be used for the single document summarization task, as the extractive summaries are commonly composed of the sentences of the original text and term counts proved to be a reliable method for the summary sentence identification. Roughly based on [1] the ISF and ITF measures were calculated the following way:

$$isf_{tf} = \log \frac{|S|}{|\{s \in S : t \in s\}|} \tag{2.2}$$

where $|S|$ is the total number of sentences in the document; $|\{s \in S \ t \in s\}|$ number of sentences where the term t appears.

$$itf = \log \frac{|V|}{|\{t \in V : t \in d\}|} \tag{2.3}$$

where $|V|$ is the vocabulary size of the document; $|\{t \in V \ t \in d\}|$ number of times where the term t appears in the document, i.e. term frequency.

2.3   Word Sense Disambiguation

Although the approaches based on TF result in a rather competitive baseline, they fail to capture the semantics of a document. The cases of synonymy like between the nouns "a ship" and "a boat" can be captured employing a Word Sense Disambiguation (WSD) module and using concepts instead of terms. The TF is substituted by the concept frequency and Formulas (2.1), (2.2) and (2.3) will be modified the following way:

$$Sc_{cf(S_j)} = \frac{\sum_{i=1}^{m} cf_i}{n} \qquad (2.4)$$

where S_j is the j_{th} sentence $S_j = \{t_1, t_2, ..., t_m\}$ with m tokens; cf_i is the frequency of the WordNet[1] synset id in the document that the sense of the i_{th} term belongs to; n is the number of sentence tokens (stopwords removed).

$$isf_{cf} = \log \frac{|S|}{|\{s \in S : c \in s\}|} \qquad (2.5)$$

where $|S|$ is the total number of sentences in the document; $|\{s \in S \ c \in s\}|$ number of sentences s where the WordNet synset id c in the document appears, for all $c \in d$.

$$icf = \log \frac{|V|}{|\{t \in V : c \in d\}|} \qquad (2.6)$$

where $|V|$ is the vocabulary size of the document as measured in the number of different WordNet synset ids appearing in the text; $|\{c \in V \ c \in d\}|$ number of times where the concept c appears in the document, i.e. concept frequency.

A graph-based method for TS combined with WSD is described in [11]. The authors report the recall values up to 0.4651 for ROUGE-1 on the DUC 2002 data which improves significantly over the DUC baseline of 0.4113 and outperforms the best DUC 2002 system. The important implementation detail of this system is that the concepts were used only for nouns. The use of verbs in the graph showed the decrease in the system performance. The final set of features considered in the present paper includes concept frequencies both for nouns and verbs, nouns and adjectives.

2.4   Anaphora Resolution

Taking into account the significance of most frequent words for the final summary generation mentioned in the Section 2.1, it becomes particularly important to resolve anaphora in the original document. The previous work on including anaphora resolution (AR) reports some increase in performance. [13] achieve the improvement of around 1.5% over their summarization system based on the lexical LSA by incorporating anaphoric information into it. The performance was tested on the DUC 2002 data using the ROUGE evaluation toolkit. The authors also mention two strategies for including anaphoric relations: (1) addition, when anaphoric chains are treated as another kind of terms for the input matrix construction; (2) substitution, when each representative

---

[1] WordNet http://wordnet.princeton.edu/

of an anaphoric chain in the text is substituted by the chain's first representative. The evaluation results show that the substitution approach performs significantly worse than the addition approach and in some tests even worse than the same system without including anaphora resolution.

2.5 Textual Entailment

The task of Textual Entailment (TE) is to capture the semantic inference of texts. In the framework of TS it was used for different purposes: (1) to segment the input document into subtopics [14]; (2) when evaluating the set of generated summaries to identify the summary that can be best deduced from the original document [6]; (3) to eliminate redundant information from the final summary. In [8] TE was used in its latter application which yielded the increase in the ROUGE-1 values over the TF baseline on the task of single document summarization. The TE module similar to [8] and [4] was employed for the present research. The sentences of the original document are being handled sequentially by the TE module which determines whether the sentence that is currently being processed can be inferred from the stack of already processed ones.

2.6 Corpus-Tailored Stopwords

Inspecting the first summaries we discovered that the words like "just", "near", "away", "ago", "today" among others were not in the standard stopword list. It was decided to extend the 350-words stopword list by the missing words from the list of 245 most frequent words of the DUC 2002 corpus[2] that was extracted using the Lucene 3.5.0[3] indexing module.

3 Evaluation Environment

3.1 Evaluation Corpus and Metrics

For our experiments we used the Document Understanding Conferences 2002 data for single-document summarization task. The corpus consists of 567 newswire articles covering a wide range of topics. Each newswire article is accompanied by one or more abstractive model summaries. The model summaries are approximately 100 words long and were created manually by humans.

The system's summaries were evaluated against the model ones using the ROUGE metrics [7], which is a standard measure for evaluation by now. ROUGE-N is a family of metrics based on the overlap n-grams between a system summary and one or more model summaries, where N stands for the n-gram length. For the present paper we computed the ROUGE-1 recall values (see Section 4) that were shown to correlate with human assessment [13].

---

[2] DUC Conferences: http://duc.nist.gov/

[3] Apache Lucene http://lucene.apache.org/

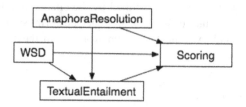

Fig. 1. Interaction of semantic components

3.2 System Settings

The core of our system is the scoring module (see Figure 1). In its basic setting it computes the score of each sentence in the original document based on the TF as in Formula (2.1). The scoring module can be set to filter out the stopwords from a custom stopword list. The second scoring strategy involves WSD and rates the sentences based on the CF as in Formula (2.4). The WSD was performed using the built-in function of Freeling[4]. Freeling package provides a number of WSD algorithms. We experimented with the most frequent sense (MFS) and the PageRank algorithm (UKB). Another setting for the WSD module is to disambiguate only nouns. This was motivated by the results provided in [11]. The more sophisticated versions of TF and CF involve computing ISF and ITF as described in Formulas (2.2), (2.3), (2.5) and (2.6). All the scoring module settings except for the stopwords, ITF and ISF can be found in the results Table 1 in the column names[5]. Before rating the sentences the original text can be processed using AR, TE, or both. When AR is applied, each anaphor is substituted by its antecedent and then the text is sent to the scoring module. The JavaRAP[6] was used to solve this task. TE is used to remove semantically redundant sentences. It in its turn can also be combined with the WSD module replacing each representative of a WordNet synset with one and the same synset member. Finally, AR can be combined together with the TE module. The AR can be applied either only for redundancy detection, i.e. the resolved text is being consumed by the TE module, but the original version of the remaining sentences sent further for scoring, or the sentences with the resolved anaphoric expressions will be submitted for scoring.

The following settings have been evaluated so far:

- **SSW**: using the standard stopword list
- **ASW**: using the standard and the extended stopword list
- **ASW ITF**: ASW combined with the ITF
- **ASW ISF**: ASW combined with the ISF
- **ASW AR**: ASW combined with the AR
- **ASW TE**: ASW combined with the TE
- **ASW WSD TE**: use ASW, replace the words of the selected parts of speech with the same member of the WordNet synset, then process the result using the

---

[4] Freeling http://nlp.lsi.upc.edu/freeling/

[5] N stands for nouns, NVA stands for nouns, verbs, adjectives – the parts of speech WSD is applied to.

[6] JavaRAP http://aye.comp.nus.edu.sg/~qiu/NLPTools/JavaRAP.html

TE module. The scoring module is applied to the original version of remaining sentences, thus making it possible to evaluate the impact of both MFS and UKB algorithms on the resulted data.

- **ASW AR WSD TE**: is similar to the previous one with the difference that AR is applied before the WSD.The scoring module is again applied to the original version of the remaining sentences.

Each of the listed settings is further combined with either TF or CF scoring strategy as shown in Table 1.

The scoring module arranges the sentences in the descending order. The top N sentences with the common length not exceeding the size of 100 words are selected for the final summary.

4   Results and Discussion

Table 1 shows the ROUGE-1 recall values for the system settings listed in Section 3.2. The system components interact in a complex way. This way the name conventions for the columns cannot be associated exclusively with the scoring module strategies. For example, the system setting ASW WSD TE uses WSD prior to scoring, so the particular WSD algorithms used during this step is also indicated in a column name. As well as the stopwords filtering was used not only during the scoring procedure, but also before applying WSD.

Table 1. ROUGE-1 Recall values

	ROUGE-1 Recall				
	TF	**CF-MFS**		**CF-UKB**	
		NVA	**N**	**NVA**	**N**
SSW	**0.40906**	0.40869	0.41717	0.41765	0.41396
ASW	0.41779	0.40976	0.41466	0.41785	0.41462
ASW ITF	0.36924	0.36828	0.36668	0.36890	0.36719
ASW ISF	0.38126	0.37985	0.37894	0.37924	0.37562
ASW AR	0.38945	0.38873	0.39077	0.39146	0.38991
ASW TE	0.41804	0.41807	0.41596	**0.41897**	0.41665
ASW WSD TE	0.41807	0.41796	0.41627	**0.41894**	**0.41843**
ASW AR WSD TE	**0.43235***	**0.43050***	**0.42963***	**0.43196***	**0.43017***

The TF combined with the SSW was selected as the baseline. This baseline is slightly lower than the DUC 2002 baseline of 0.41132 [13]. Only some of the system settings outperformed it. All of them involved TE. As expected, the best setting is the ASW AR WSD TE – the combination of all the semantic modules used sequentially applied one after the other. The scores for this setting were evaluated using the t-test and the statistically significant values are indicated with the asterisk in Table 1. As for the scoring strategies, the TE-based and the CF-UKB applied for NVA outperform the CF-MFS strategies. The worst results were obtained for the AWS ITF, AWS ISF and AWS

AR settings. For further experiments we plan to substitute the JavaRAP by another AR module and examine the difference.

The best results for the systems mentioned in Section 1 involving TE [8] and a WSD-based graph implementation of [11] are 0.4518 and 0.4651 respectively as reported in [11]. They outperform our best result of 0.43235. However, our system improved on the best score of the LeLSA+AR system [13]. This shows that combining TE, WSD and AR with graph-based method is worth experimenting with.

5   Conclusion and Future Work

This paper presented the initial results of the ongoing research on the impact of a set of statistical and semantic features on the TS task. In particular, we analyzed the impact of textual entailment, anaphora resolution, word sense disambiguation, term frequency, inverse term and sentence frequency and corpus-tailored stopword list. The results showed that each of these features alone do not improve the performance, moreover inverse sentence and term frequencies and anaphora resolution influence it negatively and we would like to further investigate these issues. But the combination of the semantic features that include anaphora resolution, textual entailment and word sense disambiguation outperforms the baseline. In the future we plan to experiment with a different anaphora resolution system, compute the scores for all the mentioned settings using the standard stopword list and extend our system to cover other languages.

Acknowledgements. This research work has been funded by the Spanish Government through the project TEXT-MESS 2.0 (TIN2009-13391-C04) and by the Valencian Government through projects PROMETEO (PROMETEO/2009/199) and ACOMP/2011/001.

References

1. Blake, C.: A Comparison of Document, Sentence, and Term Event Spaces. In: ACL-44 Proceedings of the 21st International Conference on Computational Linguistics and the 44th Annual Meeting of the Association for Computational Linguistics, pp. 601–608. ACL, Stroudsburg (2006)
2. Chuang, W.T., Yang, J.: Text Summarization by Sentence Segment Extraction Using Machine Learning Algorithms. In: Terano, T., Chen, A.L.P. (eds.) PAKDD 2000. LNCS, vol. 1805, pp. 454–457. Springer, Heidelberg (2000)
3. Fujii, Y., Kitaoka, N., Nakagawa, S.: Automatic Extraction of Cue Phrases for Important Sentences in Lecture Speech and Automatic Lecture Speech Summarization. In: INTER-SPEECH, pp. 2801–2804 (2007)
4. Ferrández, O., Micol, M., Muñoz, R., Palomar, M.: A perspective-based approach for solving textual entailment recognition. In: ACL-PASCAL Workshop on Textual Entailment and Paraphrasing, pp. 66–71 (2007)
5. Harabagiu, S., Lacatusu, F.: Topic Themes for Multi-document Summarization. In: 28th Annual International ACM SIGIR Conference on Research and Development in Information Retrieval, pp. 202–209 (2005)
6. Harabagiu, S., Hickl, A., Lacatusu, F.: Satisfying Information Needs with Multi-document Summaries. Information Processing & Management 43(6), 1619–1642 (2007)

7. Lin, C.Y.: ROUGE: A Package for Automatic Evaluation of Summaries. In: ACL Text Summarization Workshop, pp. 74–81 (2004)
8. Lloret, E., Ferrández, O., Muñoz, R., Palomar, M.: A Text Summarization Approach Under the Influence of Textual Entailment. In: 5th International Workshop on NLPCS, pp. 22–31 (2008)
9. Luhn, H.P.: The Automatic Creation of Literature Abstracts. IBM Journal of Research and Development 2(2), 157–165 (1958)
10. Nenkova, A., Vanderwende, L., McKeown, K.: A Compositional Context Sensitive Multidoscument Summarizer: Exploring the Factors That Influence Summarization. In: 29th SIGIR, pp. 573–580. ACM, New York (2006)
11. Plaza, L., Díaz, A.: Using Semantic Graphs and Word Sense Disambiguation. Techniques to Improve Text Summarization. Procesamiento del Lenguaje Natural 47, 97–105 (2011)
12. Saggion, H.: A Robust and Adaptable Summarization Tool. Traitement Automatique des Languages 49(2), 103–125 (2008)
13. Steinberger, J., Poesio, M., Kabadjov, M.A., Ježek, K.: Two Uses of Anaphora Resolution in Summarization. Information Processing and Management 43(6), 1663–1680 (2007)
14. Tatar, D., Tamaianu-Morita, E., Mihis, A., Lupsa, D.: Summarization by Logic Segmentation and Text Entailment. In: 33rd CICLing, pp. 15–26 (2008)

Using Dependency-Based Annotations
for Authorship Identification

Charles Hollingsworth[1,2]

[1] Institute for Artificial Intelligence
The University of Georgia
Boyd GSRC, Room 111
Athens, GA 30602
[2] Applied Systems Intelligence
3650 Brookside Parkway
Suite 500
Alpharetta, GA 30022
http://www.ai.uga.edu,
http://www.asinc.com

Abstract. Most statistical approaches to stylometry to date have focused on lexical methods, such as relative word frequencies or type-token ratios. Explicit attention to syntactic features has been comparatively rare. Those approaches that have used syntactic features typically either used very shallow features (such as parts of speech) or features based on phrase structure grammars. This paper investigates whether typed dependency grammars might yield useful stylometric features.

An experiment was conducted using a novel method of depicting information about typed dependencies. Each token in a text is replaced with a "DepWord," which consists of a concise representation of the chain of grammatical dependencies from that token back to the root of the sentence. The resulting representation contains only syntactic information, with no lexical or othographic information. These DepWords can then be used in place of the original words as the input for statistical language processing methods.

I adapted a simple method of authorship attribution — nearest neighbor based on word frequency rankings — for use with DepWords, and found it performed comparably to the same technique trained on words or parts of speech, even outperforming lexical methods in some cases. This indicates that the grammatical dependency relations between words contains stylometric information sufficient for distinguishing authorship. These results suggest that further research into typed-dependency-based stylometry might prove fruitful.

Keywords: stylometry, authorship attribution, syntax, dependency grammar, DepWords.

1 Introduction

Most statistical approaches to authorship attribution to date have focused on lexical features, such as word frequencies or type-token ratios. Little attention has been paid to syntactic features. This lack of attention was due in part to the lack of syntactically annotated texts on which to train classifiers. Only recently did the availability of fast and accurate natural language parsers allow for serious research into syntactic stylometry.

P. Sojka et al. (Eds.): TSD 2012, LNCS 7499, pp. 314–319, 2012.

Among those studies that do examine syntactic features, most have focused on phrase-structure grammars. This is unfortunate, because dependency grammars offer the benefit of simplicity and a smaller set of features to be analyzed. In this paper I present a novel annotation system based on typed dependencies, and investigate its usefulness for the task of authorship attribution.

1.1 Previous Work

So far, there has not been much work in authorship attribution that focused specifically on syntactic features. The overview of the field by Holmes in 1994 [3] names a number of lexical features used for authorship attribution, such as type-token ratio, vocabulary distributions, word frequencies, and hapax legomina and dislegomina. A number of "quasi-syntactic" features, such as function words and part of speech distributions, are also given, but Holmes notes the lack of analysis at syntactic or higher levels. Another survey by Juola in 2008 [4] briefly mentions syntactic features, but includes in this category word n-grams and punctuation. Explicit attention to data from parsed sentences seems to be rare.

One work that does use parse data is [1]. In this work, the authorship attribution task was performed, not on the actual words of the texts, but on "pseudo-words" representing rewrite rules extracted from syntactically annotated versions of the texts. Tests of word frequency and vocabulary richness were found to yield better results when performed on the syntactic pseudo-words than on the original words.

Other attempts at syntactic stylometry included using probabilistic context-free grammars induced from the training set to classify test samples [10], analyzing syntax subtree frequency and syntax tree depth [5], and analyzing frequency of clauses and sentence complexity [6]. Notably, all of these methods use phrase structure grammars; there does not seem to have been any serious use of dependency grammars for stylometric analysis.

2   The DepWords Representation

I investigated whether information about the grammatical dependencies within a sentence could be used for stylometric analysis. Borrowing the technique from [1] of using syntax-based pseudo-words instead of actual words as units of analysis, I devised my own dependency-based representation, which I call DepWords. The DepWords format is a concise representation of the grammatical dependencies between words in a sentence. Any consistent system of typed dependencies could be used, but the version presented here is based on the typed dependencies output of the Stanford Parser [7], which produces an output that assigns a dependency type to each governor-dependent pair in a sentence. The current version of the Stanford Parser (2.0.1 as of this writing) uses fifty-three different dependency types, including the ROOT type that indicates a word is not dependent on any other. Since the number of types is conveniently one more than twice the number of letters in the Latin alphabet, an obvious representation was to assign an arbitrary upper or lowercase letter to each dependency type, plus a pound sign (#) to represent the root.

The DepWord corresponding to a given word in a sentence is composed of the single-character representations of each dependency relation from the sentence root to the word in question. For example, the root itself would be represented as #, while a word that depends directly on the root with the nsubj relation would be represented as #c. A word dependent on this word with the nn relation would be represented as #ca, and so forth.

For example, here is the first sentence of *Dead Men Tell No Tales*, by E. W. Hornung:

> Nothing is so easy as falling in love on a long sea voyage, except falling out of love.

The first few dependencies for this sentence produced by the Stanford Parser are as follows:

```
nsubj(easy-4, Nothing-1)
cop(easy-4, is-2)
advmod(easy-4, so-3)
. . .
```

Finally, here is the DepWords representation of this sentence:

```
#c #O #D # #o #oi #oio #oioj #oio #oiojS #oiojF #oioja #oioj
#oigr #oigR #oig #oigq #oigo #oigoj #r .
```

(The end of each sentence is marked by a lone period.)

Several advantages of the DepWords representation are immediately apparent. This format presents some fairly detailed syntactic information about a text in a document roughly the same size as the original text. It uses ordinary ASCII characters, and is composed of space-separated words, and thus can be used as input to text-processing software just like any unannotated English-language text. Thus some very simple textual operations can give us detailed information about a text's syntax. By computing a word frequency histogram, for example, we can find out what the most common dependency relationships are in a document. By calculating average word lengths, we can get a rough idea of the average complexity of sentences (with more complex sentences tending to have longer chains of dependencies). If we want to collect statistics about a particular type of dependency, we can filter the text for words ending in that type's corresponding letter.

One possible disadvantage of the DepWords format is that it does not preserve information about which particular words are involved in a given dependency relation. For example, in the DepWords representation of the passage from *Dead Men Tell No Tales* given above, we can see that there are two words marked #oioj, and two words marked #oio. We know that each of the latter is a dependent of one of the former, but we have no way of knowing which. Nevertheless, the information that is communicated in the DepWords format should still allow us to draw some interesting statistical conclusions about a document. To test this hypothesis, I performed an experiment to see how DepWords compared to other types of features for the task of authorship attribution.

3   The Experiment

My experiment was based on that presented in [9]. That work used *rank distance* as a measure of similarity between texts. Rank distance consists of the sum of the absolute values of the differences between rankings for each style marker. For example, if you are comparing texts based on word frequencies, and the most common word in one text is the third most common in another, then the rank distance between those two texts is at least $|1 - 3| = 2$. In calculating rankings, ties are assigned the arithmetic mean of their rankings. If the second and third place entries both have the same frequency, then each gets a ranking of 2.5.

The experiment in [9] based rank distance on the relative frequencies of function words, using the same set of function words introduced in [8]. These rank distances were used to classify texts by author via a simple nearest neighbor algorithm. For my own experiments, I decided to base rank distances on several different types of features, both lexical and syntactic, and see how they compared to my own DepWord format. I tested the performance of frequencies of units, bigrams, and trigrams of words, parts of speech, dependency types, and DepWords. I also tested using just the function words found in [8] and [9]. Instead of a 1-nearest neighbor algorithm, I used a 3-nearest neighbor algorithm. A text was assigned to the author who wrote at least two of the three nearest texts. If all three were by different authors, the text was assigned to the author of the nearest one.

Since the number of units, bigrams, or trigrams for some of these features numbered in the hundreds of thousands, comparing the rankings of all features would have been very slow and computationally demanding. Instead, I compared only smaller sets of features, in the hopes that a more manageable number of features might still have decent accuracy. For each experiment I would determine the top n features for the training set as a whole, and each pair of texts was compared according to their relative rankings of these top n features only. Each type of feature was tested five times, with $n = 10, 50, 100, 500,$ and $1,000$. Note that in some of these cases the latter few tests were redundant; for example, since there are only 53 Stanford Dependencies, all tests in which $n > 53$ were identical. For function words, since there were only seventy such words, all tests in which $n > 70$ were identical.

My choice of texts for the authorship attribution task was inspired by Goldman [2], who investigated English-language detective novels, mostly from the 1920s, freely available from Project Gutenberg (http://www.gutenberg.org). This corpus was chosen because the texts were readily available in digital form, and because for the most part they were novel-length works of prose from a single literary genre written during the same time period, thus helping to ensure that any differences found were due to the authors' individual styles. I included eleven works by E. W. Hornung, fourteen by J. S. Fletcher, and fifteen by Sax Rohmer, for a total of forty texts. Unlike Goldman, I chose not to use the works of Arthur J. Rees, because there are only four texts by this author found on Project Gutenberg. I also omitted the works of Maurice Leblanc, who originally wrote in French, because it would be impossible to determine whether any stylistic peculiarities belonged to the author or to his translators. I manually edited the text files to remove chapter headings and other superfluous text, and used the Stanford Parser to generate part of speech tags and typed dependency parses for each text. The dependency parses were further used to generate DepWords representations of the texts.

3.1   Results

For each experiment, I performed a ten-fold cross validation. The forty texts were arbitrarily divided into ten groups of four texts each, and the experiment repeated ten times, each time using one group as the test set and the other thirty-six as the training set. I then calculated the percentage of correct classifications for all ten runs. While not every fold contained texts from all three authors, the distribution of texts guaranteed that in each run, the training set contained multiple texts from each author. The results are shown in Table 1.

Table 1. Experimental results

Feature	Top 10	Top 50	Top 100	Top 500	Top 1000
Word 1-gram	80.0	97.5	97.5	97.5	97.5
Word 2-gram	80.0	97.5	97.5	97.5	97.5
Word 3-gram	87.5	97.5	97.5	97.5	97.5
POS 1-gram	87.5	95.0	95.0	95.0	95.0
POS 2-gram	85.0	97.5	95.0	97.5	97.5
POS 3-gram	85.0	92.5	95.0	97.5	97.5
Dependency 1-gram	72.5	95.0	95.0	95.0	95.0
Dependency 2-gram	85.0	92.5	97.5	97.5	97.5
Dependency 3-gram	80.0	95.0	97.5	97.5	97.5
DepWord 1-gram	45.0	92.5	95.0	95.0	95.0
DepWord 2-gram	82.5	92.5	95.0	97.5	97.5
DepWord 3-gram	90.0	97.5	95.0	97.5	97.5
Function Words	90.0	97.5	97.5	97.5	97.5

As would be expected, the higher the number of features used to calculate rank distances, the better the classifier performed. The top 1000 of any kind of style marker gave a success rate of 97.5 percent, with the exception of part of speech, dependency, and DepWord units, all of which had a success rate of 95.0. None of the experiments had a 100% success rate, and the one text that was consistently misclassified, even when all others were classified correctly, was *In the Days of Drake* by J. S. Fletcher. This work was written in 1897, whereas most of the other works by that author were written in the 1920s; furthermore, it was not a work of detective fiction, but an historical novel set in the time of Sir Francis Drake. These two factors doubtless influenced the work's style, and it is interesting that even the purely grammatical features reflected this difference.

Also of note is the fact that there is considerable variation among the different features as to how quickly they improve when the number of rankings considered is increased. It is hardly surprising that looking at only the top ten dependency or DepWord units gives poor results, since the ten most frequent types of dependencies used in a text would be expected to be fairly constant for a given language and not good indicators of style. However, when trigrams rather than individual units are considered, the top ten DepWords perform as well as or better than the top ten of all other types of features, at 90%.

The big winner here, however, is still function words. They give the same results as DepWord trigrams, except in the case where the top 100 most frequent features are considered, in which the accuracy of DepWord trigrams actually decreased slightly.

4 Conclusions and Further Research

For the specific authorship attribution task presented here, there does not seem to be a compelling reason to adopt the DepWords approach. The generation of a DepWords representation of a text is a slow process, since it requires the entire document be parsed using the Stanford Parser. Function word frequencies are much easier and faster to calculate, and appear to yield slightly better results. However, the fact that DepWords performed as well as or better than other methods at similar tasks indicates that there is important stylometric information to be found in typed dependencies.

In the future, it would be beneficial to see how typed-dependency-based classification stacks up against other methods in a wider range of tests. This experiment only looked at a small number of authors, in a single genre, with mostly book-length works. It may be the case that typed-dependency-based methods actually offer an advantage over other methods for some other type of texts. Repeating this experiment with shorter works, larger numbers of authors, or texts from different genres might be beneficial. Furthermore, in this experiment, DepWords were only tested in isolation. It could be the case that typed-dependency-based methods combined with other methods yield better results than either alone.

References

1. Baayen, R., van Halteren, H., Tweedie, F.: Outside the cave of shadows: Using syntactic annotation to enhance authorship attribution. Literary and Linguistic Computing 11(3), 121–131 (1996)
2. Goldman, E., Allison, A.: Using grammatical Markov models for stylometric analysis. Class project, CS224N, Stanford University (2008), Retrieved from,
   http://nlp.stanford.edu/courses/cs224n/2008/reports/17.pdf
3. Holmes, D.I.: Authorship attribution. Computers and the Humanities 28(2), 87–106 (1994)
4. Juola, P.: Authorship Attribution. Now Publishers, Delft (2008)
5. Kaster, A., Siersdorfer, S., Weikum, G.: Combining text and linguistic document representations for authorship attribution. In: SIGIR Workshop: Stylistic Analysis of Text for Information Access (STYLE), pp. 27–35. MPI, Saarbrücken (2005)
6. Levitsky, V., Melnyk, Y.P.: Sentence length and sentence structure in English prose. Glottometrics 21, 14–24 (2011)
7. Marneffe, M., MacCartney, B., Manning, C.D.: Generating typed dependency parses from phrase structure parses. In: Proceedings of the 5th International Conference on Language Resources and Evaluation, pp. 449–454 (2006)
8. Mosteller, F., Wallace, D.L.: Inference and disputed authorship: The Federalist. Addison-Wesley, Massachusetts (1964)
9. Popescu, M., Dinu, L.P.: Rank distance as a stylistic similarity. In: Coling 2008: Companion Volume — Posters and Demonstrations, pp. 91–94 (2008)
10. Raghavan, S., Kovashka, A., Mooney, R.: Authorship attribution using probabilistic context-free grammars. In: Proceedings of the ACL 2010 Conference Short Papers, pp. 38–42 (2010)

A Space-Efficient Phrase Table Implementation Using Minimal Perfect Hash Functions

Marcin Junczys-Dowmunt

Faculty of Mathematics and Computer Science
Adam Mickiewicz University
ul. Umultowska 87, 61-614 Poznań, Poland
junczys@amu.edu.pl

Abstract. We describe the structure of a space-efficient phrase table for phrase-based statistical machine translation with the Moses decoder. The new phrase table can be used in-memory or be partially mapped on-disk. Compared to the standard Moses on-disk phrase table implementation a size reduction by a factor of 6 is achieved.

The focus of this work lies on the source phrase index which is implemented using minimal perfect hash functions. Two methods are discussed that reduce the memory consumption of a baseline implementation.

Keywords: Statistical machine translation, compact phrase table, minimal perfect hash function, Moses.

1 Introduction

As the size of parallel corpora increases, the size of translation phrase tables used for statistical machine translation extracted from these corpora increases even faster. The current in-memory representation of a phrase table in Moses [1], a widely used open-source statistical machine translation toolkit, is unusable for anything else but toy-size translation models or prefiltered test set data. A binary on-disk implementation of a phrase table is generally used, but its on-disk size requirements are significant.

This paper continues the research of [2] towards a compact phrase table that can be used as a drop-in replacement for both, the binary phrase table implementation and the in-memory phrase table of Moses. An important requirement is the faithful production of translations identical to translations generated from the original phrase table implementation if the same phrases, scores, and settings are provided. The phrase table design is inspired by the architecture of large-scale n-gram language models which employ minimal perfect hash functions (MPH) as their main indexing structure. In this paper we investigate two methods to reduce the size of a baseline source phrase index implementation.

2 Related Work

Zens and Ney [3] describe a phrase table architecture on which the binary phrase table of Moses is based. The source phrase index consists of a prefix tree. Memory

P. Sojka et al. (Eds.): TSD 2012, LNCS 7499, pp. 320–327, 2012.

requirements are low due to on-demand loading. Disk space requirements however can become substantial. A suffix-array based implementation of a phrase table for Moses that can create phrase pairs on-demand is introduced by Levenberg et. al [4]. While it surely is a promising alternative, we do not compare this approach with ours, as they differ in assumptions.

Other approaches, based on phrase table filtering [5], can be seen as a type of compression. They reduce the number of phrases in the phrase table by significance filtering and thus reduce space usage and improve translation quality at one stroke.

The architecture of the source phrase index of the discussed phrase table has been inspired by the efforts concerned with language model compression and randomized language models [6,7]. Guthrie et. al [7] who describe a language model implementation based on a minimal perfected hash function and fingerprints generated with a random hash function is the greatest influence.

3 Experimental Data

The presegmented version of Coppa, the Corpus Of Parallel Patent Applications [8], a parallel English-French corpus of WIPO's patent applications published between 1990 and 2010, is chosen for phrase table generation. It comprises more than 8.7 million parallel segments with 198.8 million English tokens and 232.3 million French tokens. The phrase table that is used throughout this paper has been created with the standard training procedure of Moses. Word alignment information is included. There are ca. 2.15×10^8 distinct source phrases and 3.36×10^8 phrase pairs.

4 Compact Phrase Table Implementation

Figure 1 illustrates the architecture of the discussed phrase table implementation. One important design guideline is the representation of most of the data in plain array structures that can be either fully read into memory or directly mapped from disk to memory. The phrase table consists of three main modules that are described in more detail in the following subsections.

4.1 Source Phrase Index

The source phrase index assigns integer positions to source phrases. As mentioned above, its structure is inspired by [7] who use a similar implementation for huge n-gram language models. The most important part of the index is a minimal perfect hash function (MPH) that maps a set S of n source phrases to n consecutive integers. This hash function has been generated with the CHD algorithm included in the freely available CMPH[1] library [9]. The CHD algorithm generates very small MPH (in this case 109 Mbytes) in linear time.

---

[1] http://cmph.sourceforge.net

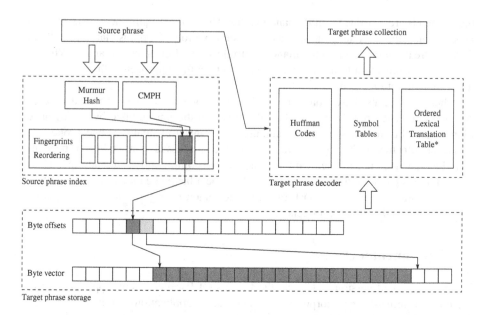

Fig. 1. Simplified phrase table implementation schema

The MPH is only guaranteed to map known elements from S to their correct integer identifier. If a source phrase is given that has not been seen in S during the construction of the MPH, a random integer will be assigned to it. This can lead to false assignments of target phrase collections to unseen source phrases, so-called *false positives*. Guthrie et. al [7] propose to use a random hash algorithm (MurmurHash3[2]) during construction and store its values as fingerprints for each phrase from S. For querying, it suffices to generate the fingerprint for the input phrase and compare it with the fingerprint stored at the position returned by the MPH function. If it matches, the phrase has been seen and can be further processed. For 32 bit fingerprints there is a probability of 2^{-32} of an unseen source phrase slipping through. Fingerprints are stored in an array at the position assigned to the phrase by the MPH. Since for each source phrase one fingerprint has to be stored, the array of fingerprints uses around 820 Mbytes.

MPHs generated using the CHD algorithm are not order-preserving, hence the original position of the source phrase is stored together with each fingerprint. Results for lexicographically ordered queries lie close or next to each other. If the data is stored on disk, this translates directly to physical proximity of the data chunks on the drive and less movement of the magnetic head. Without order-preservation the positions assigned by the MPH are random which can render a memory-mapped version of the phrase table near unusable. For each source phrase a 32 bit integer reordering position is stored, this consumes another 820 MBytes. In total the source phrase index consumes 1,750 MBytes. In this paper we examine two methods that allow to shrink memory consumption in comparison to the described baseline.

---

[2] http://code.google.com/p/smhasher/wiki/MurmurHash3

4.2   Target Phrase Storage

The target phrase storage contains collections of target phrases at position assigned by the index. It consists of a byte vector that stores target phrase collections consecutively according to the order of their corresponding source phrases. A target phrase collection consists of one or more target phrases. A target phrase is a sequence of target word symbols followed by a special stop symbol, a fixed-length sequence of scores, and a sequence of alignment points followed again by a special stop symbol.

Random access capability is added by the byte offset vector. For every target phrase collection, it stores the byte offset at which this collection starts. By inspecting the next offset the end position of a target phrase collection can be determined. While the byte vector is just a large array of bytes, the byte offset vector is a more sophisticated structure. Instead of keeping offsets as 8-byte integers[3], differences between the offsets are stored. A synchronization point with the full offset value is inserted and tracked for every 32 values.[4] This turns the byte offset vector into a list of rather small numbers, even more so when the byte array is compressed. Techniques from inverted list compression for search engine indexes are used to reduce the size further: Simple-9 encoding for offset differences and Variable Byte Length encoding for synchronization points.

Size reduction is achieved by compressing the symbol sequence of a target phrase collection, for instance using symbol-wise Huffman coding. The number of Huffman codes can become large for a particular type of symbols, e.g. there are nearly 13.6 million distinct scores (32 bit floats). Three different sets of Huffman codes are used to encode target phrase words, scores, and alignment points. In every set are as many codes as distinct symbols of the encoded type. This variant of the phrase table is denoted as "Baseline". Another encoding schema — Rank-Encoding (denoted "R-Enc") — has been proposed by [2]. Target phrase words are encoded using the positions of source phrase words they are aligned with. This reduces the entropy of target words and results in shorter Huffman codes. See Table 2 for the size characteristics of both these variants.

4.3   The Phrase Decoder

The target phrase decoder contains the data that is required to decode the compressed byte streams. It includes source and target word lists with indexes, the Huffman codes, and if Rank Encoding is used a sorted lexical translation table. Huffman codes are stored as canonical Huffman codes, a memory efficient representation.

5   Memory Requirements for Order Preservation

Order preservation comes at a cost of 4 bytes (32 bits) per source phrase. This can be significantly reduced if the order itself is exploited. For lexicographically sorted phrases, the cost for a chunk of order information can be reduced to n bits if every 2^n-th source phrase is stored explicitly. We call these phrases *landmarks*. That way

---

[3] 4-byte integers could hold byte offsets up to 4 Gbytes only.

[4] This is an arbitrarily set step size.

a partition of 2^n equally[5] sized ranges on the set of sorted source phrases S is created. For each range an MPH is generated together with corresponding fingerprint arrays and reordering arrays. Landmarks are stored in a sorted structure and retrieved by binary search.

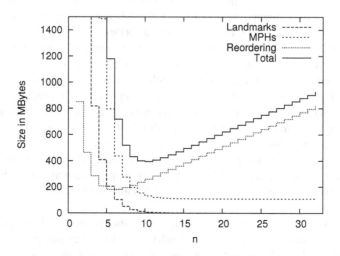

Fig. 2. Memory-requirement for chosen n (Coppa phrase-table)

Given a source phrase s located in the sorted phrase table at position p, the value $\hat{p}$ in the reordering array corresponding to s is calculated as $\hat{p} = p \bmod 2^n$. This value is always smaller than 2^n and can be stored in a n-bit sized storage.

Querying is done as follows: For a phrase s, the largest i is determined among the sorted landmark phrases, for which lexicographically $s_i \leq s$. The phrase s is hashed using the i-th MPH which assigns the precalculated fingerprint and the position $\hat{p}$ to s. Using MurmurHash3 a fingerprint is calculated for s. If fingerprints match, the original position $p = 2^n i + \hat{p}$ is returned.

Fig. 2 depicts the relation between n and memory requirements for landmark phrases, reordering information, and MPHs using the example of the Coppa phrase table. Fingerprints are omitted since their memory usage is fixed at 820 MBytes. The minimum of 393 MBytes is reached for $n = 10$ with 1024 phrases per range. For smaller n the space usage of the MPHs and the landmark phrases is growing rapidly. This is due to the exponential increase in number of landmarks, MPHs, and reordering arrays for small n. Overhead per MPH is ca. 100 bytes, per reordering array 8 bytes. For $n = 8$ memory consumption (425 MBytes) is slightly increased in comparison to $n = 10$, but full byte values are easier stored. This variant is denoted as "Ord-8". It can be safely assumed that a choice of n between 8 and 12 results in general in a significant space reduction.

---

[5] The last range may be smaller.

6 Memory-Requirements for Fingerprints

6.1 Estimating an Optimal Fingerprint Size

Fingerprint bit length is a certain candidate for optimization, but this may affect translation quality. In this section a method is described that allows to estimate a safe and space-optimal bit number *before* the construction of the compact phrase table using the text version of the phrase table.

When creating an SMT system, typically in-domain test sets or development sets distinct from the training data are available. Such a sample text can be used to create a set P of all phrases up to a fixed length[6] occurring in that text. During the translation of the sample, the phrase table is being queried with all phrases from P. As introduced above, S is the set of all source phrases present in the phrase table. The intersection $P \cap S$ is the set of phrases from the sample for which translations exist. Conversely, $P \setminus S$ is the set of unseen phrases that have no translation. The following sums are defined:

$$C_P = \sum_{p \in P} c(p), \quad C_S = \sum_{p \in P \cap S} c(p), \quad C_U = \sum_{p \in P \setminus S} c(p) = C_P - C_S \quad (1)$$

where $c(p)$ is the number of occurrences of a phrase p in the sample. C_P is the sum of occurrences of all phrases in the sample, C_S counts occurrences of all seen sample phrases, C_U is the number of occurrences of all unseen sample phrases.

$C_{\text{fp}}(b)$ is the observed number of false positives that occur during the translation of the sample text if a fingerprint size of b is chosen. Dividing $C_{\text{fp}}(b)$ by N — the number of sentences in the sample text — a more comparable observed per-sentence false positive ratio denoted as ofpr(b) is obtained. This value is not known yet, but the expected value can be calculated as follows:

$$\text{efpr}(b) = E\left[\text{ofpr}(b)\right] = E\left[\frac{C_{\text{fp}}(b)}{N}\right] = \frac{C_U}{2^b N}. \quad (2)$$

By fixing efpr(b) at a desired value f, the number of fingerprint bits b is given as:

$$\text{bits}(f) = \left\lceil \text{efpr}^{-1}(f) \right\rceil = \left\lceil \log_2 \frac{C_U}{f \cdot N} \right\rceil. \quad (3)$$

WIPO released a distinct test set[7] for the COPPA corpus, for which the following values are obtained: $N = 7{,}122$, $C_P = 919{,}755$, $C_S = 487{,}137$, and $C_U = 432{,}618$. For this test set and a bit size of $b = 32$, a per-sentence false positive rate efpr(b) = 1.414×10^{-8} is estimated, i.e. a false positive is expected to occur once every 71 million sentences. It seems reasonable to choose a smaller b, for instance, by fixing efpr(b) = 0.001 (one false positive occurring every 1,000 sentences) which results in bits(0.001) = 16. A 16 bit fingerprint reduces the size requirements by half.

---

[6] Standard settings in the Moses training procedure limit source and target phrases to 7 symbols. The decoder should be configured to use the same phrase length limit.

[7] http://www.wipo.int/patentscope/translate/coppa/testSet2011.tmx.gz

6.2 Fingerprint Size and Translation Quality

Table 1 compares the epfr(b) for chosen fingerprint sizes with observed false positive rates for the WIPO test set. Additionally, the obtained BLEU scores are given. The column "Used" contains the total number of sentences in which false positives actually surfaced causing an unwanted translation result.

Table 1. BLEU score in relation to expected and observed false positive rates

b	efpr(b)	ofpr(b)	Used	BLEU
32	$1.41 \cdot 10^{-8}$	0	0	46.86
24	$3.62 \cdot 10^{-6}$	0	0	46.86
20	$5.79 \cdot 10^{-5}$	$1.42 \cdot 10^{-4}$	1	46.86
16	$9.27 \cdot 10^{-4}$	$5.68 \cdot 10^{-4}$	3	46.86
12	$1.48 \cdot 10^{-2}$	$1.42 \cdot 10^{-2}$	36	46.79
8	0.24	0.25	494	46.02
4	3.80	3.78	3438	37.35

As expected, 32 bit down to 24 bit fingerprints are very successful in prohibiting false positives. For 20 bits a single false positive appears as a translation option and also surfaces in one translation (out of over 7,000). Similarly, 16 bit fingerprints offer an acceptable protection, only 3 sentences have been translated differently. BLEU scores remain unchanged for 20 bit and 16 bit fingerprints. For 12 bits there is a small decrease in BLEU that might still be acceptable for some applications. Bit lengths of 8 and lower have a serious effect on BLEU and should not be used. We choose 16 bit fingerprints for the Coppa phrase table. The translation quality seems acceptable and again full byte values can be used. This variant is denoted as "Fprt-16".

7 Summary and Future Work

Table 2 summarizes the memory consumption for variants of the presented phrase table. Variants to the right include the memory reduction methods of the previous variants. Figures for the Moses binary phrase table are given for comparison. The memory usage of the source phrase index was reduced by 52 percent without sacrificing translation quality in terms of BLEU. If target phrase compression techniques are applied (described in more depth in [2]), the whole phrase table is reduced by 34 percent compared to the baseline. The final variant of the described phrase table implementation uses only 17 percent disk space of the Moses binary phrase table.

Future work will address the problem of scores in the phrase table which currently consume the most space. For information on performance see [2]. First experiments suggest that the described phrase table allows much faster decoding than the Moses binary phrase table, especially on machines where disk I/O is a bottle neck. Also, the described source index implementation seems to be a good starting point for a distributed implementation.

Table 2. Comparison of phrase table implementations

	Moses	Baseline	R-Enc	Ord-8	Fprt-16
Total size in Mbytes :	29,418	7,681	5,967	5,463	5,052
Ordered source phrase index (Mbytes):	5,953	1,750	1,750	1,246	835
Bytes per source phrase:	29.1	8.5	8.5	6.1	4.1
Target phrase storage (Mbytes):	23,441	5,873	4,127	4,127	4,127
Target phrase decoder (Mbytes):	—	59	90	90	90

Acknowledgements. This paper is based on research funded by the Polish Ministry of Science and Higher Education (Grant No. N N516 480540).

References

1. Koehn, P., Hoang, H., Birch, A., Callison-Burch, C., Federico, M., Bertoldi, N., Cowan, B., Shen, W., Moran, C., Zens, R., Dyer, C., Bojar, O., Constantin, A., Herbst, E.: Moses: Open Source Toolkit for Statistical Machine Translation. In: Annual Meeting of the Association for Computational Linguistics (ACL). The Association for Computer Linguistics, Prague (2007)
2. Junczys-Dowmunt, M.: A Phrase Table without Phrases: Rank Encoding for Better Phrase Table Compression. In: Proc. of the 16th Annual Conference of the European Association for Machine Translation (EAMT), pp. 241–252 (2012)
3. Zens, R., Ney, H.: Efficient Phrase-table Representation for Machine Translation with Applications to Online MT and Speech Translation. In: Proc. of the Human Language Technology Conference of the North American Chapter of the Association for Computational Linguistics (2007)
4. Levenberg, A., Callison-Burch, C., Osborne, M.: Stream-based Translation Models for Statistical Machine Translation. In: Human Language Technologies: The 2010 Annual Conference of the North American Chapter of the Association for Computational Linguistics, pp. 394–402 (2010)
5. Johnson, J.H., Martin, J., Fost, G., Kuhn, R.: Improving translation quality by discarding most of the phrasetable. In: Proc. of EMNLP-CoNLL 2007, pp. 967–975 (2007)
6. Talbot, D., Brants, T.: Randomized Language Models via Perfect Hash Functions. In: Proc. of ACL 2008: HLT, pp. 505–513. Association for Computational Linguistics, Columbus (2008)
7. Guthrie, D., Hepple, M., Liu, W.: Efficient Minimal Perfect Hash Language Models. In: Proc. of the 7th Language Resources and Evaluation Conference (2010)
8. Pouliquen, B., Mazenc, C.: COPPA, CLIR and TAPTA: three tools to assist in overcoming the language barrier at WIPO. In: MT-Summit 2011 (2011)
9. Belazzougui, D., Botelho, F.C., Dietzfelbinger, M.: Hash, Displace, and Compress. In: Fiat, A., Sanders, P. (eds.) ESA 2009. LNCS, vol. 5757, pp. 682–693. Springer, Heidelberg (2009)

Common Sense Inference Using Verb Valency Frames

Zuzana Nevěřilová and Marek Grác

NLP Centre, Faculty of Informatics, Masaryk University
Botanická 68a, 602 00 Brno, Czech Republic
{xpopelk,xgrac}@fi.muni.cz

Abstract. In this paper we discuss common-sense reasoning from verb valency frames. While seeing verbs as predicates is not a new approach, processing inference as a transformation of valency frames is a promising method we developed with the help of large verb valency lexicons. We went through the whole process and evaluated it on several levels: parsing, valency assignment, syntactic transformation, syntactic and semantic evaluation of the generated propositions.

We have chosen the domain of cooking recipes. We built a corpus with marked noun phrases, verb phrases and dependencies among them. We have manually created a basic set of inference rules and used it to infer new propositions from the corpus. Next, we extended this basic set and repeated the process. At first, we generated 1,738 sentences from 175 rules. 1,633 sentences were judged as (syntactically) correct and 1,533 were judged as (semantically) true. After extending the basic rule set we generated 2,826 propositions using 276 rules. 2,598 propositions were judged correct and 2,433 of the propositions were judged true.

1 Introduction

The cookbook "story" *fry the onion till it looks glassy* means peel a fresh, uncooked onion, chop it, put grease into a cooking pot and heat it, put the onion into the pot and wait until the onion looks glassy. In NLP systems we have to deal with implicit information to resolve "stories" such as: fry the onion till it looks glassy, reduce heat and cover. Where the heat comes from? What to cover?

Texts in natural languages usually contain "facts" (also known as common sense propositions or common sense facts) that are considered to be true in "normal" situations (also referred as stereotypical information [1]), e.g. fried onion looks glassy. This information is obvious for humans therefore rarely mentioned. The problem of the implicit information has been recognized since the beginning of the AI research. There are many approaches including frames [2] or scripts [3].

In this paper we concentrate on inferring new propositions from verb valency frames. The technique is based on transformations[1] on syntactic level and evaluation

---

[1] The word *transformation* is not linked to Chomsky's transformational grammar, but to Sowa's broader definition of logic as "any precise notation for expressing statements that can be judged true or false". In the same context an inference rule is defined as "a truth-preserving transformation: when applied to a true statement, the result is guaranteed to be true" [4].

P. Sojka et al. (Eds.): TSD 2012, LNCS 7499, pp. 328–335, 2012.
© Springer-Verlag Berlin Heidelberg 2012

on semantic level. The system works with syntactic units – noun phrases (NP)[2] and verb phrases (VP), but during the evaluation the meaning of the proposition is examined. A corpus of cooking recipes was created and the work is thus related to the cooking domain. This domain is quite strictly delimited and moreover it contains verbs describing mostly actions (fry, pour, cut etc.).

The aim of this inferencing prototype is to answer questions such as "do this meal contain gluten?" or "do I need a blender to cook this meal?" While the answer to the former question is not yet reachable, the answer to the latter can be provided already.

The paper structure follows: Section 2 depicts using verbs as predicates w.r.t. Czech. Section 3 focuses on recognizing textual entailment. In Section 4 we describe the whole process of inferring new propositions in detail. We start with describing the nature of the cooking recipes language, the processes of annotating the corpus and building inference rules. Afterwards, the inference algorithm is provided with example outputs. The number of rules increased automatically using verb valency lexicon VerbaLex. Section 5 provides evaluation and discussion respectively. Section 6 proposes further development directions.

2 Verb Frames and Semantics

Verbs mostly describe an action or state. Since the verb "is the hook upon which the rest of a sentence hangs" [5], it is often seen as a predicate (for example $tastelike(x, y)$ means that x tastes like y). *Verb valency* then refers to the number of arguments of a verbal predicate. *Syntactic valencies* describe the syntactic properties (such as subject or object) of an argument. In Czech (as well as most other Slavic languages) syntactic properties are expressed by the case and possibly a preposition (e.g. syntactic subject is in nominative).

Semantic valencies assign semantic roles to arguments of a verbal predicate. "A semantic role is the underlying relationship that a participant has with the main verb in a clause" [6].

VerbaLex [7] is a large valency lexicon of Czech verbs and their arguments (in frame lexicons often called slots). It captures the syntactic information (prepositions and cases of the arguments in VerbaLex) as well as semantics (reference to semantic roles and Princeton WordNet [8] (PWN) hypernym.

3 Relationship to Recognizing Textual Entailment

Recognizing Textual Entailment (RTE) is a sub-problem of NLP. Its focus is in determining if a statement (called *hypothesis*) can be inferred from a given text. RTE systems consist of syntactic parsing, role labeling, named entities recognition, logical representation and other modules.

Apart from the ad-hoc and shallow approaches the sound approaches (e.g. [9]) use tree transformation operations that generate the hypothesis from the given text and knowledge based operations.

---

[2] For the purpose of this paper we take noun phrases and prepositional phrases together and later call it NPs.

Since our work is linguistically-motivated we use syntactic parsing but the result is not a complex dependency tree but rather one or more small trees or better a *bush*. This notion is discussed in Section 4.2.

At the first stage we did not use a *knowledge base* for recognizing the entities in the text. However the assumption that knowledge base will improve the results is expressed at the end of this paper. On the other hand the set of manually created inference rules itself is a knowledge base. It adds information about arguments of the verbs and relations between different verb frames. It also introduces new entities such as ingredients and cookware.

4   Common Sense Inference in Cooking Recipes

4.1   The Language of Cooking Recipes

The language of cooking recipes differs from the general language in the following attributes:

- use of imperative. In Czech cooking recipes most cooking recipes authors use first person plural (literally "we fry the onion...") instead of imperative. Sometimes, infinitive or imperative forms are used. In all verbs occurring in cooking recipes 6 % were imperatives, 51 % were indicatives in first person plural, 11 % were infinitives (some of these infinitives are bare, i.e. are together with another verb such as "let the onion fry"). The remaining verbs were 3rd person indicatives (such as "the onion looks glassy").
- frequent use of phrasal coordinations of NPs and of VPs: in cooking recipes corpus there are approx. three times more coordinations than in a corpus of blog texts.

4.2   Building Annotated Corpus

The annotation method was that of the BushBank project [10]. The corpus was annotated on several language levels: tokens (words and sentence boundary marks), morphology (lemma and morphologic tag for tokens), syntactic structures (NPs, VPs, coordinations and clauses), relations between syntactic structures (dependencies). The annotation of tokens was done purely by annotators' intuition since it is straightforward for humans to detect word and sentence boundaries. Detecting boundaries of NPs and VPs (or better detecting errors in NPs' and VPs' boundaries) was also quite an easy task. For searching dependencies verb valency lexicon VerbaLex was used. However, we did not find a way of using VerbaLex automatically because of high semantic ambiguity of verbs. Therefore annotators only consulted VerbaLex during their work.

Data for annotation were obtained purely by automatic tools (desamb [11], SET [12]) and validity of syntactic structures and their relations were confirmed during manual annotation. This means that structures that were not identified by automatic tools could not be added by annotators.

This was done contrary to traditional requirements in which we tried to obtain *completeness* of the annotation. BushBank ideas put greater emphasis on *simplicity* of the annotation (without definition of all border-line cases), *usability* (proved by the evaluation described in Section 5) and *rapid-development* (annotation itself was done in 40 (wo)man-hours). As we are working on a concept, data were manually checked by just one annotator. We plan using at least two annotators in the future with measuring their agreement.

4.3 Inference Rules

The inference rule for a particular sentence contains an input verb phrase *Vinput*, the output verb phrase *Voutput*, information about the grammatical polarity[3] preservation n, information on how the arguments participate in the inference process (syntactic rules S) and inference type t (see below), in short the rule I is a tuple $I = (Vinput, Voutput, n, S, t)$. The grammatical polarity preservation allows to formulate rules that result in sentences with opposite grammatical polarity (e.g. "*x* cooks *y*" effects in "*y* is *not* raw").

Each syntactic rule $S \in S$ is a pair of syntactic properties of the *Vinput* dependent *SPinput* and syntactic properties of the *Voutput* dependent *SPoutput*, in short $S = (SPinput, SPoutput)$. Syntactic property is a pair of the appropriate case of the dependent and a preposition. Prepositions can be either none (direct case) or prepositions agreeing to a case. Case is marked by a number[4].

The inference itself is a process of filling up an output verb frame with definite arguments and creating another verb frame with some of the previous arguments. The algorithm is described in Section 4.4.

```
<inference type="effect" verb="dochutit" mean="to_flavour">
  <ruleset id="taste_like" inf_verb="chutnat" negation="False">
    <rule case="c4" prep="" inf_case="c1" inf_prep=""/>
    <rule case="c7" prep="" inf_case="c6" inf_prep="po"/>
  </ruleset>
</inference>
```

Fig. 1. Example of the inference rule notation: *toflavour*(x, y, z) (has effect) *totaste*(y, z) means that "*x* flavours *y* with *z*" has effect "*y* tastes like *z*" (*x* is not part of the inference and therefore is not mentioned). Translation: *dochutit*=to flavour, *chutnat*=to taste, *po*=preposition (here meaning "like").

The system covers the following inference types t: effect (66 rules), precondition (47 rules), near synonymy (75 rules), conversion between active and passive verb forms (24 rules).

The inference rules were created by expert linguists according to their introspection and experience. Figure 1 shows an example of the inference rule description.

After creation the inference algorithm (described below) was applied and the resulting propositions were evaluated syntactically. This evaluation lead back to *a*) inference rules correction *b*) improvements of syntactic rules for sentence generation. Table 1 shows progressive improvements after each such cycle.

---

[3] The distinction of affirmative and negative.

[4] 1 – nominative, 2 – genitive, 3 – dative, 4 – accusative, 6 – locative, 7 – instrumental.

Table 1. Manual creation and evaluation of inference rules. After 3 cycles the system did significantly improve its outputs. At the same time the number of applied rules increased.

cycle	# of generated propositions	# of correct propositions	# of rules used
1	1,792	1,016	168
2	1,734	1,415	174
3	1,783	1,633	175

4.4   The Inference Algorithm and Example Outputs

For the verb phrase present in the sentence $Vinput \in Sinput$ find all inference rules that contain $Vinput$ as a input verb phrase.

For each $I = (Vinput, Voutput, n, S, t)$:

1. find all dependents $D_1, \ldots, D_n$ of $Vinput$ in $Sinput$.
2. transform $Vinput$ to $Voutput$. Since the Czech language uses declination we have to find the right form of the output verb. This is done by a) determining the morphological tag of $Vinput$ (person, number, tense, polarity), b) generating an appropriate verb form for $Voutput$. For this purpose we have used the morphological analyzer/generator majka [13]. In case of passive verbs we have to transform the grammatical categories of the passive participle (number and gender) as well. This transformation depends on grammatical categories of the inferred subject ($SPoutput_i$ where case is nominative).
3. determine all dependents $\{D_1, \ldots, D_k\}$ that have syntactic properties of $SPinput$
4. transform all the dependents selected in the previous step according to their corresponding rule $S = (SPinput_i, SPoutput_i)$. The NP has to change its form according to the new case. The case was determined during the annotation, the preposition is changed (to that of $SPoutput_i$) and a new form of the NP is generated using the morphological analyzer/generator majka [13].
5. generate a sentence $Soutput$ from $Voutput$ and transformed dependents. The new proposition is generated as a sequence of $SPoutput_1, \ldots, SPoutput_l$ and $Voutput$. Since Czech is a nearly free word order language, at least the NPs' order can be interchanged without worries.

The ideal case happens when all dependents in $Sinput$ are transformed to dependents of *output*. Some inferences were incomplete. Examples of both types are shown in Table 2.

4.5   Adding New Rules Using VerbaLex

It is obvious that manual creation of inference rules cannot lead to significant results in short-term. We extended the existing rule set using VerbaLex. We extracted all verb frames that contain verbs $Vinput$ and one of their functors is SUBS(tance). Afterwards, we have created rules by replacing $Vinput$ by its synonyms in the particular verb frame. Using this procedure we added aspectual pairs, writing variants and synonyms with the same syntactic valencies.

The number of inference rules increased from 212 to 599. Since in VerbaLex, only 3% of verbs are biaspectual, about half of the new rules employ different aspect of the

Table 2. Example (syntactically and semantically correct) outputs

Sinput	*t*	*Soutput*
Rozpustíme máslo (Melt the butter)	effect	máslo může být tekuté (butter can be liquid)
Kuřecí prsa naklepeme natenko (Tenderize chicken breast thinly)	precondition	kuřecí prsa jsou druh masa (Chicken breast is a meat)
Nahrubo nastrouháme všechny sýry (Grate coarsely all sort of cheese)	precondition	na všechny sýry vezmeme struhadlo (We take a grater suitable for all sort of cheese)
Občas trochu podlít vodou (Occasionally baste with water)	equals	vodu nalíti (Pour water)

same verb (e.g. vysušit/perfective – vysoušet/imperfective – both meaning to dry up). The rest of the rules was generated from synonyms.

Afterwards, we did the same evaluation as described in Table 1. N.B. that the number of rules used in inference did not increase accordingly to the increase of the rules. It is caused by the fact that some of the late rules lead to sentences that are correct but unnatural, e.g. *tvarovat těsto (to shape a pastry)* is usual but *modelovat těsto (to form a pastry)* is not.

5 Evaluation

We have picked up 164 verbs occurring in Czech cooking recipes. For these verbs 212 inference rules were manually created. The inference process was tested on a corpus of 37 thousands tokens (2,400 sentences). As the result 1,783 new sentences were generated and evaluated.

Evaluation proceeded on two levels: syntactic and semantic. At each level annotators had to decide whether or not the new sentence is correct. We then observed and classified the types of errors.

From 1,783 sentences 1,633 were evaluated syntactically correct (using 175 rules). Afterwards, within 1,633 sentences 1,533 was evaluated semantically correct (the new proposition was judged true given the original proposition).

Next step went with extending the rule set automatically. By adding synonyms from VerbaLex, we increased the number of inference rules nearly three times – up to 599 rules. From these 599 rules, 276 were used, 2,826 propositions were generated and 2,598 of them were evaluated syntactically correct. From these 2,598 propositions 2,443 were evaluated as semantically correct.

Inference most often results in incomplete propositions. This is caused by frequent occurrence of verb coordinations and other ellipses. A working solution would be to generate sentences for the verb coordinations, e.g. from "chop onions, stir and cover" generate "chop onions", "stir onions" and "cover (the pot containing) onions". Ellipses could also be solved by anaphora resolution, e.g. "chop onions and put it into the pot" will lead to "chop onions" and "put onions into the pot". These features should be implemented as a preprocessing module to the parser.

Second, parsing was not successful on NPs or VPs containing unknown words. For this reason we plan to use named entity taxonomies (e.g. from Wikipedia pages) as a preprocessing prior to parsing. Unknown words were exclusively from the food domain, e.g. amasaké (amazake drink), feta (feta cheese), mascarpone, žervé (fresh cheese). These words very often are not inflected.

6   Conclusion and Future Work

We made a prototype application for automatic inference. We went the whole way from syntactic parsing, NPs and VP detection, verb valency assignment and generating new sentences in Czech.

The results can be improved by a good preprocessing tools such as clause generator from verb coordinations and (domain specific) named entities recognizer. These tools should improve the syntactic parsing significantly but currently are not ready to use with Czech language. Involving taxonomies (such as Czech WordNet [14]) in the process seems to be a good choice, however currently we do not have one that is rich enough and contains Czech literals at the same time.

Second, we plan to create even more rules since we know that the coverage even on the restricted domain is still low. To obtain more inference rules automatically we plan to use verb grouping according to semantic role patterns (e.g. verbs such as pour, spill, sprinkle have the same semantic roles in their slots) [15].

The results have shown that we have to concentrate on automatic valency assignment instead of manual annotation. This next step can lead to the last one on a way from a prototype to a working application – an user interface that will allow users to ask questions on cooking recipes.

Acknowledgments. This work has been partly supported by the Ministry of Education of CR within the LINDAT-Clarin project LM2010013 and by the Czech Science Foundation under the project P401/10/0792.

References

1. Lehmann, D.J.: Stereotypical reasoning: Logical properties. CoRR cs.AI/0203004 (2002)
2. Minsky, M.: A framework for representing knowledge. Technical report, Massachusetts Institute of Technology, Cambridge, MA, USA (1974)
3. Schank, R.C., Abelson, R.P.: Scripts, Plans, Goals, and Understanding: An Inquiry Into Human Knowledge Structures (Artificial Intelligence), 1st edn. Lawrence Erlbaum Associates, Hardcover (July 1977)
4. Sowa, J.: Fads and fallacies about logic. IEEE Intelligent Systems, 84–87 (2007)
5. Kingsbury, P., Kipper, K.: Deriving verb-meaning clusters from syntactic structure. In: Proceedings of the HLT-NAACL 2003 workshop on Text meaning, vol. 9. HLT-NAACL-TEXTMEANING 2003, pp. 70–77. Association for Computational Linguistics, Stroudsburg (2003)
6. Loos, E.E., Anderson, S., Dwight, H., Day, J., Jordan, P.C., Wingate, J.D.: Glossary of linguistic terms (January 2004),
   http://www.sil.org/linguistics/GlossaryOfLinguisticTerms/

7. Hlaváčková, D., Horák, A.: Verbalex – new comprehensive lexicon of verb valencies for Czech. In: Proceedings of the Slovko Conference (2005)
8. Fellbaum, C.: WordNet: An Electronic Lexical Database (Language, Speech, and Communication). MIT Press, Hardcover (May 1998)
9. Stern, A., Dagan, I.: A confidence model for syntactically-motivated entailment proofs. In: Proceedings of the International Conference Recent Advances in Natural Language Processing 2011, RANLP 2011 Organizing Committee, Hissar, Bulgaria, pp. 455–462 (September 2011)
10. Grác, M.: Case study of BushBank concept. In: PACLIC 25th Pacific Asia Conference on Language, Information and Computation, pp. 353–361 (2011)
11. Šmerk, P.: Towards Morphological Disambiguation of Czech. Ph.D. thesis proposal, Faculty of Informatics, Masaryk University (2007)
12. Kovář, V., Horák, A., Jakubíček, M.: Syntactic Analysis Using Finite Patterns: A New Parsing System for Czech. In: Vetulani, Z. (ed.) LTC 2009. LNCS, vol. 6562, pp. 161–171. Springer, Heidelberg (2011)
13. Šmerk, P.: Fast morphological analysis of Czech. In: Proceedings of the RASLAN Workshop 2009, Masarykova univerzita (2009)
14. Pala, K., Smrž, P.: Building Czech Wordnet (7), 79–88 (2004)
15. Nevěřilová, Z.: Semantic Role Patterns and Verb Classes in Verb Valency Lexicon. In: Sojka, P., Horák, A., Kopeček, I., Pala, K. (eds.) TSD 2010. LNCS, vol. 6231, pp. 150–156. Springer, Heidelberg (2010)

A Genetic Programming Experiment
in Natural Language Grammar Engineering

Marcin Junczys-Dowmunt

Faculty of Mathematics and Computer Science
Adam Mickiewicz University
ul. Umultowska 87, 61-614 Poznań, Poland
junczys@amu.edu.pl

Abstract. This paper describes an experiment in grammar engineering for a shallow syntactic parser using Genetic Programming and a treebank. The goal of the experiment is to improve the *Parseval* score of a previously manually created seed grammar. We illustrate the adaptation of the Genetic Programming paradigm to the problem of grammar engineering. The used genetic operators are described. The performance of the evolved grammar after 1,000 generations on an unseen test set is improved by 2.7 points F-score (3.7 points on the training set). Despite the large number of generations no overfitting effect is observed.

Keywords: Shallow parsing, genetic programming, natural language grammar engineering, treebank.

1 Introduction

This paper describes an experiment in grammar engineering for a shallow syntactic parser using Genetic Programming and a treebank. The goal of the experiment is to improve the *Parseval* [1] score of a previously manually created seed grammar. The shallow parser cannot be easily trained on a treebank using classical grammar extraction methods like the creation of a probabilistic context-free grammar. Neither does the parser support weights nor is the grammar formalism context-free. Parsing rules are applied in the same order as they appear in the rule set, no search is carried out during parsing. This requires a human grammar engineer to tune the grammar very carefully.

The Genetic Programming [2,3] approach has been chosen because we regard it to be similar to the grammar engineering process as it is performed by a human grammar engineer, based on a trial-and-error search roughly guided by a treebank. We illustrate the adaptation of the Genetic Programming paradigm to the problem of grammar engineering by treating grammars as programs.

2 Related Work

Various flavors of genetic algorithms have been widely used for the induction of context free grammars (e.g. [4,5]), but only few (e.g. [6,7]) focus on natural language grammars.

P. Sojka et al. (Eds.): TSD 2012, LNCS 7499, pp. 336–344, 2012.

These works use Genetic Programming or other genetic algorithms for the unsupervised inference of context free grammars from unannotated corpora with rather unsatisfactory results.

For the training on annotated data like treebanks, direct grammar extraction approaches are dominant. Probabilistic context free grammars can be read out from the given tree structures, probabilities are based on frequency of the extracted structure occurring in the treebank, see for instance [8] for experiments on the Penn Treebank or [9] for results for the German TüBa D/Z treebank. We know of no work that uses treebanks to automatically improve manually created grammars.

3   The Seed Grammar

3.1   The Shallow Parser

The shallow parser used in our experiments is a component of the *PSI-toolkit* [10]. It is based on the *Spejd* [11] shallow parser and employs a similar rule formalism. It has been used as a parser for French, Spanish, and Italian in the syntax-based statistical machine translation application *Bonsai* [12] and in other applications.

The parsing process relies on a set of string matching rules constructed as regular expressions over single characters, words, part-of-speech tags, lemmas, grammatical categories etc. Apart from the matching portion of a rule, matching patterns for left and right contexts of the main match can be defined. The parse tree construction process is linear, matching rules are applied deterministically. The first match is chosen and a spanning edge is added to the result. No actual search is performed during parsing. The parsing process for a sentence is finished if during an iteration no rule can be applied. The parser cannot make use of weights or other disambiguation methods.

3.2   Manual Grammar Engineering

For this kind of parser the use of a treebank is limited, for instance, it is not possible to use a PCFG. Acceptable parsing quality is achieved by the careful manual construction of the parsing grammar. The order of the rules in a grammar determines the order of application to a sentence.

However, a treebank in combination with an evaluation metric like *Parseval* can be used to guide the grammar engineer through a trial-and-error process. A human grammar designer seeks to improve the grammar in such a way that the score is improved. If the amount of work that needs to be invested into the grammar is not reflected in further improvement, the construction can be regarded as finished.

Relying on *The French Treebank* [13], a basic French grammar has been created for the PSI-toolkit shallow parser which is also used for French-Polish translation in the mentioned *Bonsai* MT system. The manual work on that grammar was stopped at an F-score level of 76.55% on the development set (76.20% on the test set). This grammar serves as the starting point for the experiment.

4   Genetic Programming

Genetic programming (GP) [2,3] is a genetic algorithm flavor that evolves computer programs. A population of programs is optimized according to a fitness function that measures the performance of an individual on a training set. Programs that perform better than others are allowed to produce off-springs which are added to the next generation of computer programs. Repeating this process for a large number of iterations is expected to result in a population of programs that perform reasonably well at the task the fitness function was based on.

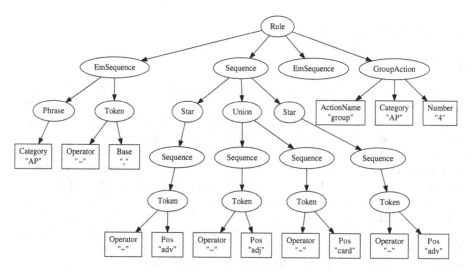

Fig. 1. Chromosome tree representation of rule `Example 1`

4.1   Chromosome Representation

Genetic programming evolves computer programs, which are most commonly represented as tree structures. Child nodes contain arguments to the functions represented by their respective parent nodes. Running a program is equivalent to a recursive evaluation of the tree. In the terminology of genetic algorithms such a tree representation is called a *chromosome*.

If the shallow parser is treated as a programming language interpreter, a grammar is nothing else but an interpretable program. This assumption makes it straightforward to apply GP directly to grammar engineering. The grammar only needs to be represented as a tree structure that can be modified by genetic operators. The original grammar is a text file consisting of rules similar to the following:

```
Rule "Example 1"
Left:   [type=AP] [base~","];
Match:  ([pos~"adv"])* ([pos~"adj"]|[pos~"card"]) ([pos~"adv"])*;
Eval:   group(AP,4);
```

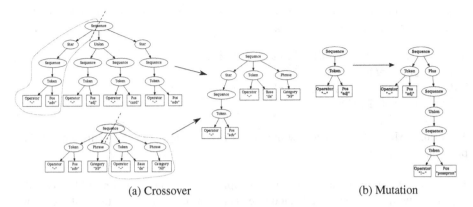

(a) Crossover   (b) Mutation

Fig. 2. Variable-arity genetic operators

This rule creates an adjective phrase from an adjective or a cardinal number that can be preceded or followed by any number of adverbs (including none). The rule is only allowed to create the adjective phrase if its immediate left context is another adjective phrase followed by a comma. Fig. 1 gives the tree representation of this rule. A full grammar consists of many such trees, the roots of which are children of a single RuleSet node. We differentiate between fixed-arity nodes (fixed number of children) and variable-arity nodes (variable number of children) for which specialized genetic operators are implemented.

4.2   Genetic Operators

The main genetic operators used in GP — and other genetic algorithms — are crossover and mutation. In our case two variants exist for each operator depending on the arity of the nodes involved. The genetic operators are type-aware and guaranteed to generate a correct grammar.

Crossover.   Crossover creates a new individual by combining genetic material of two individuals from the previous generation that have been selected for reproduction. A node of the first individual is randomly selected and based on its type a node of the same type is randomly selected from the second individual.

Fixed-arity Crossover.   Fixed-arity crossover simply replaces all child nodes (and attached subtrees) of the first chosen node with the children of the second node.

Variable-arity Crossover.   Variable-arity crossover is a variant of "cut-and-splice" crossover. The number of children for both chosen nodes may vary, therefore two random crossover points are chosen for each sequence of children. The new individual contains all children of the first node left of the first crossover point and all children of the second node right of the second crossover point. The example below is a possible result of the crossover of rule `Example 1` with another rule. Fig. 2a illustrates the operation for fragments of both rules.

```
Rule "Example 1 variable-arity crossover"
Left:  [type=AP] [base~","];
Match: ([pos~"adv"])* [base~"de"] [type=NP];
Eval:  group(AP,4);
```

Mutation. Mutation forms a new individual from a single selected individual of the previous generation.

Fixed-arity Mutation. Fixed-arity mutation replaces a randomly chosen child of a fixed-arity node with a randomly generated tree. The root of this tree has to be of the same type as the replaced node.

Variable-arity Mutation. Variable-arity mutation simply adds a randomly generated branch to the list of children of the mutated variable-arity node. The rule below is the result of the mutation of a node of rule `Example 1` illustrated in Fig. 2b.

```
Rule "Example 1 - variable-arity mutation"
Left:  [type=AP] [base~","];
Match: ([pos~"adv"])* ([pos~"adj"]
       (([pos!~"posspron"]))+|[pos~"card"]) ([pos~"adv"])*;
Eval:  group(AP,4);
```

4.3   Fitness Function

The measure how well an individual is doing compared to other individuals in its generation is called *fitness*. The labeled version of the *Parseval* [1] metric is a natural candidate for a fitness function in the case of grammar engineering. Therefore, the *Parseval* F-score is chosen as the first component of our fitness function. Obviously, the greater the F-score the fitter is the evaluated individual.

Bloat — a rapid increase in the size of evolved trees — is a common phenomenon for genetic programming. Many strategies have been proposed to counteract the bloat effect [3, p. 78]. We choose to use a multi-objective fitness function: apart from optimizing F-score, tree complexity measured as the number of nodes is also optimized by applying Ockham's razor: if two individuals have the same F-score, the smaller tree is considered fitter than its larger competitor.

4.4   Algorithm

Algorithm 1 contains the pseudo-code of the main evolution procedure. The parameter `seedGramar` is the seed grammar the algorithm is supposed to improve. The first population consists of `populationSize` copies of that grammar. We set `populationSize` to 500 individuals per generation and limit the evolutionary process to 1,000 generations (`maxGenerations`). The parameters `eliteFraction` and `selectFraction` are

---

Algorithm 1. Main evolution procedure

Require: seedGrammar, trainingSet, populationSize, eliteFraction, selectFraction, crossProb, mutateProb, maxGenerations, maxComplexity, order

```
seedScore ← EVAL(seedGrammar, trainingSet)
seedNodes ← COMPLEXITY(seedGrammar)
population ← {(seedGrammar, seedScore, seedNodes) | 1, 2, ..., populationSize}
for i ← 1 to maxGenerations do
    nextPopulation ← BESTN(population, order, populationSize × eliteFraction)
    selection ← SELECT(population, order, populationSize × selectFraction)
    for k ← 1 to popsize do
        offspring ← GENOP(selection, crossProb, mutateProb)
        offsprComplexity ← COMPLEXITY(offspring)
        if offsprComplexity > maxComplexity then
            score ← 0
        else
            score ← EVAL(offspring, trainingSet)
        end if
        nextPopulation ← nextPopulation ∪ {(offspring, score, offsprComplexity)}
    end for
    population ← nextPopulation
end for
return BESTN(population, order, 1)
```

---

both set to 0.2. This means that the first 100 best individuals of a population are copied unaltered to the next population and are the only ones allowed to reproduce. The total number of nodes in a tree is limited by `maxComplexity` which is set to 25,000. The two parameters `crossProb` and `mutateProb` determine the probabilities of the respective genetic operators. For our experiment, a mutation probability of 0.2 and a crossover probability of 0.8 were chosen. The sort order in the population is determined by the parameter `order` which implements Ockham's razor for the fitness function as described in the previous section.

5 Evaluation

5.1 Training and Test Data

The French Treebank [13] comprises ca. 21,100 syntactically annotated sentences. Originally, the first 15,000 sentences were set apart as a training set, but for this experiment only the first 2,500 sentences are used in order to make calculations feasible. For 1,000 iterations of the algorithm with 400 new individuals per iteration the training set needs to be parsed 400,000 times. The performance of the best individual in each generation is evaluated on the last 6,000 sentences of the treebank which have not been seen during training.

Table 1. Performance for chosen generations

		Training data			Test data		
Generation	Complexity	Precision	Recall	F-score	Precision	Recall	F-score
0	2,672	77.29	75.83	76.55	76.90	75.52	76.20
50	4,235	78.80	76.87	77.83	78.13	76.31	77.21
100	6,834	79.46	77.40	78.42	78.70	76.78	77.73
500	14,256	81.08	78.38	79.71	80.02	77.26	78.61
1,000	10,395	81.49	79.03	80.24	79.99	77.74	78.85

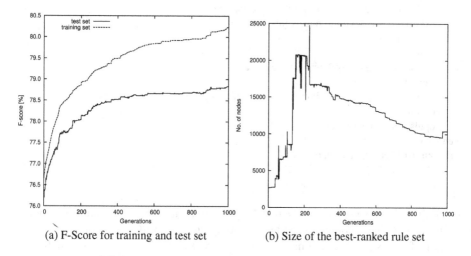

(a) F-Score for training and test set   (b) Size of the best-ranked rule set

Fig. 3. Progression of F-Score and Complexity

5.2 Results

Tab. 1 and Fig. 3 summarize the results for the best individual of each generation. The 0th generation represents the seed grammar. As shown on Fig. 3a, the increase in quality progresses slower on the test set than on the training data. However, during the first 1,000 iterations no overfitting seems to occur. From the 400th generation on, little improvement on the test data can be observed, but the curve follows the shape of the curve for the training data. The performance of the evolved grammar after 1,000 generations on the unseen test set is improved by 2.7 points F-score (3.7 points on the training set).

The changes in size of the best rule set in each generation are presented in Fig. 3b. At first, increasing the size of the rule set seems to be a good strategy to increase parse quality. A cut-off size of 25,000 nodes is applied and one might expect the grammars to reach this maximum size and to stay there for a longer time ([14]). Interestingly, apart form a single peak in the 226th generation this does not happen. Starting from this generation, the second optimization objective, rule set complexity, plays a greater role.

Size decreases steadily and compared to the rather dynamic first phase of the process no significant jumps in size occur.

6   Conclusions and Future Work

We demonstrated that a Genetic Programming algorithm can improve our complex, manually created natural language grammar with the help of a treebank by around 2.7 points *Parseval* F-score on an unseen test set. The grammar is non-probabilistic and not context-free. It can be assumed that the presented method could be used to any type of grammar that can be represented as a tree structure. The process is however quite resource-intensive.

Future research needs to focus on larger training data sets and better parameter settings for the presented algorithm. Another important direction will include the full extraction of a grammar from treebanks without a seed grammar. In that case a more popular treebank shall be used to make results comparable to traditional approaches to treebank grammars.

Acknowledgements. This paper is based on research funded by the Polish Ministry of Science and Higher Education (Grant No. N N516 480540).

References

1. Abney, S., Flickenger, S., Gdaniec, C., Grishman, C., Harrison, P., Hindle, D., Ingria, R., Jelinek, F., Klavans, J., Liberman, M., Marcus, M., Roukos, S., Santorini, B., Strzalkowski, T.: A Procedure for Quantitatively Comparing the Syntactic Coverage of English Grammars. In: Proceedings of a Workshop on Speech and Natural Language, San Francisco, pp. 306–311 (1991)
2. Koza, J.R.: The Genetic Programming Paradigm. In: Dynamic, Genetic, and Chaotic Programming, New York, pp. 203–321 (1992)
3. Poli, R., Langdon, W.B., McPhee, N.F.: A Field Guide to Genetic Programming (2008), http://www.gp-field-guide.org.uk
4. Dunay, B.D., Petry, F.E., Buckles, W.P.: Regular Language Induction with Genetic Programming. In: Proc. of the 1994 IEEE World Congress on Computational Intelligence, Orlando, pp. 396–400. IEEE Press (1994)
5. Keller, B., Lutz, R.: Learning Stochastic Context-Free Grammars from Corpora Using a Genetic Algorithm. University of Sussex (1997)
6. Smith, T.C., Witten, I.H.: A Genetic Algorithm for the Induction of Natural Language Grammars. In: Proc IJCAI 1995 Workshop on New Approaches to Learning for Natural Language Processing, pp. 17–24 (1995)
7. Korkmaz, E.E., Ucoluk, G.: Genetic Programming for Grammar Induction. In: 2001 Genetic and Evolutionary Computation Conference, San Francisco (2001)
8. Klein, D., Manning, C.D.: Accurate Unlexicalized Parsing. In: Proc. of the 41st Annual Meeting of the Association for Computational Linguistics, pp. 423–430 (2003)
9. Kübler, S., Hinrichs, E.W., Maier, W.: Is it really that difficult to parse German. In: Proc. of the Conference on Empirical Methods in Natural Language Processing, pp. 111–119 (2006)
10. Graliński, F., Jassem, K., Junczys-Dowmunt, M.: PSI-toolkit: A Natural Language Processing Pipeline. In: To appear in: Computational Linguistics — Applications. SCI. Springer

11. Przepiórkowski, A., Buczyński, A.: ♠: Shallow parsing and disambiguation engine. In: Proceedings of the 3rd Language & Technology Conference, Poznań (2007)
12. Junczys-Dowmunt, M.: It's all about the Trees — Towards a Hybrid Syntax-Based MT System. In: Proceedings of IMCSIT, pp. 219–226 (2009)
13. Abeillé, A., Clément, L., Toussenel, F.: Building a Treebank for French. In: Treebanks: Building and Using Parsed Corpora, pp. 165–188. Springer (2003)
14. Crane, E.F., McPhee, N.F.: The Effects of Size and Depth limits on Tree Based Genetic Programming. In: Genetic Programming Theory and Practice III, pp. 223–240. Springer (2005)

User Modeling for Language Learning in Facebook

Maria Virvou[1], Christos Troussas[1], Jaime Caro[2], and Kurt Junshean Espinosa[2]

[1] Department of Informatics, University of Piraeus, Piraeus, Greece
[2] Department of Computer Science, University of the Philippines Diliman
Quezon City, Philippines
{mvirvou,ctrouss}@unipi.gr,
jdlcaro@dcs.upd.edu.ph, kpespinosa@up.edu.ph

Abstract. The rise of Facebook presents new challenges for matching users with content of their preferences. In this way, the educational aspect of Facebook is accentuated. In order to emphasize the educational usage of Facebook, we implemented an educational application, which is addressed to Greek users who want to learn the Conditionals grammatical structure in Filipino and vice versa. Given that educational applications are targeted to a heterogeneous group of people, user adaptation and individualization are promoted. Hence, we incorporated a student modeling component, which retrieves data from the user's Facebook profile and from a preliminary test to create a personalized learning profile. Furthermore, the system provides advice to each user, adapted to his/her knowledge level. To illustrate the modeling component, we presented a prototype Facebook application. Finally, this study indicates that the wider adoption of Facebook as an educational tool can further benefit from the user modeling component.

Keywords: User Modeling, Facebook, Social Networking Sites, Intelligent Tutoring Systems, Computer Assisted Language Learning, Initialisation.

1 Introduction

In the last decade, we have witnessed major improvements in the area of information and communications technology, which induced changes in pedagogy. Currently, social networks are being adopted rapidly by millions of users worldwide, most of whom are students [19]. Social network tools support educational activities by making feasible the interaction, collaboration, active participation, information and resource sharing, and critical thinking between users [20]. Social networks may offer students the possibility of increasing avenues for learning. Hence, the use of educational applications along with the opportunity of communicating with peers in order to achieve a common purpose has become significant. Using social networks, such as Facebook, in educational and instructional contexts can be considered as a potentially powerful idea simply because students spend anyway a lot of their spare time on these online networking activities [21].

On the other hand, this rapid development accentuates the learning of multiple foreign languages, a phenomenon which is responsible for joining different cultures from all over the world. Considering the scientific area of Intelligent Tutoring Systems

P. Sojka et al. (Eds.): TSD 2012, LNCS 7499, pp. 345–352, 2012.

(ITSs), there is an increasing interest in the use of computer-assisted foreign language instruction. The need for tutoring systems that may provide user interface friendliness and also individualized support to errors via a student model are even greater when students are taught a foreign language through a social network. Student modeling may include modeling of students' skills and of declarative knowledge and can perform individualized error diagnosis of the student.The use of social networks in educational contexts is a burning issue in recent scientific literature. However, the integration of social networks in the learning environment along with the modeling of users who are trying to attain more robust learning opportunities is not yet well studied.

In view of the above, we propose an educational application in Facebook for learning the grammatical phenomenon of conditionals. This application is addressed to Greek students who want to learn conditionals in Filipino and vice versa. The users are modeled so that they can receive advice from the system. The prototype system combines an attractive multimedia interface and adaptivity to individual student needs in the social network. The communication between the system and its potential users as students is accomplished through the use of web services.

2   Related Work

In this section we present the scientific work for student modeling, related firstly with Social Networking Sites(SNSs) and secondly with Computer Assisted Language Learning (CALL).

The advent of SNSs has made an impact in various areas including educational context. Many researchers investigated the use of SNSs such as Facebook in the context of learning, specifically language learning. Piriyasilpa [10] discussed the effects of the inclusion of a Facebook activity as part of the language classroom. Ho-Abdullah et al. [7] also found out that the informal setting of a web-based social networking site has provided students opportunity for improving in their English language skills. Ota [12] investigated the benefits of SNS communities (i.e. Facebook) for learning Japanese as a second language (L2). They found out that SNSs has provided a portal for L2 learners to access other information and sources. Hiew [6] studied the perceptions of English as a Second Language (ESL) learners' in the use of Facebook journal as a channel for language learning outside the classroom. Furthermore, Promnitz-Hayashi [11] has found out that simple activities in Facebook has helped the less language-proficient students to become more actively involved in the language learning process.

On the other hand, Antal and Koncz [2] reviewed the student modeling problem for computer-based test systems and proposed a novel method for the graphical representation of student knowledge. Moreover, Ferreira and Atkinson [5] presented a model of corrective feedback for an ITS for Spanish as a foreign language and proposed the design of a component of effective teaching strategies into this ITS. Dickinson et. al. [3] designed a paper-based system that provides feedback on particle usage for first-year Korean learners, who learn a second language. Amaral et. al. [1] analyzed student input for different activities and motivated a broader perspective of student models. Tsiriga and Virvou [16] presented a framework for the initialization of student models in web-based educational applications. Finally, Virvou and Troussas [18] described a ubiquitous e-learning tutoring system for multiple language learning in which students "naturally" interact with the system in order to get used to electronically supported computer-based learning via user modeling and error proneness.

However, after a thorough investigation in the related scientific literature, we came up with the result that there was no implementation of language learning application in social networks and specifically in Facebook that incorporates user modeling. Hence, we implement a prototype application, which provides intelligence in its diagnostic component and offers advice based on students' performance.

3   General Architecture of the System

This section describes the general architecture of the user modeling for language learning in Facebook.

3.1   Description

Our system's architecture (Fig. 1) is an Intelligent Language Tutoring System (ILTS) that runs on the Facebook platform [4]. It takes advantage of the provided social networking APIs of the Facebook platform to handle basic web application tasks such as user account authentication, data persistence through the Graph API, and social analytics.

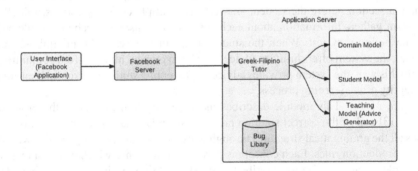

Fig. 1. Architecture of ILTS

3.2   Architecture

Our system is a multi-tiered client-server architecture which consists of the user inter-face (the Facebook application), the Facebook server, and the application server which hosts the ILTS modules: Intelligent Tutoring/Teaching Model, the Student Model, and the Domain Model.

1. User Interface. The User Interface is the Facebook application. In the Facebook platform [4], the application runs on a Canvas Page, which contains the HTML,

JavaScript, and CSS. When the user requests for a page, that request is sent through the Facebook server and is loaded back and rendered on an iframe on that page for user viewing. Then the user interface renders an adapted content generated by the tutoring module strategy for the said user.

2. Facebook Server. The Facebook Server acts as an intermediary of the application with the user. The user interacts with the Facebook application, which sends the request to the App server through the Facebook server. And the feedback is sent through the same path as discussed in [14].

3. Application Server. This hosts the core modules of the proposed system, namely the Tutoring/Teaching Model, Domain Model, Student Model. The Tutoring/Teaching Model includes the Advice Generator Component. All the core modules interact with a database server, which contains the domain representation, the student models, and the bug libraries.

a. Teaching Model. This module determines the pedagogical strategies that structure the interaction with the student [13]. It decides what item to teach next, what problem to give next and how to correct the bugs. Firstly, the student may start by studying the domain theory and then proceeding with the evaluation section, which consists of exercises in the form of multiple-choice questions. In the evaluation section, the system provides advice to students on preventing errors. The system evaluates the student's answer and delivers the appropriate feedback, namely a proposal of revising the theory. In each cycle, the said student model is dynamically updated. This process is supported by the main component of ILTSs, which is the intelligent error diagnosis [17] which makes use of the bug library.

b. Advice Generator Component. The advice generator component is activated in the evaluation section before the student answers the multiple choice questions. Initially, the system gathers information about each student from his/her Facebook profile and from the preliminary test. When the student is about to answer the multiple choice exercises, the system offers him/her advice in order to prevent him/her from committing error. In this way, the system provides advice in preventing students' errors and ensures the integrity of the learning process.

c. Domain Model. This module describes the necessary knowledge for the problem domain [13] and of the correct solution process [8]. In language learning, this module defines all the grammatical structures through what we can call a concept, which we can write as production rules. Each concept is part of a course unit, which can be a theory, an example, or an exercise as described in [9]. In addition, each course unit includes meta-descriptions such as level of difficulty and student cognitive level. Aside from the core concepts, this module also contains the bug library component, which is described next.

d. Bug Library Component. This is the library of the mal-rules in the chosen domain topic [13,8]. In the application domain, this refers to the common or possible mistakes in learning conditional grammatical structures in both language domains. This can also be expressed in production rules and is used by the tutoring module particularly in the intelligent error diagnosis.

e. Student Model. There are two phases of the student modeling: initial phase and the subsequent phase. The student model is initialized by gathering information about the user and by taking the preliminary test and is updated through all the interactions in the learning process. In the application domain, the Facebook user profile is the basis of the initial student model. Information concerning the user, such as the age, the country of origin and residence, the knowledge background and the number of spoken languages, are drawn from his/her Facebook profile. Each student will also take a preliminary test to assess the level of knowledge in the application domain. The aforementioned preliminary test will use the English language so that each student will be evaluated solely on his/her knowledge in the application domain. For this application, it is assumed that users are well-versed in the English language. Based on the student's performance in the preliminary test and the information gathered from his/her Facebook profile, the system will assign him/her into one of the four categories, namely: novice, intermediate, advanced, expert. Hence, at the start the system is able to gather personal data, knowledge of the domain if any, and student characteristics that will be helpful in initializing each student's model.

The stereotype and overlay techniques in classifying users are also applied [15] because it is taken in one single observation (as the case in the initialization phase), represents a model of the user as a subset of the expertise model.

Subsequent phase includes the multiple observations obtained during the interaction process. This helps in constructing a more reliable model using inductive methods like Bayesian networks. During the learning process, this module keeps a historical log of the student's weakness and progress [17].

4   Objective and Educational Usage of Facebook

Social networks, particularly Facebook, attracted great number of users in a short span of time [22]. Furthermore, it is highly considered as an educational tool because of its beneficial features such as enabling peer feedback and collaboration or either interactivity and active participation. It can enhance informal learning and support social connections within groups of learners and with those involved in the support of learning. The following model (Fig. 2) is constructed to shed light on the adoption of Facebook platform for our application along with its educational usage, by providing [21]:

a. Ease of access: The ease of use of many social networking services such as Facebook, can provide benefits to users by simplifying access to other tools and applications.
b. Familiarization: A significant factor, which influenced the adoption of Facebook, is the little technical knowledge that a user should possess to use it due to its widespread use and common acceptance.
c. Usefulness: Facebook enhances the individuals' productivity. Moreover, various opportunities, among which information sharing, collaboration and entertainment, influence the adoption of Facebook.
d. Social influence: Given that Facebook is a social utility used by many people worldwide, social norms must have a significant role in individuals' use of this tool [21]. People join Facebook in order to connect with several social environments or to keep the communication with their existing friends, while others become members of a group

upon their friends' invitation. Hence, this fact accentuates the perception that social influence plays a crucial role in people's decision to take part in social networking.

e. Peer feedback: Educational usage of Facebook for peer feedback consists of the enabling of communication among users/students so that they stay aware about significant information shared by others related to the curriculum.

f. Cooperation: The idea of collaborative learning can undoubtedly be expressed through the use of Facebook. In this way, students can exchange ideas, help their peers and work together in order to enhance the educational experience.

g. Knowledge sharing: A crucial aspect incorporated in the educational usage of Facebook is the exchange of resources, documents and useful knowledge concerning the curriculum taught. Furthermore, Facebook provides the additional possibility of multimedia sharing so that students may share with their peers audio, video, images, and other materials related to their curriculum.

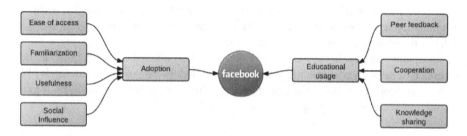

Fig. 2. Adoption and Educational usage of Facebook

5   General Overview of the System

The following figure (Fig. 3) are snapshots of the preliminary test, which aims to evaluate the student's understanding and mastery of the subject matter and the exercise part of the lesson. In the preliminary test (left), the English language is used as the medium of instruction so that the student will be evaluated solely on his/her knowledge state in the application domain. Particularly, in this figure the student is asked to identify the conditional type of the sentence structure presented. Correspondingly, he/she then selects the correct answer from the given choices. In this way, the system attempts to cluster the students in order to better assist them in the educational process. The exercise part of the lesson (right) basically contains three fundamental components, namely the question, the choices, and the system's advice. In this case, the student is asked to translate a Filipino conditional structure into Greek. S/he picks from the choices given the best translation of the given conditional. The system's advice that appears, highlighted at the bottom, guides the student on what to be careful about. The aforementioned system advice is generated dynamically based on the student's previous interaction with the application.

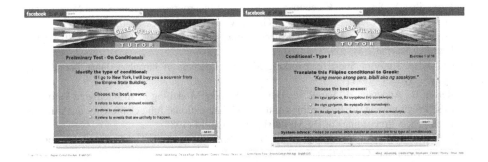

Fig. 3. Overview of the system

6   Conclusions and Future Work

We have presented a user modeling architecture for language learning in Facebook. The related scientific literature showed that the use of SNSs, such as Facebook, as a platform for learning may further benefit the educational process, while user modeling offers personalization in language learning. Hence, we introduced a prototype Facebook application that can model the users for adaptive and individualized learning experience.

Future plans include system evaluation to examine the degree of usefulness of the user modeling and error proneness components of our Facebook application.

References

1. Amaral, L., Meurers, D.: Conceptualizing Student Models for ICALL. In: Conati, C., McCoy, K., Paliouras, G. (eds.) UM 2007. LNCS (LNAI), vol. 4511, pp. 340–344. Springer, Heidelberg (2007)
2. Antal, M., Koncz, S.: Student Modeling for a Web-Based Self-Assessment System. Expert Systems with Applications 38(6), 6492–6497 (2011)
3. Dickinson, M., Eom, S., Kang, Y., Lee, C.M., Sachs, R.: A Balancing Act: How Can Intelligent Computer-Generated Feedback be Provided in Learner-To-Learner Interactions. Computer Assisted Language Learning 21(4), 369–382 (2008)
4. Facebook Platform. Facebook Developers (February 5, 2012), http://developers.facebook.com/
5. Ferreira, A., Atkinson, J.: Designing a Feedback Component of an Intelligent Tutoring System for Foreign Language. Knowledge-Based Systems 22(7), 496–501 (2009)
6. Hiew, W.: English Language Teaching and Learning Issues in Malaysia: Learners' Perceptions via Facebook Dialogue Journal. Journal of Arts, Science and Commerce (ResearchWorld), 11–19 (2011)
7. Ho-Abdullah, I., Hashim, R.S., Jaludin, A., Ismail, R.: Enhancing Opportunities for Language Use through Web-based Social Networking. In: 2011 International Conference on Social Science and Humanity, pp. 136–139. IACSIT Press, Singapore (2011)
8. Mizoguchi, R.: Student Modeling in ITS. Institute of Scientific Research and Industrial Research, Osaka University, Japan (1994)
9. Naganathan, E.R., Maheswari, N.U., Venkatesh, R.: Intelligent Tutoring System Using Hybrid Expert System with Speech Model in Neural Networks. International Journal of Computer Theory and Engineering 2, 1793–8201 (2010)

10. Piriyasilpa, Y.: See you in Facebook: The Effects of Incorporating Online Social Networking in the Language Classroom. Journal of Global Management Research, 67–80 (2011)
11. Promnitz-Hayashi, L.: A Learning Success Story Using Facebook. Studies in Self-Access Learning Journal 2(4), 309–316 (2011)
12. Ota, F.: A Study of Social Networking Sites for Learners of Japanese. Japan Foundation Sydney, Sydney (2011)
13. Reiser, B., Anderson, J., Farell, R.: Dynamic Student Modelling in an Intelligent Tutor for LISP Programming. In: Proceedings of the Ninth International Joint Conference on Artificial Intelligence, Proceedings of IJCAI, Pittsburgh, pp. 8–14 (1985)
14. Stallons, J.: The Architecture of Applications Built on the Flash And Facebook Platforms. Adobe Developer Connection (2011)
15. Tsiriga, V., Virvou, M.: Web Passive Voice Tutor: An Intelligent Computer Assisted Language Learning System over the WWW. In: Proceedings of IEEE International Conference on Advanced Learning Technologies, Madison, WI, pp. 131–134 (2001)
16. Tsiriga, V., Virvou, M.: A Framework For The Initialization Of Student Models In Web-Based Intelligent Tutoring Systems. User Modelling and User-Adapted Interaction 14(4), 289–316 (2004)
17. Virvou, M., Troussas, C.: Personalized Teaching of Multiple Languages through the Web. International Journal for e-Learning Security 1(12), 52–59 (2011)
18. Virvou, M., Troussas, C.: CAMELL: Towards A Ubiquitous Multilingual E-Learning System. In: CSEDU 2011 – Proceedings of the 3rd International Conference on Computer Supported Education, vol. 2, pp. 509–513 (2011)
19. Lenhart, A., Madden, M.: Teens, privacy, and online social networks. Pew Internet and American Life Project Report (2007)
20. Ajjan, H., Hartshorne, R.: Investigating faculty decisions to adopt Web 2.0 technologies: Theory and empirical tests. The Internet and Higher Education 11(2), 71–80 (2008)
21. Mazman, S., Usluel, Y.: Modeling educational usage of Facebook. Computers and Education 55(2), 444–453 (2010)
22. Boyd, D.M., Ellison, N.B.: Social Network Sites: Definition, History, and Scholarship. Journal of Computer-Mediated Communication 13(1), 210–230 (2007)

Detection of Semantic Compositionality
Using Semantic Spaces

Lubomír Krčmář[1], Karel Ježek[1], and Massimo Poesio[2]

[1] Faculty of Applied Sciences, University of West Bohemia, Czech Republic
{lkrcmar,jezek_ka}@kiv.zcu.cz
[2] University of Essex, United Kingdom
poesio@essex.ac.uk

Abstract. Any Natural Language Processing (NLP) system that does semantic processing relies on the assumption of semantic compositionality: the meaning of a compound is determined by the meaning of its parts and their combination. However, the compositionality assumption does not hold for many idiomatic expressions such as "blue chip". This paper focuses on the fully automatic detection of these, further referred to as non-compositional compounds.

We have proposed and tested an intuitive approach based on replacing the parts of compounds by semantically related words. Our models determining the compositionality combine simple statistic ideas with the COALS semantic space. For the evaluation, the shared dataset for the Distributional Semantics and Compositionality 2011 workshop (DISCO 2011) is used. A comparison of our approach with the traditionally used Pointwise Mutual Information (PMI) is also presented. Our best models outperform all the systems competing in DISCO 2011.

Keywords: DISCO 2011, compositionality, semantic space, collocations, COALS, PMI.

1   Introduction

In NLP, representing word meanings in vector spaces has become very common. The underlying idea, popularized by Firth [1], is that "a word is characterized by the company it keeps". Recently, new attempts to characterize the meanings of compounds have been presented.

The proposed approaches rely on the compositionality assumption. However, this assumption does not hold for many compounds. Compare e.g. compositional "short distance" or "students learn" with non-compositional "blue chip" or "beg the question". Non-compositional compounds make it harder for distribution-based systems to succeed. Therefore, and also because they may be handled in a special way in other NLP applications, it is useful to detect them.

The rest of this paper is organized as follows. Section 2 briefly introduces related work. Section 3 describes the task and the evaluation methods. The proposed system, including its relation to PMI, is described in Section 4. The results and their analysis are provided in Section 5. Section 6 concludes by summarizing and proposing future directions.

P. Sojka et al. (Eds.): TSD 2012, LNCS 7499, pp. 353–361, 2012.
© Springer-Verlag Berlin Heidelberg 2012

2   Related Work

Our paper is based on the results of the DISCO 2011 workshop presented in [2], where the task, evaluation methods, results and a comparison of all the proposed systems are described. We employ semantic spaces, more generally referred to as Vector Space Models (VSM) of semantics, introduced in [3]. Specifically, we work with the COALS[1] semantic space presented in [4] and implemented as a part of the S-Space Package described in [5]. We exploit the freely available POS[2]-tagged and lemmatized ukWac corpora presented in [6].

The approach most similar to ours is described in [7], where COALS and correlation measures are also used. However, the results of the presented experiments based on replacing the parts of compounds are poor. We believe this might be caused by inappropriately applied clustering.

Another approach close to ours can be found in [8]. Similarly to us, Lin in [8] replaces the parts of compounds with words having the same distributional characteristics. Unlike us, Lin does not directly use the numbers of occurrences of alternative compounds but explores the differences between the values of mutual information counted for the original compounds and their alternatives.

Approaches different from ours are based on traditional statistical measures of association such as PMI or T-score. This is the case of [9], where these and other measures are described. However, the results described in [2] show that statistical approaches achieve worse results in comparison to VSM-based approaches. We believe that the drawback of statistical approaches applied to the DISCO 2011 task is that they reflect the abnormal occurrences of the compounds in comparison to the occurrences of their constituents rather than the non-compositionality characteristic of compounds.

3   Task Description

The task of the DISCO 2011 workshop was to build a system capable of assigning scores to the given sets of compounds according to their compositionality. The proposed systems were evaluated using provided datasets scored by humans.

Table 1. The sample ordered compounds together with their numerical, and coarse scores from the TrainD dataset

Type	compound	Coarse	Score	Ordering
EN_V_OBJ	beg question	low	18	1
EN_V_OBJ	pull plug	low	21	2
EN_V_SUBJ	company take	medium	50	3
EN_ADJ_NN	hard work	medium	51	4
EN_ADJ_NN	short distance	high	97	5
EN_V_SUBJ	student learn	high	98	6

[1] Correlated Occurrence Analogue to Lexical Semantics
[2] Part Of Speech

Table 2. Number of compounds (with coarse scores)

dataset	AN	SV	VO	Sum
TrainD	58 (43)	30 (23)	52 (41)	140 (107)
ValidationD	10 (7)	9 (6)	16 (13)	35 (26)
TrainValD	68 (50)	39 (29)	68 (54)	175 (133)
TestD	77 (52)	35 (26)	62 (40)	174 (118)

The compounds scored by humans were obtained from the ukWac corpora. They were provided without their contexts, although they had been scored by humans for whom the contexts had been available. Compounds of the types: adjective-noun (AN), verb-subject (SV), and verb-object (VO) were investigated. The data were split into TrainD, ValidationD, and TestD datasets. All the provided compounds were assigned numerical scores from 0 to 100 by humans. The compounds in the range of 0–25 were further labeled as "low" (denoting that the assumption of compositionality does not hold for them), those between 38–62 as "medium", and the ones from 75 to 100 as "high", thus giving the coarse scores. The sample data are illustrated in Table 1. All the provided datasets are summarized in Table 2, where TrainValD denotes the dataset originating from the concatenation of TrainD and ValidationD.

The proposed systems were evaluated using the following methods: Average Point Difference (APD), Spearman's (ρ) and Kendall's (τ) correlations for numerical scores, and Precision (P) for coarse scores. APD is defined as the average difference between the system's and human's scores for the same compound:

$$APD(S, H) = \frac{1}{N} \sum_{i=1..N} |h_i - s_i| \ .$$

The precision is measured as the number of correct coarse predictions:

$$P(S, H) = \frac{1}{N} \sum_{i=1..N} \begin{cases} s_i == h_i & 1 \\ otherwise & 0 \end{cases} \ .$$

We argue that the best evaluation methods are the measures of correlation. There are at least two reasons for this. Firstly, it is the ordering of compounds that the systems should be able to imitate. Secondly, it is not possible to "cheat" the measures of correlation since there is no average, or high value which could improve the results, as is the case with numerical, and coarse scores. For these reasons we employed the correlation measures and created the *hard_baseline* system which assigns the average scores obtained from TrainValD.

4   System Description

For our system, the COALS semantic space S_{coals} was built from the whole ukWac corpora, but only for the words occurring at least 50 times. As an input for the COALS, every lemma of word token was concatenated with its POS tag.[3]

---

[3] Only the first two letters of categories were used. The "book", for example, could become the "book_NN", or "book_VV" after our preprocessing.

Next, the module capable of returning the most similar words of the same POS category from S_{coals} to the given word was implemented. The quality of S_{coals} was tested by looking at the words returned as the most similar to the words randomly chosen from TrainValD. Then, the numbers of occurrences in the ukWac corpora for every compound of the adjective-noun, verb-(determiner)-noun, and noun-verb types[4] were counted.

Having prepared the modules, we explored our three hypotheses:

1. H_o: The compositionality is determined by the comparison of occurrences of the original compounds and their alternatives.
2. H_w: The occurrence of any alternative is more significant than the number of its occurrence.
3. H_t: For different types, different numbers of alternative compounds taken into account can better imply the compositionality.

For the following description, let $\langle h, m \rangle$ be the number of occurrences of an original compound composed of the head[5] word h and modifying word m. Let a_i^h be the ith word most similar to h in S_{coals}. Then, $\langle a_i^h, m \rangle$ denotes the number of occurrences of the alternative compound which originates from the original one by replacing its h part by a_i^h (h's ith alternative). Similarly, $\langle h, a_j^m \rangle$ denotes the number of occurrences of the compound originating by replacing the modifying part of the original compound, e.g., in the case of the "blue chip" compound, the third alternative compound could be the "yellow chip", if the third closest word in S_{coals} to "blue" was "yellow".

Further, let p be the number of used alternatives created by replacing the h part of the original compound by p words most similar in S_{coals}, and q the number of used alternatives created by replacing the original's m part by q words most similar in S_{coals}. Then, let c_b be the basic compositionality score, determining the extent to which the compound is *compositional*, defined as follows:

$$c_b = \frac{\sum_{i=1}^{p} \langle a_i^h, m \rangle + \sum_{j=1}^{q} \langle h, a_j^m \rangle}{\langle h, m \rangle} .$$

We define our basic model for scoring compounds $B_{p,q}$ as the model assigning c_b to every scored compound. We find the appropriate values for p and q with the help of TrainValD.

Similarly, let c_l be the logarithmic compositionality score defined as follows:

$$c_l = \frac{\sum_{i=1}^{p} \log(\langle a_i^h, m \rangle + 1) + \sum_{j=1}^{q} \log(\langle h, a_j^m \rangle + 1)}{\log(\langle h, m \rangle + 1)} ,$$

where adding one ensures non-negativity of the logarithmic values. Then, $L_{p,q}$ denotes our logarithmic model assigning c_l to every scored compound.

---

[4] Corresponding to the AN, VO and SV compound types.
[5] The word that is superior from the syntactic point of view.

In the same manner, we define the $R_{p,q}$ model assigning c_r to the scored compounds defined as follows:

$$c_r = \frac{\sum_{i=1}^{p} \langle a_i^h, m \rangle * \sum_{j=1}^{q} \langle h, a_j^m \rangle}{\langle h, m \rangle} .$$ (1)

The $R_{p,q}$ model is inspired by PMI since PMI, used for determining the *non-compositionality* of compounds, assigns the S_{PMI} score defined as follows:

$$S_{PMI} = \log \frac{\langle h, m \rangle / N}{\langle ., m \rangle / N * \langle h, . \rangle / N} ,$$

where $\langle h, . \rangle$ denotes the sum of occurrences of compounds with the particular head word h as its left part, and N is the number of all compounds. Once we order the compounds by S_{PMI}, we can omit the logarithm and constant N and obtain the ordering PMI score. To predict compositionality of compounds instead of non-compostionality, we swap the numerator for the denominator in the obtained expression. Further, if we take into account only the alternative compounds whose head words are close to the original head word, and the alternative compounds whose modifying words are close to the original modifying word, we derive the aforementioned (1).

Finally, we find the best model for every compound type and combine the three simple models that arise into one. We combine these models by mapping them into the order of compounds obtained from TrainValD. In this manner, we create our combined models trained on TrainValD which are able to assign the ordering scores to every[6] compound. Thus, the ordering scores reflect the degree of compositionality similarly as the compositionality scores do in applying the simple models. Due to a lack of space, we do not describe the way we combine the three models in detail. We denote the combined models according to the names of the simple models which they are composed of. For example, $B_{0,4}L_{300,24}L_{26,22}$ denotes the combined model composed of the $B_{0,4}$ simple model for the AN compound type, $L_{300,24}$ for SV, and $L_{26,22}$ for VO.

To compare our system with the systems of others, we create mapping to the numerical (s_n) and coarse (s_c) scores. At first, we map all the compositionality scores obtained from every model to the ordering scores (s_o) obtained from TrainValD. Next, we map s_o to s_{tmp} by applying the $s_{tmp} = 7.301 s_o^{0.506}$ regression function which originates from mapping the order of TrainValD to the human numerical scores. Finally, we transform s_{tmp} to s_n by pulling up s_{tmp} to the average numerical score obtained from TrainValD using the $s_n = 66.39 - 0.8(s_{tmp} - 66.39)$ formula.[7] We further split all the numerically scored compounds into three groups to obtain the s_c. We used 31.5 and 68.5 thresholds since these values result from [2].

5 Results

We used TrainValD to find the best p, q values from these numbers: $0, 1, \ldots, 40, 50, \ldots,$ $100, 200, \ldots, 1000, 2000, \ldots, 4000$. We evaluated our models according to ρ. Our results

---

[6] Even to any compound which does not occur in TrainValD.

[7] The s_n are pulled up to 66.38 by 20%, which is the first and only tried value.

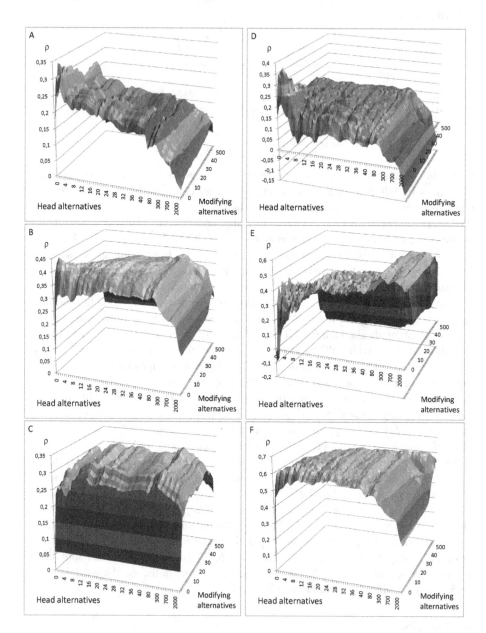

Fig. 1. Graphs depicting the dependency of ρ on p, q values. Graphs A, B, and C depict the results of the $B_{p,q}$, $L_{p,q}$, and $R_{p,q}$ models in order applied to TrainValD. Graphs D, E, and F depict the results of $L_{p,q}$ restricted to the AN, SV, and VO compound types in order.

Table 3. ρ, and τ; APD (d), and precision (p) scores for the best (highest ρ) three simple and combined models chosen from all the tested models on TrainValD applied to TrainValD (matching number(coarse) scored compounds: 175(133))

	ρ_all	ρ_AN	ρ_SV	ρ_VO	τ_all	τ_AN	τ_SV	τ_VO
$L_{1,5}$	0,444	0,370	0,332	0,605	0,302	0,247	0,193	0,443
$L_{2,5}$	0,438	0,336	0,346	0,596	0,295	0,227	0,215	0,440
$L_{3,5}$	0,434	0,325	0,334	0,602	0,296	0,225	0,220	0,437
$B_{0,4}L_{300,24}L_{26,22}$	**0,537**	**0,385**	**0,592**	**0,660**	0,380	**0,264**	**0,417**	0,486
$B_{0,4}L_{300,24}L_{25,20}$	**0,537**	**0,385**	**0,592**	0,659	**0,382**	**0,264**	**0,417**	**0,491**
$B_{0,4}L_{300,23}L_{26,22}$	**0,537**	**0,385**	0,589	0,659	0,380	**0,264**	0,409	**0,491**

	d_all	d_AN	d_SV	d_VO	p_all	p_AN	p_SV	p_VO
hard_baseline	19,12	18,21	18,06	20,64	0,571	0,600	0,586	0,537
$L_{1,5}$	16,78	18,38	17,40	14,83	0,647	0,560	**0,690**	**0,704**
$L_{2,5}$	16,84	18,87	16,97	14,74	0,632	0,560	0,621	**0,704**
$L_{3,5}$	16,76	18,71	17,03	14,66	0,617	0,560	0,655	0,648
$B_{0,4}L_{300,24}L_{26,22}$	14,56	**16,98**	**12,63**	13,26	**0,677**	**0,640**	**0,690**	**0,704**
$B_{0,4}L_{300,24}L_{25,20}$	**14,50**	**16,98**	**12,63**	13,10	**0,677**	**0,640**	**0,690**	**0,704**
$B_{0,4}L_{300,23}L_{26,22}$	14,52	**16,98**	12,70	**13,10**	**0,677**	**0,640**	**0,690**	**0,704**

Table 4. ρ, and τ; APD (d), and precision (p) scores for the best (highest ρ) three simple and combined models in TrainValD applied to TestD (matching number(coarse) scored compounds: 174(118)). Best_score presents the best results in DISCO 2011.

	ρ_all	ρ_AN	ρ_SV	ρ_VO	τ_all	τ_AN	τ_SV	τ_VO
best_scores	0,350	-	-	-	0,240	-	-	-
$L_{1,5}$	0,358	0,376	0,140	0,635	0,245	0,254	0,133	0,466
$L_{2,5}$	0,360	0,386	0,170	0,630	0,250	0,261	0,153	0,445
$L_{3,5}$	0,347	0,372	0,220	0,622	0,243	0,266	**0,170**	0,443
$B_{0,4}L_{300,24}L_{26,22}$	0,414	**0,417**	0,238	0,673	0,289	**0,281**	0,140	0,494
$B_{0,4}L_{300,24}L_{25,20}$	0,421	**0,417**	0,238	0,690	0,294	**0,281**	0,140	0,505
$B_{0,4}L_{300,23}L_{26,22}$	**0,423**	**0,417**	**0,240**	0,690	0,295	**0,281**	0,143	**0,505**

	d_all	d_AN	d_SV	d_VO	p_all	p_AN	p_SV	p_VO
hard_baseline	16,87	17,75	**15,52**	16,54	0,585	0,654	0,346	0,650
best_scores	16,19	**14,62**	15,72	14,66	0,585	**0,731**	**0,577**	0,650
$L_{1,5}$	16,57	18,44	22,08	**11,14**	0,568	0,539	0,269	**0,800**
$L_{2,5}$	16,66	17,67	22,35	12,21	0,559	0,539	0,269	0,775
$L_{3,5}$	16,56	17,57	22,52	11,94	0,576	0,519	0,346	**0,800**
$B_{0,4}L_{300,24}L_{26,22}$	15,83	17,35	20,29	11,42	**0,593**	0,615	0,308	0,750
$B_{0,4}L_{300,24}L_{25,20}$	15,79	17,35	20,29	11,32	**0,593**	0,615	0,308	0,750
$B_{0,4}L_{300,23}L_{26,22}$	**15,78**	17,35	20,22	11,32	**0,593**	0,615	0,308	0,750

in TrainValD presented by graphs A, B, and C in Fig. 1 confirm our H_o hypothesis. Graph A shows that we should favor values close to five for p, q in our $B_{p,q}$ models. Graph B confirms our H_w hypothesis since the results of $L_{p,q}$ models are better for all the non-marginal[8] p, q values. Graph B also shows that $L_{p,q}$ models are more resistant to bad choices of p, q. Still, the best results are achieved by favoring the values around five. Graph C shows that our $R_{p,q}$ models achieve reasonable results for all the possible non-marginal choices of p, q. However, they don't achieve such high correlations as the other two models.

We chose the three simple models which had achieved the best correlation values in TrainValD to be tested in TestD. Further, we found the best models for every compound type and combined them. We used three combined models which had achieved the best correlation values and were chosen from eight models originating by combining the best two models for every compound type. The results of the best models applied to TrainValD are presented in Table 3.

As for the H_t hypothesis, this is confirmed by the results presented by graphs D, E, and F in Fig. 1, where the results of $L_{p,q}$ models are depicted. The results show that for the AN compounds, only low numbers of p, q should be used. On the other hand, all the models with non-marginal values of p, q work well for VO. Graph E shows that for the SV type it might be good to use more than 100 alternative compounds where the head word is being replaced[9].

Our results in TestD, which are depicted in Table 4, confirm the success of our models. The results are even better for the AN and VO compound types. On the other hand, the results for SV are bad. We hypothesize that we should extract compounds occurrences from a bigger window than from that of size two and thus get more proper counts of the alternative compounds for the SV type.[10]

6   Conclusion

Our approach turned out to be successful. We showed its relevance to the traditionally applied PMI. We further showed an interesting approach for combining the best models for different compound types which helped us to achieve the best results in all the main evaluation methods available for the DISCO 2011 task. Our system is very successful when being applied to the VO compound type.

For future work, we would also like to explore the applicability of different semantic spaces from COALS, employ clustering on semantic spaces, or try different numbers of alternative words for every compound by specifying a similarity threshold.

Acknowledgments. We wish to thank Dr Paul Scott and the Language and Computation group from the University of Essex for their valuable advice. Also, the access to computing and storage facilities owned by parties and projects contributing to the National Grid Infrastructure MetaCentrum, provided under the programme "Projects of Large Infrastructure for Research, Development, and Innovations" (LM2010005) is highly appreciated.

---

[8] Not 0 or higher than 2,000.

[9] However, this small hypothesis was not confirmed in TestD.

[10] We believe that the SV type is the least resistant one to this inaccuracy.

References

1. Firth, J.R.: A synopsis of linguistic theory 1930–1955. Studies in Linguistic Analysis (special volume of the Philological Society) 1952-59, 1–32 (1957)
2. Biemann, C., Giesbrecht, E.: Distributional Semantics and Compositionality 2011: Shared Task Description and Results. In: Proceedings of the Workshop on Distributional Semantics and Compositionality, pp. 21–28 (2011)
3. Turney, P.D., Pantel, P.: From Frequency to Meaning: Vector Space Models of Semantics. Artificial Intelligence Research 37, 141–188 (2010)
4. Rohde, D.L.T., Gonnerman, L.M., Plaut, D.C.: An Improved Model of Semantic Similarity Based on Lexical (2005) (unpublished manuscript)
5. Jurgens, D., Stevens, K.: The S-Space Package: An Open Source Package for Word Space Models. In: Proc. of the ACL 2010 System Demonstrations, pp. 30–35 (2010)
6. Baroni, M., Bernardini, S., Ferraresi, A., Zanchetta, E.: The WaCky wide web: a collection of very large linguistically processed web-crawled corpora. Language Resources and Evaluation 43, 209–226 (2009)
7. Johannsen, A., Martinez, H.: Rishøj, C., Søgaard, A.: Shared task system description: Frustratingly hard compositionality prediction. In: Proceedings of the Workshop on Distributional Semantics and Compositionality, pp. 29–32 (2011)
8. Lin, D.: Automatic Identification of Non-compositional Phrases. In: Proceedings of the 37th Annual Meeting of the ACL on Computational Linguistics, vol. 37, pp. 317–324 (1999)
9. Chakraborty, T., Pal, S., Mondal, T., Saikh, T.: Shared task system description: Measuring the Compositionality of Bigrams using Statistical Methodologies. In: Proc. of the Workshop on Distributional Semantics and Compositionality, pp. 38–42 (2011)

User Adaptation in a Hybrid MT System

Feeding User Corrections into Synchronous Grammars and System Dictionaries

Susanne Preuß[1], Hajo Keffer[1], Paul Schmidt[1],
Georgios Goumas[2], Athanasia Asiki[2], and Ioannis Konstantinou[2]

[1] GFAI – Gesellschaft zur Förderung der Angewandten Informationsforschung
http://www.iai-sb.de/iai/
[2] National Technical University of Athens
School of Electrical and Computer Engineering
Computing Systems Laboratory
http://www.cslab.ece.ntua.gr

Abstract. In this paper we present the User Adaptation (UA) module implemented as part of a novel Hybrid MT translation system. The proposed UA module allows the user to enhance core system components such as synchronous grammars and system dictionaries at run-time. It is well-known that allowing users to modify system behavior raises the willingness to work with MT systems. However, in statistical MT systems user feedback is only 'a drop in the ocean' in the statistical resources. The hybrid MT system proposed here uses rule-based synchronous grammars that are automatically extracted out of small parallel annotated bilingual corpora.

They account for structural mappings from source language to target language. Subsequent monolingual statistical components further disambiguate the target language structure. This approach provides a suitable substrate to incorporate a lightweight and effective UA module. User corrections are collected from a post-editing engine and added to the bilingual corpus, whereas the resulting additional structural mappings are provided to the system at run-time.

Users can also enhance the system dictionary. User adaptation is organized in a user-specific commit-and-review cycle that allows the user to revise user adaptation input. Preliminary experimental evaluation shows promising results on the capability of the system to adapt to user structural preferences.

1 Introduction

It is well-known that allowing users modify system behavior raises their willingness to work with MT systems. Thus, User Adaptation (UA) has received remarkable attention in recent years. The PACO report [1] provides a survey on post editing tools and argues that UA is an important feature for increasing the acceptance of MT systems. Most MT systems, however, have conceptual problems to provide UA. Statistical MT (SMT) systems face the problem that small-scale user feedback has insignificant effect on system resources that are based on large data sets. Therefore, SMT systems restrict their UA functionality to shallow string manipulation.On the other hand, rule-based MT

P. Sojka et al. (Eds.): TSD 2012, LNCS 7499, pp. 362–369, 2012.

(RBMT) systems face the problem that the rule components are often manually written which requires sophisticated linguistic skills that cannot be expected by the end-users. Therefore, classical RBMT also provides only shallow string manipulation operations as UA or restricted facilities for bilingual lexicon extension.

The FAUST project explicitly addresses the problem of user feedback. The project's goal is to develop interactive MT systems which adapt rapidly (!) and intelligently to user feedback. A major issue is to identify noise that users tend to produce when giving feedback to MT systems. The major achievement planned is to "develop mechanisms for instantaneously incorporating user feedback into the machine translation engines" [2]. For SMT this amounts to specially weighting user feedback. Whether this is different to string replacement remains to be seen.

In this paper we present a UA mechanism built on top of a novel, Hybrid MT system that makes use of already available tools and lexica, automatically built large monolingual corpora and synchronous grammas, and easily assembled small bilingual corpora. A key goal of the MT system is to provide a full translation path for new language pairs with significantly reduced implementation effort. Adhering to this philosophy, the UA module is capable of providing a lightweight and effective mechanism that allows users modify system components towards a system adapted to their needs.

2   A Hybrid MT System

Design Philosophy. The key idea of the hybrid MT system is to use publicly available tools and resources so that development time and costs are kept low and the system can be easily extended to new language pairs without trading translation quality. Such tools are statistical taggers and lemmatizers and statistical chunkers. Large monolingual corpora for target languages (TL) are automatically built out by properly mining the world wide web. The only bilingual resources needed are small parallel bilingual corpora consisting of a few hundred sentences and bilingual dictionaries which are often available from publishers, at least if the system is not used commercially. Any other linguistic resources are derived from the resources listed above, mainly including SL models for statistical chunkers, synchronous grammars, statistical models for TL lemma disambiguation and TL token generation tables. For a detailed description see [3].

Translation Process. The translation process consists of three steps:

1. *SL annotation*: The incoming SL sentences are annotated with the help of the statistical taggers, chunkers and clause chunkers[1].
2. *Bilingual structure generation*: The synchronous grammar takes the annotated SL structures as input and generates a set of TL structures, including TL lemmas and tags.
3. *TL disambiguation*: Monolingual statistical models disambiguate the TL structures. For the best TL structure, tokens are generated.

---

[1] A pattern-based clause chunker has been developed that recognizes SVO and SOV patterns and can be easily adapted to new languages by specifying tag classes such as the set of finite verb tags and the set of nominal tags.

Thus the synchronous grammar component is sandwiched between two statistical components that do SL disambiguation and TL disambiguation. This architecture makes it viable to work with relatively small bilingual corpora for synchronous-grammar generation. The resulting grammar need not be specific enough to do SL or TL disambiguation.

Synchronous Grammar Component. The limited size of the bilingual corpus does not allow statistically derived synchronous grammars as in [4]. Also, unlike other tree-to-tree translation approaches [5,6,7], the system proposed does not use deep syntactic processing of the corpus data. Instead, shallow parsers are used to annotate the small bilingual corpus. Thus, the challenge is to extract the richness of cross-linguistic rules and generalizations out of a very limited set of flatly structured and monolithically annotated bilingual sentence pairs.

To extend the coverage of the synchronous grammar beyond the patterns found in the bilingual corpus, the sentential chunk and tag alignments are broken down into the smallest self-contained alignments[2] and then converted into productions[3]. As an example for structural divergences between languages, consider the case of separable prefix verbs in German. The German verb 'annehmen' ('accept') is split into 'nehmen' and 'an' in certain constellations. Thus there is a structural change in the translation from English to German:

SL: John accepted the offer from Mary.

TL: John nahm das Angebot von Mary an.

Gloss: 'John took the offer of Mary on.'

The diverging structures can be represented schematically as shown in Fig. 1 where B represents the verbal constituents.

The minimal self-contained alignments are converted into productions. Productions have the following format: 'SL rule(s)' ⇔ 'TL rule(s)'. Subscripts indicate alignments.

---

[2] Definition of minimal self-contained alignment:

If $SL_k \ldots SL_i$ is a sequence of source language elements and $TL_x \ldots TL_y$ is a sequence of target language elements, with k, i, x, y, n and $m \in Na$ and $k \leq i$ and $x \leq y$, then $SL_k \ldots SL_i$, $TL_x \ldots TL_y$ is a minimal self-contained alignment for the elements SL_k and TL_x if i and y are the smallest natural numbers for which the following holds:

1. each element in $SL_k \ldots SL_i$ is aligned to an element in $TL_x \ldots TL_y$ and
2. each element in $TL_x \ldots TL_y$ is aligned to an element in $SL_k \ldots SL_i$ and
3. no element SL_{k-n} is aligned to an element TL_{y+m} and
4. no element SL_{i+n} is aligned to an element TL_{x-m}

[3] The system employs two more strategies to extend the coverage of the synchonous grammar beyond the patterns found in the bilingual corpus:

Tag reduction: The complex tags of morphologically rich languages such as German are reduced to forms that contain only the information that is relevant for the translation process. Thus otherwise derivable agreement information is deleted.

Equivalence classes of tags: The productions are multiplied out according to equivalency classes of tags. I.e. finite verb tags form an equivalence class. Thus, if a production exists for a 3. person singular present tense tag, the corresponding productions for all other persons, numbers and tenses are generated.

SL: A   B   A   C   D

TL: A   B   A   C   B   D

Fig. 1. Alignments

$$MO_1 \Rightarrow A_2 MO_3 \Leftrightarrow MO_1 \Rightarrow A_2 MO_3$$
$$MO_1 \Rightarrow B_2 A_3 C_4 MO_5 \Leftrightarrow MO_1 \Rightarrow B_2 A_3 C_4 B_2 MO_5$$

Fig. 2. Productions for isomorphic and complex alignments

MO is the mother node. If the sequences consist of chunks, then MO is a clause node, if the sequences consist of tags, then MO is a chunk node. For each production a recursive and non-recursive variant is generated. For example, the isomorphic alignment of A – A and the complex alignment of BAC – BACB are converted into the recursive productions shown in Fig. 2.[4] Thus, structural identity between SL and TL results in hierarchical trees, whereas structural divergence results in flat trees.

In order to process the synchronous grammars, a simple Earley chart parser is adopted. The Earley chart parser takes the SL side of the bilingual productions to build up SL tree structures. The Earley chart parser also does a lexicon lookup in the bilingual dictionary. The TL side of the productions and the TL lemmas found in the lexicon are stored in specific features in the SL trees. Then the information on TL structure and TL lemmas is unpacked and a set of TL structures including TL lemmas is generated.

3   User Adaptation in the Hybrid MT System

To address the challenges of UA, we first need to set a solid basis of key decisions, then provide a system design, and finally implement an effective and easy-to-use to system.

User Adaptation Philosophy. When it comes to designing a module with a goal to adjust a complex system to the user needs, a lot of subtle issues arise. Thus, a number of high-level decisions, forming a design philosophy, need to be taken. In the following paragraphs we discuss a number of them.

Make the frequent case fast: As mentioned above the ultimate goal of the UA module is to save user's time. This is achieved by exploiting previous corrections to avoid future ones.

Involve the user: Ideally, the system should work transparently to the user. However, this cannot be the case for a complex, multi-user system. Wrong, funny, or malevolent corrections may contaminate the system and greatly degrade performance. Thus, it is advisable to involve the user in the adaptation process.

Keep the core system intact: The proposed system is implemented utilising a number of critical linguistic resources, such as language models, monolingual corpora and resources derived thereof, parallel corpora and bilingual lexica. A key decision in the

---

[4] In order to cut down the number of generated SL and TL tree structures, tag and lemma disjunctions are represented locally. Tag disjunctions contain frequencies of occurrences which are converted into probabilities that are taken into account by the TL disambiguation components. The frequencies account for minor tagging and alignment errors and foreground the most frequent cases.

UA approach is to keep these core system resources unaffected by user corrections in order to maintain the system's initial functionality.

User Adaptation Design. The proposed UA system targets adaptation regarding both word selection and general structural aspects of the translation. Users may correct single word translations or freely post-edit the sentence. The module design is based on the following key ideas:

- Freely post-edited translations affecting the parallel corpora and the 2nd translation step (bilingual structure generation)
- Word-level adaptation affecting the lexica and the 3rd translation step (TL disambiguation)

The information collected from users' post-editing is stored in two types of resources, which are separate for each user:

- *User lexica*: When the user changes the translation of a single word, a new lexicon entry for the specific pair of SL and TL words will be added to the system per user. If the user decides that the new translation should always be used, then the existing entry will be replaced by the new one.
- *User parallel corpora*: The changes made by the user in the structure of a sentence are stored as a new pair of SL and TL sentence in a user parallel corpus. The UA module comprises components that interact with the system's post-processing module (and GUI in general), and components that are incorporated in the main translation engine of the system. Figure 3 shows the general design of the UA module and how user corrections are collected from the post-editing module and enhance user lexica and parallel corpora.

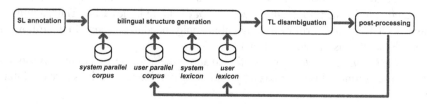

Fig. 3. The proposed UA design

Supported Functionality. *Word replacement*: In the post-editing module users are able to replace a single word either by an alternative word provided by the system lexicon, or enter a new word. Users can freely modify the lemma form provided by the post-editing module into any inflected form. This action is logged by the UA module as a candidate for user adaptation in subsequent translations.

Free text editing: Post-editing offers the functionality of free text editing. These changes are also logged by the UA module as a triplet: SL sentence, system TL sentence, and user TL sentence. The original system TL sentence is stored in order to facilitate quality control of the system translations after UA has been applied. Pre- and post-UA system translations can be compared.

Enable/disable UA: User adaptation is offered to registered users that can enable/disable the UA feature at will.

Review/commit changes: With UA enabled, and after a series of corrections, users are able to review corrections made since UA was enabled. In the case of replacing a word with an existing translation alternative, they are provided with the option to inform the system that the correction made was temporary (it affected only one translation) or that this replacement should replace subsequent occurrences of the SL word from now on and make it the preferred translation.

Versioning: Users are provided with the functionality to create "snapshots" of their adapted environment. Starting from the initial, pure system state (called "vanilla") committed changes start to build the adapted environment, called "working". After a period of satisfactory system behavior, the users may decide to backup the working adaptation environment into a version called "stable".

Feeding User Input Back to the System. The additional sentence pairs are added to a user-specific copy of the bilingual corpus. The corresponding synchronous grammar productions and the dictionary update are done at run-time. The lemmas of the new lexical entries are determined by lemmatizing them in the sentence they occur in. Further linguistic annotation such as tags is not needed in the lexicon.[5]

4 Experimental Evaluation Scores

For an initial experiment, 265 sentences German - English have been taken from a multi-lingual EU web site. The sentences have been minimally adapted in order to get a sentence-aligned parallel corpus. Then the corpus has been split into a training corpus of 225 sentence pairs and a test corpus of 40 sentence pairs. The training corpus has been further split into randomly selected subsets consisting of 25 sentence pairs which have been subsequently added to the bilingual corpus of the MT-system thus simulating structural user adaptation. Figure 4 visualizes the evaluation scores in relation to corpus size.

The evaluation scores confirm the contention that increasing the bilingual corpus improves the evaluation scores. However, the different development of BLEU and NIST scores needs further explanation. The development of the BLEU scores is expected, namely that the scores of the training set and the independent test set increase in the same order of magnitude, since in the present concept of structural user adaptation the training corpus is mainly used for mining grammatical structures which are expected to be roughly the same in the test set and the training set. The fact that the NIST scores of the training set increase to a much larger extent than the NIST scores of the test set is unexpected at first. However, considering that the NIST scores rank matches of content words higher than matches of grammatical function words, the hypothesis is that the content words of the training set are better translated because test and training set

---

[5] If the TL lemmas in the new lexical entries are not part of existing lexical entries, the token generation table and the TL lemma disambiguation models have to be updated as well since for efficiency reasons both are restricted to the TL lemmas found in the dictionary. The update of the statistical models is not done at run-time.

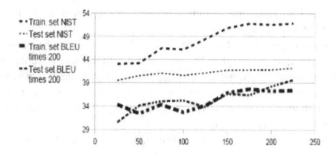

Fig. 4. Results for SL=German and TL=English translations

differ on the lexico-syntactic level. The German SL sentences contain many compounds and other complex lexico-syntactic expressions which translate into complex syntactic expressions in English, which have to be instantiated by bilingual productions. Thus, the difference in NIST scores is due to a difference in lexico-syntactic structures. This hypothesis is supported by the fact that allmost all of the corpus sentences provide new productions, namely 231 out of the 265. Further experiments are needed to examine whether the difference in NIST scores levels out at some point. In an additional experiment, 200 sentences randomly taken from large web corpora and translated by professional translators have been added to the corpus. They yield the same curves. While the additional experiment confirms that the domain of the corpus is irrelevant, the convergence point has not been reached yet.

In order to provide not only scores but also real system translations, consider the following example from the test set: The sentence *"Diese Übereinkunft könnte implizit von der Partei angenommen werden."* has as reference translation: *"This agreement could be implicitly adopted by the party."*. The system translations with 25, 50–100, 125 and 150–225 training sentences are: 25: *"This instrument knows implicitly of the party taken will."* (BLEU: 0.0000 NIST: 1.6610), 50–100: *"This instrument know implicitly of the party will be accepted."* (BLEU: 0.0000 NIST: 1.8120), 125: *"This implicitly arrangement may be taken from the party."* (BLEU: 0.0000 NIST: 1.9932), 150–225: *"Implicitly this agreement could be taken from the party."* (BLEU: 0.3928 NIST: 2.6575). Structurally the final translation is correct but the verb and preposition disambiguation does not provide the intended reading. Also, the position of the adverb is different than in the reference translation. Thus adding more reference translations will improve the evaluation scores.

Human evaluation of the translations has revealed that the translations of some German compounds are missing and that the word order of verbs in long sentences and sentences with complex sentential embeddings tends to be incorrect. The two phenomena have an interesting connection. Both can be improved by lexical user adaptation. Adding new lexical entries improves not only the translation of words, it also improves the alignment of the bilingual parallel corpus which has effects on the productions. In particular long-distance phenomena such as the differing word order of verbs in German and English can only be accounted for if the chunks intervening between the verbs are aligned. Thus adding new lexical entries also improves the treatment of word order phenomena.

5 Conclusions

First experiments show that the proposed user adaptation concept successfully allows the user to improve system behavior and that the system is robust enough to take 'naturally occuring', freely translated sentences as input.[6]

References

1. Paco-Project: Deliverable 5.1 Report on State of the art in Post-Editing Interfaces, http://www.ccl.kuleuven.be/~vincent/ccl/papers/paco_d5.1.pdf
2. Byrne, B.: FAUST - Feedback Analysis for User Adaptive Statistical Translation. In: Proceedings EAMT 2010 (2010), http://www.faust-fp7.eu/faust/Main/FaustPapers
3. PRESEMT Deliverable D2.1: System Specifications
4. Chiang, D.: Hierarchical phrase-based translation. Computational Linguistics 33(2), 201–228 (2007)
5. Eisner, J.: Learning non-isomorphic tree mappings for machine translation. In: Proc. of the 41st Annual Meeting of the Assoc. for Computational Linguistics (2003)
6. Cowan, B., Kucerova, I., Collins, M.: A discriminative model for tree-to-tree translation. In: Proc. of the 2006 Conference on Empirical Methods in Natural Language Processing, pp. 232–241 (2006)
7. Zhang, M., Jiang, H., Aw, A.T., Sun, J., Li, S., Tan, C.L.: A Tree-to-Tree Alignment-based Model for Statistical Machine Translation. In: MT-Summit 2007, pp. 535–542 (2007)

[6] The research leading to these results has received funding from the European Community's Seventh Framework Programme (FP7/2007–2013) under grant agreement n° 248307 (PRESEMT project).

Using Cognates to Improve Lexical Alignment Systems

Mirabela Navlea and Amalia Todirascu

LILPA, Université de Strasbourg
22 rue René Descartes, BP 80010, 67084 Strasbourg Cedex, France
{navlea,todiras}@unistra.fr

Abstract. In this paper, we describe a cognate detection module integrated into a lexical alignment system for French and Romanian. Our cognate detection module uses lemmatized, tagged and sentence-aligned legal parallel corpora. As a first step, this module apply a set of orthographic adjustments based on orthographic and phonetic similarities between French - Romanian pairs of words. Then, statistical techniques and linguistic information (lemmas, POS tags) are combined to detect cognates from our corpora. We automatically align the set of obtained cognates and the multiword terms containing cognates. We study the impact of cognate detection on the results of a baseline lexical alignment system for French and Romanian. We show that the integration of cognates in the alignment process improves the results.

Keywords: cognate detection and alignment, lexical alignment.

1 Introduction

We present a lexical alignment system for French and Romanian integrating a new cognate identification method. The system uses parallel law corpora. Our goal is to improve the results of the French - Romanian lexical alignment system by using cognates. The evaluation shows that the integration of the cognate detection module significantly improves the overall system's performance.

We define cognates as translation equivalents having identical forms or sharing orthographic or phonetic similarities (common etymology, borrowings). Cognates are very frequent between close languages, such as French and Romanian. Generally, cognates are used to detect sure translation equivalents. When they are integrated into lexical alignment tools, cognates can improve statistical translation models [1]. Cognate identification methods adopt statistical techniques. These methods might be improved by the use of various linguistic information.

As far as we know, no specific cognate identification method is available for French and Romanian, except the multilingual system developed by [2] for Romance languages.

The task of cognate identification is difficult, especially to distinguish false friends (bilingual pairs of words having different meanings, but high orthographic similarities) from real cognates. [3] develop classifiers for French and English cognates based on several dictionaries and manually built lists of cognates. [3] distinguish between:

1. cognates (*texte* (FR) - *text* (EN), *sens* (FR) - *sense*(EN));
2. false friends (*blesser* 'to injure' (FR) - *bless* (EN));

P. Sojka et al. (Eds.): TSD 2012, LNCS 7499, pp. 370–377, 2012.

3. partial cognates, depending on the context (*facteur* (FR) - *factor* or *mailman* (EN));
4. genetic cognates, sharing common origins (*chef* (FR) - *head* (EN));
5. unrelated pairs of words (*glace* (FR) - *ice* (EN) and *glace* (FR) - *chair* (EN)).

Our detection method identifies cognates, partial and genetic cognates. This method is used to improve a French - Romanian lexical alignment system. So, we aim to obtain a high precision for the cognate identification method, by combining n-gram methods and linguistic information (lemmas, POS tags). In addition, our method detects and eliminates false friends and unrelated pairs of words using the frequencies of the cognate candidates in the studied corpus. We use a lemmatized, tagged and sentence-aligned parallel corpus. Unlike [3], we do not use other external resources (dictionaries, lists of cognates) for our cognate identification method. These ressources are not available for French and Romanian. Since we do not use such external resources, our method could easily be extended to other Romance languages.

One of the main techniques used by cognate identification methods is to detect the orthographic similarities between two words of a bilingual pair. An efficient method is the 4-gram method [4]. This method considers two words as cognates if their length is greater than or equal to 4 and at least their first 4 characters are common. Other methods apply Dice's coefficient [5] or a variant of this coefficient [6]. Dice's coefficient computes the ratio between the number of common character bigrams and the total number of bigrams in both words. Also, some methods use the Longest Common Subsequence Ratio (LCSR) [7,8]. LCSR is computed as the ratio between the length of the longest common substring of ordered (and not necessarily contiguous) characters and the length of the longest word. Thus, two words are considered as cognates if LCSR value is greater than or equal to a given threshold.

Similarly, other methods estimate the distance between two words as the smallest sequence of transformations (substitutions, insertions and deletions) used to obtain one word from another [9]. These language independent methods use exclusively statistical techniques. Linguistic methods use the phonetic distance between two words belonging to a bilingual pair [10].

In our method, we use a specific combination of cognate identification techniques. So, we define a set of orthographic adjustments between French - Romanian pairs of words, based on their orthographic and phonetic similarities. Then, statistical methods are combined with linguistic information (lemmas, POS tags), to detect cognates.

In the next section, we describe the parallel corpora used for our experiments. In Section 3, we present the lexical alignment method and we explain our cognate identification module in Section 4. We present the evaluation of our method in Section 5. Our conclusions and further work figure in Section 6.

2 The Corpus

Our cognate identification method requires sentence-aligned, tagged and lemmatized parallel corpora. We use a freely available legal parallel corpus (*DGT-TM*) [1] which is based on the *Acquis Communautaire* corpus. *DGT-TM* contains 9,953,360 tokens

---

[1] http://langtech.jrc.it/DGT-TM.html

in French and 9,142,291 tokens in Romanian. It is sentence-aligned. We extracted a test corpus of 1,000 sentences (the sentence length being at most 80 words) for the evaluation purposes. This test corpus contains 33,036 tokens in French and 28,645 tokens in Romanian and it is manually word aligned.

Our method for cognate identification exploits linguistic information such as POS tags and lemmas. We then tag and lemmatize our corpora with TTL tagger available for French [11] and Romanian [12]. The tagger provides detailed information about lexical categories, morpho-syntactic properties (number, gender for nouns and adjectives, time, mood for verbs) (*ana* attribute in Fig. 1) and lemmas. In addition, the tagger includes the identification of simple noun, prepositional and verb phrases (represented in the *chunk* attribute).

We give an example of TTL's output for French in Fig. 1 below:

```
<seg lang="fr"><s id="ttlfr.4">
<w lemma="il" ana="Pp3fs" chunk="Vp#1">Elle</w>
<w lemma="poursuivre" ana="Vmip3s" chunk="Vp#1">poursuit</w>
<w lemma="un" ana="Da-ms" chunk="Np#1">un</w>
<w lemma="but" ana="Ncms" chunk="Np#1">but</w>
<w lemma="non" ana="R" chunk="Np#1,Ap#1">non</w>
<w lemma="lucratif" ana="Af-ms" chunk="Np#1,Ap#1">lucratif</w>
<c>.</c></s></seg>
```

Fig. 1. A sample of TTL's output for French

3   Lexical Alignment System

Our lexical alignment system uses GIZA++ [13] and it applies several heuristic rules, a collocation dictionary [17] and a cognate identification module to complete the lexical alignment. GIZA++ implements IBM models [14]. These models build word-based alignments from aligned sentences. Indeed, each source word has zero, one or more translation equivalents in the target language.

We use the lemmatized, tagged and sentence-aligned parallel corpus described in the previous section. We use lemmas followed by the first two letters of the morpho-syntactic tag [15] to avoid data sparseness. For example, the same French lemma *communiqué* 'communicated' can be a common noun or an adjective: *communiqué_Nc* vs. *communiqué_Af*. This operation morphologically disambiguates the lemmas and improves the GIZA++'s overall performance.

We improve the results provided by GIZA++. First, we intersect the results of the bidirectional alignments to keep only sure alignments [16]. Then, we apply our cognate identification method (described in the next section) to complete the list of cognates obtained by GIZA++.

Also, GIZA++ does not provide many-to-many alignments. Thus, to complete word alignments, we apply a French - Romanian dictionary of verbo-nominal collocations [17]. Collocations are multiword expressions, composed of words related by

Table 1. Orthographic adjustments for the [k] phoneme

French	Romanian	Examples
ck	c [k]	sto**ck**age - sto**c**are
cq	c [k]	gre**cq**ue - gre**c**
q (final)	c [k]	cin**q** - cin**c**i
qu(+i) (medial)	c [k]	é**qu**ilibre - e**ch**ilibru
qu(+e) (medial)	c [k]	mar**qu**er - mar**c**a
qu(+a)	c(+a) [k]	**qu**alité - **c**alitate
que (final)	c [k]	prati**qu**e - practi**c**ă

lexico-syntactic relations [17]. The dictionary contains the most frequent verbo-nominal collocations (*faire usage* vs. *a face uz 'make use'*) extracted from legal corpora. Finally, we apply a set of heuristic rules exploiting linguistic information [15]:

1. we define POS affinity classes, explaining translation rules (a noun might be translated by a noun, a verb or an adjective);
2. we align content words according to the POS affinity classes;
3. we align similar chunks (noun phrase with noun phrase);
4. we identify null aligned words and we propose translation equivalents inside the chunks aligned in a previous step. To this purpose, we apply a set of morpho-syntactic contextual heuristic rules [18]. These rules are based on morpho-syntactic differences between French and Romanian.

4   Cognate Detection Method

In our approach, cognates fulfill the linguistic constraints below:

1. their lemmas are translation equivalents in two parallel sentences;
2. they have identical lemmas or have orthographic/phonetic similarities between lemmas;
3. they are content-words (nouns, verbs, adverbs, etc.) having the same POS tag or belonging to the same POS affinity class.

Firstly, we apply a set of empirically established orthographic adjustments based on orthographic and phonetic similarities between French - Romanian pairs of words. We present some orthographic adjustments in the Table 1.

Indeed, as French has an etymological writing and Romanian generally use a phonetic writing, we detect phonetic mappings between lemmas and we apply a set of 18 orthographic adjustments from French to Romanian [19]. For example, cognates *équilibre* vs. *echilibru* 'balance' become *ecuilibre* vs. *echilibru*. We use here two adjustments: **é** is replacing by [e] and **q** (qu(+i) medial) become [k] (as in Romanian). These rules are context-sensitive (*équilibre* vs. *echilibru* 'balance') or context-independent (*stockage* vs. *stocare* 'storage'). This step improves significantly the recall of the method.

Secondly, we iteratively extract cognates by categories [19]:

1. cross-lingual invariants (numbers, acronyms and abbreviations, punctuation signs);
2. identical cognates (*principal* 'main' vs. *principal*);
3. similar cognates:
   (a) 4-grams [4]; The first 4 characters of lemmas are identical. The length of these lemmas is greater than or equal to 4 (*texte* vs. *text* 'text').
   (b) 3-grams; The first 3 characters of lemmas are identical and the length of the lemmas is greater than or equal to 3 (*date* vs. *dată* 'date').
   (c) 8-bigrams; Lemmas have a common sequence of characters among the first 8 bigrams. At least one character of each bigram is common to both words. This condition allows the jump of a non identical character (*versement* vs. *vărsământ* 'payment'). This method applies only to long lemmas (length greater than 7).
   (d) 4-bigrams; Lemmas have a common sequence of characters among the 4 first bigrams. This method applies for long lemmas (length greater than 7) (*fourniture* vs. *furnizare* 'provision') but also for short lemmas (length less than or equal to 7) (*entrer* vs. *intra* 'enter').

In addition, we use two input data disambiguation strategies:

1. we compute frequency for ambiguous candidates (the same source lemma occurs with several target lemmas *autorité – autoritate|autorizare*) and we keep the most frequent candidate. Thus, we partially eliminate false friends and unrelated pairs of words ;
2. we delete reliable considered cognates (high precision) from the input data at each iteration.

5   Evaluation

In this section, we evaluate the impact of cognate detection method on a baseline lexical alignment system for French and Romanian. This baseline system uses the legal parallel corpus described in section 2. It is based on bidirectional alignments performed by GIZA++ [13] and their intersection [16].

We automatically align the set of extracted cognates and of multiword terms containing cognates. Then, we add these alignments to the baseline system. We evaluate the resulting alignment in terms of precision, recall, F-measure and AER score [13] by comparing it to our manually aligned reference corpus. In addition, we ignore null words in both alignments. Since there is no distinction between sure and possible links in our reference alignment, we consider all alignments as sure, in order to calculate the AER score. Then, it is obtained as follows:

$$AER = 1 - \text{F-measure}$$

In a previous work [19], we evaluate our cognate detection method against a list of cognates manually built from the test corpus. This list contains 2,034 pairs of cognates.

Our method extracted 1,814 correct cognates out from 1,914 provided candidates, obtaining a precision of 94.78%, a recall of 89.18% and a F-measure of 91.89%. In addition, we compare our method with pure statistical methods such as LCSR, DICE's coefficient and 4-grams. We show that our method improves the f-measure with 41.42% (compared to LCSR), with 33.28% (compared to DICE's coefficient) and with 11.02% (compared to 4-grams).

To find the best way to exploit the extracted cognates during the alignment process, we perform experiments in two directions (see Table 2):

1. filtering the baseline system by the set of cognates;
2. enriching the baseline system with the cognate alignments;

Table 2. Evaluation of the built alignments

Systems	Precision	Recall	F-measure	AER
1. baseline	95.56%	52.91%	68.11%	31.89%
2. baseline + cognate filter	96.51%	42.31%	58.83%	41.17%
3. baseline + cognate alignment	95.13%	54.53%	69.32%	30.68%
4. baseline + cognate alignment + multiword terms alignment	95.23%	55.72%	70.30%	29.70%

We first filter the baseline system by the set of cognates to increase the precision of the system. In this case, we keep only correct cognate links provided by the baseline (*sens - sens* 'meaning'; *améliorer - ameliora* 'improve'). The precision increases from 95.56% to 96.51%, but we observe a significant decrease of the recall from 52.91% to 42.31%, which augments the AER score with 9.28% (from 31.89% to 41.17%). This is mainly due to the high frequency of synonyms in the corpus (*sens - sens* vs. *sens - înțeles* 'meaning'; *améliorer - ameliora* vs. *améliorer - îmbunătăți* 'improve'), but our system does not yet include techniques to handle synonyms.

In order to augment the recall of the results, we enrich the baseline system with cognate alignments. In this case, the recall increases with 1.62% (from 52.91% to 54.53%) and the AER score decreases with 1.21% (from 31.89% to 30.68%).

In addition, we exploit the extracted cognates to capture alignments of multiword terms containing cognates. This step concerns terms provided by the TTL tagger [12] in Romanian. Indeed, this tagger recognizes a set of Romanian multiword terms (*cooperare_europeană* 'European cooperation', *autorizație_de_transport* 'transport authorisation'), while in French these terms are tokenized. We show an example of multiword terms' alignment in Figure 2.

We obtain 477 correct supplementary links which improve the recall with 1.19% (from 54.53% to 55.72%) and decrease the AER score with 0.98% (from 30.68% to 29.70%), compared to the previous system adding only 1:1 cognate alignments.

The integration of the cognate list into the lexical alignment process improves the overall results. Thus, the recall augments with almost 3% (from 52.91% to 55.72%) and the AER score decreases with about 2% (from 31.89% to 29.70%), compared to the baseline system.

Fig. 2. Example of multiword terms alignment recognized by TTL in Romanian

6   Conclusion and Further Work

We present a cognate detection module integrated by a French - Romanian lexical alignment system. This method uses lemmatized, tagged and sentence-aligned legal parallel corpora. It uses a new combination of orthographic adjustments, statistical techniques and linguistic information which significantly improves the module's overall results. However, the provided results are dependent on the studied languages, on the corpus domain and on the data volume. We plan more experiments using other corpora from other domains. The extracted cognates are exploited to improve a baseline alignment system. Thus, we study the impact of cognate detection on a baseline lexical alignment system for French and Romanian. We align cognates and multiword terms containing cognates. We show that the integration of cognates in the alignment process improves its results.

References

1. Kondrak, G., Marcu, D., Knight, K.: Cognates Can Improve Statistical Translation Models. In: Human Language Technology Conference of the North American Chapter of the Association for Computational Linguistics (HLT-NAACL 2003) Companion volume, Edmonton, Alberta, pp. 46–48 (2003)
2. Bergsma, S., Kondrak, G.: Multilingual Cognate Identification using Integer Linear Programming. In: RANLP 2007, Borovets, Bulgaria, pp. 11–18 (2007)
3. Inkpen, D., Frunză, O., Kondrak, G.: Automatic Identification of Cognates and False Friends in French and English. In: RANLP 2005, Bulgaria, pp. 251–257 (2005)
4. Simard, M., Foster, G., Isabelle, P.: Using cognates to align sentences. In: Proceedings of the Fourth International Conference on Theoretical and Methodological Issues in Machine Translation, Montréal, pp. 67–81 (1992)
5. Adamson, G.W., Boreham, J.: The use of an association measure based on character structure to identify semantically related pairs of words and document titles. Information Storage and Retrieval 10(7-8), 253–260 (1974)
6. Brew, C., McKelvie, D.: Word-pair extraction for lexicography. In: Proceedings of International Conference on New Methods in Natural Language Processing, Bilkent, Turkey, pp. 45–55 (1996)

7. Melamed, D.I.: Bitext Maps and Alignment via Pattern Recognition. Computational Linguistics 25(1), 107–130 (1999)
8. Kraif, O.: Identification des cognats et alignement bi-textuel: une étude empirique. In: Actes de la 6éme conférence annuelle sur le Traitement Automatique des Langues Naturelles, TALN 1999, Cargése, pp. 205–214 (1999)
9. Wagner, R.A., Fischer, M.J.: The String-to-String Correction Problem. Journal of the ACM 21(1), 168–173 (1974)
10. Oakes, M.P.: Computer Estimation of Vocabulary in Protolanguage from Word Lists in Four Daughter Languages. Journal of Quantitative Linguistics 7(3), 233–243 (2000)
11. Todiraşcu, A., Ion, R., Navlea, M., Longo, L.: French text preprocessing with TTL. In: Proceedings of the Romanian Academy, Series A: Mathematics, Physics, Technical Sciences and Information Science, vol. 12(2), pp. 151–158. Romanian Academy Publishing House, Bucharest (2011)
12. Ion, R.: Metode de dezambiguizare semantică automată. Aplicaţii pentru limbile engleză şi română. Ph.D. Thesis, Romanian Academy, Bucharest, 148 p. (May 2007)
13. Och, F.J., Ney, H.: A Systematic Comparison of Various Statistical Alignment Models. Computational Linguistics 29(1), 19–51 (2003)
14. Brown, P.F., Della Pietra, V.J., Della Pietra, S.A., Mercer, R.L.: The mathematics of statistical machine translation: Parameter estimation. Computational Linguistics 19(2), 263–312 (1993)
15. Tufiş, D., Ion, R., Ceauşu, A., Ştefănescu, D.: Combined Aligners. In: Proceedings of the Workshop on Building and Using Parallel Texts: Data-Driven Machine Translation and Beyond, pp. 107–110. Michigan, Ann Arbor (2005)
16. Koehn, P., Och, F.J., Marcu, D.: Statistical Phrase-Based Translation. In: Proceedings of Human Language Technology Conference of the North American Chapter of the Association of Computational Linguistics, Edmonton, pp. 48–54 (May-June 2003)
17. Todiraşcu, A., Heid, U., Ştefănescu, D., Tufiş, D., Gledhill, C., Weller, M., Rousselot, F.: Vers un dictionnaire de collocations multilingue. Cahiers de Linguistique 33(1), 161–186 (2008)
18. Navlea, M., Todiraşcu, A.: Linguistic Resources for Factored Phrase-Based Statistical Machine Translation Systems. In: Proceedings of the International Workshop on Exploitation of Multilingual Resources and Tools for Central and (South-) Eastern European Languages, 7th International Conference on Language Resources and Evaluation (LREC 2010), Malta, pp. 41–48 (2010)
19. Navlea, M., Todiraşcu, A.: Using Cognates in a French - Romanian Lexical Alignment System: A Comparative Study. In: Proceedings of RANLP 2011, pp. 247–253. INCOMA Ltd., Bulgaria (2011)

Disambiguating Word Translations
with Target Language Models

André Lynum, Erwin Marsi, Lars Bungum, and Björn Gambäck

Department of Computer and Information Science
Norwegian University of Science and Technology (NTNU)
Sem Sælands vei 7–9, NO–7491 Trondheim, Norway
{andrely,emarsi,larsbun,gamback}@idi.ntnu.no

Abstract. Word Translation Disambiguation is the task of selecting the best translation(s) for a source word in a certain context, given a set of translation candidates. Most approaches to this problem rely on large word-aligned parallel corpora, resources that are scarce and expensive to build. In contrast, the method presented in this paper requires only large monolingual corpora to build vector space models encoding sentence-level contexts of translation candidates as feature vectors in high-dimensional word space. Experimental evaluation shows positive contributions of the models to overall quality in German-English translation.

Keywords: word translation disambiguation, vector space models.

1 Introduction

Word-for-word translation is in itself not a sufficient approach to machine translation (MT), as there are many aspects of translation that cannot be captured in this way, such as non-compositional meaning and syntactic structure, especially if the languages involved are typologically far apart. Nonetheless word translation is a common challenge for MT systems that partly rely on lexical resources such as translation dictionaries. Apart from lexical coverage, the major challenge in word translation is that, according to a bilingual dictionary or some other word translation model, a word often has multiple translations. For instance, the English word *bank* may be translated into German as *Bank* in a financial setting, but as *Ufer* in the context of a river. Determining the correct translation in a given context is called *Word Translation Disambiguation* (WTD). Full word translation is a two-step process involving (1) generation of translation candidates, followed by (2) translation disambiguation. Although the work here only addresses WTD (i.e., step 2), this still contributes to resolving full word translation as a common subtask in MT.

The work on WTD described here has been carried out as a part of the PRESEMT project on MT [1]. One of the features that distinguishes PRESEMT from mainstream statistical MT is that it tries to avoid relying on large parallel text corpora for training purposes, resources that are both scarce and expensive to build. Instead, the PRESEMT approach aims at learning patterns in the source and target language, and the mapping between them, from a large annotated target language corpus in combination with a very

P. Sojka et al. (Eds.): TSD 2012, LNCS 7499, pp. 378–385, 2012.

small parallel corpus. In a similar vein, most empirical approaches to WTD crucially depend on word-aligned parallel text, while our goal is to develop data-driven methods for WTD that rely solely on bilingual dictionaries and large-scale monolingual corpora.

The core idea underlying our approach follows. Suppose we have the English sentence *They pulled the canoe up on the bank* and want to translate the word *bank* into German. According to an English-German dictionary, the translation is either *Bank* or *Ufer* (for sake of simplicity, assume there are just two translations). Furthermore, suppose sentences containing either *Bank* or *Ufer*, such as *Die Bank mit den zufriedensten Kunden*, *Wird das gekenterte Kanu ans Ufer gebracht*, etc., can be retrieved from a large corpus of German text. We then look for the German sample sentence which most closely matches the English sentence, or more precisely, the German sample of which the context of *Bank/Ufer* most closely matches the context of the English *bank*. Obviously, directly matching English and German contexts is not going to work, so first the English context is translated to German word-for-word by dictionary look-up, giving *Sie zogen die Kanu am Bank/Ufer*. We can now infer that the second German sample is more similar to the English input than the first, because it shares the word *Kanu* with the translated context. As it is a sample for translation candidate *Ufer*, it supports *Ufer* rather than *Bank* as the correct translation in the given context.

Evidently the outline above abstracts away from a number of important issues. A major issue is how to determine similarity between a given input context and the collected sample contexts for each translation candidate. The approach to measuring similarity taken here is that of Vector Space Models (VSM), originally developed in the context of Information Retrieval to measure similarity between user queries and documents [2]. A VSM for words is a high-dimensional vector representation resulting from a statistical analysis of the contexts in which the words occur. Similarity between words is then defined as the similarity between their context vectors, usually measured by cosine similarity. A major advantage of the VSM approach is the balance of reasonably good results with a simple mathematical model. It is particularly attractive in our setting since it does not require any external knowledge resources besides a large target language corpus and is fully unsupervised (i.e., no need for expensive annotation).

2   Related Work

WTD has been regarded as one of the core problems in MT, both in SMT systems such as the phrase-based Moses system and traditional knowledge-based systems where lexical translation ambiguities are one of the main sources of candidate translations. In phrase-based SMT in particular several experiments have shown improved translation quality through incorporation of a dedicated WTD model extending the limited context normally taken into consideration by the decoder [3]. In this paper we incorporate a WTD model within a hybrid MT system where such a component is an explicit part of the pipeline. Such systems often integrate a WTD-like empirical model, but in that context the models tend to be tightly connected to the linguistic knowledge part of the system.

As a task in itself, WTD has primarily been based on parallel text, including several submissions to the Semantic Evaluation (SemEval-2) 2010 Cross-Lingual Word Sense

Disambiguation (CL-WSD) workshop [4,5,6]. However, there have been some notable efforts on building WTD systems without parallel corpora, initially by estimating word translation probabilities using Expectation Maximization (EM) and a language model [7], and then comparing this and the more naïve approach of using the most frequent translation from a bilingual dictionary with methods based on parallel text [8]. Similarly, EM has been used with lexical association metrics based on target language corpora in order to estimate lexical translation probabilities for Cross Language Information Retrieval [9].

VSMs have been used previously for WTD, but only for the purpose of extracting ranked lists of translation candidates for extending a dictionary [10]. The present work is to the best of our knowledge the first utilizing such models for WTD within an MT system.

3   Implementation

The present WTD approach comprises two parts: (1) an off-line processing step in which translation ambiguities in the dictionary are identified, context samples from a target language corpus are collected and VSMs are constructed; (2) online application of VSMs in a full-fledged MT system. These steps are described below with regard to German-English translation in the PRESEMT system.

3.1   Model Construction

Model contruction relies on two types of resources: a translation dictionary and text corpora for the source and target languages. An existing proprietary German-English translation dictionary was reused within the PRESEMT project. It is lemma-based, includes multi-word expressions, provides part-of-speech (POS) tags on both source and target side, and contains over 576k entries. The share of ambiguous entries is over 26% with an average ambiguity of 3.32 translations. Furthermore, a number of very large corpora of text from the web were collected [11]. The English corpus contains over 3.6B words, the German corpus over 2.8B words. All text was lemmatized and POS tagged with the TreeTagger [12], and subsequently indexed with Manatee, a corpus management tool that provides fast querying and basic statistical measures such as word/lemma counts [13].

First, the translation dictionary is searched for translation ambiguity: source language entries having at least two possible translations. Once these ambiguities have been identified, samples for the target words can be retrieved from a target language corpus and used to build word translation disambiguators. Disambiguation is not worth the effort for very infrequent source words, as those are unlikely to have much impact on the translation quality, so words below a certain threshold (10k) are disregarded. Likewise, disambiguation is infeasible for very infrequent translation candidates, because of a lack of samples, so these are filtered as well (threshold 10k). For German-English, this resulted in selection of 5,409 entries, which is 8.6% of all single-word ambiguous entries.

Second, for each target lemma plus POS combination, usage samples (full sentences) are retrieved from the target language corpus. Very short samples (< 5 tokens) or very long (> 100 tokens) are removed, as are stopwords and remaining "garbage" like separator lines consisting of dashes (through regular expression filters). The maximum number of samples per translation candidate defaults to 10k. For efficient storage, samples are vectorized using a vocabulary mapping lemmata to numerical IDs and collectively stored as sparse matrices. The most frequent target language lemmata (frequency > 100) form the basis of the vocabulary. The resulting vocabularies are large (> 182k lemmata for English), but used only for context sampling; the vocabularies are further pruned during model creation.

Third, the vectorized samples are combined into VSMs. The vocabulary is then further pruned to reduce space and processing requirements. Currently all target lemmas lacking from the translation dictionary are removed, as these can never appear in (potential) translations of the context produced by the PRESEMT system. Other more sophisticated pruning approaches that can reduce the model size substantially have been explored and may be incorporated in the future. After pruning, a matrix consisting of all context vectors for a single translation candidate is reshaped by applying one or more transformations to it. Transformations can, e.g., reduce dimensionality or data sparseness, or promote generalization. Earlier work explored different forms of feature weighting (e.g., TF*IDF) and dimensionality reduction (e.g., Latent Semantic Indexing) and showed that simply calculating the centroid vector for each candidate provides a good trade-off between complexity and performance [14]. As a result, each translation candidate is represented by a single prototype vector in the VSM.

Finally, for each of the selected ambiguous dictionary entries, a corresponding VSM is created containing the prototype vectors for all translation candidates. For German-English this amounts to 5,409 individual models The resulting sparse matrices, along with various mappings such as target lemmas to matrix indices needed during decoding, are stored in a directory structure representing the complete model.

3.2 Model Application

The WTD models are by themselves not sufficient for complete disambiguation, since they do not fully cover all possible word translation ambiguities, only the most frequent ones involving content words (the models are currently used for nouns, adverbs and adjectives). The WTD models are therefore combined with a standard statistical n-gram language model (LM) in order to account for local word ordering and translation candidates not covered by the WTD models. Intuitively we expect the LM and WTD models to complement each other with the first representing local lexical relations and word order, while the second contributes word order independent global lexical relations.

The WTD model scoring is done in a preliminary processing sequence. First, a vector for the target sentence is constructed. This involves looking up all possible translations for the source lemmas in the dictionary and adding their counts to the sentence vector, as far as the they are part of the vocabulary. Next, the appropriate WTD model is retrieved for each ambiguous lemma, and the cosine similarity of the sentence vector to each candidate centroid is computed. The similarities are normalized to a probability measure over the translation candidates for each ambiguous word, to avoid scaling problems with similarity scores from different models. Dynamic

Table 1. NIST and BLEU scores on German-English development data

System	NIST	BLEU
PRESEMT	5.55	0.230
PRESEMT + MF	5.57	0.238
PRESEMT + VSM	5.60	0.230

programming through Viterbi decoding constrained with beam search finds the optimal lemma sequence from the ordered sequence of word translation candidate sets. This is similar to decoding the optimal sequence from Hidden Markov Models except that there are no distinct "observation probabilities". The WTD model probability is factored during the calculation of the probabilities at each point in the trellis to optimize the formula:

$$p(w_1^n) \approx \arg\max_{w_1^n} \prod_{i=1}^{n} p_{wtd}(w_i) p_{lm}(w_i|w_{i-1}, w_{i-2})$$

Calculation of similarities in sparse high-dimensional vector spaces may result in a set of similarities not differing appreciably numerically [15]. In order to accentuate small differences in similarity, a non-linear scaling is applied to the similarity scores to increase contrast, applying sigmoid or polynomial transformations before vector normalization. Currently the two factors used during decoding, the n-gram transition probabilities and the WTD similarities are combined as simple factors. A linear weighting of the factors optimized globally on end-to-end translation quality can potentially even out discrepancies in the perplexity of the models, avoiding the situation where one model overruns the other.

4   Evaluation

The impact of the WTD models on the overall translation quality was measured on a development data set consisting of 50 German sentences comprising 447 tokens in total. There are five reference translations in English, produced by a native speaker of German and checked by a native speaker of English. Performance was assessed through two automatic evaluation measures, NIST [16] and BLEU [17]. Scores were calculated using the scoring script from the NIST Open Machine Translation evaluation series. Table 1 presents scores for three different systems: (1) the PRESEMT system in its default setting, which relies on an n-gram language model for WTD; (2) with a WTD module always selecting the most frequent (MF) translation according to the corpus; (3) with WTD using VSMs with polynomial scaling as described in this paper.

The results show a small improvement of 0.05 NIST points of the VSM-based system over the baseline PRESEMT system, but the difference with the most frequent-based system (PRESEMT + MF) is negligible. Likewise differences in terms of BLEU scores are minimal. Despite the marginal nature of the improvements, the WTD models in general have a positive impact on translation quality. In total there were nine differences

Table 2. Sentences for which the VSM translation differs from the baseline system

#	Source sentence
1	Es ist wichtig, das kulturelle Vakuum zwischen unseren beiden **Regionen** zu füllen.
2	Dieser Vorschlag stammt von den **Organisatoren** des Forums der Völker.
3	Ich unterstütze die **Erweiterung** der Europäischen Union.

#	REF1	REF2	REF3	REF4	REF5	BASE	VSM
1	regions	regions	regions	regions	regions	locations	regions
2	organisers	organisers	organisers	organisers	organisers	organisers	promoters
3	enlargement	extension	expansion	enlargement	expansion	stretch	addition

in word translation between the baseline PRESEMT system (BASE) and the VSM approach, as illustrated in Table 2, comparing these to each other and five reference translations. Five of the VSM-based translations are probably better than the base translations because they correspond to at least one reference translation (such as sentence 1 in Table 2), three of them are likely to be worse (e.g., sentence 2), whereas the last one (sentence 3) seems equally bad. Looking at the WTD similarities (sentence 2), we see that the model distinguishes between "organisers" and "organizers" which may unduly depress the similarity to these two candidates. For sentence 3 all reference terms have much higher similarity than "stretch", but are lower than the top ranked "addition". These examples indicate that models with more sophisticated prototype vectors are necessary for the WTD system to further improve the performance of the MT system.

5   Discussion and Future Work

The results for the VSM models are promising, but appear to have limited impact when combined with LM Viterbi decoding. One factor may be that the current WTD models only cover the the most frequent ambiguities; another that the combination of the LM and VSM models as independent factors is probably too simplistic. At the very least these two models should be weighted in a manner that offer proper influence of the VSM over the LM decoding, not allowing the LM to dominate the selection of translation candidates. Currently the two factors used during decoding, the n-gram transition probabilities and the WTD similarities are combined as simple factors. A linear weighting of the two factors optimized globally on end-to-end translation quality can potentially even out discrepancies in the perplexity of the two models, avoiding the situation where one model overruns the other.

The experiments used centroids as a prototype vector for a given translation candidate, which has given better results than k-means classification against all the contexts. Intuitively, the prototype represents the word sense of the translation candidate and an interesting extension of the model would be to use multiple such prototype vectors for each candidate. In this way the model can represent a range of word sense semantics or language usages where a single prototype is not sufficient since they may not form a coherent group in the vector space. Data mining such relations is a challenging task though, and further complicated by the extreme sparsity of the sentence context, although this sparsity can be reduced by inducing latent relations in the data (e.g., by using Latent Semantic Indexing), or by adding second-order co-occurrence relations either directly or through Reflective Random Indexing [18].

Another avenue for future research is extending the approach to translation of compounds and multi-word expressions. Current efforts focus on integrating changes in word order into the translation candidate decoding process and exploiting the multi-word content in bilingual dictionaries in combination with monolingual corpora.

Acknowledgements. This research has received funding from the European Community's 7th Framework Programme under contract nr 248307 (PRESEMT). Thanks to the other project participants and the anonymous reviewers for several very useful comments.

References

1. Tambouratzis, G., Sofianopoulos, S., Vassiliou, M., Simistira, F., Tsimboukakis, N.: A resource-light phrase scheme for language-portable MT. In: Forcada, M.L., Depraetere, H., Vandeghinste, V. (eds.) Proceedings of the 15th Conference of the European Association for Machine Translation, Leuven, Belgium, pp. 185–192 (2011)
2. Salton, G.: Automatic Text Processing: The Transformation, Analysis and Retrieval of Information by Computer. Addison-Wesley, Reading (1989)
3. Carpuat, M., Wu, D.: Improving statistical machine translation using word sense disambiguation. In: Proceedings of the 2007 Joint Conference on Empirical Methods in Natural Language Processing and Computational Natural Language Learning (EMNLP-CoNLL), pp. 61–72. ACL, Prague (2007)
4. van Gompel, M.: UvT-WSD1: A cross-lingual word sense disambiguation system. In: Proceedings of the 5th International Workshop on Semantic Evaluation, pp. 238–241. ACL (2010)
5. Vilariño Ayala, D., Balderas Posada, C., Pinto Avendaño, D.E., Rodríguez Hernández, M., León Silverio, S.: FCC: Modeling probabilities with GIZA++ for Task 2 and 3 of SemEval-2. In: Proceedings of the 48th Annual Meeting of the Association for Computational Linguistics, 5th International Workshop on Semantic Evaluation, pp. 112–116. ACL, Uppsala (2010)
6. Silberer, C., Ponzetto, S.P.: UHD: Cross-lingual word sense disambiguation using multilingual co-occurrence graphs. In: Proceedings of the 48th Annual Meeting of the Association for Computational Linguistics, 5th International Workshop on Semantic Evaluation, pp. 134–137. ACL, Uppsala (2010)
7. Koehn, P., Knight, K.: Estimating word translation probabilities from unrelated monolingual corpora using the em algorithm. In: National Conference on Artificial Intelligence (AAAI 2000), Langkilde, pp. 711–715 (2000)
8. Koehn, P., Knight, K.: Knowledge sources for word-level translation models. In: Lee, L., Harman, D. (eds.) Proceedings of the 2001 Conference on Empirical Methods in Natural Language Processing, pp. 27–35. ACL, Pittsburgh (2001)
9. Monz, C., Dorr, B.J.: Iterative translation disambiguation for cross-language information retrieval. In: Proceedings of the 28th Annual International ACM SIGIR Conference on Research and Development in Information Retrieval, SIGIR 2005, pp. 520–527. ACM, New York (2005)

10. Rapp, R.: Automatic identification of word translations from unrelated english and german corpora. In: Proceedings of the 37th Annual Meeting of the Association for Computational Linguistics, ACL 1999, pp. 519–526. ACL, College Park (1999)
11. Pomikálek, J., Rychlý, P., Kilgarriff, A.: Scaling to billion-plus word corpora. Advances in Computational Linguistics 41, 3–13 (2009)
12. Schmid, H.: Probabilistic part-of-speech tagging using decision trees. In: Proceedings of International Conference on New Methods in Language Processing, Manchester, UK, vol. 12, pp. 44–49 (1994)
13. Rychlý, P.: Manatee/Bonito – a modular corpus manager. In: 1st Workshop on Recent Advances in Slavonic Natural Language Processing, Brno, Masaryk University, pp. 65–70 (2007)
14. Marsi, E., Lynum, A., Bungum, L., Gambäck, B.: Word translation disambiguation without parallel texts. In: International Workshop on Using Linguistic Information for Hybrid Machine Translation (LIHMT 2011), Barcelona, Spain, pp. 66–74 (2011)
15. Weber, R., Schek, H.J., Blott, S.: A quantitative analysis and performance study for similarity-search methods in high-dimensional spaces. In: Gupta, A., Shmueli, O., Widom, J. (eds.) Proceedings of 24rd International Conference on Very Large Data Bases, VLDB 1998, August 24-27, pp. 194–205. Morgan Kaufmann, New York City (1998)
16. Doddington, G.: Automatic evaluation of machine translation quality using n-gram co-occurrence statistics. In: Proceedings of the Second International Conference on Human Language Technology Research, pp. 138–145. Morgan Kaufmann Publishers Inc. (2002)
17. Papineni, K., Roukos, S., Ward, T., Zhu, W.: BLEU: a method for automatic evaluation of machine translation. In: Proceedings of the 40th Annual Meeting on Association for Computational Linguistics, ACL, pp. 311–318 (2002)
18. Cohen, T., Schvaneveldt, R., Widdows, D.: Reflective random indexing and indirect inference: A scalable method for discovery of implicit connections. Journal of Biomedical Informatics 43(2), 240–256 (2010)

English-Vietnamese Machine Translation
of Proper Names
Error Analysis and Some Proposed Solutions

Thi Thanh Thao Phan and Izabella Thomas

Centre de Recherche Lucien Tesnière, Université de Franche-Comté, France
`thao.phan_thi_thanh@edu.univ-fcomte.fr`,
`izabella.thomas@univ-fcomte.fr`

Abstract. This paper presents some problems involved in the machine translation of proper names (PNs) from English into Vietnamese. Based on the building of an English-Vietnamese parallel corpus of texts with numerous PNs extracted from online BBC News and translated by four machine translation (MT) systems, we implement the PN error classification and analysis. Some pre-processing solutions for reducing and limiting errors are also proposed and tested with a manually annotated corpus in order to significantly improve the MT quality.

Keywords: proper names, machine translation, pre-processing.

1 Introduction

Proper names are of great importance to both linguistics and NLP domains due to their frequency and their particular functions. In fact, PNs have been used in all kinds of written texts such as newspapers, articles, books, etc. Interestingly, they constitute more than 10% of newspaper texts [2], which indicates a high rate of PNs' occurrences in this type of document. PNs also make a significant contribution to various NLP applications such as MT, CLIR, IE and QA. The crucial role of PNs is to provide the specific identification of persons, places, organizations, and artifacts. Nevertheless, it is a thorny task to deal with PNs in general and with PN translation from a source language (SL) into a target language (TL) due to their peculiar features. The number of words used in constructing PNs is potentially infinite [9]. PNs belong to an open class of words; hence, it is complicated to formulate and analyze PNs based on "pure" linguistic rules. Moreover, PNs also constitute a considerable portion of "unknown words" in corpora [13], which cannot be stored in dictionaries because of their uniqueness and their quantity. As a result, PN translation often requires different approaches and methods than translation of other types of words [1].

Despite a variety of important studies on PN translation for several pairs of languages such as English, Chinese, French, Japanese, etc., which report good results [10,12], there is little research on the English-Vietnamese (E-V) language pair. Many computational linguists in Vietnam's NLP community have considered and worked on MT and Named Entity Recognition [3,4,5,8], but none of them considers MT of PNs. There exist a large number of linguistics problems in E-V MT of PNs. Some personal names,

P. Sojka et al. (Eds.): TSD 2012, LNCS 7499, pp. 386–393, 2012.

which should be kept unchanged, are incorrectly translated into Vietnamese, e.g., *Vince Cable, Mark Lowen, Jack Warner*, etc. The same errors occur with geographic names, product names (e.g., *Victoria Embankment, Turkey, Apple*, etc.). Mistakenly translating those proper nouns as if they were common nouns often leads to incomprehensibility or ambiguities that requires extensive post-editing work.

Our study addresses the problematic issue of PN MT and is divided into 3 main parts: the first part concerns the building of an E-V comparable corpus dedicated to PN analysis; the second part introduces the corpus-based linguistic error classification and analysis of PN errors made by E-V MT systems; the last part details some preprocessing solutions for reducing PN errors.

2 Methods

2.1 Building a Bilingual Parallel Corpus of Texts Dedicated to PN Processing

Due to the benefits of corpora in supplying training materials for NLP applications (e.g., training of statistical POS taggers and parsers, training of example-based and SMT systems, etc.), a large number of corpora have been built and analyzed with satisfactory achievements. So far there have been a few studies on building E-V bilingual corpora [3,12], which can be used efficiently in many NLP applications. For example, EVC-an annotated E-V parallel corpus of over 5 million words is used for Vietnamese-related NLP tasks such as Vietnamese word segmentation, POS-tagger, word sense disambiguation in E-V MT [11]. These corpora are mostly collected from various bilingual texts (books, dictionaries, available corpora) in several chosen domains (science, technology and daily conversation). Yet, none of them seems appropriate for the study of PN translation.

Our corpus consists of 1,500 original English texts of approximately 110,000 words (i.e. 50–70 words per text) with a great number of PNs translated by 4 E-V MT systems including Vietgle, Google Translate, Bing and EVtran [14], which are widely used in Vietnam. All the texts extracted from online BBC News and related to various topics provide authentic and diverse training data for MT systems. Our method of building this corpus consisted of following steps:

1. Collecting and selecting texts for the E-V corpus: resources of raw or authentic texts from online BBC News on a variety of popular topics.
2. Arrangement of the texts according to topics and chronology to achieve the frequency of PNs in different domains.
3. Automatic translation of the texts by 4 E-V MT systems mentioned above to provide the corresponding versions in the SL and the TL.
4. Manual annotation of PNs was the final step for finishing this E-V corpus.

2.2 Corpus-Based Error Classification

MT of PNs from English into Vietnamese mainly makes mistakes due to the non-translation (so-called NT) and wrong translation (WT). Errors of NT often happen to abbreviations, professional titles, organization names, weekdays, etc. Meanwhile, the WT errors are analyzed on the basis of various linguistics criteria as follows: 1. Graphics; 2. Lexis; 3. Syntax; 4. Transcription/transliteration.

For the analysis of the E-V parallel corpus of texts with PNs described above, we propose our typology of E-V PN translation errors (shown in Fig. 1). As regards NT errors, we mention 7 sub-classes (NT1 to NT7) including NT for abbreviations or acronyms, professional titles, human titles, organization names, etc. Concerning WT errors, we classify them into 4 categories and 23 sub-types.

The graphic errors (GE) caused by wrong capitalization of PNs is divided into 12 GE sub-classes (GE1 to GE12). There is a great difference in capitalization between English and Vietnamese. In Vietnamese, all PNs that are personal names, geographic names, professional names, organization names, names of nationalities, etc. should be capitalized with the initial letters; however, months (*January, June*, etc.), weekdays (*Monday, Tuesday*, etc.) and human titles (*Mr., Mrs., Ms.*, etc.) are capitalized in English, but they are not in Vietnamese (except at the beginning of a sentence) [7].

Concerning lexical errors (LE), there are 4 LE sub-classes resulting from WT by incorrect words, missing words, redundant words and unknown words. Since the polysemy of vocabulary remains problematic in MT, it is difficult to avoid LE in E-V MT of PNs. The WT of English personal names into Vietnamese, e.g., *Sepp Blatter, Spin Boldak, Bin Hammam*, etc. (which should be kept unchanged) usually gives rise to the incomprehensibility of the texts.

In view of syntactic errors (SY), we mention three sub-classes: 1. Incorrect translation of possessive structures (SY1); 2. Incorrect order of words in noun phrases (SY2); 3. Wrong order of personal names (SY3). Regarding SY1, the singular possessive form with apostrophe s ('s) after a PN in English is sometimes mistranslated as if it were the contraction of the verb "to be" conjugated with the third person singular personal pronouns, e.g. *"Apple's success"* is automatically translated by *"Táo là thành công"* (Apple fruit is success) instead of *"thành công của thương hiệu Apple"* (the success of Apple trade-mark).

The PN transcription errors (TE) are caused by incorrect transcription or transliteration of PNs in Vietnamese and divided into 4 classes: TEs for geographic names, TEs for monetary units, TEs for nationalities and TEs for languages. Since there do not exist any consistent regulations for writing and transliteration/transcription of foreign PNs into Vietnamese [6], both linguists and computational linguists are confused with plenty of ways in PN writing and transliteration, e.g., *"Washington"* (English) is written in Vietnamese like *"Oa-sinh-tơn"* or *"Hoa Thịnh Đốn"* (Sino-Vietnamese) or *"Washington"* (current usage). Obviously, it is impossible to avoid TEs in E-V automatic translation. For instance, *"Burundian soldiers"* is translated by *"lính Burundian"* or *"lính Burund"*; in reality, it should be translated by *"lính Burundi"* because of *"Burundi"* in current use.

2.3   Error Analysis

On the basis of corpus-based error classification described above, we analyzed the results of the four machine PN translations of the E-V parallel corpus. According to our results, there are 8,549 PNs in 1,500 texts with approximately 6 PNs per text of

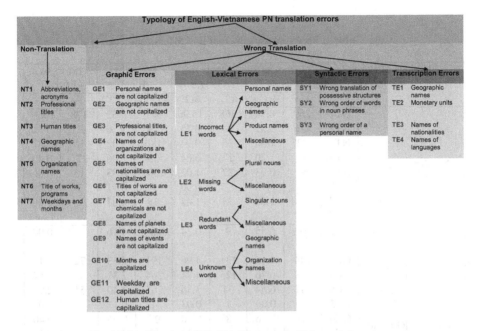

Fig. 1. Classification of English-Vietnamese PN translation errors

50–70 words. It shows that the rate of PNs in online BBC News is rather high (8.3–13.6% of the total words in a text), which emphasizes the important role of a correct PN translation for information access.

Our results (shown in Table 1) reveal a large number of errors made by 4 MT systems. There are 3224 errors/8549 PNs (37.71%) made by Vietgle, 2,636 errors/8,549 PNs (30.83%) made by Google, 3,579 errors/8,027 PNs (44.58%) by Bing and 1726 errors/3654 PNs (47.23%) by EVtran. Thus, the rate of MT errors of PNs is quite high.

We find each class of errors in the translations of all 4 MT systems. In particular, the number of LEs made by 4 translation engines is high, e.g., 32.87% by Vietgle, 35.96% by Google, 36.18% by Bing and 29.54% by EVtran. Consequently, it is necessary to pay close attention to this typology of errors. Similarly, the rate of SYs caused by these MT systems is relatively high, specifically, 19.94% by Vietgle, 17.62% by Google, 16.65% by Bing and 36.84% by EVtran. EVtran has made the highest rate of SY errors among the 4 translation engines. In brief, in comparison with the total of GEs and TEs, the number of SYs and LEs is relatively high. Hence, it is an essential task to reduce these kinds of errors in MT quality improvement.

2.4   Pre-processing

Pre-processing plays a significant role in ameliorating MT systems [7] as there remain a great number of unsolved problems concerning the adjustment of input texts to achieve better outputs through an automatic translation process. Consequently, we apply the following pre-processing steps to our corpus:

Table 1. PN errors made by four E-V MT systems

Typology of Errors		Errors by Vietgle		Errors by Google		Errors by Bing		Errors by EVtran	
		(1)	% (1)'	(2)	%(2')	(3)	% (3')	(4)	% (4')
Non-Translation	NT1	269	8.34	368	13.96	341	9.52	178	10.31
	NT2	116	3.59	20	0.75	79	2.20	2	0.11
	NT3	42	1.30	25	0.94	180	5.02	10	0.57
	NT4	30	1.20	17	0.64	38	1.06	7	0.40
	NT5	50	0.93	27	1.02	30	0.83	15	0.86
	NT6	4	0.12		0.00	4	0.11	7	0.40
	NT7	33	1.02	38	1.44	35	0.97		0.00
	Total	523	16.50	495	18.75	707	19.71	219	12.65
Wrong translation	GE1	18	0.55	9	0.34	6	0.16		0.00
	GE2	144	4.46	131	4.96	191	5.33	72	4.17
	GE3	218	6.76	125	4.74	264	7.37	44	2.54
	GE4	211	6.54	94	3.56	265	7.40	51	2.95
	GE5		0.00	1	0.03	4	0.11		0.00
	GE6	15	0.46	8	0.30	25	0.69	7	0.40
	GE7	5	0.15	6	0.22	12	0.33	4	0.23
	GE8	8	0.24	17	0.64	28	0.78	8	0.46
	GE9	6	0.18	7	0.26	23	0.64	2	0.11
	GE10	19	0.58	40	1.51	27	0.75	24	1.39
	GE11	96	2.97	111	4.21	24	0.67	41	2.37
	GE12	120	3.72	60	2.22	6	0.16	31	1.79
	Total	860	26.67	609	23.10	875	24.44	284	16.45
	LE1	805	24.95	402	15.23	556	15.53	395	22.86
	LE2	222	6.87	271	10.27	229	6.39	62	3.58
	LE3	65	2.01	86	6.23	70	1.95	22	1.27
	LE4	187	5.80	189	7.17	440	12.29	31	1.79
	Total	1092	33.87	948	35.96	1295	36.18	510	29.54
	SY1	248	7.69	264	10.01	364	10.17	287	16.62
	SY2	290	8.99	117	4.43	147	4.10	247	14.31
	SY3	105	3.25	81	3.07	85	2.37	102	5.90
	Total	643	19.94	462	17.52	596	16.65	636	36.84
	TE1	55	1.70	79	2.99	74	2.06	54	3.12
	TE2	11	0.34	19	0.72	20	0.55	10	0.57
	TE3	37	1.14	21	0.79	10	0.27	13	0.75
	TE4	3	0.09	3	0.11	2	0.05		0.00
	Total	106	3.28	122	4.62	106	2.96	77	4.46
Total of Errors/ Total of PNs		3224/ 8549	37.71	2636/ 8549	30.83	3579/ 8027	44.58	1726/ 3654	47.23

(Row groups within "Wrong translation": GE1–GE12 = Graphic Errors; LE1–LE4 = Lexical Errors; SY1–SY3 = Syntactic Errors; TE1–TE4 = Transcription Errors)

1. Manual annotation of corpus.

   An annotated corpus offers good training data of both SL and TL, which are useful for improving MT quality. The annotation of a corpus with PNs is an essential task to reduce PN translation errors. We use the tags shown in Table 2 to annotate our corpus of English texts. Here is an example of annotated texts:

   <NE>CIA<PER><Professional Title>director</Professional Title><PERS>Leon Panetta</PERS></PER></NE>should take over as the next defence secretary when<PER><PERS>Robert Gates</PERS></PER>retires in late June.

Table 2. Tag elements for English corpus annotation

1	<NE>	Named Entity boundary
2	<GEO>	Geographic names
3	<ORG>	Organization names
4	<SCI>	Scientific names
5	<TWP>	Titles of Works and Publications
6	<TIME>	Calendar and Time designations
7	<MISC>	Other proper names
8	<PER>	Personal names including full names & human titles or professional titles
9	<Human Title>	e.g., Mr., Mrs., Ms, etc.
10	<Professional Title>	e.g. Pilot, Secretary, Doctor, etc.
11	<PERS>	Personal names

2. Automatic simplification of certain syntactic structures in original texts.

   The program is written to search and change automatically the English possessive structure "PN's noun/noun phrase" =>"the noun/noun phrase + of + PN", since the second one is less ambiguous when translating to Vietnamese. This can limit a large number of SY1, e.g. *"Gen Petraeus's job"* =>*"the job of Gen Petraeus"*, which is then correctly translated by all the E-V MT systems.

3. Automatic marking of PNs, which should not be translated, from the English annotated corpus. This is necessary to correct LE1, e.g., all the personal names tagged by <PERS></PERS>are marked by <DNT_xxx>(xxx =0001, 0002,...). These will then not be translated by the MT systems.

4. Automatic replacement of the text marked <DNT_xxx>with the original text to retrieve the expected result.

Our demo MT pre-processing program built for correcting the PN translation errors consists of the five following steps:

Step 1: Enter the input (tagged)text(a text file or richtext document).

Step 2: Parse the input text (the text will be parsed and all the occurrences of PN possession structures are found and replaced by the new ones).

Step 3: Output of the marked text (all the occurrences of PNs, which will not be translated, marked with <DNT_xxx>).

Step 4: Paste the marked text after being translated with MT systems.

Step 5: Replace the text marked with the original text to retrieve the result.

3   Testing and Results

The pre-processing program presented above has been implemented with our English corpus, which was translated once more by the four MT systems. The results shown in Table 3 indicate the total PN errors corrected, including 340 LEs and 349 SYs by Vietgle, 236 LEs and 339 SYs by Google, 269 LEs and 438 SYs by Bing, 207 LEs and 383 SYs by EVtran. It corresponds to the reduction of 689/3,224 errors by Vietgle, 575/2,636 errors by Google, 707/3,579 errors by Bing and 670/1,726 errors by EVtran. Due to the simplification of possessive structures, all the syntactic errors resulting from the wrong interpretation of the "s" particle have been corrected by 4 MT systems. In addition, the WT errors of personal names, geographic names, product names can be reduced.

Table 3. Total errors corrected in the pre-processed corpus

Typology of Errors		Errors by Vietgle		Errors by Google		Errors by Bing		Errors by EVtran	
		(1)	% (1)'	(2)	% (2')	(3)	% (3')	(4)	% (4')
Lexical Errors(LE)	LE1	340/ 1092	10.54	236/ 948	8.95	269/ 1295	7.51	207/ 510	11.99
Syntactic Errors(SY)	SY1	246	7.63	261	9.90	361	10.08	284	16.45
	SY3	103	3.19	78	2.85	77	2.15	99	5.73
Total		349/643	10.82	339/462	12.86	438/596	12.23	383/636	22.19
TOTAL (LE+SY)		689/ 3224	21.37	575/ 2636	21.81	707/ 3579	19.75	670/ 1726	38.81

4   Conclusion

In this paper, we discussed the importance of PNs in NLP applications and some problems relating to PN translation. Being aware of the vital role of bilingual corpora in solving those problems, we have built an E-V parallel corpus of texts with numerous PNs. The error analysis results motivated us to search for pre-processing solutions for the reduction of PN errors. Our pre-processing approach consists of three main steps: pre-processing of the source texts, manual annotation of the corpus and extraction of annotated PNs. Our method has been tested and achieved certain noticeable results. The rate of errors made by 4 MT systems has been reduced, specifically, 21.37% for Vietgle, 21.81% for Google, 19.75% for Bing and 38.81% for EVtran.

Our study has contributed to the improvement of MT quality in Vietnam's NLP community. Our future work will explore different solutions for correcting other PN errors such as graphic errors and transcription errors. It is expected that our research will be applied to both English-Vietnamese and French-Vietnamese translation engines in the coming years.

References

1. Babych, B., Hartley, A.: Improving MT Quality with Automatic Named Entity Recognition. Centre for Translation Studies, University of Leeds, UK (2003)
2. Coates-Stephens, S.: The Analysis and Acquisition of Proper Names for the Understanding of Free Text. Computers and the Humanities 26(4), 441–456 (1993)
3. Dinh, D., Nguyen, L.T.N., Do, X.Q., Van, C.N.: A hybrid approach to word order transfer in the English to Vietnamese machine translation (2004)
4. Ho, T.B., Ha, N.K., Nguyen, T.P.T.: Issues and First Development Phase of the English-Vietnamese Translation System EVSMT 1.0. (2007)
5. Nguyen, H.D.: Vietnamese-English Cross-Language Information Retrieval (CLIR) using bilingual dictionary. Hewlett-Packard Company, USA (2008)
6. Nguyen, V.K.: Principle draft of writing and pronouncing foreign PNs in the government management protocols. Journal of Linguistics and Society No. 6(128), 1–6 (2006)
7. Phan, T.T.T.: Proper Name Errors in Online Translation Texts from English to Vietnamese: an Analysis and a Proposed Solution. In: Bulag NLP and HLT, International Review 2010, No. 34, pp. 111–133. Presses UFC (2010)
8. Tran, Q.T., Pham, T.X.T., Ngo, Q.H., Dinh, D., Collier, N.: Named Entity Recognition in Vietnamese Document. Program in Informatics (4), 5–13 (2007)
9. Wolinski, F., et al.: Automatic Processing of Proper Names in Texts (1995)
10. Dinh, D., Hoang, K.: POS-Tagger for English-Vietnamese Bilingual Corpus. In: Proceedings of Human Language Technology- North American Chapter of the Association for Computational Linguistics (2003)
11. Do, D., et al.: Word alignment in English-Vietnamese bilingual corpus. In: Proceedings of EALPIIT 2002, Hanoi, Vietnam, pp. 3–11 (2002)
12. Do, T.N.D., Le, V.B., Besacier, L., Bigi, B.: Mining a comparable text corpus for a Vietnamese-French statistical MT system. In: Proceedings of the 4th Workshop on SMT, pp. 165–172. ACL, Athens (2009)
13. Krstev, C., Vitas, D., Maurel, D., Tran, M.: Multilingual Ontology of Proper Names. In: Vetulani, Z. (ed.) Proceedings of 2nd Language & Technology Conference, Poznan, Poland, pp. 116–119 (2005)
14. Website links of E-V MT systems,
    http://tratu.vietgle.vn/hoc-tieng-anh/dich-van-ban.html,
    http://translate.google.com,
    http://www.microsofttranslator.com/,
    http://vdict.com/#translation

Improved Phrase Translation Modeling Using MAP Adaptation

A. Ryan Aminzadeh[1], Jennifer Drexler[2], Timothy Anderson[3], and Wade Shen[2,*]

[1] US Department of Defense
ryan.aminzadeh@gmail.com
[2] Massachusetts Institute of Technology, Lincoln Laboratory
{j.drexler,swade}@ll.mit.edu
[3] Air Force Research Laboratory
timothy.anderson@wpafb.af.mil

Abstract. In this paper, we explore several methods of improving the estimation of translation model probabilities for phrase-based statistical machine translation given in-domain data sparsity. We introduce a hierarchical variant of *maximum a posteriori* (MAP) adaptation for domain adaptation with an arbitrary number of out-of-domain models. We note that domain adaptation can have a smoothing effect, and we explore the interaction between smoothing and the incorporation of out-of-domain data. We find that the relative contributions of smoothing and interpolation depend on the datasets used. For both the IWSLT 2011 and WMT 2011 English-French datasets, the MAP adaptation method we present improves on a baseline system by 1.5+ BLEU points.

1   Introduction

Real-world performance of statistical MT models is often limited by the availability of training bitext. Performance of SMT models is sensitive not only to the amount of training data, but also to the domain from which these data are drawn. For optimal performance, developers of SMT systems will typically make significant investments to acquire bitexts *in the domain of interest*, which can be difficult and expensive.

In this paper, we describe efforts to improve SMT performance through better use of out-of-domain bitexts. We do this in the context of *maximum a posteriori* (MAP) estimation, a well-known approach for adaptation of statistical models that has been applied to the related problems of speech recognition [1] and language modeling [2,3], and which we apply here to the phrase translation tables used during SMT. We extend this method, which we developed previously for two phrase tables [4], to an arbitrary number of models.

In MAP adaptation, translation probabilities from the in-domain model are backed-off to out-of-domain estimates when the phrase pair occurs rarely in the in-domain data. When the phrase pair does not occur in the out-of-domain model, this back-off results in

---

* This work is sponsored by the Air Force Research Laboratory under Air Force contract FA8721-05-C-0002. Opinions, interpretations, conclusions and recommendations are those of the authors and are not necessarily endorsed by the United States Government.

a smoothing effect. Linear interpolation, another domain adaptation technique, also has a smoothing effect for phrase pairs that are not contained in all of the models being interpolated. When applied to a single phrase table, smoothing has been shown to improve SMT performance [5]. This observation leads us to investigate the extent to which the gains produced by domain adaptation are a result of smoothing.

In this paper, we empirically explore the relationship between phrase table smoothing and interpolation methods. We compare the results of MAP adaptation with the results of two other domain adaptation methods, linear interpolation and phrase table fill-up, and test these techniques on both smoothed and unsmoothed phrase tables.

2   Prior Work

2.1   Domain Adaptation

The phrase translation models used in SMT define a feature vector, $\phi(s, t)$, for each source/target phrase pair extracted during training. MAP adaptation and linear interpolation are applied to all features in that vector, but for conciseness we describe them here only for a single feature, $p(s|t)$.

MAP adaptation is a Bayesian estimation method that attempts to maximize the posterior probability of a model given the data, as in [1]. The standard MAP formulation defines two probability distributions: a prior distribution ($p(s|t, \lambda_0)$) and a distribution estimated over the adaptation data ($p(s|t, \lambda_{adapt})$). For phrase table estimation, the in-domain corpus is treated as adaptation data for estimating $p(s|t, \lambda_{adapt})$ and the out-of-domain corpus is assumed to be $p(s|t, \lambda_0)$.

$$\hat{p}(s|t, \lambda) = \frac{N_{adapt}(s, t)}{N_{adapt}(s, t) + \tau} p(s|t, \lambda_{adapt}) + \frac{\tau}{N_{adapt}(s, t) + \tau} p(s|t, \lambda_0) \quad (1)$$

where $N_{adapt}(s, t)$ is the joint count of s and t in the adaptation data and τ is the MAP relevance factor. We previously used this method to effectively make use of out-of-domain data to improve performance of phrase tables trained with limited amounts of data [4]. Foster et al. [6] also used a MAP-based approach for phrase table interpolation, improving performance over a baseline system for a number of different datasets.

Linear interpolation of translation probabilities, with fixed weights for each corpus, has also been shown to improve SMT performance [7]. The basic formulation is:

$$\hat{p}(s|t, \lambda) = \sum_{i=1}^{M} \alpha_i(s, t) p(s|t, \lambda_i) \quad (2)$$

where M is the number of corpora and α_i is the interpolation coefficient for the ith corpus. Foster and Kuhn [7] test several strategies for determining the best mixture weights, comparing uniform weights with TF/IDF-, perplexity-, and EM-based techniques. Although they obtain small improvements using the more complex techniques, all variations yield performance gains of about 1 BLEU point on the NIST 2006 Chinese dataset.

Phrase table fill-up has also recently been used for domain adaptation [8]. Fill-up is initialized with all phrase pairs from the in-domain phrase table. Phrase pairs from the out-of-domain table are then added only if they are not in the in-domain table. The probabilities associated with phrase pairs added to the table are not changed. An extra binary feature is added to the table, indicating which model each phrase came from; that feature can then be used during optimization to penalize out-of-domain phrase pairs relative to those from the in-domain table. Additional phrase tables can be "cascaded" together, with phrase pairs from a new table added only when they are not already contained in the filled-up table. Another binary feature is added to the final phrase table for each additional model. On the 2011 IWSLT English-French dataset, fill-up adaptation improves performance by 0.7 BLEU points, and is comparable to linear interpolation with uniform weights.

2.2 Phrase Table Smoothing

Foster et al. [5] review a large number of phrase table smoothing techniques, and find that all of them significantly improve the performance of a baseline SMT system. Kneser-Ney [9] and modified Kneser-Ney [10] smoothing have the best performance overall, producing gains of almost 1.5 BLEU points on an English-French task. The phrase table formulation of Kneser-Ney smoothing is [5]:

$$p(s|t) = \frac{N(s,t) - D}{N(t)} + \alpha(t)P_b(s) \tag{3}$$

$$\alpha(t) = Dn_{1+}(*, t) / \sum_{\tilde{s}} N(\tilde{s}, t) \tag{4}$$

$$P_b(s) = n_{1+}(s, *) / \sum_{\tilde{s}} n_{1+}(\tilde{s}, t) \tag{5}$$

where $n_{1+}(s, *)$ is the number of target phrases aligned with source phrase s, and $n_{1+}(*, t)$ is the reverse. In modified Kneser-Ney, the discount D is replaced by an empirically derived discount D_i, which is dependent on the joint count $N(s, t) = i$ of the source-target phrase pair.

Chen et al. [11] introduce an enhanced low frequency (ELF) feature designed to penalize phrase pairs with low joint counts during optimization. The feature is: $h_{elf}(s, t) = e^{(-1/N(s,t))}$, which is a $1/N(s, t)$ penalty in a log-linear model. Using this feature on the WMT 2010 French-English dataset, the authors report a gain of 0.55 BLEU points over a baseline phrase table. When added to modified Kneser-Ney, this feature produces an additional gain of .07 BLEU points.

3   Methods

3.1   MAP

We extend MAP to an arbitrary number of corpora, M, in a hierarchical fashion, as shown in Figure 1. To build this hierarchy, the training corpora are sorted based on their distance from a development dataset that matches the test domain. Models trained on corpora that are more distant from the test domain are successively MAP-adapted

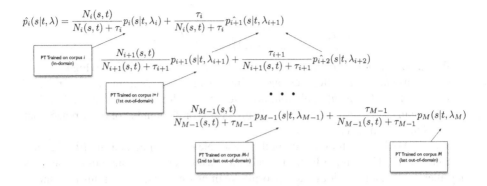

Fig. 1. MAP with multiple corpora

with models estimated from less distant corpora. The formulation of MAP for multiple corpora used in this paper is shown below:

$$\hat{p}_i(s|t, \lambda) = \frac{N_i(s,t)}{N_i(s,t) + \tau_i} p_i(s|t, \lambda_i) + \frac{\tau_i}{N_i(s,t) + \tau_i} \hat{p_{i+1}}(s|t, \lambda_{i+1}) \qquad (6)$$

The final probability estimate for the given phrase pair is $\hat{p}_1(s|t)$.

We experiment with two methods of constructing the MAP hierarchy. First, we determine the similarity between each corpus and the test domain. Similarity scores are computed using 1-gram document vectors created from source language data, with all words with *count* < 5 and all stop words removed. Background statistics are derived by concatenating all corpora used in these experiments, and used to compute a TFLLR weighting [12] which is applied to the document vectors. The score for a given model is the cosine similarity between the training data for that model and the development data. These similarity scores are then sorted to produce an ordering for MAP. Second, we use each phrase table individually to translate the development dataset and create a MAP hierarchy from the resulting BLEU scores.

We present a method for determining τ_i based on the similarity scores described above. We define τ_i as $\frac{\bar{N}_i(1-s_i)}{s_i}$, so that the interpolation coefficient for each model is, on average, equal to the similarity score for that corpus, s_i. As described in 3.2, these normalized scores are also used as corpus weights for linear interpolation. With this choice of τ_i, the smoothing effect of MAP on the in-domain model is, on average, the same as the smoothing effect of linear interpolation with those weights.

3.2   Linear Interpolation

There are many ways to determine the coeffients used for linear interpolation, and no consensus on the best method of doing so [8]. For this paper, we compare the similarity scores described above with uniform interpolation weights.

In both cases, linear interpolation has a smoothing effect similar to that produced by MAP estimation, but with a key difference. When a phrase exists only in the in-domain model, MAP reduces to:

$$\hat{p}(s|t, \lambda) = \frac{N_{adapt}(s, t)}{N_{adapt}(s, t) + \tau} p(s|t, \lambda_{adapt}) \tag{7}$$

MAP assumes that low-count phrase pairs are poorly estimated relative to higher-count pairs, and this smoothing effect removes more probability mass from the least reliable estimates. During linear interpolation, however, probabilities are trusted solely based on the domain used to estimate them, with no regard for the number of occurrences that produced that estimation.

Like the MAP smoothing effect, both Kneser-Ney smoothing and the ELF feature are count-based. Combining linear interpolation with these other smoothing methods may, therefore, result in additional performance improvements. We test this hypothesis by interpolating smoothed phrase tables. When combining linear interpolation with the ELF feature, we add one ELF feature to the phrase table for each corpus.

3.3 Phrase Table Fill-Up

The rankings used to determine the MAP hierarchies are also used as the cascade orderings for phrase table fill-up. Fill-up interpolation has no smoothing effect, so we test it both with and without smoothing. When combining this interpolation method with the ELF feature, we use a slight modification of the original phrase table fill-up method. We use that ELF feature in place of the binary fill-up feature, and include an ELF feature for the in-domain model, to allow the optimizer to penalize low frequency words from that model. For comparison, we include a binary feature for the in-domain model in our baseline fill-up implementation.

4   Experiments

4.1   Data

All experiments described in this paper test English-French translation using four corpora: Europarl (EP), Gigaword, News Commentary (NC), and TED. The first three came from WMT 2011 [13], and the last from IWSLT 2011[14]. The News Commentary and TED datasets are relatively small at just over 100K sentences each, while Europarl contains 600K sentences, and Gigaword 22.5M sentences.

We run all experiments on both the WMT 2011 and IWSLT 2011 test datasets. When testing on the WMT data, we use the WMT 2010 test set as development data for optimization. We consider the TED data to be the in-domain set for the IWSLT tests, and experiment with treating either the NC or EP data as in-domain for WMT.

4.2   System

Our baseline system uses a standard SMT architecture and has performed well on past evaluations [4,15].

We use interpolated Knesser-Ney n-gram language models built with the MIT Language Modeling Toolkit [16]. Additional class-based language models were also

Table 1. Smoothing Results

Dataset	IWSLT		WMT			
Phrase Table	TED	TED	NC	EP	NC	EP
Language Models	TED	all	NC	EP	all	all
Baseline	28.63	30.75	21.48	23.49	24.94	25.69
ELF	**30.01**	30.91	21.87	23.29	24.30	26.16
KN	29.21	30.71	21.83	**23.96**	24.67	**26.34**
KN + ELF	29.91	**31.35**	**22.01**	23.95	**25.05**	26.18

Table 2. Phrase Table Fill-Up Results

Dataset	IWSLT		WMT	
Ordering	Similarity	BLEU	Similarity	BLEU
Fill-up	30.70	**31.43**	25.77	26.52
Fill-up + ELF	31.02	30.45	25.95	25.97
Fill-up + KN	30.63	31.19	25.93	**27.3**
Fill-up + KN + ELF	30.73	31.09	26.10	26.66

trained on the TED data and used for rescoring when translating the IWSLT dataset. All phrase tables were created with IBM Model 4 alignments [17]; alignments extracted using the Berkeley Aligner and competitive linking algorithm (CLA) were added to the NC and TED phrase tables [18].

Our translation model assumes a log-linear combination of phrase translation models, language models, etc. We optimize the combination weights over a development set using minimum error rate training with a standard Powell-like grid search [19]. We use the Moses decoder [20].

All scores reported here are average BLEU scores obtained from three rounds of optimization.

4.3   Results

We ran four experiments exploring the effect of smoothing on a single phrase table: baseline, ELF, KN, and ELF + KN. We did two sets of these experiments: one in which only the in-domain phrase table and language model were used, and one in which the in-domain phrase table was paired with language models from all available corpora. Results are shown in Table 1. As expected, KN with the ELF feature has the best overall performance. The ELF feature seems to be particularly useful on the IWSLT dataset; the results are less consistent for WMT. These results suggest that the EP model should be considered in-domain for WMT. The extra language models provide a boost of 2 to 2.5 BLEU points across the board.

We experimented with all of the above smoothing options in combination with both fill-up and distance-based interpolation. We used all four language models for all domain adaptation experiments.

Fill-up results are in Table 2. For the WMT dataset, fill-up provides a clear improvement over the baseline. Performance on the WMT dataset is strongly impacted

Table 3. Linear Interpolation Results

Dataset	IWSLT		WMT	
Weights	Uniform	Similarity	Uniform	Similarity
Original	31.82	31.83	**27.71**	27.6
KN	31.56	31.76	27.61	27.65
ELF	31.78	**32.00**	27.66	27.67
KN + ELF	31.62	31.64	27.70	27.58

Table 4. MAP Results

Dataset	IWSLT	WMT
Similarity-based τ	32.06	27.05
$\tau = 4$	**32.42**	27.00
$\tau = 12.5$	32.33	**27.26**
$\tau = 25$	31.75	27.11
$\tau = 50$	32.37	27.12

by the fill-up cascade ordering, most likely because the similarity-based ordering treats the NC model as in-domain, while the BLEU ordering treats the EP model as in-domain.

Fill-up order has less of an impact on the IWSLT dataset; the TED model is considered in-domain for both. Regardless of order, fill-up performs poorly on the IWSLT dataset. The TED training data is very well-matched to the IWSLT test domain, and it may be that some additional phrases from the out-of-domain corpora hurt rather than help performance in this scenario.

Linear interpolation results are in Table 3. For both datasets, the choice of weights had little impact on performance; smoothing also does not provide any additional improvement on top of interpolation. While linear interpolation is better than the baseline for both datasets, the gain is much larger for WMT.

MAP using the BLEU ordering outperformed MAP with the similarity-based ordering for both datasets. Using the BLEU ordering, we experimented with several constant values of MAP τ, as well as the similarity-based values described in Section 3.1. The results are shown in Table 4. Overall, MAP performance is relatively insensitive to the value of τ chosen. On the IWSLT dataset, MAP equals or outperforms linear interpolation for all values of τ, while linear interpolation consistently outperforms MAP on the WMT dataset.

The two datasets used here have very different characteristics. The TED training data is well-matched to the IWSLT test domain, but contains relatively few sentences. As a result, the model is poorly estimated, and benefits greatly from count-based smoothing. Data from other domains, however, does not consistently improve IWSLT performance. In this situation, MAP adaptation outperforms the other interpolation and smoothing techniques tested here.

The EP model, on the other hand, is trained on a large amount of data, but that data does not come from the WMT test domain. WMT performance is very much improved by the incorporation of out-of-domain data, and is less impacted by smoothing techniques. Linear interpolation performs better than MAP adaptation in this case.

Table 5. Results Summary

dataset	IWSLT	WMT
baseline, single LM	28.63	23.49
best smoothed baseline, single LM	30.01	23.95
baseline, all LMs	30.75	25.69
best smoothed baseline, all LMs	31.35	26.34
fill-up, BLEU ordering	31.43	26.52
fill-up + KN, BLEU ordering	31.19	27.30
uniform linear interpolation	31.82	**27.72**
linear interpolation, SIM weights	31.83	27.6
linear interpolation, SIM + KN	31.76	27.65
MAP, similarity-based τ	32.06	27.05
MAP, best constant τ value	**32.42**	27.26

5   Conclusion

In this paper, we examine several methods of phrase table interpolation and compare them with the hierarchical MAP adaptation technique that we present. We also explore the relative contributions of smoothing and the addition of out-of-domain data to the performance gains achieved through phrase table interpolation.

For both the WMT 2011 and IWSLT 2011 datasets, phrase table fill-up performs worse than both linear interpolation and MAP adaptation. Linear interpolation is more effective than MAP on the WMT dataset, while the reverse is true for IWSLT.

These results can be explained by the nature of the datasets themselves. Count-based smoothing, and thus MAP adaptation, has a greater impact when the in-domain model is poorly estimated but well-matched to the test domain, as in the IWSLT dataset. When the converse is true, as in the WMT dataset, the incorporation of additional data provides the greatest performance improvement.

Our results impact MT applications where training bitext that is well-matched to the domain of interest is not available in sufficient quantities. In such applications, our results show that hierarchical MAP-based adaptation of the in-domain phrase translation model to general-purpose models allows for more reliable MT performance measured at 1.5+ BLEU.

This work has shown the merit of hierarchical MAP-based adaptation; future research will aim to optimize the way in which it is applied. Further experiments can be done to determine whether development set BLEU score is the best metric for generating the adaptation ordering, and, if so, whether it can be approximated by another, simpler metric that does not require building a full translation system for each corpus.

References

1. Gauvain, J.L., Lee, C.H.: Maximum a posteriori estimation for multivariate Gaussian mixture observations of Markov chains. IEEE Transactions on Speech and Audio Processing 2, 291–298 (1994)

2. Federico, M.: Bayesian estimation methods for n-gram language model adaptation. In: Proceedings of International Conference on Spoken Language Processing, pp. 240–243 (1996)
3. Bacchiani, M., Riley, M., Roark, B., Sproat, R.: Map adaptation of stochastic grammars. Computer Speech & Language 20, 41–68 (2006)
4. Shen, W., Delaney, B., Aminzadeh, A.R., Anderson, T., Slyh, R.: The MIT-LL/AFRL IWSLT-2009 System. In: Proc. of the International Workshop on Spoken Language Translation, Tokyo, Japan, pp. 71–78 (2009)
5. Foster, G., Kuhn, R., Johnson, J.H.: Phrasetable smoothing for statistical machine translation. In: Conference on Empirical Methods in Natural Language Processing, Sydney, Australia (2006)
6. Foster, G., Goutte, C., Kuhn, R.: Discriminative instance weighting for domain adaptation in statistical machine translation. In: Proceedings of the 2010 EMNLP, Cambridge, MA, pp. 451–459 (2010)
7. Foster, G., Kuhn, R.: Mixture-model adaptation for SMT. In: ACL Workshop on Statistical Machine Translation, Prague, Czech Republic (2007)
8. Bisazza, A., Ruiz, N., Federico, M.: Fill-up versus interpolation methods for phrase-based smt adaptation. In: International Workshop on Spoken Language Translation (2011)
9. Kneser, R., Ney, H.: Improved backing-off for m-gram language modeling. In: International Conference on Acoustics, Speech, and Signal Processing, ICASSP 1995, vol. 1, pp. 181–184 (1995)
10. Chen, S.F., Goodman, J.: An empirical study of smoothing techniques for language modeling. Computer Speech and Language 13, 359–393 (1999)
11. Chen, B., Kuhn, R., Foster, G., Johnson, H.: Unpacking and transforming feature functions: New ways to smooth phrase tables. In: Proceedings of MT Summit XIII, Xiamen, China (2011)
12. Campbell, W., Campbell, J., Gleason, T., Reynolds, D., Shen, W.: Speaker verification using support vector machines and high-level features. IEEE Transactions on Audio, Speech, and Language Processing 15, 2085–2094 (2007)
13. Callison-Burch, C., Koehn, P., Monz, C., Zaidan, O.: Findings of the 2011 workshop on statistical machine translation. In: Proceedings of the 6th Workshop on Statistical Machine Translation, Edinburgh, Scotland, pp. 22–64 (2011)
14. Federico, M., Bentivogli, L., Michael Paul, S.S.:: Overview of the iwslt 2011 evaluation campaign. In: Proceedings of the International Workshop on Spoken Language Translation, San Francisco, CA (2011)
15. Shen, W., Anderson, T., Slyh, R., Aminzadeh, A.: The MIT-LL/AFRL IWSLT 2010 MT system. In: Proc. of the International Workshop on Spoken Language Translation, Paris, France (2010)
16. Hsu, B.J., Glass, J.: Iterative language model estimation: Efficient data structure and algorithms. In: Proc. Interspeech (2008)
17. Brown, P.F., Della Pietra, S.A., Della Pietra, V.J., Mercer, R.L.: The Mathematics of Statistical Machine Translation: Parameter Estimation. Computational Linguistics 19, 263–311 (1993)
18. Chen, B., Cattoni, R., Bertoldi, N., Cettolo, M., Federico, M.: The ITC-irst SMT System for IWSLT-2005. In: Proceedings of the IWSLT 2005 (2005)
19. Och, F.J.: Minimum Error Rate Training in Statistical Machine Translation. In: Proceedings of ACL (2003)
20. Koehn, P., Hoang, H., Birch, A., Burch, C.C., Federico, M., Bertoldi, N., Cowan, B., Shen, W., Moran, C., Zens, R., Dyer, C., Bojar, O., Constantin, A., Herbst, E.: Moses: open source toolkit for statistical machine translation. In: Proceedings of the ACL, ACL 2007, Stroudsburg, PA, USA, pp. 177–180 (2007)

Part III

Speech

"**Speech:** The expression of or the ability to express thoughts and feelings by articulate sounds: *he was born deaf and without the power of speech.*"
NODE (The New Oxford Dictionary of English), Oxford, OUP, 1998, page 1788, meaning 1.

A Romanian Language Corpus
for a Commercial Text-To-Speech Application

Mihai Alexandru Ordean[1], Andrei Şaupe[1], Mihaela Ordean[1],
Gheorghe Cosmin Silaghi[2], and Corina Giurgea[1]

[1] iQuest Technologies, Str. Motilor 6–8, 400001, Cluj-Napoca, Romania
{Mihai.Ordean,Andrei.Saupe,
Mihaela.Ordean,Corina.Giurgea}@iquestint.com
[2] Babeş-Bolyai University, Business Information Systems Department,
Str. Theodor Mihali 58–60, 400591, Cluj-Napoca, Romania
Gheorghe.Silaghi@econ.ubbcluj.ro

Abstract. Text and speech corpora are a prerequisite for the development of an effective commercial text-to-speech system, using the concatenative technology. Given that such a system needs to synthesize both common and domain-specific discourses, the considered corpora are of main importance. This paper presents the authors' experience in creating a corpus for the Romanian language, designed to support a concatenative TTS system, able to reproduce common and domain-specific sentences with naturalness.

1 Introduction

Text and spoken language corpora of appropriate size and quality are prerequisites for the development of a commercial text-to-speech (TTS) system. When such corpora, covering not only the common vocabulary of a certain language but also the specific vocabulary of the target domains, are not fully available, they should be built from scratch.

A successful commercial TTS system needs to achieve a high speech quality. Speech quality can be characterized by intelligibility, naturalness, and clarity [1]. Natural sounding speech – the target of a TTS system, is defined as speech which sounds as if a native speaker would produce it. The intelligible speech represents the utterance that just can be understood and clear speech is the one that can be understood without difficulty. *Naturalness* is a multifaceted characteristic and the most important aspect of the naturalness of synthetic speech is how easy it is for a human to understand it [2]. Assessing the naturalness in synthetic speech is an intractable problem in information technology today [3], but naturalness is important because it opens the possibility of scientific evaluation for the competing hypothesis using perceptual tests [2].

Our main long-term goal is to create a text-to-speech system for the Romanian language using the concatenative approach, which can be easily adapted to a given domain of discourse. For achieving this goal, in this paper we present our experience in building a Romanian language corpus for our commercial TTS system. In this respect, we will use facets of naturalness in order to evaluate the effectiveness of the constructed corpus, used within our commercial TTS system, adapted for a specific domain: public

P. Sojka et al. (Eds.): TSD 2012, LNCS 7499, pp. 405–414, 2012.

administration. Our created corpus should have the minimal size and should give a good naturalness to the synthesized speech. The final goal of the corpus creation process is to allow a minimum *time-to-market* for commercial applications.

This paper evolves as follows. Section 2 presents other efforts to create text and speech corpora for the Romanian language. Section 3 introduces our corpus. We first define the properties our corpus should have, and next, we present the steps we performed in order to create the corpus. Section 4 evaluates the corpus, showing how inclusion of different parts improves or does not improve the naturalness of the TTS system for the common speech and for the specialized speech for public administration.

2   Related Work

In the last twenty years, especially in the last decade, a lot of efforts have been made by different research teams in order to develop Romanian linguistic resources. A review of the main achievements of the research actions addressing the computational aspects of the Romanian language could be found in [4].

Presented in the aforementioned paper as an objective for future work, the Web-portal hosting both Romanian language resources - corpora, dictionaries, collection of linguistic data in both symbolic and statistical form, and processing tools for application development are available on sections [5]. The SRoL database includes files with basic sounds of the Romanian language, dialectal voices, pathological voices and gnathosonic and gnathophonic sounds. It represents the first Internet-based annotated database of emotional speech for the Romanian language and contains more than 1500 recordings in different coding formats (.wav, .ogg, .txt, 22 kHz sampling rate, 24 bit or 16 bit precision).

Burileanu et al. [6,7] reported a relatively medium-size database they have created in order to build a robust spontaneous speech recognition system for the Romanian language. The database has the following features: (i) the signals originate mainly from the acquisition of TV streaming in Romanian, broadcasted on the Internet; (ii) duration of the recordings is around 4 hours of speech; (iii) number of speakers is 12 – 8 women and 4 men; (iv) number of sessions per speaker range from 3 to 20; (v) the intervals between sessions range from one day to a couple of weeks; (vi) the number of word types is around 3,000 (medium vocabulary); and (vii) speech register is spontaneous speech with sampling frequency of the signal 8 kHz.

3   The Corpus

The corpus we aim to create should be an integral part of a concatenative TTS system, with a high degree of naturalness. To accomplish this objective, we considered the following directions:

- minimizing the number of concatenations.
- maximum coverage of diphones. A diphone is considered to be the transition between two phones that also contains half of each phone. Diphones are also differentiated by their phones properties (phonetic processes), syllable properties

(phones syllable position and syllable stress) and word properties (phones word position)
- enlarging the selection database for acoustic units
- recording the smallest audio corpus possible, fitting the properties enumerated above.

Such a corpus should assure the best possible coverage of the Romanian common vocabulary. It should contain audio recordings without intonation which would make it possible to apply intonations at a later stage, using algorithms (e.g. during speech synthesis). It must be mentioned that based on our previous experience in creating a restricted corpus, we have set an additional objective: to bound the total size of recordings to 5 hours. Reaching this target involves limitations which need to be taken into consideration when establishing the size of the corpus.

In this section we present our efforts to create the text and speech corpora for the Romanian language, fitting the characteristics just presented above.

3.1 Text Corpus

From the point of view of the covered vocabulary, the Text corpus is represented in Figure 1. As shown, the text corpus has two major components:

- CORE, consisting of (i) the Basic component which contains a part of the common vocabulary of the Romanian language which needs to be as large as possible and (ii) Extra words (EW) component which contains all the words covering the diphones of the Romanian language and that are not found in the other components of the text corpus
- Extensions of the vocabulary from specific fields – terms from the target domain vocabulary (the legal domain in our case) and the vocabulary used in the written press in the target domain

Previous to the text corpus creation, a text selection was built, containing: (i) belletristic and religious literature (size 50 Mb, sources: around 125 tomes), (ii) legal documents (size 178 Mb), and (iii) texts from newspapers (size 0.5 Mb, sources: local, regional and national newspapers). Starting with the existing text selection, using distinct selection

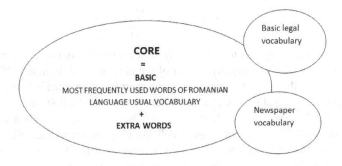

Fig. 1. Text corpus structure

criteria for the `Basic + Extensions` part and the EW part, the `Text Corpus` was built in several steps as follows. First, for the `Basic` part of the `CORE` as for both Basic legal and newspaper `Extensions` – sentences were selected directly using an algorithm based on sentence frequency cost defined in Eq. (1):

$$SFreq_{cost} = -\sum_{i=1}^{n} word_{i_{Freq}} \qquad (1)$$

where $word_{i_{Freq}}$ represents the frequency of $word_i$ if $word_i$ was not previously selected and 0 if $word_i$ was previously selected. Sentences longer than 50 phones or shorter than 3 words were omitted for the selection. The limits were set so that sentences are short enough but not too short to be easily pronounced with no intonation or breathing pause. The upper limit was introduced in such a manner that no breathing pause needs to be taken as it introduces prosodic phrase intonation. Also sentences with less than 3 words seemed to produce a voice inflection because of their abrupt end. We notice that the sentence frequency cost is the negative sum over the frequencies of the sentence words that were not previously selected.

Second, if there are more candidates for the current selection step we use the sentence diphone coverage cost of Eq. (2) to filter them.

$$SDiphone_{cost} = \frac{1}{k} \sum_{i=1}^{n} diphone_{i_{Freq}} \qquad (2)$$

where $diphone_{i_{Freq}}$ is the count of the ith sentence diphone in the current sentence, if the $diphone_i$ was not previously selected and 0 if $diphone_i$ was previously selected. k represents the number of selected diphones. We notice that the sentence diphone coverage cost is the mean of all the unseen sentence diphones weighted by their in-sentence frequency, as selecting a sentence that repeats the same diphone was not desired. If even after the diphone cost calculation there are more candidates, the first shortest sentence is selected to save recording space.

For `Extra words EW` – we selected the words that cover diphones of the Romanian language and that are not found in other components of the `Text Corpus`. To perform the selection, we used an algorithm based primary on diphone coverage cost, computed as in Eq. (3).

$$WDiphone_{cost} = -\sum_{i=1}^{n} cost \qquad (3)$$

where $cost$ is 1 if $diphone_i$ was not previously selected and 0 otherwise. We notice that the word diphone cost is the negative count of the unseen word diphones. If there are more candidates for the current selection step, the most frequent word is chosen. Similar to the sentence selection algorithm, the last filter criterium is the first shortest word.

After word selection done with the help of the previosly described procedure, sentences were built by a team of linguist collaborating on the project. The inclusion of as many exception words as possible in each sentence was taken into consideration in order to reduce the duration of the audio recordings. The current linguistic corpus covers a number of sentences/words as follows:

- `Basic` (the common vocabulary) contains 938 sentences and 8,354 words selected, estimated to covering almost 1h 30' audio recording. In fact, the recorded time is 1h 22'
- EW contains 2,934 words initially selected with an estimated recording time of 2h 51'. This estimation was based on the hypothesis that each word is part of a different sentence. From the processing done by the linguist team, it resulted a total of 1,229 sentences consisting of 8,646 words. Because the linguists managed to encapsulate more extra words in the same sentence, the audio recording time was re-estimated to 1h 33'. Finally, the audio recording time was 1h 45'.
- `Extension`: the legal domain extension contains 672 sentences and 5,238 words, the estimated audio recording time was 1h. Finally, the audio recording time was 1h 01'
- `Extension`: the press extension contains 183 sentences and 1,495 words; estimated audio recording time: 15'. Finally, the audio recording time was 14'

We should mention that, the core can be upgraded based on the available resources. The size of the `text corpus` was estimated by the experience with the previous audio corpus that consisted of 411 sentences and 2,219 words, which covered 23'41" of audio recording. Based on this previous experience we computed the audio recording times as follows:

- Phoneme mean recording time: 0.118273645"
- Sentence mean recording time: 2.75"
- Break between sentences mean recording time: 0.75"

This leads to a final audio corpus of 4h 22', which fits the length limitation of 5 hours.

3.2 Speech Corpus

Recordings were done using a female voice. Recordings were done in a flat tone of voice; the same timbre and diction was maintained during the entire session. They took place over six days, in four hour-long sessions. The result was an audio corpus long of 4h22'.

After recording, the speech corpus underwent an annotation process, on two levels:

A. on a textual level. The first 200 sentences were annotated manually using PRAAT, based on the textual prosody described in the phonetic transcript files. Textual prosody included information about phonemes and their properties, syllables and the accents on syllables for recorded words. The automated annotation of the rest of the corpus was done using the HTK speech recognition software, trained with the 200 manually annotated files and phonetic transcripts.

B. on an audio level. Using MATLAB, a wavelet D Mayer analysis was carried out in order to select the fundamental frequency (under 150 Hz). Based on signals resulted from MATLAB processing and using PRAAT, we conducted an automated annotation and analysis of glottal impulses and the amplitude of the vocal signal.

4 Corpus Evaluation

To show the validity of the created corpus, we analyzed how the following properties are fulfilled: (i) the influence of the corpus on the number of concatenations on an unknown text, (ii) diphone coverage and (iii) creating a larger diphone database leads (or not) to an improvement of the unit selection in the TTS system. The corpus was evaluated on unseen text from a common language domain (literature) and from a legal domain (public administration), in parallel. For both domains, the texts were extracted from the same sources as the ones used for the training stage. The first test, dedicated to the common language domain, consisted of 4,457 words and the second test, dedicated to the legal domain, consisted of 737 words. Based on the above criteria, the gain in naturalness brought on by the corpus can be deducted. Based on the different component gains, we argue whether the cost of fully annotating a specific part of the corpus is motivated. We mention that the target domain of the TTS application plays a big role in evaluating the gain in naturalness brought by the corpus components.

4.1 The Influence of the Corpus on the Number of Concatenations

In our perspective, the most important criteria in obtaining naturalness is the amount of continuous context found in the current corpus for an unseen text. The smaller number of concatenations made on a text, the closer the resulting sound will be to natural spoken language. Fig. 2 presents the rise in continuous output brought by adding different corpus parts. The first component added is the core of our corpus (Basic and EW) as it is designed to be the foundation of our TTS system. After adding the core, the effects of adding the Extension components (legal and press) are evaluated in parallel. While both parts bring similar gain for the common domain, it can be seen that for the public administration domain the gain of the legal component is clearly higher (right graphic). For the last part the gain of adding the newspaper component as a continuation of the legal one was considered. Fig. 2 also presents the continuous percentage of the audio output in relation with two thresholds. The lower one (1 concatenation / 1 word) is

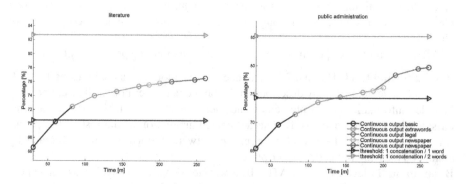

Fig. 2. Continuous output for unseen text

where the percentage of continuous audio output would be if one concatenation would be considered after each word, while the upper threshold (1 concatenation / 2 words) is for one concatenation after every other word. One can notice that these thresholds differ for each text, the reason being that the average word in the public administration domain is longer than the ones in literature.

Evaluated by this metric, the legal component increases the continuous audio output percentage for the public administration domain to a roughly one concatenation after a word and a half from one concatenation for every word (~3% more continuous output), making it an important part for that market segment, while the newspaper component falls behind. We conclude that, for the common domain, the paper and legal components of the corpus bring little gain and they can be excluded from the final annotation with little loss in naturalness.

4.2 Diphone Coverage

Another way for the synthesized speech to be closer to human spoken language is to have as little distortions brought on by concatenation points. For our system this means good diphone coverage as the phone concatenation quality is worse. This is due the fact that the natural transition between the concatenated phones is lost and the synthesized one do not rise to the quality of a natural transition. Fig. 3 shows the diphone coverage brought on by adding different corpus parts, similar to the previous evaluation. It begins with the diphone coverage of the core component and continues in parallel with the newspaper and legal extension components. The right-side of Fig. 3 highlights the legal component as an important asset to the TTS system for the public administration domain. In the left-side of Fig. 3, the legal and newspaper corpus parts bring little to no diphone coverage for the literature (common) domain, while in the public administration domain only the newspaper component does not increase the metric. From this perspective, even the last part of the EW corpus, which was added for the purpose of rare diphone coverage, is not justified.

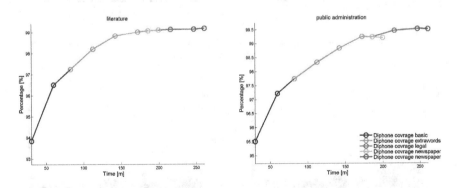

Fig. 3. Diphone coverage for unseen text

4.3 Diphone Database Enhancement

The least important abstraction of naturalness, in our perspective, is the diphone database growth. This is a plus due to the fact that it permits the unit selection algorithms to select the best fit for a concatenation point. On the other hand, any newly added component will increase the diphone database. Therefore, instead of simply counting the diphones of a component, which would be directly proportional to its size, we considered the frequency of a diphone in the tested text. Fig. 4 presents the histograms made after adding the core component (upper slice) of 186 minutes, core-legal components respectively (lower slice) of 261 minutes. The histograms present how many selection candidates are available for each concatenation point. By doing this, we emphasize the usefulness of the diphones brought on by the components, not their count. Also, for the candidate counts that were larger than 500 (the right histogram limit), we considered them as being equal to 500, but only for presentation purposes.

The correct concatenation candidate count is used to compute the histogram mean value. Taking into consideration the mean points of these histograms, which is actually the frequency weighted diphone mean, we created a metric that shows the diphone database enhancement. Fig. 5 depicts the histogram means and shows the diphone gain brought by different corpus components. Similar to the other evaluations it starts with the core component followed by both newspaper and legal corpus parts, than presenting

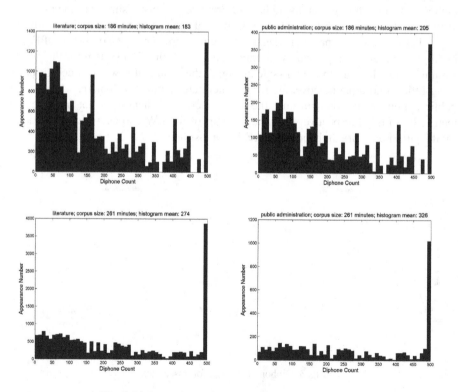

Fig. 4. Diphone histogram for two distinct corpus splits

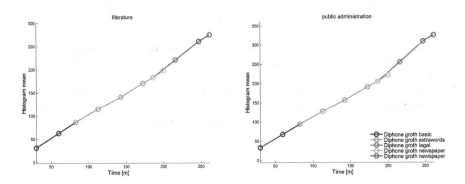

Fig. 5. Diphone database enhancement for unseen text

the newspaper component as an extension to the core-legal corpus. We notice that this is a weak metric, as all the components have similar weights. Even so, it can still be used to assess decisions regarding which corpus component is best to continue with in case the other metrics are not conclusive.

5   Conclusion

This paper presents our efforts to create a corpus for the Romanian language, with the purpose of developing a concatenative TTS system with a higher degree of naturalness. Our corpus is composed of a basic vocabulary, containing common words, together with extensions for a specific domain – the public administration. Using various statistics, we show how the inclusion of different parts of the corpus affects or not the naturalness of the synthesized speech.

We obtained a satisfying degree of naturalness by processing only the core of the corpus. Adding domain-specific components has a bigger impact on the speech quality than continuing the annotation of the core after a certain point. Therefore, to deploy the TTS system for various specific domains, we need to develop specific vocabularies, rather than further processing the existing recordings.

References

1. Sanders, W.R., Gramlich, C., Levine, A.: Naturalness of synthesized speech. In: University-Level Computer-Assisted Instruction at Stanford: 1968-1980, pp. 487–502 (1981), http://suppes-corpus.stanford.edu/pdfs/CAI/II-8.pdf
2. Hawkins, S., Heid, S., House, J., Huckvale, M.: Assessment of naturalness in the prosynth speech synthesis project. In: IEE Colloquium on Speech Synthesis (2000), http://www.phon.ucl.ac.uk/home/mark/papers/iee00hawkins.pdf
3. Keller, E., Bailly, G., Monagham, A., Terken, J., Huckvale, M.: Improvements in Speech Synthesis: Cost 258: The Naturalness of Synthetic Speech. Wiley (2011)
4. Cristea, D., Forascu, C.: Linguistic resources and technologies for romanian language. Computer Science Journal of Moldova 14, 34–73 (2006)

5. Feraru, S., Teodorescu, H., Zbancioc, M.: SRoL – Web-based Resources for Languages and Language Technology e-Learning. International Journal of Computers Communications & Control 5, 301–313 (2010)
6. Burileanu, C., Popescu, V., Buzo, A., Petrea, C.S., Ghelmez-Hanes, D.: Spontaneous speech recognition for romanian in spoken dialogue systems. Proceedings of the Romanian Academy, Series A 11, 83–91 (2010)
7. Burileanu, C., Buzo, A., Petre, C.S., Ghelmez-Hanes, D., Cucu, H.: Romanian spoken language resources and annotation for speaker independent spontaneous speech recognition. In: Fifth International Conference on Digital Telecommunications, pp. 7–10. IEEE Press (2010)

Making Community and ASR Join Forces
in Web Environment

Oldřich Krůza and Nino Peterek

Charles University in Prague
Faculty of Mathematics and Physics
Institute of Formal and Applied Linguistics
Malostranské nám. 25, Prague, Czech Republic
{kruza,peterek}@ufal.mff.cuni.cz

Abstract. The paper presents a system for combining human transcriptions with automated speech recognition to create a quality transcription of a large corpus in good time. The system uses the web as interface for playing back audio, displaying the automatically-acquired transcription synchronously, and enabling the visitor to correct errors in the transcription. The human-submitted corrections are then used in the statistical ASR to improve the acoustic as well as language model and re-generate the bulk of transcription. The system is currently under development. The paper presents the system design, the corpus processed as well as considerations for using the system in other settings.

1 Introduction

For a setting where recorded speech data are available, it is often desirable to obtain a transcribed version of the spoken words because processing digital text is much simpler than processing digital audio. Human transcription is costly and no automatically acquired transcription is perfect, so human revision of ASR output is a common scenario.

In our setting, where an uncatalogized spoken corpus and a community of lay volunteers are available, we feel a need for a system that would employ the modern speech-recognition technology and thoroughly exploit the precious brain cycles of the involved humans. The web with its ever-growing possibilities offers a neat platform for creating such a system.

1.1 Spoken Corpus of Karel Makoň

Karel Makoň (1912–1993) was a Czech mystic who authored 27 books on spirituality and Christian symbolism, and translated and commented 28 others[1]. Aside from his writing, he was giving talks in a close, private circle of friends as the topics he was covering were not safe to express openly under the socialist regime. His friends and apprentices recorded most of his lectures on magnetic tapes. The recording started in

---

[1] http://www.makon.cz/

P. Sojka et al. (Eds.): TSD 2012, LNCS 7499, pp. 415–421, 2012.

late 60's using reels, then switched to cassettes in the course of the 70's and went on until Makoň ceased to give lectures in 1992, one year before his death.

The recordings were archived in the homes of their makers, losing the quality as the magnetic signal was slowly fading over time. Attempts have been made to digitize the material, but using little to no automation only a fraction has been converted, leaving the rest of the material at time's mercy. A systematic digitization took place only from 2010 to 2012 when most of the material was converted to wav files.

The complete corpus, as of the time of writing, comprises about 1,000 hours and is available on-line[2]. Our work deals with processing this corpus, although the system under development is intended to work on any data and we hope it will be useful in a broader spectrum of settings.

2   Motivation for Transcription

The digitized corpus is largely unstructured. The only metadata we have, are years and sequential numbers in case of the cassettes. For some reels, there are handmade indexes listing the topics covered and the corresponding time positions on the counter of the device. These are typed on paper and have been photographed. The indexes, however, cover only 76 of the 1,096 total files.

This is a similar setting to the Malach [1] project. Malach was approached by doing automatic speech recognition and building a search engine on the ASR output. Our setting differs from that of Malach significantly, as we merely have to deal with material by one single speaker, and we just have to handle about one thousand hours in comparison with Malach's stunning one hundred sixteen thousand hours. This plus the fact that there is an active community around the legacy of Karel Makoň motivates us to attempt a full, high-quality, verbatim transcription of Makoň's talks.

3   Architecture of the System

Obviously, there are two basic ways to do speech-to-text transcription:

– manually,
– automatically.

Manual transcription usually delivers very high quality, but it is unbearably time-consuming. Automatic transcription can be quite fast, but the amount of errors leaves much to be desired. Hence, we propose and are developing a system to draw on the power of both to converge quickly to a good transcription.

Figure 1 outlines the system architecture:

1. An initial transcription is obtained by means of an automatic speech recognition system.
2. The resulting transcription is transformed into a JavaScript data structure suitable for synchronous displaying on the web interface.

---

[2] http://www.makon.fm/

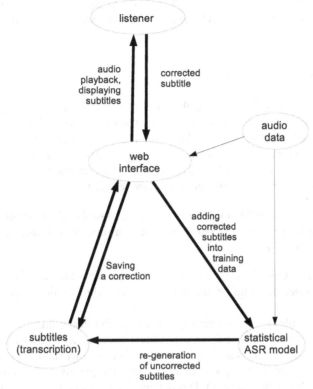

Fig. 1. Schema of the system architecture

3. Visitors use the web interface to listen to the recordings, at the same time seeing the synchronized transcription.
4. When encountering an error in the transcription, the visitor selects the erroneous text and edits it, entering the correct transcription.
5. The corrected subtitle is immediately saved, so that it is shown on subsequent requests of that passage.
6. The corrected subtitle is added to the training data of the ASR system.
7. The ASR model is re-trained with the new training data.
8. The re-trained ASR model is applied and the bulk of the transcription that has not yet been human-corrected, is re-generated.

3.1   First Transcription

We did some first experiments to obtain the initial transcription. We tried to train an acoustic model on 15 minutes of manually transcribed material from one of the recordings. The training was executed using HTK [2]. We also adapted the acoustic model from CUCFN [3] to Makoň's speech using the mentioned 15 minutes of manual transcription.

For language modeling, we trained a bigram model on Karel Makoň's books and one on PDT [4]. We're searching for a fitting ASR configuration to use the models. The current results on ASR for Czech [5] give us certain expectations with respect to accuracy.

3.2   Web Interface

The web interface where people can collaborate on speech transcription is a cornerstone of the whole project. The audio is played back by browser's native HTML5 `<audio>` element, falling back to flash[3] where HTML5 is not supported[4]. The subtitles are shown as three lines of text with the current word being highlighted. The lines scroll down as the recording is played back. This was chosen for ergonomic reasons. Two other possible implementations that came into mind were rejected:

1. a horizontally-scrolling line of text, like a `<marquee>` HTML element,
2. subtitles like in a movie: one phrase shown at a time, which disappears and is replaced by a new one.

Using a marquee-like solution (point 1) would cause constant scrolling in a non-constant speed, since the rate of the speech varies. This would quickly become annoying.

Using a movie-subtitle-like solution (point 2) has the disadvantage of making it impossible to select (and thus correct) a span of text reaching across the given phrases.

In our solution, several lines are shown, they are horizontally static, and the current word is highlighted, always being kept on the middle line (unless the beginning or end of the recording is being played back). This delivers the reader enough context and feels similar to reading static text. A drawback of this method is that showing many lines gets computationally intense. We wrap each word into a dedicated HTML element to be able to keep track of the currently played word and to be able to track down the corresponding subtitles when the user selects and corrects some text chunk. A good trade-off seems to be using three lines. There is always enough context, it is never hard to spot the highlighted word, and the browser stays responsive.

The highlighting of the current word still has a small difficulty. Since the time-update event, which triggers re-drawing of the highlighted word, is quite coarsely granular – i.e. it fires only about 4 times per second –, it happens that a word is highlighted with a noticeable delay. This reduces the comfort slightly but can be avoided using look-ahead methods.

As hinted at in point 2 of the architecture outline, the subtitles are stored in a JavaScript structure. To be more precise, we're storing it in the JSONP [6] format. This allows us to have the subtitles on an external CDN[5] and dedicate the web server to processing the corrections. Another obvious advantage is that the browser can parse the subtitles rapidly, not losing responsivity for a long period.

---

[3] `http://www.adobe.com/products/flashplayer.html`

[4] jPlayer was used for this; `http://jplayer.org/`

[5] Content-delivery network.

4   Dealing with Submitted Corrections

Processing the submissions from the visitors is probably the most important point of the work. Working collaboratively on large data over web is a very popular approach, most notably exploited by Wikipedia[6]. Our setting shares some notions with that of Wikipedia, but is obviously quite different. Not only because our data are acoustic but mainly because in our case, there is one correct output that we hope the users to provide, whereas in case of Wikipedia, the users are creative. This makes our task much easier in respect of conflict resolution. When two people edit the same Wikipedia article at the same time, an interactive edit-conflict procedure has to be employed[7].

To explain our situation, we'll first outline what happens when a correction is sent by a visitor.

1. The visitor selects the subtitles where the error is contained. The selection is automatically padded to whole words.
2. The selected subtitles are replaced by a textarea where the words are copied.
3. When the textarea loses focus, and if it had changed, then the new text is sent to the server using an AJAX [7] request. Along with the text, the timestamp of the first word and of the one after the last word is sent.
4. On the server, the correction is saved into the database.
5. The corresponding audio sequence is extracted and matched against the provided text.

   If the forced alignment fails, the process ends and no further modifications take place; an error response is sent to the client.

   If, on the other hand, the forced alignment of the provided subtitles to the audio succeeds, then the individual words get timestamps and the process continues.
6. The corrected span of subtitles is replaced with the submitted aligned text and the transcription is saved.
7. The submitted transcription is added to training data for the acoustic model.
8. The submitted transcription is added to the training data for the language model.
9. Re-training of the model is scheduled.[8]

Notice that if two people send overlapping corrections, they are simply pasted onto each other. The intersection is saved from the latter of the concurring submissions. Assuming they were both correct, there is no conflict.

If we give up the assumption of the correctness of the submissions, we have two ways of dealing with that. Firstly, notice point 5 of the correction sequence: If the alignment fails, then the subtitle is not saved. This protects the system from outright vandalism and spam, since transcriptions that are too far from the audio will be automatically rejected.

Other, more subtle forms of errors introduced by the visitors will naturally be harder to compensate. One thing we might do, is to track which correction comes from which user. If a user tends to consistently make a certain kind of mistake, it may be possible

---

[6] http://www.wikipedia.org/

[7] http://en.wikipedia.org/wiki/Help:Edit_conflict

[8] Actually re-training the model and re-recognizing the whole corpus after every submissions is, of course, infeasible.

to revert it, or to set up a group of proof-readers. This is pure theory though. We'll have to deal with these problems when they come.

We were making a simplification up to this point: Upon submission of a corrected subtitle, we cannot just add this to the training data. We also have to remove previous corrections of the same passage, so that previously introduced errors can be compensated.

4.1   Foreign Words

Another scenario that brings in complications is, when foreign words or any words with non-standard pronunciation are a part of the corrected subtitle. Suppose that the name *George* appears in the audio and is not correctly recognized. Also suppose the name is not present in the dictionary of the speech recognition system. Since the mechanism for accepting or rejecting a submission is based on acoustic fit and we use a simple rule-based word-to-phonemes converter[9], if the user enters *George*, the word-to-phonemes converter will assume pronunciation /gɛorgɛ/ instead of the correct /dʒɔːɹdʒ/ – and thus will likely be rejected as non-matching.

We approach the problem by instructing the users to type in foreign words in their phonetic transcription. In this case, *džordž*. After the subtitle is saved, the words become data structures with independent word forms and pronunciations. Then, the proper spelling *George* can be introduced.

5   Using the System in Other Settings

As much as our primary aim is to process this specific corpus, we hope the functionality developed to be applicable in a wide range of settings. Applying the system for a different corpus should be straight-forward. Complications could arise if the corpus were stored in smaller files since we don't take care of playing one file after another: we assume that they are separate and the user is only ever interested in one specific file at a time.

Using the system for a corpus in another language would also make modifications necessary. Mainly the word-to-phonemes conversion would have to be implemented in a way that is suitable for the given language. Of course the whole ASR can be replaced – and we plan to do this to try out different ones.

One part that lends itself neatly for modifications and re-use is the web interface. For data, where audio, or video for that matter, and aligned transcription are available, the interface can be used as a generic subtitle displayer.

In music, the audio-text synchronous display could be used for collaborative transcription of lyrics or even for karaoke. One possible serious drawback in this would be the difficult matching of lyrics to audio. The music mixed into the words, artistic interpretation of the vocalist and a variety of speakers (singers actually) would likely make the alignment much less robust. In a setting though, where the aligned lyrics with

---

[9] We use phonetic alphapet designed by Psutka et al. 1997 [8] based on Czech phonology as described by Palková 1992 [9].

potential errors are on input, the system could still be well used to for display and gathering corrections.

In language teaching, students could use the system to match spoken words with written ones to learn pronunciation of their language of interest. Introducing deliberate errors and expecting corrections could be a way of examination.

6   Conclusion

The described system is under development. Much of the functionality has already been implemented, although the current form of the user interface is crude at best. Our work on this project has merely begun. We are thrilled to develop a system that on one hand will systematize and make available maybe the most comprehensive opus on Christian mystic in Czech, on the other hand will open new possibilities in collaborative transcription of speech.

Acknowledgements. The research was supported by GACR project number 201/09/H057. This work has been using language resources stored by the LINDAT-Clarin project of the Ministry of Education of the Czech Republic (project LM2010013).

References

1. Byrne, W., Doermann, D., Franz, M., Gustman, S., Hajič, J., Oard, D., Picheny, M., Psutka, J., Ramabhadran, B., Soergel, D., Ward, T., Zhu, W.-J.: Automatic recognition of spontaneous speech for access to multilingual oral history archives. In: Proc. Eurospeech 2003, Geneva, Switzerland (2003), http://www.ee.umd.edu/~oard/pdf/tsap04.pdf
2. Young, S., Evermann, G., Gales, M., Hain, T., Kershaw, D., Liu, X.A., Moore, G., Odell, J., Ollason, D., Povey, D., Valtchev, V., Woodland, P.: The HTK Book (2006), http://htk.eng.cam.ac.uk/
3. Byrne, W., Hajič, J., Ircing, P., Jelinek, F., Khudanpur, S., McDonough, J., Peterek, N., Psutka, J.: Large Vocabulary Speech Recognition for Read and Broadcast Czech. In: Matoušek, V., Mautner, P., Ocelíková, J., Sojka, P. (eds.) TSD 1999. LNCS (LNAI), vol. 1692, pp. 235–240. Springer, Heidelberg (1999)
4. Böhmová, A., Hajič, J., Hajičová, E., Hladká, B.: The Prague Dependency Treebank: A Three-Level Annotation Scenario (2007), http://www.scientificcommons.org/43211198
5. Psutka, J., Hajič, J., Byrne, W.: The development of ASR for Slavic languages in the MALACH project in Proc. ICASSP (2004), http://svr-www.eng.cam.ac.uk/ wjb31/ppubs/ icassp04-malach-final.pdf
6. Ippolito, B.: Remote JSON – JSONP (2005), http://bob.pythonmac.org/ archives/2005/12/05/remote-json-jsonp/
7. Garrett, J.J.: Ajax: A new approach to web applications (2005), http://mmccauley.net/com601/material/ajax_garrett.pdf
8. Nouza, J., Psutka, J., Uhlíř, J.: Phonetic alphabet for speech recognition of Czech Radioengineering, vol. 6, pp. 16–20 (1997), http://www.radioeng.cz/fulltexts/1997/97_04_04.pdf
9. Palková, Z.: Fonetika a fonologie Češtiny Univerzita Karlova, Praha (1992)

Unsupervised Synchronization of Hidden Subtitles with Audio Track Using Keyword Spotting Algorithm

Petr Stanislav, Jan Švec, and Luboš Šmídl

Department of Cybernetics, Faculty of Applied Sciences
University of West Bohemia, Czech Republic
{pstanisl,honzas,smidl}@kky.zcu.cz

Abstract. This paper deals with a processing of hidden subtitles and with an assignment of subtitles without time alignment to the corresponding parts of audio records. The first part of this paper describes processing of hidden subtitles using a software framework designed for handling large volumes of language modelling data. It evaluates characteristics of a corpus built from publicly available subtitles and compares them with the corpora created from other sources of data such as news articles. The corpus consistency and similarity to other data sources is evaluated using a standard Spearman rank correlation coefficients. The second part presents a novel algorithm for unsupervised alignment of hidden subtitles to the corresponding audio. The algorithm uses no prior time alignment information. The method is based on a keyword spotting algorithm. This algorithm is used for approximate alignment, because large amount of redundant information is included in obtained results. The longest common subsequence algorithm then determines the best alignment of an audio and a subtitle. The method was verified on a set of real data (set of TV shows with hidden subtitles).

Keywords: keyword spotting, text alignment, subtitles.

1 Introduction

The need of large text corpora is common in many tasks including optical character recognition, machine translation or automatic speech recognition. In automatic speech recognition there is a need of data which are very close to the target domain. This request is often hard to fulfill especially in the spontaneous or conversational speech recognition tasks. Some tasks allow to prepare language model in advance from shorthand records such in the task of automatic subtitling of the Czech Parliament Meetings [1,2]. Other tasks allow to use topic dependent language model focused on the discussed topics [3].

This paper describes the use of movie and television subtitles for creating conversational speech language models that are then used in an automatic subtitling system. The Section 2 describes the corpora created from available subtitles as well as a number of statistics of these data. The Section 3 presents a novel robust method for unsupervised synchronization of a reference text and a corresponding audio. Section 4 concludes the paper.

P. Sojka et al. (Eds.): TSD 2012, LNCS 7499, pp. 422–430, 2012.

2   Subtitles Corpora

The advance in a digital multimedia broadcasting and sharing of multimedia across the Internet causes a huge number of community-generated content to be available. One of the largest sources of such data are publicly available subtitles for digital movies and television shows. Although these data are community-generated they comprise a number of valuable data suitable for many linguistic tasks. Another source of language data are the hidden subtitles broadcast by public televisions. Such data have a much better quality because they contain only a very few typos and the language structure of these subtitles is mostly correct.

This paper deals with the analysis of available subtitle data in Czech language with the focus on language modelling. First, the corpus built from community generated subtitles is described. Then we compare this corpus with the collection of hidden subtitles broadcast by Czech Television, the public television broadcaster. The corpora were integrated into an existing language modelling system and database described in [4] and the similarity and consistency of the parts with respect to the rest of the database were evaluated. In this analysis we used the Spearman rank correlation coefficient of the distance of ranks of 500 most frequent words. We also present other properties of the corpus such as average length of the sentence or average number of tokens per show. We analysed the duplicates in each of the subtitles source and the statistics were computed with duplicates removed to minimize the effect of duplicated texts on the presented statistics.

2.1   Community Generated Subtitles

We used a collection of 32,643 documents containing subtitles in three different formats - SubRip (.SRT extension) which is the most common subtitle format in the movie fans community, VobSub (.SUB extension) is related to a DVD format and SubStation Alpha (.SSA extension) which is considered an obsolete format used mainly in the anime genre.

The processing of community generated subtitles starts with automatic detection of character encoding which is very important for further automatic processing. Because the collection contains a small number of subtitles in Slovak and English, we used simple rule-based language detection. The methods uses the relative frequency of language specific characters for three languages occurring in the subtitles: Czech, Slovak and English. Using this method we got 29,560 documents classified as containing Czech subtitles, 2,806 Slovak and 148 English.[1] In the rest of the paper we will denote the part of community generated subtitles with an abbreviation SUB.

2.2   Hidden Subtitles

The collection of hidden subtitles contains 42,094 documents with hidden subtitles prepared and broadcast by Czech Television from January 2000 till the end of February

---

[1] In fact, the subtitles labeled as English can also contain Czech and Slovak subtitles written without diacritics but this causes no problems because subtitles with language other than Czech are removed from further processing.

2012. The subtitles are always in known encoding and in Czech language. The length of television shows which are subtitled vary from a few minutes to a few hours. We classified the 50 most frequent TV shows with a topic. The six most frequent topics were *News, Travelling, Magazine, Education, Journalism* and *Talk show*. In the following text this part of corpus containing hidden subtitles will be denoted with the abbreviation IVY.

2.3   Subtitles Processing

To allow subsequent processing of the subtitles, we first stripped the subtitle timing information. Although the subtitle timing can be useful in some tasks, the timing is designed to ease the reading and understanding of the subtitle and therefore the subtitle timing does not necessary correspond with an audio track of the show. We have designed an algorithm which performs unsupervised synchronization of subtitles with the audio. The algorithm is described later in this paper.

The text of subtitles stripped from timing information is then processed with a language modelling tool which is built to support language processing of large text corpora. It uses a large set of text processing algorithms such as text normalization and tokenization. The data are stored in an SQL database. The tool is able to process corpora with billions of tokens separated into millions of documents [4,3].

First the knowledge based cleaning algorithm is performed to remove a large number of metadata occurring in the subtitles – for example information about audio and video codecs, the series of the TV show, version information etc. Then the numerals are expanded into word forms. The stream of words is then tokenized using a generic knowledge based tokenizer. The tokenized text is then processed using a set of token normalization rules (manually and semi-automatically generated) which replaces sequences of tokens with another sequence of tokens and possibly with a so-called multi-words (a token formed from adjacent words). The rules mainly unify different spelling and common typos in names of corporations, proper names, geographic names etc. The whole set contains about 196k normalization rules from which there is about 70k rules resulting in a multi-word.

Then we applied two high-level text processing algorithms – a duplicity detection and topic identification. The duplicity detection algorithm is based on the shingling method and uses a symmetrized similarity measure [4]. The topic identification algorithm uses a language-model-based approach to topic identification and each document is labelled with three topics selected from a topic hierarchy containing approx. 500 different topics [3]. The topic identification is performed only on the IVY part of the corpus. The SUB part contains subtitles with topics which are very inconsistent with the designed topic tree. On the opposite site the topics of IVY subtitles (Sec. 2.2) strongly correspond with the predefined topics.

2.4   Corpora Analysis

First, we analysed the number of duplicates in both subtitles corpora and we compared the results with the rest of the language modelling database. The results are shown in Tab. 1. The community created subtitles (SUB) contain a large number of duplicated

Table 1. Number of documents, number of originals, number of duplicates and the duplicity rate depending on the part of the corpus. The part *rest* denotes the whole language modelling database without the SUB and IVY parts.

	# doc	# origs	# dupl	dup. rate
SUB	29,560	17,747	11,813	40.0%
IVY	45,049	42,538	2,511	5.6%
rest	3,534,320	2,941,700	592,620	16.8%

Table 2. Number of raw tokens (including punctuation marks), tokens (excluding punctuation marks) and the size of the vocabulary for different data sources

	# raw tokens	# tokens	vocab. size
SUB	88.2M	67.6M	973k
IVY	101.7M	84.9M	961k
rest	1,040.5M	876.4M	3,700k

Table 3. Mean and standard deviation of lengths of sentences and documents depending on the part of language modelling database. The total numbers of sentences (# sent) and documents (# doc) are also shown. The row *all* includes also the IVY and SUB corpora.

	# sent	E{len($sent$)}	σ\{len($sent$)\}	# doc	E{len(doc)}	σ\{len(doc)\}
IVY	9.2M	9.26	7.24	42.5k	1,994.26	1,324.39
SUB	14.4M	4.68	4.09	17.7k	3,806.09	1,941.96
all	117.5M	8.77	7.54	2,997.8k	343.56	545.77

Table 4. Similarity and consistency of different parts of language modelling database.

	ANP	CNO	IDS	LID	OVM	PAL	**IVY**	**SUB**
ANP	0.937	0.907	0.976	0.975	0.670	0.858	0.868	0.667
CNO		0.930	0.915	0.946	0.511	0.897	0.653	0.421
IDS			0.954	0.974	0.600	0.826	0.842	0.658
LID				0.950	0.632	0.903	0.800	0.594
OVM					0.911	0.633	0.726	0.626
PAL						0.950	0.629	0.409
IVY							0.987	0.880
SUB								0.994

documents (about 40%) because in the corpus there are many versions of subtitles for a given show. They only differ in the timing or the subtitles are corrected by another author and repeatedly published. Therefore the following statistics are computed only on *originals*. As an original we denote the document which does not duplicate any other document.

The total amount of data available in both subtitles corpora is described in Tab. 2. The table shows that the subtitles corpora contain about 15% of tokens in the whole language modelling database. The community created subtitles (SUB) have a richer vocabulary than the hidden subtitles (IVY). In other words the same vocabulary size is achieved with a smaller number of tokens.

The next analysis compares the length of sentences and the length of whole documents across different parts of the language modelling database. The length of a sentence (len($sent$)) is a number of tokens between two punctuation marks. The length of a document (len(doc)) is a number of tokens in the whole documents excluding the punctuation. The Tab. 3 shows that the SUB data have much shorter sentences than the IVY data and whole language modelling database. On the opposite side the documents from the SUB corpus have significantly larger number of tokens. These statistics show that the conversational style of speech contained in the SUB data substantially differ from the other parts of the database even from the IVY data.

Our last analysis evaluates the similarity and consistency of different parts of the language modelling database (Tab. 4). We used a standard Spearman rank correlation coefficient of the distance of ranks of 500 most frequent words [5,6]. The scores were evaluated against the other data sources included in the language modelling database [4].[2] The consistency (shown in Tab. 4 on the diagonal) of all parts is relatively high. The similarity scores shows that both subtitles corpora (SUB and IVY) are very usable as a part of a language modelling database because they contain data which are not similar to other sources already present in the database. The SUB subtitles have the largest similarity score if evaluated against the IVY subtitles. This is caused by the use of spontaneous speech in the shows and movies in both subtitle corpora. The higher similarity between IVY and other sources is caused by the presence of similar topics (see Sec. 2.2).

3   Unsupervised Alignment of Text to the Audio

Large amount of annotated speech data is needed to build an acoustic models for an automatic speech recognition systems, spoken term detection and so on. Many sources of data are not suitable for training because of missing or incorrect time alignment between annotation and the matching audio. Therefore a novel algorithm was designed for the alignment of the text and the corresponding audio. The main assumption of the algorithm is the existence of a consent of sequence of utterances in audio data and a sequence of text references.

In the Fig. 1, the processing of audio data A and text reference R that are on the input of the system is illustrated. These text references R are preprocessed (for example the punctuation is erased). Then the control keywords C are chosen from text reference R and searched in audio data A. Then the audio is splited into $m+1$ smaller segments, thus $A = \{a_1, a_2, \ldots, a_{m+1}\}$, where m is number of the found control keywords. Further, each segment a_i is processed individually in the algorithm section called alignment,

---

[2] In shorthand, the CNO, IDS, LID and PAL sources contain data from different Internet news servers. The ANP source is mixed from printed news and transcriptions of television and radio broadcasts. The OVM source contain transcriptions of political discussion shows.

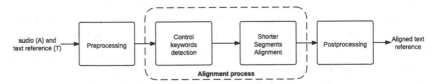

Fig. 1. Scheme of an alignment system

see Fig. 1. After that, the obtained segment alignments are connected into aligned text reference R_a with corresponding audio file A.

Modified Longest Common Subsequence – Falsely detected keywords may be found in the results of keyword spotting algorithm. Therefore the cleaning algorithm is implemented. The principle is based on searching for the longest common subsequence by using combination dynamic programming technique with a classic divide and conquer approach, see [7]. This algorithm is widely used, but in this case one of disadvantages is that only one longest common subsequence is found. For example, if two strings $s_1 = ABCD$ and $s_2 = ACBD$ are the input of the LCS algorithm, then the result is $LCS(s_1, s_2) = ABD$ or $LCS(s_1, s_2) = ACD$[3]. Therefore the modified LCS algorithm is used. Modified algorithm compares keywords with result from keyword spotting algorithm. Unlike standard algorithm all longest common subsequences are found. For each LCS the combined score is computed as the sum of each word scores. Then the output of the algorithm is the longest common subsequence with the lowest combined score.

Data Preprocessing – One of the crucial requirements for the alignment method is the independece on a data source. Therefore it is necessary to perform standardization of the input data before the alignment process. The standardization is also useful for comparison of the results of the alignment from different data sources. The standard format of audio files $\tilde{A}$ is obtained by converting any audio format to the mono pulse-code-modulated stream of samples that is of sampling frequency $f_s = 8,000\,\mathrm{Hz}$ and of sample size 16 bit. This stream is the input for the further data processing. Text references conversion into a standardized format $\tilde{R}$ is performed in several steps. Firstly all punctuations and all symbols that are neither letters nor Arabic numbers are erased, then all Arabic numbers are replaced by a corresponding word that is defined in a so-called 'dictionary'.

Alignment Algorithm – The algorithm of alignment uses a keyword spotting technique [8] to find a specific time boundary of a particular word C. The keyword spotting technique is based on a statistical approach and is speaker independent.

Data processing procedure is as follows. Audio data are split into segments. Then the text reference are detected in audio segments a these results are improved by using the Longest Common Subsequence (LCS) method.

---

[3] Result depends on the exact implementation of algorithm.

The *audio data segmentation* using control keywords C is the first part of the alignment process. The control keywords C are obtained from text reference $\tilde{R}$ that is being processed. The frequency of each included word is counted. Then all words are sorted by frequency from the smallest to the largest one and the stop words are erased. Then n (value is specified experimentally as $n = \frac{duration(\tilde{A})}{20}$) longest words are chosen as control keywords $C = \{c_1, c_2, \ldots, c_n\}$. Then obtained control keywords are searched in audio data using the keyword spotting algorithm. But a false detection can occur in the process of the control keyword detection (CKD) in audio data $\tilde{A}$. The modified longest common subsequences algorithm is used for false detection elimination. Then the cleaned results of CKD $\tilde{C} = \{\tilde{c}_1, \tilde{c}_2, \ldots, \tilde{c}_m\}$, $m \leq n$ are used for division of the preprocessed audio signal into shorter segments $\tilde{a}_i, i = 1, \ldots, m + 1$. Moreover, it is required that each segment $\tilde{a}_i, i > 1$ begins by the control keyword $\tilde{c}_{i-1}$ and is of minimal required duration t_d. After segmentation of audio data $\tilde{A}$ into parts, every part a_i has the start time t_{i-1} and end time t_i in ideal situation. But the ideal segmentation process is not suitable for further processing. Therefore the parameter $\varepsilon = 2.0s$ is defined as set-up value of overlap of neighboring segments. Then the segment $\tilde{a}_{i-1}$ ends at time $\tilde{t}_i = t_i + \varepsilon$ and the next segment $\tilde{a}_i$ starts at $\tilde{t}_i = t_i - \varepsilon$. Then probability of the searched words occurrence is increased on the border of segments.

The principle of *alignment* is based on processing the segments chronologically. It means that we start by $i = 1$ corresponding with audio data segment $\tilde{a}_i$ and text reference r_i and continue with processing $\tilde{a}_{i+1}$, r_{i+1} until $i = m$. We assume that every segment starts with known control keyword. Then we can suppose that the corresponding part of text reference r_i is contained in processed audio segment $\tilde{a}_i$ but the each word position is unknown. Therefore the current part of text reference r_i is used as a keywords K and the keyword detection process is performed on the current audio part $\tilde{a}_i$. Analogously, the results $\tilde{K}$ of this process are the score $s_{\tilde{k}_i}$ of each keyword $\tilde{k}_i$ and its time occurrence. Naturally, false detections can occur. In this case the LCS algorithm is used analogically. The keyword detection is based on the statistical approach, therefore the control keywords $\tilde{C}$ need not be placed necessarily to the point where they were detected. The wrong detection results in *appearance of controversial words*. The controversial word can not be clearly assigned to neither the end of previous segment $\tilde{a}_{i-1}$ nor start of currently processed segment $\tilde{a}_i$. In this case the controversial parts of two consecutive segments are connected into one new segment and the alignment process is repeated. Note that the controversial segment starts at the end of the last detected word from previous part and ends at the beginning of first detected word of current part. The inclusion of found word in specific segment is realised on the basis of its start time being included in obtained results. If the non-empty set of controversial words remains after processing, then all the words are considered as a part of previous segment.

Postprocessing – In the last phase of this process, all single segments are connected back in one segment R_a. As a result of detection, the parts of text reference $\tilde{R}$ were found in segments and then in a whole audio file, too. If any unknown word exists between any of two detected words, then all unknown words can be assigned to the

Table 5. Mean and variance of durations of segments with errors

Class	OK	X	B	M	E	Σ
Count	1,912	31	19	72	22	2,056
E{dur} [s]	4.0	13.7	6.9	21.1	10.3	4.9
σ{d} [s]	4.7	12.4	4.6	9.6	7.2	6.2

section defined by these two words. On the other hand, if only two words are detected, then the area is considered as non-transcribed part of the audio data.

4   Conclusions

We evaluated the presented algorithm on a real task – alignment of hidden subtitles of main television news called *Události* broadcast by Czech Television at the beginning of February 2012. Since no precisely aligned data were available we manually rated every aligned segment. Each segment was classified as *OK* for correctly aligned segment, *X* for completely bad segment, *B, M, E* for small error in alignment at the beginning resp. middle and end of the segment. We used approximately 2.8 hours of recordings in our evaluation which were divided into 2,056 segments. Our evaluation shows that 93% of segments are correctly aligned (with 0.95 confidence interval 91.96%, 94.02%). We also evaluated the mean and standard deviation of duration of segments manually classified into the mentioned classes. The results are presented in Tab. 5.

This paper describes the subtitles corpora as a source of data for language modeling of conversational speech. The algorithm presented in the second part of the paper also allows automatic alignment of the subtitles with the underlaying audio track. The algorithm can be used in an automatic dubbing system [9] or in acoustic modeling task, for example as a source of data for supervised adaptation of acoustic model.

Acknowledgments. This work was supported by the European Regional Development Fund (ERDF), project "New Technologies for Information Society" (NTIS), European Centre of Excellence, ED1.1.00/02.0090, by the grant of the University of West Bohemia, project No. SGS-2010-054, and by the Technology Agency of the Czech Republic, project No. TE01020197.

References

1. Pražák, A., Psutka, J.V., Hoidekr, J., Kanis, J., Müller, L., Psutka, J.: Automatic Online Subtitling of the Czech Parliament Meetings. In: Sojka, P., Kopeček, I., Pala, K. (eds.) TSD 2006. LNCS (LNAI), vol. 4188, pp. 501–508. Springer, Heidelberg (2006)
2. Ircing, P., Psutka, J., Psutka, J.V.: Using Morphological Information for Robust Language Modeling in Czech ASR System. IEEE Transactions on Audio Speech and Language Processing 17, 840–847 (2009)
3. Skorkovská, L., Ircing, P., Pražák, A., Lehečka, J.: Automatic Topic Identification for Large Scale Language Modeling Data Filtering. In: Habernal, I., Matoušek, V. (eds.) TSD 2011. LNCS, vol. 6836, pp. 64–71. Springer, Heidelberg (2011)

4. Švec, J., Hoidekr, J., Soutner, D., Vavruška, J.: Web Text Data Mining for Building Large Scale Language Modelling Corpus. In: Habernal, I., Matoušek, V. (eds.) TSD 2011. LNCS, vol. 6836, pp. 356–363. Springer, Heidelberg (2011)
5. Spoustová, D., Spousta, M., Pecina, P.: Building a Web Corpus of Czech. In: Proceedings of the Seventh conference on International Language Resources and Evaluation (LREC 2010), Valletta, Malta (2010)
6. Kilgarriff, A.: Comparing corpora. Intl. Journal of Corpus Linguistics 6, 97–133 (2001)
7. Hirschberg, D.S.: A linear space algorithm for computing maximal common subsequences. Communications of the ACM 18, 341–343 (1975)
8. Šmídl, L., Trmal, J.: Keyword spotting result post-processing to reduce false alarms. Recent Advances in Signals and Systems 9, 49–52 (2009)
9. Matoušek Jindřich, V.J.: Improving Automatic Dubbing with Subtitle Timing Optimisation Using Video Cut Detection. In: Proceedings of the ICASSP, Kyoto, Japan (2012)

Did You Say What I Think You Said?

Towards a Language-Based Measurement of a Speech Recognizer's Confidence

Bernd Ludwig and Ludwig Hitzenberger

University Regensburg, Chair for Information Science
Universitätstraße 31, D-93047 Regensburg
{bernd.ludwig,ludwig.hitzenberger}@ur.de

Abstract. In this paper we discuss the problem that in a dialogue system, speech recognizers should be able to guess whether the speech recognition failed, even if no correct transcription of the actual user utterance is available. Only with such a diagnosis available, the dialogue system can choose an adequate repair strategy and try to recover from the interaction problem with the user and avoid negative consequences for the successful completion of the dialogue. We present a data collection for a controlled out-of-vocabulary scenario and discuss an approach to estimate the success of a speech recognizer's results by exploring differences between the N-gram distribution in the best word chain and in the language model. We present the results of our experiments that indicate that differences can be found to be significant if the speech recognition failed severely. From these results, we derive a quick test for failed recognition that is based on a *negative* language model.

1 Introduction

Miscommunication is a major problem for the success of natural language dialogue system in every application scenarios [1]. One source of failed interactions is the lack of a robust methodology for the NLU component of a dialogue system to detect that result of the speech recognition process is incorrect for some reason. One important cause for this problem is the role of the language model employed by a speech recognizer. It heavily influences the way in which the acoustic hypotheses in a word lattice are disambiguated.

The standard way to build language models is to collect a corpus of typical utterances in the intended application domain. However, there is neither a well-established methodology nor an accepted set of criteria to determine which utterances should be contained in a language model and which not. The only rule widely accepted is Mercer's thesis[1] "There is no data like more data" [2], which is definitely not wrong, but nevertheless insufficient. While current work on language models (e.g. [3,4]) focusses on improving the mathematical models [5], we also take into account common patterns of spoken language usage.

---

[1] In http://www.lrec-conf.org/lrec2004/doc/jelinek.pdf, Jelinek credited Mercer for this quote.

P. Sojka et al. (Eds.): TSD 2012, LNCS 7499, pp. 431–437, 2012.
© Springer-Verlag Berlin Heidelberg 2012

In our opinion, there is a lack of research work on the issue on whether and how a language model should be constructed. An answer to this open question would be important to understand the interplay between speech recognition, dialogue processing, and problem solving (the actual reason of why dialogues are conducted). Therefore, in order to improve the current state-of-the-art of keyword or keyphrase spotting as a mean of extracting meaning from speech recognizer output, it might be a good idea to develop an approach how to estimate the probability of a word chain not being a typical language pattern. With this knowledge, the dialogue manager – i.e. the reasoning component responsible for automatically deciding how to continue a dialogue – could apply much more sophisticated decision strategies than those currently used to recover from misunderstandings between user and system [6].

The example below from a scenario of controlling electronic devices in a household illustrates this problem:

> wie an Wetter heute ich Vorschau   *how on weather today I preview*

has been recognized for the actual user utterance

> was kommt heute im Fernsehen?   *what will be on TV today?*

In order to avoid that key phrases of the word chain will be processed by the dialogue manager with a high risk of the dialogue to fail, a "quick" test without deep analysis is required. This test should predict that the chain is highly irregular with respect to the language model used by the speech recognizer.

We observe that this differs significantly in what is modelled from state-of-the-art approaches to confidence scoring in ASR. While these are commonly based on so-called *back-off models* [7,8,9,10] our approach does not aim at smoothing the probability by the following approximation[2]:

$$
P(w_i|w_{i-n+1}\cdots w_{i-1}) \approx
\begin{cases}
d_{w_{i-n+1}\cdots w_i} & \text{if } C(w_{i-n+1}\cdots w_i) > k \\
\dfrac{C(w_{i-n+1}\ldots w_{i-1}w_i)}{C(w_{i-n+1}\cdots w_{i-1})} & \\
\alpha_{w_{i-n+1}\cdots w_{i-1}} & \text{otherwise} \\
\cdot P(w_i|w_{i-n+2}\cdots w_{i-1}) &
\end{cases}
$$

w_i constitute the tokens in the utterance $w_1 \ldots w_n$, $d_{w_{i-n+1}\cdots w_i}$ is the weight of the conditional probability of w_i given the preceeding word chain $w_{i-n+1} \cdots w_{i-1}$, while $\alpha_{w_{i-n+1}\cdots w_{i-1}}$ is the weight of the back-off value if the token sequence $w_{i-n+1} \cdots w_i$ occurs less the k times in the training corpus (see [11] for details).

However, the case $C(w_{i-n+1} \cdots w_i) \leq k$ indicates a severe mismatch between the language model and the hypothesis that the sequence $(w_{i-n+1} \cdots w_i)$ actually has been uttered by the user.

Therefore, each such observation of a mismatch calls for diagnosis on whether the recognizer can find more plausible alternatives in the word lattice or misunderstanding occured to due e.g. acoustic problems or the user uttering words not contained in the vocabulary.

---

[2] $C(w_{i-n+1} \cdots w_i)$ is the absolute frequency of the sequence $(w_{i-n+1} \cdots w_i)$ in the language model.

2 Detecting Failed Speech Recognition

In this paper, we report about a series of experiments with language models for a home device domain. With these experiments, we provide evidence for the hypothesis that the frequency of N-grams not observed in the training corpus significantly correlates with the irregularity of the word chain. In such case, this correlation may be used as a predictor for the irregularity of an utterance and provide the necessary data for the quick test introduced above.

2.1 The Data Set

For our experiments, we worked with the IISAH corpus described in [12]. It contains 2,758 dialogue turns to control home devices. A standard way to construct a language model for a speech recognizer is to use these turns directly as a supervised data set. However, utterances collected in user studies or Wizard-of-Oz experiments are full of irregular phenomena of spoken language: incomplete utterances, interruptions, and self-repairs. Our hypothesis is that a language model that can be used to predict the irregularity of a word chain

1. should not contain disfluencies (as typically found in irregular word chains)
2. should not contain turns with more than one language pattern as at the boundary of two patterns irregular N-grams are generated.

In order to support our research hypothesis, we split the corpus into segments of 200 utterances and segmented the utterance whereever a disfluency had been detected. For example, the original transcribed turn

bitte SMS ach so das geht ja nur aufs Handy Moment zurück
text message please, oh only works with cell phones, one moment, back

has been segmented into five turns:

bitte SMS | ach so | das geht ja nur aufs Handy | Moment | zurück

2.2 The Experiment

The subcorpora were built in order to allow several annotators to label and transcribe the data. From the resulting subcorpora, we selected six by random. From each of them, we drew 30 utterances and recorded them with a smart phone. They served as control group as they basically repeat the language model.

To test the capability of a language model to generalize to unseen sequences of words, we constructed six other samples of 30 utterances consisting of words in the language model only, but without repeating complete patterns. In order to find out whether the irregularity of a word chain can be predicted by analyzing how many N-grams in the word chain are not included in the language model, we generated a third set of sample utterances with words not contained in the language model (except inevitable function words). We observe that for these 30 utterances the speech recognition will fail necessarily – in this case we are not interested in the recognition

Table 1. Comparison of mean error rates for the original language model. The p-values are computed using the two sample Welch test.

Subcorpus	Control Group	Known Vocabulary	p-value Control vs. Known	Unknown Vocabulary	p-value Control vs. Unknown
C_1	0.354	0.346	0.910	0.586	0.006
C_2	0.266	0.451	0.010	0.732	2.619e-10
C_3	0.232	0.370	0.018	0.698	2.594e-08
C_4	0.198	0.500	2.804e-06	0.697	1.023e-09
C_5	0.329	0.492	0.020	0.672	2.978e-06
C_6	0.309	0.648	4.213e-05	0.600	9.85e-05

rate, but we want to know whether we can observe significant differences between the output of the recognizer and the language model that encodes typical patterns of language use for the application domain. Overall the corpus for the experiment contains 3,056 words in 542 utterances (5.64 words per utterance).

The speech recognition process was conducted twice for all six subcorpora: first using the original language model, and then using the language model with all disfluencies eliminated. Table 1 shows the mean error rates of the first run for the control group, the second group with vocabulary taken from the language model, and the third group with unknown vocabulary. For comparing the error rates in each group statistically, we performed Welch tests for metric features. With this test, we assess statistically significant differences in the mean error rate for each group in each corpus compared to the control group.

3   Results of the Corpus Analysis

Table 1 presents the resulting p-values; the smaller the value, the more certain we can accept the hypothesis that with respect to the word error rate the two groups are actually different.

Table 1 and Table 2 reveal several interesting facts about the impact of the different language models on the word error rate: With the original model, the average error rate for the control group of all six subcorpora is 0.281 while for the model without disfluencies the error rate is 0.262 on the average. This is a first indication that the model without disfluencies does not affect the disambiguation process as negatively as the model with all disfluencies from the transcriptions. We note that the second group (*known vocabulary*) gives further empirical evidence for our hypothesis: in this group, the original model produces an average error rate of 0.468, while for the model without disfluencies, the average error is 0.338. In the first case, the error rate increases by 66.3%, while without disfluencies the average error increases just by 29.1%. We conclude that the disfluencies lead to wrong disambiguations.

On the other hand, the disfluencies do not impede the diagnosis of recognition failures severely: in the group *unknown vocabulary* the error rate is several orders of magnitude higher compared to the control group. In this case the differences between

Table 2. Comparison of mean error rates for the language model with all disfluencies eliminated. The p-values are computed using the two sample Welch test.

Subcorpus	Control Group	Known Vocabulary	p-value Control vs. Known	Unknown Vocabulary	p-value Control vs. Unknown
C_1	0.285	0.276	0.900	0.823	2.494e-11
C_2	0.207	0.355	0.034	0.856	< 2.2e-16
C_3	0.153	0.246	0.079	0.746	< 2.2e-16
C_4	0.241	0.410	0.012	0.850	3.135e-16
C_5	0.313	0.333	0.758	0.880	2.274e-12
C_6	0.370	0.405	0.619	0.781	2.070e-08

the original model and the model without disfluencies may result from the fact that the vocabulary of the original model is larger, as it contains more function words. However, in both cases, the word error rate is a highly reliable indicator for the speech recognition to perform poorly for some reason.

4 A Quick-Test Procedure for Failed Speech Recognition

As far as the development of a quick test for failed speech recognition is concerned, we performed some experiments, in which we counted how many bigrams in a word chain occur in the language model as well, and how many bigrams have never been observed in the training of the speech recognizer. We have done this because we assume that the oracle in a labelled corpus (i.e. the transcription of the actual user utterance) may be approximated by computing the relative frequency of unseen bigrams in the word chain: if this frequency is very low, most bigrams can be found in the language model. In such a case, the disambiguation of the word lattice is very close to the patterns of language codified in the language model. Therefore the probability for the word chain to be faulty is quite low. On the other hand, with many unseen bigrams in the word chain, only very few typical patterns could be found in the word lattice and therefore it is plausible to assume that the utterance diverges quite drastically from the language model or there are severe anomalies in the speech signal.

Table 3 indicates that this hypothesis can be validated statistically on the basis of the six subcorpora introduced earlier. In the group with *unknown vocabulary* the frequency of unseen bigrams is different from the control group with much higher significance than from the group *known vocabulary*. In Table 4 we present experiments with a language model showing a structure that is very different from the original one: beyond eliminating all disfluencies, in this language model we segmented the utterances into grammatical chunks. This approach produced many segments consisting of one or two words only. In our opinion, such a language model is an extreme opposite to the original model as it almost coincides with applying no language model at all and relying on the acoustic models only. This observation explains the high frequency of unseen bigrams in Table 4. However, again we note that in the *unknown vocabulary* group the p-values are much higher than in the comparison between the control group and the *known vocabulary* group.

Table 3. Comparison of average probability of an unseen bigram to appear for the original language model. The p-values are computed using the two sample Welch test.

Subcorpus	Control Group	Known Vocabulary	p-value Control vs. Known	Unknown Vocabulary	p-value Control vs. Unknown
C_1	0.372	0.439	0.510	0.675	0.007
C_2	0.194	0.336	0.083	0.720	3.974e-08
C_3	0.242	0.427	0.025	0.748	1.733e-08
C_4	0.279	0.541	0.005	0.662	2.545e-05
C_5	0.201	0.418	0.013	0.637	3.312e-06
C_6	0.229	0.631	1.473e-05	0.744	4.494e-08

Table 4. Comparison of average probability of an unseen bigram to appear for the language model with all disfluencies eliminated

Subcorpus	Control Group	Known Vocabulary	p-value Control vs. Known	Unknown Vocabulary	p-value Control vs. Unknown
C_1	0.608	0.694	0.257	0.934	8.154e-06
C_2	0.477	0.716	0.001	0.911	4.827e-10
C_3	0.554	0.661	0.081	0.920	4.769e-09
C_4	0.668	0.732	0.333	0.832	0.009743
C_5	0.712	0.639	0.219	0.908	0.0005325
C_6	0.861	0.914	0.270	0.940	0.08607

5   Conclusion and Further Work

From our experiments, we conclude that it is possible to estimate the probability of a word chain resulting from failed speech recognition by counting bigrams in the chain that do not occur in the language model. This is what we call *negative language model*. It is guided by the idea that the language model contains representative patterns of language use specific for an application domain. Our experiments also indicate that it is unclear what makes a language model representative: from our analyses we conclude that it should not contain disfluencies. In our future work, we will try to develop a best practice between the extreme approaches to constructing language models as described above. Furthermore, we will investigate word chains into more details. Our aim is not only to establish a quick test, but also to develop methods based on linguistic information for localizing errors in the chain. We assume that the integration of such methods in dialogue systems could further enhance their performance due to improved recovery strategies for human-computer-interaction.

References

1. Chang, J.C., Lien, A., Lathrop, B., Hees, H.: Usability evaluation of a volkswagen group in-vehicle speech system. In: Schmidt, A., Dey, A.K., Seder, T., Juhlin, O. (eds.) Automotive UI, pp. 137–144. ACM (2009)

2. Chelba, C., Jelinek, F.: Structured language modeling. Computer Speeech & Language 14, 283–332 (2000)
3. Bocchieri, E., Dimitriadis, D.C.D.: Speech recognition modeling advances for mobile voice search. In: Proceedings of Acoustics, Speech and Signal Processing (ICASSP 2011), Prague, pp. 4888–4891 (2011)
4. Chen, L., Chin, K.K., Knill, K.: Improved language modelling using bag of word pairs. In: Proceedings of Interspeech 2009, Brighton, pp. 2671–2674 (2009)
5. Jurafsky, D., Martin, J.H.: Speech and Language Processing. Prentice Hall (2009)
6. Hacker, M.: Context-aware speech recognition in a robot navigation scenario. In: Proceedings of the 2nd Workshop on Context Aware Intelligent Assistance, pp. 4–15 (2012)
7. Katz, S.M.: Estimation of probabilities from sparse data for the language model component of a speech recogniser. IEEE Transactions on Acoustics, Speech, and Signal Processing 35, 400–401 (1987)
8. Chelba, C., Brants, T., Neveitt, W., Xu, P.: Study on interaction between entropy pruning and kneser-ney smoothing. In: Proceedings of Interspeech 2010, pp. 2242–2245 (2010)
9. Uhrik, C., Ward, W.: Confidence Metrics Based on N-Gram Language Model Backoff Behaviors. In: Fifth European Conference on Speech Communication and Technology. ISCA (1997)
10. Jiang, H.: Confidence measures for speech recognition: A survey. Speech Communication 45, 455–470 (2005)
11. Katz, S.: Estimation of probabilities from sparse data for the language model component of a speech recogniser. IEEE Transactions on Acoustics, Speech, and Signal Processing 35, 400–401 (1987)
12. Spiegl, W., Riedhammer, K., Steidl, S., Nöth, E.: FAU IISAH Corpus – A German Speech Database Consisting of Human-Machine and Human-Human Interaction Acquired by Close-Talking and Far-Distance Microphones. In: Proceedings of the Seventh Conference on International Language Resources and Evaluation (LREC 2010), pp. 2420–2423. ELRA (2010)

Dealing with Numbers
in Grapheme-Based Speech Recognition*

Miloš Janda, Martin Karafiát, and Jan Černocký

Brno University of Technology, Speech@FIT and IT4I Center of Excellence
Czech Republic
{ijanda,karafiat,cernocky}@fit.vutbr.cz

Abstract. This article presents the results of grapheme-based speech recognition
for eight languages. The need for this approach arises in situation of low
resource languages, where obtaining a pronunciation dictionary is time- and cost-
consuming or impossible. In such scenarios, usage of grapheme dictionaries is the
most simplest and straight-forward. The paper describes the process of automatic
generation of pronunciation dictionaries with emphasis on the expansion of
numbers. Experiments on GlobalPhone database show that grapheme-based
systems have results comparable to the phoneme-based ones, especially for
phonetic languages.

Keywords: LVCSR, ASR, grapheme, phoneme, speech recognition.

1 Introduction

With fast spread of speech processing technologies over the last decade, there is a
pressure to speech processing community to build Large Vocabulary Continuous Speech
Recognition (LVCSR) systems for more and more different languages. One of essential
components in the process of building speech recognizer is pronunciation dictionary, that
maps orthographic representation into a sequence of phonemes — the sub words units,
which we use to define acoustic models during the process of training and recognition.

The acquisition of quality hand-crafted dictionary requires linguistic knowledge
about target languages and is time- and money-consuming, especially for rare and low-
resource languages. For these, several approaches for automatic or semi-automatic generation of dictionaries have been introduced, typically based on contextual pronunciation rules [1], neural networks [2] or statistical approaches [3].

The most straightforward method is to generate pronunciation dictionary as sequence
of graphemes and thus to directly use orthographic units as acoustic models (see [4,5]).
This approach is suitable for phonetic languages, where relation between the written and
the spoken form is reasonably close. The most widely used phonographic writing script
is the Roman script, so it is not surprising, that grapheme-based speech recognition
(GBSR) has been extensively tested on Western languages using this script. Later

---

* This work was partly supported by Czech Ministry of Trade and Commerce project
No. FR-TI1/034, by Czech Ministry of Education project No. MSM0021630528 and by
European Regional Development Fund in the IT4Innovations Centre of Excellence project
(CZ.1.05/1.1.00/02.0070).

P. Sojka et al. (Eds.): TSD 2012, LNCS 7499, pp. 438–445, 2012.

Table 1. Numbers of speakers, amounts of audio material (hours) and sizes of dictionary (words)

Lang.	Speakers	TRAIN (h)	TEST (h)	DICT
CZ	102	27	1.9	33k
EN	311	15	1.0	10k
GE	77	17	1.3	47k
PO	102	27	1.0	56k
SP	100	21	1.2	42k
RU	115	20	1.4	29k
TU	100	15	1.4	33k
VN	129	16	1.3	8k

experiments and results in this paper show, that the grapheme-based approach is also suitable for Cyrillic [6] or for the tonal languages like Vietnamese or Thai [7].

2 Experimental Setup

This section presents the data corpus and details the generation of grapheme based dictionaries with two possibilities (with and without expansion of numbers).

2.1 Data

GlobalPhone [8] was used in our experiments. The database covers 19 languages with an average of 20 hours of speech from about 100 native speakers per language. It contains newspaper articles (from years 1995–2009) read by native speakers (both genders). Speech was recorded in office-like environment by high quality equipment. We converted the recordings to 8 kHz, 16 bit, mono format.

The following languages were selected for the experiments: Czech (CZ), German (GE), Portuguese (PO), Spanish (SP), Russian (RU), Turkish (TU) and Vietnamese (VN). These languages were complemented with English (EN) taken from Wall Street Journal database. See Table 1 for detailed numbers of speakers, data partitioning and vocabulary sizes. Each individual speaker appears only in one set. The partitioning followed the GlobalPhone recommendation (where available).

When preparing the databases for baseline phoneme-based systems, several problems were encountered. The biggest issue was the low quality of dictionaries with many missing words. The Vietnamese dictionary was missing completely. The typos and miss-spelled words were corrected, numbers and abbreviations were labeled and missing pronunciations were generated with an in-house grapheme-to-phoneme (G2P) tool trained on existing pronunciations from given language. The dictionaries for Vietnamese and Russian were obtained from Lingea[1]. The CMU dictionary[2] was used for English. Each language has its own phoneme set and for better handling with different locales, all transcripts, dictionaries and language models (LMs) were converted to Unicode (UTF-8).

---

[1] http://www.lingea.com

[2] http://www.speech.cs.cmu.edu/cgi-bin/cmudict

Table 2. OOV rates, dictionary sizes, LM sizes and sources for individual languages

Lang	OOV	LM Dict Size	LM Corpus Size	WWW Server
CZ	3.08	323k	7M	www.novinky.cz
EN	2.30	20k	39M	WSJ - LDC2000T43
GE	1.92	375k	19M	www.faz.net
PO	0.92	205k	23M	www.linguateca.pt/cetenfolha
SP	3.10	135k	18M	www.aldia.cr
RU	1.44	485k	19M	www.pravda.ru
TU	2.60	579k	15M	www.zaman.com.tr
VN	0.02	16k	6M	www.tintuconline.vn

The data for LM training were obtained from Internet newspaper articles using RLAT and SPICE tools from the KIT/CMU[3]. The sizes of corpora gathered for LM training, and the sources are given in Tab. 2. Bigram LMs were generated for all languages except Vietnamese — a syllable language — for which a trigram LM was created.

2.2   Grapheme-Based Dictionaries

As proposed in the Introduction, the conversion of dictionaries to grapheme form was done. Word lists were obtained from current pronunciation dictionaries. An alternative would be to derive lists of words directly from transcripts, but we wanted to guarantee the same size of vocabulary in both (phoneme and grapheme) dictionaries and thus guarantee the same OOV rate for both systems and comparable results.

Prior to dictionary conversion to grapheme form, the word-lists were pre-processed: special characters like asterisk, brackets, colons, dashes, dollar symbols, etc. were removed. In the first version of grapheme dictionaries, we also removed all marked numbers from the vocabulary. After these operations, the grapheme based dictionary was obtained by simple splitting the words to letters, and finally, all graphemes were converted to lowercase (e.g. WORD → w o r d).

The transcripts of CZ, EN, VN did not contain any numbers, but we had to investigate how to deal with them for GE, SP, PO, RU, and TU. With deletion of numbers from dictionaries, we had to adequately change the transcripts to be consistent. One option was to remove all utterances, where a number is spoken (*grap_v0*). Another option was to map missing numbers in transcript into "unknown" *<UNK>* symbol (*grap_v1*).

The above mentioned processing of numbers however led to significant loss of acoustic data available for training (see Table 3). In average, we lose about 3.4 hours of data for the first variant, which represents about 17% on 20 hours of speech. The rate of numbers in the original dictionaries is about 3%. These differences can produce large degradation of recognition accuracy in the final results, so we decided to make another versions of grapheme dictionaries using numbers expansion.

---

[3] http://i19pc5.ira.uka.de/rlat-dev, http://plan.is.cs.cmu.edu/Spice

Table 3. Amount of audio data in different setups (with and without numbers)

	With numbers		Without numbers		Difference
Lang	[hours]	[utts]	[hours]	[utts]	[hours in %]
GE	16.37	9034	14.96	8390	-8.6%
PO	16.75	7350	12.33	5805	-26.3%
SP	15.36	5227	10.77	4064	-29.8%
RU	19.49	9771	16.73	8822	-14.1%
TU	14.49	5988	10.75	4775	-25.8%

2.3 Grapheme-Based System with Number Expansion

From the previous analysis, it is obvious that numbers need to be processed in a less aggressive way. Then second version of dictionaries (*grap_v2*) with number expansion were generated. For number expansion we used standard ICU library[4], which can be used for most languages and supports large variety of locales. With number expansion, we obtained complete dictionaries with all words including numbers and all acoustic data, without any loss of information, could be used.

We observed that a number in dictionary can have two meanings. One as normal word — cardinal number (e.g. 911 → n i n e h u n d r e d a n d e l e v e n), and another as a sequence of digits, i.e. for phone numbers, credit card numbers, etc. (e.g. 911 → n i n e o n e o n e). In *grap_v2* version, we did not use any variants and transcribed numbers in the first mentioned way (as cardinal number, e.g. 911 → n i n e h u n d r e d a n d e l e v e n).

Then, another version of dictionaries was produced (*grap_v3*), where we combined both of these variants and all numbers were expanded as in version *grap_v2* plus as a sequence of digits (so each number exists in dictionary two times with different pronunciations). In fact, pronunciation of numbers in form of single spoken digits is not frequent in GlobalPhone data, so we did not expect any substantial improvement with *grap_v3*.

2.4 Number Expansion for Spanish with Gender Dependency

For Spanish, a more sophisticated expansion of numbers was tested. Here, the gender of noun following a number can change the expansion of the number, for example:

un lápiz (one pencil)
una pluma (one pen)
uno (one - as single number)

cincuenta y un lápices (fifty-one pencils)
cincuenta y una plumas (fifty-one pens)
cincuenta y uno (fifty-one - as single number)

---

[4] http://site.icu-project.org/

```
doscientos dos coches (202 cars)
doscientas dos casas (202 houses)
```

The underlined numbers vary according to gender. When a number ends in '-uno' (one), the form '-un' is used before masculine nouns, and '-una' before feminine nouns. The 'uno' form is used only in counting. The hundreds portions of numbers change in gender even when other parts of the number intervene before the noun[5].

The limitation of this expansion is in fact, that whole transcription is needed for dictionary generation, as we first need to obtain pairs of numbers and corresponding nouns. With such a list of pairs, we can expand numbers in correct way, according to noun gender. As will be seen in the results, this method significantly improves accuracy, even over results of phoneme-based system. On the other hand, this approach uses morphology information about target language, which can be also used in standard phoneme dictionary and thus accuracy of phoneme-based system could also be improved.

3   Experimental Framework

The KALDI toolkit[6] was used for all recognition experiments [9]. We setup five systems:

- **Phon**: phoneme-based, which is set as a baseline.
- **Grap_v0** - grapheme-based, without numbers (with reduced acoustic data)
- **Grap_v1** - grapheme-based, without numbers (no reduction of data, numbers mapped to <*UNK*> symbol in transcripts)
- **Grap_v2** - grapheme-based with expanded numbers (no reduction of data). All numbers expanded as cardinal number
- **Grap_v3** - grapheme-based with expanded numbers (no reduction of data). All numbers expanded in two meanings — as cardinal number and in form of single spoken digits.

As features, we extract 13 Mel-frequency cepstral coefficients (MFCCs) and compute delta and delta-delta features. For all four setups, we first train a monophone system (*mono*) with about 10k diagonal Gaussians. Next, we train initial triphone system with about 50k diagonal covariance Gaussians (5000 states). This system is retrained into triphone system (*tri2c*) with the same number of parameters, and per-speaker cepstral mean normalization applied. The last system — *SGMM* — is built on subspace-GMMs [10] modeling of triphones. This system uses 400-component full-covariance Gaussian background model, about 6,000 tree leaves and 22k Gaussians in total.

4   Results

All results are given in terms of word accuracy. Table 4 presents the results for monophone system, the second column shows numbers of phonemes, resp. graphemes for different languages. Last column gives absolute improvement in accuracy between phoneme (phon) and grapheme system (grap_v2).

---

[5] http://spanish.about.com/cs/forbeginners/a/cardinalnum_beg.htm
[6] http://kaldi.sourceforge.net

Table 4. Accuracy of monophone system for different languages

	Count	MONO					ACC (Diff)
Lang	phon/grap	phon	grap_v0	grap_v1	grap_v2	grap_v3	phon/grap_v2
CZ	41/44	**64.2**	62.7				−1.5%
EN	40/27	**71.1**	43.9				−27.2%
GE	42/31	**51.9**	42.8	42.2	43.1	43.3	−8.8%
PO	34/40	**54.1**	48.0	47.6	48.3	47.7	−5.8%
SP	36/34	**61.5**	58.5	59.7	59.5	59.6	−2.0%
RU	54/34	**50.5**	47.1	47.4	47.3	47.2	−3.2%
TU	30/33	46.9	46.4	48.0	47.1	**48.1**	0.2%
VN	85/94	**61.1**	55.7				−4.2%

Table 5. Accuracy of triphone GMM system for different languages

		TRI2c				ACC (Diff)
Lang	phon	grap_v0	grap_v1	grap_v2	grap_v3	phon/grap_v2
CZ	**76.0**	75.9				−0.1%
EN	**82.6**	76.0				−6.6%
GE	**71.0**	70.2	70.5	70.7	70.8	−0.3%
PO	**72.9**	70.3	69.5	71.8	71.8	−1.1%
SP	75.4	74.5	**75.6**	75.4	75.4	0%
RU	**65.2**	63.3	63.9	64.1	64.1	−1.1%
TU	66.0	63.9	65.7	**66.1**	**66.1**	0.1%
VN	71.1	**71.6**				0.5%

As we can see, baseline phoneme-based systems have the best results in monophone training for almost all languages, the grapheme-based systems are about 2–8% absolutely worse. Only Turkish is an exception, with 0.2% better accuracy. On the other hand, the biggest hit is observed for English. Here, the results and big reduction of the number of acoustic units (from 40 phonemes to 27 graphemes) are related to the fact, that English spelling is not phonetically-based.

Table 5 shows the results for triphone GMM system. Here, grapheme-based setups have about 0.1–2% worse accuracy than phonemes, for EN, the degradation is about 6% against the baseline. These improvements are caused by possibility of triphone system to model wider context of graphemes. For some languages (SP, TU, VN), triphone grapheme based system works even better than phoneme one, this fact could indicate poor quality of the original dictionaries.

Table 6 presents the results for SGMM systems. Although we did not trained jointly shared parameters on all languages, we could see improvement in accuracy, obtained by simple usage of Sub-space Gaussian Mixture Models on each language. SGMMs have about 3–6% better results for phonemes and about 4–7% for graphemes against similar systems in triphone GMM training. The gap between phoneme and grapheme systems is further decreased.

The last Table 7 summarizes the result for Spanish, where three columns represent different systems (monophone, triphone GMM, SGMM) and rows present the results

Table 6. Accuracy of SGMM system for different languages

Lang	phon	SGMM				ACC (Diff)
		grap_v0	grap_v1	grap_v2	grap_v3	phon/grap_v2
CZ	**79.0**	78.5				−0.5%
EN	**85.4**	79.7				−5.7%
GE	**76.2**	75.5	75.1	75.6	75.1	−0.6%
PO	**76.3**	74.2	74.8	75.6	75.8	−0.7%
SP	78.7	78.6	78.9	**79.4**	79.2	0.7%
RU	68.5	68.2	68.4	**69.1**	68.3	0.6%
TU	70.3	69.7	70.3	70.0	**70.8**	−0.3%
VN	77.9	**78.6**				0.7%

Table 7. Accuracy of SGMM system for different languages

SP	MONO	TRI2c	SGMM
Phon	**61.5**	75.4	78.7
Grap_v2	59.5	75.4	79.4
Grap_v4	59.7	**76.0**	**79.6**

of baseline phoneme system, grapheme system with number expansion (grap_v2) and results after gender dependent expansion of numbers (grap_v4) as described in section 2.4.

Advanced expansion outperforms baseline in triphone GMM (0.6% absolute better) and SGMM system (+0.2%). Monophone result is still under baseline, but there is an improvement over the system with basic expansion of numbers (grap_v1), thus we can claim, that improved expansion of number gives better results than basic one. On the other hand, for rare languages we mostly know nothing about language morphology and rules, so advanced expansion of numbers is out of question.

5   Conclusion

We have shown that grapheme-based speech recognition, that copes with the problem of low-quality or missing pronunciation dictionaries, is applicable for phonetic languages and also tonal languages like Vietnamese. For class of non-phonetic languages, like English, using of models with wider context gives also comparable results and grapheme based approach can be, with small limitation, usable also for these languages. Improved expansion of numbers for Spanish also answered the question, whether we are able to do the number expansion better in situation, where we have information about target language and its morphology and rules. Grapheme-based straightforward approach, supported by the expansion of numbers in dictionaries, is advantageous especially in situation of low-resource languages and could be successfully used in building speech recognizers for rare languages.

References

1. Black, A., Lenzo, K., Pagel, V.: Issues in building general letter to sound rules. In: Proceedings of the ESCA Workshop on Speech Synthesis, Australia, pp. 77–80 (1998)
2. Fukada, T., Sagisaka, Y.: Automatic generation of multiple pronunciations based on neural networks. Speech Communication 27(1), 63–73 (1999)
3. Besling, S.: Heuristical and statistical Methods for Grapheme-to-Phoneme Conversion, Konvens, Wien, Austria, pp. 23–31 (1994)
4. Killer, M., Stüker, S., Schultz, T.: Grapheme Based Speech Recognition. In: Proceedings of the EUROSPEECH, Geneve, Switzerland, pp. 3141–3144 (2003)
5. Schillo, C., Fink, G.A., Kummert, F.: Grapheme Based Speech Recognition For Large Vocabularies. In: Proceedings of ICSLP 2000, pp. 129–132 (2000)
6. Stüker, S., Schultz, T.: A Grapheme Based Speech Recognition System for Russian. In: Specom 2004 (2004)
7. Charoenpornsawat, P., Hewavitharana, S., Schultz, T.: Thai grapheme-based speech recognition. In: Proceedings of the Human Language Technology Conference of the NAACL, Stroudsburg, PA, USA, pp. 17–20 (2006)
8. Schultz, T., Westphal, M., Waibel, A.: The globalphone project: Multilingual lvcsr with janus-3. In: Multilingual Information Retrieval Dialogs: 2nd SQEL Workshop, Plzeň, Czech Republic, pp. 20–27 (1997)
9. Povey, D., Ghoshal, A., et al.: The Kaldi Speech Recognition Toolkit. In: Proceedings of the ASRU, Hawaii, US (2011)
10. Povey, D., Burget, L., et al.: The subspace Gaussian mixture model – A structured model for speech recognition. Computer Speech and Language 25(2) (2011)

Discretion of Speech Units for the Text Post-processing Phase of Automatic Transcription (in the Czech Language)

Svatava Škodová[1], Michaela Kuchařová[2], and Ladislav Šeps[2]

[1] Department of the Czech Language and Literature
[2] Institute of Information Technology and Electronics, Technical University of Liberec
Studentská 2, 461 17, Liberec, Czech Republic
{svatava.skodova,michaela.kucharova1,ladislav.seps}@tul.cz

Abstract. In this paper we introduce an experiment leading to the improvement of the text post-processing phase of automatic transcription of spoken documents stored in the large Czech Radio audio archive of oral documents. This archive contains the largest collection of spoken documents recorded during the last 90 years. The underlying aim of the project introduced in the paper is to transcribe a part of the audio archive and store the transcription in the database, in which it will be possible to search for, and retrieve information. The value of the search is that one can find the information on the two linguistic levels: in the written form and the spoken form. This doubled information-storage is important especially for the comfortable retrieval of information and it diametrically extends the possibilities of work with the information. One of the important issues of the conversion of spoken speech to written texts is the automatic delimitation of speech units and sentences/clauses in the final text processing, which is connected with the punctuation important for convenient perception of the rewritten texts. For this reason we decided to test Czech native speakers' perception of speech and their need of punctuation in the rewritten texts. We compared their results with the punctuation added by an automaton. The results should serve to train a program for automatic discretion of speech units and the correct supplying of punctuation. For the experiment we prepared a sample of texts spoken by typologically various speakers (the amount of speech was 30 minutes; 5,247 words), these automatically rewritten texts were given to 59 respondents whose task was to supply punctuation to the automatically rewritten texts. We used two special tools to run this experiment; *NanoTrans* – this tool was used by respondents for supplying the punctuation. The other tool for viewing and comparing the respondents' and machine performance, especially written for the probe, was *Transcription Viewer*. In the text we give detailed information about these comparisons. In the final part of the paper we propose further improvements and ideas for future research.

Keywords: audio archive processing, audio search, punctuation, sentence delimitation, speech-to-text, speech units.

1 Introduction

One of the most prospective application fields for the automatic speech recognition is the processing of large audio or multimedia archives. The goals of these applications

P. Sojka et al. (Eds.): TSD 2012, LNCS 7499, pp. 446–455, 2012.

are usually to automatically create databases of document transcriptions, to index them and to provide some tools for the database searching in order for anyone to be able to deploy a full-text search of the archive along with some means of playing the corresponding parts of the documents. Such systems have been designed for broadcast archives [1], for national spoken language [2, 3] or for some special-purpose document collections [4]. Our laboratory has been working on these tasks since 2005 [5] and some of the developed tools have been already deployed in broadcast data mining [6]. The research presented in this paper is a part of a large applied-research project supported by the Czech Ministry of Culture. It aims at processing the archive of historical and contemporary recordings of the Czech Radio and at making its content available for search. In this paper we deal with one of the last stages of the whole project - with the post-processing of recognized texts. Currently, large volumes of audio data can be transcribed automatically with reasonable accuracy. The problem is that standard ASR systems produce only a raw stream of words, leaving out important information, such as sentences boundaries [7]. These boundaries are clearly visible in written texts by means of punctuation and capitalization, but hidden, if present at all, in speech.

Nevertheless, for the final transcriptions, it is essential to present them in a form comfortable for the reader. This involves the automatic identification of sentences and adding punctuation into them. Such a completed text is easier to read and to understand [8]. Even though the contemporary version of our program developed for the transcription uses formal criteria for sentence delimitation and succeeds in this task, we still suppose that the human necessity for unit delimitation in text is much higher. Therefore, we suggested a simple experiment in order to determine the human need of the delimitation of text units.

2   Experiment

For the experiment we prepared a selection of texts spoken by typologically various speakers; the total amount of speech was 30 minutes (5,247 words). These texts were transcribed by the automatic speech recognition system (the current version of our system is presented in [5]) and the transcripts were given to 59 correctors[1] to check all the errors and to add punctuation. The correctors both saw the texts automaticaly rewritten and they listened to the speech. We provided them with the both language resources, because in their task they combined their ability to work with both these part of language: the primary source of language was spoken but they corrected its rewritten form and they placed sentence boundaries to this form[2].

---

[1] We distinguish the terms *transcriber*, an automatic tool for transcribing texts from spoken to written form, and *corrector*, a human correcting the written form transcribed automatically.

[2] At the same time we run another experiment showing differences in the amount of used punctuation depending on the source of language. One group of correctors placed sentence boundaries to the text according to the sound, the other group used only the written form. The data of this experiment are not included into this study, it is just important to stress that the correctors who dealt only with the transcription used more punctuation than the correctors who dealt with the sound. The data of this experiment are available in [10].

At the same time we used an automaton [9] to do the same task and we compared its outcomes with the correctors; namely the amount of punctuation added by human correctors in comparison with the automaton. We used two special tools to run this experiment; *NanoTrans* – this tool was used by respondents to punctuate the text and *Transcription Viewer* – this tool was used for evaluation of the outcomes.

2.1 Material and Participants

For the experiment test we selected radio recordings of several qualitatively different speakers as follows:

i. Trained professional speakers.

ii. Speakers who are not trained professionals but their verbal expression is very cultivated.

iii. Speakers whose verbal expression is fuzzy and they violate the standard syntactic rules for Czech sentences.

The total number of speakers was ten; five of whom were professional reporters; five men, five women.

2.2 Text Editing and Correction

The punctuation system of written Czech undergoes clearly defined rules; in our project the main problem is that it is sometimes difficult to follow these rules while transcribing speech. As far as there are no codified rules for using punctuation in spoken type of language, the correctors were instructed to follow the current orthographic codification of Czech [11] and to add all the types of punctuation possible. We did not create any special manual for this experiment.

The experiment was also intended to investigate the agreement among the correctors on the delimitation of utterance boundaries[3]. We do not discuss the presuppositions concerning the cues according to which the correctors decided on sentence boundaries (i.e. speakers' intonation, pauses, meaning of utterances etc.), if it was not possible to use standard rules for written language. This will undergo the future research.

3 Transcription Tools

Transcribing audio recordings into time-aligned text units by humans is a straightforward but tiring procedure consisting of repetitive listening to short pieces sections (several dozens of seconds at maximum) of original recordings, writing down their contents, marking at least the beginning and end of the section with appropriate time, and selecting the speaker.

To simplify the transcription process, recordings are processed by a speech recognition system first. Therefore, instead of manual transcribing, a user simply corrects the

---

[3] We do not discuss the ability of the correctors to use the punctuation as far as they are linguists. To ensure the quality of correction, we chose students and specialists of the Department of the Czech Language of the Technical university of Liberec.

wrongly recognized parts. The output is a time-aligned text. The next step is the automatic splitting of the time-aligned text into short parts (the target length is between 10 and 30 seconds) called paragraphs, delimited by non-speech events, such as breathing sounds or pauses. After that, the preprocessed transcription is distributed to correctors.

There are very few publicly available applications used for audio transcription tasks. Probably the most known one is *Transcriber* and *TranscriberAG* [12, 13]. *Transcriber* is a powerful tool but it lacks some features the experiment required, for instance simultaneous support for a text and phonetic transcriptions, time-alignment at a word level (generated by a speech recognizer) and several other useful features and utilities. Adding them to an existing tool like *Transcriber* would be troublesome; therefore we decided to create a new tool using up-to-date technologies. We named it *NanoTrans*. The user interface of this program can be seen in Fig. 1.

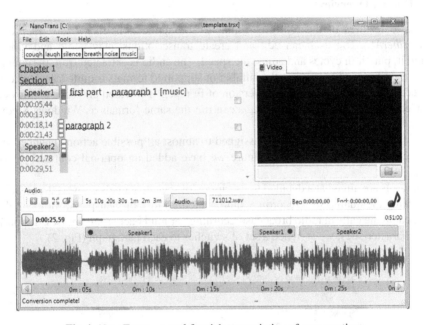

Fig. 1. *NanoTrans* – a tool for rich transcription of conversations

3.1   User Interface

NanoTrans is a Microsoft Windows application based on Microsoft .NET framework 4.0 and Windows Presentation Foundation for the user interface. Unlike *Transcriber*, a version for Linux has not been developed.

Its user interface consists of four parts – Main menu and toolbar, Transcription, Waveform, and Video frame – see Fig. 1.

NanoTrans is still in development, minor user interface changes are likely to be made prior to its official release.

The transcription frame is a part where text and phonetic transcriptions are displayed and edited. It is represented by three nested layers. The two topmost layers are called

Chapter and Section; they are intended to split a long transcription into shorter and more manageable parts. The third layer is called Paragraph. Each Paragraph contains a transcription of an utterance spoken by one speaker; so it is the counterpart of *Transcriber's* segment.

The waveform frame contains a graphical representation of the audio signal, paragraphs aligned to the audio signal and controls to adjust wave rendering and to play audio. Also paragraph positions can be created and edited from here.

Optionally, the user can also watch a video if he/she wants to transcribe audio-visual recordings. In this case, the video playback is synchronized with the audio signal. This can make the identification of speakers more convenient by providing visual information to correctors.

3.2 Feature Overview

The basic features of the tool are similar to the applications like *Transcriber* and *TranscriberAG*. The user can edit and create transcriptions, play audio and video, rewind it, play it in cycles and adjust the speed. The audio is extracted from the source file by the FFmpeg library, thus the number of supported formats is quite large and can be modified by building individual version of FFmpeg. The video playback employs a .NET component, therefore *NanoTrans* can run the same formats as Windows Media Player.

A rich set of keyboard shortcuts assigned to almost all possible action enhances the user's comfort. As an additional feature we have added an optional control by foot pedals.

Our tools are aimed at a detailed transcription of speech, namely conversations, such as marking the quality of a recording and/or intelligibility of their parts; for example, recorded phone calls in radio broadcast recordings or loud background noises. To mark these characteristics, we assigned a set of switches to each paragraph (Fig. 2).

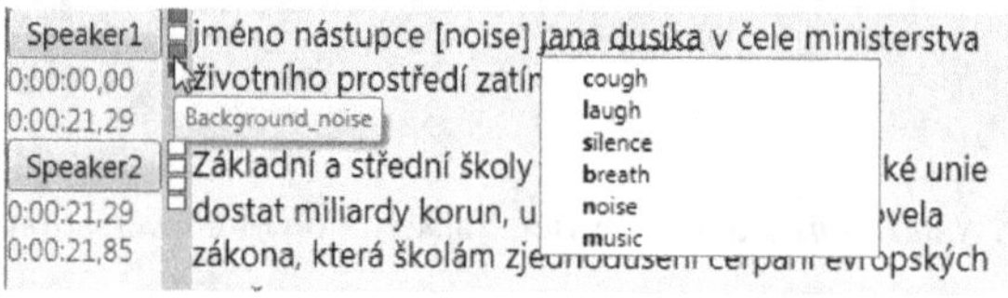

Fig. 2. A snapshot showing how signal quality can be easily labeled and how a non-speech sound can be added to speech transcription

To identify a short non-speech event or a short audio noise inside a paragraph we introduced special labels. They have a form of a text description surrounded by square brackets (e.g. [breath]). In a text editor they cannot be edited after insertion, only deleted. For the insertion of these labels, one can use either the toolbar or keyboard shortcuts or the suggestion window.

If the transcription contains time-alignment data (i.e. words with their time stamps provided by the speech recognizer), during playback the currently played words are highlighted by green background. This synchronization between the playback and editing makes the process of correcting the transcriptions faster and more comfortable.

In order to eliminate potential typing and grammar errors, a customizable spell-checking option was added. It is implemented in a standard way – by underlining the words or strings that are not in the vocabulary list by a red wavy line. The spell-checking feature is supported by open source *Hunspell* platform known from *Open Office, Mozilla Firefox* or *Google Chrome*.

4   Evaluation of Transcripts

To evaluate the transcripts processed within the *NanoTrans*, we created a special program called *Transcription Viewer*, designed to make statistical linguistic analyses. This program displays each transcription separately and aligns it, so that we could easily compare the differences between these transcriptions.

4.1   Description of the Program

Transcription Viewer allows us to observe differences between individual transcriptions. The program also computes statistics as a count of punctuation marks or a count of all words for one block. Fig. 3 shows the main window of the program. At the top of the program window, there is shown the number of the actual transcript block and the total count of transcript blocks. There is the name of the speaker as well. Below, there is a table with individual transcripts and statistics. In the left column of the table, there are the names of the correctors and the basics statistics. In the center section, there are individual transcripts aligned. Under the individual transcriptions, there is a line with the statistically derived, most probable transcription created by the program on the basis of the individual transcripts shown in the central section. Under this line, there is another one with the count of occurrences of a word from statistical sentence in individual transcripts.

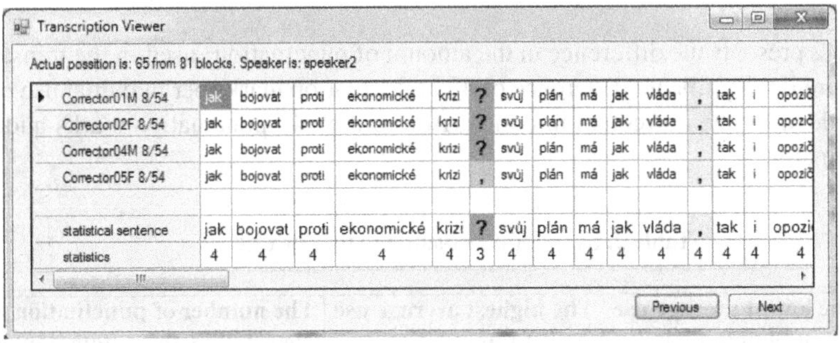

Fig. 3. Program *Transcription Viewer* used for comparing the transcripts provided by different correctors

The colors in the program window in the central section are used for better distinction between the differences and for special marks. In the recently conducted experiments it focuses on a statistical analysis of punctuation marks, therefore program highlights them. This is done by enlarging the font and by using colors. The punctuation marks indicating the end of a sentence (e.g. '.', '?', '!') are highlighted by red color. The marks which indicate a partition of a sentence (e.g. ',', ';', ':') are labeled using blue color. Next, the words in which the correctors differ from the statistical sentence are enhanced by a green color, as shown in Fig. 3. These tools make the evaluation of experiments faster and more accurate.

5   Evaluation of Speech Units Delimitation via Punctuation Use

In the final part of the text, we present the comparison of punctuation amount placed by the correctors and the automaton. Apart from that, we compare the correctors' agreement on the delimitation of sentences through *Transcription Viewer*.

5.1   An Overview of Language Material

The overview in Table 1 shows the complete number of positions (words and punctuation) transcribed in the corrected text.

For the analysis presented in this text, we employed only the highest and lowest average of used positions[4] added by all 59 correctors and the automaton. There is the variety of 681 positions per the maximum amount of 4,802 positions which can be caused by punctuation by avoiding/adding words in the stream of recognized speech.

Table 1. Score of positions (words and punctuation) in the corrected text

The lowest average use of positions	The highest average use of positions	The number of positions detected by the automaton
4,479	4,802	4,121

Table 2 presents the difference in the amount of punctuation[5] used in the transcribed text. There is a significant difference of 489 punctuation marks per maximal use of 832 marks added by the correctors in comparison to the 343 punctuation marks added by the automaton.

Table 2. Score of punctuation in the corrected text

The lowest average use of punctuation	The highest average use of punctuation	The number of punctuation detected by the automaton
456	832	343

---

[4] The number of positions includes words and punctuation marks.

[5] The number includes only the punctuation marks.

Table 3. Score of punctuation in the text

	Punctuation Min 456	Punctuation Max 832	Punctuation Automaton 343
Min Positions 4479	**10.18%**	**18.58%**	
Max Positions 4802	**9.50%**	**17.33%**	
Positions Automaton 4121			**8.32%**

Table 4. Absolute agreement on punctuation use

	Full stop	Comma	Question mark	Exclamation mark
Total agreement	37	247	19	1
Punct. Min : 456	8.11%	54.17%	4.17%	0.22%
Punct. Max: 832	4.45%	29.69%	2.28%	0.12%

In the case of automaton, the punctuation formed only 8.32% of the whole amount of positions in the text. In the case of correctors, the punctuation was much more frequent. Considering the lowest and the highest average use of positions captured by the correctors and the lowest and highest average use of the punctuation, we get an approximate idea of the human readers' need of text boundaries delimitation, as shown in Table 3.

It is necessary to mention that some punctuation added by some correctors was incorrect. Because the automaton uses only formal lexical markers for adding the punctuation, it was nearly 100% correct as regards the use of commas (i.e. commas before conjunctions); only 2 occurrences of punctuation were erroneous. The correctors did not achieve this level of accuracy in the use of commas.

Table 4 below introduces the correctors' absolute punctuation agreement on particular punctuation. We did not compare this use to the automaton, because it is not programmed to use all the types of punctuation marks available to the correctors.

The highest agreement was observed in the area of the use of commas; it is the highest one because the borders of sentences in this category are delimited by a formal element of a lexical character – by a conjunction. The correctors absolutely agreed on commas before the conjunctions že/that, ale/but, který/which, aby/to. All of the other marks exhibit much lower agreement. Comparing the results of absolute punctuation agreement to the whole amount of the punctuation (to the highest and the lowest total use of punctuation), there was only 4.45% ÷ 8.11% agreement on the use of a full stop; 29.69% ÷ 54.17% of a comma; 2.28% ÷ 4.17% of a question mark and 0.12% ÷ 0.22% of an exclamation mark.

Our presupposition concerning the delimitation of sentence borders was that the professional speakers who follow syntactic rules for written language and follow explicit intonation scheme will exhibit a higher agreement score in punctuation placed by the correctors. This presupposition was confirmed, but it is necessary to stress that the absolute agreement on punctuation marks use was much lower than expected. The delimitation of sentence borders due to the type of speaker is the aim of further research.

6   Conclusion and Future Work

National archives of spoken documents represent considerable corpora for linguistics research. So far, most of the linguistic research concerned with the written data, because it was not possible to manipulate the spoken material. The project with its tools referred to in the text opens the possibilities to examine spoken data, reconsider the current generally accepted descriptions of language and compare the results to the rules for written language. *NanoTrans*, which was used for this specific research, has the advantage of parallel storage of written and spoken varieties of texts and so it allows for valid comparisons.

We have explored the ability of human transcribes in detection sentence boundaries of speech transcribed automatically into written texts; we compared the agreement of humans on adding punctuation marks and we compared the average number of added punctuation marks with the amount of punctuation added by automaton. The experiment is illustrative of several facts about sentence delimitation in transcription of primary spoken data: (1) the automaton adds correctly nearly all punctuation following a formal marker (conjunction); people do not achieve such a level of accuracy and sometimes do not add punctuation connected with a formal marker; (2) the absolute agreement on the sentence boundaries in transcribed speech is very low – correctors agreed on the sentence boundaries but they did not agree on the type of punctuation, (3) the experiment results confirmed the presupposition that correctors require a higher degree of structuration of a text than it is supplied by the automaton. Compared to the automaton, the correctors added more than twice as much punctuation.

Currently we are planning to use the material for analyzing the formal character of positions where correctors agreed on adding punctuation, and to define formal rules for punctuation by the automaton to fill the need of users for more detailed structuration of a text via larger amount of punctuation.

Acknowledgments.   This work was supported by project no. DF11P01OVV013 provided by Czech Ministry of culture in research program NAKI.

References

1. Hayashi, Y., et al.: Speech-based and Video-supported Indexing Multimedia Broad-cast News. In: Proc. ACM SIGIR (2003)
2. Ordelman, R., de Jong, F., Huijbregts, M., van Leeuwen, D.: Robust Audio Indexing for Dutch Spoken Word Collections. In: 16th Int. Conference of the Association for History and Computing, Humanities, Computers and Cultural Heritage, Amsterdam, pp. 215–223 (2005)
3. Hansen, J.H.L., Huang, R., Zhou, B., Seadle, M., Deller, J.R., Gurijala, A.R., Kurimo, M., Angkititrakul, P.: SpeechFind: Advances in Spoken Document Retrieval for a National Gallery of the Spoken Word. IEEE Trans. on Speech and Audio Processing 13(5), 712–730 (2005)
4. Byrne, W., et al.: Automatic Recognition of Spontaneous Speech for Access to Multilingual Oral History Archives. IEEE Transactions on Speech Audio Processing 12(4), 420–435 (2004)
5. Nouza, J., Žďánský, J., Červa, P., Kolorenč, J.: A System for Information Retrieval from Large Records of Czech Spoken Data. In: Sojka, P., Kopeček, I., Pala, K. (eds.) TSD 2006. LNCS (LNAI), vol. 4188, pp. 485–492. Springer, Heidelberg (2006)

6. Nouza, J., Blavka, K., Bohac, M., Cerva, P., Zdansky, J., Silovsky, J., Prazak, J.: Voice Technology to Enable Sophisticated Access to Historical Audio Archive of the Czech Radio. In: Grana, C., Cucchiara, R. (eds.) MM4CH 2011. CCIS, vol. 247, pp. 27–38. Springer, Heidelberg (2012)

7. Kolář, J.: Automatic Segmentation of Speech into Sentence-like Units. University of West Bohemia, Pilsen (2008)

8. Müllerova, O.: Mluvený text a jeho syntaktická výstavba. Academia, Praha (1994)

9. Bohac, M., Blavka, K., Kuchařová, M., Škodová, S.: Post-processing of the Recognized Speech for Web Presentation of Large Audio Archive. In: International Conference on Telecommunications and Signal Processing, Czech Republic (2012)

10. Plocova, H.: Comparison of the Boundaries of Sentence Units Depending on the Form of Realisation. Bachelor thesis. Technical university of Liberec, Liberec (2012)

11. Pravidla çeského pravopisu. Členicíu znaménka, pp. 58–75. Academia, Praha (1993)

12. Barras, C., Geoffrois, E., Wu, Z., Liberman, M.: Transcriber: Development and Use of a Tool for Assisting Speech Corpora Production. Speech Communication Special Issue on Speech Annotation and Corpus Tools 33(1-2) (2000)

13. Geoffrois, E., Barras, C., Bird, S., Wu, Z.: Transcribing with Annotation Graphs. In: Second International Conf. on Language Resources and Evaluation (LREC), pp. 1517–1521 (2000)

On the Impact of Annotation Errors on Unit-Selection Speech Synthesis*

Jindřich Matoušek, Daniel Tihelka, and Luboš Šmídl

University of West Bohemia, Faculty of Applied Sciences, Dept. of Cybernetics
Univerzitní 8, 306 14 Plzeň, Czech Republic
{jmatouse,dtihelka,smidl}@kky.zcu.cz

Abstract. Unit selection is a very popular approach to speech synthesis. It is known for its ability to produce nearly natural-sounding synthetic speech, but, at the same time, also for its need for very large speech corpora. In addition, unit selection is also known to be very sensitive to the quality of the source speech corpus the speech is synthesised from and its textual, phonetic and prosodic annotations and indexation. Given the enormous size of current speech corpora, manual annotation of the corpora is a lengthy process. Despite this fact, human annotators do make errors. In this paper, the impact of annotation errors on the quality of unit-selection-based synthetic speech is analysed. Firstly, an analysis and categorisation of annotation errors is presented. Then, a speech synthesis experiment, in which the same utterances were synthesised by unit-selection systems with and without annotation errors, is described. Results of the experiment and the options for fixing the annotation errors are discussed as well.

Keywords: speech synthesis, unit selection, annotation errors.

1 Introduction

Nowadays, two approaches to speech synthesis, unit selection and HMM-based speech synthesis, are very popular and are almost exclusively used to synthesise speech. Both approaches are *corpus-based* in that they utilise large speech corpora, though each of them in a different manner. The principle of *unit-selection-based speech synthesis*, a signal-based approach, is to select the largest suitable segment of natural speech from the source speech corpus according to various phonetic, prosodic and positional criteria, in order to prevent potential discontinuities in the synthesised speech, and to smoothly concatenate them. On the other hand, in *HMM-based speech synthesis*, a model-based approach, the speech corpus is used as a basic material for statistical modelling of speech by means of hidden Markov models (thereof the abbreviation HMM). In this case, the resulting speech is generated from the HMMs. Both approaches applied to Czech were described e.g. in [1] or [2], respectively. This paper deals with the unit-selection approach.

* Support for this work was provided by the TA CR, project No. TA01030476, and by the European Regional Development Fund, project "New Technologies for Information Society", European Centre of Excellence, ED1.1.00/02.0090.

P. Sojka et al. (Eds.): TSD 2012, LNCS 7499, pp. 456–463, 2012.

The unit selection technique is known for its ability to produce nearly natural-sounding synthetic speech, but, at the same time, also for its need for very large speech corpora. Speech corpora containing tens of hours of speech are not rare in this technique. In addition, since the quality of unit-selection synthesis system crucially depends on the quality of the speech corpus and its textual, phonetic and prosodic annotations and indexation, special attention should be paid to the whole process of speech corpus preparation, including the very precise annotation. The problems with the accurate annotation lie in the fact that, due to huge corpora, the annotation process is time-consuming and, thus, costly. Although some attempts were made to annotate the corpora automatically, or semi-automatically [3,4,5], the automation is still error-prone. From this point of view, it is still advantageous, when annotation is made by human experts.

The process of recording and annotation of speech corpus suitable for speech synthesis of Czech speech was described in [6]. In this paper, the impact of annotation errors on the quality of unit-selection-based synthetic speech is analysed.

The paper is organised as follows. In Section 2, an analysis and categorisation of annotation errors is presented. A speech synthesis experiment, in which the same utterances were synthesised by systems with and without annotation errors, is described in Section 3. Results of this experiment and options for fixing the annotation errors are discussed in Section 4. Finally, conclusions are drawn in Section 5.

2   Annotation Errors

In the first phase, the "prescribed" sentences (ANN0) selected by a sentence-selection algorithm were used as patterns. Following the annotation rules described in [6], the annotation ANN0 was modified by the first annotator. As a result, ANN1 annotation was obtained. Approximately 28% of all transcribed sentences and 4% of all words contained errors and were corrected in ANN1 (see the column ANN1-ANN0 in Table 1).

Being aware of the importance of the precise annotation of the source speech data for corpus-based speech synthesis, all annotations were subject of a revision in the 2nd phase. The revision ANN2 was made by another annotator—she used ANN1 annotations and corrected them if needed. As shown in Table 1 (column ANN2-ANN1), approximately 4% of all sentences and 1% of all words were found to be misannotated by the first annotator and were corrected by the second annotator.

The typical annotation errors were: *different words* (words which are different in the original and revised annotations), *extra words* (words which were present in original annotations but actually were not spoken; thus these words were deleted in the revised annotations), and *missing words* (words which were missing in original annotations and added to revised annotations).

The frequency of the annotation errors within the different annotation phases is shown in Table 1. Some examples of the annotation errors are presented in the left part of Fig. 1.

Both the words missed in ANN1 (missing words) and deleted in ANN2 (extra words) were mostly monosyllabic words (e.g. "se", "si", etc.). The differences between words

Table 1. Comparison of the pattern (ANN0), initial (ANN1) and revised (ANN2) annotations (relative frequency in percents) and percentage of words and sentences which contained errors

Differences	ANN0-ANN1	ANN0-ANN2	ANN1-ANN2
Different words	0.79	0.88	0.09
Extra words	0.07	0.08	0.08
Missing words	0.05	0.05	0.03
Word errors	3.51	3.78	0.38
Sentence errors	27.68	29.27	3.76

Table 2. Differences between words and their examples as annotated in ANN1 and revised in ANN2. Percentage is shown within all differences.

Typo	Perc. [%]	ANN1	ANN2
TYPO1	47.37	blondýna	blondýnka
LAST	19.30	jak	jako
LENGTH	14.04	benzinů	benzínů
TYPO2	10.53	pude	bude
MISP	8.77	Jankulovski	Jarkulovski

in both annotations (different words) are summarised in Table 2. They typically consist in:

- the last letter of a word was missing or extra (LAST);
- a vowel letter was shortened or lengthened (LENGTH);
- a word was mistyped in ANN1 as another meaningful word (TYPO1);
- a word was mistyped in ANN1 as a non-sense word (TYPO2);
- a word had been pronounced as a non-sense word but the original transcription from ANN0 was left in ANN1, or ANN2, respectively (MISP).

As can be seen in Fig. 1, annotation errors cause serious phonetic segmentation errors. As the task of phonetic segmentation is to delimit speech segments (usually phone-like units or diphones) in the source recordings, wrongly segmented segments can be concatenated during unit-selection speech synthesis. The impact of the annotation/segmentation errors on synthetic speech is further examined in Section 3.

3   Speech Synthesis Experiment

To evaluate the impact of annotation errors on unit-selection-based synthetic speech, the following speech synthesis experiment was carried out:

1. Annotation errors were identified in a Czech male speech corpus prepared specially for the purpose of unit-selection speech synthesis [6]. From 615,915 phone tokens present in the corpus 2,371 tokens (0.38% of all tokens) corresponded to the errors.
2. Two acoustic unit inventories were built from the speech corpus [7]:

(a) One inventory (INV1) was built from the speech corpus with the annotation errors.

(b) The other inventory (INV2) was built from the speech corpus without the annotation errors.

3. Representative annotation errors were then selected as "reference" errors.

4. 1,043,938 phrases were synthesised by a Czech speech synthesis system [1] with the INV1 inventory, and the phrases, in which speech segments from the representative misannotated (and badly segmented) words were used, were stored. It was found that 49,196 synthesised phrases (almost 5% of all phrases) contained at least one segment from a misannotated word. The examples of such synthetic phrases are shown in the right part of Fig. 1.

5. Selected phrases were then synthesised with the same synthesis system again, this time with the INV2 inventory. The phrases were selected to contain both major (segmentation of more consecutive segments was wrong like in Fig 1b,c,e) and minor (few segments were badly segmented like in Fig 1a,d) annotation errors introduced in Section 2.

6. Informal listening tests were carried out to compare the selected phrases synthesised from both INV1 and INV2 inventories.

Five listeners experienced with synthetic speech participated in the listening tests. All of them unmistakably evaluated the corrected inventory INV2 as significantly better. Moreover, they reported that the annotation errors have a drastic impact on the quality of synthetic speech. This is not surprising in the case of synthetic speech generated from the major annotation errors (like in Fig. 1b,c,e). Due to the major errors, the synthetic speech contained an extra word(s) (as in the case of Fig. 1b where an extra word "toho" appeared in synthetic speech) or parts of words (as in the case of Fig. 1c,e), which made the synthetic speech unintelligible.

As for the minor annotation errors like the one in Fig. 1d (where the word "piktogramy" was misannotated as "piktografy", causing e.g. the word "žirafy" being synthesised as a nonsense word "žiramy") or the one in Fig. 1a (where the word "je" was misannotated as "byla" causing the word pair "by jí" being synthesised as "je jí"), such phone-level misplacement caused local unintelligibility and were assessed as very disruptive as well.

4   Discussion and Future Work

As described in the previous section, annotation errors have a drastic impact on the quality of synthetic speech, devaluing thus the quality of text-to-speech (TTS) systems, especially in real-time applications like a railway information service [8] or a talking head [9]. It is clear that annotation phase is a very important part of TTS system design and should receive appropriate attention. However, as mentioned earlier, due to enormous size of current speech corpora it is almost impossible to have the annotation free of errors. Therefore, methods for automatic detection of annotation errors would be very welcome and crucial for the quality of unit-selection-based synthetic speech. As shown in Section 3, fixing only 0.38% of all segments in speech corpus increases quality of almost 5 % of all phrases.

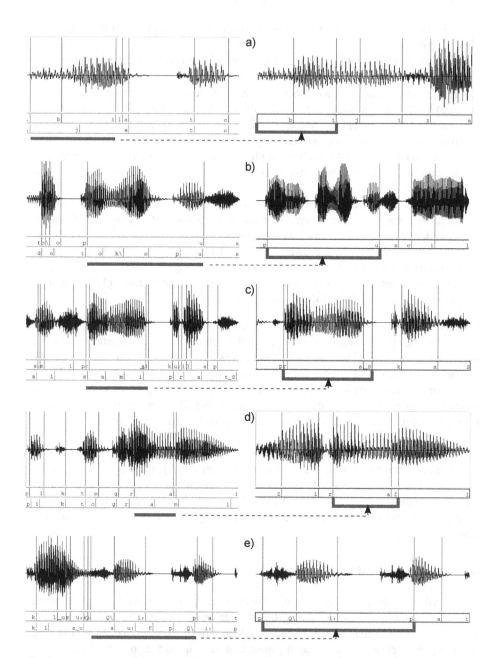

Fig. 1. Examples of annotation errors and their impact on unit-selection-based synthetic speech. Misannotated source recordings are depicted on the left. The upper line under each waveform contains the wrong annotation/segmentation, the lower line contains the correct annotation/segmentation. Synthetic speech waveforms with the corresponding misplaced segments are shown on the right. The thick solid lines show the misannotated segments.

One possibility is to fix the annotation errors manually "on the fly" as the errors are reported during the verification phase of the system design. Another possibility is to detect the annotation errors directly in the speech corpora. To do that, several methods can be proposed:

– **Segmentation control.** As segmentation errors are a good signal of bad annotation, the automatically segmented boundaries can be checked for errors. The badly segmented boundaries are typically those which delimit speech segments with very short or very long duration. Such an approach is intuitive and fast, but efficient rules have to be proposed in order not to have too many false alarms. Word-level errors could be fixed this way except the words with similar length.

– **Speech recognition.** An automatic speech recognition (ASR) system [10] can be utilised to recognise words and/or phones automatically from speech signals in speech corpus. Both word- and phone-level recognisers can be used. The language model for such recogniser is created from the reference annotations and must be accompanied by a general language model. The combination with the general model ensures the ability to detect non-standard pronunciation and annotation errors.

– **Keyword spotting.** Keyword spotting technique could be used to check the words in the annotation [11]. Only different words and missing words (as introduced in Section 2) can be detected in this way. The main advantage is that keyword spotting technique uses only acoustic model. This makes it possible to discover such errors that are caused by such non-standard combination of words. In addition, the acoustic model can be easily trained on speaker's voice. The main disadvantage is the relatively large number of false alarms.

– **Comparison with synthetic speeech.** Based on its textual representation, each utterance from the speech corpus can be synthesised by another TTS system. Subsequently, the synthesised utterance can be compared with the original utterance [12]. If the signals are different (in terms of word deletions, insertions or substitutions), the utterance can be considered to contain annotation errors.

– **Classification.** The problem of detection of annotation errors can be viewed as a task of automatic classification. Each word (or each utterance) can be described by appropriate features (like number of short or long segments in a word/utterance, number of phones in a word, number of words in an utterance, context of neighbouring words, etc.). Two-class classification can be carried out in a word-by-word or utterance-by-utterance manner, and each token (word or utterance) can be then classified as containing or not an annotation error. Different kinds of classifiers like support vector machines (SVM), classification and regression trees (CART), artificial neural networks (ANN), etc. could be utilised [13].

– **Combination of more approaches.** The above approaches can be combined together in order to achieve better performance. One possibility is to propose a "meta classifier", and results from other approaches to use as features for such a meta classifier.

In our work, preliminary experiments with keyword-spotting and segmentation control techniques were carried out. Both techniques were able to detect a majority of errors. On the other hand, they produced many false alarms, i.e. many words were detected as erroneous, but, in fact, there were no errors in these words at all. Another sort of errors than the basic ones presented in Section 2 were found during these experiments. These errors related to prosodic expression of an utterance (e.g. emphasised or emotionally expressed words, not typical for "neutral" style used in unit-selection speech synthesis [14,15]), the presence of atypical Czech sounds—"parasitic" glottalization phenomena or schwa-like segments [16]. So, first of all, we plan to focus on the detection of gross annotation errors. Other detection techniques mentioned above will also be examined and the accuracy of (gross) annotation errors detection will be compared.

5   Conclusion

In the presented paper, the impact of annotation errors on the quality of unit-selection-based synthetic speech was analysed. Based on the source speech data, a categorisation of annotation errors into several classes was carried out. Although the likelihood of using badly annotated/segmented speech unit during synthesis is relatively small, listening tests have shown the annotation errors have a drastic impact on the quality of synthetic speech. Therefore, any error found and removed is important to the overall impression of the quality of synthetic speech. Several methods have been proposed for automatic detection of annotation errors; two of them, keyword-spotting and segmentation control techniques, were preliminarily carried out. Both techniques were able to detect a majority of errors, but, on the other hand, they also produced a relatively large number of false alarms. In our future work, other techniques (and their combination) will be experimented with in order to detect gross annotation errors with a high accuracy and, at the same time, to eliminate the large number of false alarms.

References

1. Tihelka, D., Kala, J., Matoušek, J.: Enhancements of Viterbi Search for Fast Unit Selection Synthesis. In: Proc. Interspeech, Makuhari, Japan, pp. 174–177 (2010)
2. Hanzlíček, Z.: Czech HMM-Based Speech Synthesis. In: Sojka, P., Horák, A., Kopeček, I., Pala, K. (eds.) TSD 2010. LNCS, vol. 6231, pp. 291–298. Springer, Heidelberg (2010)
3. Cox, S., Brady, R., Jackson, P.: Techniques for Accurate Automatic Annotation of Speech Waveforms. In: Proc. ICSLP, Sydney, Australia (1998)
4. Tachibana, R., Nagano, T., Kurata, G., Nishimura, M., Babaguchi, N.: Preliminary Experiments Toward Automatic Generation of New TTS Voices from Recorded Speech Alone. In: Proc. Interspeech, Antwerp, Belgium, pp. 1917–1920 (2007)
5. Aylett, M.P., King, S., Yamagishi, J.: Speech Synthesis Without a Phone Inventory. In: Proc. Interspeech, Brighton, England, pp. 2087–2090 (2009)
6. Matoušek, J., Romportl, J.: Recording and Annotation of Speech Corpus for Czech Unit Selection Speech Synthesis. In: Matoušek, V., Mautner, P. (eds.) TSD 2007. LNCS (LNAI), vol. 4629, pp. 326–333. Springer, Heidelberg (2007)
7. Matoušek, J., Tihelka, D., Psutka, J.V.: Experiments with Automatic Segmentation for Czech Speech Synthesis. In: Matoušek, V., Mautner, P. (eds.) TSD 2003. LNCS (LNAI), vol. 2807, pp. 287–294. Springer, Heidelberg (2003)

8. Švec, J., Šmídl, L.: Prototype of Czech Spoken Dialog System with Mixed Initiative for Railway Information Service. In: Sojka, P., Horák, A., Kopeček, I., Pala, K. (eds.) TSD 2010. LNCS, vol. 6231, pp. 568–575. Springer, Heidelberg (2010)
9. Železný, M., Krňoul, Z., Císař, P., Matoušek, J.: Design, Implementation and Evaluation of the Czech Realistic Audio-Visual Speech Synthesis. Signal Processing 12, 3657–3673 (2006)
10. Müller, L., Psutka, J.V., Šmídl, L.: Design of Speech Recognition Engine. In: Sojka, P., Kopeček, I., Pala, K. (eds.) TSD 2000. LNCS (LNAI), vol. 1902, pp. 259–264. Springer, Heidelberg (2000)
11. Šmídl, L., Trmal, J.: Keyword Spotting Result Post-processing to Reduce False Alarms. In: Recent Advances in Signals and Systems, vol. 9, pp. 49–52. WSEAS Press, Budapest (2009)
12. Malfrere, F., Deroo, O., Dutoit, T., Ris, C.: Phonetic Alignment: Speech Synthesis-Based Vs. Viterbi-Based. Speech Communication 40, 503–515 (2003)
13. Lu, H., Wei, S., Dai, L., Wang, R.-H.: Automatic Error Detection for Unit Selection Speech Synthesis Using Log Likelihood Ratio Based SVM Classifier. In: Proc. Interspeech, Makuhari, Japan, pp. 162–165 (2010)
14. Grůber, M.: Acoustic Analysis of Czech Expressive Recordings from a Single Speaker in Terms of Various Communicative Functions. In: Proc. ISSPIT, Bilbao, Spain, pp. 267–272 (2011)
15. Přibil, J., Přibilová, A.: An Experiment with Evaluation of Emotional Speech Conversion by Spectrograms. Measurement Science Review 10, 72–77 (2010)
16. Matoušek, J., Skarnitzl, R., Machač, P., Trmal, J.: Identification and Automatic Detection of Parasitic Speech Sounds. In: Proc. Interspeech, Brighton, England, pp. 876–879 (2009)

Analysis of the Influence of Speech Corpora in the PLDA Verification in the Task of Speaker Recognition

Lukáš Machlica and Zbyněk Zajíc

University of West Bohemia in Pilsen
Faculty of Applied Sciences, Department of Cybernetics
Univerzitní 22, 306 14 Pilsen
{machlica,zzajic}@kky.zcu.cz
http://www.kky.zcu.cz/en

Abstract. In the paper recent methods used in the task of speaker recognition are presented. At first, the extraction of so called i-vectors from GMM based supervectors is discussed. These i-vectors are of low dimension and lie in a subspace denoted as Total Variability Space (TVS). The focus of the paper is put on Probabilistic Linear Discriminant Analysis (PLDA), which is used as a generative model in the TVS. The influence of development data is analyzed utilizing distinct speech corpora. It is shown that it is preferable to cluster available speech corpora to classes, train one PLDA model for each class and fuse the results at the end. Experiments are presented on NIST Speaker Recognition Evaluation (SRE) 2008 and NIST SRE 2010.

Keywords: PLDA, latent space, fusion, supervector, FA, i-vector.

1 Introduction

A major progress in the task of speaker recognition was done introducing supervector based techniques. Supervector is in fact a high dimensional feature vector obtained by the concatenation of lower dimensional vectors containing speaker dependent parameters – the most effective turned out to be the parameters related to Gaussian Mixture Models (GMMs) [1]. First attempts to incorporate supervectors into the speaker recognition task utilized Support Vector Machines (SVMs) along with distinct kernel functions [2]. Since GMMs belong to the class of generative models, whereas SVMs are based on the discrimination between classes, the techniques comprising both methods are also known as *hybrid modelling*. Subsequently, additional techniques were added to solve the problem of the change of operating conditions, namely Nuisance Attribute Projection (NAP) [3]. NAP is based on an orthogonal projection, where directions most vulnerable to environment changes are projected out. Since supervectors are of substantially high dimension (tens of thousands), which is often higher than the number of supervectors provided for the training, it is obvious that a lot of dimensions will be correlated with each other, and that the information about the identity will be contained in a subspace of a much lower dimension. This idea was incorporated into the principle of Joint Factor Analysis (JFA) [4], where the word *joint* refers to the fact that not only the speaker, but also the channel variabilities are treated in one JFA model. However, since experiments

P. Sojka et al. (Eds.): TSD 2012, LNCS 7499, pp. 464–471, 2012.

in [5] showed that the channel space obtained by JFA does still contain some information about the speaker's identity, JFA was slightly adjusted giving rise to *identity vectors*, or *i-vectors* [6]. The main difference between JFA and i-vectors is that i-vectors do not distinguish between speaker and channel space. They work with a *total variability space* containing simultaneously speaker and channel variabilities, whereas JFA treats both spaces individually.

Parallel to JFA a very similar approach was introduced in the image recognition called Probabilistic Linear Discriminant Analysis (PLDA) [7]. The only difference from JFA is that in PLDA ordinary feature vectors are used instead of GMM based supervectors (for details on the treatment of supervectors in JFA see [4]). Since PLDA is a generative model, it allows to compute the probability that several i-vectors originate from the same source, and thus it is well suited as a verification tool for a speaker recognition system [8]. System presented in this paper will utilize i-vectors (described in Section 3) based on GMM supervectors (see Section 2) with a PLDA model (refer to Section 4) used in the verification phase.

The crucial problem when proposing a speaker verification system composed of modules (e.g. JFA, PLDA) described above is that data from a lot of speakers are required, moreover several sessions have to be available for each speaker in order to train a reliable i-vector extractor and a PLDA model. The problem faced in this paper will address the question whether distinct speech corpora (e.g. Switchboard 1, Switchboard 2, NIST SRE 2004, NIST SRE 2006, etc.) should be pooled together and used to train one PLDA model, or if each corpus should be used individually to train a separate PLDA model. In the latter scenario the results are fused at the end. Experiments can be found in Section 5.

2 Supervector Extraction Based on GMMs

At first a Universal Background Model (UBM) has to be trained. UBM is in fact a Gaussian Mixture Model (GMM), however it is trained from a set containing a lot of speakers. The speakers data should match all the conditions, in which the recognition system is going to be used. UBM consists of a set of parameters $\lambda_{\text{UBM}} = \{\omega_m, \mu_m, C_m\}_{m=1}^M$, where M is the number of Gaussians in the UBM, ω_m, μ_m, C_m are the weight, mean and covariance of the mth Gaussian, respectively. Let $O_s = \{o_{st}\}_{t=1}^{T_s}$ be the set of T_s feature vectors o_{st} of dimension D belonging to the sth speaker, and

$$\gamma_m(o_{st}) = \frac{\omega_m \mathcal{N}(o_{st}; \mu_m, C_m)}{\sum_{m=1}^M \omega_m \mathcal{N}(o_{st}; \mu_m, C_m)} \tag{1}$$

be the posterior probability of mth Gaussian given a feature vector o_{st}. And let $m_0 = [\mu_1^T, \ldots, \mu_m^T, \ldots, \mu_M^T]^T$ be the supervector composed of UBM means. Then, for each speaker two supervectors are extracted

$$n_s = \sum_{t=1}^{T_s} \left([\gamma_1(o_{st}), \ldots, \gamma_m(o_{st}), \ldots, \gamma_M(o_{st})]^T \otimes 1_D\right) \text{ of size } DM \times 1,$$

$$b_s = \sum_{t=1}^{T_s} \left[\gamma_1(o_{st})o_{st}^T, \ldots, \gamma_m(o_{st})o_{st}^T, \ldots, \gamma_M(o_{st})o_{st}^T\right]^T \text{ of size } DM \times 1, \tag{2}$$

where $\otimes$ is the Kronecker product, and $\mathbf{1}_D$ is a D dimensional vector of ones. Note that n_s is the supervector containing "soft" counts of feature vectors aligned to Gaussians $1, \ldots, M$, and denoting N_s a diagonal matrix containing n_s on its diagonal, $m_s = N_s^{-1} b_s$ is the new Maximum Likelihood (ML) estimate of supervector m_0 given the dataset O_s. At last note that the Maximum Aposteriory Probability (MAP) adaptation [9] of means of the UBM according to the given data set O_s expressed in the supervector notation is given as $m_{\text{MAP}} = \tau m_s + (1 - \tau) m_0$ for some relevance factor τ.

3 i-Vector Extraction

The concept of the i-vectors extraction is based on Factor Analysis (FA) extended to handle session and speaker variabilities of supervectors to Joint Factor Analysis (JFA) [4]. Contrary to JFA, different sessions of the same speaker are considered to be produced by different speakers [5]. The generative i-vector model has the form

$$\psi_s = m_0 + T w_s + \epsilon, \quad w_s \sim \mathcal{N}(0, I), \quad \epsilon \sim \mathcal{N}(0, \Sigma) \tag{3}$$

where T (of size $D \times D_w$) is called the total variability space matrix (it contains both the variabilities between speakers and the channel variabilities between distinct sessions of a speaker), w_s is the sth speaker's i-vector of dimension D_w having standard Gaussian distribution, m_0 is the mean vector of ψ_s, however often the UBM's mean supervector is taken instead as a good approximation (therefore the same notation m_0 is used), and ϵ is some residual noise with a diagonal covariance Σ constructed from covariance matrices $C_1, \ldots, C_m$ of the UBM ordered on the diagonal of Σ.

To train the matrix T two steps are iterated in a sequence. Given a training set of S couples of supervectors b_s, n_s, and the diagonal matrix N_s containing n_s on its diagonal, these steps are:

1. use previous estimate of T to extract new i-vectors for all speakers $1, \ldots, S$

$$w_s = (I + T^{\text{T}} \Sigma^{-1} N_s T)^{-1} T^{\text{T}} \Sigma^{-1} \bar{b}_s, \tag{4}$$

2. according to the new i-vectors compute block-wise a new estimate of T

$$T_m = \left(\sum_{s=1}^{S} \bar{b}_{sm} w_s^{\text{T}} \right) \left(\sum_{s=1}^{S} N_{sm} \left(w_s w_s^{\text{T}} + \left(I + T^{\text{T}} \Sigma^{-1} N_s T \right)^{-1} \right) \right)^{-1}, \tag{5}$$

where $\bar{b}_s = b_s - N_s m_0$ is the centered version of b_s around the mean m_0, and the index m in $T_m, \bar{b}_{sm}, n_{sm}$ (and N_{sm}) refers to blocks of $T, \bar{b}_s, n_s$ (and thus to N_{sm}) of sizes $D \times D_w, D \times 1, D \times 1$ so that $T^{\text{T}} = [T_1^{\text{T}}, T_2^{\text{T}}, \ldots, T_M^{\text{T}}], \bar{b}_s^{\text{T}} = [\bar{b}_{s1}^{\text{T}}, \bar{b}_{s2}^{\text{T}}, \ldots, \bar{b}_{sM}^{\text{T}}], n_s^{\text{T}} = [n_{s1}^{\text{T}}, n_{s2}^{\text{T}}, \ldots, n_M^{\text{T}}]$, respectively. In fact, also Σ may be updated in each iteration, for details see [10].

4 Probabilistic Linear Discriminant Analysis (PLDA)

Let us assume that the i-vector extractor (4) was already trained, and that for each feature set O_s of a speaker s one i-vector w_s was extracted. Further, let us assume that several sessions $\{O_{sh}\}_{h=1}^{H_s}$ of a speaker s are available, and that for each set of feature vectors O_{sh} of each session $h = 1, \ldots, H_s$ one i-vector w_{sh} was extracted. Since in the i-vector extraction phase no distinction between session space and speaker space were made a new model in the total variability space will be now described that is going to utilize also the session variabilities.

PLDA is a generative model of the form

$$w_{sh} = m_w + Fz_s + Gr_{sh} + \epsilon, \ \epsilon \sim \mathcal{N}(0, S) \tag{6}$$

where m_w is the mean of w_{sh}, columns of F span the speaker identity space, z_s of dimension D_z are coordinates in this space and they do not change across sessions of one speaker, columns of G span the channel space, r_{sh} of dimension D_r are the session dependent speaker factors, and ϵ is some residual noise with diagonal covariance S and a zero mean. Further restrictions are put on distributions of latent variables z_s and r_{sh}, namely that both follow standard Gaussian distribution $\mathcal{N}(0, I)$. Hence, $w_{sh} \sim \mathcal{N}(m_w, FF^T + GG^T + S)$. It is common and reasonable assumption that $D_z \ll D_w$ and that $D_z + D_r \approx D_w$. To train the model parameters F, G and S one has to solve the system of equations [7]

$$\begin{bmatrix} w_{s1} - m_w \\ w_{s2} - m_w \\ \vdots \\ w_{sH_s} - m_w \end{bmatrix} = \begin{bmatrix} F & G & 0 & \ldots & 0 \\ F & 0 & G & \ldots & 0 \\ \vdots & \vdots & \vdots & \ddots & \vdots \\ F & 0 & 0 & \ldots & G \end{bmatrix} \begin{bmatrix} z_s \\ r_{s1} \\ r_{s2} \\ \vdots \\ r_{sH_s} \end{bmatrix} + \begin{bmatrix} \epsilon_1 \\ \epsilon_2 \\ \vdots \\ \epsilon_{H_s} \end{bmatrix}, \ \hat{\Sigma}_{H_s} = \begin{bmatrix} S & 0 & \ldots & 0 \\ 0 & S & \ldots & 0 \\ \vdots & \vdots & \ddots & \vdots \\ 0 & 0 & \ldots & S \end{bmatrix}, \tag{7}$$

and when rewritten to a compact form we get

$$\hat{w}_s = A_{H_s} \hat{z}_s + \hat{\epsilon}, \tag{8}$$

where $\hat{\epsilon} \sim \mathcal{N}(0, \hat{\Sigma}_{H_s})$. Matrices A_{H_s}, $\hat{\Sigma}_{H_s}$ depend on s through the number of their row- and column-blocks given by the number of sessions H_s of speaker s. Problem (8) is a standard FA problem, for details on how to solve it see the appendix in [7].

4.1   Verification

In the verification phase two hypotheses are tested [7], namely

- hypotheses $\mathcal{H}_s$ that two i-vectors w_1 and w_2 share the same identity,
- hypotheses $\mathcal{H}_d$ that the identity of two i-vectors w_1 and w_2 differs.

The log-likelihood ratio is given as

$$\text{LLR}(w_1, w_2) = \log \frac{p(w_1, w_2 | \mathcal{H}_s)}{p(w_1 | \mathcal{H}_d) p(w_2 | \mathcal{H}_d)} =$$

$$= \log \mathcal{N} \left(\begin{bmatrix} w_1 \\ w_2 \end{bmatrix}; \begin{bmatrix} m_w \\ m_w \end{bmatrix}, \begin{bmatrix} C_w & C_F \\ C_F & C_w \end{bmatrix} \right) - \log \mathcal{N} \left(\begin{bmatrix} w_1 \\ w_2 \end{bmatrix}; \begin{bmatrix} m_w \\ m_w \end{bmatrix}, \begin{bmatrix} C_w & 0 \\ 0 & C_w \end{bmatrix} \right), \tag{9}$$

where $C_w = FF^T + GG^T + S$ and $C_F = FF^T$. Note that in this verification scenario we do not care about the form of the decomposition of w_1 or w_2 (latent variables z_s, r_{sh} stay unknown). The question stated is whether two vectors share the same identity given the subspaces generated by F and G.

5   Experiments

The question raised is whether all the available data from several distinct corpora should be pooled and used to train one PLDA model (thus find a decomposition of the total variability space based on all the available data), or if it would be more efficient to find a characteristic decomposition of the total variability space for each corpus individually (hence train several PLDA models), score each pair of vectors in relation to each total space decomposition, and finally fuse the obtained scores (we will use linear combination). We believe that the latter case makes the verification more robust since possible undesirable deviations in acoustic conditions of distinct corpora may become less evident. However, individual corpora still have to contain enough data to be able to train a reliable PLDA model.

5.1   Used Corpora

In order to be able to perform reliable tests we utilized corpora: NIST SRE 2004, NIST SRE 2005, NIST SRE 2006, Switchboard 1 Release 2 and Switchboard 2 Phase 3 for development purposes, and NIST SRE 2008, NIST SRE 2010 were used for calibration of Fusion Coefficients (FCs), and for the evaluation of generality of obtained FCs, respectively. We used only those speakers from development corpora who had more than 4 recorded sessions. Further, the development corpora were divided into 3 classes:

1. NIST040506 – containing 3787 recordings of 465 males of approximately 8 sessions for each male speaker,
2. SW1 – containing 2342 recordings of 211 males of approximately 11 sessions for each male speaker,
3. SW2 – containing 2183 recordings of 216 males of approximately 10 sessions for each male speaker,

Each of the recordings had approximately 5 minutes in duration including the silence. The division into the classes was made in relation to the similarity of corpora determined according to recording conditions given in the LDC Corpus Catalog[1].

In order to train the FCs "short2-short3 trials" from NIST SRE 2008 [11] were utilized, only telephone speech from males was used (648 target speakers and 1,535 test speakers) yielding 16,968 trials in total. To test the validity of learned FCs "core-core trials" from NIST SRE 2010 [12] were used, and again only telephone speech from males was used (1,394 target speakers and 2,474 test speakers) yielding 74,762 trials in total. The duration of all the test and target recordings in both corpora was approximately 5 minutes including the silence.

---

[1] http://www.ldc.upenn.edu/Catalog/index.jsp

5.2   Feature Extraction

The feature extraction was based on Linear Frequency Cepstral Coefficients (LFCCs), Hamming window of length 25 ms was used, the shift of the window was set to 10 ms. 25 triangular filter banks were spread linearly across the frequency spectrum, and 20 LFCCs were extracted, delta coefficients were added leading to a 40 dimensional feature vector. Also the Feature Warping (FW) normalization procedure was applied utilizing a sliding window of length 3 seconds. Right before the FW Voice Activity Detector (VAD), based on detection of energies in filter banks located in the frequency domain, was used in order to discard the non-speech frames. All the feature vectors were at the end down-sampled by a factor of 2.

The number of Gaussians in the UBM was set to 1024. The size of the total variability space matrix T in the i-vector extraction was set to $1024 \times 40 \times 800$, thus the latent dimension (dimension of i-vectors) was $D_w = 800$. At last, the dimension of the speaker identity space in the PLDA model was set to $D_z = 100$ and the dimension of the session/channel space was set to $D_r = 800$, thus F was of size 800×100, and the channel matrix was a square matrix of size 800×800. The disproportion between dimensions of speaker and channel subspaces was adopted from [8].

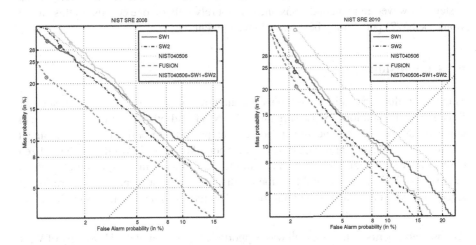

Fig. 1. DET curves for NIST SRE 2008 and NIST SRE 2010. Circles denote points where minDCF occurred.

Table 1. Results are given as EER [%] / minDCF. In the last column of the table also results for PLDA trained on pooled corpora NIST040506 + SW1 + SW2 is given

	SW1	**SW2**	**NIST040506**	**FUSION**	**pooled**
NIST 2008	9.87/0.034	8.47/0.041	8.83/0.043	**7.06/0.031**	8.78/0.045
NIST 2010	9.89/0.050	8.11/0.046	11.58/0.057	**7.58/0.043**	9.26/0.051

5.3 Results and Analysis

UBM and the i-vector's extractor described in Section 3 were trained on the pooled dataset NIST040506 + SW1 + SW2. Next, three PLDA models were trained utilizing subsequently each of the 3 corpora classes. Trials from NIST SRE 2008 were then scored using all 3 PLDA models, and the scores were used to train the fusion coefficients via the linear logistic regression from the FoCal toolkit [13]. Finally, in order to test the validity of learned FCs the same approach was performed with trials from NIST SRE 2010, but the already learned FCs were used in the linear combination of obtained scores. Results are shown in Figure 1 and Table 1, also minimum of the Decision Cost Function (DCF) is reported. In order to compute the value of DCF the cost of missing a target was set to 10, the cost of the false alarm was set to 1, and the probability of seeing a true trial was set to 0.01. These values are adopted from the NIST Speaker Recognition Evaluation (SRE) 2008 [11].

We have trained one PLDA model also from polled corpora NIST040506 + SW1 + SW2 (this was not used in the fusion). Best results are obtained for the fused system in both NIST SRE 2008 and NIST SRE 2010. Note that PLDA trained only on SW2 outperforms all the other PLDA models trained on other corpora (even on the pooled corpora), but the fusion still increases the performance of the speaker verification system. However, in real conditions one can not rely only on one corpus (in this case it would be SW2) performing best on the development set.

6 Conclusion

Since often the verification conditions are unknown in advance (e.g. in the Speaker Recognition Evaluations (SREs) organized by NIST and other institutions) we cannot count on the use of one specific speech corpus performing best on the development set. It is more convenient to utilize several corpora. We have shown that if the utilized corpora have sufficient amount of data to train reliable PLDA models, it is preferable to train several PLDA models and fuse the results. The verification becomes more robust since the deviations in acoustic conditions of distinct corpora become less evident.

Acknowledgments. This research was supported by the grant of the University of West Bohemia, project No. SGS-2010-054.

References

1. Campbell, W., Sturim, D., Reynolds, D.: Support Vector Machines Using GMM Supervectors for Speaker Verification. IEEE Signal Processing Letters 13, 308–311 (2006)
2. Longworth, C., Gales, M.: Parametric and Derivative Kernels for Speaker Verification. In: Interspeech 2007, pp. 310–313 (2007)
3. Solomonoff, A., Quillen, C., Campbell, W.: Channel compensation for SVM speaker recognition. In: Odyssey, pp. 57–62 (2004)
4. Kenny, P.: Joint Factor Analysis of Speaker and Session Variability: Theory and Algorithms. Tech. report, Centre de Recherche Informatique de Montral (2006)

5. Dehak, N.: Discriminative and Generative Approaches for Long- and Short-term Speaker Characteristics Modeling: Application to Speaker Verification. Ph.D. thesis, École de Technologie Supérieure, Université du Québec (2009)
6. Dehak, N., Kenny, P., Dehak, R., Dumouchel, P., Ouellet, P.: Front-End Factor Analysis For Speaker Verification. IEEE Transactions on Audio, Speech and Language Processing (2010)
7. Prince, S., Elder, J.: Probabilistic Linear Discriminant Analysis for Inferences About Identity. In: IEEE 11th International Conference on Computer Vision, pp. 1–8 (2007)
8. Matějka, P., Glembek, O., Castaldo, F., Alam, J., Plchot, O., Kenny, P., Burget, L., Černocký, J.: Full-covariance UBM and Heavy-tailed PLDA in I-Vector Speaker Verification. In: ICASSP 2011, pp. 4828–4831 (2011)
9. Reynolds, D.A., Quatieri, T.F., Dunn, R.B.: Speaker Verification Using Adapted Gaussian Mixture Models. Digital Signal Processing 10, 19–41 (2000)
10. Patrick, K., Pierre, O., Najim, D., Vishwa, G., Pierre, D.: A Study of Interspeaker Variability in Speaker Verification. IEEE Transactions on Audio, Speech and Language Processing 16, 980–988 (2008)
11. The NIST Year, Speaker Recognition Evaluation Plan (2008),
    http://www.itl.nist.gov/iad/mig/tests/spk/2008/
    sre08_evalplan_release4.pdf
12. The NIST Year, Speaker Recognition Evaluation Plan (2010),
    http://www.itl.nist.gov/iad/mig/tests/
    spk/2010/NIST_SRE10_evalplan.r6.pdf
13. Brummer, N.: FoCal: Tools for fusion and calibration of automatic speaker detec- tion systems (2006), http://sites.google.com/site/nikobrummer/focal

Adaptive Language Modeling
with a Set of Domain Dependent Models

Yangyang Shi, Pascal Wiggers, and Catholijn M. Jonker

Interactive intelligence Group, Delft University of Technology
Mekelweg 4, 2628CD, Netherlands
shiyang1983@gmail.com

Abstract. An adaptive language modeling method is proposed in this paper. Instead of using one static model for all situations, it applies a set of specific models to dynamically adapt to the discourse. We present the general structure of the model and the training procedure. In our experiments, we instantiated the method with a set of domain dependent models which are trained according to different socio-situational settings (ALMOSD). We compare it with previous topic dependent and socio-situational setting dependent adaptive language models and with a smoothed n-gram model in terms of perplexity and word prediction accuracy. Our experiments show that ALMOSD achieves perplexity reductions up to almost 12% compared with the other models.

1 Introduction

A language model judges whether a sequence of words is a fluent sentence in a language or not. Statistical language models do so by modeling probability distributions over word sequences. This paper focuses on language models that adapt to the domain to which they are applied.

Current state of the art statistical language models are smoothed n-grams, maximum entropy models, and more recently recurrent neural networks [1]. All of these models rely on a relatively small history of previous words to predict the next word in a sentence and do not take into account the larger context of the conversation at hand. In other words, these models assume that the same word distributions can be used in all situations.

However, when dealing with different tasks or situations, people cognitively adapt their language [2]. For example, consider the syntax and lexicon employed in a formal article, versus a casual conversation. In this paper, we study models that take into account the diversity and variability of language over context.

Taking contextual information into account is beneficial to language models, as is shown by topic dependent and socio-situational setting dependent language models [3,4,5,6,7]. Such models typically interpolate between multiple domain specific language models. This leads to robust models that favor coherent discourses. The price for this robustness is the fact that through interpolation the results of component models that match the current discourse may be weighed down by non-matching components. Therefore, a challenge is to create statistical language models that can dynamically select the domain specific models that best fit the current fragment of discourse. In this way, it can avoid the impact of non-matching component.

P. Sojka et al. (Eds.): TSD 2012, LNCS 7499, pp. 472–479, 2012.

Theoretically, better performance can be achieved if a specific model would be available that matches the statistical regularities in the current discourse [8]. For example, a model trained on texts on economics and marketing will get better results in predicting texts from the Wall Street Journal than a general model.

Clearly, the chance of having a model that exactly matches the current discourse is small, but what if we have available a rich set of models that correspond to different topics and/or different types of discourse, can we then dynamically select one model among those models most suited for the current discourse? And would this outperform an interpolation approach? These are the questions investigated in this paper.

The paper is organized as follows. In the next section, related work is discussed. In section 3, we give the theoretical background and motivation of our ALMOSD model, and discuss the structure of the model as well as the procedures we used for training a specific instance of this adaptive language modeling approach. In Section 4, we compare a smoothed n-gram model, a topic dependent and a socio-situational setting dependent language model with ALMOSD in terms of perplexity and word prediction accuracy. Finally, based on the results, conclusions are drawn in Section 5.

2   Related Work

Several adaptive language modeling approaches have been proposed in the literature. [9] classifies the adaptation methods into three categories according to the underlying philosophy: model interpolation, constraint specification and meta-information (e.g. semantic knowledge, topic knowledge, syntactic infra-structure) extraction.

In model interpolation, [10] proposed the class based n-gram language models. Their adaptation strategy is to assign words with similar meaning and syntactic function into one class.

In constraint specification adaptive language models should satisfy the features extracted from the adaptation data. Exponential models trained using a maximum entropy approach, separately assign different weights for each feature [11].

[6] and [12] proposed a mixture language models adaptation method. A collection of sub-models are trained on separate pre-defined domains. Mixture language models linearly interpolate these sub-models. However, as discussed in [13], in actual usage, the mixture language model doesn't work well, partly because of the complicated smoothing. [7,14] explicitly model the domain as a variable in their language models. In this case, it avoids complicated smoothing, as only one model needs to be trained.

In meta-information extraction, there has been much previous research in applying topic information in language modeling [4,6].

All these adaptation methods finally generate one general language model to capture the diversity of natural language. In ALMOSD, we use different models to represent different domains of natural language.

3   The Model

As discussed by [15], psychophysical evidence for the existence of parallel processing channels in human processing of speech has been found, especially in dealing with

Table 1. Perplexity results of specific sub-models (SM), a smoothed trigram (ST), a topic dependent (TM) and socio-situational setting dependent (SSM) model

comp	socio-situational setting	SM	ST	TM	SSM
a	Spontaneous conversations ('face-to-face')	220	222	226	221
b	Interviews with teachers of Dutch	196	214	213	212
c&d	Spontaneous telephone dialogues	188	191	192	190
e	Simulated business negotiations	110	153	154	152
f	Interviews/ discussions/debates	274	284	283	281
g	(political) Discussions/debates/ meetings	287	372	366	349
h	Lessons recorded in the classroom	298	315	314	313
i	Live (eg sports) commentaries (broadcast)	275	425	402	369
j	Newsreports/reportages (broadcast)	345	369	367	361
k	News (broadcast)	366	573	560	563
l	Commentaries/columns/reviews (broadcast)	425	440	435	434
m	Ceremonious speeches/sermons	398	444	436	434
n	Lectures/seminars	-	-	-	-
o	Read speech	573	705	682	695

unexpected words. This evidence inspired us to design an adaptive language modeling which can automatically select one from a set of domain dependent models.

The other source of inspiration is the behavior of language models. It is well known that domain-specific language models perform much better on a given domain than general-purpose models. This is illustrated in Table 1, where perplexity results are shown for data from 14 different domains. Perplexity is a measurement of the performance of language models. A better model returns lower perplexity on the same test data set. The perplexity is calculated according to:

$$PP = 2^{-\frac{1}{t} \log P(w_1 w_2 ... w_t)},$$
(1)

where $w_1 w_2 \ldots w_t$ is the data in the test set.

The results in the column marked SM in Table 1 are obtained by domain specific models, trained on similar data from the same domain, whereas all results the column marked ST are obtained by a smoothed trigram trained on data from all domains. The columns marked TM and SSM show the results obtained with two sentence-level mixture models the first of which contains topic-specific submodels found by automatic clustering and the second of which uses the domain-specific models of column SM as components (the details of these models will be explained in the next section).

These results suggest that a model that predicts the next word according to a distribution that fits that of a domain specific model will outperform a static model as well as mixture models. So, if k is the current domain and P_k a corresponding model (e.g. on of the models listed in the third column of Table 1), then for an adaptive model $\hat{P}$ the following should hold:

$$\hat{P}(w_i|w_1 \ldots w_{i-1}) = P_k(w_i|w_1 \ldots w_{i-1}),$$
(2)

where w_i is the word at position i. Note that this formulation allows the selection of a different k for every word in a discourse.

The question then is how to select the right specific model P_k at every point in time. For this purpose we introduce the function $\Phi()$ that predicts the domain k based on the word history:

$$\hat{P}(w_i|\Phi(w_1 \dots w_{i-1}) = k) = P_k(w_i|w_1 \dots w_{i-1}). \qquad (3)$$

We chose to use the set of specific models themselves to implement this function, i.e. the model that best matches the history seen sofar, is the one used to predict the next word:

$$\Phi(w_1 \dots w_{i-1}) = \arg\max_k P_k(w_1 \dots w_{i-1}). \qquad (4)$$

An alternative way to think of this model is as a model that puts all word histories that best match a particular model k in one equivalence class. This implies that in general, there is no need to use the sub-models themselves to select the appropriate distribution for prediction, any suitable function of the word history will do. Also, there is no restriction on the submodels that can be used, one could for example include a general purpose model or mixture models as components as well to deal with those cases in which no appropriate specific model is available.

3.1 Models Training

All sub-models used in this paper are interpolated trigrams trained with a two phase procedure. For all models the same vocabulary is used. The process of training of the sub-models is as follows:

Initial Training. Initially, a unigram, bigram and trigram model are trained on all training data using MLE. Next these models are interpolated:

$$\hat{p}(w_i|w_{i-2}w_{i-1}) = \lambda_1 p(w_i|w_{i-2}w_{i-1}) + \lambda_2 p(w_i|w_{i-1}) + \lambda_3 p(w_i), \qquad (5)$$

where w_i is the i-th word in a sentence. The interpolation weights λ_1, λ_2 and λ_3 are estimated using a held-out data set. This model is used as our baseline smoothed trigram model (ST) and as a basis for all other models.

Sub-model Training. In the second phase, the complete data set is partitioned into a set of sub-domains (Table 1). Each domain specific model is trained on the corresponding subset. The final component based model is obtained by interpolating this model with the general model of phase 1:

$$P(w_i|w_{i-1}\dots w_{i-n}) = \theta_C P_C(w_i|w_{i-1}\dots w_{i-n}) + \theta_S P_S(w_i|w_{i-1}\dots w_{i-n}), \qquad (6)$$

where $\theta_C + \theta_S = 1, \theta_C, \theta_S \geq 0$, P_C, P_S represent the probability learned in complete training and subset training, respectively.

In initial training the complete data trained models actually are the average of the distributions of each subset. They avoid overfitting in some degree, but they also ignore the characteristics of each subset. The sub-models which are the interpolation of the complete data and subsets data, highlight the distribution of each subset. At the same time, they are controlled by the complete data distribution to avoid overfitting.

4   Experiments

4.1   Data

The Corpus Spoken Dutch (Corpus Gesproken Nederlands; CGN) [16,17], an 8 million word corpus of contemporary Dutch spoken in Flanders and Netherlands is used in our experiments. This data set is made up of 15 components, each related to a socio-situational setting. The socio-situational settings used in this paper are shown in Table 1.

In the experiment, 80% of the data in every component was randomly selected for training, 10% for development testing and 10% for evaluation. A vocabulary with 44,368 words was created, which contains all unique words that occur more than once in the training data. All words in the test data that are not in the vocabulary were replaced by an out-of-vocabulary token.

4.2   Topic and Socio-situational Setting Dependent Language Models

We compared our model with a baseline smoothed trigram, but also with two mixture models: one based on topic-dependent models (TM) that were found by automatic clustering of the data and one based on the same set of socio-situational models that make up the components of our model (SSM). Both are sentence level mixture models [6]:

$$p(w_{1,N}) = \sum_T p(T) \prod_{i=1}^{N} p(w_i|w_1 \ldots w_{i-1}, t_i), \qquad (7)$$

and

$$p(T) = p(t_1) \prod_{i=2}^{N} p(t_i|t_{i-1}), \qquad (8)$$

where t_i represents the topics or socio-situational settings at time i.

In these models, the current topic or socio-situational setting is dependent on the previous one; the current word is dependent on a history of two words and the previous topic or socio-situational setting. For the details of combining topic information in dynamic language models see [4,6].

4.3   Results

Table 2 shows the performance of the models on the entire test set in terms of perplexity. The interpolated models perform only slightly better than the smoothed trigram. AL-MOSD clearly outperforms the three other models with a perplexity reduction of 11.91%.

In addition to perplexity, we also use word prediction accuracy to measure the performance of the language models. Word prediction has many applications in natural language processing, such as augmentative and alternative communication, spelling correction, word and sentence auto completion, etc. Typically word prediction provides one word or a list of words which fit the context best. This function can be realized by statistical language models as a side product. Looking at this from the other side, word prediction accuracy actually provides a measurement of the performance of language models [18].

Table 2. Comparison of the models in terms of perplexity (ppl)

model	perplexity
smoothed n-gram ST	277
topic-based mixture model TM	274
socio-situational setting mixture model SSM	272
ALMOSD	244

Table 3. Comparison of the models in terms of word prediction accuracy (wpa) per component of the data set and for the entire test set

comp	ALMOSD	ST	TM	SSM
a	16.14	15.35	15.36	15.37
b	14.90	14.83	14.83	14.87
cd	18.10	17.40	17.42	17.41
e	19.07	18.02	18.02	17.97
f	13.98	14.22	14.25	14.27
g	14.93	14.00	14.14	14.23
h	13.42	13.41	13.40	13.40
i	15.70	13.44	13.48	13.53
j	13.88	12.75	12.95	13.17
k	18.26	15.54	15.51	15.62
l	12.38	12.86	12.86	12.98
n	12.61	12.48	12.51	12.61
o	12.64	11.85	11.95	11.98
overall	16.09	15.36	15.38	15.39

Table 3 compares ALMOSD, the smoothed trigram model ST, the topic dependent adaptive language models (TM) and the socio-situational setting dependent adaptive language model (SSM) in terms word prediction accuracy. The models are compared on the entire test sets as well as per component of the CGN listed in Table 1. It can be seen that one the entire data set, the ALMOSD model outperforms the three other models. The model also performs best on most individual components. It does especially well on component k that contains broadcast news (a word prediction accuracy of 18.26% vs a prediction accuracy of 15.62% by the socio-situation setting mixture model).

For these results it should be noted that our models were trained on data from the same set of domains that they were tested on, in case of out-of-domain data, a difference between our model and mixture based models is to be expected. However, note that there is no restriction on the set of models that can be included as components. To handle out-of-domain data, one could include for example a mixture model component.

5 Conclusion

Arguing that domain specific language models perform better than general purpose models, we propose an adaptive language modeling method called ALMOSD that combines a set of domain dependent models with a function that selects the most appropriate domain dependent model for the current situation. In particular, for the task of word prediction the function selects that component model that fits the word history better than the other component models. The prediction of the selected model is chosen as the next word prediction of ALMOSD. At every point in time the model that assigns the highest probability to the word history is chosen.

Our experiments we show that ALMOSD is able to reduce perplexity by 11.91% compared to a smoothed n-gram model. It also outperforms the other models tested on a word prediction task.

The architecture of ALMOSD makes it easy to experiment with other functions to select the appropriate component model for the current situation. For example, we will experiment with a limited history horizon, e.g., looking back at most 20 sentences. Other ideas are to experiment with (combinations of) dynamic classification functions of domains.

Furthermore, the architecture of ALMOSD also makes it easy to plug in other models, e.g., richer or models for other specific domains than used in our experiments.

References

1. Mikolov, T., Karafiát, M., Burget, L., Cernocký, J., Khudanpur, S.: Recurrent neural network based language model. In: INTERSPEECH, pp. 1045–1048 (2010)
2. Foster, P., Skehan, P.: The influence of planning and task type on second language performance. Studies in Second Language Acquisition 18, 299–323 (1996)
3. Wiggers, P.: Modelling Context in Automatic Speech Recognition. Ph.D. thesis, Delft University of Technology (2008)
4. Wiggers, P., Rothkrantz, L.: Combining Topic Information and Structure Information in a Dynamic Language Model. In: Matoušek, V., Mautner, P. (eds.) TSD 2009. LNCS, vol. 5729, pp. 218–225. Springer, Heidelberg (2009)
5. Iyer, R., Ostendorf, M.: Modeling long distance dependencies in language: Topic mixtures versus dynamic cache models. IEEE Trans. Speech Audio Process. 7, 236–239 (1999)
6. Iyer, R., Ostendorf, M., Rohlicek, J.R.: Language modeling with sentence-level mixtures. In: HLT 1994: Proceedings of the Workshop on Human Language Technology, pp. 82–87. Association for Computational Linguistics, Morristown (1994)
7. Shi, Y., Wiggers, P., Jonker, C.M.: Language modelling with dynamic bayesian networks using conversation types and part of speech information. In: The 22nd Benelux Conference on Artificial Intelligence, BNAIC (2010)
8. Shi, Y., Wiggers, P., Jonker, C.M.: Combining Topic Specific Language Models. In: Habernal, I., Matoušek, V. (eds.) TSD 2011. LNCS, vol. 6836, pp. 99–106. Springer, Heidelberg (2011)
9. Bellegarda, J.: Statistical language model adaptation: review and perspectives. Speech Communication 42, 93–108 (2004)
10. Brown, P.F., Pietra, V.J.D., de Souza, P.V., Lai, J.C., Mercer, R.L.: Class-based n-gram models of natural language. Computational Linguistics 18, 467–479 (1992)
11. Rosenfeld, R.: A maximum entropy approach to adaptive statistical language modelling. Computer Speech & Language 10, 187–228 (1996)
12. Seymore, K., Rosenfeld, R.: Using story topics for language model adaptation. In: Kokkinakis, G., Fakotakis, N., Dermatas, E. (eds.) EUROSPEECH. ISCA (1997)

13. Adda, G., Jardino, M., Gauvain, J.L.: Sixth European Conference on Speech Communication and Technology, Eurospeech 1999, budapest, Hungary, September 5-9. ISCA (1999)
14. Wiggers, P., Rothkrantz, L.J.M.: Topic-based language modeling with dynamic bayesian networks. In: Proceedings of the Ninth International Conference on Spoken Language Processing, pp. 1866–1869 (2006)
15. Hermansky, H.: Dealing with Unexpected Words in Automatic Recognition of Speech. In: Habernal, I., Matoušek, V. (eds.) TSD 2011. LNCS, vol. 6836, pp. 1–15. Springer, Heidelberg (2011)
16. Hoekstra, H., Moortgat, M., Schuurman, I., van der Wouden, T.: Syntactic annotation for the spoken dutch corpus project (cgn). Computational Linguistics in the Netherlands 2000, 73–87 (2001)
17. Nelleke, O., Wim, G., Frank Van, E., Louis, B., Jean-pierre, M., Michael, M., Harald, B.: Experiences from the spoken dutch corpus project. In: Proceedings of the Third International Conference on Language Resources and Evaluation, pp. 340–347 (2002)
18. van den Bosch, A.: Scalable classification-based word prediction and confusible correction. Traitement Automatique des Langues 46, 39–63 (2006)

Robust Adaptation Techniques Dealing with Small Amount of Data*

Zbyněk Zajíc, Lukáš Machlica, and Luděk Müller

University of West Bohemia in Pilsen, Univerzitní 22, 306 14 Pilsen
Faculty of Applied Sciences, Department of Cybernetics
{zzajic,machlica,muller}@kky.zcu.cz

Abstract. The worst problem the adaptation is dealing with is the lack of adaptation data. This work focuses on the feature Maximum Likelihood Linear Regression (fMLLR) adaptation where the number of free parameters to be estimated significantly decreases in comparison with other adaptation methods. However, the number of free parameters of fMLLR transform is still too high to be estimated properly when dealing with extremely small data sets. We described and compared various methods used to avoid this problem, namely the initialization of the fMLLR transform and a linear combination of basis matrices varying in the choice of the basis estimation (eigen decomposition, factor analysis, independent component analysis and maximum likelihood estimation). Initialization methods compensate the absence of the test speaker's data utilizing other suitable data. Methods using linear combination of basis matrices reduce the number of estimated fMLLR parameters to a smaller number of weights to be estimated. Experiments are aimed to compare results of proposed basis and initialization methods.

Keywords: ASR, adaptation, fMLLR, robustness, initialization, basis.

1 Introduction

For state of the art Automatic Speech Recognition (ASR) systems it is usual to implement the speaker adaptation of their acoustic models. A popular way is to use a linear transformation, especially the method of feature Maximum Likelihood Linear Regression (fMLLR), which transforms acoustic features in order to improve the fit of transformed features and the Speaker Independent (SI) acoustic model. fMLLR tries to find a linear transformation ($d \times (d + 1)$ matrix, where d is the features dimension) of an acoustic space, which maximizes the probability of test data given a SI model.

In the case where small amount of adaptation data is available the number of free parameters is too high to estimate these parameters reliably. Transformation matrix becomes ill-conditioned, and can lead to poor recognition rates. To avoid this problem various solutions have been proposed, e.g. lower the number of free parameters using diagonal or block-diagonal matrices, proper initialization of transformation matrices [8] or the eigenspace approach [7].

---

* This research was supported by the Ministry of Culture Czech Republic, project No. DF12P01OVV022.

P. Sojka et al. (Eds.): TSD 2012, LNCS 7499, pp. 480–487, 2012.

Initialization methods compensate the absence of test speaker data utilizing other suitable data. Possibilities how to chose the initialization data are presented in Section 3. We present two different methods of initialization distinguished upon the source providing the initialization data – data derived from SI model [8] and data collected from nearest speakers [1].

Another discussed solution solving the problem of small amount of adaptation data is based on the estimation of a subspace of the acoustic space, in which we try to estimate the parameters for speaker adaptation. Hence, the number of free parameters to be estimated decreases. Our aim is to restrict the form of the fMLLR transformation matrix to a weighed linear combination of some basis matrices. The method how to find these basis matrices is described in Section 4. Given appropriate basis matrices the fMLLR problem degrades to the problem of estimation of weights of the linear combination. If low amounts of adaptation data are available significantly smaller number of weights is chosen. Hence, only a small portion of the amount of former parameters have to be estimated.

Experiments, see Section 5, are mainly aimed to compare distinct choices of basis matrices in method of a linear combination of basis, and to explore their contribution to the adaptation with limited amount of data. For comparison the results of initialization methods are presented too.

2   Adaptation

The adaptation adjusts the SI model so that the probability of the adaptation data would be maximized. Instead of storing a huge amount of data to estimate the adaptation formulas, adaptation methods need only following statistics:

$$\gamma_{jm}(t) = \frac{\omega_{jm} p(o_t|jm)}{\sum_{m=1}^{M} \omega_{jm} p(o_t|jm)} \tag{1}$$

denoting the mth mixture component's posterior of the jth state of the HMM,

$$c_{jm} = \sum_{t=1}^{T} \gamma_{jm}(t), \tag{2}$$

representing the "soft" count of feature vectors aligned to mixture component m,

$$\varepsilon_{jm}(o) = \sum_{t=1}^{T} \gamma_{jm}(t)o_t , \quad \varepsilon_{jm}(oo^{\text{T}}) = \sum_{t=1}^{T} \gamma_{jm}(t)o_t o_t^{\text{T}} \tag{3}$$

denoting the first and the second moments of features aligned to mixture component m in the jth state of the HMM. Output probabilities of HMM states are given by Gaussian Mixture Models (GMMs) described by mean μ_{jm}, covariance matrix C_{jm} and weight ω_{jm} of each mixture component $m = 1, \ldots, M$.

2.1   Feature Maximum Likelihood Linear Regression (fMLLR)

fMLLR approach tries to find a linear transformation in order to increase the probability of adaptation data given an acoustics model. Contrary to other adaptation methods, fMLLR can adapt more model components at once using the same transformation (e.g. only one matrix for all the model means), thus they require lower amount of adaptation data since number of free parameters to be estimated is low. Similar model components are clustered into clusters $K_n, n = 1, \ldots, N$ in order to lower the number of adapted parameters. Advantage of fMLLR is that it transforms directly acoustics features instead of an acoustics model (this is the case for MLLR), what is less time-consuming. fMLLR transforms features o_t according to the formula

$$\bar{o}_t = A_{(n)} o_t + b_{(n)} = W_{(n)} \xi(t) , \tag{4}$$

where $W_{(n)} = [A_{(n)}, b_{(n)}]$ represents the adaptation transformation matrix corresponding to the cluster K_n, and $\xi(t) = [o_t^T, 1]^T$ stands for the extended feature vector. The estimation formulas for rows of $W_{(n)}$ are given as

$$w_{(n)i} = G_{(n)i}^{-1} \left(\frac{v_{(n)i}}{\alpha_{(n)}} + k_{(n)i} \right) , \tag{5}$$

where $v_{(n)i}$ is the ith row vector of cofactors of matrix $A_{(n)}$, $\alpha_{(n)}$ can be found as a solution of a quadratic function defined in [9],

$$k_{(n)i} = \sum_{m \in K_n} \frac{\mu_{mi} \varepsilon_m(\xi)}{\sigma_{mi}^2} , \quad G_{(n)i} = \sum_{m \in K_n} \frac{\varepsilon_m(\xi \xi^T)}{\sigma_{mi}^2} , \tag{6}$$

where $G_{(n)i}, k_{(n)i}$ are matrices of accumulated statistics (3) for all mixture components m contained in a given cluster $K_{(n)}$, and

$$\varepsilon_m(\xi) = \left[\varepsilon_m^T(o), c_m \right]^T , \quad \varepsilon_m(\xi \xi^T) = \begin{bmatrix} \varepsilon_m(oo^T) & \varepsilon_m(o) \\ \varepsilon_m^T(o) & c_m \end{bmatrix} . \tag{7}$$

Equation (5) is the solution of the minimization problem with auxiliary function [9]

$$Q(W_{(n)}) = \text{tr}(K_{(n)}^T W_{(n)}) - \sum_i w_i^T G_i w_i + \beta_{(n)} \log |\det A_{(n)}| , \tag{8}$$

where

$$\beta_{(n)} = \sum_{m \subset K_{(n)}} c_m . \tag{9}$$

Matrices $W_{(n)}$ are estimated iteratively, thus they have to be suitably initialized (if enough data are available the initialization can be random, otherwise see Section 3).

3   Initialization of the Adaptation Matrix

The matrices of accumulated statistics (6) are dense and have a lot of parameters to be estimated. One of the problem arises in cases when low amount of adaptation data is available. Such situations can lead to ill-conditioned transformation matrices W_n and to the degradation of system's performance. Therefore, it is suitable to initialize matrices (6) with proper values in order to increase the robustness of the estimation process. In following sections two principles of initialization are introduced – data derived from the SI model are used, or data from similar speakers from a development database are utilized. However, as have been already mentioned in Section 2 we do not need the data directly, we need only their statistics (zero, first and second moments). These (initialization) statistics are then added to the given speaker's test statistics.

3.1 Initialization Derived from SI Model

The idea proposed in [8] is to utilize data that mostly match the SI model (when none new adaptation data are available, the estimated transformation matrix should equal the identity matrix). For this purpose we can use directly the SI model parameters. Now the initialization of (6) takes the form

$$k_{(n)i} = \sum_{m \in K_{(n)}} p_m \frac{\mu_{mi}}{\sigma_{mi}^2} \begin{bmatrix} \mu_m \\ 1 \end{bmatrix}, \quad G_{(n)i} = \sum_{m \in K_{(n)}} p_m \frac{1}{\sigma_{mi}^2} \begin{bmatrix} \mu_m \mu_m^T + C_m & \mu_m^T \\ \mu_m & 1 \end{bmatrix}, \quad (10)$$

where μ_m, C_m are mean and covariance of a mixture component from cluster $K_{(n)}$, and p_m is a smoothing weight. Higher values of p_m indicate a higher influence of the initialization, the adaptation is less effective since the estimates are more restricted. In our case we set $p_m = \omega_m$ equal to the weight of a mixture component. In the initialization step given actual test speaker's statistics are added to these initialization statistics.

3.2 Sufficient Statistics from Closest Speakers

A subset of speakers from development set is selected, who are close in the acoustic space to the given test speaker whose speech is going to be recognized. Transformation matrix for the test speaker is then estimated using the already stored HMM data statistics of selected speakers [1]. The principle can be summarized to three steps:

– **Accumulation of sufficient statistics** – for each speaker s from the development set matrices $k_{(n)i}^s$ and $G_{(n)i}^s$ given in (6) are accumulated and stored off-line.
– **Selecting a cohort of speakers** – N-best speakers are selected from the development database according to their closeness to the test speaker. Test speaker's data are scored on-line against each of the development speakers' GMM. N-best speakers with highest verification scores are added to the cohort.
– **Summation of cohort statistic** – matrices of accumulated statistics (6) are initialized as a sum of all statistics from speakers in the cohort:

$$k_{(n)i} = \sum_{s=1}^{N} k_{(n)i}^s, \quad G_{(n)i} = \sum_{s=1}^{N} G_{(n)i}^s, \quad (11)$$

for each cluster n and each row i of $W_{(n)}$. At the end, statistics of the actual test speaker are added to these initialization statistics.

4   Linear Combination of Basis Matrices

Another way how to deal with the problem of insufficient amount of data when estimating transformation matrices containing many free parameters ($D = d \times (d + 1)$, where d is dimension of feature space) is to restrict the parameter space to a subspace [5]. The speaker dependent transformation matrix is given as a linear combination of the form

$$W = W_0 + \sum_{b=1}^{B} \alpha_b W_b, \tag{12}$$

where W_b are basis matrices, $W_0 = [I; 0]$, α_b are speaker-specific weights, and B in range $1 \leq B < D$ is the rank of the subspace. Basis matrices are estimated on the development database beforehand.

The only parameters estimated in the adaptation process are the weights α_b. These weights are found by maximizing the auxiliary function (8) w.r.t. these weights [6]. Only test speaker's statistics (6) and (9) are used in the estimation process. We use the method of Newton's gradient adopted from [6]. Just one coefficient has to be determined empirically during the adaptation, namely the number of basis matrices B. We determine B with respect to the amount of adaptation data.

A naive approach for finding basis matrices is to use directly transformation matrices of all speakers from the development database. The speakers are clustered, and for each cluster a transformation matrix is trained from data of all clustered speakers. Obtained matrices (one matrix for each cluster) are than used as the basis matrices. Since no prior information on the significance of individual matrices is available, the data have to be clustered into the different number of final clusters for each number B. Loosely speaking, having K matrices obtained from K clusters, and if $B < K$ one can not decide which of the K matrices to take as the basis matrices. In following subsections more sophisticated methods how to find the basis matrices will be presented.

4.1   Eigen Decomposition (ED)

Eigen decomposition was introduced in [2]. It finds eigenvectors of a covariance matrix constructed from all supervectors $w = \text{vec}(W)$ of T transformation matrices W from development speakers, where the operator vec concatenates the rows of a matrix so that a high dimensional vector – supervector – is formed. The dimension of the supervector is D, usually $T \ll D$. In the speaker recognition these eigenvectors are called eigenvoices. The term eigenvoice was originally related to supervectors based on GMM means, however we will use this term also when adaptation matrices W are utilized as supervectors since these matrices do carry speaker dependent information.

In order to compute the eigen decomposition of the supervector covariance matrix $Z^T Z$ is utilized (see e.g. [7]), where Z is the $T \times D$ data matrix containing supervectors stored in its rows normalized to have zero mean. Only B eigenvectors related to highest eigenvalues are used in the linear combination (12). These chosen eigenvectors define directions with most variability of input data.

4.2 Factor Analysis (FA)

FA is an alternative to ED. FA [3] is a statistical method used to describe feature vectors from an input set by low B dimensional latent representations. It is based also on the analysis of the variability in the input space. The components of the latent representation/vectors are called factors. The main difference from EV is that FA is represented as a generative model that involves also a noise component ε. Given the data matrix Z described in previous subsection, the FA model is given as

$$Z = LF + \varepsilon, \tag{13}$$

where $\varepsilon \sim \mathcal{N}(0, \Psi)$ is the noise term with diagonal covariance matrix

$$\Psi = \mathrm{Cov}(\varepsilon) = \mathrm{Diag}(\psi_1, \ldots, \psi_K), \tag{14}$$

L is of size $D \times B$ and is called the factor loading matrix, and F of size $B \times T$ is the matrix of factors. Note that each column f_n of F is the latent representation of the supervector w_n stored as a column of Z. Additionally, f_n is assumed to have standard normal distribution with zero mean and unit covariance matrix.

In this scenario the basis matrices W_b in (12) are represented by the columns of the factor loading matrix L. Note that an individual matrix L have to be computed for distinct numbers B of basis matrices. One cannot take the first B columns of the L matrix since the columns in L are in any way sorted according to their "relevance" (no eigenvalue counterparts in FA). However, these matrices are computed off-line.

4.3 Independent Component Analysis (ICA)

ICA [4] is a method of blind source separation. The ICA model is given as

$$Z = AS, \tag{15}$$

where $S = [s_1, \ldots, s_T]$ is the matrix of independent components. The crucial assumption is that s_t are independent and that their distribution is *not* Gaussian. The Gaussionality involved in previous methods does not allow to discover the rotation of the latent space [12]. Hence, ICA is less restricted than EV or FA. Again, for each number B of basis vectors individual matrix A have to be estimated off-line.

4.4 Maximum Likelihood (ML) Estimation

In [6] authors have shown, involving several approximations, the connection between Maximum Likelihood (ML) estimation and eigen decomposition. Taking a second-order Taylor expansion of the auxiliary function (8) for $w = \mathrm{vec}(W)$, taken around $w = w_0$ we get

$$Q^s(w) = (\Delta w)^T p^s - \frac{1}{2}(\Delta w)^T H^s (\Delta w), \tag{16}$$

where $\Delta w = w - w_0$, and p^s, H^s are computed from matrices of accumulated statistics k_i^s, G_i^s and β^s for speaker s. It can be shown [6] that when limiting w to the form of a linear combination (12), the objective (16) is maximized when the basis vectors $w_b = \mathrm{vec}(W_b)$ are taken to be the eigenvectors of $M = \sum_s \frac{1}{\beta^s} p^s p^{sT}$.

5 Experiments

5.1 SpeechDat-East (SD-E) Corpus

For experiment purposes we used the Czech part of SpeechDat-East corpus (see [11]). In order to extract the features Mel-frequency cepstral coefficients (MFCCs) were utilized, 11 dimensional feature vectors were extracted each 10 ms utilizing a 32 ms hamming window, Cepstral Mean Normalization (CMN) was applied, and Δ, Δ^2 coefficients were added. A 3 state HMM based on triphones with 2,105 states total and 8 GMM mixture components with diagonal covariances in each of the states was trained on 700 speakers with 50 sentences for each speaker (cca 5 sec. on a sentence). Using the same data UBM containing 256 mixture components was trained, and subsequently all the GMMs of individual development speakers were MAP adapted. To test the systems performance different 200 speakers from SD-E were used with 50 sentences for each speaker, however a maximum of 12 sentences was used for the adaptation. A language model based on trigrams used in the recognition [10]. The vocabulary consisted of 7,000 words.

5.2 Adaptation Setup

In our experiments we utilized fMLLR adaptation. Before the own adaptation statistics (6) for all development speakers were precomputed (see Section 3.2), also supervectors $w^s = \text{vec}(W^s)$ based on fMLLR matrices were extracted from each speaker from the development set (see Section 4). In the case of the basis method only one (global transformation) fMLLR matrix was used (all the mixture components share the same cluster), but when sufficient statistics from closest speaker are used for the initialization we decided to utilize multiple clusters. The clustering of model components (GMM means) was performed via a regression tree [13]. The occupation threshold for nodes in the regression tree was set to 1,000. Only one iteration of fMLLR was carried out.

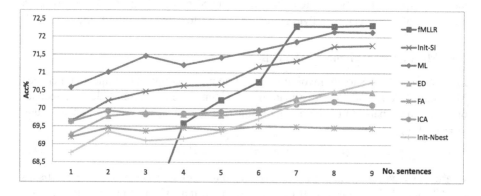

Fig. 1. Accuracy (Acc)[%] of recognition utilizing described methods in dependence on the number of adaptation sentences

5.3 Results

All methods were tested on varying number of adaptation sentences. The graph in Figure 1 depicts results of ASR with fMLLR adaptation. As expected, basic fMLLR (fMLLR without any initialization) performs poor on small amount of adaptation data. Individual initializations are denoted as Init-SI (init. derived from SI), Init-Nbest (init. based on sufficient statistics from N-best speakers), and techniques based on linear combination of basis are denoted according to the method used, namely: ED, FA, ICA and ML approach.

6 Conclusion

The presented experiments proved the urgency of additional solution for fMLLR adaptation in the task of extremely small data sets. All methods proposed in this work show the improvement in comparison with the fMLLR approach. The best results can be obtained by a linear combination of basis found by maximum likelihood estimation.

References

1. Zajíc, Z., Machlica, L., Müller, L.: Initialization of fMLLR with Sufficient Statistics from Similar Speakers. In: Habernal, I., Matoušek, V. (eds.) TSD 2011. LNCS, vol. 6836, pp. 187–194. Springer, Heidelberg (2011)
2. Westwood, R.: Speaker Adaptation Using Eigenvoices. Ph.D. thesis on Cambridge University Engineering Department (1999)
3. Gorsuch, R.L.: Factor Analysis, 2nd edn. Psychology Press (1983)
4. Comon, P., Jutten, C.: Handbook of Blind Source Separation: Independent Component Analysis and Applications. Elsevier (2010)
5. Visweswariah, K., Goel, V., Gopinath, R.: Maximum Likelihood Training of Bases for Rapid Adaptation (unpublished manuscript)
6. Povey, D., Yao, K.: A Basis Representation of Constrained MLLR Transforms for Robust Adaptation. In: Computer Speech & Language, pp. 35–51 (2012)
7. Chen, K., Liau, W., Wang, H., Lee, L.: Fast speaker adaptation using eigenspace-based maximum likelihood linear regression. In: International Conference on Spoken Language Processing, Beijing, China, pp. 742–745 (2000)
8. Li, Y., et al.: Incremental on-line feature space MLLR adaptation for telephony speech recognition. In: International Conference on Spoken Language Processing, Denver, pp. 1417–1420 (2002)
9. Povey, D., Saon, G.: Feature and model space speaker adaptation with full covariance Gaussians. In: Interspeech, pp. 1145–1148 (2006)
10. Pražák, A., Psutka, J.V., Hoidekr, J., Kanis, J., Müller, L., Psutka, J.: Automatic Online Subtitling of the Czech Parliament Meetings. In: Sojka, P., Kopeček, I., Pala, K., et al. (eds.) TSD 2006. LNCS (LNAI), vol. 4188, pp. 501–508. Springer, Heidelberg (2006)
11. Pollak, P., et al.: SpeechDat(E) – Eastern European Telephone Speech Databases, XLDB – Very Large Telephone Speech Databases (ELRA), Paris (2000)
12. Tipping, M.E., Bishop, C.M.: Mixtures of Probabilistic Principal Component Analysers. Neural Computation 11, 443–482 (1999)
13. Gales, M.J.F.: The generation and use of regression class trees for MLLR adaptation. Cambridge University Engineering Department (1996)

Language Modeling of Nonverbal Vocalizations
in Spontaneous Speech

Dmytro Prylipko[1], Bogdan Vlasenko[1], Andreas Stolcke[2], and Andreas Wendemuth[1]

[1] Cognitive Systems, Otto-von-Guericke University, 39016 Magdeburg, Germany
{dmytro.prylipko,bogdan.vlasenko,andreas.wendemuth}@ovgu.de
[2] Conversational Systems Lab, Microsoft, Mountain View, CA, USA
andreas.stolcke@microsoft.com

Abstract. Nonverbal vocalizations are one of the characteristics of spontaneous speech distinguishing it from written text. These phenomena are sometimes regarded as a problem in language and acoustic modeling. However, vocalizations such as filled pauses enhance language models at the local level and serve some additional functions (marking linguistic boundaries, signaling hesitation). In this paper we investigate a wider range of nonverbals and investigate their potential for language modeling of conversational speech, and compare different modeling approaches. We find that all nonverbal sounds, with the exception of breath, have little effect on the overall results. Due to its specific nature, as well as its frequency in the data, modeling of breath as a regular language model event leads to a substantial improvement in both perplexity and speech recognition accuracy.

1 Introduction

The conversational speech we produce every day is remarkably different from the types of speech for which language technology is successfully applied [1]. Campbell argues that spontaneous speech encodes two distinct streams of information, linguistic and interpersonal [2]. While the former carries the verbal meaning of a message, the latter accompanies it and enriches our speech with paralinguistic messages and cues. They coexist and merge into a common information stream which is called human spoken speech. For recognition and synthesis of spontaneous speech, special attention must be paid to conversational speech phenomena, both in acoustic and language modeling.

An example of such phenomena are nonverbal and nonspeech sounds. Nonverbals produced with human vocal apparatus can provide valuable paralinguistic cues, but do not have any linguistic meaning (filled pauses, clicks and lip smacks, cough, laughter). They occur frequently in spontaneous speech and have been investigated since 1990s.

Schultz and Rogina in [3] took into consideration different human and non-human noises present in transcriptions and found that modeling them as regular words improves recognition performance compared with treating them as silences, i.e. non-linguistic events expose some regularities well-captured by standard N-grams.

Further research has been devoted mostly to the use of filled pauses and linguistic *disfluencies* (DFs) such as false starts, repetitions, or repairs. Stolcke and Shriberg introduced a language model (LM) that explicitly models these phenomena [4]. That work confirmed that DFs have a systematic and nonrandom nature that can be captured

P. Sojka et al. (Eds.): TSD 2012, LNCS 7499, pp. 488–495, 2012.

Table 1. Distribution and type of the nonverbals in the Verbmobil corpus

Filled pauses	Breath	Human noise	Laugh	Throat clean	Swallow	Lip smack	**Total**
8 464	21 149	1 710	231	93	228	5 844	**37 719**

by LMs and contain information for predicting following words. They also argue that one function of filled pauses is marking the beginning of the linguistic segment. Further research confirmed importance of keeping filled pauses in the LM conditioning context and their role as linguistic boundaries [5], [6]. These articles address two of the four fundamental challenges outlined by Shriberg in [7], namely, recovering hidden punctuation (such as sentence and phrase boundaries) and coping with disfluencies.

The results with filled pauses show us the potential of using spontaneous speech characteristics (usually regarded as a source of problems) for modeling more natural human speech. However, among all nonverbal sounds, only the role of filled pauses has been investigated extensively for LM purposes.

Modern transcriptions of spontaneous speech corpora contain a broad range of different nonverbal and nonspeech markers such as breath, laughter, cough, lip smack, verbal and nonverbal noises, throat cleaning etc. But it is not obvious whether it is worth modeling them as LM events or would be better to eliminate them from the model. Furthermore, state-of-the-art language modeling tools provide a wide range of modeling techniques, some specially designed for speech disfluencies. In this paper we investigate these techniques applied to several classes of nonverbal sounds, in order to determine which nonverbals have the potential to improve language models of conversational speech.

2   Speech Corpus Overview

We performed our experiments on the Verbmobil corpus (German part) [8]. This corpus represents the data collected within the Verbmobil project, which had as its task domain the negotiation of meetings and trip planning between two speakers.

The corpus consists of 1,658 spontaneous dialogs with 13,890 turns produced by 655 speakers. The text data comprises about 350K words, with a vocabulary of 6,680 unique words. The total duration of the recorded speech is 33:51:42 h.

Three kinds of filled pauses (*äh*, *ähm*, *hm*) and hesitation were regarded as the same event (filled pause) representing the same linguistic function. For detailed information about the nonverbals present in the corpus see Table 1.

In this paper we consider only nonverbal events; nonspeech sounds like paper rustle or squeak were eliminated from the textual data. All experiments were conducted using 10-fold speaker independent cross-validation.

3   Approach to Modeling of Nonverbals within Conversational Speech

As the baseline model we utilized a backoff trigram trained and tested using the SRILM toolkit [9]. In this study we evaluated four modeling approaches:

1. **Clean model:** Nonverbals are excluded both from data and language model.

2. **Regular model:** Nonverbals are modeled as regular words. Inclusion of nonverbal markers into the test set causes an issue: increased number of tokens to be predicted makes the direct perplexity comparison with the previous model inconsistent. In order to overcome this issue we calculated two values for this kind of model: 'all tokens' and 'verbal only'. The former was calculated as usual on the full test set. For the latter we took into account only probabilities of the verbal tokens, thereby we estimate this value on the same amount of data as the perplexity value obtained with the 'clean' model.

3. **Omitted event model:** Nonverbals are modeled, but omitted from context. In this model we assume that the fluent context is a better predictor for the following word, thus the nonverbal token is excluded from the N-gram context and the preceding word is included instead. For instance, for the sentence with medial filled pause *ähm*:

```
Ja wunderbar <ähm> wiederhören
```

the following trigrams are generated:

```
Ja wunderbar <ähm>       Ja wunderbar wiederhören
```

but not the following trigram:

```
wunderbar <ähm> wiederhören
```

The nonverbal itself is considered as a usual word conditioned on the corresponding cleaned-up context. This is the 'cleanup model' in [4]. When evaluated only on the verbal tokens, this model becomes equivalent to the 'clean' model; therefore, only results including nonverbals are presented.

4. **Hidden event model:** modeling of nonverbals as hidden events. In this case speech is considered as a word stream with probabilistic events between words, some of which are hidden from direct observation, but are included in predicting context. The language model is trained on data containing the nonverbal tokens, but tested on data containing only verbal tokens. This approach is used for modeling of disfluencies and sentence or topic boundaries.

4   Experiments and Results

Our experiments show that including nonverbal tokens lowers the perplexity of the full test data (see Table 2). At the same time, including nonverbals into the model as regular words increases the perplexity of the verbal tokens. However, experiments including individual nonverbal types let us see that their effects differ. The overall gain is mostly due to the low local perplexity of the most frequent nonverbals, namely filled pauses, breath, and lip smack. For these event types the perplexity of the verbal tokens is higher than for the full test data. The difference is proportional to the frequency of the nonverbal type.

Table 2. Perplexity of the test data including different nonverbals. Highlighted values show an improvement compared with the baseline model with removed nonverbals.

Configuration	Perplexity			
Excluding all	66.85			
Including:	as regular word		as omitted event	as hidden event
	all tokens	verbal only		
– filled pauses	67.20	68.28	67.28	**66.27**
– breath	**62.59**	70.15	**63.22**	**65.88**
– laughter	67.02	66.86	67.04	**66.83**
– verbal noise	67.82	67.36	67.58	**66.74**
– throat clean	66.97	66.87	66.96	66.85
– swallow	67.08	66.89	67.06	**66.84**
– lip smack	67.07	67.93	67.06	**66.54**
Including all	**63.80**	73.88	**65.01**	**66.65**

As seen in Table 2, including any event type as a word into the language model actually slightly increases the perplexity, except for breath, which has a remarkable influence on the overall result. Breath is unlike the other nonverbals we consider here. Including it into the model as a regular word gives a major reduction in perplexity, while modeling it as an unobserved item reduces the beneficial effect.

The special status of breath events can be explained by their nature: breathing tends to occur at the end of full phrases within long sentences. The following sentence illustrates this fact:

Ja und dann bin ich im Dezember ab siebzehnten weg <BREATH> bis Silvester dann. (*Yeah so I will be away in December from the 17th <BREATH> see you on the New Year's Eve then.*)

We also found that breaths often occur before or next to the other nonverbals, especially filled pauses or lip smacks. Other modeling approaches for breath are the hidden event model or omitting them from context. However, experimental results show that the lowest perplexity is obtained with breath as a regular event within N-grams (see Tables 2, 3).

Table 3. Local perplexity at breath markers

Model	Breath		Breath+1		Breath+2	
	bigram	trigram	bigram	trigram	bigram	trigram
Regular	13.84	14.04	131.54	110.70	59.66	51.15
Omitted event	15.70	16.28	148.74	150.72	59.44	54.49
Hidden event	–	–	172.43	170.76	56.67	49.62
Clean	–	–	178.87	183.18	57.03	52.46

Filled pauses (FP) are of interest due to their importance for natural language understanding and their role as markers of linguistic and prosodic boundaries. Results presented in Table 4 confirm that FPs are better predictors for the following words and omitting them from context makes local perplexity significantly worse. At the same time, including filled pauses as words in the model does not improve the overall perplexity, but actually increases it (cf. Table 2).

Table 4. Local perplexity at filled pause positions

Model	FP		FP+1		FP+2	
	bigram	trigram	bigram	trigram	bigram	trigram
Regular	35.21	36.98	299.87	280.37	84.11	76.69
Omitted event	36.68	38.80	355.37	370.07	83.26	78.92
Hidden event	–	–	339.38	351.38	81.63	75.41
Clean	–	–	353.26	369.42	81.72	77.56

Previous research has shown that cutting context before medial filled pauses (using bigrams instead of trigrams) can improve language models on acoustically segmented sentences [5]. This can be explained by the role of medial FPs as linguistic segment boundaries, thus previous context often belongs to another phrase. However, our experiments on Verbmobil show that regular trigrams still provide lower local perplexity after FPs than regular bigrams. In order to check this we conducted an additional experiment just with those filled pauses which appear in the middle of sentences. The results are presented in Table 5.

Table 5. Local perplexity at medial filled pauses

Model	FP-M		FP-M+1		FP-M+2	
	bigram	trigram	bigram	trigram	bigram	trigram
Regular	41.70	44.16	337.43	319.59	88.41	80.35
Omitted event	43.19	46.08	394.28	413.20	88.39	84.98
Hidden event	–	–	380.44	392.36	86.51	80.74
Clean	–	–	393.98	414.48	86.75	83.46

As seen in Table 5, local perplexity after medial FPs in general corresponds to the values for all FPs (Table 4). An interesting result is that bigrams provide lower perplexity for the filled pauses themselves. Also, when FPs are removed from context (in omit and 'clean' models), bigrams predict following words slightly better than trigrams.

Experiments with Speech Recognition. A certain correlation exists between language model perplexity and word error rate (WER) [10]. However, even substantial reductions in perplexity do not always provide a marked benefit in word accuracy. To check to what extent breath markers can benefit recognition performance, we performed several experiments with speech recognition. The LM setup for speech recognition was slightly different from that for perplexity evaluation. We included into the model only those nonverbals which are most relevant for natural language processing and understanding, namely, filled pauses and laughter. Other nonverbal and nonspeech sounds were modeled acoustically, but with no output into the final recognition hypothesis ('invisible' event), so as to avoid insertions caused by misrecognition of noises as short words. In order to evaluate the role of breath specifically as a linguistic boundary marker, we removed initial and final breath markers in both hypotheses and references (i.e., they did not affect word accuracy).

Speech recognition was performed using the HTK toolkit [11]. For acoustic modeling we employed three-state left-to-right hidden Markov models/Gaussian mixture models (HMM/GMMs) with 32 mixtures. Sixteen nonverbal and nonspeech sounds were modeled with nine-state GMMs due to their longer average duration. In contrast to language models, we created separate acoustic models for three types of filled pauses (*äh*, *ähm*, *hm*) and hesitation to account for the different acoustic realizations. The feature set consisted of the typical 13 Mel-frequency cepstral coefficients (MFCC) including the 0th coefficient (energy) and delta and acceleration coefficients. Features were extracted from frames of 25 ms length sampled every 10 ms.

Table 6. Perplexity and word error rate for different language models

Model	Perplexity	WER	Nonverbal WER
Trigram excluding breath	67.38	30.21	47.89 †
Breath as hidden event	66.39	30.17	45.22 †
Breath as omitted event	65.54	29.03	42.23
Breath as regular word	64.16	28.97	42.43
Breath as word, eliminated	–	27.72	30.31 †

† *For these setups breath markers were excluded from both output and reference labels, thus the values were obtained just on two nonverbal tokens: filled pauses and laugh.*

Results presented in Table 6 show a 4.3% relative reduction of the word error rate in comparison to the setup where breath is modeled only as an acoustic event. Since breath markers convey little linguistic meaning, they can be removed from the output using post-processing. In this case the relative WER improvement increases to 8.2%, in spite of the higher perplexity of the verbal tokens alone for this language model type (see Table 2). Modeling breath as a hidden event allows us to apply the model on unmarked data, possibly from other corpora. In this case we also see a small improvement in both perplexity and word accuracy.

Analysis of the recognition rates of nonverbals shows high insertion rates for those nonverbals. Moreover, breath is inserted much more often than laughter or filled pauses. For example, when modeling breath as a regular word, relative to a total token count of 24,284, 21,203 are recognized correctly, and 7,223 are inserted incorrectly. When breath is excluded from the model, the proportions are roughly the same: 8,436 tokens in total, 6,797 are recognized correctly, with 2,401 insertions. However, when modeling breath as a regular word with eventual removal from the hypothesis, the number of insertions falls dramatically: 6,721 correctly recognized, with only 842 insertions. We surmise that breath is often substituted for short noises and verbal sounds, which in other setups would be recognized as filled pauses. Thus, breath can 'eat up' a substantial number of false insertions of other nonverbals; eliminating it from the output thus removes all these errors.

5   Conclusions and Future Work

Previous research showed that nonverbal sounds have rather strong local effect and do not influence the overall perplexity and speech performance result much [4,5,6]. Our experiments show that despite an evident gain in local predictions, increased perplexity of the verbal content almost nullifies this advantage or even leads to higher total perplexity. There is also a remarkable difference between the perplexities of the verbal tokens alone and the full test data. The effect correlates with the frequency of a certain nonverbal in the corpus. We can assume there is a threshold (about 1.5% for Verbmobil) after which perplexity of the full test data improves compared to the value obtained on the verbal part alone. If the frequency of a nonverbal type relative to all tokens is greater than 4–5%, including it in the model as a regular word may lead to a substantial improvement in both perplexity and word error rate. For the Verbmobil corpus, breath is an example of such a nonverbal type.

Other than benefiting perplexity and accuracy, including nonverbal tokens into the language model can enrich transcriptions with paralinguistic information, which may be important for natural speech processing and understanding. The most significant events for this purpose are laughter and filled pauses. Our experiments on the Verbmobil database show that modeling laughter and filled pauses as regular words is preferable compared to omitting these events from context or modeling them as hidden events.

For future work, we would like to employ the language modeling techniques considered in this paper together with detection methods based on the acoustic signal (e.g., as presented in [12]). This approach could further improve spontaneous speech recognition using some of the same phenomena that have been considered troublesome in the past.

Acknowledgments. The authors acknowledge the support provided by the federal state Sachsen-Anhalt with the Graduiertenförderung (LGFG scholarship). We also acknowledge the German Research Foundation (DFG) for financing our computing cluster used for parts of this work.

References

1. Ostendorf, M., Shriberg, E., Stolcke, A.: Human language technology: Opportunities and challenges. In: Proceedings of ICASSP, vol. 5, pp. 949–952. IEEE (2005)
2. Campbell, N.: On the Use of NonVerbal Speech Sounds in Human Communication. In: Esposito, A., Faundez-Zanuy, M., Keller, E., Marinaro, M. (eds.) Verbal and Nonverbal Commun. Behaviours. LNCS (LNAI), vol. 4775, pp. 117–128. Springer, Heidelberg (2007)
3. Schultz, T., Rogina, I.: Acoustic and language modeling of human and nonhuman noises for human-to-human spontaneous speech recognition. In: Proceedings of ICASSP, vol. 1, pp. 293–296. IEEE, Detroit (1995)

4. Stolcke, A., Shriberg, E.: Statistical language modeling for speech disfluencies. In: Proceedings of ICASSP, Atlanta, GA, pp. 405–408 (1996)
5. Siu, M., Ostendorf, M.: Modeling disfluencies in conversational speech. In: Proceedings of ICSLP, vol. 1, pp. 386–389. IEEE, Philadelphia (1996)
6. Siu, M., Ostendorf, M.: Variable N-grams and extensions for conversational speech language modeling. IEEE Transactions on Speech and Audio Processing 8, 63–75 (2000)
7. Shriberg, E.: Spontaneous speech: How people really talk and why engineers should care. In: Proceedings of EuroSpeech, pp. 1781–1784 (2005)
8. Burger, S., Weilhammer, K., Schiel, F., Tillmann, H.G.: Verbmobil Data Collection and Annotation. In: Verbmobil: Foundations of Speech-to-Speech Translation, pp. 537–549. Springer (2000)
9. Stolcke, A.: SRILM – an extensible language modeling toolkit. In: Proceedings of ICSLP, vol. 2, pp. 901–904 (2002)
10. Dietrich, K., Peters, J.: Testing the correlation of word error rate and perplexity. Speech Communication 38, 19–28 (2002)
11. Young, S., Evermann, G., Gales, M., Hain, T., Kershaw, D., Liu, X., Moore, G., Odell, J., Ollason, D., Povey, D., Valtchev, V., Woodland, P.: The HTK book (for HTK Version 3.4). Cambridge University Press, Cambridge (2000)
12. Fukuda, T., Ichikawa, O., Nishimura, M.: Breath-detection-based Telephony Speech Phrasing. In: Proceedings of INTERSPEECH, pp. 2625–2628 (2011)

Acoustic Segmentation Using Group Delay Functions and Its Relevance to Spoken Keyword Spotting

Srikanth R. Madikeri and Hema A. Murthy

Indian Institute of Technology Madras, Chennai 600036, India

Abstract. In this paper, a new approach to keyword spotting is presented that uses event based signal processing to obtain approximate locations of sub-word units. A segmentation algorithm based on group delay functions is used to determine the boundaries. units. These sub-word units are used as individual inputs to an unconstrained endpoints dynamic time warping-based (UE-DTW) template matching algorithm. Appropriate score normalisation is performed using scores of background words. The technique is tested using MFCC and Modified Group Delay features. Performance gains of 13.7% (relative) improvement over the baseline for clean speech is observed. Further, for noisy speech, the degradation is graceful.

1 Introduction

Keyword spotting is the task of finding the occurrences of a given keyword K, if any, in a speech utterance $\mathbf{O}$. Keyword spotting is required for audio indexing of large databases with lecture materials or broadcast news, behavior modelling of speakers, etc. In this respect, several key approaches have been explored. The use of continuous density HMMs (Hidden Markov Models) for modelling garbage/filler and keyword models [1] have gained prominence [2,3] in the literature. These approaches, however, require large amounts of data in order to obtain good estimates. Out Of Vocabulary (OOV) words, however, lead to poor performance of the system. Alternate approaches have also been considered. In [4], a segmental DTW (Dynamic Time Warping) algorithm [5] is used to detect keywords in an unsupervised fashion. A large-margin based discriminative training technique [6] has proven to be effective for this purpose as well.

The core part of a keyword spotting relies on segmenting the input speech utterance. In the HMM-based technique utilising keyword and garbage models, Viterbi decoding is applied based on the estimated models. In ASR (Automatic Speech Recognition)-based techniques, language models are used while decoding the utterance [7]. These methods require a large amount of data for training the HMM models. In this paper, a data-independent, group-delay based algorithm is used to segment the speech utterance. The segmentation algorithm gives acoustic boundaries whose resolution can be varied from phone to word level. These acoustic boundaries can be used to match with one or multiple spoken references of a keyword. Segments obtained are used as test sequences to a *unconstrained end-points* DTW (UE-DTW) algorithm with the spoken keyword as a reference. This method requires very little data and is also vocabulary

P. Sojka et al. (Eds.): TSD 2012, LNCS 7499, pp. 496–504, 2012.

independent. Before aligning the reference with test segments, boundaries are expanded to accomodate for the errors owing to hard boundaries produced by segmentation. To provide robustness to the system, a novel score normalization technique is developed.

The rest of the paper is organized as follows: Section 2 describes the method of automatic segmentation for continuous speech using group delay functions. In Section 3, the UE-DTW alignment algorithm used for keyword spotting is described. Section 4, describes the experiments conducted showing the accuracy of the technique.

2   Group-Delay Based Segmentation of Continuous Speech

The importance of phase-based features for speech recognition has already been discussed in detail in [8]. Further developments ([9]) have also re-iterated the importance of phase in processing of speech.

To segment a continuous speech utterance into acoustic units, the group-delay function is used as follows: a smoothed short term energy (STE) contour is used to locate syllable centres using the peaks in the STE function. The STE is symmetrized to resemble magnitude spectrum, where the valleys of the STE function characterize boundaries. Zero crossings in the group delay function of the corresponding minimum phase signal corresponds to the boundaries. This technique is delved with in detail in [10].

The resolution of the boundaries to fit to sub-word units is adjusted using the following expression

$$N_c = L_f / W \tag{1}$$

where N_c is the cepstral lifter parameter for group delay function, L_f is STE sequence length and W is called the *window scale factor*. N_c controls the segment boundary resolution. Since, L_f is fixed, W acts as liftering parmeter changing the resolution at which group delay segments boundaries. Typically, a low W produces phone-level segments while higher values increase the semantic context of the acoustic units produced. Segment boundaries correspond to events in speech.

2.1   Advantages

There are many advantages to using event detection algorithms for speech segmentation: first, any technique on the acoustical properties of the speech signal does not require prior knowledge. Second, the algorithms based on signal characteristics are fast. In the proposed technique, the computational complexity is $O(N_c \log N_c)$ – the time required to compute FFT and a hidden linear cost to search for zero-crossings. Whereas, decoding in HMM framework takes quadratic time. Also, in the group-delay based segmentation algorithm, the segment boundaries can be adjusted to different degrees of resolution by adapting W.

In the context of keyword spotting there are further benefits in using this technique. For example, number of searches for word boundaries that need to be performed in the case of template matching algorithms is approximately the number of words in the test utterance. This is a significant advantage compared to keyword-filler models based system in the context where the decoding time is quadratic in the number of models.

Another significant advantage is the seamless operation on OOV words. An instance of the keyword to be matched is adequate to spot a keyword. The method is also, ideally, language and text independent.

2.2 Handling Misalignments

Although the segmentation algorithm can resolve boundaries quite accurately, there can be misalignments. Further, even if the boundaries are approximately correct, they are hard. Boundaries produced by the segmentation algorithm are, therefore, expanded by a few frames on either sides. Spurious boundaries would contain more acoustic information. A leeway of 0.1s is used in the experiments. To handle missed boundaries and spurious ones in the context of keyword spotting we use a soft-ended dynamic time warping algorithm. This is discussed in the next section.

3 Template Matching with DTW

In the context of speech, dynamic time warping was initially used for isolated word recognition [11]. Its inability to perform well on longer segments and ability of HMMs to do better in such cases have led to the widespread use of HMMs for continuous speech recognition. Moreover, in the HMM-framework, accurate speech segmentation is not required as the goal is primarily transcription of the entire utterance rather than the segmentation of the utterance into words.

Each segment is passed through a template matching algorithm. DTW based algorithm is used to align each reference keyword K with $\mathbf{O}^{(K)}$. Keywords are detected if the distance satisfies a threshold. The hypothesis being that a segment is rejected when the distance is greater than the threshold.

3.1 Unconstrained End Points DTW

Keywords when present in the expanded test segments could be only a part of it. Thus an *unconstrained end-point* DTW is used for template matching with references. Here, only a subsequence of the test segment is matched with the reference by removing the end point constraints from the conventional DTW technique. The resulting distance measure is asymmetric (this is admissible in our case).

Once the constraints are chosen, the UE-DTW distance between the keyword $\mathbf{O}^K$ and a segment $\mathbf{O}^P$ in the test sequence is computed as

$$DTW(\mathbf{O}^K, \mathbf{O}^P) = \arg \min_{1 \le a \le b \le t_2} DTW(\mathbf{O}^K, \mathbf{O}^P_{a,b}) \qquad (2)$$

where $\mathbf{O}^P_{a,b}$ is the subsequence of $\mathbf{O}^P$ from index a to b.

A sample DTW-match is shown in Fig. 1. It can be seen that the entire reference is matched with only a subsequence of the test segment.

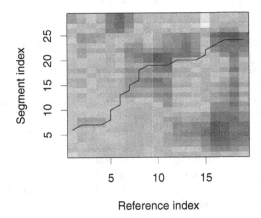

Fig. 1. Depicted above is the behavior of UE-DTW alogrithm for a reference against a segment in test utterance. The greener regions correspond to lesser distance. Note: In this example, the segment does not contain the keyword.

4   Experiments

In Figure 1 the regions that are dark green correspond to a smaller distance. In this example, the segment does not contain the keyword. Independent experiments are conducted on TIMIT ([12] and NTIMIT corpora [13]). The choice of databases is justified as follows: TIMIT is clean speech and NTIMIT is noisy. Performance on TIMIT is like a control experiment to evaluate the benefits of using this approach vis-a-vis state-of-the-art system. The performance on NTIMIT would demonstrate the robustness of this technique. For DTW alignment, DTW library [14] for R statistical package [15] is used.

The keywords for the experiments are chosen such that there are at least 3 test cases containing each keyword in both test sets. 40 keywords are chosen. These are given in Table 1. At most one keyword is present in the test cases. That is, test utterances that do not contain any keyword are also considered. The gender information of reference or test cases are ignored. There are a total of 259 test cases that contain only one keyword. There are 80 randomly chosen test cases that do not contain any keyword. There are a total of 565 words in all of them put together generating 28,250 test cases.

For acoustic segmentation, 25 ms frames are used with 15 ms shifts to generate the STE. Silence regions are removed and each of the remaining segments are processed separately to get segment boundaries. W is set to 3.

For reference matching, two different features are considered – MFCC and Modfied Group Delay (ModGD) [16]. For MFCC, 12 cepstral coefficents are extracted with a window size of 25 ms at a rate of 18 ms. Velocity and acceleration parameters are appended. For ModGD, $\gamma = 0.9$, $\alpha = 0.4$ are used [16]. Velocity parameters are appended.

Table 1. List of keywords used with total no. of references

age(2)	autumn(7)	bright(3)	bring(3)
broken(7)	charge(8)	development(8)	destroy(9)
done(3)	feel(4)	felt(9)	forces(10)
form(3)	garbage(8)	gets(7)	gives(9)
history(7)	intelligence(7)	itself(4)	leaves(8)
lines(10)	meet(8)	meeting(3)	necessary(4)
nice(7)	nobody(4)	organizations(7)	others(8)
paper(8)	perhaps(2)	played(3)	project(8)
provides(3)	quick(8)	quite(9)	rare (9)
related(4)	seemed(7)	sense(4)	something(3)
street(2)	subway(7)	thing(9)	thus(2)
tomorrow(3)	upon(3)	woman(10)	working(3)
yes(3)	yourself(3)		

4.1 Baseline System

The baseline system uses the HMM-based Keyword/Filler model approach to keyword spotting. The keyword models and filler models form their own respective network of HMMs so that a decoded keyword may or may not be a keyword. The triphone models themselves are trained from WSJ0 database's training set (SI_TR_S). It should be noted that the training data set used here is comparably large compared to that used in the proposed system. The triphone HMMs have 5 states with 16 mixtures for the GMM for each state. Only MFCC (with velocity and accelaration parameters) features are used in the baseline system. The HTK toolkit ([17]) is used to build these models and for Viterbi decoding during testing. The filler models consist of approximately 100 frequently occuring words (excluding the keywords) in the training set of the TIMIT database.

4.2 Score Normalization

The length-normalized distance scores obtained for each segment in the test utterance have considerable variability contributing to too many false alarms and misses. To suppress the variabilities in the scores, a normalization technique inspired by [18] is developed. Scores are normalized as follows: a random set of sentences that do not belong to any of the test sentences are chosen. Words from these sentences are extracted based on labeling information already available. These words form a *background set* for every keyword being spotted. Each keyword in the background set is aligned against the expanded segments in the test case using UE-DTW algorithm. Mean of these scores for each segment are used to normalise the score of the keyword for that segment. Variance normalization did not improve the performance.

Formalising score normalization:

$$s'_{i,j} = s_{i,j} - \frac{1}{|C|} \sum_{c \in C} s_{i,j}^{(c)} \tag{3}$$

(a)

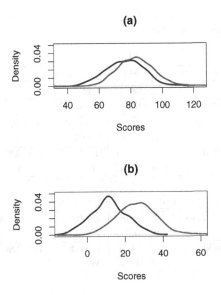

(b)

Fig. 2. Illustration of the effect of score normalization on true (blue) and impostor (red) scores. A shift is observed reducing false alarms and misses.

where $s_{i,j}$ is the DTW distance of the reference with the jth segment of ith test case, C refers to the background set, $|C|$ refers to the size and $s_{i,j}^{(c)}$ refers to the DTW score for $c \in C$ against the same segment in the test case. 10 random sentences containing 71 words are picked from the train set in TIMIT such that, they do not contain the the keywords. The score normalization was particularly effective in accomplishing two things: first, all scores are normalized irrespective of the mismatch between reference and test envirnoments. Second, it proved very useful in reducing the number of false alarms. Fig. 2 illustrates the separability of the score distributions of the keywords and non-keywords achieved by score normalization.

4.3   Results

The results are presented as ROC (Receiver Operating Characteristics) curves and their AUC (Area Under the Curve) value (following the pattern in [6]). Experiments are conducted to reduce the bias (if any) of the reference samples used on the results. Also, the number of references to be used requires careful consideration. Cases of one, two and all available references are considered. These are represented as NOR=1, NOR=2 and NOR=all, repsectively, in the Table presented (Tab. 2). To reduce the influence of the choice of references for the first two cases, references are randomly picked over several iterations. Both the best and the worst case results are given. The results on TIMIT and NTIMIT are summarised and compared with the HMM baseline in Table 2.

From Table 2, following observations are made: (i) the approach performs pretty well even when the number of references used is as minimal as a single utterance, (ii) more

Table 2. Results for experiments conducted on TIMIT and NTIMIT datasets in terms of AUC (NOR: Number of references, B=best, W=worst)

TIMIT			
HMM Baseline (MFCC+vel+accl) = 0.86			
Proposed System			
Feature	NOR=1	NOR=2	NOR=all
MFCC+vel+accl	B=0.881	B= 0.885	B = **0.92**
	W= 0.807	W= 0.833	W= -
ModGd+vel	B=0.886	B= 0.9055	B = **0.977**
	W= 0.847	W= 0.873	W= -
NTIMIT			
HMM Baseline (MFCC+vel+accl) = 0.78			
MFCC+vel+accl	B=0.806	B= 0.814	B = **0.840**
	W= 0.743	W= 0.766	
ModGd+vel	B=0.834	B= 0.839	B = **0.861**
	W=0.785	W=0.786	

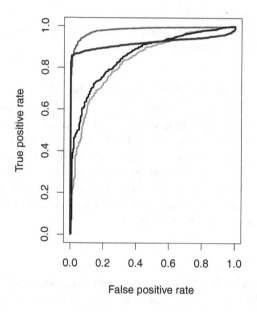

Fig. 3. Comparison of systems that use all available references for spoken keyword spotting. Legends- Red: ModGD on TIMIT, Blue: MFCC on TIMIT, Black: ModGD on NTIMIT, Green: MFCC on NTIMIT.

the number of references, better the performance (iii) ModGD consistently performs better than MFCC (iv) in case of using only one utterance, the worst case performance was heavily influenced by the choice of reference. The overall performance on NTIMIT when compared to TIMIT decreased as expected. Overall, for the case when all available references of a keyword were used, a relative improvement of 13.7% is observed over that of an HMM-based baseline. In the case where at most 2 references were used, a relative improvement of 29% is observed. Also, in the worst case for NOR=2, the degradations observed is acceptable for NOR=2 despite having used little data.

All systems are compared together in the ROC plot in Fig. 3. In summary, ModGD-based systems performed better than MFCC counterpart. Overall, the proposed approach is a very good alterantive to conventional approaches to spoken keyword spotting.

5 Summary

The problem of spotting a spoken keyword in a speech utterance is addressed. The test utterance is segmented using a group delay-based acoustic segmentation algorithm. The output of this gives acoustic evidence for possible subword boundaries. The boundary information is used by a UE-DTW algorithm with references of keywords to be spotted. Score normalisation is performed to compensate for variabilities in scores due to test segment length and mismatch conditions. The overall approach, has superior performance and is also robust while have minimal data-dependency. Results on TIMIT and NTIMIT corroborate these claims. In the best case, a relative improvement of 13.7% is observed compared to the Keyword/Filler model-based system.

References

1. Rose, R., Paul, D.: A Hidden Markov Model based keyword recognition system 1, 129–132 (1990)
2. Benayed, Y., et al.: Confidence measures for keyword spotting using support vector machines. In: Proc. ICASSP, vol. 1, pp. 588–591 (2003)
3. Szöke, I., Schwarz, P., Matějka, P., Karafiát, M.: Comparison of keyword spotting approaches for informal continuous speech (2005)
4. Zhang, Y., Glass, J.: Unsupervised spoken keyword spotting via segmental DTW on Gaussian posteriorgrams. In: IEEE Workshop on ASRU, pp. 398–403 (2009)
5. Park, A., Glass, J.: Unsupervised pattern discovery in speech. IEEE Trans. on Audio, Speech, and Language Processing 16, 186–197 (2008)
6. Keshet, J., et al.: Discriminative keyword spotting. Speech Communication 51, 317–329 (2009)
7. Vergyri, D., Shafran, I., et al.: The SRI/OGI 2006 Spoken Term Detection system. In: Proc. of Interspeech, pp. 2393–2396 (2007)
8. Nagarajan, T., Murthy, H.A., Hegde, R.M.: Segmentation of speech into syllable-like units. In: Proceedings of EUROSPEECH, pp. 2893–2896 (2003)
9. Shi, G., et al.: On the importance of phase in human speech recognition. IEEE Transactions on Audio, Speech, and Language Processing 14, 1867–1874 (2006)
10. Prasad, V.K., et al.: Automatic segmentation of continuous speech using minimum phase group delay functions. Speech Communication 42, 429–446 (2004)
11. Rabiner, L., Juang, B.H.: Fundamentals of speech recognition. Prentice-Hall, Inc., Upper Saddle River (1993)

12. Fisher, W.M., Doddington, G.R., Goudie-Marshall, K.M.: The DARPA speech recognition research database: Specifications and status. In: Proceedings of DARPA Workshop on Speech Recognition, pp. 93–99 (1986)
13. Jankowski, C., et al.: NTIMIT: A phonetically balanced, continuous speech telephone bandwidth speech database. In: Proc. ICASSP, vol. 1 (1990)
14. Giorgino, T.: Computing and Visualizing Dynamic Time Warping Alignments in R: The dtw Package
15. R Development Core Team: R: A Language and Environment for Statistical Computing. R Foundation for Statistical Computing, Vienna, Austria (2011) ISBN 3-900051-07-0
16. Hegde, R.M., et al.: Significance of joint features derived from the modified group delay function in speech processing. In: EURASIP 2007 (2007)
17. Woodland, P., Odell, J., Valtchev, V., Young, S.: Large vocabulary continuous speech recognition using HTK. In: Proc. of ICASSP, pp. 125–128 (1994)
18. Auckenthaler, R., et al.: Score normalization for text-independent speaker verification systems. Digital Signal Processing 10, 22–54 (2000)

An In-Car Speech Recognition System
for Disabled Drivers

Jozef Ivanecký and Stephan Mehlhase

European Media Laboratory
Schloss-Wolfsbrunnenweg 35, 69118 Heidelberg, Germany
{jozef.ivanecky,stephan.mehlhase}@eml.org

Abstract. Automatic Speech Recognition (ASR) is becoming a standard in nowadays cars. However, ASR in cars is usually restricted to activities not directly influencing the driving process. Thus, the voice-controlled functions can rather be classified as comfort functions, e. g. controlling the air condition, the navigation and entertainment system or even the mobile phone of the driver. Obviously this usage of an ASR system could be extended in two directions: On the one side, the speech recognition system could be used to control secondary functions in the car like lights, windscreen wipers or windows. On the other side, the comfort functions could be enriched by utilizing services like weather inquiries, SMS dictation or online traffic information. Compared to todays usage these extensions require a different approach than the one employed today. Controlling secondary functions in the car by voice demands the usage of a very reliable, real-time, local ASR. At the same time a large vocabulary ASR system is required for comfort functions like dictation of messages.

In this paper, we describe our efforts towards a hybrid speech recognition system to control secondary functions in the car. We also provide an extended comfort functionality to the driver. The hybrid speech recognition system contains a fast, grammar-based, embedded recognizer and a remote, server-based, LM-based, large vocabulary ASR system. We will analyze different aspects of such a design and the integration of it into a car. The main focus of the paper will be on maximizing the reliability of the embedded recognizer and designing an algorithm for switching dynamically between the embedded recognizer and the server-based ASR system.

1 Introduction

Automatic speech recognition (ASR) is becoming more and more common in todays cars [2]. The used ASR systems can be classified in two distinct classes: On the one hand there are integrated ASR systems, which control basic comfort functions like air conditioning, radio, or navigation system, e. g. to enter the address. On the other hand, todays upper class cars are utilizing speech recognition system running on a server which is accessed through the Internet. This allows for more complex tasks, e. g. supporting inquiries for weather or traffic information.

Irrespective of the used ASR technology, in general the set of controlled in-car devices and functions does not expand to the secondary functions (e. g. lights or windscreen wipers). The driver can reach those without having to stop focusing on the

P. Sojka et al. (Eds.): TSD 2012, LNCS 7499, pp. 505–512, 2012.

driving process itself. Pressing a switch is in general, faster and more natural than to use a spoken command for such a task. However, controlling comfort functions is a more complicated process. Complex tasks like music selection require a significant amount of the driver's attention. Therefore, the driver benefits from controlling these functions by voice. In cases where the driver has to use a joystick instead of a steering wheel, e. g. due to a disability, controlling the secondary functions takes significant additional effort. Therefore, it makes sense to expand the voice control to include the secondary functions as well. The requirements for controlling secondary and comfort functions differ: On one hand a reliable, real-time speech recognition system with a safety model for incorrectly recognized commands is required for secondary functionalities. On the other hand controlling the comfort functions by voice, does not require real-time speech recognition. Also, a mis-recognized comfort function does not directly influence safety.

In this paper we describe our effort towards the implementation of a hybrid ASR system. A real-time, grammar-based embedded recognizer is used to recognize secondary functions commands directly in the car. A large vocabulary, LM-based recognizer connected via the Internet is used for advanced comfort functionality. We investigate different methods for dynamically switch between those recognizers, which is an important step towards reaching the aforementioned goals.

Remaining parts of the paper are organized as follows: In Section 2 we define secondary and comfort functions of a car. In Section 3 we describe the design of the two different ASR systems used for in-car speech recognition. Section 4 describes experiments used to evaluate the in-car speech recognition. In Section 5 a brief summary is provided.

2   Secondary and Comfort Functions

We define 3 classes of functions available in a car. They differ in terms of availability, simplicity of usage and required promptness of the reaction.

1. *Secondary functions:* Obligatory functionality of each car which does not belong to the primary functions (accelerator, breaks, steering wheel, . . .). Examples are the different kind of lights, car horn or windscreen wipers. They are easily accessible and intuitively to operate. The reaction time of all these devices is instant and reliability is very high.
2. *Basic comfort functions:* Optional equipment of a car related to driving comfort, e. g. air conditioning or radio. They are usually easily accessible but not always intuitively to operate. As before, the reaction time is instant. Malfunctioning is not significantly influencing car usability.
3. *Advanced comfort functions:* Optional equipment of a car related to driving comfort, e. g. navigation system or traffic information systems. In general, they are rather complex to operate and the reaction time is not instant. Some of these functions require Internet access. Malfunctioning affects only the comfort of the driver.

Secondary functions are easily accessible in any car and there is seemingly no need to use voice control. However, the situation is fundamentally different in cars modified to be used by disabled driver. Depending on the level of disability, controlling secondary functions with ordinary control levers may vary from easy to impossible. In the latter situation, speech recognition might be a more natural way to control the secondary functions of a car.

3   Hybrid Speech Recognition

Because of the different requirements for the aforementioned in-car functions, it is difficult to use a single ASR system. For the secondary and basic comfort functions it is necessary to use a real-time local ASR system with very high recognition accuracy. This is achieved by a small vocabulary grammar-based system directly integrated into the car. The advanced comfort functions often require a large vocabulary, but do not require as high accuracy and low latency as ASR for the secondary functions. We are using a LM-based recognition server accessed through the Internet to provide this functionality. Finally, we designed a system which dynamically switches between the two recognition systems to provide a uniform interface to the user.

In the literature the term *Hybrid Speech Recognition* is used to describe a combination of HMM and ANN-based recognizers. In this paper however, we use it to refer to the combination of a grammar-based, real-time recognizer with a server-based, large vocabulary recognizer.

3.1   ASR for Secondary and Basic Comfort Functions

Embedded recognizers were originally designed to run on significantly slower hardware than available today. Therefore, in case of a small grammar the real-time requirement is easily satisfied. The main challenge for such a system is to meet the very low error rate requirements. An incorrect recognition can trigger an unwanted action, which, in a certain ill-timed moment, can lead to dangerous situations, e. g. switching off the lights during the night or switching on the opposite turning signal. Therefore, a robust safety model in case of an incorrect recognition is needed.

We are using commercially available embedded recognizers (Loquendo and SVOX). To run the recognizer we used the same platform as in [1]. We were focusing mainly on grammar and application design to achieve maximal accuracy and reliability. Usually if the grammar offers a big variety of commands the error rate of the recognition increases. Therefore, we tried to minimize the grammar size and avoid acoustic similarities between the commands. As there are many ways to toggle specific devices, we focused on the most common short and long forms. For instance, for turning on the high beams the short form is "*Fernlicht an*" whereas the long form is "*Das Fernlicht einschalten*"[1]. The vocabulary size of the resulting grammars is only around 30 words.

The system is operating in *Push-to-Talk* (P2T) mode, which means that the system is only listening while a button is pressed. The *Push-to-Activate* (P2A) mode, in which the user only pushes the button once to indicate the start of the utterance, could be easier to use. However, we decided for the P2T system for accuracy reasons. Especially at high

---

[1] German terms to switch on the high beams.

speeds the automatic end-pointing needed in the P2A system poses a problem due to the environmental noise.

Irrespective of the activation mode, the button used is serving also safety purposes. If the user presses the button again shortly after the recognition finished, he cancels the initiated action. Such a behavior should avoid unwanted situations caused by incorrect speech recognition and consecutive actions.

3.2   ASR for Advanced Comfort Functions

In order to provide the user with the comfort functions as defined in Section 2, the speech recognition system must be able to deal with a large vocabulary. Therefore, it is no longer feasible to use a grammar-based recognition system. We decided to use a server-based, large vocabulary speech recognition system. It is located in a computing center and consequently requires an in-car Internet connection to be available. Some cars already come with support for mobile network connectivity, it is possible to place UMTS routers in the car. If that is not possible the tethering capabilities of smart-phones could be used.

Regarding the recognition time, there are two considerations to take into account: On one hand, in case of accessing the advanced comfort functions it is no longer necessary to provide the user with recognition results in real-time. On the other hand, it is also important that processing is not taking too long as the driver gets distracted from driving when the system is not working as he expects to, i. e. not reacting to his voice input promptly. Given that the audio data needs to be transferred to the server which in turn sends back the recognition result using a possibly slow and unreliable mobile Internet connection, it was necessary to build a robust system which can handle outages in a non-disruptive way.

In order to decrease the recognition time, the service uses a custom network protocol to transfer the audio data in small chunks. The protocol allows the server to send back partial results as soon as they are available. With this protocol it was possible to create a service already starting to process the audio data while sending. Optimizing the server-side processing of the received audio signal allows to further decrease the perceived decoding time. Using this technique, we were able to reduce the perceived recognition time factor from around 3 down to around 1. The *perceived recognition time* specifies the time the user perceives as waiting time from finishing to speak until the system reacts to his input. The *actual recognition time* can differ, mainly due to the time needed to transfer the data to the server.

The recognition system we are using is working with a language model with a vocabulary size of over 1 million words, specifically tailored for mobile search and dictation applications. The server-based system is designed to be highly scalable and can serve many clients at the same time without performance degradation.

3.3   Which One to Use?

The audio signal is always processed by the in-car recognition system. A control application has to decide if the command was aimed at the secondary or basic comfort functionality or whether it is part of the advanced comfort functions. We evaluated 3 different approaches on how to distinguish between them:

1. *Confidence score:* Only the confidence score of the recognized utterance is taken into account. If the score is below a certain threshold, the audio signal is sent to the server-based recognizer.
2. *Out of grammar model:* If the recognition result is tagged as out of grammar (OOG), the audio signal is sent to the server-based recognizer. The confidence score is not taken into account.
3. *OOG model with trigger word:* As the previous method, but a special key word has to precede the "out of grammar" part.

If the decision algorithm decides that the utterance has to be sent to the server-based recognizer, the application informs the user about it and waits for a reply from the server. This kind of functionality assumes a working Internet connection as explained in Section 3.

4   Evaluation

The evaluation is split into two major aspects. The first aspect is to examine the speech recognition accuracy for different grammars and noise levels. The second aspect is evaluating the switching between the local and the remote recognizer. In order to evaluate our system we recorded a test set. For data collection the P2T mode was used and the microphone was at a distance of $20 - 30$ cm to the speaker. The recorded data consists of 10 speakers (4 female and 6 male voices) of which 2 were non-native German speakers. For each we recorded 2×30 commands, containing

- 10 long commands for controlling secondary functions (*den Blinker links ausschalten, die Lichthupe einschalten, . . .*),
- 10 short commands for controlling secondary functions (*Blinker links an, Lichthupe, . . .*),
- 5 commands controlling comfort functions with a trigger word (*Komfortfunktion: Wettervorhersage für Heidelberg, Komfortfunktion: Radio: SWR3 wählen, . . .*), and
- 5 commands controlling comfort functions without a trigger word (*Wettervorhersage für Heidelberg, Radio: SWR3 wählen, . . .*).

The recording took place in 2 different environments: A quiet office environment and a noisy environment with in-car noise up to 80 dB, responsible for low SNR and the Lombard effect during the recording.

4.1   Speech Recognition

For the recognition accuracy test we created two different grammars. The first grammar is covering only the long forms of the commands and was designed to be used only with the first 10 test sentences recorded by each speaker. The second grammar is covering all commands for the secondary functions. The second one was used for all recorded commands to examine whether the error rate is getting worse with bigger

Table 1. Speech recognition and action accuracy (SER – Sentence Error Rate, SA – Sentence Accuracy, AER – Action Error Rate, AA – Action Accuracy, ASCF – Average Sentence Confidence Score)

	SER	SA	AER	AA	ASCF
	Quiet environment				
Long form – reduced grammar	2%	84%	2%	84%	84.51%
Long form – full grammar	15%	80%	1%	94%	84.56%
Short form – full grammar	9%	91%	3%	97%	84.01%
	Noisy environment				
Long form – reduced grammar	0%	94%	0%	84%	81.88%
Long form – full grammar	13%	86%	1%	98%	81.38%
Short form – full grammar	11%	88%	6%	93%	77.36%

command variety in the recognition grammar as expected. However, more important than the speech recognition accuracy is the accuracy of the actions triggered by the voice command. Even an incorrectly recognized command can trigger the correct action. Therefore we examined action accuracy as well as recognition accuracy.

In Table 1 the results for the sentence accuracy and the action accuracy obtained on the test set are shown. From the speech recognition point of view the most important results are the sentence accuracy (SA) and sentence error rate (SER). It is difficult to decide which combination of grammar and set of commands to use based on these results alone. In the quiet environment the short form commands with the full grammar give the best accuracy, whereas in the noisy environments the long forms with the reduced grammars give the best results.

Taking the action accuracy (AA) and more importantly the action error rate (AER) into account, Table 1 gives a better indication which is the safest grammar and commands combination. The smallest AER and biggest AA are always achieving using the long form of commands. Whether the grammar should also contain the short forms is subject to practical testing we plan to do in the future.

The table shows also the average sentence confidence scores[2]. We did not take into account the confidence score during the evaluation. However, using also such an information is an option how to further eliminate incorrect actions caused by an incorrect recognition result. On the other side the result rejection based on the confidence score will decreased the action accuracy. The number of commands from the recognition test with confidence score below 50% was 5. In 4 of these 5 cases the recognition was incorrect. Therefore, if we used a minimum sentence confidence score for the secondary functions of 50%, it would further reduce SER or AER but AA as well.

4.2   Speech Recognizer Selection

The recognizer selection tests included all three approaches described in Section 3.3. For the confidence score approach we re-used the grammars used for the tests in

---

[2] Confidence score of a particular recognizer was scaled into to the range 0 to 100.

Table 2. Maximal sentence confidence score for the comfort function commands with the secondary function grammars

	With trigger word	Without trigger word
	Quiet environment	
Reduced grammar	36%	59%
Full grammar	42%	60%
	Noisy environment	
Reduced grammar	42%	46%
Full grammar	42%	61%

Table 3. Out of grammar (OOG) recognized for secondary function commands

	Quiet env.	Noisy env.
OOG w/o trigger word	76%	84%
OOG with trigger word	0%	0%

Section 4.1. With those grammars we tried to recognize the recorded commands aimed at the comfort functions. Of course the recognizer produced a recognition result containing a sentence from the grammar. But now the sentence confidence score is taken into account as well. Therefore, we examined the maximum score a sentence for a comfort function would gain, which are listed in Table 2. Comparing these values with the sentence confidence scores reported in Table 1, in all cases we observe a satisfactory difference. The lowest confidence scores were achieved for commands containing a trigger word. The best result was achieved with the combination of using such a trigger word and the grammar containing only the long forms.

For the garbage-based experiments, we modified the recognition grammar to include also an out of grammar (OOG) model. In the experiment with garbage preceded by a trigger word a command "<Trigger word> OOG;" was added. In the other experiment just the command "OOG;" was added. We were observing how many times the result "OOG" appeared among the recognized commands for secondary functions and how many times "OOG" did not appear among the comfort functions commands.

Table 3 shows how often "OOG" was returned when feeding secondary function commands into the speech recognition engine. We did the experiment with and without the trigger word "Komfortfunktion" which is not part of the remaining grammar. The results indicate, that for a reliable separation of secondary and comfort functions, the usage of some kind of trigger word is necessary. In the following experiment we used the grammar containing the trigger word and used the comfort function commands as input for the recognizer. In nearly all cases (98% in quiet, 100% in noisy environment) the recognizer returned the "OOG" indicator.

In case of the comfort functions the error rate, i. e. cases in which the output should be "OOG" but was not, is more important than the accuracy. A comfort function command which is accepted by a secondary function grammar could trigger an unwanted action on the secondary functionality in the car. The error rate measured in quiet and noisy environment was 0%. Consequently, the results are confirming the previous indication, that the usage of an adequate trigger word is a reliable way to determine which recognizer to use.

5 Summary

In this paper we described various aspects of the development of an in-car speech recognition application for disabled drivers. We analyzed the usage options for disabled drivers and identified possible risks that need to be minimized. First experiments indicate the feasibility of our approach, but also unveiled the need for further work and massive testing in a real-life environment in order to maximize safety of the driver. The speech recognition accuracy in this case is only of secondary importance.

References

1. Ivanecký, J., Mehlhase, S., Mieskes, M.: An Intelligent House Control Using Speech Recognition with Integrated Localization. In: Proc. of Ambient Assisted Living – 4th AAL Congress, Berlin (2011)
2. Heisterkamp, P.: Linguatronic product-level speech system for Mercedes-Benz cars. In: Proc. of the First International Conference on Human Language Technology Research, San Diego (2001)

Captioning of Live TV Programs through Speech Recognition and Re-speaking*

Aleš Pražák, Zdeněk Loose, Jan Trmal, Josef V. Psutka, and Josef Psutka

University of West Bohemia, Department of Cybernetics
Plzeň, Czech Republic

Abstract. In this paper we introduce our complete solution for captioning of live TV programs used by the Czech Television, the public service broadcaster in the Czech Republic. Live captioning using speech recognition and re-speaking is on the increase and widely used for example in BBC; however, many specific issues have to be solved each time a new captioning system is being put in operation. Our concept of re-speaking assumes a complex integration of re-speaker's skills, not only verbatim repetition with fully automatic processing. This paper describes the recognition system design with advanced re-speaker interaction, distributed captioning system architecture and neglected re-speaker training. Some evaluation of our skilled re-speakers is presented too.

Keywords: live captioning, speech recognition, re-speaking.

1 Introduction

As the public service television companies are pushed by law and by the society of deaf and hard of hearing to provide closed captions to all live TV programs, large vocabulary continuous speech recognition (LVCSR) systems are demanded as a solution. The advantage of using an automatic speech recognition is a high transcription speed in comparison with human typing speed and a cheaper operation. On the other hand, automatic speech recognition systems are in general more erroneous than human.

There are many laboratory recognition systems that are being developed on synthetic tasks to get the best recognition accuracy, but almost none of them can be used for live captioning. A task of live captioning requires low-latency real-time recognition, so many speech processing methods described in scientific papers, which are designed for multi-pass recognition with enormous time requirements, cannot be used. However, the recognition accuracy is not the only quality criterion for the task of live captioning. The readability and overall intelligibility of the resulted captions is very important for deaf and hard of hearing viewers too. This comprises issues such as reduction and rectification of the original message if necessary, a segmentation of recognition output or its instant correction in case of crucial misrecognitions.

To overcome problems arising from acoustics and speaker variability in the original acoustic track a re-speaker (or shadow-speaker) approach was introduced [1,2]. This

---

* This paper was supported by the Technology Agency of the Czech Republic, project No. TA01011264 and by the grant of The University of West Bohemia, project No. SGS-2010-054.

P. Sojka et al. (Eds.): TSD 2012, LNCS 7499, pp. 513–519, 2012.

simplifies the recognition task significantly, so fine-tuned speaker-dependent acoustics is used and a well-qualified re-speaker controls the whole caption creation process. On the other hand, this approach for captioning of live TV programs brings up a problem of re-speaker training, because it can be a long and expensive process.

During our cooperation with the Czech Television, the public service broadcaster in the Czech Republic, we have developed and implemented a system for automatic live captioning of the meetings of the Parliament of the Czech Republic directly from the real acoustic track [3] (furthermore, our speech synthesis was used for automatic dubbing from subtitles for hard of hearing [4]). We operate such a system over four years with more than 600 captioned hours. The recognition accuracy reaches 90% (depending on the topic discussed). Other systems using text-to-speech alignment were introduced recently [5,6]. Since this approach is not suitable for vast majority of live TV programs, we have enhanced the recognition system for a re-speaker operation and designed the captioning system topology as distributed.

This paper describes our complete solution for live TV captioning from recognition system preparation to the distributed captioning system architecture and one chapter is dedicated to the re-speaker training methodology. Some evaluation of our skilled re-speakers is presented too.

2   Speech Recognition System

To achieve the best recognition performance a language model of the recognition system should be specific for each type of a live program (e.g. political debates, TV shows, sport programs). Thus we need a huge amount of language model training data from different sources of different natures. We have collected a large corpus containing the data from newspapers (520 million tokens), web news (350 million tokens), subtitles (200 million tokens) and transcriptions of some TV programs (175 million tokens) [7]. These data are automatically processed using topic detection and daily updated, so we have topic specific language models with up to date vocabularies.

Now we use trigram back-off language models with mixed-case vocabularies with more than one million words; however, even such models cannot cover all the words of the language with a high degree of inflection. To cover common out-of-vocabulary (OOV) words we have prepared special large lists of named entities. These lists include for example Czech surnames, Czech villages and streets, world countries and cities in all grammatical cases and other word forms with pronunciation variants. These words are not present in the active vocabulary, but they can be added to the vocabulary by the re-speaker during recognition with full trigram statistics based on classes trained for each list and word form. The idea is if a named entity was misrecognized during re-speaking, a re-speaker types the named entity, so the word and all its word forms with pronunciation variants are added to the vocabulary. Then the re-speaker erases misrecognized words and re-speaks named entity once more, so from now on the named entity in all its word forms will be recognized correctly.

To allow manual sentence segmentation by the re-speaker, the language model considers punctuation marks as words. The re-speaker uses his/her hands to press punctuation marks on keyboard during inter-word pauses, so they can be processed by the recognition system that presents the punctuation marks directly in its result. This approach allows the recognition system to benefit from extra information, because punctuation marks are then used as a context words for subsequent recognitions.

Since a re-speaker's voice is being recognized instead of real acoustic track of a TV program, speaker-dependent acoustic model should be trained. We train such a model for each re-speaker in two steps. In the first step, we use a gender-dependent acoustic model which is automatically adapted to the speaker's voice characteristics just during recognition. An unsupervised incremental fMLLR adaptation is carried out in the background based on recognized words and their confidence scores [8]. This approach can improve recognition performance until enough speech data (30 hours at minimum) is gathered. In the second step, a speaker-dependent acoustic model is trained unsupervised almost from scratch based on gathered data, their recognized transcriptions and word confidence scores. We use triphone acoustic models with PLP parameterization (see [9] for optimization methodology) on 22 kHz acoustic signal.

We employ one-pass in-house LVCSR system optimized for real-time operation. The recognition system is based on three-state Hidden Markov Models and lexical tree structure of a recognition network. To enable recognition of more than one million words in real-time, we use triphone in-word and triphone left cross-word context, but open right cross-word context. This approach reduces decoding time significantly with minimum impact on recognition performance. The implementation is focused on low-latency real-time operation with very large vocabularies on multi-core systems. Due to high efficient decoder parallelization and a graphic processor unit (GPU) utilization, we can recognize more than one million words in real-time on modern four-core laptop computers.

3   Captioning System Architecture

It comes from the principle of recognition system, that a caption can be created only after a last word of the caption was re-spoken and assumed not to change by recognition of the following speech signal. Considering this, about three last so-called "pending" words of the recognized text are not immediately sent to the captioning server, but they can be displayed to the re-speaker highlighted, so the re-speaker knows, which words can be eventually corrected. To be able to quickly correct any of the pending words (not only the last one) in case of misrecognition, the best method is to erase them all (by a key press) and re-speak consequently. This feature can dramatically decrease the error rate of the final captions. On the other hand, a time lag of the final caption can be reduced if the re-speaker manually dispatches the pending words. This is very important when the re-speaker does not speak for a longer time (because of TV jingle or he/she is listening ahead) and the pending words remain unsent. It can significantly reduce the time lag between the words uttered in the original speech and corresponding words in the captions.

A time lag of the final caption dictated by the length of the caption and the pending words is about 3–5 seconds, moreover the transmission of the caption to the viewer takes another 3 seconds. To avoid the time lag completely, the broadcaster would have to hold up the broadcasting up to 10 seconds to provide precisely-timed captions. Since the Czech Television is not ready to do such a thing, we receive the real acoustic track of the TV program ahead of its transmission, so we can eliminate above mentioned 3 seconds of the time lag.

Because of the strict electronic security policies at the Czech Television (CTV) in Prague, the captioning system architecture was designed as highly distributed with the centre at the University of West Bohemia (UWB) in Pilsen (see Figure 1). The interconnection between CTV and UWB is done by a point-to-point connection over the ISDN network. To be able to carry both audio signal to UWB and generated captions to CTV, the ISDN is used only as a full-duplex data carrier with bandwidth of 128 kbit/s. Specialized terminal adapters commonly used in CTV with transparent low-latency compression are used. The captioning server at UWB distributes audio signal by VoIP service to the re-speakers to arbitrary locations with internet connection. Since a visual component of a TV program is not delivered, common DVB signal is displayed to the re-speaker. The synchronization delay of 3 seconds between audio and video is not crucial, because the visual component is intended only for re-speaker's overview about the situation.

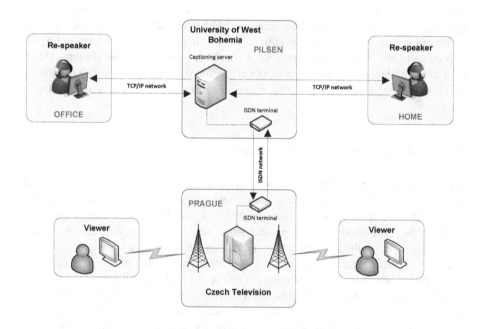

Fig. 1. Captioning system architecture

The whole recognition process with keyboard interaction with the re-speaker is carried out on a laptop computer. Since boxed pop-on closed captions are preferred by deaf and hard of hearing, formatting of captions is done automatically at the captioning server. To trade-off between the caption time lag and its length, one row captions are preferred, but other requirements relating the readability of the final captions are considered too. Based on speaker change markers provided by the re-speaker, an automatic speaker coloring of final captions is performed – blue for the host and alternating white and yellow colors for guests.

4 Re-speaker Training

It is impractical to bring a re-speaker to the real captioning system and let him/her do the real job of a skilled re-speaker throwing away the resulting captions for a few months. More effective is to use a training system that shortens (and thus cheapens) the training process, so we have developed a special training system for re-speakers that provides gradual training process under surveillance of a skilled supervisor. Since the re-speaking in our concept is not only verbatim repetition with no interaction, a re-speaker candidate should train his/her skills gradually from the basics to the most demanding tasks. To become accustomed to the real captioning software, the whole training process should be a part of such a software. Thus we have developed captioning software with four training phases, the fifth phase represents the real captioning process.

The first training phase is intended to train re-speaker's skill to listen and speak simultaneously. The re-speaker opens prepared video file and practices speaking while playing any part of the video. The aim is not to re-speak word-by-word, but become accustomed to speaking meaningfully while listening to and perceiving the original acoustic track. This phase does not employ recognition system, but all utterances are recorded for later playback by the supervisor. The first phase helps to sort out re-speaker candidates which do not have basic prerequisites to do the job.

The second phase of the training system assists in optimizing the re-speaker's utterance to the recognition system demands, so this phase integrates the LVCSR system and displays its output to the re-speaker. The main objective of the re-speaker is to re-speak original speech word-by-word so that the recognition accuracy is as high as possible. The re-speaker just mechanically re-speaks what he/she listens to, so he/she can focus on altering the utterance (mainly the pronunciation) and its influence on the recognition results. Since a transcription of demanded re-speaker utterance is known in advance, the misrecognitions in the recognized text can be highlighted based on Levenshtein alignment. In addition, overall recognition accuracy is displayed, so the candidate's training progress can be monitored.

In the third phase of the training system a trainee learns rephrasing and simplifying when it is required. A re-speaker has to check instantly recognized text and perform potential corrections of pending words. Other keyboard commands are used to dispatch pending words, to add new words to the vocabulary (if OOV words cannot be paraphrased) and to mark speaker changes too. The fourth phase simulates the real captioning system with all the features needed during the real caption generation including punctuation marks insertion.

The overall duration of a re-speaker training is very individual, but according to our expertise, an average time of intensive training plan is from 2 to 3 months (75 training hours at minimum).

5 Evaluation of Re-speakers

Now we have five skilled re-speakers and three re-speaker candidates in different phases of the training process. We produce captions for two live TV programs regularly. The first one is weekly political debate over two hours long, so two re-speakers relay each other in the middle of the program. The second one is one hour daily talk show on different topics that is captioned by one re-speaker.

Some statistics from live captioning of 50 minute talk show about economics for three skilled re-speakers are presented in Table 1. The recognition accuracy is computed including re-speaker's slips of the tongue not considering the corrections made by the re-speaker, so the raw audio recording of the captioning session was recognized. On the other hand, the caption accuracy represents the accuracy of final captions that were transmitted to the captioning server and displayed to viewers.

Table 1. Evaluation of re-speakers

Re-speaker	EK	NZ	DT
Recognition accuracy	96.93%	95.91%	93.36%
Caption accuracy	98.42%	97.20%	97.21%
Words	4 572	4 717	5 082
Word additions	3	2	4
Word corrections	63	50	128
Word dispatches	103	35	24
Commas	390	359	438
Full stops	377	402	392
Question marks	88	89	100
New speakers	231	192	252

6 Conclusion and Future Work

We have presented our complete solution for captioning of live TV programs including the recognition system, captioning system architecture and re-speaker training methodology. Although we do live captioning only for a small portion of live TV programs broadcasted by the Czech Television now, according to the experience in the world, speech recognition application with the re-speaking is the only feasible solution for 24/7 captioning of live TV programs.

Since major sporting events will be held this year, we prepare the language models for ice-hockey and football TV programs according to the demands of the society of deaf and hard of hearing. Each sport has quite a large number of active and retired sportsmen, but all their names cannot be present in the active vocabulary of the recognition system. Thus a method for automatic vocabulary and language model adaptation should be developed.

References

1. Evans, M.J.: Speech Recognition in Assisted and Live Subtitling for Television. R&D White Paper WHP 065, BBC Research & Development (2003)
2. Marks, M.: A distributed live subtitling system. R&D White Paper WHP 070, BBC Research & Development (2003)
3. Pražák, A., Psutka, J.V., Hoidekr, J., Kanis, J., Müller, L., Psutka, J.: Automatic Online Subtitling of the Czech Parliament Meetings. In: Sojka, P., Kopeček, I., Pala, K. (eds.) TSD 2006. LNCS (LNAI), vol. 4188, pp. 501–508. Springer, Heidelberg (2006)
4. Matoušek, J., Vít, J.: Improving automatic dubbing with subtitle timing optimisation using video cut detection. In: IEEE International Conference on Acoustics, Speech, and Signal Processing, Kyoto, pp. 2385–2388 (2012)
5. Bordel, G., Nieto, S., Penagarikano, M., Rodriguez-Fuentes, L.J., Varona, A.: Automatic Subtitling of the Basque Parliament Plenary Sessions Videos. In: 12th Annual Conference of the International Speech Communication Association, pp. 1613–1616. Causal Productions (2011)
6. Ortega, A., Garcia, J.E., Miguel, A., Lleida, E.: Real-Time Live Broadcast News Subtitling System for Spanish. In: 10th Annual Conference of the International Speech Communication Association, pp. 2095–2098. Causal Productions (2009)
7. Švec, J., Hoidekr, J., Soutner, D., Vavruška, J.: Web Text Data Mining for Building Large Scale Language Modelling Corpus. In: Habernal, I., Matoušek, V. (eds.) TSD 2011. LNCS, vol. 6836, pp. 356–363. Springer, Heidelberg (2011)
8. Zajíc, Z., Machlica, L., Müller, L.: Initialization of fMLLR with Sufficient Statistics from Similar Speakers. In: Habernal, I., Matoušek, V. (eds.) TSD 2011. LNCS, vol. 6836, pp. 187–194. Springer, Heidelberg (2011)
9. Psutka, J.V., Šmídl, L., Pražák, A.: Searching for a robust MFCC-based parameterization for ASR application. In: International Conference on Signal Processing and Multimedia Applications, pp. 196–199. INSTICC Press, Lisbon (2007)

Investigation on Most Frequent Errors
in Large-Scale Speech Recognition Applications

Marek Boháč, Jan Nouza, and Karel Blavka

SpeechLab, Faculty of Mechatronics, Technical University of Liberec
Studentská 2, 46117 Liberec, Czech Republic
{marek.bohac,jan.nouza,karel.blavka}@tul.cz

Abstract. When automatic speech recognition (ASR) system is being developed
for an application where a large amount of audio documents is to be transcribed,
we need some feedback information that tells us, what the main types of errors
are, why and where they occur and what can be done to eliminate them. While
the algorithm commonly used for counting the number of word errors is simple,
it does not care much about the nature and source of the errors. In this paper,
we introduce a scheme that offers a more detailed insight into analysis of ASR
errors. We apply it to the performance evaluation of a Czech ASR system whose
main goal is to transcribe oral archives containing hundreds of thousands spoken
documents. The analysis is performed by comparing 763 hours of manually
and automatically transcribed data. We list the main types of errors and present
methods that try to eliminate at least the most relevant ones. We show that the
proposed error locating method can be useful also when porting an existing ASR
system to another language, where it can help in an efficient identification of
errors in the lexicon.

Keywords: speech recognition, transcription error detection.

1 Introduction

Every system for automatic speech recognition (ASR) produces a certain amount
of transcription errors. They occur mainly due to a) the system's imperfect lexicon
(e.g. missing words or incorrect pronunciations), b) inadequate acoustic and language
models (causing confusions between acoustically similar words and homophones), c) a
speaker and his/her style of speaking (namely non-standard or sloppy pronunciation),
and d) noisy acoustic environment.

The performance of an ASR system is often evaluated in terms of a word-error-rate
(WER). During the system development, this single measure is a good indicator of the
performance progress. However, when the system reaches a certain mature level, it may
happen that additional improvements will be reflected only by a negligible reduction
of the WER. In this case, it is important to conduct a more detailed analysis of the
recognition errors and focus on eliminating those error types that occur most frequently,
as shown e.g. in [1,2].

When working with inflective languages, a thorough analysis of ASR errors is even
more essential. It should help to answer questions, like: How large the lexicon should

P. Sojka et al. (Eds.): TSD 2012, LNCS 7499, pp. 520–527, 2012.

be? Will a very large lexicon containing many acoustically similar inflected word-forms guarantee a lower WER? How much care should be paid to generating (and checking) basic and alternative pronunciations for these large lexicons? What percentage of errors is caused by omitted or inserted short words? Do all the detected errors have the same importance? Etc.

As our research is focused particularly on Czech and other Slavic languages, which are known for their rich morphology with many inflected word-forms, learning the answers on the above questions is very important [3]. From subjective evaluation of our Czech ASR system we knew that a large portion of errors was related to the shortest words, to verbs in the past tense with -*li* and -*ly* endings, or to minimal word-pairs that differ only in the length of one vowel. Yet, we believed that an objective, large-scale investigation could reveal some other sources of frequent errors. For that purpose we have developed a scheme that can locate and identify ASR errors in a more precise way than the standard WER evaluation method based on the Minimum Edit Distance (MED, [4]) algorithm (e.g. that implemented in the HTK toolkit [5]). Our method takes into account not only the orthography but also pronunciation, which leads to a more accurate alignment between the reference text and the recognized one and helps to locate errors even on a sub-word level.

2   Research Motivation

Since 2011 our team has been involved in a large applied-research project supported by the Czech Ministry of culture, whose aim is to transcribe the audio archive of the Czech Radio and to make its content publically available and searchable. The archive is huge and contains more than 100,000 hours of spoken documents recorded during the last nine decades. Our solution is based on the ASR technology we have been developing for more than 10 years [6]. As most of the archive data will be processed automatically, we need feedback information on the precision and readability of the transcriptions. Therefore, during the development, we regularly test the performance of the ASR module, search for the most frequent and most relevant errors and, consequently, we try to eliminate them. As we have human-made transcriptions to some 2,000 archive documents (almost 1,000 hours of audio), we can use these recordings as a large resource of development and test data. Recently, they have been used for the detailed error analysis proposed and described in this paper. The same scheme can be utilized also for several other tasks, e.g. during porting our ASR system to other Slavic languages, where it helps in an efficient identification of lexicon errors.

3   Scheme for Error Detection and Location

The standard procedure for the WER calculation (e.g. program HResult in [5]) is based on the MED alignment of the reference and recognized texts. It employs dynamic programming to match the two texts with the aim of achieving maximum number of hits (H). The non-matched words are marked as substitutions (S), insertions (I) or deletions (D), which is done without considering any measure of similarity between them. So, for example, when Czech word '*nato*' is recognized as two-word string '*na to*', the procedure will report one substitution and one insertion. If we used the direct output of the MED alignment for generating statistics of errors, we could arrive at wrong conclusions, such as word '*nato*' is often confused with '*na*' and word '*to*' is frequently omitted. Therefore, we propose a more accurate error detecting and locating scheme. It is derived from the MED algorithm, but it takes more information into account – the similarity in orthography and in pronunciation. The scheme is depicted in Fig. 1.

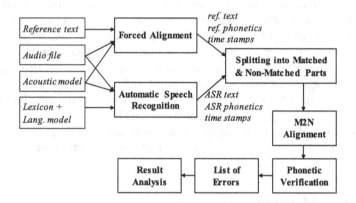

Fig. 1. Overall structure of the proposed scheme

It utilizes three inputs: the audio file, its reference transcription and the output of the recognizer. The recognizer processes the audio file in two different manners: first in the forced-alignment (FA) mode and then in the standard ASR mode. In both the cases, it produces rich information for each processed word, namely its pronunciation (which is important especially for those words that have several alternative phonetic forms) and a so called time stamp (word beginning and ending times). The ASR and FA outputs are time aligned (using these time stamps) and compared in order to delimit the matched parts from those that do not match. The latter contain errors caused either by the recognizer or by the human transcriber. The next step is the analysis of these non-matched parts by an M-word-to-N-word (M2N) alignment procedure (it takes pronunciation and other aspects into account) and verification of the reference transcription.

3.1 Employed System, Modules and Methods

Here we describe the employed ASR system, supporting modules and methods.

Automatic Speech Recognition System. Since 2006, when we launched our ASR platform for Czech broadcast monitoring and transcription [7], the system has been significantly improved. Its recent version operates with a 483K word lexicon (the lexicon size as well as the number of pronunciation forms gradually increases). The lexicon and the language model have been optimized for the task mentioned in section 2, using a 30 GB corpus made of contemporary and historical newspapers texts and broadcast transcriptions. The acoustic model is based on context-dependent HMMs (with 5,562 physical states and 180,864 gaussians) trained on almost 320 hours of annotated speech recordings provided by more than 5,000 speakers. The system operates with the WER in range from 5 % (for professionally read speech) to 20 % (conversational speech). More details about the system can be found in [6] and [8].

Forced Alignment Module. The above system can be set up to run in the forced-alignment mode. In this case, it takes an audio recording and its text transcriptions and tries to assign the best fitting pronunciations and time stamps to all words. The task can be accomplished even if the transcription is not accurate – if there are some missing, substituted or extra words in the text [9].

Grapheme to Phoneme Conversion. The Grapheme to Phoneme (G2P) conversion module employs two knowledge sources to assign the pronunciation to a word. The first one is the lexicon (usually with multiple phonetic forms per word) and the second one is a set of rules (based on [10]) together with a list of exceptions (mostly for foreign words) that is used to generate pronunciations to out-of-vocabulary (OOV) terms. Note that in this article we use the Czech phonetic alphabet [11] for phonetic transcriptions.

Maximum Alignment Score. The MAS is a method for comparing two strings derived from the MED. MAS utilizes dynamic programming to align two strings by maximizing the score given as a sum of weights ($H_w = 1$; $S_w=D_w=I_w = 0.7$). Our implementation can process whole words as well as single characters.

Splitting into Matched and Non-matched Parts. This is done to segment a recording into parts whose FA and ASR transcriptions are identical and those that do not match (and probably contain errors). It is done by aligning the two transcriptions using their time stamps. (To ease the procedure, each word is represented by its time center.) We take the first word in the FA text and find the ASR word with the closest time center. If they are identical, we set a new start of the matched part, otherwise we proceed with next FA word. Then we try to append the following words and compare them. If they are same we continue in searching the end of the matched part. This is done until we reach the last transcription word. The words or their sequences that stay outside the matched parts are labeled as non-matched parts. These enter the next procedure.

M2N Alignment. Here, the goal is to align two text sequences (ref_ortho, asr_ortho) using their phonetic forms (ref_phon, asr_phon). We want to determine which M words in the recognized sequence correspond to N words in the reference. We start with computing the MAS of both phonetic transcriptions. If there is a hit ($H_s = 1$) or a substitution ($S_s = 0.7$), we raise the value in the SCORE matrix (of the word pair to which the phonemes belong). Deletions and insertions have zero score ($D_s=I_s = 0$). Maximum values of rows and columns of the SCORE matrix mean the words belong together. By evaluation of these pairs, we get the (phonetically) aligned word segments. An example is shown in Tables 1, 2 ('*' marks the segment borders and '‖' marks word borders). We compared ref: '*se také bavili muži z náměstí*' to the asr: '*sedákem balily můžeš na místě*'.

Phonetic Verification. The reference data gained via the FA method can contain some transcription errors. To detect and eliminate them, we verify the phonetic sequence to the reference transcription. All possible phonetic transcriptions of the reference sequence are generated by G2P. Then the phonetic score S_p is computed (1), where ph_{ref} stands for the reference phonetics, ph_i is the ith phonetics generated by the G2P. N_{hits} is the number of hits and N_{subs} stands for number of substitutions in the MAS output. The best S_p reached over all phonetics is the segment score. If the S_p reaches the acceptance threshold $S_p^* = 0.6$, the sequence is processed as correctly aligned. The acceptance threshold was set experimentally on 300 error-sequences from an independent development set.

$$S_p(ph_{ref}, ph_i) = \frac{1 \cdot N_{hits}(MAS(ph_{ref}, ph_i)) + 0.7 \cdot N_{subs}(MAS(ph_{ref}, ph_i))}{length(MAS(ph_{ref}, ph_i))} \quad (1)$$

Table 1. Example: The values of the SCORE matrix

	sedákem	balili	můžeš	na	místě
se	**2.0**	0.0	0.0	0.0	0.0
také	**2.4**	0.7	0.0	0.0	0.0
bavili	0.0	**5.7**	0.0	0.0	0.0
muži	0.0	0.0	**3.4**	0.0	0.0
z	0.0	0.0	**0.7**	0.0	0.0
náměstí	0.0	0.0	0.0	**1.7**	4.4

Table 2. Example: Results of the M2N alignment (for used phonetic alphabet see [11])

ref_ortho	s	e	t	a	k	é		*	b	a	v	i	l	l	i	*	m	u	ž	i	z	*	n	á	m	ě		s	t	í
ref_phon	s	e	t	a	k	é		*	b	a	v	i	l	l	i	*	m	u	ž	i	z	*	n	á	m	ň	e	s	ť	í
MED	H	H	S	S	H	I		*	H	H	S	H	H	H	*	H	S	H	S	S	*	H	S	H	D	S	H	H	S	
asr_phon	s	e	d	á	k	e	m	*	b	a	l	i	l	i		*	m	ú	ž	e	š	*	n	a	m		í	s	ť	e
asr_ortho	s	e	d	á	k	e	m	*	b	a	l	i	l	y		*	m	ů	ž	e	š	*	n	a	m		í	s	t	ě

4   Experimental Data

We conducted two large-scale experiments. In the first one (Exp1), we utilized 23.6 hours of speech data. It was recordings of individual sentences (5 to 20 words long) read by several tens of speakers (students, their friends or their relatives). The reference texts in this set contained 67,838 words in 5,745 files. The second experiment (Exp2) was focused on typical broadcast speech. We used 763 hours of recordings of radio programs (mostly news and debates), for which we had transcriptions provided by a media monitoring company. These transcriptions contained the total number of 3,849,129 words in 1,868 files. It should be noted that they were not always exactly verbatim, as some parts (e.g. jingles, headlines, greetings as well as repeated words and hesitations) were often omitted. These inconsistences were identified and eliminated during the FA pass. When we applied the proposed error analysis scheme, 12,336 errors were detected in Exp1 and 526,065 errors in Exp2. There were 523 errors/hour in Exp1 and 689 errors/hour in Exp2. The higher error rate in the latter case was mainly due to the spontaneous character of speech.

5 Error Analysis

The results from both the experiments have form of detailed lists of all the errors. The lists can be filtered and sorted according to various criteria. We can see, for example, which word substitutions are the most frequent ones, which word endings contribute most to confusions, which compound (or quasi-compound) words are recognized as de-compound ones (and vice versa), which lexicon entries may have incorrect and inappropriate phonetic forms.

5.1 Word-Ending Errors

One of the most frequent source of transcription errors are the confusions caused by acoustically very similar (or even the same) word endings. This type of error represents 24.5 % (Exp1) and 20.4 % (Exp2) of all the errors detected by our system. The most frequent ones are listed in Table 3A. We can see, that often it is just the length of the final vowel that cause many confusions (e.g. í <=> i, é <=> e). Frequent are also substitutions between final vowels or diphthongs (e.g. e <=> i, ou <=> u). Let us note that these confusions may be strongly affected by speakers with sloppy pronunciation. As expected, the homophone verb forms with *-li/-ly* ending are among the top 5 word-ending errors.

5.2 Word Substitutions

The second type of very frequent errors are substitutions, deletions and insertions of short words. In Czech, there are 8 single-phoneme words (conjunctions and prepositions *a, i, o, u, k, s, v, z*) and all belong to the 50 most frequent lexemes. These and several other short and often confused words are listed in Table 3B.

5.3 Other Findings

The detailed inspection of the error lists helped us to identify possible mistakes in the phonetic part of the lexicon. It was mainly in the *di, ti, ni* syllables whose correct pronunciation depends on the (Czech or foreign) word origin. We have also learned how many errors are results of just one confused phoneme (7.6% in Exp1 and 3.7% in Exp2). Another type of statistics revealed the most frequently confused homophones.

Even though the two experiments used different types and size of speech recordings, the achieved results are very similar. We found that 93 of the top 100 errors in Exp1 were among the top 250 errors in Exp2 and, seen from the other side, 87 of top 100 errors from Exp2 occurred among top 250 ones in Exp1.

Table 3. The most frequent ending and word substitutions (symbol ø marks omission)

A - ending substitutions			B - word substitutions		
ending	Exp1	Exp2	error	Exp1	Exp2
e <=> i	200	8996	a <=> ø	602	14992
ou <=> u	216	8811	je <=> ø	171	6472
o <=> u	131	7230	v <=> ø	254	4384
í <=> i	199	6407	i <=> ø	137	3693
li <=> ly	142	6321	si <=> se	95	1524
é <=> e	89	5027	to <=> ø	18	1385
á <=> a	126	4924	o <=> ø	65	1269
ú <=> u	154	4145	se <=> ø	90	1048
a <=> ø	131	3209	k <=> ø	56	737
ím <=> í	90	2723	ale <=> ø	13	732
o <=> ou	43	2676	z <=> ø	65	723
ení <=> í	72	2521	na <=> ø	32	693
o <=> ø	95	2256	ale <=> a	32	669
á <=> é	29	2169	je <=> i	17	637
í <=> i	41	2141	tak <=> pak	25	627
a <=> y	25	1915	že <=> ø	22	533
í <=> ích	43	1899	na <=> a	12	486

5.4   How to Eliminate Frequent Errors

Having the statistics on the most frequent errors, we can try to find ways to eliminate them. We could see that many errors were related to very short words. Therefore, we have adopted two solutions: 1) We have significantly increased the number of multi-word lexicon entries. There are 5,120 multi-words in the current lexicon and most of them consist of a short conjunction or preposition followed by a frequently co-occurring word. 2) All the short (and other often confused) words were assigned multiple alternative phonetic forms. These try to cover voiced as well as unvoiced pronunciation, presence or absence of the glottal stop, influence of stress, etc.

6   Conclusions

In this study we present a method that allows detailed analysis of most frequent errors in ASR applications. The method is based on the alignment of the ASR output with the reference transcription that goes into a sub-word level and utilizes also phonetic information. We have applied the proposed scheme in two large-scale experiments (using 763 and 23 hours of speech) and obtained rather representative statistics on errors occurring at the output of an advanced Czech ASR system. The identification of the most frequent error types and their partial elimination contributed to the performance improvement of the archive transcription system by 1.5% [8]. The scheme also helps us in porting ASR systems to other Slavic languages (e.g. Croatian [12]), where it facilitates the detection of the most frequent lexicon errors caused by incorrect phonetic forms.

Acknowledgment. The work was supported by project no. DF11P01OVV013 provided by the Ministry of Culture of the Czech Republic (program NAKI).

References

1. Goldwater, S., Jurafsky, D., Manning C.D.: Which Words are Hard to Recognize? Prosodic, Lexical and Disfluency Factors that Increase ASR Error Rates. In: Proc. of ACL/HLT 2008, Columbus (OH), pp. 380–388 (2008)
2. Nemoto, R., Vasilescu, I., Adda-Decker, M.: Speech Errors on Frequently Observed Homophones in French: Perceptual Evaluation vs. Automatic Classification. In: Proc. of LREC 2008, pp. 2189–2195. Marrakech, Morocco (2008)
3. Nouza, J., Žd'ánský, J., Červa, P., Silovský, J.: Challenges in Speech Processing of Slavic Languages (Case Studies in Speech Recognition of Czech and Slovak). In: Esposito, A., Campbell, N., Vogel, C., Hussain, A., Nijholt, A. (eds.) Second COST 2102. LNCS, vol. 5967, pp. 225–241. Springer, Heidelberg (2010)
4. Wagner, R.A., Fischer, M.J.: The String-to-String Correction Problem. Journal of the ACM 21(1), 168–173 (1974)
5. HTK Documentation, http://htk.eng.cam.ac.uk/docs/docs.shtml
6. Nouza, J., Blavka, K., Boháč, M., Červa, P., Žd'ánský, J., Silovský, J., Pražák, J.: Voice Technology to Enable Sophisticated Access to Historical Audio Archive of the Czech Radio. In: Grana, C., Cucchiara, R. (eds.) MM4CH 2011. CCIS, vol. 247, pp. 27–38. Springer, Heidelberg (2012)
7. Nouza, J., Žd'ánský, J., Červa, P., Kolorenč, J.: A System for Information Retrieval from Large Records of Czech Spoken Data. In: Sojka, P., Kopeček, I., Pala, K. (eds.) TSD 2006. LNCS (LNAI), vol. 4188, pp. 485–492. Springer, Heidelberg (2006)
8. Nouza, J., Blavka, K., Červa, P., Žd'ánský, J., Silovský, J., Boháč, M., Pražák, J.: Making Czech Historical Radio Archive Accessible and Searchable for Wide Public. Journal of Multimedia 7(2), 159–169 (2012)
9. Boháč, M., Blavka, K.: Automatic Segmentation and Annotation of Audio Archive Documents. In: Proc. of ECMS 2011, pp. 61–66 (2011)
10. Palková, Z.: Fonetika a Fonologie Češtiny. Karolinum, Prague (1997) (in Czech)
11. Nouza, J., Psutka, J., Uhlíř, J.: Phonetic Alphabet for Speech Recognition of Czech. Radioengineering 6(4), 16–20 (1997)
12. Nouza, J., Červa, P., Žd'ánský, J., Kuchařová, M.: A Study on Adapting Czech Automatic Speech Recognition System to Croatian Language. In: Proc. of 54th International Symposium, ELMAR 2012, Zadar, Croatia (September 2012)

Neural Network Language Model with Cache

Daniel Soutner, Zdeněk Loose, Luděk Müller, and Aleš Pražák

University of West Bohemia, Faculty of Applied Science, Dept. of Cybernetics
Univerzitní 8, 306 14 Plzeň, Czech Republic
{dsoutner,zloose,muller,aprazak}@kky.zcu.cz

Abstract. In this paper we investigate whether a combination of statistical, neural network and cache language models can outperform a basic statistical model. These models have been developed, tested and exploited for a Czech spontaneous speech data, which is very different from common written Czech and is specified by a small set of the data available and high inflection of the words. As a baseline model we used a trigram model and after its training several cache models interpolated with the baseline model have been tested and measured on a perplexity. Finally, an evaluation of the model with the lowest perplexity has been performed on speech recordings of phone calls.

Keywords: neural networks, language modelling, automatic speech recognition.

1 Introduction

Statistical language models (LM) play an important role in the state-of-art large vocabulary continuous speech recognition (LVCSR) systems. Statistically computed n-gram models and class-based LMs are the main models used in LVCSR systems. However, in recent years, some other types of LMs have also attracted a lot of attention; among the most successful examples of such models are neural network based language model [2,5] (NNLM). The main reason for their attractiveness is their ability to model a natural language in another way than the standard n-gram models, so that their combination with standard back-off n-gram models increases the speech recognition accuracy. Another approach to language modelling are the cache models (CacheLM) [6] – kind of language model which reflects short-term patterns of word use. We focused on cache models based on the words and a cache based on the word classes. In this article we are describing our experiments with combinations of neural network models and cache models to get better performance.

Our data for training and measuring of performance were the Czech phone calls with a small corpus size and specific type of language (an unusual word inflection and pronunciation, a unique vocabulary), so different attitude of language modelling is needed.

The rest of the paper is organised as follows: Our task is specified and data are described in Section 2, used NNLMs and CacheLMs are reviewed in Section 3. In Section 4 various combinations of LMs are evaluated, first on perplexity and finally on phone records.

P. Sojka et al. (Eds.): TSD 2012, LNCS 7499, pp. 528–534, 2012.

2   Data Description

As a training and test data we used Czech spontaneous speech which was recorded from phone calls. These calls were acquired as "Free calls" where people could phone for free while giving permission to use anonymously their calls for the speech recognition experiments.

We had to deal with the task where recorded data were very different from a common written Czech language. This is not a trivial task, as shown in [8] or in [11], where the records with spontaneous speech were also processed. The data are specified by:

– a high inflection of Czech language (cases, various verb forms,...)
– word inflection is partially different from written Czech language
– unusual words used by speakers (slang, diminutives,...)
– only a small set of data available (about 1.5M words)
– the records contain a lot of non-speech events
– the sentences are relatively short
– the vocabulary is relatively small (about 65k words)

The statistics of the used corpus is shown in Table 1; the corpora was divided into tree parts: training, development and test set. The characteristics of the test phone records used for models evaluation on speech recognition are shown in Table 2.

Table 1. BH text corpus

	Text data		
	Sentences	Words	OOVs
Train	397219	1580254	–
Dev	3236	13250	350
Test	3236	14477	385
Sum	403691	1607981	

Table 2. BH test records

	Records
Length h:mm	2:16
Files/Sententces	4115
Speakers	50

3   Language Models

Given the data described in Section 2, we decided to use not only a standard statistical back-off model, but also the neural network model and the cache model. We expect that NNLM will do better smoothing on such a small data that we have. The Cache LM we are testing because we suppose that they are able to model a topic of the conversation.

3.1   Neural Network Language Model

The architecture of used NNLM is based on the approach used in [5]. As it could be seen in Figure 1, we separated the desired 4-gram LMs into two neural networks, with one hidden layer each.

The first NN is a bigram neural network. The task of training this NN is relatively simple: given the word w from a vocabulary V, estimate the probability of the next word in the text. For that purpose a neural network with input and output layer of the size $|V|$ is used, where the word w in the input layer means that all input neurons are set to 0 except the one that corresponds to word w which is set to 1. Further, one hidden layer of a small size (10–50 neurons) is used to ensure that similar n-grams are somewhat clustered in the multi-dimensional space, thus giving similar outputs. The hidden layer uses a sigmoid activation function. The output layer uses a softmax function that normalizes output neurons values to sum up to 1.

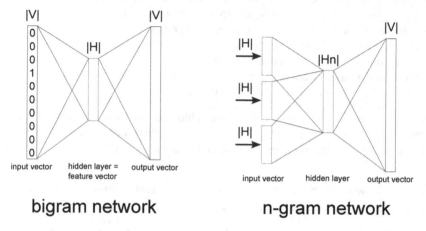

bigram network   **n-gram network**

Fig. 1. Schema of NNLM used

The second NN is an n-gram neural network (in our case 4-gram). Its structure is similar to the structure of the bigram neural network described in the previous step. But instead of $n-1$ word indexes (each of a size $|V|$) it uses as its input the hidden layer of the first NN of these $n-1$ words, thus greatly reducing the size of the input vector (from $(n-1)|V|$ to $(n-1)|H|$). This hidden layer effectively reduces the size of the input vector to a lower dimension without a significant loss of information; the input of the hidden layer is called a feature vector (FV).

The separation of the whole NNLM on a bigram NN and the n-gram NN with a low dimensionaly input has a very little impact on accuracy of the resulting NN, but it brings high benefit of speed and memory-saving.

The speed of NN is further increased by reducing the size of a vocabulary by merging the low-count words (i.e. less than 2 occurrences) in one token; we call them *RARE* words.

3.2 Cache Language Model

There are many ways how LMs could adapt themselves and one of them is a cache model [6]. Another approach should be for example the topic detection of the conversation or text [9] or unsupervised adapting of whole language model [10]. The cache LM uses a window of the n most recent words to determine the probability distribution of the next word. Because of a small corpus we decided to deal only with unigram cache models.

Simple Word Cache. Standard n-gram LMs will assign a low probability to the words that have been observed rarely. The cache models consider that words that showed in recent past are more likely to appear again. The base cache model (*linear*) [6] could be formulated as

$$P_{CA}(w_i|w_{i-M}^{i-1}) = \frac{1}{M} \sum_{m=1}^{M} \delta(w_i, w_{i-m}), \tag{1}$$

where δ means Kronecker delta: $\sum_{w_i \in V} \delta(w_i, w_l) = 1$, $\delta(w_i, w_l) = 0$ for $i = l$ and $\delta(w_i, w_l) = 0$ otherwise.

Modifications of this concept can be obtained when we consider a model with forgetting - the weights of words appeared in the past are lowered linearly (*triangle model*) or exponentially (*exponential model*).

The right choice of the cache length is also needed; with a respect to our data, we were testing 25, 50, 75 and 100 words.

Cache Based on Feature Vectors. We tried to deal with the inflection of Czech language with clustering similar words to word classes [4]. The Class cache model determines the probability of the next word given the last recent word classes, instead of words.

We also performed clustering based on the neural networks, more specifically on feature vectors (FV). From input data we trained feature vectors as mentioned in Section 3. Based on this FV (after training various lengths of vector, we finally decided to choose 10) we clustered words with **K-means** algorithm into 700–2,000 classes. Again we tested a linear cache model, triangle and exponential.

4   Experiments

The experiments were divided into two phases: first we choose the best models according to their perplexity and after that we evaluate their performance on phone calls with LVCSR system.

4.1   Perplexity Results

The perplexity was computed with and without start-of-sentence and end-of-sentence tokens (<s> and </s>) noted as PPL (resp. PPL1). The sentences in corpus are short (average 4.47 words/sentence) so, as we show further, it matters if start- and end-of-sentence tokens are used. The word list has 65k words, for NNLMs is shortened to 30k words.

Perplexity performance of standard n-gram LM is shown in Tab. 3 – we received the same results for 3-gram and 4-gram. With respect to our 4-gram based NNLM we use a as baseline model a 4-gram model with Kneser-Ney smoothing (KN4). These n-gram models were created using SRILM toolkit [1]. The best results measured on perplexity for different types of our proposed models are shown in Tab. 4, 5 and 6.

Various NNLM were tested (10–50 hidden layers, different vocabulary size). Some of results are in Tab. 4. Models with the best performance are ANN10 (with 10 hidden neurons and 30k vocabulary) and ANN50 (50 hidden neurons and 20k vocabulary). The results for combination of n-gram model combined with Class cache are in Tab. 5; the best performance have LMs with triangle forgetting and without forgetting, a better one is the cache LM with the length 50 and the optimal number of classes is about 1,600.

Tab. 6 shows results of Word cache LMs. The models with different word cache length were tested (25–100 words) and 3 types of forgetting as described in Section 3.2. For this type of LMs in general it seems to be better cache without forgetting (*linear*). The best result was achieved on the model with the cache length of 50 words.

Finally, Tab. 7 summarize the best partial results. As we expected, the combination of all three models (KN, NNLM and CacheLM) performs best, the model KN4 + ANN + Word cache performs even better than model with a Class cache. We observed, that this LMs decreases perplexity to the 61% resp. 45% of the original value.

Table 3. Baseline statistical models

	Statistical model		
Model	type	PPL	PPL1
KN2	2-gram	160.7	500.2
KN3	3-gram	150.7	462.3
KN4	4-gram	**150.7**	**462.4**

Table 4. Combined model with 4-gram and NNLM

KN4 + NNLM			
hidden layer	vocabulary	PPL	PPL1
10	20k	155,5	333,2
20	30k	181,8	332,8
10	30k	**93,5**	318,1
50	20k	150,0	**312,5**

Table 5. Combined models 4-gram, NNLM and Class cache

KN4 + Class cache			
Classes	Cache	PPL	PPL1
1600	50, exponential	100.0	310.6
1600	25, triangle	95.3	305.4
700	50, linear	93.1	299.7
1300	50, linear	92.8	297.9
1600	50, linear	**92.7**	**297.1**

Table 6. Combined models with 4-gram, NNLM and Word cache

KN4 + Word cache		
Cache model	PPL	PPL1
50 words, exponential	75.8	235.6
50 words, triangle	70.7	231.7
25 words, exponential	73.3	225.9
50 words, linear	**68.8**	**212.2**

4.2 System Evaluation

The evaluation of our models was done with our own LVCSR decoder and output word lattices were rescored with our proposed language models. An acoustic model was trained from approximately 100 hours of the phone calls with the same nature as the test set. Non-speech events were deleted from both refernece and recognized files. The out-of-vocabulary rate is 2.93%.

Table 7. Best LMs evaluation on perplexity

Model	PPL	PPL1
KN4	150.7	462.4
KN4 + ANN10	93.5	318.2
KN4 + ANN50	150.1	297.1
KN4 + Class cache 1600 lin-50	87.7	293.5
KN4 + ANN50 + Class cache 1600 lin-50	95.0	285.6
KN4 + ANN10 + Word cache lin-50	**68.8**	**211.5**

Table 8. Best LMs evaluation on recognition

Model	Accuracy (%)	Correctness (%)
KN4 + ANN10	44.07	46.47
KN4 + Class cache 1600 lin-50	44.26	46.49
KN4 + ANN10 + Class cache 1600 lin-50	44.27	46.51
KN4 + ANN10 + Word cache lin-50	44.27	46.51
KN4 + Word cache 1600 lin-50	44.53	46.86
KN4 + Class cache 1600 lin-50	44.66	47.15

5 Conclusion

In this paper we investigated the performance of NNLM with cache LM used on Czech spontaneous phone calls. Consistent accuracy improvements obtained on a state-of-the-art LVCSR system suggest that our combined ANN and cache models are useful on this task. Future research will focus on different types of cache (i.e. various types of clustering into classes) and improving the robustness of model.

Acknowledgments. This work has been supported by the Ministry of Industry and Trade, project No. MPO FR-TI1/486, and by the grant of The University of West Bohemia, project No. SGS-2010-054.

References

1. Stolcke, A.: SRILM – an extensible language modeling toolkit. In: INTERSPEECH (2002)
2. Mikolov, T., Kopecký, J., Burget, L., Glembek, O., Černocký, J.: Neural network based language models for highly inflective languages. In: ICASSP, pp. 4725–4728 (2009)
3. Schwenk, H., Gauvain, J.: Training Neural Network Language Models on Very Large Corpora. In: HLT/EMNLP (2005)
4. Brown, P.F., Pietra, V.J.D., Souza, P.V.D., Lai, J.C., Mercer, R.L.: Class-Based n-gram Models of Natural Language. Computational Linguistics, 467–479 (1992)
5. Bengio, Y., Ducharme, R., Vincent, P., Janvin, C.: A neural probabilistic language model. J. Mach. Learn. Res. 3, 1137–1155 (2003)

6. Kuhn, R., De Mori, R.: A Cache-Based Natural Language Model for Speech Recognition. IEEE Transactions on Pattern Analysis and Machine Intelligence, 570–583 (June 1990)
7. Trmal, J., Zelinka, J., Müller, L.: Adaptation of a Feedforward Artificial Neural Network Using a Linear Transform. In: Sojka, P., Horák, A., Kopeček, I., Pala, K. (eds.) TSD 2010. LNCS, vol. 6231, pp. 423–430. Springer, Heidelberg (2010)
8. Pavel, I., Josef, P., Psutka Josef, V.: Using Morphological Information for Robust Language Modeling in Czech ASR System. IEEE Transactions on Audio Speech and Language Processing 17, 840–847 (2009)
9. Skorkovská, L., Ircing, P., Pražák, A., Lehečka, J.: Automatic Topic Identification for Large Scale Language Modeling Data Filtering. In: Habernal, I., Matoušek, V. (eds.) TSD 2011. LNCS, vol. 6836, pp. 64–71. Springer, Heidelberg (2011)
10. Bacchiani, M., Roark, B.: Unsupervised language model adaptation. In: Proceedings of the IEEE International Conference on Acoustics, Speech, and Signal Processing (ICASSP), pp. 224–227 (2003)
11. Psutka, J., Švec, J., Psutka, J.V., Vaněk, J., Pražák, A., Šmídl, L., Ircing, P.: System for Fast Lexical and Phonetic Spoken Term Detection in a Czech Cultural Heritage Archive. EURASIP Journal on Audio, Speech, and Music Processing (2011)

TENOR: A Lexical Normalisation Tool
for Spanish Web 2.0 Texts

Alejandro Mosquera* and Paloma Moreda

DLSI-Universidad de Alicante, Alicante, Spain
{amosquera,moreda}@dlsi.ua.es
http://www.dlsi.ua.es

Abstract. The lexical richness and its ease of access to large volumes of information converts the Web 2.0 into an important resource for Natural Language Processing. Nevertheless, the frequent presence of non-normative linguistic phenomena that can make any automatic processing challenging. We therefore propose in this study the normalisation of non-normative lexical variants in Spanish Web 2.0 texts. We evaluate our system by restoring the canonical version of Twitter texts, increasing the F1 measure of a state-of-the-art approach for English texts by a 10%.

Keywords: Web 2.0, lexical variants, normalisation.

1 Introduction

Since the first appearance of social media, Web 2.0 applications have gained popularity on the Internet. Collaborative encyclopedias like Wikipedia, micro-blogging sites like Twitter and social networks like Facebook are among the top ranked websites by number of visits[1]. These applications have changed the flow of on-line information. This paradigm shift focuses on users who generate, share and consume this information. The informal nature of this exchange and the geographical and social diversity of users are reflected in their written language, with frequent occurrence of non-normative linguistic phenomena such as emoticons, slang and non-standard contractions among others.

For example, in the particular case of Twitter[2], the maximum number of characters per message is limited to 140, so it is common to find abbreviations and non-standard contractions in order to save space. Thus, as in SMS messages, some words or syllables can be represented by letters or numbers that have the same pronunciation but whose size is smaller. For example, the Spanish word *cansados (tired)* has a pronunciation equivalent to *cansa2*. Likewise, the syllable or conjunction *que (that)* can be replaced by *k* or *q*. Another way of shortening words is to omit certain letters, usually vowels.

---

* This paper has been partially supported by Ministerio de Ciencia e Innovación – Spanish Government (grant no. TIN2009-13391-C04-01), and Conselleria d'Educació – Generalitat Valenciana (grant no. PROMETEO/2009/119, ACOMP/2010/286 and ACOMP/2011/001).
[1] http://www.alexa.com/topsites
[2] http://www.twitter.com

P. Sojka et al. (Eds.): TSD 2012, LNCS 7499, pp. 535–542, 2012.

For example, the Spanish word *trabajo (job)* can be abbreviated as *trbj*. Moreover, the expression of emotions or moods is usually performed with emoticons, a combination of alphanumeric characters and punctuation symbols.

Because its lexical richness and the large volume of texts available, the Web 2.0 can be considered an important resource for natural language processing (NLP). However, the informal features present in these texts can complicate their automatic processing. Among the studies that address this problem we highlight those using normalisation techniques. Understanding the concept of normalisation as a process designed to "clean" texts by transforming non-standard lexical variants into their canonical forms. However, to our knowledge, the vast majority of the work along this topic have been focused on English texts. For this reason, this study proposes the normalisation of non-standard lexical variants in Web 2.0 Spanish texts. The implementation of the proposed approach (TENOR) will be used to restore the canonical version of Spanish texts extracted from Twitter.

The main contributions in this paper are:

(1) We have implemented a phonetic indexing algorithm adapted to the peculiarities of Spanish language.
(2) An unsupervised lexical normalisation of Spanish Web 2.0 texts has been performed using similarity measures, dictionaries, heuristic rules and language models have been developed.
(3) We demonstrate that the implementation of the proposed system obtains results comparable to the current state of the art for other languages, such as English.

The paper is organised as follows: Section 2 describes the state of the art. In Section 3 we explain our normalisation approach. The obtained results are evaluated in Section 4. Finally, Section 5 discusses the conclusions and future work.

2   Related Work

We have identified three major trends to tackle text normalisation. The first one relies on machine translation, the second is based on orthographic correction and the third one applies automatic speech recognition (ASR) techniques.

Machine translation has been proved useful for normalizing SMS texts [1] taking non-normative texts as the source language and its normalised equivalent as target language. This system has also been used to translate Spanish SMS texts [2] using MOSES [3] statistical translation engine. However, translation approaches need relatively large, previously normalised and aligned corpora to obtain good results [4]. Shannon's noisy channel model [5] is often used in automatic orthographic correction and spell checking systems [6]. However, there are intentional misspellings with aim to express emphasis or feelings *(goooooooooool!!)* and non-standard homophone contractions to save space *(knsado - cansado)*. In both cases, and when dealing with texts of limited length, performing text normalisation using this model can be challenging, since the context can not play a key role.

Finally, ASR techniques are based on the assumption that most non-standard lexical variants have a standard homophone equivalence *(ksa (house) - casa)*. Using phonetic coding algorithms in order to obtain a list of homophone candidates it is possible to perform text normalisation by extracting the most appropriate candidate with the help of language models [7].

3 TENOR Text Normalisation Proposal

Our text normalisation proposal follows a similar strategy to the successfully used in Web 2.0 and SMS English texts [8], using ASR techniques but adapted to the peculiarities of the Spanish language. In order to evaluate our normalisation process, a set of Spanish texts extracted from Twitter have been used.

First, we define the scope of our approach in Section 3.1. In Section 3.2 we explain our methodology.

3.1 Scope of Work

In this study we will focus on perform a word-level normalisation of lexical variants in Spanish texts. As a first approximation only the words marked as out-of-vocabulary (OOV) will be processed (see Section 3.2.1).

3.2 Methodology

The text normalisation process used by TENOR consists of two steps: In the first one, we used a classification method in order to detect non-standard lexical variants or OOV words. In the second one, the selected words in the previous step are normalised and replaced with their original canonic form.

Detection of OOV Words. In this study we refer to OOV words as those that are not part of the standard Spanish vocabulary and for this reason they need to be standardised. However, the detection of such words is not a trivial task: The presence of proper names, location names, neologisms and acronyms makes difficult to know whether a word belongs to the language or otherwise is a lexical variant. To carry out this process we have analysed two different classification methods: the first one makes use of dictionaries and heuristics while the second one performs a classification based on syllable n-grams.

Dictionaries. The expanded GNU Aspell[3] dictionary has been used and augmented with a list of country names, city names, abbreviations and common names extracted from Wikipedia[4]. The resulting lexicon of 931,435 words includes different verb tenses in both upper and lower case forms. Some limitations of using this lexicon are the absence of imperative tenses and the difficulty of detecting diminutive or augmentative words using heuristic rules, as these tend to be irregular. For simplicity, compound words have been treated as a single word.

---

[3] http://aspell.net
[4] http://es.wikipedia.org

Heuristics. Heuristics based on word capitalisation have been used in order to identify named entities and acronyms. Likewise, some special Twitter tags were used to make a slight syntactic disambiguation, such as: @ (User Name) # (Tag), RT (Retweet) and TT (Trending Topic), thus avoiding the processing of such elements.

Syllable N-gram Models. The use of syllabic and pseudo-syllabic language models have the ability to discriminate between languages, identify misspellings and lexical variants [9]. For example, the segmentation of the Spanish word *vosotrxs* would produce the following sequence: *vo-so-trxs*, where the syllable *trxs* is not present in the standard Spanish language.

Two trigram syllable models have been trained, the first on the CESS-ESP corpus [10] in order to detect normative Spanish and second on a set of 100 texts from Twitter selected by hand for their high level of informal vocabulary and non-normative expressions. A classifier was built using the two trained syllable models, marking words as OOV if the proportion of common syllabic trigrams with the Twitter model are higher than with the CESS-ESP.

Lexical Variant Substitution. This section will discuss the different steps that are performed in order to replace the lexical variants classified as OOV in the previous section for their normalised form. First, several filtering techniques employed to "clean" noisy texts will be introduced. The next step details the process of replacing abbreviations and transliterations. Then, the phonetic indexing algorithm implemented in TENOR for obtaining lists of words with equivalent pronunciation will be discussed. Subsequently, this method is applied in order to identify possible substitution candidates to replace the non-normative lexical variants. Finally, we explain how the use of word similarity algorithms and language models can help to select the most appropriate from the list of candidate words.

Filtering. First, all non-printable characters and non-standard punctuation symbols were eliminated with the exception of emoticons. While these may be beyond the scope of the study to not be considered lexical variants, its filtering could impact other NLP tools and applications such as sentiment analysis or opinion mining.

Abbreviations and Transliterations. The second step checks if the analysed OOV word is an abbreviation, in which case it gets replaced by its equivalent standard form. Moreover, heuristic rules are used in order to reduce letter repetition within words *(nooo!, gooooolll)*. Furthermore, we check for the presence of phonetic transliterations using numbers whose pronunciation is often used to shorten the length of the message *(separa2, ning1)* or combination of letters and numbers *(c4s4)* thus being replaced by using a simple set of conversion rules. Finally, emoticons have been grouped in two categories (happy and sad) in order to be replaced by their textual equivalence using heuristic rules based on regular expressions.

In addition, a table of equivalence with 46 of the most common multi-word expression abbreviations *(qtal - qué tal, xfa - por favor)* has been manually compiled and used in order to improve the normalisation results.

Phonetic indexing. A phonetic index with the entries of the lexicon expanded in 3.2.1. have been constructed. These words were grouped by their approximate pronunciation with the metaphone algorithm [11] by using a set of rules. To our knowledge, there is not a specific metaphone version for Spanish language, for this reason we have developed an adaptation based on the original English implementation (see Table 1). For example, the metaphone *(JNTS)* would index the words *gentes, gentíos, jinetas, jinetes, juanetes, juntas y juntos* between others.

Table 1. Spanish metaphone transformation rules, exceptions and examples

Substitutions	Exceptions and	Examples
á → A	C → X	(acción)
ch → X	C → Z	(césar, cien)
ç → S	C → K	(casa)
é → E	G → J	(gente, ecología)
í → I	G → G	(gato)
ó → O	H → mute	(hola)
ú → U	Q → K	(queso)
ñ → NY	W → U	(whisky)
gü → W		
ü → U		
b → V		
ll → Y		

In the next part of the process TENOR obtains the metaphone for each analysed OOV word, checking for their presence in the previously built phonetic index. Then, with the entries of matching metaphones a list of substitution candidates is generated.

Lexical Similarity. The Gestalt algorithm [12] which is based on the principle of maximum common subsequence, produces a similarity score between two strings with values between 0 and 100, where 100 is maximum similarity and 0 is the absence of similarity. TENOR calculates the similarity of OOV words against each of the phonetic candidates obtained in the previous step with this algorithm. Then, candidates with a similarity value below 60 have been discarded empirically, because after this threshold have not been observed reliable results.

Language Models. Finally, when two or more candidate words share the same similarity score a language model is used for obtaining the most appropriate substitution. A trigram language model has been trained on the CESS-ESP [10] corpus using smoothing techniques [13]. We have used NLTK NgramModel [14] implementation in order to determine the replacement that minimises the model perplexity, taking the latter as a measure of model quality.

4 Evaluation

We have used the measures of precision, recall and F1 defined in [15] for the evaluation of the obtained results.

4.1   Used Corpus

Because this study performs a non-supervised lexical normalisation we do not need any training corpus. However, an annotated data set has been used to evaluate TENOR performance. In order to do this, 1000 Spanish texts extracted from the Twitter network have been tagged by hand, marking the words individually as OOV or IV. In the special case of OOV words their normalised form have been provided.

4.2   Results

The distribution of non-standard linguistic phenomena is shown in Table 2-a. The most frequent observed language deviation is the emphatic letter repetition within a word.

TENOR evaluation was conducted in two phases: First we have considered the detection of OOV words and then analysed the results of OOV normalisation. Both the F1 obtained in the OOV classification (89%) and normalisation (83%) show an improvement over state-of-the-art approaches for the English language (see Table 2-b).

Table 2. a) Results of OOV classification using language models (FV-M), dictionaries (FV-D), OOV normalization (NRM), OOV normalization using table of common abbreviations (NRM-A) and Han2011 results for OOV classification (FV) and normalisation (NRM) of English Tweets. *P* precision, *R* recall and *F1* the harmonic mean of precision and recall b) Lexical variant distribution.

	(a)					(b)	
Task	**P**	**R**	**F1**	**Type**		**Ratio**
TENOR(FV-M)	58.4%	81.5%	68%	Letter substitution		7,55%
TENOR(FV-D)	**82.7%**	**98%**	**89.7%**	Letter omission		14,66%
Han'11(FV)	61.1%	85.3%	71.2%	Emphatic repetition		36%
TENOR(NRM)	94.1%	56%	70.2%	Absence of orthographic accents		7,55%
TENOR(NRM-A)	**96.1%**	73%	**83%**	Other		34,22%
Han'11(NRM)	75.3%	**75.3%**	75.3%			

4.3   Analysis

We have noticed the difficulty of distinguish between unintentional and intentional spelling errors. While unintentional errors are usually caused by typing mistakes or lack of Spanish knowledge (see Table 3-d), the use of SMS-style contractions (see Table 3-a), deliberate omission of orthographic punctuations or making use of expressions in other languages such as English, Catalan, Basque, Galician and Valencian (see Table 3-e) are intentional lexical variants typical of informal language registers. Moreover, the use of language models for candidate selection is not always effective due to the lack of context in very small texts (see Table 3-f).

Table 3. Examples of Spanish texts from Twitter before and after the normalisation process

	Raw Text	Normalised
a)	tdo StO no s cierT, stams caNsa2	todo esto no es cierto, estamos cansados
b)	xfa apoyo xa 1 niño d 3 añits	por favor apoyo para 1 niño de 3 añits
c)	mal momemto para sufrur!	mal momento para sufrir!
d)	bamos a x ellos nesecitamos el apollo!!!!!	vamos a por ellos necesitamos el apoyo!
e)	amunt! valencia, visca el barça!	aumento! Valencia, busca el F.C. Barcelona!
f)	el no aprobara	el no aprobara

5   Conclusions and Future Work

In this paper we have undertaken a study on lexical normalisation of Spanish Web 2.0 texts using the TENOR tool. By evaluating its validity using Twitter texts, the obtained results are comparable to the state of the art for the English language, improving its results by a 26% in OOV detection and a 10% in text normalisation (F1). In order to do this, we investigated word similarity measures and phonetic indexing algorithms, adapting and porting the metaphone algorithm from English to Spanish.

In a future work we plan to develop an English version of TENOR. Moreover, the application of language detection techniques adapted to micro-texts, such as those extracted from Twitter and instant messaging applications, could be useful for detecting the presence of mixed languages. Finally, the use of named entity recognition systems and Web 2.0 metadata can provide additional information such as geographic location or user language, what will allow us to improve the obtained results further.

References

1. Aw, A., Zhang, M., Xiao, J., Su, J.: A phrase-based statistical model for SMS text normalization. In: Proceedings of the COLING/ACL, pp. 33–40 (2006)
2. López, V., San-Segundo, R., Martín, R., Echeverry, J.D., Lutfi, S.: Sistema de traducción de lenguaje SMS a castellano. In: XX Jornadas Telecom I+D, Valladolid, Spain (2010)
3. Hoang, H., Birch, A., Callison-burch, C., Zens, R., Aachen, R., Constantin, A., Federico, M., Bertoldi, N., Dyer, C., Cowan, B., Shen, W., Moran, C., Bojar, O.: Moses: Open source toolkit for statistical machine translation, pp. 177–180 (2007)
4. Kaufmann, J.: Syntactic Normalization of Twitter Messages. REU Site for Artificial Intelligence Natural Language Processing and Information Retrieval Research Project 2 (2010)
5. Shannon, C.E.: A mathematical theory of communication. The Bell Systems Technical Journal 27, 379–423 (1948)
6. Choudhury, M., Saraf, R., Jain, V., Sarkar, S., Basu, A.: Investigation and modeling of the structure of texting language. In: Proceedings of the IJCAI-Workshop on Analytics for Noisy Unstructured Text Data, pp. 63–70 (2007)
7. Gouws, S., Metzler, D., Cai, C., Hovy, E.: Contextual Bearing on Linguistic Variation in Social Media. In: ACL Workshop on Language in Social Media (LSM) (2011)

8. Han, B., Baldwin, T.: Lexical normalisation of short text messages: Makn sens a #twitter. In: Proceedings of the 49th Annual Meeting of the Association for Computational Linguistics: Human Language Technologies, pp. 368–378. Association for Computational Linguistics, Portland (2011)

9. Garcia, R.G., Dimitriadis, Y., Merino Pastor, F., Coronado, J.L.: Error detection in character recognition using pseudosyllable analysis. In: International Conference on Document Analysis and Recognition, vol. 1, p. 446 (1995)

10. Martí, M.A., Taulé, M.: Cess-ece: corpus anotados del español y catalán. Arena Romanistica. A New Nordic Journal of Romance Studies 1 (2007)

11. Philips, L.: The double metaphone search algorithm. C/C++ Users Journal 18, 38–43 (2000)

12. Ratcliff, J.W., Metzener, D.E.: Pattern matching: The gestalt approach. Dr. Dobb's Journal 13, 46–72 (1988)

13. Chen, S.F., Goodman, J.: An empirical study of smoothing techniques for language modeling. In: Proceedings of the 34th Annual Meeting of the Association for Computational Linguistics (ACL 1996), pp. 310–318 (1996)

14. Bird, S.: Nltk: the natural language toolkit. In: Proceedings of the COLING/ACL on Interactive presentation sessions, COLING-ACL 2006, pp. 69–72. Association for Computational Linguistics, Stroudsburg (2006)

15. Tang, J., Li, H., Cao, Y., Tang, Z.: Email data cleaning. In: KDD 2005: Proceeding of the Eleventh ACM SIGKDD International Conference on Knowledge Discovery in Data Mining, pp. 489–498. ACM Press, New York (2005)

A Bilingual HMM-Based Speech Synthesis System for Closely Related Languages

Tadej Justin[1], Miran Pobar[2], Ivo Ipšić[2], France Mihelič[1], and Janez Žibert[3]

[1] Faculty of Electrical Engineering, University of Ljubljana
Tržaška 25, 1000 Ljubljana, Slovenia
http://www.luks.fe.uni-lj.si

[2] Department of informatics, University of Rijeka
Omladinska 14, 51000 Rijeka, Croatia
http://www.inf.uniri.hr

[3] Faculty of Mathematics, Natural Sciences and Information Technologies,
University of Primorska, Glagoljaška 8, 6000 Koper, Slovenia
http://www.famnit.upr.si/en

Abstract. In this paper we investigate a bilingual HMM-based speech synthesis developed for Slovenian and Croatian languages. The primary goals of this research are to investigate the performance of an HMM-based synthesis build from two similar languages and to perform a comparison of such synthesis system with standard monolingual speaker-dependent HMM-based synthesis. The bilingual HMM synthesis is built by joining all the speech material from both languages by defining proper mapping of Slovenian and Croatian phonemes and by adapting acoustic models of Slovenian and Croatian speakers. Adapted acoustic models are then served as basic building blocks for speech synthesis in both languages. In such a way we are able to obtain synthesized speech of both languages, but with the same speaker voice. We made the quantitative comparison of such kind of synthesis with monolingual counterparts and study the performance of the synthesis in a relation to the amount of data, which is used for building the synthesis system.

Keywords: bilingual speech synthesis, HMM speech synthesis, Slovenian synthesis, Croatian synthesis.

1 Introduction

There is quite a large number of languages with small database resources and they are staying behind the more developed languages regarding speech-technologies point of view. On the other hand many of these languages have their own specialities and all these technologies need to be applied to those languages with special care. Recently, the techniques for developing HMM-based polyglot synthesis were introduced that could be also used for such kind of languages [1]. The main goal of such a synthesis systems is to combine data from multiple speakers in different languages into the same HMM-based synthesis capable of speaking all of the languages used in training.

In this research we investigate the bilingual HMM-based speech synthesis of Slovenian and Croatian language. Similar HMM-based syntheses with limited domain

P. Sojka et al. (Eds.): TSD 2012, LNCS 7499, pp. 543–550, 2012.

databases for both languages were already presented in the past [2,3]. In this paper we explore the possibilities of improving the acoustical models for Slovenian and Croatian HMM-based speech synthesis with adaptive training of target speaker voices from bilingual voice models trained on both Slovenian and Croatian speakers data. To obtain such bilingual speaker independent model the manual mapping of similar phonemes in both languages was performed. Such synthesis is capable to synthesize texts in both languages, which enabled us to perform quantitative comparison of such kind of synthesis with monolingual counterparts and study the performance of the synthesis in a relation to the amount of data, which is used for building the synthesis system.

2   HMM-Based Bilingual System

With the use of the HMM-based synthesis it is possible to built synthesis systems by applying adaptation of speaker voice to models trained with monolingual or multilingual data [1]. For the latter, the task of phone mapping between languages is required. To synthesize speech in the other language, cross-language speaker adaptation should be used where the adaptation is calculated for phonemes common to both languages and is applied to speaker independent models of the other language. Adapted models can be used to synthesize speech in both languages with common speaker characteristic of the target voice.

In our experiments we performed speaker adaptive training [4] for building different target speaker synthesis voices from basic multilingual multi-speaker models. The adaptation was done by using constrained maximum linear likelihood regression method [5] and fed by different target speaker's data of both evaluated languages.

2.1   Datasets for Speech Synthesis

We used the data from the Croatian VEPRAD [6] radio news and weather report corpus and the Slovenian VNTV [7] database of the same domain. From each corpus speech data from three male speakers were used for building the speaker-independent models and the fourth voice was used as a target voice for adaptation and also for building monolingual speaker-dependent model.

The total amount of data for each voice was randomly selected from speech database in a way that selected utterances represented all phones that were in databases and that the total amount of each voice's speech data was nearly the same. For the speaker dependent model it was used one selected speaker's voice of each language in total amount of 41 min and 43 min for Slovenian and Croatian languages, respectively. For the training of basic HMMs the speech data from three different speakers were used in total amount of 45 and 46 minutes of both languages separately and then the target speaker data was applied in total amount of 41 and 43 minutes in the Slovenian and in the Croatian case, respectively.

2.2   Phoneme Mapping

Since the databases used in our experiments already had phonetic dictionaries closely related to SAMPA phonetic alphabets of respective languages, we decided to map the phonemes with a help of SAMPA alphabet [8]. The task of joining phonemes was done with a phonetic expert, who has knowledge about both languages.

In the Croatian database the vowels were marked as stressed or unstressed presenting the 10 different vowels. That is in contrast with the Slovenian database, where the vowels are divided on long and short ones and stressed and unstressed. Since the Slovenian language consists of 8 marked vowels and Croatian of only 5 vowels the mapping was performed by listening the words that contain the short and long vowels in the VNTV database and comparing them to words in the VEPRAD database that contain stressed and unstressed vowels. We decided to map the short Slovenian vowels with the Croatian unstressed vowels and long Slovenian vowels with the stressed Croatian vowels. Additionally the Croatian syllable-forming /r/ and Slovenian unstressed schwa /@/ were mapped. All other Slovenian vowels, stressed schwa /@/, short stressed /O/ and short stressed /u/, stayed unmapped. Consonants mappings were performed easier by joining phonemes which corresponds to the same letter in both alphabets. Nevertheless there were some special consonants, who did not find its pair. Those were for Croatian /nj/, /lj/,/dz/, /dz'/ and /ts'/. The Slovene unmapped consonants were only /nj/ and /lj/. Finally, the joined phone set consisted of 44 phonemes, of which 33 appeared in both languages, 7 appeared only in Slovenian and 4 phonemes appeared only in Croatian speech data.

With prepared audio datasets and language dependent grapheme to phoneme trans-formation we generated the language dependent labels files with EHMM Labeler in-cluded in FestVox toolkit [9]. The phone mapping of phones in each language was conducted on each of such prepared label files.

2.3   HMM Training and Synthesis

For all realized synthesis systems we used the HTS synthesis toolkit [10].

First the speaker dependent acoustic HMMs models were built for selected Slovenian and Croatian speakers. For training the multilingual average model we used speaker adapted training procedure [4] to adapt the average model to target voice of speakers in both languages. In both speaker dependent and speaker adaptive training we used only one decision tree per state for all the phonemes. The questions of the decision trees refer only to the phonetic features of the phoneme and its immediate left and right context. All the HMMs were constructed as five-state left-to-right models.

To synthesize text to speech, first the grapheme to phoneme conversion had to be performed. This was done in our case with appropriate language dependent words-to-phoneme dictionary. Further on the transcriptions were mapped to a joined representation of phonemes. Such transcription was then converted into a sequence of HMM states.

3   Evaluation Experiments

We tested six different synthesizers. Two of them were strictly language and voice dependent, i.e., the HMMs for each language were built from the data of individual speakers of both languages, respectively. Other two speech-synthesis systems were

Table 1. Summary of speech data for training HMM models. The amount of speech material for each model in minutes are shown in () brackets.

synthesis system	SD data	ID data	adaptation data
SD(CRO)	VEPRAD (41)		
SD(SLO)	VNTV (45)		
ID(CRO)-ADAPT(SLO)		VEPRAD (45)	VNTV (43)
ID(SLO)-ADAPT(CRO)		VNTV (46)	VEPRAD (41)
ID(SLO+CRO)-ADAPT(CRO)		VEPRAD + VNTV (91)	VEPRAD (41)
ID(SLO+CRO)-ADAPT(SLO)		VEPRAD + VNTV (91)	VNTV (43)

developed to test impacts of cross-language training of HMMs. This means, that we built synthesis systems by performing adaptation of base HMMs, which were trained on one language, from the data of the other language. The last two synthesizers were built with speaker adaptive training to target language voice, but the base HMMs were trained from multi-speaker data of both languages, in this case it was used six different voices of both languages.

Altogether we built six different speech synthesis systems: SD(SLO) is a Slovenian speaker dependent speech synthesis, ID(CRO)-ADAPT(SLO) corresponds to basic HMM-synthesis trained on Croatian data and adapted from the Slovenian speaker data, and analogous is the ID(SLO)-ADAPT(CRO) synthesis, where the languages were switched; the ID(SLO+CRO)-ADAPT(SLO) and the ID(SLO+CRO)-ADAPT(CRO) synthesis systems represent the HMM synthesis systems, where the basic HMMs were trained on Slovenian and Croatian speakers data and adapted to Slovenian and Croatian speaker voice, respectively. The data that were used for building each individual speech synthesis system are summarized in Table 1.

We were mainly interested in two questions when building such systems:

- How does the adaptation of one target language speaker from another language monolingual average model impact the final synthesis in each language?
- Can we obtain better synthesized voice with adaptation of a target speaker from monolingual or multilingual basic voice models?

All the synthesizers were used for synthesizing texts in Slovenian and Croatian language. In total we trained six synthesizers with different sets of training data. Two of them were build as speaker dependent synthesizer and other four were trained with speaker adaptive training in order to perform adaptation of the Slovenian or the Croatian target voice from different basic models.

3.1 Evaluation Tests

To evaluate the performance of all synthesis systems a listening test was conducted using 5-point MOS scale [11]. Both Slovenian and Croatian native speakers participated in the test, mostly students of electrical engineering or informatics at the universities of Ljubljana, Slovenia, and Rieka, Croatia. Subjects were evaluating the tests using the headphones by filling out the web based forms for each of the synthesizers. One test

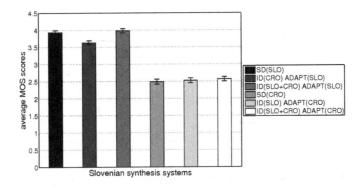

Fig. 1. MOS evaluation of Slovenian synthesis systems, number of evaluators was 31

form was used for one synthesized voice. The listening tests were randomly chosen, so each listener evaluated different sequence of speech synthesis systems. Subjects always rated only speech in their native language.

Factors considered in the evaluation were: speech overall quality of heard audio, comprehension and articulation of synthesized speech, pronunciation anomalies, naturalness of speech, and how native the speech sounds. All factors were evaluated in scores from bad (1) to excellent (5).

Each listener evaluated 6 sets of two samples of synthesized speech from 8 to 10 seconds, from each of the models in 1. Both samples were listened one after another. The first sample was obtained by synthesizing the same testing sentence for all synthesis systems, which was tested against the second sentence, which was different and obtained by different synthesizer. The test sentences were selected from the utterances in the VEPRAD and VNTV databases, but from the speakers that were not used in process of building the synthesis systems. With that kind of selection we obtained test sentences from the same domain. Further on, the selection was performed automatically following the requirement that the text data forming the evaluation sentences for each synthesis system had to include at least one occurrence of phoneme from all the phonemes used in the realized systems. In such way we obtained about 30 test samples for each language with approximate length between 8 to 10 seconds.

The evaluation results measured the overall quality of synthesis systems by collecting evaluation scores of all questions together and the average scores were computed. The significance of average scores was tested by performing non-parametric analysis-of-variance (ANOVA) test by using Kruskal-Wallis tests with multi comparison analysis [12].

3.2 Evaluation Results

The evaluation results from Slovenian native speakers is presented in Fig. 1 and an analysis of significant differences in average evaluation scores are shown in Fig. 2. The results shows all evaluated synthesis systems in Slovenian language.

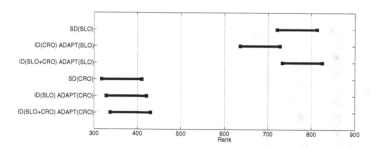

Fig. 2. The ANOVA test of significant differences between speech synthesis systems in the Slovenian MOS evaluation. Significant differences between scores are present in cases when 95% confidence intervals around each median value of different synthesis systems do not intersect with each other.

The analysis of variance tests in 2 show that Slovenian speaker dependent model SD(SLO) and other two models with speaker characteristics of Slovenian target speaker the ID(CRO)-ADAPT(SLO) and the ID(SLO+CRO)-ADAPT(SLO) were evaluated significantly different by Slovenian listeners to models SD(CRO), ID(SLO)-ADAPT(CRO) and ID(SLO+CRO)-ADAPT(CRO). Even though the synthesizer ID(SLO+CRO)-ADAPT(SLO) built with multilingual basic model and adapted to the Slovenian target speaker achieved overall the best average MOS score, there are no significant difference to the other two Slovenian target speaker models. On the other hand, the model ID(CRO)-ADAPT(SLO) achieved the lowest average MOS score among all trained models with the Slovenian speaker characteristics. The Slovenian MOS test was conducted by 31 native Slovenian listeners.

The Croatian evaluation was performed by 20 native Croatian listeners. The results are shown in Fig. 3 and 4. The Croatian listeners evaluated the model, which was built with multilingual average model and adapted to the Croatian target speaker, ID(SLO+CRO)-ADAPT(CRO), is significantly better than other models. The other models are not significantly different to each other, even though that the average evaluation scores of synthesis systems with the Croatian target performed are slightly better than the scores of the Slovenian target speakers.

Both evaluations showed expected performances of tested speech synthesis systems. In general, the syntheses with the target speakers of the same language as was evaluators' language were evaluated significantly better than those, which were in the other languages. These was proved by performing ANOVA tests in both evaluation cases. In the Slovenian evaluation all synthesis systems with the Slovenian target speakers performed significantly better than with the Croatian target speaker, while in the Croatian evaluations this was true only for the case with synthesis based on HMMs trained on the multilingual data and adapted to the Croatian target speaker. On the other hand, increasing the speech data material by mixing data from speakers of both languages increased the evaluation scores in Croatian case, while such synthesis was not significantly better in the Slovenian evaluation.

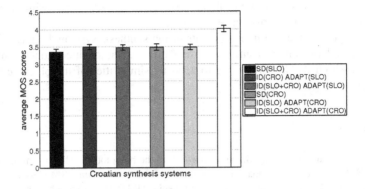

Fig. 3. MOS evaluation of Slovenian synthesis systems, number of evaluators was 20

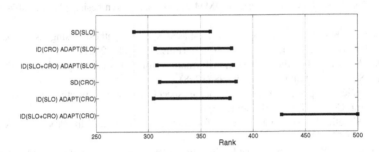

Fig. 4. The ANOVA test of significant differences between speech synthesis systems in the Croatian MOS evaluation. A significant differences between scores are present in cases when 95% confidence intervals around each median value of different synthesis systems do not intersect with each other.

4   Conclusion

The paper presents joined works of two research groups, who aimed to develop an HMM-based speech synthesis system of Croatian and Slovenian language. We built several versions of speech synthesis systems in order to evaluate the correspondence of the training material and adaptation data to final quality of the speech synthesis systems in both tested languages. With the phoneme mapping based on SAMPA phonetic alphabet we joined phonemes of Croatian and Slovenian languages, which enabled us to use the same speaker voice for making synthesis in both languages. Each synthesizer could produce Slovenian and Croatian syntheses of limited domain texts. The evaluations of different synthesis systems with two groups of Croatian and Slovenian listeners showed expected performances of the tested systems. It was shown that the syntheses with the target speakers of the same language as was evaluators' language were evaluated significantly better than those, which were in other languages. On the other hand, increasing the speech data material by mixing data from speakers of both languages increased the evaluation scores in Croatian case, while it was not significantly better in the Slovenian evaluation.

Future work will include improvements of the synthesis systems by increasing the training and adaptation data and by testing different automatic phoneme mapping techniques, which would better suite to acoustic similarities of the phonetic units in the source and target languages.

References

1. Latorre, J., Iwano, K., Furui, S.: New approach to the polyglot speech generation by means of an HMM-based speaker adaptable synthesizer. Speech Commun. 48, 1227–1242 (2006)
2. Vesnicer, B., Mihelič, F.: Evaluation of the Slovenian HMM-Based Speech Synthesis System. In: Sojka, P., Kopeček, I., Pala, K. (eds.) TSD 2004. LNCS (LNAI), vol. 3206, pp. 513–520. Springer, Heidelberg (2004)
3. Martinčić-Ipčić, S., Ipčić, I.: Croatian HMM-based speech synthesis. CIT 14, 307–313 (2006)
4. Yamagishi, J., Kobayashi, T.: Average-voice-based speech synthesis using hsmm-based speaker adaptation and adaptive training. IEICE Trans., 533–543 (2007)
5. Gales, M.: Maximum likelihood linear transformations for HMM-based speech recognition. Computer Speech and Language 12, 75–98 (1998)
6. Martinčić-Ipčić, S., Ipčić, I.: Veprad: a croatian speech database of weather forecasts. In: Proc. 25th Int. Conf. ITI, pp. 321–326 (2003)
7. Žibert, J., Mihelič, F.: Slovenian weather forecast speech database. In: Proc., SoftCOM, vol. 1, pp. 199–206 (2000)
8. Wells, J.C.: SAMPA computer readable phonetic alphabet. In: Handbook of Standards and Resources for Spoken Language Systems. Mouton de Gruyter, Berlin (1997)
9. Prahallad, K., Black, A.W., Mosur, R.: Sub-phonetic modeling for capturing pronunciation variation in conversational speech synthesis. In: Proc. of IEEE Int. Conf. Acoust., Speech, and Signal Processing (2006)
10. Zen, H., Oura, K., Nose, T., Yamagishi, J., Sako, S., Toda, T., Masuko, T., Black, A.W., Tokuda, K.: Recent development of the HMM-based speech synthesis system (HTS). In: Proc. APSIPA 2009, Sapporo, Japan (2009)
11. International Telecommunication Union: ITU-T Recommendation P.800.1: Mean Opinion Score (MOS) terminology. Technical report (2006)
12. Hochberg, Y., Tamhane, A.C.: Multiple Comparison Procedures. Wiley, New York (1987)

The Role of Nasal Contexts on Quality of Vowel Concatenations

Milan Legát[1] and Radek Skarnitzl[2,*]

[1] University of West Bohemia in Pilsen, Faculty of Applied Sciences
Department of Cybernetics, Czech Republic
legatm@kky.zcu.cz
[2] Charles University in Prague, Faculty of Arts
Institute of Phonetics, Czech Republic
radek.skarnitzl@ff.cuni.cz

Abstract. This paper deals with the traditional problem of occurrence of audible discontinuities at concatenation points at diphone boundaries in the concatenative speech synthesis. We present results of an analysis of effects of nasal context mismatches on the quality of concatenations in five short Czech vowels. The study was conducted with two voices (one male and one female), and the results suggest that the female voice vowels /a/, /e/ and /o/ are inclined to concatenation discontinuities due to nasalized contexts.

Keywords: speech synthesis, unit selection, concatenation cost, nasality, phase mismatch, pitch marks.

1 Introduction

Despite the increasing popularity of HMM based and hybrid speech synthesis methods, the unit selection concatenative systems still represent the mainstream in many real life applications, especially in limited domains where synthesized chunks are combined with pre-recorded prompts. In such applications, the ability of the unit selection to deliver highly natural and to the recordings well fitting output are the key factors. Not surprisingly, the unit selection also remains the first choice for eBook reading applications, which have been acquiring a lot of interest over the recent years.

Among the unit selection related issues that continue to be non-resolved, the audible discontinuities appearing at concatenation points play an important role. Many studies have been published over the last one and a half decades, dealing with the design of concatenation cost functions [1,2,3], to name but a few. Despite the considerable amount of efforts, none of them unfortunately succeeded in providing a clear answer on how to measure the concatenation discontinuities. The presented results have even sometimes been in contradiction.

A few papers, addressing the issue of phonetic effects on vowel concatenations [4], phonetic effects on discontinuity detection in general [5], or using phonetic features as a support for acoustic measures [6], have also been published.

* This research was supported by the Technology Agency of the Czech Republic, project No. TA01030476, and by the grant of the University of West Bohemia, project No. SGS-2010-054.

P. Sojka et al. (Eds.): TSD 2012, LNCS 7499, pp. 551–558, 2012.

We believe that the main contribution of this paper consists in a detailed analysis of one particular example of concatenation artifacts, which shows how complex the problem is, and that generalizations are difficult to make in this area. This could be one of the explanations for the failing of most of the traditional concatenation cost functions, whose common feature is speaker and phoneme generalization, to reliably reflect the human perception of the concatenation artifacts.

The rest of this paper is organized as follows. The next section gives a motivation for choosing nasals as the context of interest of this study. Section 3 describes our experimental set up, including a data collection and a listening test design. In Section 4, we present the results of our experiment and form a hypothesis to explain the obtained results. Section 5 describes more detailed analysis of the concatenation points with respect to the hypothesis, and finally, in Section 6, we draw conclusions and outline some future work intentions.

2   Motivation

From the articulatory perspective, nasalized vowels are quite simple—they differ from their oral counterparts only in lowering the velum. However, a large and complex resonance space emerges as a result of opening the nasal cavity, which is why nasalized vowels and vowels in the context of nasals in general represent, from the acoustical perspective, probably the most complicated sounds of human speech. This simple articulatory gesture is acoustically manifested in various ways, depending especially on the quality of the vowel and also on the degree of acoustic coupling between the two resonance chambers. For these reasons—and also because the nasal cavity and the paranasal cavities of every speaker are different—there are only few universal acoustic correlates of nasality.

The acoustic complex of nasalized vowels consists of nasal formants (formants of the pharyngonasal tract), oral or vocalic formants (whose frequency may, however, be shifted compared to non-nasal vowels), and antiformants which frequently appear in pairs with nasal formants [7].

The most important acoustic features responsible for the perceptual impression of nasality appear in low frequencies. One of the main correlates of nasality, regardless of vowel quality, is the relative lowering of the intensity of F1, specifically by 6–8 dB according to [9]. The second "universal" feature related to the nasality is the presence of a spectral peak around 250 Hz, which corresponds to the first formant of the pharyngonasal tract, typically marked as N1. The presence of antiformants due to coupling of the nasal cavity is also universal, but their specific frequencies differ in various studies (see [10] for a review).

For the purposes of concatenating units in a speech synthesis system, the presence of nasality in only one of two concatenated diphones may lead to perceived discontinuities, which was indeed observed in our informal experiments. A nasalized vowel may, on the one hand, manifest higher intensity in low frequencies (around 250 Hz, the N1) and, on the other hand, the energy roll-off above this peak is likely to be considerably greater due to the weaker F1 and generally stronger spectral slope. Our hypothesis is that it may be undesirable to concatenate a nasalized vowel with a nonnasal vowel or vice versa, since the energy difference in specific frequency bands may cause the impression of discontinuity. The aim of this paper is thus to examine whether controlling for the contextual nasality conflicts will lead to better continuity of concatenations in vowels.

3   Experimental Setup

3.1   Test Material

Recordings covering five Czech short vowels—/a/, /e/, /i/, /o/ and /u/(Czech SAMPA notation)—in all consonantal contexts were made in an anechoic room by two professional speakers—male and female. The recorded script was composed of three word sentences containing CVC word in the middle each, e.g. /kra:lofski: **kat** konal/. Recorded data were re-synthesized using the "half sentence" method [11]. This method consists in cutting the sentences in the middle of the vowels in the central words and combining the left and right parts, which results in a large set of sentences containing only one concatenation point in the middle of the central CVC word each and covering the vowels in all possible consonantal contexts. Note that the concatenations were done pitch synchronously using a simple overlap-and-add and weighting by the Hanning window, but no smoothing algorithm was applied for the reasons explained in [12].

3.2   Definition of Nasalization Mismatch

Since this work deals with the nasal contexts, a selection was made that only contained synthetic sentences containing a *nasalization mismatch*, henceforth referred to as the NAMI set (NAMI stands for the NAsalization MIsmatch). The rest of the sentences formed a set NOMI (NO MIsmatch). The *nasalization mismatch* was defined as a disagreement between an original context of a vowel and its target context, no matter if the disagreement was on the left or on the right side of the synthesized vowel. As an example, let us take two words /t_San/ and /t_Sas/, and create a synthetic word /t_Sa-as/ (dash symbol marks the concatenation point) using the left part of the first word and the right part of the second one. In our analysis, this synthetic word would be considered as containing the *nasalization mismatch*, because the original right context of the vowel /a/ in the first word was the nasal /n/, whereas in the synthetic word, the right context is the fricative /s/.

3.3   Mitigating the Role of F0 Discontinuities

F0 discontinuities are unquestionably a significant source of concatenation artifacts [12]. In order to be able to analyze the effect of nasalization, it was needed to factor the F0 discontinuities out. In most related studies, the standard procedure is to smooth the concatenation points with respect to differences in pitch and energy to make sure that any perceived discontinuity is not due to pitch or energy "jumps" at the concatenation points. As mentioned above, we have decided not to apply any pitch smoothing algorithm during concatenation, which was mainly to avoid the risk of introducing any sort of F0 smoothing artifacts that could influence listeners' ratings.

We have shown that clustering of pitch contours and concatenating words whose pitch contours fall within the same clusters is not a reliable way of predicting F0 concatenation artifacts [13]. Still, the information contained in the pitch contours can be leveraged for predicting concatenation discontinuities with a high accuracy using machine learning techniques [14].

For this work, we applied the SVM models trained on fine grained pitch contours extracted from a vicinity of concatenation points (see details in [14]) to identify sentences of the NAMI set that were supposed to be smooth, i.e. not containing an audible discontinuity at the concatenation points, according to the models' prediction. Let us further refer to this set as NAMI-S (NAsalization MIsmatch Smooth). The same models were analogically applied to obtain a set NOMI-S (NO MIsmatch Smooth).

3.4   Listening Test

Test Stimuli. As the next step, we randomly selected pairs of sentences—one sentence from the NAMI-S set and one from the NOMI-S set—containing the same word in the middle. Each vowel was represented by 15 pairs of sentences, resulting in the total number of 150 audio samples per voice presented to listeners. The sequence of pairs of samples of different vowels was also randomized. Two listening tests were organized— one for the male voice and one for the female voice.

Subjects. The subjects taking part in the listening tests were TTS experts and students working on TTS related projects. There were 9 and 10 subjects who finished the male and the female voice listening tests, respectively. Most of the subjects participated in both listening tests.

Procedure. The listeners were presented with the pairs of audio samples in a randomized order. Their task was to indicate whether or not they heard a concatenation discontinuity in any of the two samples, and which of the two samples they found better. It was also possible to say that none of the samples in a pair was better.

Both listening tests were conducted using a web interface allowing the listeners to work from home. It was, however, stressed in the test instructions that the tests shall be done in a silent environment and using headphones. As a preparation, the participants were presented (prior to each listening test) with a couple of samples containing audible discontinuities. There were no restrictions on how many times the listeners could play each sentence before providing their answers.

4   Listening Test Evaluation

Since the reliability of results obtained in any listening test—no matter if the participants are experts or not—is always an issue, a listeners ratings analysis was conducted in line with the procedures described in [15]. Based on the analysis results, one listener was excluded from each listening test.

As the next step, we have collected two sets of "facts". The first—discontinuity detection—set of "facts" was composed of audio samples for which more than or equal to 80% of listeners indicated that they heard a discontinuity (henceforth refferred to as

Table 1. *Counts of the discontinuity detection "facts" (DISC_FACT)*

	Female		Male	
	NAMI-S	**NOMI-S**	**NAMI-S**	**NOMI-S**
/a/	14	0	0	0
/e/	15	1	1	0
/i/	6	0	0	0
/o/	9	0	2	0
/u/	1	1	0	0

Table 2. *Counts of the preference "facts" (PREF_FACT). None stands for no preference "facts".*

	Female			Male		
	NAMI-S	**NOMI-S**	**None**	**NAMI-S**	**NOMI-S**	**None**
/a/	0	15	0	0	0	3
/e/	0	11	0	0	1	4
/i/	1	5	0	0	1	4
/o/	0	10	0	0	2	6
/u/	2	3	0	0	0	6

DISC_FACT). The second set of "facts" was based on preference scores, i.e. stimuli for which more than or equal to 80% of listeners expressed their preference for one of the samples in a pair or indicated no preference (henceforth referred to as PREF_FACT). Note that the majority threshold was set ad hoc to obtain a reasonable compromise between robustness and quantity of the "facts".

The results of the "facts" collection are summarized in Tab. 1 and Tab. 2. It is obvious from both tables that the *nasalization mismatches* at the concatenation points do not matter for the male voice, which was used in our study. This is in contrast to the female voice where we can see that especially for the vowels /a/, /e/ and /o/, there is a strong impact of the *nasalization mismatch* on the perceived quality of concatenations. To less extend, the effect can also be found for the vowel /i/.

It was interesting to speculate about the reasons for the obtained results from the perspective of the theory of the speech production. Since the problems were observed mainly in the female voice vowels /a/, /e/ and /o/, one interesting hypothesis that arose, was that the perceptual effects were due to a complex interaction of spectral peaks in a low frequency range.

For the female speakers' high vowels (/i/ and /u/), there are three spectral peaks in the frequency range between 200 and 500 Hz—fundamental frequency (F0), as well as the first oral and nasal formants (F1 and N1, respectively); see, for instance, [9]. It is well known that frequency components lying within 3 to 3.5 Bark from each other tend to be perceptually integrated into one broader peak [16]. That is exactly the case with the spectral peaks mentioned above.

It was therefore possible that the concatenation of an oral and a vowel from a nasal context could result in some sort of discontinuity acoustically, but since it is the energy within this 3.5-Bark band, which is relevant perceptually, the discontinuity could be inaudible. Supposing that this was true, it would have explained why /i/ and /u/ behave differently—the other vowels' F1 lies in higher frequencies and therefore falls outside of the 3.5-Bark range of the perceptual integration. It would have also explained why we did not find a similar situation for the male voice—his F0 lies much lower than the peak complex of F1 and N1.

5   Analysis of Discontinuities

To verify the hypothesis formulated in the previous section, we more closely investigated the concatenation areas in both time and frequency domains, and we got an intriguing finding. As shown in Fig. 1, the reason for the perceived discontinuities was a phase mismatch at the concatenation points. The phase mismatch appeared as a consequence of misplacement of pitch marks by our pitch marking algorithm [17], which got confused by strengthening of a harmonic signal component the peaks of which were in close vicinity of the F0 peaks. This strengthening of the harmonic component appears when a vowel (/a/, /e/ or /o/, to be more precise) stands in a context of a nasal consonant. For the vowel /a/, all audible discontinuities can be fixed by manual relabeling of the pitch marks in the concatenation areas, which was confirmed by an informal listening test.

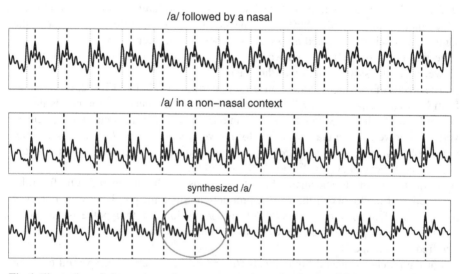

Fig. 1. Illustration of phase mismatch at a concatenation point due to mislabeling of pitch marks. Dashed lines show the positions of pitch marks as originally given by the automatic pitch marking algorithm. The dotted lines show the manual corrections of the pitch marks' positions. The circled area contains a pointer to an artificial signal peak due to the phase mismatch at the concatenation point.

The situation was however more complicated for the vowels /e/ and /o/, for which the perceived discontinuities could not be removed in this way in all sentences. The reason was that the harmonic component mentioned above got not only stronger, but also interfered with the F0 peaks. In such cases, it seems to be more advisable to avoid concatenations at all.

Interestingly enough, for the male voice we did not observe this phenomenon, which also explains why the *nasalization mismatch* does not matter for this particular voice.

Regarding the high vowels of the female voice, the discussed harmonic component is very weak in both nasal and non-nasal contexts. More important factor seemed to be energy differences at the concatenation points (especially for /i/). It appears to be, however, perceptually of less importance, as the listeners found the discontinuity "facts" in a smaller number of sentences.

6   Conclusion and Future Work

In this paper, we have closely investigated the effects of nasal context mismatches on the quality of concatenations in vowels for two Czech speakers—male and female. The results clearly showed that the *nasalization mismatches* have a strong effect on perceived quality of concatenations in vowels /a/, /e/ and /o/ for our female speaker. Upon closer inspection of spectrograms and oscilograms of the concatenation points, it was found out that the concatenation discontinuities were due to strengthening of a harmonic component of the vowels in nasalized contexts, which in most cases resulted in phase mismatches at the concatenation points. In some cases, when the harmonic component interferes with the F0 peaks, the concatenation artifacts can not be removed by fixing the phase at the concatenation points and it is therefore better to completely avoid such concatenations. For the male voice, no impact of the *nasalization mismatches* was found.

Since the study was conducted for two voices only, it needs to be repeated using various speakers, and possibly also in more languages, in order to draw some more general conclusions. Still, we believe that the findings shown in this paper are an important indication of what we have so far been missing when measuring quality of concatenations in the diphone based concatenative speech synthesis. We have made similar observations for different consonantal phonetic context mismatches, and we plan to publish the results of these analyses in near future.

References

1. Klabbers, E., Veldhuis, R.: Reducing audible spectral discontinuities. IEEE Transactions on Speech and Audio Processing 9, 39–51 (2001)
2. Bellegarda, J.R.: A novel discontinuity metric for unit selection text-to-speech synthesis. In: SSW5 2004, Pittsburgh, PA, USA, pp. 133–138 (2004)
3. Vepa, J.: Join cost for unit selection speech synthesis. Ph.D. thesis, University of Edinburgh (2004)
4. Syrdal, A.K.: Phonetic effects on listener detection of vowel concatenation. In: EURO-SPEECH 2001, Aalborg, Denmark, pp. 979–982 (2001)

5. Syrdal, A.K., Conkie, A.: Perceptually-based data driven join costs: comparing join types. In: INTERSPEECH 2005, Lisbon, Portugal, pp. 2813–2816 (2005)
6. Kawai, H., Tsuzaki, M.: Acoustic measures vs. phonetic features as predictors of audible discontinuity in concatenative speech synthesis. In: ICSLP 2002, pp. 2621–2624. Denver, Colorado (2002)
7. Fujimura, O., Lindqvist, J.: Sweep-tone measurements of vocal-tract characteristics. J. Acoust. Soc. Am. 49, 541–558 (1971)
8. Fant, G.: Acoustic theory of speech production. Mouton, The Hague (1960)
9. House, A.S., Stevens, K.N.: Analog studies of the nasalization of vowels. J. Speech Hearing Disorders 21, 218–232 (1956)
10. Hawkins, S., Stevens, K.N.: Acoustic and perceptual correlates of the non-nasal–nasal distinction for vowels. J. Acoust. Soc. Am. 77, 1560–1575 (1985)
11. Legát, M., Matoušek, J.: Design of the Test Stimuli for the Evaluation of Concatenation Cost Functions. In: Matoušek, V., Mautner, P. (eds.) TSD 2009. LNCS, vol. 5729, pp. 339–346. Springer, Heidelberg (2009)
12. Legát, M., Matoušek, J.: Analysis of Data Collected in Listening Tests for the Purpose of Evaluation of Concatenation Cost Functions. In: Habernal, I., Matoušek, V. (eds.) TSD 2011. LNCS, vol. 6836, pp. 33–40. Springer, Heidelberg (2011)
13. Legát, M., Matoušek, J.: Identifying Concatenation Discontinuities by Hierarchical Divisive Clustering of Pitch Contours. In: Habernal, I., Matoušek, V. (eds.) TSD 2011. LNCS, vol. 6836, pp. 171–178. Springer, Heidelberg (2011)
14. Legát, M., Matoušek, J.: Pitch contours as predictors of audible concatenation artifacts. In: Proceedings of the World Congress on Engineering and Computer Science, San Francisco, USA, pp. 525–529 (2011)
15. Legát, M., Matoušek, J.: Collection and Analysis of Data for Evaluation of Concatenation Cost Functions. In: Sojka, P., Horák, A., Kopeček, I., Pala, K. (eds.) TSD 2010. LNCS, vol. 6231, pp. 345–352. Springer, Heidelberg (2010)
16. Chistovich, L.A.: Central auditory processing of peripheral vowel spectra. J. Acoust. Soc. Am. 77, 789–805 (1985)
17. Legát, M., Matoušek, J., Tihelka, D.: On the detection of pitch marks using a robust multi-phase algorithm. Speech Communication 53, 552–566 (2011)

Analysis and Assessment of State Relevance in HMM-Based Feature Extraction Method

Rok Gajšek, Simon Dobrišek, and France Mihelič

Faculty of Electrical Engineering, University of Ljubljana,
Tržaška 25, SI-1000 Ljubljana, Slovenia
{rok.gajsek,simon.dobrisek,france.mihelic}@fe.uni-lj.si
http://luks.fe.uni-lj.si/

Abstract. In the article we evaluate the importance of different HMM states in an HMM-based feature extraction method used to model paralinguistic information. Specifically, we evaluate the distribution of the paralinguistic information across different states of the HMM in two different classification tasks: emotion recognition and alcoholization detection. In the task of recognizing emotions we found that the majority of emotion-related information is incorporated in the first and third state of a 3-state HMM. Surprisingly, in the alcoholization detection task we observed a somewhat equal distribution of task-specific information across all three states, resulting in constantly producing better results if more states are utilized.

Keywords: emotion recognition, HMM, HMM adaptation, MAP.

1 Introduction

Modern systems for paralinguistic information analysis from speech such as, speaker's gender or age, emotion or other psychophysical effects, are mostly based on a procedure of Universal Background Model (UBM) adaptation [1,2]. Usually the UBM is represented as a Gaussian Mixture Model (GMM) and the adaptation is done following the maximum a posteriori (MAP) adaptation criteria. In our previous work [3] we presented a similar adaptation scheme where instead of the GMM, an HMM likelihood function is used as an UBM. The adaptation criteria remains MAP. We have shown that through the use of an HMM, whose components represent particular phones, the paralinguistic information can be better represented in the values of the adapted means of Gaussian distributions. In this paper we present an evaluation and assessment of the distribution of the paralinguistic information across the states of the HMM. Specifically, we evaluate the recognition accuracy of systems based on the mean vectors from different combinations of HMM states.

In order to reliably assess the importance of different HMM states for paralinguistic recognition tasks we selected two corpora with different types of speech. For the emotion recognition evaluation we used the FAU Aibo Emotion Corpus (FAU-Aibo) [4] which consists of spontaneous speech uttered by German children. For the second task of alcohol intoxication detection a Slovenian corpus of alcoholized speech is used [5].

The reminder of the paper is organized as follows. In Section 2 a description of FAU-Aibo and Vindat databases is given. Section 3 provides an overview of the HMM-based

P. Sojka et al. (Eds.): TSD 2012, LNCS 7499, pp. 559–565, 2012.
© Springer-Verlag Berlin Heidelberg 2012

feature extraction method, proposed for the various tasks of paralinguistic information recognition, as well as our reasoning for evaluating the impact of using different combinations of HMM states. Experiments and results are discussed in Section 4 and in Section 5 our final conclusions are given.

2 Databases

FAU-Aibo corpus [4,6] contains spontaneous children's speech labeled according to the emotional states being expressed in the recordings. The Wizard-of-Oz type of scenario was designed, where children utter commands to the little robot dog Aibo, while a person in the back is actually controlling the motion of the robot. This setup is effective for inducing an emotional response from the participant since the controller can intentionally disobey the participant's commands. In order to have the same behavior of the robot for all the participants, its movement and reactions to certain tasks were predefined. The session included five object localization tasks with children directing the robot to the defined spot, and the main task of leading the robot through a predefined path and at certain stops commanding the robot to perform a particular action. On one hand, the predefined and disobedient actions of the robot dog lead to frustration and other negative emotions, while the actions, compliant with the children's commands, induce positive emotions.

The FAU-Aibo database contains the recordings of 51 children between the ages of 10 and 13, of which 30 were female and 21 were male. In our experiments we follow a 2-class protocol of the Interspeech 2009 Emotion Challenge described in [1]. The class of negative emotions (NEG) consists of emotional labels *angry*, *touchy*, *reprimanding* and *emphatic*, while the idle class (IDL) consists of all non-negative emotional states. The sessions took place at two schools and according to the location of the recording the corpus is split into the training (school #1) and test (school #2) sets. The number of utterances per class is shown in Table 1, where an imbalance towards the idle class (IDL) can be observed. The number of samples for the negative class (NEG) is approximately half of the number of samples for the IDL class, which is usually the case in databases of spontaneous emotions.

Table 1. Number and distribution of samples in (a) the FAU-Aibo and in (b) the Vindat

(a) FAU-AIBO 2-class: negative vs. idle

#	IDL	NEG	$\sum$
Train	6601	3358	9959
Test	5792	2465	8257
$\sum$	12393	5823	**18216**

(b) VINDAT 2-class: non-alcoholized vs. alcoholized

#	NON-ALCO	ALCO	$\sum$
All	450	421	**871**

2.1 VINDAT

The VINDAT database [5] contains the recordings of people speaking at different levels of alcohol intoxication. Ten Slovene speakers, five men and five women, took part in the recording sessions. The average age of the adults was 35 years. The recording session for each speaker consisted of two parts. For the first part, fourteen Slovenian words were selected based on their demanding pronunciation (for example, the Slovene translation of otorhynolaryngologist – "otorinolaringologinja"). The selected words formed the center of a sentence, with meaningful words added left and right, in order to avoid the changes in speed and intonation that are usually found at the beginning and the end of spoken utterances. In the second part, the speakers repeated sentences previously read to them by the operator in an attempt to record speech closer to the natural speaking style.

The participants were recorded in three sessions based on the amount of the consumed alcohol. The alcohol levels were measured by a hand-held indicator, usually employed by the police for the inspection of drivers on the roads. The device measures the level of intoxication as the amount of milligrams per liter of exhaled air ($^0/_{00}$). In the first session, the participants were sober with $0^0/_{00}$. Before the second and the third session, each speaker consumed a selected amount of alcoholic beverage and measured the level of alcoholization prior to recording session. Understandably, at least 15 minutes passed since the last drink was consumed, before each measurement was made in accordance with the instructions of the alcohol-level indicator. Half of the participants exhibited $0.5^0/_{00}$ (which is a legal limit for driving in Slovenia) prior to the second recording, and by the last session, all participants had an alcohol level above $0.5^0/_{00}$.

For our task of alcoholization recognition, all the utterances labeled as less than $0.5^0/_{00}$ were assigned to the non-alcoholized class (NON-ALCO) and the rest (equal or more than $0.5^0/_{00}$) were assigned to the alcoholized class (ALCO). The threshold was set in accordance with the local legal limit for driving, as well as in accordance with the Intoxication Sub-challenge in [2] and the comparative database of [7]. In the lower part of Table 1, the distribution among the NON-ALCO and the ALCO classes is presented and it can be seen that the classes are almost balanced. While the text uttered in the database recordings was predefined and should as such match the content of the recordings, after listening to all the utterances, approximately 10% were not in accordance with the proposed text. These exceptions were carefully corrected enabling the use of transcriptions in our HMM adaptation.

3 HMM-Based Feature Extraction Method

HMM-based feature extraction method [3] is a modification of the standard GMM-UBM adaptation [8]. In the classification scheme based on GMM-UBM adaptation, the first step is to build the UBM model using all available training data. Once the UBM is acquired a set of adapted GMMs is estimated based on the UBM and each particular sample in the corpus. From the set of newly adapted models the means of Gaussian distributions are extracted separately for each model and transformed into a vector, sometimes referred to as a super-vector. The set of super-vectors is then used in the selected classifier. In [3] we proposed to use an HMM model instead of a GMM to represent the UBM. The idea is based on the fact that the components in the GMM are constructed based on statistical information whereas for the monophone HMM the components represent individual allophone. Hence, the super-vector's elements, which are derived from the adaptation procedure, represent utterance specific information seperatly for each allophone contained in the utterance. The elements that correspond to the phonemes missing in the specific utterance are not updated and keep the value from the UBM. The procedure is described in detail below.

The adaptation formula for the mean vector of the mixture component i in the HMM state s is

$$\hat{\mu}_{si} = \alpha_{si} \mathbf{E}_{si}(X) + (1 - \alpha_{si})\mu_{si}^{UBM},$$ (1)

where μ_{si}^{UBM} is the mean of the HMM-UBM s state and i-th mixture, $\mathbf{E}_{si}(x)$ is the mean of the observed adaptation data and α_{si} is the adaptation parameter. Similarly to the case of GMM adaptation, it is defined as

$$\alpha_{si} = \frac{n_{si}}{n_{si} + \tau},$$ (2)

where τ is the relevance factor and n_{si} is the occupation likelihood of the adaptation data, defined as

$$n_{si} = \sum_{t=1}^{T} Pr_{si}(t).$$ (3)

The mean of the observed adaptation data $\mathbf{E}_{si}(x)$ from Eq. 1 is calculated as

$$\mathbf{E}_{si}(X) = \frac{\sum_{t=1}^{T} Pr_{si}(t)\mathbf{x}_t}{n_{si}},$$ (4)

where Pr_{si} denotes the probability of occupying the mixture i of state s at time t, and $\mathbf{x}_t$ is the adaptation data feature vector at time t. One can see that the MAP adaptation of the means of the HMM (Eq. (1-4)) is an extension of the MAP adaptation of the GMM, hence, if there is only one state in the HMM, the equations become identical.

The adaptation formula for means (Eq. (1)) is made up of two terms, the UBM value of mean μ_{si}^{UBM} and the calculated mean of the observed adaptation data $\bar{\mu}_{si}$, both weighted by the combination of the occupation likelihood n_{si} and the relevance factor τ. The distance between the new MAP estimated mean and the UBM initial mean is determined by the occupation likelihood n_{si} for the particular component i of state s. The higher the value of n_{si}, the more influence the adaptation data has on the new adapted mean. And vise-versa, if the value of the n_{si} for the i component is small, the value of the mean vector will remain similar to the initial UBM mean. Likewise, the relevance factor τ controls the weighting of the prior mean and the influence of the adaptation data as well. However, it is independent of the adaptation data, and thus controls mostly the speed of the convergence. It should be noted that equations 1-3 are run iteratively until either the predefined number of iterations is reached, or the difference between the new and the old means is smaller than a predefined threshold. Thus, the notation μ_{si}^{UBM} is appropriate only for the first iteration, when the value of the mean vector from the UBM is used, and should be, in later iterations, better denoted by μ_{si}, symbolizing the current value of the mean being updated. However, the symbol μ_{si}^{UBM} is left in the equations since it better represents the idea of adapting the same HMM-UBM for each particular utterance.

Although the MAP adaptation can be used to update all the parameters of the HMM-UBM, we leave the weights, covariances, transition probabilities and state probabilities intact. Hence, all utterance-specific information or differences from the "general" speech, represented by the HMM-UBM, is captured in the new vectors of means. In this way, the exploitation of the means as a classification feature is enabled.

Once the above-described adaptation of the HMM-UBM for all train and test samples is finished, a set of new HMMs is obtained. These HMMs share the same values for all the parameters, except for the means of the Gaussian densities in the states. Next, for every utterance-specific HMM, the mean vectors are pooled together and combined in a new super-vector. The size of the super-vector is the number of Gaussians in the HMM-UBM times the dimension of the front-end acoustic feature vector. The set of super-vectors is obtained for the training and the test sets. These sets represent a suitable input to the classification, which is in our case realized by the SVM classification, like in the case of the GMM adaptation. Hence, the improvement of our HMM-UBM-MAP against the GMM-UBM-MAP modeling can be easily demonstrated.

3.1   Analysis of the HMM States

The HMM-based feature extraction method, described in the previous section, updates means in all 3 states of the HMM according to the MAP criteria. We want to evaluate if the paralinguistic information is equally distributed across all three states. In the case of stationary allophones the central state represents the segment of the phone were the speech signal exhibits as quasi-stationary, and the first and third state capture transitions to and from the allophone. Therefore, we devised an experimental protocol where would compare using only the means form the central states of HMM, only the means from the first and the third state against using means from all three states. This way we could reliable assess the importance of particular states to the information about the speaker's emotional state or intoxication level.

4   Experiments and Results

Both systems, for emotion recognition and intoxication detection, were based on MFCCs 0–12 and their Δs, calculated on 20 millisecond frames, with a delay of 10 milliseconds between the frames. The filter bank consisted of 22 Mel-spaced filters. The 0-th cepstral coefficient corresponds to the energy of the frame.

The UBM training and the adaptation process was done using the HTK toolkit [9] with the MAP relevance factor τ set to 16 and the number of the MAP iterations limited to 5. After the set of super-vectors was constructed from the means of the adapted GMMs a SVM classification was employed. A linear kernel was trained using sequential minimal optimization algorithm (SMO) was used for the training of the

Table 2. Comparison of emotion recognition performance using different combinations of HMM states for the FAU-Aibo corpus

# Gaussians	Number of used HMM states		
	States 1,2 and 3	State 2	States 1 and 3
120	70.3%	69.3%	70.6%
200	69.2%	68%	69.5%

SVM classifier as implemented in the Weka toolkit [10]. The same SVM classification parameters were used in all the tests.

In emotion recognition experiments, the total number of allophones used was 40. Consequently, the total number of means in the adapted GMMs equalled 120. Next, we evaluated the recognition accuracy for 3 cases: *(i)* using all three states, *(ii)* using only the central state and *(iii)* using only the first and the third state. In the first case with all three states used the dimension of the super-vector constructed from the means of the GMM was 3,120. The length of the supervector is a product of 40 allophones $\times$ 3 states $\times$ 26 acoustic features = 3,120. If we omitted the central state and only used the transition states the dimension of the super-vector lowered to 2,080, and finally if only the central state was used the dimension further reduced to 1,040.

In order to robustly evaluate the importance of different HMM states we conducted the same experiments as described above for the case of using a total of 200 Gaussians. To increase the number of Gaussians we split the distributions consisting of a single Gaussian in states that had the highest variance to the combination of Gaussians. Again, three tests were conducted using the above described combinations of states. The results of all experiments in emotion recognition using the FAU-Aibo corpus are shown in Table 2. In both cases of different number of Gaussians the results drop if only the central state is used however, if only the transition states are considered the result is slightly better. This observation leads us to believe that the first and the last state of the HMM capture more paralinguistic information than the central state of the monophone model.

The Vindat corpus transcript contains 41 unique allophones. The number is similar to the number of allophones in the FAU-Aibo corpus, hence the supervector dimensions are similar. The FAU-Aibo corpus has a defined set of training and test samples where as for the Vindat database an experimental protocol is not defined. Due to the smaller size of the corpus we conducted a 5-fold speaker-independent cross-validation. In each fold 20% of the samples was used for testing and 80% for training plus, the speakers from the test set were not included in the training set.

Table 3. Comparison of alcoholization recognition performance using different combinations of HMM states for the Vindat corpus

# Gaussians	Number of used HMM states		
	States 1,2 and 3	State 2	States 1 and 3
123	70.9%	65.6%	69%
200	69.2%	64.4%	67.6%

Similarly to the emotion recognition tests, we conducted experiments employing a different number of Gaussians. When only one Gaussian per state is used the total number of Gaussians equals 123. For the second evaluation the number of Gaussians was increased to 200. The results for the intoxication recognition are presented in Table 3. Opposite to the results in emotion recognition, the recognition rates for the alcoholization detection task are best if all HMM states are considered. The recognition accuracy drops in both attempts to omit some of the HMM states. Result is consistent regarding the complexity of the likelihood function, which is controlled by the number of Gaussians used.

5   Conclusion

We evaluated and assessed the importance of different HMM states in an HMM-based feature extraction method used to model paralinguistic information. Specifically, we evaluated the distribution of the paralinguistic information across different states of the HMM in two classification tasks: emotion recognition and alcoholization detection. In the task of recognizing emotions we found that the majority of emotion-related information is incorporated in the first and third state if standard 3-state HMMs are used. Surprisingly, in the alcoholization detection task we observed a somewhat equal distribution of information across all states, constantly producing better results if more states are utilized.

References

1. Schuller, B., Steidl, S., Batliner, A.: The INTERSPEECH 2009 Emotion Challenge. In: INTERSPEECH 2009. ISCA, pp. 312–315 (2009)
2. Schuller, B., Steidl, S., Batliner, A., Schiel, F., Krajewski, J.: The INTERSPEECH 2011 Speaker State Challenge. In: INTERSPEECH 2011. ISCA (2011)
3. Gajšek, R., Mihelič, F., Dobrišek, S.: Speaker state recognition using an hmm-based feature extraction method. Computer Speech and Language (to be published, 2012)
4. Steidl, S.: Automatic Classification of Emotion-Related User States in Spontaneous Children's Speech. Logos Verlag, Berlin (2009)
5. Mihelič, F., Gros, J., Dobrišek, S., Žibert, J.: Spoken language resources at luks of the university of ljubljana. International Journal of Speech Technology 6, 221–232 (2003)
6. Batliner, A., Steidl, S., Hacker, C., Nöth, E.: Private emotions vs. social interaction – a data-driven approach towards analysing emotion in speech. User Modeling and User-Adapted Interaction 18, 175–206 (2008)
7. Schiel, F., Heinrich, C., Barfüsser, S.: Alcohol language corpus: the first public corpus of alcoholized german speech. Language Resources and Evaluation (to appear, 2012)
8. Reynolds, D.A., Quatieri, T.F., Dunn, R.B.: Speaker Verification Using Adapted Gaussian Mixture Models. Digital Signal Processing 10, 19–41 (2000)
9. Young, S.J., Evermann, G., Gales, M.J.F., Hain, T., Kershaw, D., Moore, G., Odell, J., Ollason, D., Povey, D., Valtchev, V., Woodland, P.C.: The HTK Book, version 3.4.1. Cambridge University Engineering Department, Cambridge, UK (2009)
10. Hall, M., Frank, E., Holmes, G., Pfahringer, B., Reutemann, P., Witten, I.H.: The WEKA data mining software: an update. SIGKDD Explor. Newsl. 11, 10–18 (2009)

On the Impact of Non-speech Sounds
on Speaker Recognition

Artur Janicki*

Institute of Telecommunication, Warsaw University of Technology
ul. Nowowiejska 15/19, 00-665 Warsaw, Poland
A.Janicki@tele.pw.edu.pl

Abstract. This paper investigates the impact of non-speech sounds on the performance of speaker recognition. Various experiments were conducted to check what the accuracy of speaker classification would be if non-speech sounds, such as breaths, were removed from the training and/or testing speech. Experiments were run using the GMM-UBM algorithm and speech taken from the TIMIT speech corpus, either original or transcoded using the G.711 or GSM 06.10 codecs. The results show a remarkable contribution of non-speech sounds to the overall speaker recognition performance.

Keywords: speaker recognition, GMM-UBM, non-speech sounds, TIMIT.

1 Introduction

Non-speech sounds are often perceived as something unwanted, disturbing the normal process of speech analysis. This is the case indeed when dealing with automatic speech recognition, where e.g., a breath can be confused with the phoneme [x] or [h], what can obviously lead to an incorrect word recognition. But there are situations when non-speech sounds can play positive role. E.g., they are added to the speech signal during speech synthesis, to increase the naturalness of speech.

The aim of this work is to check if non-speech sounds, such as breath sounds, are helpful or disturbing in speaker recognition. The results of experiments with speaker classification will be presented, in which training and testing speech will contain or will be deprived of paralinguistic cues, such as breathing. The impact of non-speech sounds on the classification performance will be shown and discussed. In this paper we will concentrate mostly on breath sounds, as they are commonly present in the recordings.

The paper is organized as follows: first, selected other studies on non-speech sounds and speaker recognition will be presented. Then, the experiment is described in details, followed by presentation of the results and the discussion. Conclusions are drawn at the end.

1.1 Non-speech Sounds

Non-speech sounds are naturally present in recordings containing speech, what can be easily explained by the human physiology and articulation mechanism. Humans

* The calculations were made in the Interdisciplinary Centre for Mathematical and Computational Modeling (ICM) of the University of Warsaw (computational grant No. G46-2).

P. Sojka et al. (Eds.): TSD 2012, LNCS 7499, pp. 566–572, 2012.

breathe, cough, sneeze, yawn, smack, etc., with or without controlling it. Sounds of hesitation, such as filled pauses, are met especially in conversational speech.

Many scientists address the analysis of non-speech sounds by investigating negative aspects of them. In [1] the authors aim at detecting and removing breath sounds from recordings containing speech and singing, in order to improve the aesthetics of the recorded voice. In [2] the researcher tries to make a speech recognition system robust against paralinguistic cues such as filled pauses or lip smacking.

Other scientists try to recognize paralinguistic clues in order to draw conclusions out of their number, quality, location, etc. In [3] the authors detect breath sounds and based on them they diagnose psychiatric diseases, such as schizophrenia. In [4] artificial neural networks are used to detect inspiration and expiration events for automated analysis of respiratory data acquired during sleep for medical purposes. In [5] non-speech sounds are automatically classified. The researchers use support vector machines (SVMs) and multivariate adaptive regression splines (MARS) to classify sneezing, screaming, laughter and snoring, based on spectral and MFCC parameters.

1.2 Speaker Recognition

Voice can be treated as one of biometric features. Even though robustness of speaker recognition is not as high as, e.g., the one of iris recognition, but its huge advantage is that it can be performed remotely, e.g., over the phone during client authorization by the bank. What is more, acquiring a speech signal is not so intrusive as, e.g., acquiring somebody's fingerprints.

Speaker recognition can be text-dependent, when the system is expecting an agreed sentence, or text-independent, when the recognition process does not rely on a specific phonetic order. The most common technique for the former case are Hidden Markov Models (HMMs), while Gaussian Mixture Models (GMMs) are usually used in the latter case. GMMs are usually created based on the MFCC parameters. Parameters of an ith Gaussian component: λ, μ, and Σ (weight, mean values and covariance matrix, respectively) are often derived from the so called Universal Background Model (UBM), using the MAP algorithm [6].

Support Vector Machines (SVMs) have also been successfully employed for speaker recognition. In [7] the authors proposed using an SVM machine to classify supervectors containing GMM parameters (more precisely: Gaussian mixture mean values). Similar approach was described in [8], where a multi-class SVM classifier was used.

In speaker recognition usually non-speech frames are removed, assuming, among others, that they contain unwanted noise. E.g., in [9] the authors designed a voice activity detector (VAD) to be used in speaker verification, which aimed at removing all non-speech parts because, as the authors claim: "including non-speech frames in the modeling process would bias the resulting model, especially if the number of non-speech frames is significant".

The following sections will try to address the problem of impact of non-speech sounds on speaker recognition performance.

2   Experiments

In this section the experiment setup is described, including description of the speech data, its labeling, and methodology used for testing.

2.1   Speech Data

The TIMIT speech corpus [10] was used as the database of recordings. Although it was originally designed for studies of speech recognition, this corpus has been used as well for a number of studies on speaker recognition (e.g. [11],[12], and [13]), as it contains recordings of 630 speakers, which is a relatively large number. The drawback of the TIMIT corpus is that it contains only single-session recordings, so the problem of the speaker's inter-session variability was not investigated in this study.

Table 1. Assignment of the TIMIT phonemes to the classes considered in this study. S denotes silence, V – vowels, C – consonants, N – non-speech sounds.

class	TIMIT phonemes
S	sil (always), h# pau epi (if below an energy threshold)
V	iy ih eh ey ae aa aw ay ah ao oy ow uh uw ux er ax ix axr ax-h
C	m n ng em nx en eng p t k b d g dx s sh ch f th hh z zh jh v dh hv el l r w y dcl gcl bcl tcl kcl pcl
N	h# pau epi (if over an energy threshold)

Each of the speakers utters ten sentences, each one lasting 3.2 s on average. The audio material per single speaker is relatively short: ca. 32 s, in total for training and for testing.

The TIMIT database is annotated, using, among others, phonetic transcription. Thanks to that it was possible to take a subset of the speech signal and test its impact on the speaker recognition performance. Non-speech sounds are not specifically annotated, but they are mostly found in segments marked with the labels: "h#" ("begin/end marker"), "pau" ("pause"), or "epi" ("epenthetic silence"). Additional energetic condition was checked to distinguish the frames containing non-speech sounds from the ones with silence. Speech files with concatenated non-speech sounds from all speakers were generated, to verify the correctness of labeling. The non-speech sounds in TIMIT were mostly sounds of breathing, with some lip smacking and single cases of laughter. Assignment of the TIMIT phonemes to the studied classes is presented in Table 1, their contribution to the training and testing data is shown in Table 2. Breathing frequency was not analyzed as the recordings were too short.

2.2   Testing Methodology

Performance of speaker recognition was tested on the text-independent speaker classification task, using speech data from the TIMIT corpus. Three types of speech quality were used in the experiments:

Table 2. Class distribution in training and testing data for S – silence, V – vowels, C – consonants, N – non-speech sounds (in % of time)

class	training	testing
S	12.77	11.89
V	39.89	42.92
C	47.33	45.19
N	5.81	6.50

- original speech, sampled with 16 kHz;
- speech with emulated fixed-telephony quality, sampled with 8 kHz. The speech signal was transcoded with the G.711 codec, i.e., the codec used in fixed telephony (also in VoIP). G.711 is in fact an application of the logarithmic quantizer. A-law option was used in the experiments;
- speech with emulated mobile phone quality, sampled with 8 kHz. In this case the speech signal was transcoded using the GSM 06.10 codec, used in the GSM telephony. Even though newer (and better) versions of this codec already exist (e.g., GSM 06.60), we wanted to check the performance when using a moderate-quality codec, such as GSM 06.10.

The speech data was parameterized using 19 MFCC parameters (plus the 0th one), with a frame length of 30 ms and a 10 ms analysis step. The UBM models with 256 Gaussian components were trained using the GMM EM-ML algorithm, separately for each type of speech quality, using 200 speakers. The remaining 430 speakers were used for classification tests, analogously to [8] and [11].

Speaker classification was realized using the GMM-UBM method. Speaker models were created by adapting the UBM model using the MAP algorithm with reference factor $RF = 1$. The training was based on the five SX sentences, using feature vectors corresponding to:

- vowels (hereinafter denoted as V), or:
- consonants (hereinafter denoted as C), or:
- solely non-speech sounds (hereinafter denoted as N), or:
- vowels and consonants (VC), or finally:
- vowels, consonants and non-speech sounds (VCN);

The same was done for the testing recordings. In addition the following combinations of testing data were evaluated:

- vowels and non-speech sounds (VN);
- consonants and non-speech sounds (CN).

Each of the SA and SI sentences were used for testing, so there were five test recordings for each of the 430 speakers. Even though the SA sentences were the same for every speaker, they were used in the testing part only, so the text-independence was preserved.

The experiments were run in the Matlab environment using the h2m toolbox [14], using parallel computing on 21 nodes of a computational cluster. The results were assessed by counting the ratio between the number of correct classifications against the total number of 2,150 test recordings.

3 Results

The results of classification accuracy are presented in Table 3. It shows that if non-speech sounds were *not* present during training (configurations: V,C,VC) then adding them to testing data (configurations: VN,CN,VCN) can sometimes decrease the classification accuracy, e.g. in configuration V/VN (training/testing) or C/CN, both for G.711, or C/CN for original speech. In the latter case it decreased from 92.74% to 88.98%, what is a remarkable loss.

Table 3. Classification accuracy [%] for systems trained (in rows) and tested (in columns) with various content of speech, for three types of speech quality. V denotes vowels, C – consonants, N – non-speech sounds.

speech	train	V	C	VC	VN	CN	VCN	N
					test			
16 kHz	V	93.63	36.33	86.98	93.16	34.56	86.33	2.75
	C	48.23	92.74	81.95	45.12	88.98	80.00	4.52
	VC	95.86	95.44	98.42	94.70	92.74	98.42	5.68
	VCN	95.53	95.58	98.51	97.26	96.70	98.98	36.08
	N	4.74	5.26	5.02	28.05	19.67	18.47	30.68
G.711	V	60.88	10.28	44.37	60.05	9.21	43.86	1.40
	C	11.02	71.72	35.40	11.07	69.40	35.21	3.65
	VC	66.33	81.21	84.47	66.28	79.49	84.19	4.25
	VCN	67.44	82.00	84.79	71.77	84.47	87.44	21.93
	N	1.63	2.33	2.00	12.28	6.51	5.86	16.83
GSM 06.10	V	50.98	6.70	32.98	49.35	5.72	31.95	1.40
	C	6.28	63.63	27.72	5.95	60.93	28.05	2.57
	VC	61.86	74.33	82.37	62.33	72.88	83.30	2.95
	VCN	63.02	75.21	82.60	68.33	77.95	86.09	17.11
	N	1.49	0.74	0.65	10.37	4.09	3.16	13.04

When the speaker models were trained using the complete speech signal with non-speech sounds (VCN), testing them with the VCN test data always yielded better results than testing them with the VC data only. The results of VCN/VCN experiments were always better than the ones for VC/VC, see Fig. 1 for comparison. The highest gain was for the worst-quality speech, i.e., GSM 06.10-transcoded signal - after adding non-speech sounds it increased from 82.37% to 86.09%. The lowest benefit was for original speech (increase from 98.42% to 98.98%), but here the recognition was close to perfect anyway.

A relatively high accuracy for systems trained and tested on non-speech sounds only is somewhat surprising: it turned out that the system was able to classify correctly 30.68% of test recordings (for the original, 16 kHz-sampled speech), relying only on non-speech sounds. The fact that the TIMIT speech corpus is unisessional (i.e., all recordings from a given speaker were made during one recording session) could

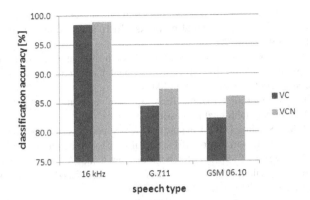

Fig. 1. Classification accuracy without (dark) and with (light) non-speech sounds, for three types of speech quality

contribute to this high results. It was investigated if potential imperfections of the TIMIT annotation (such as misplaced boundaries of "#h") could explain that score; but when listening to all speech data classified as N we could here not much apart from breathing (inspirations/expirations), some smacking etc. Apparently such a high score confirms that non-speech sounds can be of a remarkable value in speaker recognition.

4 Conclusions

The experiments showed that non-speech sounds carry information which can be beneficial during speaker recognition. The improvement is visible especially well for lower-quality signal, such as telephony speech.

Advantageous impact of non-speech sounds on speaker recognition is known from every-day experience: humans can easily recognize the people they know based only on the breath sound or the way he or she sneezes. The GMM-UBM technique proved to be able to take advantage of the acoustic information contained, e.g., in the breath sound. Therefore, it is reasonable to include non-speech sounds in the training data when creating speaker models; voice activity detectors used with speaker recognition systems, should be designed to treat non-speech sounds as a useful part of data.

References

1. Ruinskiy, D., Lavner, Y.: An effective algorithm for automatic detection and exact demarcation of breath sounds in speech and song signals. IEEE Transactions on Audio, Speech, and Language Processing 15(3), 838–850 (2007)
2. Rajnoha, J.: Speaker non-speech event recognition with standard speech datasets. Acta Polytechnica 47(4-5/2007), 107–111 (2008)
3. Rapcan, V., D'Arcy, S., Reilly, R.B.: Automatic breath sound detection and removal for cognitive studies of speech and language. In: IET Irish Signals and Systems Conference (ISSC 2009), pp. 1–6 (2009)

4. Sa, R.C., Verbandt, Y.: Automated breath detection on long-duration signals using feedforward backpropagation artificial neural networks. IEEE Transactions on Biomedical Engineering 49(10), 1130–1141 (2002)
5. Liao, W.H., Lin, Y.K.: Classification of non-speech human sounds: Feature selection and snoring sound analysis. In: IEEE International Conference on Systems, Man and Cybernetics, SMC 2009, pp. 2695–2700 (2009)
6. Reynolds, D.A., Quatieri, T.F., Dunn, R.B.: Speaker verification using adapted gaussian mixture models. In: Digital Signal Processing (2000)
7. Campbell, W.M., Sturim, D.E., Reynolds, D.A.: Support vector machines using gmm supervectors for speaker verification. IEEE Signal Processing Letters 13, 308–311 (2006)
8. Janicki, A., Staroszczyk, T.: Speaker Recognition from Coded Speech Using Support Vector Machines. In: Habernal, I., Matoušek, V. (eds.) TSD 2011. LNCS (LNAI), vol. 6836, pp. 291–298. Springer, Heidelberg (2011)
9. Hautamäki, V., Tuononen, M., Niemi-Laitinen, T., Fränti, P.: Improving speaker verification by periodicity based voice activity detection. In: Proc. 12th International Conference on Speech and Computer, SPECOM 2007, pp. 645–650 (2007)
10. Garofolo, J., Lamel, L., Fisher, W., Fiscus, J., Pallett, D., Dahlgren, N., Zue, V.: Timit acoustic-phonetic continuous speech corpus. Linguistic Data Consortium, Philadelphia (1993)
11. Besacier, L., Grassi, S., Dufaux, A., Ansorge, M., Pellandini, F.: Gsm speech coding and speaker recognition. In: Proc. ICASSP, pp. 1085–1088 (2000)
12. Jiang, T., Gao, B., Han, J.: Speaker identification and verification from audio coded speech in matched and mismatched conditions. In: Proc. of the IEEE International Conference on Robotics and Biomimetics, ROBIO 2009, pp. 2199–2204 (2009)
13. Yu, E.W.M., Mak, M.-W., Kung, S.-Y.: Speaker Verification from Coded Telephone Speech Using Stochastic Feature Transformation and Handset Identification. In: Chen, Y.-C., Chang, L.-W., Hsu, C.-T. (eds.) PCM 2002. LNCS, vol. 2532, pp. 598–606. Springer, Heidelberg (2002)
14. Cappe, O.: h2m toolkit, http://www.tsi.enst.fr/~cappe/

Automatic Rating of Hoarseness by Text-based Cepstral and Prosodic Evaluation

Tino Haderlein[1,2], Cornelia Moers[3], Bernd Möbius[4], and Elmar Nöth[1]

[1] University of Erlangen-Nuremberg, Pattern Recognition Lab (Informatik 5)
Martensstraße 3, 91058 Erlangen, Germany
Tino.Haderlein@informatik.uni-erlangen.de
http://www5.informatik.uni-erlangen.de
[2] University of Erlangen-Nuremberg, Department of Phoniatrics and Pedaudiology
Bohlenplatz 21, 91054 Erlangen, Germany
[3] University of Bonn, Department of Speech and Communication
Poppelsdorfer Allee 47, 53115 Bonn, Germany (now with Max Planck Institute for
Psycholinguistics, Wundtlaan 1, 6525 XD Nijmegen, The Netherlands)
[4] Saarland University, Department of Computational Linguistics and Phonetics
Postfach 151150, 66041 Saarbrücken, Germany

Abstract. The standard for the analysis of distorted voices is perceptual rating of read-out texts or spontaneous speech. Automatic voice evaluation, however, is usually done on stable sections of sustained vowels. In this paper, text-based and established vowel-based analysis are compared with respect to their ability to measure hoarseness and its subclasses. 73 hoarse patients (48.3 ± 16.8 years) uttered the vowel /e/ and read the German version of the text "The North Wind and the Sun". Five speech therapists and physicians rated roughness, breathiness, and hoarseness according to the German RBH evaluation scheme. The best human-machine correlations were obtained for measures based on the Cepstral Peak Prominence (CPP; up to $|r| = 0.73$). Support Vector Regression (SVR) on CPP-based measures and prosodic features improved the results further to $r \approx 0.8$ and confirmed that automatic voice evaluation should be performed on a text recording.

1 Introduction

Evaluation of voice distortions is still mostly performed perception-based. Perception of voice qualities, however, is too inconsistent among single raters to establish a standardized and unified classification. The average opinion of a panel of raters is more consistent, but this approach is not suitable for clinical application. The ideal solution would be objective, automatic assessment.

The perception experiments are applied to spontaneous speech, read-out standard sentences, or standard texts. In contrast, already used methods of automatic analysis rely mostly on sustained vowels [13]. The advantage of speech recordings, however, is that they contain phonation onsets, variation of F_0 and pauses [16]. Furthermore, they allow to evaluate speech-related criteria, such as intelligibility [5]. For this reason, also for the automatic evaluation, speech recordings should be used. This paper focuses on

P. Sojka et al. (Eds.): TSD 2012, LNCS 7499, pp. 573–580, 2012.

the automatic assessment of hoarseness and its subclasses by means of prosodic and cepstral analysis.

Hoarseness is a psycho-acoustically defined measure which was originally believed to be distinct of the other two categories roughness (or harshness) and breathiness. Nowadays, hoarseness is often seen as the superclass of these categories [1]. The Roughness-Breathiness-Hoarseness (RBH) evaluation scheme [15] takes this into account. It is an established means for perceptual voice assessment in German-speaking countries and serves as the reference for the automatic analysis presented in this paper.

Most studies on automatic voice evaluation use perturbation-based parameters, such as jitter, shimmer, or the noise-to-harmonicity ratio (NHR, [13]). However, perturbation parameters have a substantial disadvantage. They require exact determination of the cycles of the fundamental frequency F_0. In severe dysphonia it is difficult to find an F_0 due to the irregularity of phonation. This drawback can be eliminated by using the Cepstral Peak Prominence (CPP) and the Smoothed Cepstral Peak Prominence (CPPS) which represent spectral noise. They do not require F_0 detection and showed high human-machine correlations in previous studies [2,6,9].

The questions addressed in this paper are the following: How do cepstral- and text-based prosodic measures perform in comparison to established, vowel-based measures? How well does cepstral-based analysis correspond to perception-based RBH evaluation when it is supported by prosodic analysis?

In Sect. 2, the audio data and perceptual evaluation will be introduced. Section 3 will give some information about the cepstral analysis, Sect. 4 describes the vowel analysis with the Praat software. An overview of the prosodic analysis and Support Vector Regression will be presented in Sect. 5 and 6, and Sect. 7 will discuss the results.

2   Test Data and Subjective Evaluation

73 German subjects with chronic hoarseness (24 men and 49 women) between 19 and 85 years of age participated in this study. The average age was 48.3 years with a standard deviation of 16.8 years. Patients suffering from cancer were excluded. Each person uttered the vowel /e/ and read the text "Der Nordwind und die Sonne" ("The North Wind and the Sun", [11]), a phonetically balanced standard text which is frequently used in medical speech evaluation in German-speaking countries. It contains 108 words (71 distinct) with 172 syllables. The data were recorded with a sampling frequency of 16 kHz and 16 bit amplitude resolution using an AKG C 420 microphone (AKG Acoustics, Vienna, Austria).

The text recordings were evaluated perceptually by 5 speech therapists and physicians according to the German Roughness-Breathiness-Hoarseness (RBH) scale [15]. Each of the three criteria can be evaluated on a 4-point scale where '0' means "absent" and '3' means "high degree". In order to capture the fact that hoarseness is the superclass, the H rating must have either the same or a higher rating than R or B. RBH represents a short version of the GRBAS scale [10], with the categories "asthenia" and "strain" omitted.

3 Cepstral Analysis

The Cepstral Peak Prominence (CPP) is the logarithmic ratio between the cepstral peak and the regression line over the entire cepstrum at this quefrency. A strongly distorted voice has a flat cepstrum and a low CPP due to its inharmonic structure. The computation of CPP and the Smoothed Cepstral Peak Prominence (CPPS) was performed by means of the free software "cpps" [8] which implements the algorithm introduced by Hillenbrand and Houde [9]. The cepstrum was computed for each 10 ms frame, CPPS was averaged over 10 frames and 10 cepstrum bins. The vowel-based results will be denoted by "CPP-v" and "CPPS-v". For the automatic speech evaluations ("CPP-NW" and "CPPS-NW"), the first sentence only (approx. 8–12 seconds, 27 words, 44 syllables) of the read-out text was used. Sections in which the patients laughed or cleared their throat were removed from the recording for this pilot experiment by hand.

4 Analysis of Sustained Vowels with Praat

The automatic analysis of the sustained vowels (/e/) with respect to established irregularity measures was performed using the software Praat 5.1 [4]. An overview of the features is given in Table 1. For this vowel analysis, sections of at least 0.5 seconds duration of stable phonation excluding onset and offset were evaluated. From 17 speakers, a section of 0.7 seconds could be extracted; from 36 speakers, a full second was available. Although several measures are F_0-based, men and women were not analyzed separately for this study. The reason is that the goal of the analysis was to find measures which can be used independently of the speaker's gender, just like the human raters do not need different evaluation methods for men and women.

5 Prosodic Features

In order to find automatically computable counterparts for the RBH criteria, also a "prosody module" was used to compute features based upon frequency, duration and speech energy (intensity) measures.

The prosody module processes the output of a word recognition module [5] and the speech signal itself. Hence, it can use the time-alignment of the recognizer and the information about the underlying phoneme classes. For each speech unit of interest (here: words), a fixed reference point has to be chosen for the computation of the prosodic features. This point was chosen at the end of a word because the word is a well-defined unit in word recognition, it can be provided by any standard word recognizer, and because this point can be more easily defined than, for example, the middle of the syllable nucleus in word accent position. For each reference point, 28 prosodic features are computed which refer to a single word or the pause between two words. Ten of these features are additionally computed for a word-pause-word interval. A full description of the features used is beyond the scope of this paper; details and further references are given in [3].

In addition to the 38 local features per word position, 15 global features were computed from jitter, shimmer and the number of voiced/unvoiced decisions for each 15-word interval. They cover the means and standard deviations for jitter and shimmer, the number, length and maximum length each for voiced and unvoiced sections, the ratio of the numbers and ratio of the durations of voiced and unvoiced sections, the ratio of length of voiced sections to the length of the signal, and the same for unvoiced sections. The last global feature is the standard deviation of F_0.

6 Support Vector Regression (SVR)

In order to find the best subset of the prosodic features and cepstral measures to model the subjective ratings, Support Vector Regression (SVR, [17]) was used. The general idea of regression is to use the vectors of a training set to approximate a function which tries to predict the target value of a given vector of the test set. Here, the training set comprises the automatically computed measures, and the test set consists of the subjective RBH scores. For this study, the sequential minimal optimization algorithm (SMO, [17]) of the Weka toolbox [18] was applied in a 10-fold cross-validation manner.

7 Results and Discussion

For all three perceptual RBH criteria, the entire range between 0 and 3 was covered. The average values were 1.56 (standard deviation: 0.83) for R, 1.19 (0.81) for B, and 1.84 (0.84) for H (see also [14]).

The correlations between the perceptual evaluation and the single automatic measures are given in Table 1. For all evaluated criteria, the best results for single objective measures were obtained for CPPS-NW. The text-based CPP-NW and CPPS-NW perform remarkably better than all the vowel-based measures, including CPP-v and CPPS-v.

The SVR on all available measures, i.e. the Praat and CPP-based features, and the prosodic features, revealed higher correlations than on the single measures. Two feature sets were identified, where one was optimal for R and H and the other one for B (Table 2). For all criteria, CPPS-NW, the normalized energy in word-pause-word intervals (EnNormWPW), the average minimal F_0 of each word (F0MinWord), and the average F_0 at voice offset (F0OffWord) were in the respective set. Additionally, the duration of all unvoiced sections in the signal (Dur–Voiced) was part of the best set for breathiness. None of the vowel-based measures contributed to the best feature sets. This clearly indicates the need for text-based automatic evaluation.

The energy value EnNormWPW may contribute strongly to the best feature sets, because it was normalized with respect to healthy speakers [3]. Loudness effects are removed by the normalization, but if a person has a hoarse and irregular voice, then the energy level especially in the high frequency portions is higher than for normal speakers. The impact of the F_0 values F0MinWord and F0OffWord can be explained by the noisy speech that causes octave errors during F_0 detection, i.e. instead of the real fundamental frequency, one of its harmonics is found. With more "noisy speech", this may influence the F_0 trajectory and hence the correlation to the subjective results. It is not clear so far,

Table 1. Correlation *r* between perceptual and automatic evaluation (**: significant on the 0.01 level, *: significant on the 0.05 level); the perceptual result was the mean value of all raters. The names of the features in the first part of the table follow the names used in Praat [4]; APQ11 was computed for 72 patients only due to an invalid value for one patient.

data	feature(s)	R	B	H
vowel	Jitter local	0.33**	0.54**	0.51**
vowel	Jitt local absolute	0.33**	0.28*	0.34**
vowel	RAP (Rel. Avg. Perturb. Quotient), jitter of 3 periods	0.26*	0.39**	0.38**
vowel	PPQ5 (Pitch Perturb. Quotient), jitter of 5 periods	0.24*	0.32**	0.33**
vowel	Shimmer local	0.38**	0.56**	0.58**
vowel	Shimmer local absolute	0.39**	0.56**	0.59**
vowel	APQ11 (Amplitude Perturb. Quotient of 11 periods)	0.34**	0.41**	0.47**
vowel	NHR (N of [1500;4500] Hz / H of [70;4500] Hz)	0.34**	0.54**	0.53**
vowel	HNR (Mean harmonicity-to-noise ratio)	−0.40**	−0.57**	−0.59**
vowel	CPP-v (Cepstral Peak Prominence, vowel-based)	−0.25*	−0.60**	−0.53**
vowel	CPPS-v (smoothed CPP, vowel-based)	−0.17	−0.52**	−0.44**
text	CPP-NW (Cepstral Peak Prominence, text-based)	−0.47**	−0.69**	−0.69**
text	CPPS-NW (smoothed CPP, text-based)	−0.52**	−0.69**	−0.73**
v.+t.	SVR on all features: best set for B	0.72**	**0.82****	0.77**
v.+t.	SVR on all features: best set for R and H	**0.74****	0.79**	**0.79****

Table 2. SVR regression weights for the best feature subsets when predicting the perceptual RBH scores

	best set for R, H			best set for B		
predicted score	R	B	H	R	B	H
EnNormWPW	−0.063	0.565	0.372	0.014	0.589	0.389
F0MinWord	−0.678	0.076	−0.431	−0.562	−0.562	−0.408
F0OffWord	−0.223	−0.410	−0.274	−0.261	−0.261	−0.269
Dur–Voiced		—		0.470	0.262	0.044
CPPS-NW	−0.444	−0.612	−0.615	−0.215	0.641	−0.631

Table 3. Correlation *r* of the feature values of the best feature sets

feature	F0MinWord	F0OffWord	Dur–Voiced	CPPS-NW
EnNormWPW	0.12	−0.03	0.09	−0.56
F0MinWord		0.11	−0.56	0.37
F0OffWord			−0.04	0.26
Dur–Voiced				−0.49

however, why in the case of F0OffWord only the end of the voiced sections causes a noticeable effect. It may reflect changes in the airstream between the beginning and the end of words or phrases. High speaking effort leads to more irregularities especially in these positions, but this has to be confirmed by more detailed experiments. Note that the prosody module computes the F_0 values only on sections which it has previously identified as voiced. It may be this property of the software that makes F_0-based values so important for the analysis of distorted voices, although the purpose of adding cepstral measures to the feature set was to become independent of them.

The duration of the voiceless sections in the signal (Dur–Voiced) is comparable among all speakers since they read the same text. Hence, a higher duration indicates a higher percentage of voiceless sections and thus an irregular voice.

The correlations between the feature values of the best subsets are given in Table 3. A low EnNormWPW correlates significantly with a high CPPS-NW, because both indicate a high-quality voice. Likewise, CPPS-NW and Dur–Voiced are negatively correlated. A large duration of unvoiced sections correlates negatively with the average F_0 minimum. The reason may be the predefined F_0 threshold of the prosody module. The lowest F_0 that will be returned is 50 Hz, lower values are classified as voiceless. Hence, a very low voice will result in a low minimal F_0 of about 50 Hz and in a higher amount of unvoiced sections caused by F_0 values below the threshold.

The average inter-rater correlation between one rater and the average of the other ones was $r = 0.76$ for R, $r = 0.70$ for B, and $r = 0.82$ for H. For breathiness, the text-based CPP values ($r = 0.69$) alone almost reached the human reference, the SVR results even outperformed it. For roughness and hoarseness, the SVR almost reached the human inter-rater correlation.

The results on human-machine correlation with cepstral parameters confirmed some findings of other studies. Hillenbrand and Houde [9] found a significant correlation between these parameters and the perceived degree of breathiness for sustained vowels and speech recordings. This was confirmed in our study, but only for speech recordings. Heman-Ackah et al. [7] reported a correlation of the total degree of dysphonia and CPPS of $r = -0.80$ on stable vowel sections and $r = -0.86$ on sentence recordings. Their correlation between vowel- and sentence-based CPPS and the breathiness rating was $r = -0.70$ and $r = -0.71$, respectively. Our best result for a single feature for breathiness was $r = -0.69$ for both CPP-NW and CPPS-NW. The vowel-based measures, however, reached just $r = -0.60$. Nevertheless, breathiness was better modeled by the vowel-based CPP-v than roughness or hoarseness. The most probable reason for the differences in vowel analysis among the studies is that it requires stable phonation. Often a frame of one second of the vowels /a/ (predominantly), /e/, or /i/ is chosen. Other vowel segment durations from 0.1 seconds up to 3 seconds have been reported [13]. For our study, the minimum duration of stable phonation was set to 0.5 seconds, because some patients were not able to phonate longer without too much irregularity. Our subjects uttered /e/, sometimes shifted towards /ɛ/ which is the adjacent phoneme in the German vowel space. Therefore, the results may not be completely comparable to other studies. On the other hand, these variations in duration and vowel quality show that there will always be inconsistencies in the data obtained from a representative group of patients. If their influence deteriorates the evaluation results that much, then the method cannot be used for clinical purposes. This is another important argument against vowel-based perturbation analysis for voice evaluation.

CPP and CPPS cannot differentiate between different voice qualities [2]. Even with support by prosodic features, this is not possible with the available feature set, because the best sets for modeling R, B, and H are too similar. When these sets were applied to predict all of the rating criteria (Table 2), still a few clear differences in the weighting factors for the prediction formulae were revealed. The normalized energy EnNormWPW, for instance, could also be left out of the set for R. The results change only marginally then. This feature is obviously more important for the evaluation of breathiness and overall hoarseness.

The similarity of the best feature sets for all rating criteria is consistent with the interaction between different dimensions of human perception: the presence of roughness in a voice does not influence the perception of breathiness. However, the perceived degree of roughness is strongly influenced by the presence of breathiness [12]. Additionally, dysphonic voices with lower fundamental frequencies are perceived as being more rough than those with higher F_0 [19]. Our results, however, confirm the assumption that roughness and breathiness are perceived as separate dimensions and that hoarseness is the superclass of both [1]. R and H showed a better correlation with each other in the human results ($r = 0.81$) than B and H ($r = 0.76$). The correlation of R and B was only $r = 0.36$. It is remarkable that breathiness and hoarseness were better mapped by the automatically obtained measures than roughness, even more so since the optimal feature sets for R and H were the same. It will be one of the most important aspects in future work to teach the automatic analysis to tell apart roughness, breathiness, and hoarseness as well as human listeners do.

Some aspects for enhancing the human-machine correlations have not been tested in this study. Human perception is often non-linear, such as indicated by the Bark scale for pitch, for instance. Physical scales are often linear, such as the frequency measured in Hertz. Better human-machine correlations may be found with non-linear mappings between the two modalities.

8 Conclusion

The results obtained in this study allow for the following conclusions: There is a significant correlation between the subjective rating of roughness, breathiness, hoarseness, on the one hand, and the automatic evaluation, on the other. However, the three criteria cannot be rated separately with the available set of features. The human-machine correlation is about as good as the average inter-rater correlation among speech experts. Cepstral-based measures improve the human-machine correlation, but only when they are computed from a speech recording and not from a sustained vowel only. The method can serve as the basis for an automatic, objective system that can support voice rehabilitation.

Acknowledgments. This work was partially funded by the Else Kröner-Fresenius-Stiftung (Bad Homburg v.d.H., Germany) under grant 2011_A167. The responsibility for the contents of this study lies with the authors. We would like to thank Dr. Hikmet Toy for acquiring and documenting the audio data.

References

1. Aronson, A., Bless, D.: Clinical Voice Disorders. Thieme, 4th edn. (2009)
2. Awan, S., Roy, N.: Outcomes Measurement in Voice Disorders: Application of an Acoustic Index of Dysphonia Severity. J. Speech Lang. Hear. Res. 52, 482–499 (2009)
3. Batliner, A., Buckow, J., Niemann, H., Nöth, E., Warnke, V.: The Prosody Module. In: Wahlster, W. (ed.) Verbmobil: Foundations of Speech-to-Speech Translation, pp. 106–121. Springer, Berlin (2000)
4. Boersma, P., Weenink, D.: Praat: Doing phonetics by Computer, Version 5.1.33, http://www.fon.hum.uva.nl/praat (accessed May 21, 2012)
5. Haderlein, T., Moers, C., Möbius, B., Rosanowski, F., Nöth, E.: Intelligibility Rating with Automatic Speech Recognition, Prosodic, and Cepstral Evaluation. In: Habernal, I., Matoušek, V. (eds.) TSD 2011. LNCS, vol. 6836, pp. 195–202. Springer, Heidelberg (2011)
6. Halberstam, B.: Acoustic and Perceptual Parameters Relating to Connected Speech Are More Reliable Measures of Hoarseness than Parameters Relating to Sustained Vowels. ORL J. Otorhinolaryngol. Relat. Spec. 66, 70–73 (2004)
7. Heman-Ackah, Y., Michael, D., Goding Jr., G.: The Relationship Between Cepstral Peak Prominence and Selected Parameters of Dysphonia. J. Voice 16, 20–27 (2002)
8. Hillenbrand, J.: cpps.exe (software), http://homepages.wmich.edu/~hillenbr (accessed May 21, 2012)
9. Hillenbrand, J., Houde, R.: Acoustic Correlates of Breathy Vocal Quality: Dysphonic Voices and Continuous Speech. J. Speech Hear. Res. 39, 311–321 (1996)
10. Hirano, M.: Clinical Examination of Voice. Springer, New York (1981)
11. International Phonetic Association (IPA): Handbook of the International Phonetic Association. Cambridge University Press, Cambridge (1999)
12. Kreiman, J., Gerratt, B., Berke, G.: The multidimensional nature of pathologic vocal quality. J. Acoust. Soc. Am. 96, 1291–1302 (1994)
13. Maryn, Y., Roy, N., De Bodt, M., Van Cauwenberge, P., Corthals, P.: Acoustic measurement of overall voice quality: A meta-analysis. J. Acoust. Soc. Am. 126, 2619–2634 (2009)
14. Moers, C., Möbius, B., Rosanowski, F., Nöth, E., Eysholdt, U., Haderlein, T.: Vowel- and Text-based Cepstral Analysis of Chronic Hoarseness. J. Voice 26, 416–424 (2012)
15. Nawka, T., Anders, L.C., Wendler, J.: Die auditive Beurteilung heiserer Stimmen nach dem RBH-System. Sprache - Stimme - Gehör 18, 130–133 (1994)
16. Parsa, V., Jamieson, D.: Acoustic discrimination of pathological voice: sustained vowels versus continuous speech. J. Speech Lang. Hear. Res. 44, 327–339 (2001)
17. Smola, A., Schölkopf, B.: A Tutorial on Support Vector Regression. Statistics and Computing 14, 199–222 (2004)
18. Witten, I., Frank, E.: Data Mining: Practical Machine Learning Tools and Techniques, 2nd edn. Morgan Kaufmann, San Francisco (2005)
19. Wolfe, V., Martin, D.: Acoustic Correlates of Dysphonia: Type and Severity. J. Commun. Disord. 30, 403–416 (1997)

Improving the Classification of Healthy and Pathological Continuous Speech

Klára Vicsi, Viktor Imre, and Gábor Kiss

Dept. of Telecommunications and Media Informatics
Budapest University of Technology and Economics
vicsi@tmit.bme.hu
http://alpha.tmit.bme.hu/speech/

Abstract. A number of experiments were made in the field of speech diagnostic analysis in which researchers wanted to examine whether it was the acoustic characteristics of sustained voice or continuous speech that were more appropriate for distinguishing healthy from pathological voice. Since in phoniatric practice, doctors mainly use continuous speech, we also wanted to concentrate on the examination of continuous speech. In this paper we present a series of classification experiments showing how it is possible to separate healthy from pathological speech automatically, on the basis of continuous speech. It is demonstrated that the results of the automatic classification of healthy vs. pathological voice improved to a large extent by a multi-step processing methodology, in which most examples in which uncertainties occurred in the measurement of the acoustic parameters can be accounted for separately. That multi-step processing could be especially useful when pathological data is not sufficient for statistical point of view.

Keywords: pathological voice, voice disorder detection, automatic speech recognition, support vector machine.

1 Introduction

Generally in voice production, there is a close connection between variation in the voice generation organs (differences in size, in tissue flexibility, etc.) and the measurable acoustic parameters (fundamental frequency, sound pressure, spectrum, etc.) of the speech product generated.

In vocal diagnostic analyses, several examinations were made in connection with the question whether it is sustained voice or continuous speech that is more effective in distinguishing healthy voice from pathological voice ([1,2,3,4,5,6]).

In phoniatric practice, mainly continuous speech is applied by phoniatry specialists for the classification of voice quality based on hearing. This is no accident: for generating speech the cooperation of other important articulatory functions is necessary besides vibration of the vocal cords, and thus in the case of any disorder clearness of the voice can be easily disrupted. On the other hand, there is a possibility in continuous speech for the observation of suprasegmental characteristics: emphasis, intonation, and the duration of sonorants.

Let us examine the possibilities of analysis of continuous speech and sustained voice from the viewpoint of acoustic measurements. According to Rabinov et al. [7], the most

P. Sojka et al. (Eds.): TSD 2012, LNCS 7499, pp. 581–588, 2012.

reliable "tool" for the evaluation of voice quality is the human ear, after all. This can be explained by the fact that in the measurement of oscillation of amplitude and frequency of vibration of vocal cords, these parameters do not take the shape of the generated voice waves into consideration, and that the vibration of the vocal cords is accompanied by a frictional noise. These issues may contain relevant information, mainly in the case of pathological voice.

Titze and colleagues [8] suggest that acoustic measurements (jitter, shimmer) can only provide reliable information in the examination of sustained voice, because the characterization of periodicity can be determined easily due to the quasi-periodicity of the signal, and sustained voice contains enough periods for an authentic calculation of oscillations. On the other hand, in the analysis of continuous speech, where the length of examined sections is very short because of the quick voice transitions, jitter and shimmer results are less reliable.

Zhang & Jiang [1] examined the acoustic characteristics of sustained voice and continuous speech for distinguishing healthy from pathological voice. Acoustic parameters: jitter, shimmer and HNR values were taken into consideration. The authors demonstrated that continuous speech is less suitable for making a distinction between healthy and pathological voice.

While in phoniatric practice doctors mainly apply continuous speech, we also wanted to examine which kind of sound material, continuous speech or sustained vowels, are the best material for the acoustic parameters and for the automatic separation of normal and pathological speech. First, a detailed statistical analysis of acoustic parameters of vowels in continuous speech and sustained voice databases were examined, and the results were compared in healthy vs. pathological speech ([9,10]). In this paper we present our classification experiments on how it is possible to separate healthy from pathological speech automatically on the basis of continuous speech.

2   Classification Experiments

The construction of a well-designed pair of pathological and healthy speech databases was necessary for the examination and for the automatic separation of healthy and pathological voice.

2.1   Pathological and Healthy Speech Databases

The sound recordings were made in a consulting room at the Out-patients' Department of Head and Neck Surgery of the National Institute of Oncology. The following diseases occurred in the recorded database: functional dysphonia, recurrent paresis, tumours at various places of the vocal tract, gastro-oesophageal reflux disease, chronic inflammation of larynx, bulbar paresis, its symptoms(paralysis of lips, tongue, soft palate, pharynx and the muscles of larynx), amyotrophic lateral sclerosis, leukoplakia, spasmodic dysphonia and glossectomia. Recordings, for comparison, were also prepared with absolutely healthy patients who had gone to the consulting hours only for control examinations.

Speech samples were recorded by close field microphone (Monacor ECM-100), with Creative Soundblaster Audigy 2 NX: an outer USB sound card with 44,100Hz, at a 16-bit sampling rate.

The following tasks were recorded from each patient: 3 [e] vowels sustained for a long time, with a deep breath taken before the utterance of each of them, and reading out a folk tale, frequently used in the phoniatric practice, "The North Wind and the Sun".

The recorded sound samples were classified by a leading phoniatry specialist by the sound perception evaluation scale RBH, a popular scale in the practice of phoniatry (RBH stands for "Rauhigkeit" (=roughness), "Behauchtkeit" (=breathiness) and "Heiserkeit" (=hoarseness)); 0 = normal voice quality, 3 = heavy huskiness). The scale classifies the voice samples into four classes (0..3) on the basis of subjectively felt parameters provided by the RBH code. This scale was used to differentiate the degree of voice generation disorders in the database. Speech samples of the patients were labelled on the basis of this numerical scale.

Since we intended to process predefined voiced sequences of the continuous speech material, phoneme-level segmentation of voice files was necessary. It was made in a semi-automatic way, using our own automatic speech recognizer.

The continuous speech (folk tale) samples of 59 speakers were used for the classification experiment (33 pathological and 26 healthy speakers).

2.2 Measured Acoustic Parameters

Earlier we examined [10] which acoustic analyzing methods best reflected the degree of voice generation disorders (or which fitted the RBH scale of sound perception evaluation the most closely). Statistical distributions of the acoustic parameters were examined by measuring these parameters in sustained vowels and at the middle of vowels [e] in continuous speech. These vowels were extracted from the segmented and labelled voiced files. In this experiment it was found that the selected acoustic parameters in the quasi-stationary part of the vowels in continuous speech could replicate the perceptual classification of experts much better than those in the traditionally used steady state sounds. The measured and analyzed parameters were used for the classification experiments, too:

Jitter: this is the average absolute difference between consecutive time periods (T) in speech, divided by the average time period. Generally two forms of jitter are in use:

$$jitter_{local} = \frac{\sum_{i=1}^{N-1} |T_i - T_{i+1}|}{\sum_{i=1}^{N-1} T_i} \cdot 100\% \tag{1}$$

$$jitter_{ddp} = \frac{\sum_{i=2}^{N-1} |2T_i - T_{i-1}T_{i+1}|}{\sum_{i=2}^{N-1} T_i} \cdot 100\% \tag{2}$$

where N is the number of periods, and T is the length of the periods.

Shimmer: this is the average absolute difference between consecutive differences between the amplitudes of consecutive periods.

$$jitter_{local} = \frac{\sum\limits_{i=1}^{N-1} |A_i - A_{i+1}|}{\sum\limits_{i=1}^{N-1} A_i} \cdot 100\% \tag{3}$$

$$jitter_{ddp} = \frac{\sum\limits_{i=2}^{N-1} |2A_i - A_{i-1}A_{i+1}|}{\sum\limits_{i=2}^{N-1} A_i} \cdot 100\% \tag{4}$$

where A is the amplitude of the period.

HNR (Harmonics-to-Noise Ratio) represents the degree of acoustic periodicity.

$$HNR = 10 \log \frac{E_H}{E_Z} dB \tag{5}$$

where E_H and E_Z are the energy of the harmonic and noise component, respectively.

2.3 Classification of Healthy and Pathological Voices

All of the sounds [e] in the reading test were used for the classification. At the middle of the [e] sound, the following acoustic parameters were measured: local jitter, ddp jitter, local shimmer, dda shimmer and HNR values. The mean, the standard deviation, minimum, maximum, and median of the measured values were calculated. These vectors were the input of the classifier. Support Vector Machine (LibSVM, http://www.csie.ntu.edu.tw/~cjlin/libsvm/) was used for the classification. Many of researchers used this classification method, but we know until now those cases, when sustained vowels were used for discrimination ([11,12]). Leave-One-Out Cross-Validation (LOOCV) techniques were used for training and testing. Of course data of a speaker chosen for testing does not appear in the training set. For the selection of the most important parameters, a series of pilot experiments were conducted. Different groups of the acoustic parameters were used and the recognition (classification) results were examined.

The best classification results of healthy and pathological speech were obtained when jitter(ddp) and shimmer(dda) means were the incoming acoustic parameters. See Table 1.

We wanted to analyse this result further, in terms of how the healthy samples were separated from the pathological cases on the basis of these two parameters. Thus we plotted the standard deviation values as a function of mean values in the case of jitter(ddp) and shimmer(dda). See Fig. 1 and Fig. 2.

There are a few salient examples in case of the standard deviation of both jitter and shimmer. Analyzing these examples one by one, it turned out that all of those salient examples originated in measuring problems. Either the fundamental frequency

Table 1. Testing the classifier with various incoming acoustic parameters (jitter = jitter(local) and jitter(ddp) together; shimmer = shimmer(local) and shimmer(dda) together)

Statistics for sound [e]		
Acoustic parameters	mean	mean, standard deviation min, max, median
jitter, shimmer, hnr, mfcc	73%	63%
jitter, shimmer, mfcc	73%	63%
jitter, shimmer	79%	79%
jitter(local), shimmer(local)	79%	79%
jitter(ddp), shimmer(dda)	**84%**	79%
jitter, shimmer, hnr	73%	73%
hnr, mfcc	68%	63%
mfcc	73%	63%

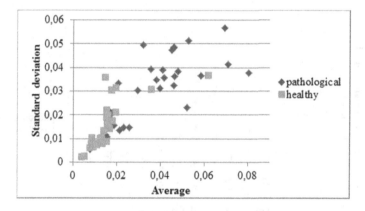

Fig. 1. Standard deviation of jitter (ddp) as a function of the mean in the case of the sound [e]

was too low or too high, or the voiced period was very short. When the fundamental frequency was too high, the fluctuation in % was smaller than expected. In cases of too low frequencies, the measurement of the fundamental frequencies became ambiguous, yielding mistakes. Both jitter and shimmer measurements are based on the measurement of fundamental frequency, thus in the cases mentioned we got nonsensical results, yielding salient data. In the case when the voiced period was very short, much shorter than the voiceless one, inadequate examples caused mistakes. Although it seems that Zhang and Jiang [1] were right when they concluded that continuous pathological speech contained too short examples for the authentic calculation of the jitter and shimmer parameters, the calculation difficulty occurred only with some of the examples, and those examples could be selected and evaluated in a different way.

2.4   Separation of the Uncertain Examples

As a first step, it was necessary to decide the threshold of the voiced/voiceless frame rate under which the examples are selected. Thus voiced/voiceless frame rates were

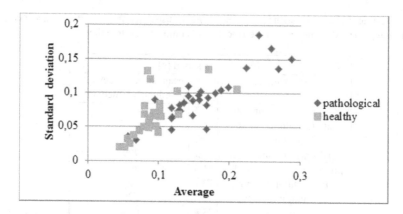

Fig. 2. Standard deviation of shimmer (dda) as a function of the mean, in the case of the sound [e]

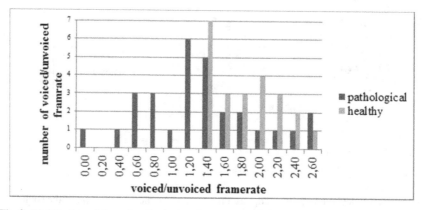

Fig. 3. Distribution of the voiced/voiceless frame rates of the healthy and pathological voices

calculated in the continuous part of the speech of the patients. A 75 ms window was used with 18.75 ms frame steps. The voiced/unvoiced frame rate was calculated as follows:

$$Fr_{v/u} = \frac{\sum\limits_{i=1}^{N} Fr(v)_i}{\sum\limits_{i=1}^{N} Fr(u)_i} \qquad (6)$$

where $Fr(v)_i$ is the number of voiced frames, and $Fr(u)_i$ is the number of voiceless frames of the i-th sound. The distribution of these voiced/unvoiced frame rates in case of healthy and pathological voices is presented in Fig. 3.

It is quite clear from the measurement that healthy speech does not exist under a 1.4 frame rate. The quality of these sounds is the worst. The error of the measurement of the fundamental frequency is high. Examples in which the voiced/voiceless frame rate was less than 0.5 were filtered out, as they are surely pathological voices. Then the classification was repeated, and now the speaker independent classification accuracy increased to 86%. As a second step, on the basis of the distribution of fundamental frequencies, those examples were separated off where the fundamental frequency was higher than 160 Hz in case of men, and 270 Hz in the case of women. These examples are also pathological voices, because in these cases the patients wanted to make their hoarse voices better by increasing the fundamental frequency. The classification was repeated again, and now its exactness increased to 88%.

3   Conclusions

In the case when pathological continuous speech database is not large enough for statistical sense, we could easily increase the exactness of the speaker independent classification of the pathological and healthy speech by the separation of uncertain examples.

In spite of the results of Zhang and Jiang [1], and in spite of the opinion of Rabinov et al. [7], it is clear for us that it is worth going further in our way and use continuous speech for the detection and classification of pathological voices. Of course, for obtaining better results, the collection of much more data is necessary. In the case when speech data are large enough may be the separation of the uncertain examples will become obsolete. What is clear, that further investigation is necessary to be undertaken for final decision which acoustic parameters are the best for the classification.

Acknowledgements.   Authors would like to thank the CESAR (http:// cesar-project.net/) project, funded under the ICT-PSP (Grant Agreement No. 271022), a partner of META-NET (http://meta-net.eu), for its support for the work done on the pathological speech database.

References

1. Zhang, Y., Jiang, J.J.: Acoustic Analyses of Sustained and Running Voices From Patients With Laryngeal Pathologies. Journal of Voice 22(1), 1–9, 0892-1997 (2006) (accepted for publication)
2. Peng, C., Chen, W., Zhu, X., Wan, B., Wei, D.: Pathological voice classification based on a single vowel's acoustic features, 0-7695-2983-6/07. IEEE (2007)
3. Parsa, V., Jamieson, D.G.: Acoustic discrimination of pathological voice: Sustained vowels versus continuous speech. Journal of Speech, Language, and Hearing Research 44, 327–339 (2001)
4. Askenfelt, A.G., Hammarberg, B.: Speech Waveform Perturbation Analysis, A Perceptual-Acoustical Comparison of Seven Measures. Journal of Speech and Hearing Research 29, 50–64 (1986)
5. Peng, C., Chen, W., Zhu, X., Wan, B., Wei, D.: Pathological Voice Classification Based on a Single Vowels's Acoustic Features. In: 7th International Conf. on Computer and Information Techn., pp. 1106–1110. IEEE (2007)
6. Ritchings, R.T., McGillion, M., Moore, C.J.: Pathological voice quality assessment using artificial neural networks. Medical Engineering & Physics 24, 561–564 (2002)

7. Rabinov, C.R., Kreiman, J., Gerratt, B.R., Bielamowicz, S.: Comparing Reliability of Perceptual Ratings of Roughness and Acoustic Measures of Jitter. Journal of Speech and Hearing Research 38, 26–32 (1995)
8. Titze, I., Wong, D., Milder, M., Hensley, S., Ramig, L.: Comparison between clinician-assisted and fully automated procedures for obtaining a voice range profile. J. Speech Hear. Res. 35, 526–535 (1995)
9. Imre, V.: Acoustical examination of pathological voices. BSc thesis, Budapest University of Technology and Economics (2009)
10. Klára, V., Viktor, I., Krisztina, M.: Voice Disorder Detection on the Basis of Continuous Speech. In: Jobbágy, Á. (ed.) 5th European IFMBE Conference. IFMBE Proceedings, vol. 37, pp. 86–89. Springer, Heidelberg (2011)
11. Everthon, S., Fonseca, J.C.: Pereira: Normal Versus Pathological Voice Signals. IEEE Engineering in Medicine and Biology Magazine (September/October 2009)
12. Markaki, M., Stylianou, Y.: Voice Pathology Detection and Discrimination based on Modulation Spectral Features. IEEE (2010)

Part IV

Dialogue

"**Dialogue:** A discussion between two or more people or groups,
especially one directed towards exploration of a particular subject
or resolution of a problem: interfaith dialogue."
NODE (The New Oxford Dictionary of English), Oxford, OUP, 1998, page 509.

Using Foot-Syllable Grammars to Customize Speech Recognizers for Dialogue Systems

Daniel Couto Vale and Vivien Mast

I5-[DiaSpace], SFB/TR8 Spatial Cognition, University of Bremen
Cartesium, Enrique-Schmidt-Straße 5, 28359 Bremen, Germany
danielvale@uni-bremen.de, viv@tzi.de

Abstract. This paper aims at improving the accuracy of user utterance understanding for an intelligent German-speaking wheelchair. We compare three different corpus-based context-free restriction grammars for its speech recognizer,
which were tested for surface recognition and semantic feature extraction on a
dedicated corpus of 135 utterances collected in an experiment with 13 participants. We show that grammars based on phonologically motivated units such as
the foot and the syllable outperform phrase-structure grammars in complex scenarios where the extraction of a large number of semantic features is necessary.

Keywords: speech recognition, dialogue system, natural language.

1 Introduction

When designing an intelligent agent for Ambient Assisted Living (AAL) environments,
automatic speech recognition is a challenging task. It is not a design option to restrict
commands to a small list of utterances with maximally different sound patterns and it
is a requirement that agents understand a large repertoire of user utterances, which may
oftentimes sound similar though realizing different meanings. On the engineering side,
this is a foreseeable catastrophe. Ontology-driven context-free restriction grammars
have been shown to give a poor quality of speech recognition for such a selection of
utterances while a more reliable corpus-driven (context-free or finite-state) restriction
grammar implies the cost of compiling a very large annotated targeted corpus [9]. For
this reason, a corpus-based (not corpus-driven) context-free restriction grammar would
be an acceptable low-cost solution if it resulted in good speech recognition.

This paper aims at improving the accuracy of speech recognition restricted by
hand-crafted corpus-based context-free grammars by developing and testing restriction
grammars with different sets of grammatical units in four human-machine interaction
scenarios. Our quest is to determine the best types of grammatical units for achieving a
good speech recognition with a low engineering cost. We conclude that phonologically
motivated units such as the syllable and the foot outperform graphologically and
ideationally motivated ones such as the word and the phrase, leading to a positive
quality-cost trade-off. We argue that the reason for this is that such units map better
onto the characteristics of situated spoken text while making a large annotated corpus
unnecessary.

P. Sojka et al. (Eds.): TSD 2012, LNCS 7499, pp. 591–598, 2012.

2   Interaction in Ambient Assisted Living

The Bremen Ambient Assisted Living Lab (BAALL, [5]) is a fully functional $60m^2$, 2 person apartment for testing and evaluating AAL technology [6]. Going way beyond automated light and temperature control, BAALL features sophisticated control of sliding doors, smart furniture, and enables higher-level services through integration [6]. Due to its complexity, intuitive and effective interaction between user and speaking agents is a priority [2]. An important factor for effectiveness of dialogue in AAL environments is that the recognizable utterances should not be restricted to lists of commands that need to be learned and memorized, since learning and memorization can be particularly demanding for elderly people.

Being tested in BAALL, the intelligent wheelchair Rolland is a modified version of the commercially available power-wheelchair Xeno [7,8,11]. It has been equipped with sensors and an onboard computer in order to be able to perceive and reason about the environment and to interact with the user in different ways. This includes a navigation assistant that enables the wheelchair to autonomously take the user to a target location [6] with qualitative spatial reasoning and a dialogue system, which is a mixed-initiative semi-autonomous agent using the DAISIE framework (DiaSpace's Adaptive Information State Interaction Executive) whose architecture was adapted to the specific needs of situated dialogue [12,13,14]. Among its components, it uses Nuance's VoCon as a speech recognizer [10] and OpenCCG as a parser [3].

3   Methodology

The German usage corpus for Rolland was collected during a Wizard-of-Oz experiment that simulates a real-life everyday scenario of wheelchair usage within BAALL [1]. In this experiment, 13 German native speakers of both sexes were told to execute 6 tasks using the intelligent wheelchair, for which they had to drive it to 9 destinations through free spoken commands. All commands were recorded and manually transcribed constituting a corpus of 135 clauses. Finally, these clauses were read out loud with pauses in between by a female German native speaker and recorded again for further processing.

Three context-free restriction grammars were manually written for the speech recognizer VoCon to cover 125 of the 135 clauses: namely Phrase-Word (PW) (see Table 1), Foot-Word (FW), and Foot-Syllable (FS) (see Table 2). The phrase-word grammar has traditional grammatical units such as Nouns(N), Prepositions(P), Noun Phrases(NP), Prepositional Phrases(PP) and Verb(V) and allows any syntactically valid structure with these units while ignoring concordance (case, gender, person and number).

The foot-word grammar has a three-rank structure starting at the lowest rank with the word (W), an intermediate unit named foot (F), and an uppermost unit named curve (C), which corresponds closely to a clause. A foot is a rhythmic unit in the compositional hierarchy of spoken language, which contains syllables as its parts and which is part of a curve [4]. In German, a foot is composed by one stressed syllable (SS) and its adjacent unstressed ones (US). The boundary of the foot was determined by analyzing frequent co-occurrences of unstressed syllables around a stressed one in our corpus. Example 1 is divided into feet, whose stressed syllables are ['kans] in the word "kannst" ['tIS] in "Tisch" and ['fa:Rn] in "fahren":

Example 1. kannst du mich / zum Tisch / fahren

In this grammar, we restricted word combinations within the foot to those found in our corpus, what eliminated non-occurring preposition-noun combinations such as "ins Sofa" (into the sofa) and verb-pronoun combinations such as "fahre ich" (do I drive?), favoring similar sounding feet such as "ans Sofa" (closer to the sofa) and "fahre mich" (take me) for speech recognition. This elimination relies solely on *de facto* occurring feet and is written as in the Foot-Word Grammar extract below. Classes of feet were created depending on their possible combinations with others inside of a curve.

```
Foot-Word Grammar
<C>  : <F1> <F2> | [...] ;
<F1> : fahre | fährst du
     | fahre mich | fährst du mich ;
<F2> : ans Sofa | an das Sofa | zum Sofa | zu dem Sofa
     | an die Couch | zur Couch | zu der Couch ;
```

The foot-syllable grammar also has the foot as its intermediate grammatical unit, but feet have syllables as their parts instead of words. By having the syllable as the lowest unit, we could fine tune the speech recognition in much detail (see below).

```
Foot-Syllable Grammar
<F2> : <USIn> <USdI> <SSkY> <USCE>
     | <USI> <USnI> <SSkY> <USCE> ;
<SSkY>    : 'kY !pronounce("'kY") ;
<USCE>    : CE !pronounce("CE") | C$ !pronounce ("C$") ;
<USI>     : I !pronounce("I")   | $  !pronounce ("$")  ;
<USIn>    : In !pronounce("In") | $n !pronounce ("$n") ;
<USnI>    : nI !pronounce("nI") | n$ !pronounce ("n$") ;
<USdI>    : dI !pronounce("dI") | d$ !pronounce ("d$") ;
```

The audio file with the spoken clauses was then played, the speech was recognized by VoCon with these three grammars, and the highest scoring recognized utterances were saved into three plain text files for further processing. In addition, two human transcriptions were made, one in words and another in feet.

Two CCG grammars were created to process VoCon's output, one for the foot transcriptions and another for the phrase transcriptions, each covering 125 of 135 expected clauses (the latter trying to come up with some feature structure for as many utterances as possible due to the low quality of the surface structure recognition). The texts were parsed with these two grammars and the feature structures were saved to a file. The features were defined in such a way that every utterance could be reverted back to its surface structure including lexical option ("Sofa" and "Couch", "Rolland" and "Roller" triggering different features), but not including details such as the position of the word "bitte" in the clause.

Table 1. Example transcriptions for the phrase-word grammar

Human Trancription	Phrase-Word
1 komme zum Bett	komm her
2 fahre in die Küche	fahre an die Küche
3 komme her	—
4 komme vor dem Sofa an	komme vor dem Sofa

Table 2. Example transcriptions for the foot grammars

Human Transcription	Foot-Word	Foot-Syllable
1 Komme zudemBett	Komme zudemBett	Komm zudemBett
2 Fahre indieKüche	Fahre indieKüche	Fahre indieKüche
3 Komme Her	Komm zudemBett	Komme Her
4 Komme vordemSofa An	Komme vordasSofa	Komme vordemSofa An

In order to evaluate the performance of the three restriction grammars, we compared the recognition results on two strata, surface structure (VoCon output) and semantic features (CCG output). For the semantic stratum, we considered the requirements of different application scenarios.

3.1 Surface Structure

In order to evaluate surface structure recognition, we compared the output of the speech recognizer to human transcriptions. Recognized utterances were only counted as correct if they completely matched their human transcription equivalent.

3.2 Semantic Features

For the parsing, we compared the semantic features extracted by OpenCCG for recognized utterance to those for the human transcription for four scenarios. Different sets of features were taken into account for comparison depending on the complexity of the scenario. Feature structure hierarchy was ignored in all scenarios.

Translation Scenario. Firstly, we tested for a very simple scenario where it is only possible to direct the wheelchair to certain destinations with limited precision, using low-level translation commands. The only relevant semantic feature is the *Relatum*, i.e. the place where the wheelchair should go to.

Pickup Scenario. If a wheelchair can distinguish whether the user is on board or not and receives a command such as "Bring mich in die Küche" when the user is not on board, it should first pick up the user before bringing him or her to the requested destination, making both the *Relata* and the *Goals* (items that should be moved) relevant.

Precision Scenario. In this scenario, we assume a wheelchair that can use all ideational features i.e. *Process-Type*, *Relative-Position*, *Place-Type*, etc. Interpersonal features such as *Politeness* and *Mood-Type* were not taken into account.

Holistic Scenario. Finally, for a wheelchair that can adjust its behavior and correct its mistakes by noticing diachronic fluctuations in politeness, interpersonal features become also relevant. In this scenario, high-level goals as well as low-level adjustments, different speech functions such as questions, and clarification dialogues can be articulated covering a broad range of tasks.

4 Results

With respect to the surface structure, the foot-syllable restriction grammar achieved 67 total matches (49.63%), the foot-word grammar 32 (23.70%), and the phrase-word grammar 16 (11.85%). Table 3 shows the results of the different grammars.

Table 3. Correct matches of the three restriction grammars for surface structures

	Correct	Wrong
FS	67	68
FW	32	103
PW	16	119

Using Pearson's Chi-squared test, this difference was shown to be highly significant ($\chi^2 = 49.57, df = 2, p < 0.0001$). The standardized residuals (Table 4) show that for the Foot-Syllable grammar, the number of correctly recognized utterances was significantly higher than avarage, whereas for the Phrase-Word grammar, it was significantly lower.

Table 4. Standardized residuals for the surface structure results

	Correct	Wrong
FS	6.70	−6.70
FW	−1.48	1.48
PW	−5.22	5.22

For the translation and pickup scenarios, Table 5 shows that all grammars reach a fairly high rate of correct feature extraction. The foot-word grammar receives the highest scores, it correctly allows correct extraction of 119 feature structures (88.15%) in the translation scenario and 104 (77.04%) in the pickup, whereas the phrase-word grammar performs worst, merely allowing respectively 83 (61.48%) and 71 (52.60%) correct feature structure extractions. The foot-syllable grammar reaches a value in between these extremes with 108 (80%) and 99 (73.33%) correct recognitions. The difference is highly significant for both scenarios (translation: $\chi^2 = 28.08, df = 2, p < 0.0001$; pickup: $\chi^2 = 21.42, df = 2, p < 0.0001$), and the standardized residuals in Table 6 show that the foot-word grammar performs significantly higher than average, while the phrase-word grammar performs significantly lower for both scenarios.

Table 5. Correct Matches

	Translation		Pickup		Precision		Holistic	
	Correct	Wrong	Correct	Wrong	Correct	Wrong	Correct	Wrong
FS	108	27	99	36	74	61	71	64
FW	119	16	104	31	36	99	34	101
PW	83	52	71	64	23	112	21	114

Table 6. Standard Residuals

	Translation		Pickup		Precision		Holistic	
	Correct	Wrong	Correct	Wrong	Correct	Wrong	Correct	Wrong
FS	1.16	−1.16	1.73	−1.73	6.66	−6.66	6.60	−6.60
FW	3.90	−3.90	2.85	−2.85	−1.87	1.87	−1.82	1.82
PW	−5.06	5.06	−4.58	4.58	−4.79	4.79	−4.78	4.78

In the precision and holistic scenarios, the results are slightly different. The foot-syllable restriction grammar retains fairly good scores: 74 correct feature structure extractions (54.81%) in the precision scenario and 71 (52.60%) in the holistic. The results of both the foot-word and the phrase-word grammar degrade to a correct feature extraction rate that is unacceptable for practical applications: the foot-word grammar only reaches 36 (26.67%) and 34 (25.19%) correct feature extractions respectively and the phrase-word grammar 23 (17.04%) and 21 (15.56%). The difference is again highly significant in both scenarios (precision: $\chi^2 = 47.18, df = 2, p < 0.0001$; holistic: $\chi^2 = 46.52, df = 2, p < 0.0001$), and the standardized residuals in Table 6 show that the foot-syllable grammar performs significantly above average, while the phrase-word grammar performs significantly below.

5   Discussion

The results point towards the superiority of the foot-syllable grammar for recognizing speech. This is particularly the case for surface structure recognition and the two most complex scenarios, where all features need to be extracted correctly. For simpler scenarios, on the other hand, the foot-word grammar performed better than the foot-syllable grammar, and the disadvantage of the phrase-word grammar was less drastic than in complex scenarios. In addition, the foot-syllable grammar was the one that best discarded the ten clauses which were not covered by the grammar, thus avoiding false recognition.

It must be noted that the good results are not only due to paradigmatic constraints on text, but also on the right choice of grammatical unit type. The foot, for instance, cooperated with our results in several ways. Firstly, it allowed creating a flatter structure that keeps information about occurring verb complements such as actors and goals (realized as pronouns), relative positions to the relatum (realized as prepositions), and place type (realized as cases or prepositions). It also allowed keeping ideational features such as five levels of politeness in a foot class, all of this keeping the restriction grammar context-free and relying neither on typologies of nouns or processes nor on delicate agreement. Finally, it helped making VoCon only output strings that could be successfully analyzed by OpenCCG. The syllable enables us to systematically increase phonological variability for unstressed syllables while enforcing exact matching for stressed ones, which usually carry the most delicate semantic content such as process types and thing types. The syllable rank also helps us allow variation while maintaining the size of the grammar constant, as syllable units can be reused more often than words.

6   Conclusion

We have used three alternative BNF grammars to restrict VoCon's speech recognition, parsed recognized strings with CCG grammars, and evaluated both the recognized surface structure and relevant semantic features for 4 concrete application scenarios. Our results shows that by using corpus data to systematically add phonological variants, and restrict syntactic structures, speech recognition results can be improved significantly. Our results have shown that, for simple scenarios, an approach that limits syntactic variability without systematically controlling the phonological stratum performs best, while for complex scenario with situated spoken interaction, a restriction grammar that additionally uses fine control over phonological variation vastly outperforms all other approaches.

We conclude that the syllable and the foot are better suited for restricting speech recognition in complex scenarios than the word and the phrase. Specially for interaction in AAL environments, where language usage is specific and large corpora are not available and very costly, a foot-syllable corpus-based context-free grammar seems to be a real alternative for both cost and quality of user utterance recognition. It remains to be shown whether this corpus-based approach can compete with corpus-driven methods in complex scenarios with large specific corpora.

Acknowledgements.   We gratefully acknowledge the support of the Deutsche Forschungsgemeinschaft (DFG) through the Collaborative Research Center SFB/ TR8 Spatial Cognition.

References

1. Anastasiou, D.: A Speech and Gesture Spatial Corpus in Assisted Living. In: Proceedings of the 8th International Conference on Language Resources and Evaluation (LREC), Istanbul (2012)
2. Augusto, J.C., McCullagh, P.: Ambient intelligence: Concepts and applications. Computer Science and Information Systems 4, 1–28 (2007)
3. Baldridge, J., Kruijff, G.-J.: Multi-Modal Combinatory Categorial Grammar. In: Proceedings of EACL (2003)
4. Halliday, M.A.K., Matthiessen, C.M.I.M.: An introduction to functional grammar, 3rd edn. Edward Arnold, London (2004)
5. Krieg-Brückner, B., Gersdorf, B., Döhle, M., Schill, K.: Technik für Senioren in spe im Bremen Ambient Assisted Living Lab (Technology for seniors-to-be in the Bremen Ambient Assisted Living Lab). In: Ambient Assisted Living – AAL: 2. Deutscher AAL-Kongress. VDE-Verlag, Berlin (2009)

6. Krieg-Brückner, B., Röfer, T., Shi, H., Gersdorf, B.: Mobility Assistance in the Bremen Ambient Assisted Living Lab. GeroPsych. 23, 121–130 (2010)
7. Krieg-Brückner, B., Shi, H., Fischer, C., Röfer, T., Cui, J., Schill, K.: Welche Sicherheitsassistenz brauchen Rollstuhlfahrer (What kind of safety assistance do wheelchair users need?). In: Ambient Assisted Living - AAL: 2. Deutscher AAL-Kongress. VDE-Verlag, Berlin (2009)
8. Mandel, C., Huebner, K., Vierhuff, T.: Toward an Autonomous Wheelchair: Cognitive Aspects in Service Robotics. In: Proceedings of Toward Autonomous Robotic Systems (TAROS 2005), pp. 165–172 (2005)
9. Martin, P.: The "Casual Cashmere Diaper Bag": Constraining Speech Recognition Using Examples. In: Interactive Spoken Dialog Systems: Bringing Speech and NLP Together in Real Applications, pp. 61–65 (1997)
10. Nuance Communications, Inc. Nuance VoCon 3200 Embedded Development System Developer's Guide (2006)
11. Otto Bock Mobility Solutions, http://www.ottobock.de
12. Ross, R.J., Bateman, J.: Daisie: Information State Dialogues for Situated Systems. In: Matoušek, V., Mautner, P. (eds.) TSD 2009. LNCS, vol. 5729, pp. 379–386. Springer, Heidelberg (2009)
13. Shi, H., Jian, C., Rachuy, C.: Evaluation of a Unified Dialogue Model for Human-Computer Interaction. International Journal of Computational Linguistics and Applications 2 (2011)
14. Shi, H., Tenbrink, T.: Telling Rolland where to go: HRI dialogues on route navigation. In: Coventry, K., Tenbrink, T., Bateman, J. (eds.) Spatial Language and Dialogue, pp. 117–216. Oxford University Press, Oxford (2009)

Coupled Pragmatic and Semantic Automata in Spoken Dialogue Management

Jolanta Bachan

Adam Mickiewicz University
ul. Wieniawskiego 1, 61-712 Poznań, Poland
jolabachan@gmail.com
http://www.bachan.speechlabs.pl

Abstract. Dialogue managers are often based explicitly on finite state automata, but the present approach couples this type of dialogue manager with a semantic model (a city map) whose traversal is also formalised with a finite state automaton. The two automata are coupled in a scenario-specific fashion within an emergency rescue dialogue between an accident observer and an ambulance station, i.e. a stress scenario which is essentially different from traditional information negotiation scenarios.

The purpose of this use of coupled automata is to develop a prototype dialogue system for investigating semantic alignment and non-alignment in a dialogue. The research on alignment of interlocutors is to improve human-computer communication in a Polish adaptive dialogue system, focusing on the stress scenario. The investigation was performed on two dialogue corpora and resulted in creating a working text-in-speech-out (TISO) dialogue system based on the two linked finite-state automata, evaluated with about 130 human users.

1 Aims and Theoretical Background

The main goal of the present investigation is to define a strategy for providing explicit models of alignment and accommodation in human-computer and human-human communication. Alignment means the adaptation of users to each other in speech style, vocabulary, pronunciation, gestures and body movements (e.g. [10,13]). Acceptable human-computer interaction is the subject of much research, but the literature on these topics does not consider alignment of synthetic speech with a human interlocutor. The present research focuses specifically on stressed (not necessarily emotional) speech in crisis situations. The models should account for speech style alignment in these situations. Speech style is a well-studied parameter, in contrast to emotion.

The research was carried in two steps: first, a preliminary study was performed on a small subset of dialogue corpus and tools for dialogue processing were created along with first dialogue automata, second, a dialogue corpus was recorded in an emergency scenario which analysis was the ground for creating a prototype dialogue system. The goal is not to provide a product or a comprehensive dialogue system but to give a proof-of-concept implementation in a new, previously unexplored semantic alignment domain in crisis scenarios and demonstrate the methodology with a Text-In-Speech-Out (TISO) approach.

P. Sojka et al. (Eds.): TSD 2012, LNCS 7499, pp. 599–606, 2012.

2   First Steps in Realistic Automata Creation: Preliminary Study

2.1   Selected Material for Analysis

For examining the details of alignment and dialogue act theory, a corpus linguistic study on a small sample was performed. For the study, the map-task dialogues from the PoInt corpus [12] were used. Two dialogues (18/,min of speech) were annotated on the dialogue act level using the selected dialogue act categories from Bunt DIT++ categories [7]. In this study the following were looked at:

1. dialogue annotation at the dialogue act level
2. annotation of turns – semantic dialogue flow
3. finite state automata of dialogue act sequences
4. most frequent dialogue acts sequences

2.2   Dialogue Act Annotation

For the preliminary analysis, two dialogues (18 min) between two females and a male and a female were annotated on the dialogue act level. The dialogue act categories for the annotations were selected from Bunt's main categories of the Dynamic Interpretation Theory, DIT [6]. More than one dialogue act category was assigned to a speaking turn, because one utterance can have more than one communicative function. In principle, multiple categories require feature-based finite state treatment [5], but the combinations were treated as atomic symbols. Abbreviations of 12 dialogue act functions chosen from Bunt's categories and used in the annotation of the selected dialogues are allo: allo-feedback, auto: auto-feedback, cnt: contact management, dir: directives, infpr: information providing, infsk: information seeking, open: open meeting, own: own commu-nication control, partner: partner communication management, social: social obligations management, time: time management, turn: turn management.

2.3   Processing of Annotations for Dialogue Analysis

To analyse the dialogue for Finite State Automata (FSA) creation the information on the dialogue act annotation and utterance transcription tiers from one map-task dialogue recording was extracted and analysed. The material was prepared using Linux scripts and was automatically divided into 49 parts and the beginnings and ends of those parts were used to determine the initial and terminal states of the automata. Those divisions into parts indicate the first silence after the last dialogue act in a sequence overlapping with the other speaker's speech, which means that the start/end of the part may occur within speaker's turn if the speaker made a pause within his turn. Fig. 1 illustrates the division of the dialogue into parts starting from the beginning of the dialogue till the 51 second.

2.4   Time Structure of the Dialogue

A computational analysis of the annotation files shows that the time relations between the utterances are very complex. The relations are not purely sequence relations but involve overlaps in time.

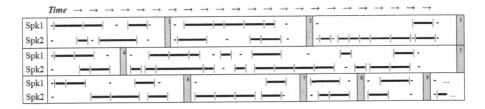

Fig. 1. Temporal sequences and overlaps in a dialogue

In Fig. 1 the temporal division of the dialogue into parts is shown. At the first silence after the last dialogue act in a sequence overlapped with the other speaker's speech, the bars represent dialogue act intervals (chunks of speech), the indices indicate indexed borders and the dash indicates the end of a turn (silence). The diagram is based on the first 51 seconds of the dialogue.

Ideally the two speakers would be modelled separately, and the parallel automata combined [11] in order to model the temporal relations. However, a simplified approach was adopted. The output dialogue act sequences underwent the following processing for the purpose of automata generalisation:

1. alphabetical sorting of dialogue act sequences,
2. reduction of multi-layered labels of dialogue acts to one-layered labels; only the first dialogue act was preserved, e.g. infpr_dir → infpr,
3. deletion of repetition of the same dialogue act,
4. re-sorting of dialogue act sequences.

The reduction process was carried out, because the fact that to the same utterance more than one communicative function could be assigned made the longer sequences unique and limited the generalisation.

2.5 Loop-Free Automata Creation

The annotation on the dialogue act tier was used to create manually a collection of loop-free automata which modelled each sequence of the dialogue acts for each of the speakers. To create the loop-free automata a matrix of dialogue acts flow for Speaker 1 and Speaker 2 with time relations was analysed separately for each speaker. Each dialogue act sequence served to define the initial node, the terminal node and the transitions between the nodes for each of the loop-free automata. Such automata were evaluated for correctness using an NDFST interpreter [9]. An example of a loop-free automaton for Speaker 1 is shown in Fig. 2.

2.6 Generalisations over Non-finite Regular Languages

Generalisations over non-finite regular languages can be expressed with FSAs with loops, and visualised with directed cyclic graphs. Altogether 22 automata with loops were created, 7 for Speaker 1 and 15 for Speaker 2.

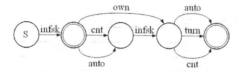

Fig. 2. Loop-free automaton for Speaker 1

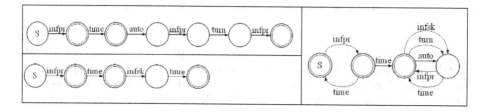

Fig. 3. Loop-free automata (left) and their counterpart with loops for Speaker 2 (right)

Analysis of the prefixes and analysis of the loop-free automata allowed the creation of a whole set of automata with loops which model non-finite regular languages. Examples of loop-free automata and their counterpart with loops for Speaker 2 are shown in Fig. 3.

2.7 Coupled Turn Automata

Coupled turn automata are based on real dialogue act sequences for both speakers. The turn automata were made by combining dialogue act automata for Speaker 1 (spk1) and Speaker 2 (spk2). Each speaker has his/her own automaton, and they are coupled in/by negotiations. An example of the coupled turn automaton is presented in Fig. 4. S is the initial node. The node more to the left shows which speaker starts the sequence. The dotted arrows show the transition of turns between the speakers.

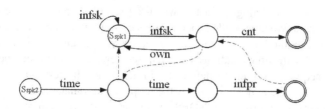

Fig. 4. Coupled turn automaton. Spk 2 starts the sequence, Spk 1 follows

2.8 Evaluation of Dialogue Act Automata

The automata were evaluated for coherence (whether the automata are syntactically correct and actually work when operational), completeness (whether the automata describe all the phenomena they are intended to describe, not necessarily only restricted to a particular corpus, but including generalisations, and possibly also judged by native speaker intuitions), soundness (whether the automata describe only the phenomena they are intended to describe not necessarily restricted to a particular corpus, but including generalisations, and possibly also judged by native speaker intuitions), consistency (whether the modelling is done in the same way for similar observations of utterances).

The Nondeterministic Finite State Transducer (NDFST) online tool [9] was used in the present study to evaluate the dialogue acts automata. All the automata underwent testing and were positively evaluated in the NDFST.

3   Corpus Linguistic Study

Two types of dialogue were recorded under laboratory conditions: a map task dialogue and a picture description dialogue, so-called diapix task [4] in stressed and neutral scenarios. For creating a dialogue model for the prototype dialogue system, only the map task dialogues in stress conditions were analysed. The map of the task is presented in Fig. 5 (left) the black circles and street names were not marked. 24 subjects were recorded in a public setting [3] in emergency scenario. Additionally, 3 pairs of close friends were recorded as a control group. For the project, 15 males and 15 females were chosen and recorded in pairs: male – male, male – female, female – female. The corpus contains 4h 12min of recordings, out of which 38min 45sec are the map-task emergency dialogues. See [1] for more details.

4   Finite-State Transducer Model of the Map

The emergency map can be represented as a finite state transducer (FST) where each junction corresponds to the transition node (Fig. 5, right). Not all the streets are open and some junctions cannot be reached by dialogue system. There is a traffic jam on the way or roadworks, and at one place the street has been blocked because of a school race. Such blockages are not taken into account when designing the FST. In the FST, $q0$ is the start node and $q13$ is the end node. Latin letters are used for transition labels.

5   Dialogue System Implementation

The prototype dialogue system is based on two FSA: Fig. 5 (right), for map traversal, and an additional FSA for the dialogue manager (Fig. 6, left). The instruction to the human caller is to direct an ambulance from the hospital to the person with a heart attack along the streets. The human user inputs chat text into the system in writing. The dialogue system communicates with the caller via audio output producing synthetic speech. The caller also has a street map on a computer screen. Either formal or informal speech style is selected by an experimenter for the dialogue. The dialogue manager schema is presented in Fig. 6 (right). The dialogue system is a TISO configuration [2]. Written input from the user is entered on the command line and the system produces synthetic speech output via loudspeakers.

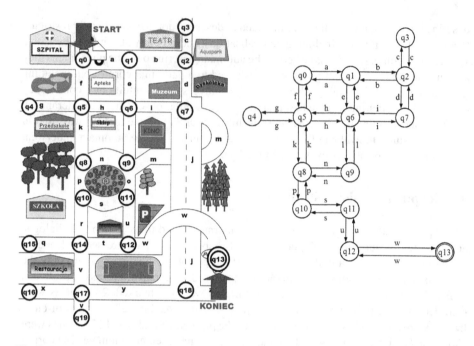

Fig. 5. Emergency map with junctions marked (left); Finite State Automaton representing the reachable junctions (right)

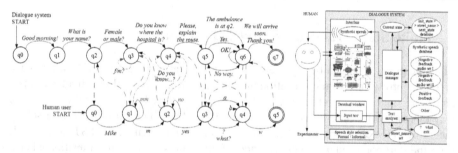

Fig. 6. Dialogue automaton with speciment utterance labels (left), dialogue manager schema (right)

5.1 Evaluation

The dialogue system was evaluated using EAGLES standards [8]. After successful diagnostic evaluation, the dialogue system faced functional testing with the human users. 52 people took part in the evaluation. In the evaluation, mainly young students took part. After each dialogue the test participant was asked to assess different domains of the system on a 5-point rating scale (1 – lowest, 5 – highest). The test participants were asked to evaluate seven categories: friendliness, speech quality, speech intelligibility, dialogue, dialogue naturalness, system attractiveness and ease of usage. Last but not least, the system was tested in field conditions at the Researchers' Night 2011 in Poznań, Poland, with about 80 people.

5.2 Results

In the laboratory test 14 females and 12 males took part to evaluate each of the two scenarios: formal and informal. The duration time of all the dialogues in formal and informal scenarios lasted about 75min 34sec and 76min 48sec respectively. The number of inputs inserted during one dialogue is almost the same and equals 20.54 inputs for the formal and 20.26 inputs for the informal scenario. All subjects accomplished the communication with the computer successfully, meaning that semantic alignment was successful despite obstacles. Misalignments happened, but the dialogue system effectively recovered from misalignments. The results of the judgement testing of the system were good: 4.11 for the formal dialogue scenario and 4.30 for the informal scenario, where 5 was the highest grade. When it comes to the field testing at the Researchers Night, all the people carried out the task successfully. The age of the youngest child was 5 years old. The best time was 55s, the longest time was 4min 30sec.

6 Conclusions

Dialogue modelling strategies were discussed, and semantic alignment was modelled. Analysis of corpora of dialogue recordings in a map task emergency scenario was used to create a finite-state automaton modelling a dialogue in a prototype dialogue system. Finally, a prototype dialogue system was developed and evaluated with human users. The prototype TISO dialogue system combined text input with speech output and its core was based on two linked finite state automata: one for the dialogue manager and one for map traversal. The laboratory setting of the evaluation task demonstrated alignment of the semantic representation of the map, as all the human users finished the task successfully.

Acknowledgments. This work was partly funded by the research supervisor project grant No. N N104 119838. The author is currently supported by grant "Collecting and processing of the verbal information in military systems for crime and terrorism prevention and control." (OR 00017012).

References

1. Bachan, J.: Developing and evaluating an emergency scenario dialogue corpus. In: Proceedings of 8th LREC 2012, pp. 1421–1427. ELRA, Istanbul (2012)
2. Bachan, J.: Modelling semantic alignment in emergency dialogue. In: Proceedings of 5th LTC, vol. 2011, pp. 324–328. Fundacja UAM, Poznań, Poland (2011)
3. Batliner, A., Steidl, S., Hacker, C., Nöth, E.: Private emotions versus social inter-action: a data-driven approach towards analysing emotion in speech. User Modelling and User-Adapted Interaction – The Journal of Personalization Research 18, 175–206 (2008)
4. Bradlow, A.R., Baker, R.E., Choi, A., Kim, M., van Engen, K.J.: The Wildcat Corpus of Native- and Foreign-Accented English. Journal of the Acoustical Society of America 121(5), pt. 2, 3072 (2007)

5. Berndsen, J.: Time Map Phonology: Finite State Models and Event Logics in Speech Recognition. Kluwer Academic Publishers, Dordrecht (1998)
6. Bunt, H.: Dialogue pragmatics and context specification. In: Bunt, H., Black, W. (eds.) Abduction, Belief and Context in Dialogue. Studies in Computational Pragmatics, pp. 81–150. John Benjamins, Amsterdam (2000)
7. Bunt, H.: DIT++ Taxonomy of Dialogue Acts (Rel. 3, ver. 2, 2008-02-08) (2008), http://let.uvt.nl/general/people/bunt/docs/dit-schema3-2.html
8. Gibbon, D., Mertins, I., Moore, R.: Handbook of Multimodal and Spoken Dialogue Systems: Terminology, Resources and Product Evaluation. Kluwer Academic Publishers, New York (2000)
9. Gibbon, D.: Nondeterministic Finite State Transducer. Version 2008-08-12 (2008), http://wwwhomes.uni-bielefeld.de/gibbon/Forms/Python/FSM/generator.html (accessed on October 12, 2011)
10. Giles, H., Coupland, N., Coupland, J.: Accomodation theory: Communication, context and consequences. In: Giles, H., Coupland, J., Coupland, N. (eds.) Contexts of Accommodation, pp. 1–68. Cambridge University Press, Cambridge (1992)
11. Kaplan, R., Kay, M.: Regular Models of Phonological Rule Systems. Computational Linguistics 20(3), 331–378 (1994)
12. Karpiński, M.: The Corpus of the Polish Intonational Database (PoInt). Investigationes Linguisticae 8, 24–25 (2002)
13. Pickering, M.J., Garrod, S.: Toward a mechanistic psychology of dialogue. Behavioral and Brain Sciences 27, 169–225 (2004)

Exploration of Metaphor and Affect Sensing Using Semantic Interpretation in an Intelligent Agent

Li Zhang

School of Computing, Engineering and Information Sciences
Northumbria University, Newcastle, UK
li.zhang@northumbria.ac.uk

Abstract. We developed a virtual drama improvisation platform to allow human users to be creative in their role-play with the interaction of an AI agent. Previously, the AI agent was able to detect affect from users' inputs with strong affect indicators. In this paper, we integrate context-based affect detection to enable the intelligent agent to detect affect from inputs with weak or no affect signals. Topic theme detection using latent semantic analysis is applied to such inputs to identify their discussions themes and potential target audiences. Relationships between characters are also taken into account for affect analysis. Such semantic interpretation of the dialogue contexts also proofs to be effective in the recognition of metaphorical phenomena.

Keywords: Affect detection, interaction contexts, metaphor and semantic interpretation.

1 Introduction

It is challenging to produce an intelligent agent which is capable of conducting drama performance, interpreting social relationships, context, general mood and emotion, reasonably sensing others' inter-conversion, identifying its role and participating intelligently in open-ended improvisational interaction. Although the first interaction system based on natural language input, *Eliza* [1], was first developed back in 1966, there were still limited attempts to enable agents to interpret open-ended literal and metaphorical inputs automatically and possess emotion intelligence. We believe it will make AI agents possess human-like behaviour and narrow the communicative gap between machines and human-beings if they are equipped to interpret human emotions during the interaction. Thus in our research, we equip our AI agent with emotion and social intelligence. According to Kappas [2], human emotions are psychological constructs with notoriously noisy, murky, and fuzzy boundaries that are compounded with contextual influences in experience and expression and individual differences. These natural features of emotion also make it difficult for a single modal recognition, such as via acoustic-prosodic features of speech or facial expressions. In this research, we intend to make our agent take multi-channels of subtle emotional expressions embedded in social interaction contexts into consideration to draw reliable affect interpretation. Thus our research focuses on the production of intelligent agents with the abilities of interpreting dialogue contexts semantically to support affect detection.

P. Sojka et al. (Eds.): TSD 2012, LNCS 7499, pp. 607–615, 2012.

Our work is conducted within a previously developed online multi-user role-play virtual drama framework, which allows school children aged 14–16 to perform drama improvisation. In this platform young people could interact online in a 3D virtual drama stage with others under the guidance of a human director. In one session, up to five virtual characters are controlled on a virtual stage by human users ("actors"), with characters' (textual) "speeches" typed by the actors operating the characters. The actors are given a loose scenario around which to improvise, but are at liberty to be creative. An intelligent agent with an affect detection component was also involved in improvisation. It was able to detect 15 emotions from human characters' each individual input, but the detection has not taken any contexts into consideration.

Moreover, the previous processing was mainly based on pattern-matching rules that looked for simple grammatical patterns partially involving specific words. It proved to be effective enough to detect affect from those inputs containing strong clear emotional indictors such as 'yes/no', 'thanks' etc. There are also situations that users' inputs do not have any obvious emotional indicators or contain very weak affect signals, thus contextual inference is needed to further derive the affect conveyed in such user inputs. Moreover, inspection of the collected transcripts also indicates that the improvisational dialogues are often multi-threaded. This refers to the situation that social responses of different discussion themes to previous several speakers are mixed up. Therefore the detection of the most related discussion themes using semantic analysis is very crucial for the accurate interpretation of the emotions implied in those with ambiguous target audiences and weak affect indicators.

2   Related Work

Tremendous progress in emotion recognition has been witnessed by the last decade. Endrass, Rehm and André [3] carried out study on the culture-related differences in the domain of small talk behaviour. Their agents were equipped to generate culture specific dialogues. There is much other work in a similar vein. Recently textual affect sensing has also drawn researchers' attention. Ptaszynski et al. [4] employed context-sensitive affect detection with the integration of a web-mining technique to detect affect from users' input and verify the contextual appropriateness of the detected emotions. However, their system targeted interaction only between an AI agent and one human user in non-role-playing situations, which greatly reduced the complexity of the modelling of the interaction context.

Moreover metaphorical language has been used in literature to convey emotions, which also inspires cognitive semanticists [5]. Indeed, the metaphorical description of emotional states is common and has been extensively studied [6], for example, "he nearly exploded" and "joy ran through me," where anger and joy are being viewed in vivid physical terms. There is also other work focusing on metaphors in affective expressions [7] useful to our application.

3   Metaphor and Affect Detection Using Latent Semantic Analysis

We noticed that the language used in the collected transcripts is often complex and invariably ungrammatical, and also contains a large number of weak cues to the affect that is being expressed. These cues may be contradictory or may work together to enable a stronger interpretation of the affective state. In order to build a reliable and robust analyser, it is necessary to undertake several diverse forms of analysis and to enable these to work together to build stronger interpretations. Thus in this work, we integrate contextual information to further derive affect embedded in contexts to provide affect detection for those without strong affect indicators.

Since human language is very diverse, terms, concepts and emotional expressions can be described in various ways. Especially if the inputs contain weak affect indicators, other approaches focusing on underlying semantic structures in the expressions should be considered. Thus latent semantic analysis (LSA) [8] is employed to calculate semantic similarities between sentences to derive discussion themes and potential target audiences for those inputs without strong affect signals.

Latent semantic analysis generally identifies relationships between a set of documents and the terms they contain by producing a set of concepts related to the documents and terms. In order to compare the *meanings or concepts* behind the words, LSA maps both words and documents into a 'concept' space and performs comparison in this space. In detail, LSA assumes that there is some underlying latent semantic structure in the data which is partially obscured by the randomness of the word choice. This random choice of words also introduces noise into the word-concept relationship. LSA aims to find the smallest set of concepts that spans all the documents. It uses a statistical technique, called singular value decomposition, to estimate the hidden concept space and to remove the noise. This concept space associates syntactically different but semantically similar terms and documents. We use these transformed terms and documents in the concept space for retrieval rather than the original terms and documents.

In our work, we employ the semantic vectors package [9] to perform LSA, analyze underlying relationships between documents and calculate their similarities. This package provides APIs for concept space creation. It applies concept mapping algorithms to term-document matrices using Apache Lucene, a high-performance, full-featured text search engine library implemented in Java [9]. We integrate this package with our AI agent's affect detection component to calculate the semantic similarities between improvisational inputs without strong affect signals and training documents with clear discussion themes. In this paper, we target the transcripts of the Crohn's disease[1] scenario used in previous testing for metaphor and affect analysis.

In order to compare the user inputs with documents belonging to different topic categories, we have to collect some training documents with strong topic themes from the Experience website (http://www.experienceproject.com/). These articles belong to 12 categories including Education, Family & Friends, Health & Wellness, etc. Since we intend to perform discussion theme detection for the transcripts of the Crohn's disease scenario, we have extracted sample articles close enough to the scenario including articles of Crohn's disease (five articles), school bullying (five articles), family care for children (five articles), food choice (three

---

[1] Peter has Crohn's disease and has the option to undergo a life-changing but dangerous surgery. He needs to discuss the pros and cons with friends and family. Janet (Mum) wants Peter to have the operation. Matthew (younger brother) is against it. Arnold (Dad) is not able to face the situation. Dave (the best friend) mediates the discussion.

articles), school life including school uniform (10 short articles) and school lunch (10 short articles). Phrase and sentence level expressions implying 'disagreement' and 'suggestion' are also gathered from articles published on the Experience website. Thus we have training documents with eight discussion themes including 'Crohn's disease', 'bullying', 'family care', 'food choice', 'school lunch', 'school uniform', 'suggestions' and 'disagreement'. Affect detection from metaphorical expressions often poses great challenges to automatic processing systems. In order to detect a few metaphorical phenomena, we include four types of metaphorical examples published on the following website: http://knowgramming.com. These include cooking, family, weather, and farm metaphors. We have also borrowed a group of 'Ideas as External Entities' metaphor examples from the ATT-Meta databank (http://www.cs.bham.ac.uk/~jab/ATT-Meta/Databank/) to enrich the metaphor categories. Individual files are used to store each type of the metaphors. All the sample documents of the above 13 categories are regarded as training files and have been put under one directory for further analysis.

We use one example interaction of the Crohn's disease scenario produced by testing subjects in the following to demonstrate how to detect the discussion themes for those inputs with weak affect indicators and ambiguous target audiences.

1. Peter: im going to *have an ileostomy* [sad]
2. Peter: *im scared* [scared]
3. Dave: *i'm ur friend* peter and *i'll stand by* you [caring]
4. Peter: yeah i know, but *the disease stuff sucks* [sad]
5. Dave: if it's what u want, you should *go for it* though [caring]
6. Janet: peter you must go throu with this operation. *Its for the best* [caring]
7. Peter: but *no one else can do* nethin [disapproval]
8. Arnold: *take it easy*, consider all your options peter [caring]
9. Matthew: u have had operations b4 I'm sure *u'll be ok* [caring]
10. Dave: what are your other options peter [*neutral: a question sentence*]
11. Peter: im trying very hard but there is too much stuff blocking my head up [*Topics: family care, ideas metaphor, bullied; Target audience: Dave; Emotion: neg.*]
12. Peter: my plate is already too full.... there aint otha options dave [*Topics: food, cooking metaphor, bullied; Target audience: Dave; Emotion: stressful*]

Affect implied by the inputs with strong affect indicators (illustrated in italics) in the above interaction is detected by the previous affect detection processing. The inputs without an affect label followed straightaway are those without strong affect indicators (10th, 11th & 12th inputs). Therefore further processing is needed to recover their most related discussion themes and identify their most likely target audiences in order to identify implied emotions more accurately. Our general regime for the detection of discussion themes is to create the 'concept' space by generating term and document vectors for all the training corpus and a test input. Then we use these transformed terms and documents in the concept space for retrieval and comparison. For example, we use the generated concept space to calculate semantic similarities between user inputs and training files with clear topic themes and search for documents including the user input closest to the vector for a specific topic theme. We start with the 11th input from Peter to demonstrate the topic theme detection. First of all, this input is stored as a separate individual test file (test_corpus1.txt) under the same folder containing all the training sample documents of the 13 categories.

As mentioned above, first of all, the corresponding semantic vector APIs are used to create a Lucene index for all the training samples and the 11th input. This generated index is also used to create term and document vectors, i.e. the concept space. Various search options could be used to test the generated concept model. In order to find out the most effective approach to extract the topic theme of the test inputs, we, first of all, provide rankings for all the training documents and the test input based on their semantic distances to a topic theme. We achieve this by searching for document vectors closest to the vector for a specific topic term (e.g. 'bullying'). The 11th input obtains the highest ranking for the topic theme, 'ideas metaphor' (top 2nd), 'cooking metaphor' (top 3rd), and 'bullied' (top 5th), among all the rankings for the 13 topics. Partial output is listed in Figure 1 for the rankings of partial training documents and the 11th input based on their semantic distances to the theme, 'ideas metaphor'.

> Found vector for 'ideas metaphor'
> Search output follows ...
> 0.9687636802981049:F:\ideas_metaphor.txt
> 0.6109025620852475:F:\test_corpus1.txt
> 0.468855438977363:F:\family_care5.txt
> 0.4384741083934003:F:\family_care2.txt
> 0.43717258989527735:F:\suggestion1.txt
> 0.43481884276082305:F:\bullied1.txt
> 0.42516120464904383:F:\bullied2.txt
> 0.42055621398181026:F:\crohn2.txt

Fig. 1. Partial output for searching for document vectors closest to the vector of 'ideas metaphor' (test_corpus1.txt containing the 11th input, ranking the top 2nd)

Semantic similarities between documents are also produced in order to further inform topic theme detection. All the training documents are taken either from articles under clear discussion themes within the 12 categories of the Experience project or the metaphor websites with clear metaphor classifications. The file titles used indicate the corresponding discussion or metaphor themes. If the semantic distances between files, esp. between training files and the test file, are calculated, then it provides another source of information for the topic theme detection. Therefore we use the CompareTerms semantic vector API to find out semantic similarities between all the training corpus and the test document. We provide the top five rankings for semantic similarities between the training documents and the 11th input in Figure 2.

The semantic similarity test in Figure 2 indicates that the 11th input is more closely related to topics of 'family care (family_care3.txt)' and 'ideas metaphor

Similarity of "family_care3.txt" with "test_corpus1.txt":
0.74371306267593305
Similarity of "ideas_metaphor.txt" with "test_corpus1.txt":
0.73447742369521 76
Similarity of "crohn3.txt" with "test_corpus1.txt":
0.71497122409146 11
Similarity of "bullied1.txt" with "test_corpus1.txt":
0.689368455185548
Similarity of "family_care2.txt" with "test_corpus1.txt":
0.67731270425495 64

Fig. 2. Part of the output for semantic similarities between training files and the 11th input

(ideas_metaphor.txt)' although it is also closely related to negative topic themes such as 'disease' and 'being bullied'. In order to identify the 11th input's potential target audiences, we have to conduct topic theme detection starting from the 10th input and retrieving backwards until we find the input with a similar topic theme or with a posed question for Peter. The pre-processing of the previous affect detection includes a syntactical parsing using a Rasp parser and it identifies the 10th input from Dave is a question sentence with the mentioning of Peter's name. Thus the syntactical processing regards the 10th input from Dave posed a question toward the target audience, Peter. We also derive its most likely topic themes for the 10th input to provide further confirmation. The processing identifies the 10th input semantically most related to the following topics, 'disagreement', 'family care' and 'suggestion'.

We also noticed that in English, the expression of question sentences is very diverse. Most of them will require replies from other characters, while there is a small group of question sentences that do not really require any replies, i.e. rhetorical questions. Such questions (e.g. "What the hell are you thinking?", "How many times do I need to tell you?", "Are you crazy?") encourage the listener to think about what the (often obvious) answer to the question must be. They tend to be used to express dissatisfaction. We especially detect such rhetorical questions using latent semantic analysis after Rasp's initial analysis. We construct two training documents for questions sentences: one with normal questions and the other with rhetorical questions. We use the topic detection to perform semantic similarity comparison between the two training document vectors and the 10th input from Dave.

It indicates that the input from Dave is more likely to be a normal question rather than a rhetorical expression. Thus it is more inclined to imply a normal discussion theme such as 'family care' than to express 'disagreement' or 'suggestion'. Thus the 10th input from Dave has the same discussion theme to one of the themes implied by the 11th input. Thus the target audience of the 11th input is Dave, who has asked Peter a question in the first place. Since the 11th input is also regarded as an 'ideas metaphor' with a high confidence score, the following processing is applied to the partial input "there is too much stuff blocking my head up" to recognize the metaphor.

1. Rasp: 'EX (there) + VBZ (is) + RG (too) + DA1 (much) + NN1 (stuff) + VVG (blocking) + APP$ (my) + NN1 (head) + RP(up)'

2. WordNet: 'stuff' -> hypernym: information abstract entity, since 'stuff' has been described by a singular after-determiner ('much'). 'Head' -> hypernym: a body part physical entity. 'Block' -> hypernyms: PREVENT, KEEP.

3. The input implies -> 'an abstract subject entity (stuff) + an action (block) + a physical object entity' (head) -> showing semantic preference violation (an abstract entity performs an action towards a physical object) -> recognised as a metaphor.

In this example, ideas are viewed in terms of external entities. They are often cast as concrete physical objects. They can move around or be active in other ways. The above processing recognises that this input shows semantic preference violation, i.e. an information abstract subject performs physical actions. Since the 11th input is also semantically close to bullied topics, it implies a 'negative' emotion.

In a similar way, the 12th input from Peter is also identified semantically most closely related to terms, 'food' and 'cooking metaphor'. The topic detection also identifies it shows high semantic similarities with training corpus under the themes of 'cooking metaphor (cooking_metaphor.txt: 0.563)' and 'being bullied (bullied3.txt: 0.513)'. Since Dave's name is mentioned in this input, the processing classifies Dave as one target audience. Thus the 12th input is regarded as a potential cooking metaphor with a negative bullying theme. The above processing using Rasp and Wordnet is also applied to it. The first part of the 12th input is interpreted as 'a physical tableware-object subject followed by a copular form and a quantity adjective'. But it does not show any semantic preference violation with only one cooking related term, 'plate'. Context information is thus retrieved for the recognition of this cooking metaphor.

We start from the 11th input to find out the topic themes of those inputs most similar to the topics of the 12th input. As discussed earlier, the 11th input is contributed by Peter as well with embedded 'family care' and 'bullied' themes, but not related to 'food'. The 10th input is a question sentence from Dave with a 'family care' theme. The 8th and 9th inputs contain strong affect indicators (see italics) implying 'family care' themes as well. The backward retrieval stops at the 7th input, the last round input contributed by Peter. Thus the 11th input shares the same 'bullied' topic with the 12th input and the 12th input contributed by the same speaker is regarded as a further answer to the previous question raised by Dave. Moreover the 11th input is recognised as an 'ideas as external entities' metaphor with a negative indication. Thus the 12th input is not really 'food' related but an extension of the ideas metaphor and more likely to indicate a physical tableware object entity, plate, is a flat, limited space for solid ideas. Therefore it is recognised as a cooking metaphor. The most recent interaction context (8th – 11th) also shares a consistent positive theme of 'family care', but not a bullying context. Thus the sick character, Peter, is thus more likely to indicate a 'stressful' emotion because of 'being bullied by disease' in the 12th input. Rule sets are generated for the metaphorical and affect reasoning using emotions embedded in social contexts and relationships between target audiences and speakers. It is envisaged the topic theme detection could also be useful to distinguish task un-related small talk and task-driven behaviours during human agent interaction.

4 Evaluation and Conclusion

We have taken previously collected transcripts recorded during our user testing to evaluate the efficiency of the updated affect detection component with contextual inference. In order to evaluate the performances of the topic theme detection and the rule based affect detection in social contexts, three transcripts of the Crohn's disease scenario are used. Two human judges are employed to annotate the topic themes of the extracted 300 user inputs from the test transcripts using the 13 topics. Cohen's Kappa is a statistical measurement of inter-annotator agreement. We used it to measure the inter-annotator agreement between human judges for the topic theme annotation and obtained 0.83. Then the 265 inputs with agreed annotations are used as the gold standard to test the performance of the topic theme detection. A keyword pattern matching baseline system was used to compare the performance with that of the LSA. We have obtained an averaged precision, 0.736, and an averaged recall, 0.733, using the LSA while the baseline system achieved an averaged precision of 0.603 and an averaged recall of 0.583. The results indicated that discussion themes of 'bullying', 'disease' and 'food choices' have been very well detected by our semantic-based analysis. The detection of 'family care' and 'suggestion' topics posed most of the challenges. Generally the LSA-based interpretation achieves promising results.

The two human judges also annotated these 265 inputs with the 15 frequently used emotions. Since 15 emotions were a big category and the annotators may not experience the exact emotions as the test subjects did, it led to the low inter-agreement between human judges. The inter-agreement between human judge A/B is 0.63. While the previous version achieves 0.45 in good cases, the new version achieves 0.57 and 0.59 respectively. Although the improvements are comparatively small due to using a large category of emotions, many expressions regarded as 'neutral' previously were annotated appropriately as emotional expressions.

Moreover, in future work, we will use articles published on the Experience website to evaluate our AI agent on metaphor recognition using contexts. We are also interested in using topic extraction to inform affect detection directly, e.g. the suggestion of a topic change indicating potential indifference to the current topic. It will also ease the interaction if our agent is equipped with culturally related small talk behavior. In the long term, we also aim to incorporate each weak affect indicator embedded in semantic analysis, speech, facial expressions and gestures to benefit affect interpretation during natural interactions. We believe these are crucial aspects for the development of intelligent agents with social and emotion intelligence.

References

1. Weizenbaum, J.: ELIZA – A Computer Program for the Study of Natural Language Communication Between Man and Machine. Communications of the ACM 9(1), 36–45 (1966)
2. Kappas, A.: Smile when you read this, whether you like it or not: Conceptual challenges to affect detection. IEEE Transactions on Affective Computing 1(1), 38–41 (2010)
3. Endrass, B., Rehm, M., André, E.: Planning Small Talk Behavior with Cultural Influences for Multiagent Systems. Computer Speech and Language 25(2), 158–174 (2011)

4. Ptaszynski, M., Dybala, P., Shi, W., Rzepka, R., And Araki, K.: Towards Context Aware Emotional Intelligence in Machines: Computing Contextual Appropriateness of Affective States. In: Proceeding of IJCAI (2009)
5. Kövecses, Z.: Are There Any Emotion-Specific Metaphors? In: Athanasiadou, A., Tabakowska, E. (eds.) Speaking of Emotions: Conceptualization and Expression, pp. 127–151. Mouton de Gruyter, Berlin and New York (1998)
6. Fainsilber, L., Ortony, A.: Metaphorical uses of language in the expression of emotions. Metaphor and Symbolic Activity 2(4), 239–250 (1987)
7. Zhang, L., Barnden, J.A.: Affect and Metaphor Sensing in Virtual Drama. International Journal of Computer Games Technology vol. 2010, Article ID 512563 (2010)
8. Landauer, T.K., Dumais, S.: Latent semantic analysis 3(11), 4356 (2008)
9. Widdows, D., Cohen, T.: The Semantic Vectors Package: New Algorithms and Public Tools for Distributional Semantics. IEEE Int. Conference on Semantic Computing (2010)

Sentence Classification with Grammatical Errors and Those Out of Scope of Grammar Assumption for Dialogue-Based CALL Systems

Yu Nagai, Tomohisa Senzai, Seiichi Yamamoto, and Masafumi Nishida

Faculty of Science and Engineering, Doshisha University
1–3 Miyakodani, Tatara, Kyotanabe-shi, Kyoto 610-0321, Japan
`{dtl0740,dtk0761}@mail4.doshisha.ac.jp,`
`{seyamamo,mnishida}@mail.doshisha.ac.jp`

Abstract. Computer Assisted Language Learning (CALL) systems are one of the key technologies in assisting learners to master a second language. The progress in automatic speech recognition has advanced research on CALL systems that recognize speech constructed by students. Reliable recognition is still difficult from speech by second language speakers, which contains pronunciation, lexical, and grammatical errors. We developed a dialogue-based CALL system using a learner corpus. The system uses two kinds of automatic speech recognizers using *ngram* and finite state automaton (FSA). We also propose a classification method for classifying the speech recognition results from the recognizer using FSA as accepted or rejected. The classification method uses the differences in acoustic likelihoods of both recognizers as well as the edit distance between strings of output words from both recognizers and coverage estimation by FSA over various expressions.

Keywords: CALL system, learner corpus, grammatical error.

1 Introduction

Language learning is a relevant topic today, where being able to communicate one's message fluently in a foreign language is important. An effective method of mastering a second language is to put oneself into a one-on-one interactive language training situation with a trained language instructor. One-on-one tutoring is usually too expensive and impractical. In reality, most students attend classes in which they have to share their teacher's attention. An automatic tutoring system may be used as a complement to the human instructor in some cases, such as in pronunciation training, and many computer assisted language learning (CALL) systems have been developed and put onto the market. One of the major problems in commonly used automated language training systems is that students are usually assigned a passive role and asked to repeat the sentence they had learned or read aloud one of the written choices. As a result, students have no opportunity to practice constructing utterances on their own.

Advances in automatic speech recognition (ASR) technologies have advanced research on dialogue-based CALL systems, which involve conversations by assigning students an active role in which they are able to construct utterances on their own.

P. Sojka et al. (Eds.): TSD 2012, LNCS 7499, pp. 616–623, 2012.

State-of-the-art ASR cannot obtain high recognition accuracy for utterances with pronunciation and grammatical errors by non-native speakers; therefore, dialogue-based CALL systems, which allow learners to construct their own utterances, direct learners to construct utterances under pre-determined conditions to achieve high recognition accuracy. For example, when FLUENCY, a CALL system developed by Eskenazi [1], asks a learner "When did you find it?", it directs him/her to construct an answer using a given keyword, e.g., "Yesterday", which makes speech recognition easier. Kweon proposed a learning system [2] in which learners complete a pre-exercise composed of vocabulary, grammar, and typical conversation examples and a real conversation exercise using the system. The pre-exercise reduces out-of-grammar utterances and makes speech recognition easier. Even after the pre-exercise, out-of-grammar utterances are produced by the learners.

These dialogue-based CALL systems use speech recognition systems that use finite state automaton (FSA) as a language model, and descriptive grammars are manually created so that FSA can accept utterances with lexical and grammatical errors. It is difficult, however, to design descriptive grammars that can accept correct expressions as well as those with various grammatical errors since the utterances the learners compose are out of the scope of these assumed descriptive grammars. To overcome this problem, dialogue-based CALL systems are designed to detect such utterances using features such as acoustic likelihood.

We developed a dialogue-based CALL system that communicates with Japanese students in English and encourages them to construct utterances on their own. The system displays an expression in Japanese that the system expects the student to utter as an answer to limit the number of hypothses in recognizing utterances. We also propose a method for detecting utterances out of the scope of the descriptive grammar assumptions using features such as acoustic likelihood and the edit-distance (ED) between strings of output words from two ASRs and the coverage of the descriptive grammars over English expressions uttered by the students.

This article is structured as follows. Section 2 gives an overview of the proposed dialogue-based CALL system. Section 3 describes a learner corpus we used to develop our dialogue-based CALL system. Section 4 describes our detection method of out-of-scope utterances. Section 5 describes experiments and results, and Section 6 concludes the article and discusses future work.

2   Overview of Proposed CALL System

Our CALL system addresses a question to learners depending on scenes of their choice, such as "Would you like to try it on?" at a shopping scene. If a learner do not understanad the system's question, he/she clicks on the repeat button. The system repeats the question and displays the question on the screen in multiple repeats. If he/she understands the question, the system displays a response in Japanese which it expects him/her to utter in English. When the learner does not think he/she is able to construct utterances corresponding to the displayed response such as "eh, soshiyoukana. shichaku-shitsuha doko desuka (Yes I would. Where is the dressing room?)", he/she can request less difficult one. Their utterances are recognized using two speech recognition systems, one of which uses FSA as the language model and the other, which is used as a verification system to detect utterances out of the scope of the descriptive grammar assumptions, uses *ngram* as the language model. The descriptive grammars from which the FSA were constructed were developed from a corpus of expressions by bilingual English/Japanese speakers and those by Japanese students speaking English as a second language. The system displays a speech recognition result and its corresponding grammatically correct expression when learners construct utterances within the assumed descriptive grammars. On the other hand, when the system detects utterances out of the scope of the assumption, the system shows a sample answer.

2.1   Recognition Engine and Acoustic Model

We used a multi-lingual speech recognition system called ATRASR [3] developed at ATR as a test-bed system of English speech recognition for Japanese. Main features of ATRASR are shortly described in the following;

(1) Speech Analysis: The speech analysis conditions were as follows; the frame length was 20 ms and the frame shift was 10 ms; 12-order MFCC, 12-order Delta MFCC, and Delta log power were used as feature parameters. Our phoneme sets consist of 44 phonemes, including silence. They are the same as those used in the Wall Street Jornal (WSJ) corpus official evaluations because in this way we could use its dictionary as a source of pronunciation base-form.

(2) Acoustic Model: We employed an gender dependent acoustic model which was trained with read speech data collected from 201 Japanese speakers (100 male and 101 female) [4]. Each speakers read out about 120 English sentences. The hidden Markov network (HMNet) [5], a kind of context-dependent phone model, was used for modeling the acoustic model. The pronunciation dictionary has about 35,000 entries mainly for vocabulary concerning travel conversation. We used a two pass decoder ATRASR developed at ATR, which uses bi-gram language model and inter-word context dependent acoustic model in the first pass, and rescores candidates obtained in the first pass with tri-gram language model and intra-word context-dependent acoustic model in the second pass.

2.2   Language Model and Recognition Accuracy

We have two language models; one is an *ngram* language model trained with Basic Travel Expression Corpus (BTEC) English data, which consists of about 500,000 sentences [6]. The other language model is an FSA model, which was mainly developed from the learner corpus created by Yasuda et al. [7]. Japanese source sentences of the learner corpus were randomly selected from the BTEC and textbooks on English for junior and senior high schools students.

Table 1 lists word accuracy for speech uttered by 14 speakers (7 females and 7 males) whose communicative skills in English were evaluated on the Test of English for International Communication (TOEIC) score [8], the most popular English assessment of communicative English skill in Japan, which ranges from 10 (lowest) to 990 (highest). Read speech denotes the speech the speakers read out that translated by a

Table 1. Word accuracy for read speech and spontaneous speech

utterance	read speech		spontaneous speech	
language model	*ngram*	FSA	*ngram*	FSA
word accuracy	81.2	92.1	74.0	62.4

bilingual English/Japanese speaker. Spontaneous speech denotes the speech which each speaker uttered by constructing the expressions on their own according to the response displayed on the screen.

3 Characteristics of Learner Corpus

Dialogue-based CALL systems allow learners to produce their own utterances, which may contain pronunciation and grammatical errors, and may produce recognition errors even for grammatically correct sentences. These CALL systems must recognize utterances with errors as it is and detect whether the error originated from abusage by the learners or from misrecognition by the speech recognizer. The abusage of the learners may depend on various factors such as proficiency of the learner and difficulty of the task; therefore, it is almost impossible to design a descriptive grammar in advance which can accept grammatically correct utterances as well as utterances with various abusage. We automatically generated a descriptive grammar from the learner corpus that also contains ungrammatical expressions.

Yasuda et al. [7] created a Japanese learner corpus of 150,000 English sentences translated by 500 Japanese speakers, of which Japanese source sentences were randomly selected from the BTEC and textbooks on English for junior and senior high schools. All 1500 Japanese sentences were translated by 100 different speakers of various TOEIC scores. Each Japanese sentence was also translated into two different ones by five bilingual English/Japanese speakers, resulting ten different correct English translations.

Some of the 100 sentences by the Japanese speakers correspond to a correct sentence generated by a bilingual speaker, even though they had grammatical errors. The learner corpus also contains sentences that do not correspond to any correct expression because only a portion of the source sentences was translated or because the source sentence was too difficult for a learner and he/she translated it to a sentence with different meaning or no meaning.

3.1 Coverage over Various Expressions

To estimate how much the learner corpus incorporates various expressions in English concerning each source sentence, we divided sentences in the learner corpus into training and evaluation subclasses, and calculated the ED between each sentence from the evaluation subclass and its most similar sentence in the training subclass. Table 2 lists the ratios of the sentence with a minimum ED of zero and less than equal 1 when n-sentences were randomly selected from the training subclass in addition to 10 correct sentences translated by bilingual speakers.

Table 2. Relation between ED and number of sentences selected as training set

No. of additional sentences (n)	0	10	20	30	40
percentage of the sentence (min. Ed = 0)	0.07	0.16	0.21	0.23	0.25
percentage of the sentence (min. Ed $\leq$ 1)	0.15	0.31	0.37	0.40	0.43

Table 2 shows that we can design descriptive grammars that can accept almost a quarter of the expressions when using expressions by the bilingual speakers and additional 40 sentences by the Japanese speakers even if the input sentence may have grammatical errors. A large portion of deleted, inserted, and substituted words are articles, prepositions, determiners, and singular/plural nouns; therefore, when we can develop descriptive grammars with less constraints such as those that can accept sentences expressed in both singular and plural nouns and in which classes of article, preposition, and determiner are used instead of each article, preposition, and determiner.

3.2 Correspondence to Grammatically Correct Expression

As described above, the learner corpus contains almost grammatically correct sentences, those with some grammatical errors, and those that are completely ungrammatical. We evaluated several sentences in the learner corpus subjectively, that is to say, a bilingual English/Japanese speaker evaluated the quality of the sentence from the viewpoint of fluency and adequacy on a 5 grade scale. We classify sentences of abusage as grammatically acceptable when a mean of score (MOS) is greater than or equal to the *Threshold*, and sentences of out-of-scope with the grammar when MOS is less than *Threshold*. We are now conducting to correspond only each grammatically acceptable sentence to a grammatically correct sentence by a bilingual English/Japanese speaker, depending on the ED between both sentences.

4   Detection of Utterances Out of Scope of Assumption

4.1 Difference of Acoustic Likelihood between Both ASRs

Watanabe et al. [9] proposed a method of using the differences of acoustic likelihood from two ASRs to determine whether recognition results from the ASR with FSA should be accepted as correct or rejected as false, and to use an ASR with a language model with larger coverage as the reference model. Classification is done according to Eq. (1) described below;

$$S = (score_{ver} - score_{sys})/T \tag{1}$$

$$S < \theta_{score}(Accept) \tag{2}$$

$$S \geq \theta_{score}(Reject) \tag{3}$$

where $score_{ver}$ and $score_{sys}$ denote the acoustic likelihood from ASR for verification and ASR with FSA, respectively, and T and θ_{score} denote utterance duration and threshold, respectively.

4.2 Other Features for Classification

We used two features, ED between strings of output words from both ASRs and estimated coverage of descriptive grammars, as shown in Fig. 1. The ASR for verification uses *ngram* as a language model, which was trained with BTEC, that covers the target topics in our CALL system. The ED between strings of output words from both ASRs described in Eq. (4) is thought to be effective as a classifier, by considering that FSA gives more strict syntactical and semantical constraints, even though both language models give similar constraints.

$$editDis(rec_{sys}, rec_{ver}) = S + I + D \qquad (4)$$

where rec_{sys} and rec_{ver} denote recognition results by ASR with FSA and ASR for verification, respectively, and S, I, and D denote numbers of substitution, insertion, and deletion, respectively, when both recognition results are optimally aligned.

The larger the descriptive grammars cover over various expressions, the fewer utterances FSA cannot accept; therefore, coverage by the descriptive grammars is thought to be an effective feature for classifying sentences into accept and reject. Coverage of the descriptive grammars cannot be directly measured in principle; therefore, we used the estimated coverage *ECov* described below;

$$ECov = (1 - aveDis/aveWord) \qquad (5)$$

$$aveDis = \sum_{y} min\{(editDis(text_x, text_y)\}/N \quad (x = 1, \dots, N) \qquad (6)$$

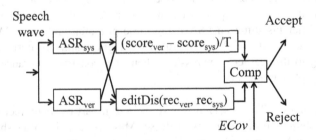

Fig. 1. Schematic classification flow of recognition results

where *aveWord* denotes the average number of words from ASR with FSA and N denotes the number of sentences with which the ED was measured.

5 Experiment and Results

We conducted classification experiments using spontaneous speech from 14 university students (7 females and 7 males). Each university students gave 68 utterances. We used

Table 3. Precision, recall, and f-measures for each combination of features

features	precision	recall	f-measure
acoustic	0.78	0.68	0.64
acoustic + edit dist.	0.81	0.81	0.81
acoustic + $ECov$	0.75	0.74	0.74
acoustic + edit dist. + $ECov$	0.81	0.81	0.81

support vector machine (SVM) as a classifier and the differences in acoustic likelihood, ED, and *ECov* as features. We used a software tool called Weka [10]. Precision, recall, and f-measures for each feature combination are listed in Table 3. Classification is best with a combination of acoustic likelihood/ED/*ECov*.

6   Conclusion and Future Work

We proposed a dialogue-based CALL system that allows learners to construct utterances on their own and detects grammatical and lexical errors from such utterances. The system uses ASR with FSA developed using sentences collected from students in English and ASR with *ngram* as a verification system. We used the differences in acoustic likelihood from both ASRs as well as the ED between strings of output words from both ASRs and *ECov* over expressions by FSA. We developed descriptive grammars using subset of sentences randomly selected from the learner corpus; however, we would like to develop descriptive grammars by only using sentences whose subjective evaluation quality is rated better than the *threshold*. We also plan to examine effects of using descriptive grammars with less constraints such as class of article, preposition, and determiner, and also effects of selecting expressions systematically selected from the learner corpus to extend coverage of the grammar over various expressions. After installing an FSA using these grammars into our CALL system, we will evaluate effects of displaying/uttering a grammatically correct expression corresponding to the learners' expression when they construct utterances within the assumed descriptive grammars.

Acknowledgements. This research was supported in part by a contract with MEXT number 22520598. The authors thank Professor Masuzo Yanagida of Doshisha University and Dr. Keiji Yasuda of NICT for their various discussions.

References

1. Eskenazi, M.: Using Automatic Speech Processing for Foreign Language Pronunciation Tutoring: Some Issues and Prototype. Language Learning and Technology 2(2), 62–76 (1999)
2. Kweon, O., Ito, A., Suzuki, M., Makino, S.: A grammatical error detection method for dialogue-based CALL system. J. Natural Language Processing 12(4), 137–156 (2005)
3. Nakamura, S., et al.: The ATR Multilingual Speech-to-Speech Translation System. IEEE Trans. ASLP 14(2), 365–376 (2006)

4. Minematsu, N., Nishina, K., Nakagawa, S.: Readspeech database for foreign language learning. JASJ 59(6), 345–350 (2003) (in Japanese)
5. Takami, J., Sagayama, S.: A successive state splitting algorithm for efficient allophone modeling. In: Proc. ICASSP, vol. 1, pp. 573–576 (1992)
6. Takezawa, T., Sumita, E., Sugaya, F., Yamamoto, H., Yamamoto, S.: Toward a broad-coverage bilingual corpus for speech translation of travel conversations in the real world. In: Proc. LREC, 27–2, pp. 147–152 (2002)
7. Yasuda, K., Kitamura, K., Yamamoto, S., Masuzo, Y.: Development and Applications of an English Learner Corpus with Multiple Information Tags. J. of NLP 16(4), 47–63 (2009) (in Japanese)
8. http://www.ets.org/toeic/
9. Watanabe, T., Tsukada, S.: Unknown Utterance Rejection Using Likelihood Normalization Based on Syllable Recognition. IEICE Trans. 75-D2(12), 2002–2009 (1992)
10. http://www.cs.waikato.ac.nz/ml/wka/

Spoken Dialogue System Design in 3 Weeks

Tomáš Valenta, Jan Švec, and Luboš Šmídl

Department of Cybernetics, Faculty of Applied Sciences
University of West Bohemia, Plzeň, Czech Republic
{smidl,honzas,valentat}@kky.zcu.cz

Abstract. This article describes knowledge-based spoken dialogue system design from scratch. It covers all stages which were performed during the period of three weeks: definition of semantic goals and entities, data collection and recording of sample dialogues, data annotation, parser and grammars design, dialogue manager design and testing. The work was focused mainly on rapid development of such a dialogue system. The final implementation was written in dynamically generated VoiceXML. The large vocabulary continuous speech recognition system was used and the language understanding module was implemented using non-recursive probabilistic context free grammars which were converted to finite states transducers. The design and implementation has been verified on a railway information service task with a real large-scale database. The paper describes an innovative combination of data, expert knowledge and state-of-the-art methods which allow fast spoken dialogue system design.

Keywords: spoken dialogue systems, language understanding, VoiceXML.

1 Introduction

We propose an innovative rapid dialogue system development approach which allows to build such a system from scratch in 3 weeks. It has been verified on a railway information service task. An expert dialogue system was chosen instead of any other (statistical) approach because the hand-crafted rules of understanding and dialogue policy are relatively easy to develop and transparent in what they do. Besides that, if we wanted to build the system from scratch in such time, we do not believe we would be able to build a statistical parser and an inference mechanism and to collect enough data to train those modules.

No dialogue design toolkit was used. For example, CSLU Toolkit [1] provides fast dialogue design as a flowchart, which is very clear, but allows to design only simple system-initiative dialogues. This toolkit is more suitable for education and practicals than building production dialogues. Moreover, integrating tools like a continuous speech recognizer or a semantic parser into an existing toolkit is more complicated then just using them from the VoiceXML interpreter or calling their existing interface (e.g. XML-RPC).

Also custom hand-crafted grammars and simple VoiceXML/PHP scripts were used rather than generated from very complex grammars like described in [2]. The approach mentioned there also involved much client-side (VoiceXML) scripting which allows almost instant validating and processing on the low level and therefore fast response

P. Sojka et al. (Eds.): TSD 2012, LNCS 7499, pp. 624–631, 2012.

and better quality results for the server side, but on the other hand it cannot take the advantage of e.g. immediate results verification in backend database. It would be also very difficult to describe the entire query sentence with garbage model (the part of an utterance with no semantic meaning, e.g. "um, eh", "could you please" etc.) by one complex grammar.

We did not design discrete components as proposed in [3], because they do not allow the flexible and complex parsing we needed. For example, initial, final and change stations look like subclasses of a parent class Station as known from object-oriented programming (inheritance), they are quite different though. Each has different activation rule (when to parse as initial), change station does not have confirm etc. Component-like behaviour, however, was implemented for particular state slots.

The knowledge and the state of the dialogue was represented in a simple structure, no sophisticated information storing and querying was needed, e.g. ontological. For such approaches there are already made tools, e.g. [4] which were successfully applied on dialogue-based retrieval of medical images or music. In our case it would mean to transform the railway connection database into a suitable structure which was unfeasible (we do not have full access to the data).

Section 2 describes the goals chosen to be the results of the work and also features of the resulting dialogue system. In Section 3 we describe the first phase of the process, which is data acquisition and processing. In Section 4 the knowledge (semantic analysis and speech grammars) incorporated into the system is described. In Section 5 we describe the dialogue manager, i.e. input (speech recognition and parsing), state transition and output as well as data access and logging. Section 6 summarizes the results we had got and mentions some tools that eased the teamwork. Finally we provide the conclusion how successful we were and how useful the final product is.

2 Goals Determination

The main goal was to design and implement a dialogue system providing users with railway connections information. It should ask mandatory questions (initial and final station) as well as allow the user to provide some complementary requests such as time of departure or arrival, whether to look for a direct train or not, query price etc. The system should accept spontaneous language requests including station names inflexions, vague expressions (e.g. afternoon) etc. The response should be sufficiently informative on one hand and brief enough on the other.

3 Data Acquisition

The most time-demanding part of the process was data acquisition. At first, possible scenarios and use-cases were determined. Then the scenarios were turned into speech grammars which allowed us to generate hundreds of unique cases. Those scenarios were then recorded by preferably distinct people, transcribed and annotated with semantic entities marked in the utterances.

```
#ABNF 1.0 UTF-8 cs;
grammar t;
root $t;
public $t = You want to (go | travel) from $city1 to $city2.;
$city1 = Pilsen | Prague;
$city2 = Olomouc | Brno;
```

Fig. 1. ABNF grammar generating "You want to go/travel from Pilsen/Prague to Olomouc/Brno"

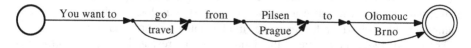

Fig. 2. Finite state automaton generated from the grammar in Fig. 1

3.1   Scenarios Generation

Several kinds of scenarios were drawn, from simple "*You want to go from Pilsen to Prague*," to "*Imagine your cousin in Brno has her birthday and you want to take there your bicycle.*" They were formulated as grammars in an ABNF (Augmented Backus-Naur Form) format (see Fig. 1) which were then turned into finite state automata (no approximation was needed because no context-free ABNF grammar recursion was used; see Fig. 2) and then arbitrary human-readable scenarios were generated using random walk algorithm.

3.2   Data Collection

After the scenarios were generated, numbered groups of four scenarios with increasing demand on speaker's verbosity and creativity were created and printed on sheets of paper. Then paid workers were sent out to record utterances from their family members, friends or even passers-by.

Each speaker could use his/her cell phone for the recording. There were five toll-free numbers available for the recording so no extra cost was required from anyone. On each of the phone lines a recording application was waiting for a call which was rejected and after few seconds, the application called back allowing us to run very simple toll-free call-back system.

Very rich variety of speakers, phones and environments is very useful not only for purposes of the dialogue system (defining semantic entities), but also for language and acoustic modelling. For a notion, 352 four-groups or 1,598 requests (some people needed more attempts) were collected during the recording phase.

3.3   Annotation

The voice audio data collected as described in previous subsection were then imported into the web application WebTransc [5] which allows transcription and semantic annotation of the utterances. The annotators inserted special tags in place of the

```
<ehm_??> <greeting_|> Hello <|_greeting> I would like to
to get <from_|> from Ostrava <|_from>
<to_|> to Beroun <|_to> be so kind and tell me the ticket
```

Fig. 3. Sample transcription and annotation

non-speech events such as `ehm_hmm`, `laugh` and additional tags which describe the semantic structure of the utterance (for example `greeting`, `from`, `to`, `time`, `train_type`). In Fig. 3 you can see a sample annotation of an utterance. Note there are exact transcription, non-speech events and semantic tags together.

4 Knowledge-Based Approach

Since we decided to build a knowledge-based system, all the modules like semantic parsers, state transition functions, actions and output generators had to be hand-crafted. Man-made grammars were created for each semantic entity. These partial grammars were then combined into one huge grammar (by simple concatenation) to parse any utterance coming to the input of the system. Out-of-grammar utterances (garbage) must have been modelled as well.

4.1 Semantic Entities Analysis

To analyse particular entities, all passages between tags (e.g. `<to_|>` and `<|_to>`) were collected from the annotations. Then a context-free speech grammar accepting those utterances was designed. It contains some other small tags allowing us to split the parsing problem into smaller pieces and to generalize classes such as *station* or *day of week*. In the understanding module only these tags are used instead of the recognized text itself. Classical parse trees are not used, the grammar serves (and in fact is implemented) as a transducer, see later.

4.2 Labelling and Semantic Interpretation

The speech grammars described in the previous paragraph are written in ABNF format. In Fig. 4 there is a sample grammar describing the station of departure (tags in curly braces), then an example input parsed by the entire grammar and finally the parser output passed to the understanding module.

Note the alternatives the user can say for indicating the initial station. In inflective languages (Czech), various inflexions of a word must also be specified, but the tags (`s_pilsen`) are left in nominative case.

4.3 Garbage Model

In order to be able to recognize spontaneous utterances, we have to model the parts with no significant semantic meaning as well and incorporate them into the grammar in the form of the garbage model.

Station of departure grammar [simplified]:

```
public $from = I am in {from} $station6 | from {from} $station2;
$station6 = Brně {s_brno} | Praze {s_prague} | ...;
$station2 = Brna {s_brno} | Prahy {s_prague} | ...;
```

Sample user input:

```
I want to go from Pilsen to Prague.
```

Parser output:

```
from s_pilsen to s_prague
```

Fig. 4. Parsing example

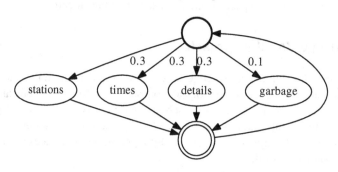

Fig. 5. Full grammar with discriminated garbage model

The garbage model in our case was generated from the transcriptions from what remained after picking out all the recognized semantic entities as described in paragraph 4.1. Some text post-processing must have been done for the recognizer to perform well. All utterances shorter than nine letters were discarded as well as the letter case and punctuation. The garbage model rule got no {tag} so it was never returned by the parser. The garbage rule expansion was also discouraged by lower prior probability parameter, otherwise almost everything would be recognized as garbage.

4.4   The Entire Grammar

In Fig. 5 the entire grammar is displayed. If we fully expanded all the nodes (i.e. *stations* → *from, to* → *station6* → ...) considering there is about 3,800 stations, each in four different cases and pretty complicated garbage model, we would get a really huge grammar. Transformed to finite state transducer, determinized and minimized gives 8,757 states and 894,742 arcs.

5   The Dialogue Manager

In the most general case, the dialogue system is modelled as a dynamic system discrete in time and variables described by equations (1):

$$
\begin{aligned}
x_{k+1} &= f(x_k, u_k) \\
y_k &= g(x_k),
\end{aligned}
\tag{1}
$$

where x_k is the state of the system in time k, u_k is the user input, $f(\cdot)$ the state transition function, $g(\cdot)$ the output function and y_k is the output. In our case, input to the dialogue manager is parser output and output is text with mark-up.

5.1 Speech Recognition and Parsing

There were two choices how to implement the dialogue manager input (i.e. the speech recognition plus the parsing). One way was to use a decoder using directly the grammar, but since it is that complex, this approach failed. Either it used vast amount of resources or (with aggressive pruning) it performed badly.

The second approach was to use an n-gram based continuous speech recognizer [6]. We used an class-based language model trained from data on a similar task. We used an orthographical transcription of words as a recognizer vocabulary and then we performed text normalization into a grammatically correct form. For example correction of words *brý en* to *dobrý den* (lit. *good morning*). The subsequent text normalization allows to map a sequence of colloquial words into a sequence of normalized words. Details are described in [7]. The normalized output of the recognizer simplifies the design of an understanding module because grammar designers do not need to take into account many colloquial forms of a given word.

Semantic tags (described in Section 4.2) produced by the parser are then passed to the dialogue manager (the transition function) for understanding and taking actions.

5.2 State Transition Function

The State transition function is the core of the dialogue system. The state contains all the information collected from every source (the user, the database) so far and is the only source for what to do or say next. The state consists of slots, which are mandatory (*initial* and *final station*) or optional (*time of departure*, *direct* etc.). The state also contains the last result of searching the database so that it can use it for requests like "next train" or "return train".

The state transition function takes the current state and the user's input (no input can be considered as a kind of input too), processes it (i.e. updates the state) and decides what to do next. It can be either a searching the database and informing the user or asking for more information if mandatory slots are empty.

Slots that hold particular piece of information can be themselves in "metastates". These metastates are 0. empty, 1. filled or 2. confirmed. The transitions are shown in Fig. 6. After some input a slot goes from empty metastate to filled metastate. Then after consistent input or implicit confirm (see later), the slot goes to confirmed metastate. The consistent input means that it is a subset of the current slot's value ($I \subseteq V$, e.g. at first the user says ambiguous station Krumlov, therefore the slot value becomes [Český Krumlov, Moravský Krumlov], the first is used. Then he/she says Moravský Krumlov and the slot becomes confirmed with value [Moravský Krumlov]).

5.3 Response Generation

The response can be either a requested piece of information or a question asking the user to fill a mandatory empty slot. The informatory response is generated entirely from the state. It is up to the transition function to gather up all the information.

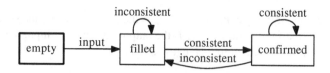

Fig. 6. The slot metastate automaton

The generated responses have increasing verbosity through the time. I.e. once a connection is found the user is told the times of departure and arrival. Then, if the user placed no other request, he/she is told where and when to change the train and then details about the connection, e.g. price, wheelchairs support etc.

Implicit Confirm is used to make sure we are searching the information the user really requested. Instead of asking the user explicitly if we understood correctly, he/she is discreetly repeated what he/she is asking for. For example:

- "I want to go to Pilsen."
- *"Where do you want to go* to Pilsen *from?"*

If the user agrees (or does not protest), the *final station* slot gets confirmed.

5.4   Implementation

The dialogue manager is implemented in PHP serving VoiceXML documents from a web server which are interpreted by a VoiceXML interpreter. The audio input received by the interpreter is passed to an automatic speech recognizer and parser and its output passed to the PHP dialogue manager. The DM output is passed via VoiceXML (along with dialogue control commands) back to the interpreter and via the Czech high quality realtime text-to-speech unit [8] played back to the user.

The data is taken from official Czech railway connection database IDOS accessed through an XML-RPC interface. This easy yet robust solution means that the information is always up-to-date and if the backend (the provider) changes, it can be easily adapted to the new conditions.

For monitoring and debugging purposes, each dialogue turn is logged into the database. The log entry contains user's input (both streams — recognized utterance and parser output), the state of the dialogue and the output of the system. Browsing the log turn by turn one can see how the state evolved which is extremely useful when debugging the state transition function.

6   Summary and Conclusion

All the steps to design and implement the system mentioned above took place in the second half of July 2011. Three researchers and five co-workers worked eight hours a day or more and went through all the stages necessary to build the dialogue given the necessary tools: VoiceXML interpreter, speech recognizer, speech synthesizer, PHP and Python scripting engines, finite state machinery (OpenFST) and cooperative and sharing tools PHPWiki and Subversion.

Table 1. Railway information service dialogue system in numbers

Recorded scenarios:	1,598
Parsing FST states:	8,757
Parsing FST arcs:	894,742
Calls to production system:	394
Distinct callers:	27
Dialogue turns:	2,029

In Table 1 you can see utilization of the system since its launch on 30 July 2011 until the end of the year 2011. It has never been thought about as a critical service nor advertised, rather used for educational and demonstration purposes.

We proved that an expert dialogue system can be built using commonly available tools and quite reasonably small amount of data in 3 weeks. Expert systems never perform as good as the statistical ones do, but they are much harder to develop from scratch and with such amount of data. The expert dialogue developed during the three weeks can be used for bootstrapping statistical dialogue's parameters as well.

Acknowledgements. This work was supported by the European Regional Development Fund (ERDF), project "New Technologies for Information Society" (NTIS), European Centre of Excellence, ED1.1.00/02.0090, by the grant of the University of West Bohemia, project No. SGS-2010-054, and by the Technology Agency of the Czech Republic, project No. TE01020197.

References

1. McTear, M.: Using the CSLU Toolkit for Practicals in Spoken Dialogue Technology. In: Proceedings of ESCA/SOCRATES Workshop on Method and Tool Innovations for Speech Science Education (1999)
2. Bringert, B.: Rapid Development of Dialogue Systems by Grammar Compilation. In: Proceedings of the 8th SIGdial Workshop on Discourse and Dialogue, Antwerp, Belgium, pp. 223–226 (2007)
3. Akolkar, R.P., Faruquie, T.A., Huerta, J., Kankar, P., Rajput, N., Raman, T.V., Udupa, R.U., Verma, A.: Reusable Dialog Component Framework for Rapid Voice Application Development. In: Heineman, G.T., Crnković, I., Schmidt, H.W., Stafford, J.A., Ren, X.-M., Wallnau, K. (eds.) CBSE 2005. LNCS, vol. 3489, pp. 306–321. Springer, Heidelberg (2005)
4. Sonntag, D., Sonnenberg, G., Nesselrath, R., Herzog, G.: Supporting a Rapid Dialogue System Engineering Process. In: Proceedings of IWSDS, Kloster Irsee, Germany (2009)
5. Šmídl, L., Valenta, T.: WebTransc – Software, Department of Cybernetics, Faculty of Applied Sciences, University of West Bohemia, Pilsen (2010), http://www.kky.zcu.cz/en/sw/wt
6. Pražák, A., Müller, L., Psutka Josef, V., Psutka, J.: Live TV Subtitling – Fast 2-pass LVCSR System for Online Subtitling. In: Proceedings of SIGMAP 2007, pp. 139–142. INSTICC PRESS, Lisabon (2007)
7. Švec, J., Šmídl, L.: Real-time Large Vocabulary Spontaneous Speech Recognition for Spoken Dialogue Systems. In: Proceedings of the 4th Int. Cong. on Image and Signal Processing, Shanghai, pp. 2458–2463 (2011)
8. Tihelka, D., Kala, J., Matoušek, J.: Enhancements of Viterbi Search for Fast Unit Selection Synthesis. In: Proceedings of Int. Conf. Interspeech 2010, pp. 174–177 (2010)

Integrating Dialogue Systems with Images

Ivan Kopeček, Radek Ošlejšek, and Jaromír Plhák

Faculty of Informatics, Masaryk University
Botanická 68a, 602 00 Brno, Czech Republic
{kopecek,oslejsek,xplhak}@fi.muni.cz

Abstract. The paper presents a novel approach, in which images are integrated with a dialogue interface that enables them to communicate with the user. The structure of the corresponding dialogue system is supported by graphical ontologies and enables the system learning from the dialogues. The Internet environment is used for retrieving additional information about the images as well as for solving more complex tasks related with exploiting other relevant knowledge. Further, the paper deals with some problems that arise from the system initiative dialogue mode and discusses the structure and algorithms of the dialogue system. Some examples and applications of the presented approach are presented as well.

1 Introduction

The idea of dialogue-based communication with images is supported by the ability of the graphics formats to encode many useful "non-graphical" pieces of information. For example, the date and time of a snapshot, GPS information, recorded sound, etc. This information can be used for image classification and semantics retrieval, see e.g. [1,2,3].

Most of the relevant information, however, is not directly feasible. Let us imagine a photo from a holiday ten years ago: the woman in the middle is my wife, but who is the guy standing behind her? It is apparently somewhere in the Alps, but what place? What is that peak in the background? Although such pieces of information are virtually inaccessible, they can be often retrieved indirectly. GPS coordinates allow us to determine where the photo has been taken. Face recognition [4,5,6] may help reveal the identities of persons. The orientation may help to determine some objects in the picture (e.g. the peak in the background).

The presented approach enables the images to gather relevant information about themselves from the users and to exploit the gathered data in the communication with the users by means of natural language.

In what follows, the term "communicative image" stands for a two dimensional graphical object (a picture, a photograph, a graph, a map,...) integrated with a dialogue interface and being equipped with an associated knowledge database. Such an image can communicate with the user and it can also learn from the communication and enlarge its knowledge database.

2 Web 2.0 Approach to Communicative Images

Since 2004, the *Web 2.0* technology [7] substitutes the static content of the web for dynamic space supporting easy information sharing and online user collaboration. User

P. Sojka et al. (Eds.): TSD 2012, LNCS 7499, pp. 632–639, 2012.
© Springer-Verlag Berlin Heidelberg 2012

collaboration together with knowledge base sharing can be considered as essential requirements for communicative images that require to handle complex and dynamic semantic knowledge. Online user cooperation helps to manage the knowledge in decentralized way. Activity of one user, e.g. publishing some historical facts about Prague castle, can be utilized by other users to improve exploration of a Prague photo downloaded somewhere from the web. This kind of distributed problem-solving is referred to as *crowdsourcing*.

In the common understanding of crowdsourcing, problems are broadcast to a group of solvers in the form of an open call for solutions. For example, a user exploring photo with a bird can ask either public or domain-specific community of ornithologists for the help with identifying and describing the bird. Positive reaction of the community members makes it possible to overcome some of the problems with understanding the visual content. Positive reactions of more members also enable to gather more precise semantic data that can be used to filter inaccurate or intentionally wrong pieces of information supplied by saboteurs.

The knowledge provided by the "crowd" can be reused by another users communicating with similar photos. However, the re-usability of the knowledge highly depends on its formal definition and precise structuring. This is why we employ ontologies in our approach.

For communicative images, crowdsourcing presents an efficient way of building up knowledge bases in long-term perspective. Image recognition and auto-detection algorithms present another facility for recognizing the graphical content. Although these techniques are still far from being able to fully describe an analyzed picture in general, specific domains, e.g. face recognition [8,6,9] or similarity search algorithms in large image collections [10,11,12], are applicable for the initial image content retrieval.

2.1   Coupling Structured Semantics with Images

To make a standard image communicative in the Web 2.0 environment, it is necessary to transform it into a format suitable for supporting structured annotation data. The SVG graphics format [13] wraps the original raster image and integrates the annotation data. The annotation data as well as general knowledge base are described by the OWL ontologies [14] that are linked from the SVG pictures via the *owl:imports* statements. In this way, the knowledge base can be shared by many pictures. On the other hand, concrete annotation data related to a concrete image, i.e. values of the properties prescribed by the ontologies, are stored directly in the SVG format in the form of XML elements. Technical details related to coupling OWL ontologies with the SVG format are discussed in [15].

Once the image is transformed into the SVG format, the system tries to acquire as much information about the image as possible, using auto-detection and image recognition techniques, similarity search algorithms searching in large collections of tagged pictures, EXIF data extraction from photos, etc. After this initial stage, the user is informed about the estimated content and asked to confirm or refute the information and to continue with questioning. New pieces of the acquired information are stored in the image ontology and reused by subsequent interactions.

2.2 Graphical Ontologies

The OWL ontologies [14] present formal specification of shared conceptualization. They describe the semantics of the data with modeling domain concepts, their relationships and attributes. In [15], a basic graphical ontology which restricts abstraction to the aspects that are suitable and utilizable for dialogue-based investigation of a graphical content is presented. Using this ontology, the annotator can express, for instance, that some object is "mostly red, oval and unusually big", etc. Moreover, the ontology supports navigation in the image, i.e. expressing either mutual or exact position of objects in the picture, with automatic reasoning.

Although graphical ontologies cover basic visual characteristics that are useful for generic dialogue interactions, verbal descriptions of domain-specific pictures require to employ specialized ontology extensions helping to generate domain-specific dialogues. For example, a domain model, *Family*, provides vocabulary and background knowledge to classify people by their family relationships. Another model, *Sights*, covers sights of interests, historical buildings, monuments, etc. Combining these two ontologies we can easily describe typical vacation photos in terms like *"my mother in law in front of the Prague castle"*. This simple information enables the communicative image to automatically infer that the annotator is married (otherwise he/she would not have a mother in law). Moreover, employing the information about Prague castle stored in the *Sights* knowledge database, the communicative image is able to provide information about historical facts associated with this sight, geographical data, etc.

3   Dialogues with Communicative Images

In this section, we discuss some dialogue-related aspects of communicative images and illustrate the presented approach by an example.

3.1   Communication Modes

The user and communicative image can communicate in the following three modes, depending on the user's choice.

1. The *information retrieval mode* (the mode with user's initiative). In this mode the user just puts questions and gets answers.
2. The *free communication mode* (the mode with mixed initiative). In this mode, there are no apriory limitations on user or image. Both user and image can ask questions and give answers.
3. The *image information supplementing mode* (the mode with image's initiative). This mode is used for supplying the image with missing pieces of information.

In each mode, the user can switch to another mode.

Let us assume, that the image in Figure 1 has been converted from classical photo into the communicative image by adding the dialogue interface, learning capability and other necessary abilities. The following example illustrates the idea of communication with images in the free communication mode (*U* stands for User, *I* for Image).

Fig. 1. A photo from a vacation

U: [activates the dialogue with the image]
I: This is a photo of a group of your friends standing before the Pernstejn castle.
/In the corresponding ontology, GROUP_OF_PEOPLE is a predefined ontology relation. FRIEND is an ontology relation related to the user (modal ontology relation)./
U: When and where this picture has been taken?
I: The picture has been taken at August 18, 2008 in the Highland, Czech Republic.
/Location has been identified using the GPS coordinates included in the metadata./
U: Who is in the picture?
/"Who" in the question refers to objects of type PEOPLE. /
I: There is your friend Michael Taylor and his wife Jane Taylor. Comparison with other photos in your album shows that there is someone called John Smith. Do you know him?
/The image took initiative./
U: Yes, he is my brother in law.
/The knowledge database is updated./
U: Is Pernstejn castle opened this Sunday?
/The request activates the search engine, which searches in the image graphical ontology, and, if this search fails, it activates real time searching Internet. If this search is successful, the relevant pieces of information are used to answer the request./
I: In April, the castle is open to visitors between 9 a.m. and 3 p.m. every Sunday.

3.2 Turn-Taking and Dealing with Contradictory Pieces of Information

An important task the communicative image has to cope with is completing the missing pieces of information and solving possible collisions if some pieces of the gathered information are contradictory. This is, of course, relevant just for the second communication mode (the *free communication mode*). Dialogue strategies for gathering information from the user in the free communication mode are based on the tradeoff between the effort to get missing pieces of information and the effort to minimize the frequency of the questions as it can be bothering for the user.

When inquiring the information from various users, it can happen that some pieces of information are contradictory. Typically, some of the objects in the image can be mistaken or wrongly recognized from some users. The contradictory pieces of information can be handled by the voting principle, considering the piece of the information to be true, when there is no contradiction, probable, when just a small number of the pieces of information is contradicting, and unsure, when there is a considerable number of contradicting pieces of information. In the case of unsure information, the dialogue strategy tries to get more information from the users, to eliminate this unfavorable state.

3.3   Communication Analysis

The interface between natural language and formalized ontology framework provides an engine transforming natural language into corresponding formal schemes. We analyzed a corpus of relevant user prompts to identify a relatively small fragment of natural language which is used by the most users. Therefore, the engine can be based on relatively simple grammars, formalized by Speech Recognition Grammar Specification [16], in combination with the frames technology and standard techniques for misunderstanding solving. For instance, the question *"How far is it from this hotel to the nearest beach?"* is resolved using the template *"How far is it from SLOT1 to SLOT2?"*. The system expects both the SLOT1 and the SLOT2 to be filled by the specific entries from the "objects" category listed below. Main principles and details of the dialogue management have been discussed in [17,18].

In the first step of the analysis, general types of expected question parameters are identified and divided into categories. This categorization is valuable for more precise identification of the possible content of the frames. The categories are as follow:

- *type_of_objects* – represents groups like "person", "thing", "animal", etc.;
- *object* - represents the specific person, dog, stone, event, etc.;
- *rng_position* – represents positions like "upside", "in the middle", "left upper corner", etc. by means of Recursive Navigation Grid [19,20];
- *position_focus* – represents photo division to "background", "main part of the photo", "foreground", etc.

In the domain of photo processing, the following categories of the users' prompts have been identified:

General Information about Image. This category covers most of the prompts that query some general knowledge about the objects in the photo. Typical prompts are: *"Describe the photo."*, *"Who are the people in the photo?"*, *"What animals are in the photo?"*, *"When this picture has been taken?"*, *"What type of camera has been used for the snapshot?"*, etc.

Search Queries cover the searching for objects both inside and outside the image. For example, the users would be curious about the specific object positions (*"Where is*

John in this picture?", *"What objects are in the upper left corner?"*), list or position of available objects (*"Where are the nearest hotels?"*), distance (*"How far is that castle (from the camera)?"*, *"How far is it from this hotel to the nearest beach?"*) or about path finding (*"How can I get from here to Matterhorn Mountain?"*).

Details about Objects. This category enables the user to inspect more specific information about the objects. The users are encouraged to ask about more detailed information, e.g. names (*"What is the name of this castle?"*), time (*"When was the castle founded?"*, personal data (*"Who was the founder of this castle?"*), type of objects (*"What is this animal?"*), etc.

Object Position category contains the users' prompts which determine the mutual position of two or more objects. *"Who is the guy standing behind my wife?"* and *"What is behind this house?"* are two examples of typical queries from this category.

Why Queries cover very common kind of questions that are usually not easy to answer unless the answer is directly encoded in the knowledge database. For example, let's assume that an ontology of birds encodes the knowledge that a young swan is grey in contract with its white adults. In this case, the question *"Why the swan is grey?"* can be easily inferred and answered by sentence like *"Because this is a young bird"*. On the contrary, it is not easy to answer question like *"Why there is no parent nearby?"* because there is probably no data in supporting ontology directly explaining absence of swan adults in the image.

Ontology Adjustment. While the previous categories represent queries, this last category contains the most complex prompts that allow modification of the knowledge base by extending the knowledge, refining ontology structure and editing or removing semantic data. For example, the *"The guy behind my wife is John Smith."* user prompt extends the knowledge, *"John is no longer my friend."* removes the semantics of John (previously marked as a friend of mine), etc.

4   Application Domains

The main benefit of the communicative images is straight availability of information about the pictures that can be otherwise either too complex or easily forgotten by the users. Users are able to explore pictures without searching in notes, databases or the Internet. Information complexity is controlled by means of dialogues in natural language, preventing the user to be overburden by provided information as well as enabling the user to drive information stream. Communicative images therefore find employment in various areas of image management and processing.

For example, communicative images are useful in the efficient processing of large collections of photos. The idea of intelligent photo album has been proposed in [21]. Chai et al. enabled the users to organize and search their collection of family photos by means of ontologies and SWRL questioning [22]. If communicative images are involved into this scheme, we see straightforward way to enhance its functionality. Because the photos in the album are organized by means of OWL ontology, it might be possible to employ the mechanism of generating dialogues from domain ontologies. In this way, the user could organize photos via dialogue as well.

The concept of communicative images seems to be also a promising approach in accessibility of graphics for people with special needs, especially for blind and visually impaired people. It provides the visually impaired users with another way to explore graphics on the Internet and gives them more detailed information about the photos in the social networks or electronic news.

Communicative maps allow the users to ask the questions about the shortest or the most convenient path to their points of interest intuitively. In the free communication mode, the users are encouraged to write their notes directly to their image. Indoor and outdoor navigation of users with visual impairment is another valuable advantage of this concept.

E-learning is another field, which can benefit from communicative images [20]. Specialized ontology can contain a lot of additional information, which is not directly presented in the picture, e.g. detailed description of objects in many languages, links to other relevant objects and terms in the ontology, etc.

The product photos in the online shops could be also presented as a communicative image. A typical user queries include comparison of two or more similar products or availability that should be answered using a natural language. Effective design of ontology of the products in the online shop supports the navigation among the different types of products. These product photos are also presented in the electronic leaflets, where all the information like number of available products, sell prize or price development should be answered.

Because the concept of communicative images is based on formal ontologies, it is fully compatible with the Semantic web paradigm and simultaneously fully supporting multilinguality. It allows the user to search inside large galleries of pictures like Flickr or Picasa and effectively resolve user's requests.

5   Conclusions and Future Work

In this paper, we have outlined basic principles of communicative images representing a new and challenging approach utilizing and integrating current technologies, especially graphical ontologies and AI-based modules. The goal is to provide users with a new dimension of exploiting images, enabling the users to communicate with them by means of natural language dialogue. Efficient images investigation is provided by the frame based dialogue management that is supported by the related graphical ontology. This approach promises important applications in many application domains, e.g. e-learning, product photos, photo albums management and assistive technologies. Our next work is aimed to enhancing dialogue strategies, inference methods, ontology management methods and testing the technologies in real online and offline environments.

Acknowledgments. This work has been supported by the European Social Fund and the The Ministry of Education, Youth and Sports, Czech Republic, under Contract No. CZ.1.07/2.2.00/15.0184 *"Innovation of the Applied Informatics Bachelor Study Programme towards the Social Informatics"*.

References

1. Sandnes, F.: Where was that photo taken? deriving geographical information from image collections based on temporal exposure attributes. Multimedia Systems, 309–318 (2010)
2. Boutell, M., Luo, J.: Photo classification by integrating image content and camera metadata. In: Proc. of the 17th Int. Conf. on Pattern Recognition, vol. 4, pp. 901–904 (2004)
3. Yuan, J., Luo, J., Wu, Y.: Mining compositional features from gps and visual cues for event recognition in photo collections. IEEE Trans. on Multimedia, 705–716 (2010)
4. Li, S.Z., Jain, A.K. (eds.): Handbook of Face Recognition. Springer (2011)
5. Wright, J., et al.: Robust face recognition via sparse representation. IEEE Trans. on Pattern Analysis and Machine Intelligence 31, 210–227 (2009)
6. Haddadnia, J., Ahmadi, M.: N-feature neural network human face recognition. In: Image and Vision Computing, pp. 1071–1082 (2004)
7. Segaran, T.: Programming Collective Intelligence: Building Smart Web 2.0 Applications. O'Reilly Media (2007)
8. Bartlett, M., Movellan, J., Sejnowski, T.: Face recognition by independent component analysis. IEEE Transactions on Neural Networks, 1450–1464 (2002)
9. Rowley, H., Baluja, S., Kanade, T.: Neural network-based face detection. IEEE Transactions on Pattern Analysis and Machine Intelligence, 23–38 (1998)
10. Batko, M., Dohnal, V., Novák, D., Sedmidubský, J.: Mufin: A multi-feature indexing network. In: SISAP 2009: 2009 Second Int. Workshop on Similarity Search and Applications, pp. 158–159. IEEE Computer Society (2009)
11. Jaffe, A., Naaman, M., Tassa, T., Davis, M.: Generating summaries and visualization for large collections of geo-referenced photographs. In: Proceedings of the 8th ACM Internat. Workshop on Multimedia Information Retrieval, pp. 89–98. ACM (2006)
12. Abbasi, R., Chernov, S., Nejdl, W., Paiu, R., Staab, S.: Exploiting Flickr Tags and Groups for Finding Landmark Photos. In: Boughanem, M., Berrut, C., Mothe, J., Soule-Dupuy, C. (eds.) ECIR 2009. LNCS, vol. 5478, pp. 654–661. Springer, Heidelberg (2009)
13. Dahlström, E., et al.: Scalable vector graphics (svg) 1.1, 2nd edn. (2011), http://www.w3.org/TR/SVG/
14. Lacy, L.W.: Owl: Representing Information Using the Web Ontology Language. Trafford Publishing (2005)
15. Ošlejšek, R.: Annotation of pictures by means of graphical ontologies. In: Proc. of Int. Conf. on Internet Computing, ICOMP 2009, pp. 296–300. CSREA Press (2009)
16. Hunt, A., McGlashan, S.: Speech recognition grammar specification version 1.0 (2004), http://www.w3.org/TR/speech-grammar/
17. Kopeček, I., Ošlejšek, R., Plhák, J.: Dialogue management in communicative images. In: Text, Speech and Dialogue – Students' section, Proceedings Addendum, University of West Bohemia in Pilsen, pp. 9–13. Publ. House (2011)
18. Kopecek, I., Oslejsek, R.: Communicative Images. In: Dickmann, L., Volkmann, G., Malaka, R., Boll, S., Krüger, A., Olivier, P. (eds.) SG 2011. LNCS, vol. 6815, pp. 163–173. Springer, Heidelberg (2011)
19. Kamel, H.M., Landay, J.A.: Sketching images eyes-free: a grid-based dynamic drawing tool for the blind. In: Proceedings of the Fifth International ACM Conference on Assistive Technologies, pp. 33–40. ACM Press (2002)
20. Kopeček, I., Ošlejšek, R.: Accessibility of graphics and e-learning. In: Proc. of the Second International Conference on ICT & Accessibility, pp. 157–165. Hammamet: Art Print (2009)
21. Chai, Y., Xia, T., Zhu, J., Li, H.: Intelligent digital photo management system using ontology and swrl. In: Proc. of the 2010 International Conference on Computational Intelligence and Security, CS 2010, pp. 18–22. IEEE Computer Society Press, Washington, DC (2010)
22. Boley, H., et al.: Swrl: A semantic web rule language combining OWL and RuleML (2004), http://www.w3.org/Submission/SWRL/

Natural Language Understanding: From Laboratory Predictions to Real Interactions

Pedro Mota, Luísa Coheur, Sérgio Curto, and Pedro Fialho

L²F / INESC-ID Lisboa
Rua Alves Redol, 9, 1000-029 Lisboa, Portugal
{pedro.mota,luisa.coheur,sergio.curto,
pedro.fialho}@l2f.inesc-id.pt

Abstract. In this paper we target Natural Language Understanding in the context of Conversational Agents that answer questions about their topics of expertise, and have in their knowledge base question/answer pairs, limiting the understanding problem to the task of finding the question in the knowledge base that will trigger the most appropriate answer to a given (new) question. We implement such an agent and different state of the art techniques are tested, covering several paradigms, and moving from lab experiments to tests with real users. First, we test the implemented techniques in a corpus built by the agent's developers, corresponding to the expected questions; then we test the same techniques in a corpus representing interactions between the agent and real users. Interestingly, results show that the best "lab" techniques are not necessarily the best for real scenarios, even if only in-domain questions are considered.

Keywords: Natural language understanding, classification, information retrieval, string similarities, laboratory experiments, real interactions.

1 Introduction

Natural Language Understanding (NLU) targets at mapping given utterances into some sort of representation that models their meaning and that a computer is able to process. Intense research has been dedicated to NLU, as it is in the basis of the development of many applications. Conversational Agents – agents that interact in natural language – are examples of such applications.

In this paper, we study the NLU process of question-answering agents, that is, agents that target to teach, ask or answer questions about a topic in which they are "experts" [2,7,14]. In addition, we focus on the particular case of the agents whose knowledge base is constituted by sets of pre-defined interactions [9,11,12]. In all these agents, the NLU task consists in, being given a question, finding the "closest" question in the knowledge base, so that its answer is appropriate to the new question. Although only appropriate in limited domains, this approach has the advantage of allowing the fast development of a prototype, even by non experts, as the knowledge sources are simple question/answers pairs [12].

In this paper, different state of the art techniques – representing what has been applied to such agents – are compared, both in a lab environment and also in a real scenario of interaction with the agent. Interestingly, results show that the best "lab" techniques

P. Sojka et al. (Eds.): TSD 2012, LNCS 7499, pp. 640–647, 2012.

are not necessarily the best for real scenarios, even if only in-domain questions are considered.

The paper is organised as follows: in Section 2 we present related work, in Section 3 we describe the surveyed NLU techniques, in Section 4 we report and discuss the results of our experiments. Finally, in Section 5 we draw some conclusions and point to future work.

2 Related Work

NLU aims at mapping utterances into some sort of representation that models their meaning and that the system knows how to interpret. The mapping must correctly capture the meaning of the given utterance and, thus, must be able to map to the same representation, utterances with the same meaning. There are many possible approaches to this task [4,15,16].

NLU is in the basis of the performance of many applications, including Conversational Agents. Although some of these are task oriented, others target to give information about a specific domain. Examples of such virtual agents are Max [6], Hans Christian Andersen (HCA) [1], Sergeant Blackwell (SB), Edgar [12] and DuarteDigital [11]. Agent Max is deployed at the Heinz Nixdorf MuseumsForum, in Germany, and its main goal is to provide explanations and comments on the museum's exhibits. HCA personifies the renown fairy tale author in a computer game and is able to talk about his work. SB plays the role of a military instructor that is prepared to talk about the US Army and answers the questions of potential recruits [7]. Both DuArte Digital and Edgar answer questions about art, in Portuguese: Edgar's topic of expertise is the Monserrate Palace and DuArte Digital's is a piece of jewelry from the 16th century, Custódia de Belém.

While Max and HCA have sophisticated NLU modules that output, respectively, conversational acts and different levels of categories (syntactic, semantic, etc.), SB, Edgar and DuArte Digital have in their knowledge sources pairs constituted by utterances (typically questions) and the respective answers. It should be clear that all utterances associated with the same answer should have the same meaning or, at least, should be related somehow so that each utterance from that set can be answered by the same sentence. Therefore, within these agents, the NLU task is to be able to map a given sentence into some "paraphrase" existing in their knowledge bases.

SB's treats NLU as a *Text Classification* problem [9], as studied in the Information Retrieval (IR) field, with a few differences. In IR the set of possible answers is perceived as a collection of documents and the strategy used consists in viewing the query and a document as samples from some probability distribution over words and make a comparison between those distributions. This approach is not suitable for SB because it is common to have question/answer pairs in which there are no common words (the vocabulary mismatch problem). Motivated by this, a new approach was developed by Leuski and his team. This approach is called Cross Language Model (CLM) and corresponds to a statistical approach in which the vocabulary mismatch problem is overcome by assuming that there are two distinct languages, one for the questions and another for the answers. Therefore, the answers need to be "translated" into the question language, or the other way around, before comparing the word distributions. A detailed description on how CLM works can be found in [9]. This same approach is also applied to these agents [10,13].

Considering Edgar's NLU capabilities, they are modelled as a classification process, which is a relatively common approach for NLU (see, for instance, [3]). As described by Edgar's authors, and considering that its knowledge base has utterances/answers pairs in the form $(\{u_{i1}, \ldots, u_{in}\}, a_i)$, where each u_{ij} represents a possible way of formulating an utterance to which a_i is an appropriate answer, the NLU process can be seen as a classification task, if we assume that each pair of the knowledge base represents a distinct category. That is, all the utterances $u_{i1}, \ldots, u_{in}$ correspond to the same category.

Finally, DuArte Digital is "modeled as a service which, for each user question, searches for the most similar question".

3   Implemented Natural Language Understanding Techniques

Being given a set of utterances and respective answers about a specific topic representing the existing knowledge base and a new (unseen) utterance, we target to identify the utterance in the knowledge base that is "closer" to the new utterance, so that we can appropriately answer to it. This corresponds to the classification task described in agent Edgar (Section 2). An example of this NLU process is to have in the knowledge base the entry *({What is your name?, Can you tell me your name?}, My name is Edgar)* and associate the input *Have you been given a name?* to it. In the end the corresponding answer is returned. As stated before, in the literature this task can be performed using different paradigms, such as Information Retrieval classification or as a lexical similarity operation. In the following we detail the methods which were used in this paper.

In the basis of our work is the platform described in [12], which is publicly available. This platform allows us to treat NLU as a classification process (as defined in Section 2), and uses a Support Vector Machines (SVM) to this end, by using the LIBSVM [5] library, with the most appropriate kernels and optimized parameters, in a one-versus-all multi-class strategy.

We have extended this platform by allowing the expansion of the training data with synonyms or paraphrases, which are defined separately. For instance, if the utterance $w_1 \ldots w_i \ldots w_j \ldots w_n$ exists in the corpus and is stated that $w_i \ldots w_j$ is the same as $s_k \ldots s_l$, then the utterance $w_1 \ldots s_k \ldots s_l \ldots w_n$ is added.

Also, we add to this platform the possibility of dealing with NLU as done in CLM and a method based on string similarity measures.

Considering the CLM approach, SB's developers have created the NPCEditor toolkit (NPC stands for Non Player Character) [8], which implements the CLM. Thus, the virtual human kit[1] containing NPCEditor was used, through the provided message API.

In what respects string similarity, we opted for three measures with distinct natures: Jaccard, Overlap and Dice. Measures such as Levenshtein were not considered as sentences with the same tokens occurring in a different order are strongly sanctioned by these.

---

[1] http://vhtoolkit.ict.usc.edu/index.php

Jaccard (Equation (1)) obtains higher scores for utterances that have similar length (a zero value means that there is nothing in common between two sentences; one is the highest possible value).

$$Jaccard(U_1, U_2) = \frac{|U_1 \cap U_2|}{|U_1 \cup U_2|} \tag{1}$$

A different philosophy is used in the Overlap measure (Equation (2)), in which sentences with different lengths are not so strongly penalised. A combination of the Jaccard and Overlap measures (Equation (3)) was implemented. In this strategy an empirical weighting factor λ is used.

$$Overlap(U_1, U_2) = \frac{|U_1 \cap U_2|}{min(|U_1|, |U_2|)} \tag{2}$$

$$JaccardOverlap(U_1, U_2) = \lambda \times Jaccard(U_1, U_2) \\ + (1 - \lambda) \times Overlap(U1, U2) \tag{3}$$

Following the perspective of trying to balance the length of the sentences, we also chose the Dice similarity measure (Equation (4)) that does not favour any particular size of utterance.

$$Dice(U_1, U_2) = 2 \times \frac{|U_1 \cap U_2|}{|U_1| + |U_2|} \tag{4}$$

In order to use either of the previously described string similarity measures for the necessary NLU task, we compute the similarity score between the input utterance and the utterances in knowledge base entries. In the end, the answer of the utterance in which the highest score was obtained is returned. Also, any n-gram version of the utterances can be used in the similarity measures.

4 Experiments

In this section we report the results of our experiments where we have tested the different approaches previously described. These experiments were made in the context of the virtual agent Edgar.

4.1 Experimental Setup

Three distinct corpora were used in the experiments (details of these corpora can be seen in Table 1):

- LAB_UNEXPANDED: built by two experts in the topic of our agent;
- LAB_EXPANDED: corresponds to an expansion of LAB_UNEXPANDED. This expansion was made using a set of 154 rules;
- REAL: gathered through a user test scenario, via web. It was told to the participants which was our agent's topic of expertise; then users were instructed to freely interact with the system. We had 6 participants in this evaluation, 3 male and 3 female, from 25 to 40 years old. No post-processing of the gathered input was made.

Table 1. Details of the different corpora

	LAB_UNEXPANDED	LAB_EXPANDED	REAL
Categories	124	124	26
Number of utterances	603	6181	112
Max utterances in a category	20	1080	-
Min utterances in a category	1	1	-
Number of words	3976	46456	611
Unique words	519	618	186

In what concerns the REAL corpus, there are 56 and 60 out of vocabulary words, concerning, respectively, the LAB_UNEXPANDED and the LAB_EXPANDED corpora. After collecting this corpus, a manual classification of the utterances was made, in order to divide individual interactions into three possible categories: in-domain, out-of-domain and context. Utterances are considered in-domain if they relate to a topic to which the agent is prepared to answer; they are out-of-domain, otherwise, except if they fall into the context category, which occurs if the utterance requires contextual information from previous interactions in order to be answered. Table 2 summarises this process.

Table 2. Characteristics of the REAL corpus

	In-domain	Out-of-domain	Context
Number of utterances	72	34	6
Number of words	390	209	12
Unique words	121	102	9

4.2   Testing with "Lab" Corpora

First experiments were done with LAB_UNEXPANDED and LAB_EXPANDED corpora. A 5-fold cross validation procedure was followed, that is, the corpora was divided in 5 random partitions and NLU techniques were trained with 4 partitions and tested with the remaining one. At the end, an average of the results was made. These results can be seen in Table 3. As expected, higher accuracy scores are obtained for all NLU techniques in the LAB_EXPANDED corpus. A reason for this is the low number of utterances in some categories in the LAB_UNEXPANDED corpus; another reason derives from the fact that many of the utterances with the same meaning were lexically very distance from each other; worse, many utterances with different meanings (and thus, quite different) were grouped together, since they could all be answered by the same utterance. Therefore, when the division of training and test corpus was made, we ended up with utterances in the test set that were completely different from the ones in the training, leading to poorer results in the LAB_UNEXPANDED corpus.

Table 3. Accuracy results of the cross validation procedure on the agent Edgar's corpus

NLU technique	LAB_UNEXPANDED	LAB_EXPANDED
SVM binary unigrams	0.71 ± 0.040	0.98 ± 0.004
CLM	0.73 ± 0.049	0.97 ± 0.008
Overlap bigrams	0.66 ± 0.051	0.96 ± 0.021
Jaccard unigrams	0.67 ± 0.021	0.97 ± 0.006
Jaccard Overlap unigrams	0.69 ± 0.034	0.97 ± 0.006
Dice bigrams	0.69 ± 0.034	0.97 ± 0.005

Table 4. Accuracy results for in-domain questions made in the user evaluation

NLU technique	LAB_UNEXPANDED	LAB_EXPANDED
SVM unigrams	0.60	0.69
CLM	0.65	0.68
Overlap bigrams	0.56	0.66
Jaccard unigrams	0.64	0.72
Jaccard Overlap unigrams	0.67	0.72
Dice unigrams	0.64	0.71

The almost perfect results in the LAB_EXPANDED corpus can be explained by the fact that there are many utterances corresponding to the same category. Thus, in this scenario, utterances in the test set are very likely to have a similar utterance in the training set.

4.3 Testing with a "Real" Corpus

For this experiment we only considered the in-domain questions of the REAL corpus. NLU techniques were trained with the LAB_UNEXPANDED and LAB_EXPANDED corpora. Here, we wanted to evaluate how the techniques under study performed in a user test scenario. Results can be seen in Table 4.

Two interesting facts that can be observed:

- In a real user scenario the NLU techniques that performed better were the string similarity ones. This was not expected because in the lab experiments these were not the best techniques;
- Results for the LAB_EXPANDED corpus were still better (a 9% increase in the best case), but the difference of performance between the techniques decreased.

4.4 Discussion

From a more detailed analysis of results, we highlight the following points:

- CLM is influenced by the number of utterances per meaning. That is, if a set U_1 of utterances with meaning M_1 share some words with another set of utterances U_2

with meaning M_2 and if the number of utterances in U_1 is (much) larger than the number of utterances in U_2, then every time these words appear, M_1 is much more likely to be chosen;

- The SVM is not sensible to the previous situation, but due to its discriminative nature, it can be biased if some word/expression only occurs in a set of utterances with the same meaning (category). In this situation, if a given utterance contains this word/expression it will most probably be mapped into that meaning;
- String similarity measures have a different problem: they just allow comparisons between utterances, two by two, and the whole picture is lost.

To conclude, lab experiments, in the case of the LAB_UNEXPANDED corpus, serve as a predictor of the NLU techniques performance in a real scenario, as the discrepancies between accuracy results of both experiments are not very high. Thus, although these observations might be biased by the low number of users, we believe that these experiments can be useful in estimating the results of NLU techniques in a user evaluation scenario.

5   Conclusions and Future Work

In this paper we tested several NLU techniques, which are applied by conversational agents that have in their knowledge base utterances/answer pairs. Two main experiments were done: one involving a lab scenario, corresponding to evaluating the system with questions selected by its developers; in the second experiment, techniques were evaluated with real users. The analysis of both experiments revealed that the best technique in lab is not necessarily the best in a real scenario and that it is possible to roughly estimate how NLU techniques will behave in a user evaluation. Drawbacks of the different techniques were also identified.

Concerning future work, since each technique has specific characteristics and different reasons for succeeding and failing, we will study ways of combining them. Dealing with out-of-domain questions is another important issue, because it is something inevitable in a real scenario. Finally, we will continue to make improvements in our prototype in order to move to other NLU approaches.

Acknowledgments. This work was supported by FCT (INESC-ID multiannual funding) through the PIDDAC Program funds, and also through the project FALACOMIGO (ProjectoVII em co-promoção, QREN n 13449) that supports Pedro Mota, Sérgio Mendes and Pedro Fialho's fellowships.

References

1. Bernsen, N.O., Dybkjær, L.: Meet hans christian anderson. In: Proceedings of the Sixth SIGdial Workshop on Discourse and Dialogue, pp. 237–241 (2005)
2. Bernsen, N.O., Charfuelan, M., Corradini, A., Dybkjær, L., Hansen, T., Kiilerich, S., Kolodnytsky, M., Kupkin, D., Mehta, M.: Conversational H.C. Andersen first prototype description. In: ADS, pp. 305–308 (2004)

3. Bhagat, R., Leuski, A., Hovy, E.: Shallow semantic parsing despite little training data. In: Proc. ACL/SIGPARSE 9th Int. Workshop on Parsing Technologies (2005)
4. Bohus, D., Rudnicky, A.I.: Ravenclaw: Dialog management using hierarchical task decomposition and an expectation agenda, Phoenix, USA, pp. 4–7 (2003)
5. Chang, C.-C., Lin, C.-J.: LIBSVM: a library for support vector machines (2001)
6. Kopp, S., Gesellensetter, L., Krämer, N.C., Wachsmuth, I.: A Conversational Agent as Museum Guide – Design and Evaluation of a Real-World Application. In: Panayiotopoulos, T., Gratch, J., Aylett, R.S., Ballin, D., Olivier, P., Rist, T. (eds.) IVA 2005. LNCS (LNAI), vol. 3661, pp. 329–343. Springer, Heidelberg (2005)
7. Leuski, A., Pair, J., Traum, D.R., McNerney, P.J., Georgiou, P.P., Patel, R.: How to talk to a hologram. In: Paris, C., Sidner, C.L. (eds.) IUI, pp. 360–362. ACM (2006)
8. Leuski, A., Traum, D.: NPCEditor: A Tool for Building Question-Answering Characters (2010)
9. Leuski, A., Traum, D.R.: A statistical approach for text processing in virtual humans, In: Proceedings of the 26th Army Science Conference (2008)
10. Leuski, A., Traum, D.R.: Npceditor: Creating virtual human dialogue using information retrieval techniques. AI Magazine, 242–256 (2011)
11. Mendes, A.C., Prada, R., Coheur, L.: Adapting a Virtual Agent to Users' Vocabulary and Needs. In: Ruttkay, Z., Kipp, M., Nijholt, A., Vilhjálmsson, H.H. (eds.) IVA 2009. LNCS, vol. 5773, pp. 529–530. Springer, Heidelberg (2009)
12. Moreira, C., Mendes, A.C., Coheur, L., Martins, B.: Towards the Rapid Development of a Natural Language Understanding Module. In: Vilhjálmsson, H.H., Kopp, S., Marsella, S., Thórisson, K.R. (eds.) IVA 2011. LNCS, vol. 6895, pp. 309–315. Springer, Heidelberg (2011)
13. Thomas, D.P.: Affect-sensitive virtual standardized patient interface system. Clinical Technologies: Concepts, Methodologies, Tools and Applications 3 (2011)
14. Pfeiffer, T., Liguda, C., Wachsmuth, I.: Living with a virtual agent: Seven years with an embodied conversational agent at the Heinz Nixdorf MuseumsForum. In: KK, pp. 273–297 (1995)
15. Rudnicky, A.: Xu W.: An agenda-based dialog management architecture for spoken language systems. In: IEEE ASRU Workshop, pp. 337–340 (1999)
16. Weizenbaum, J.: Eliza – a computer program for the study of natural language communication between man and machine. Communications of the ACM 9(1), 36–45 (1966)

Unsupervised Clustering of Prosodic Patterns in Spontaneous Speech

András Beke[1] and György Szaszák[2]

[1] Research Institute fo Linguistics, Hungarian Academy of Sciences
Beke.andras@nytud.mta.hu
[2] Dept. of Telecommunications and Media Informatics
Budapest University of Technology and Economics
szaszak@tmit.bme.hu

Abstract. Dealing with spontaneous speech constitutes big challenge both for linguistics and engineers of speech technology. For read speech, prosody was assessed as an automatic decomposition for phonological phrases using supervised method (HMM) in earlier experiments. However, when trying to adapt this automatic approach for spontaneous speech, the clustering of phonological phrase types becomes problematic: it is unknown which types can be characteristic and hence worth modelling. The authors decided to carry out a more flexible, unsupervised learning to cluster the data in order to evaluate and analyse whether some typical "spontaneous" patterns become selectable in spontaneous speech based on this automatic approach. This paper presents a method for clustering the typical prosody patterns of spontaneous speech based on k-means clustering.

Keywords: unsupervised clustering, prosody, spontaneous speech.

1 Introduction

Spontaneous speech constitutes an enormous challenge both in the linguistic and the engineering aspects of speech technology and speech understanding. Whilst the prosodic structure of read speech utterances is relatively well-explored [10], spontaneous speech seems to behave in a different manner also in terms of speech prosody [3]. The prosody of speech plays an important role in speech technology, either in speech synthesis of read and spontaneous speech, where the speech quality is highly dependent on prosody, or in speech recognition and understanding: it can be exploited in the placement of punctuation marks [8], in the improvement of speech recognition [5], in disambiguation of minimal pair sentences [13], but it also seems to be closely related to syntax [11] – especially in case of read speech, where a relatively powerful recovery of the syntax based on prosody was presented in [11]: prosody was assessed as an automatic decomposition for phonological phrases. This can be done by modelling F0 and energy contours of several phonological phrase types in form of hidden Markov models, obtained as a result of supervised machine learning. However, when trying to adapt this automatic approach for spontaneous speech, the clustering of phonological phrase types – or in a more general sense: separating typical/characteristic types or entities of prosodic units – becomes problematic: it is unknown which types can be characteristic and hence worth modelling. When using models of phonological phrases

P. Sojka et al. (Eds.): TSD 2012, LNCS 7499, pp. 648–655, 2012.

trained on read speech, recall rates fall by approx. 20–30% when tested on spontaneous speech. Some phonological phrase types seem to be missing from spontaneous speech, while others seem to be overrepresented in the automatic alignment based on read speech models. A considerable part of the prosodic patterns is most probably missed by the algorithm: these are supposed to be characteristic for spontaneous speech, but their nature is unknown. Based on these considerations, authors decided to carry out a more flexible, unsupervised learning to cluster the data in order to evaluate and analyse whether some typical "spontaneous" patterns become selectable based on this automatic approach. We present a method for clustering the typical prosody patterns of spontaneous speech based on k-means [7] clustering.

This paper is organized as follows: First, the k-means unsupervised clustering approach is presented briefly, then data processing and the clustering itself are described. Finally results are presented and discussed, final conclusions are drawn.

2   Unsupervised Clustering Methods

The goal of unsupervised learning algorithms is to allow for finding classification labels automatically. These algorithms basically operate by seeking for similarity between pieces of data, in order to determine whether some groups (termed clusters) of the data can be well characterized together. Several automatic clustering algorithms are known. K-means clustering [7] is one of the simplest and oldest unsupervised learning algorithms. Given a set of data (consisting of n different, d-dimensional observations) and the desired number of clusters (k), this algorithm clusters iteratively the data around the so called centroids. For bootstrapping, k data can be randomly chosen as centroids, then each observation is clustered to the nearest centroid. The nearest centroid is computed using some distance measure, such as the Euclidean distance or sum of squares, for example. Centroids are iteratively updated to the mean of the belonging observations, until a specified level of convergence is reached. The main drawback of the algorithm is that the number of clusters (k) has to be determined prior to the clustering itself. More on this will be explained later, in Section 4.

The clustering problem can be formulated as looking for:

$$\arg\min_{C} \sum_{i=1}^{k} \sum_{x_j \in C_i} D(x_j, \mu_i) \tag{1}$$

for clusters $C_1, C_2, \ldots, C_k$, given n observations ($k < n$), using a $D(.)$ distance function to evaluate the distance between centroid means (μ_i) and observations (x_j). Since this problem is NP hard, usually heuristic approximation is used to solve the problem.

3   Prosodic Features

The used prosodic features are based initially on fundamental frequency (F0) and energy, however, feature extraction is a two step process embedding time-warping for normalization purposes. Final feature vectors were obtained as explained in the following subsections.

3.1   Selecting the Pieces of Interest

Modern F0 extracting algorithms include post-processing methods in order to smooth and enhance the contour. As a by-product, some shorter voiceless speech segments may become voiced. Moreover, in spontaneous speech, normally unvoiced sounds may also be voiced due to coarticulation effects [4]. This means that voiced pieces in spontaneous speech usually involve segments comparable to the length of words. This is rather advantageous from our point of view, as no interpolation is carried out on unvoiced parts in order to preserve the original contour as much as possible, even if it is more fragmented. Our main goal is to cluster the voiced pieces of the speech signal in terms of their prosodic contours, hence all voiced regions are extracted. Utterances containing overlapping speech (known from speaker turn level annotation) are dropped. Each kept voiced piece is characterized with 14 dimensional feature vectors following each other every 10 ms. Basic features are F0 and energy (100 ms window). Both features have their first and second order deltas associated, whereby deltas are computed based on ±5, ±10 and ±20 frames too, yielding the final 14 dimensional feature vector. These *initial* feature vectors are redundant and serve normalization purposes as described in section 3.2, whilst unsupervised clustering will be performed on a derived feature set as described in subsection 3.3.

3.2   Normalization (Time-Warping)

Spontaneous speech shows high variability both in terms of feature space and length of the voiced pieces within the utterances. In order to eliminate the problem of unequal utterance length and to reduce dimensionality, a flexible normalization scheme was implemented: time alignment is done against a pre-trained set of background garbage models, represented as Hidden Markov Models (HMM). This ensures an equal length feature vector suitable for the unsupervised clustering. The background model is created as follows: all voiced pieces are further divided into 4 different groups: very short ($\leq$ 30 ms), short (between 30 and 100 ms), medium length (100 and 600 ms) and long ($\geq$ 600 ms). Very short pieces are excluded, because they cannot be modelled. Long pieces are also withdrawn, as they are supposed to build up from short and medium length pieces, but their exact composition is unknown yet. Short pieces are modelled with 5-state, medium length ones with 11-state left-to-right HMMs. Feature vector distributions are modelled with Gaussian mixtures up to 8 components. These two HMMs are used as background models for the alignment of voiced pieces, using again the same short and medium length pieces one-by-one (observations). The voiced frames of these pieces are aligned to HMM states by Viterbi-decoding. A normalized prosodic-acoustic match score is also available for each HMM state, interpreted as a similarity measure between the feature vector template describing the given state, and the associated frames to that state.

3.3   Computing Features for Clustering

The time-warped pieces have all their frames associated to HMM states. Final feature vectors consist of the following elements for each and every HMM state:

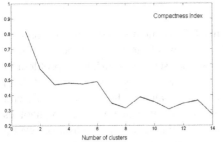

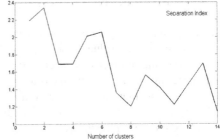

Fig. 1. Partition Index for $k = 2..14$ clusters   **Fig. 2.** Separation Index for $k = 2..14$ clusters

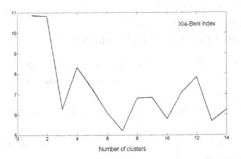

Fig. 3. Xie-Beni Index for $k = 2..14$ clusters

- Ratio of frames associated to the state vs all frames of the piece (state occupancy ratio)
- Normalized prosodic-acoustic match scores obtained from Viterbi-decoding

Adding the mean feature vectors of the initial 14 dimensional features computed from the frames associated with a given state was also considered, however, given the 14 dimensional mean feature vectors for 9 states in case of medium length pieces would have considerably augmented dimensionality, and hence these features were not added to final feature vectors. This means that *final* feature vectors were 18 dimensional for medium length pieces and 6 dimensional for short ones.

3.4   Speech Material

The material used for the experiments was the Hungarian spontaneous speech database 'BEA' [6]. This database contains five spontaneous tasks for each speaker: repeated sentences, spontaneous narration, interpreted (summarized) speech, phrasing of the subject's opinion on a given subject, multilogue. All except the repeated sentences were used, utterances were cleaned from overlapping speech. The resulting 70 hours of speech from 120 speakers formed the speech material used for the experiments.

4   Clustering of the Time-Warped Data

For clustering, the k-means approach was used. This method requires the previous specification of the value of k, that is, how many clusters should be separated. Determining the number of clusters for a given data set is a well-known problem in data clustering. A rich variety measures have been proposed so far, but none of them is capable to estimate the optimal number of clusters alone. Therefore, often several such metrics are used together and the final decision on the value of k usually involves the intervention of "human eye". Beside statistical cluster validity measures, several heuristic ones are known to evaluate cluster properties referred to as compactness and separation [2]. The term "compactness" refers to the variation among the elements belonging to the same cluster (how homogeneous, compact the cluster is), while "separation" refers to how well the clusters are isolated from one another (the better they are isolated, the more robust is the clustering). The authors used the *Fuzzy clustering and data analysis toolbox* for MATLAB [1], which implements the following metrics:

- Partition Index (PI): measures both compactness and separation. Calculated as a sum of individual cluster validity measures, normalized through division by the fuzzy cardinality of the cluster [2].
- Separation Index (SI): measures a minimum-distance separation to evaluate partition validity [2].
- Xie and Beni's Index (XB): measures a ratio of the summed variation within clusters and the separation of clusters [12].

For all of the above measures, a minimal value for the given index refers to an optimal number of clusters, however, indexes may have their minimum point for different k values, and this is the point where human judgement intervenes.

In our case, short voiced segments (S) could not been clustered, which is not surprising as they are too short to present significant changes over time, which means that short segments represent one homogeneous cluster ($k = 1$) as they are. More relevant is the determination of k for medium length segments, which are thought of as prosodic entities in our case. Results for the indexes are presented in Figures 1, 2 and 3 for $k = 2..14$. PI and SI reach a minimum at $k = 8$, however, they are sufficiently low by $k = 7$ too. The XB index reaches its local minimum at $k = 7$. Therefore the optimal number of clusters chosen is 7.

Performing the clustering for 7 clusters on the dataset yields a 7 cluster partition of the voiced segments of spontaneous speech. Clusters are represented in Figure 4.

We investigated which prosodic feature(s) had the greatest separation capability based on F-measure (see Figure 5). The greatest separation capability were found in the F0 value of the 5. and 6. HMM state of the background model.

A plot of the clusters can be obtained with an inverse time-warping of the data, tracing us back to the initial feature vectors (section 3.1) associated with it. As all medium length voiced pieces belong to a cluster, and all these pieces are aligned to the UBM, a cluster specific mean feature vector can be calculated for each state of the UBM, which contains the means of initial features, that is, F0 and energy, among others. An inverse time warping can be carried out on these, based on the averaged state occupancy ratios for the given cluster. This means that a fixed length virtual feature

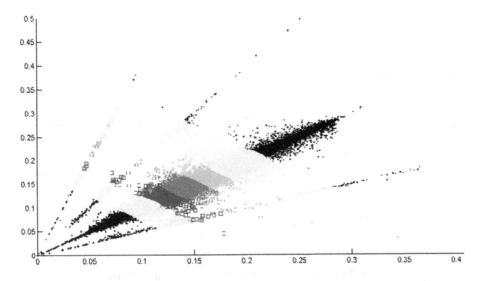

Fig. 4. Representation of the clusters

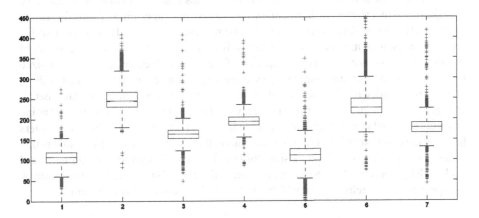

Fig. 5. Separation capability of clusters based on the sixth HMM state of F0 values

vector sequence can be generated, and F0 and energy (or other features) of this sequence can be plotted in order to follow the time course of the selected features. We generated feature sequences of 100 frames for F0 and energy, presented in Figure 6.

Based on the F0 mean contours, clusters 1, 2 and 3 can be identified with considerable F0 rise in the middle/second half of the voiced segments. Clusters 3 and 6 are close to a floating F0 contour, while clusters 4, 5 and 7 show descending F0, with cluster 5 being the most falling. The curve form is explained by the relationship between information structure and syntax. In the case of the first two clusters, we assume there was a non-initial main stress. This may explain the peaks in the middle of the curve. In this case some topic stands before the focus. This type of prosodic realization accounts

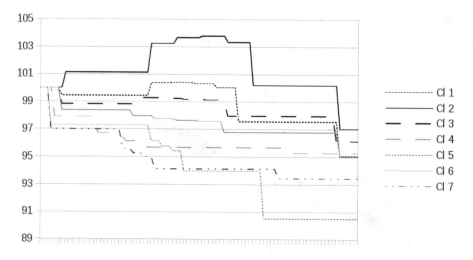

Fig. 6. Mean F0 contours for clusters (normalized to 100 Hz)

for nearly 6 percent of all type of curves. This means that in spontaneous speech, we hypothesize that utterances that contain one stress on a non-initial focused element occur rarely. The 3 clusters are assumed to be neutral utterances. This means that the utterance does not contain a single syntactically focused constituent. In this case, all of the phonological words are stressed equally (10 percent of all curves). We hypothesized that the 4–7. clusters have a single stress at the beginning of the utterance, that is these utterances started with the focused element or a quantifier and therefore are also an example of non-neutral intonation. This may explain the higher peaks at the beginning of the curve, and the continuous decrease of F0 (84% of all curves) [9]. Prosodic clusters have a relatively reduced or monotonous contour, this results from two facts: first is that means were plotted, which has a smoothing effect. The second reason can be traced back to the properties of spontaneous speech: in the case of spontaneous conversations, the speaker usually tells a story, which has a typically reduced prosodic form.

5   Conclusions

The present study aimed to explore typical prosodic patterns in spontaneous speech. In order to eliminate the problem of unequal utterance length and to reduce dimensionality, a flexible normalization scheme was implemented: time alignment is done against a pre-trained set of background garbage models, represented as a Hidden Markov Models (HMM). The ratio of frames associated to the state vs. all frames of the piece (state occupancy ratio) and the normalized prosodic-acoustic match scores obtained from Viterbi-decoding features were used for the clustering done with k-means algorithm. The optimal number of clusters was found to be 7, that is, 7 different prosodic entities could be observed in spontaneous speech. The prosodic structure of spontaneous speech seems to be composed of the following 3 main constituents: in the most common form, the utterance has a single stressed element at the beginning. In another one the utterance is realized with neutral intonation. In the third type, stress is shifted towards the middle of the utterance (normally Hungarian is a first syllable stressed language), which is supposed to indicate some short topic or unstressed conjunction words or articles at the beginning of the voiced period. The 3 main clusters can be further divided into the resulting 7 clusters, if differences in the stress and contour changes and their speed over time is also taken into account. Next steps in this research will be the training of classifiers for the obtained clusters in order to evaluate their usability in spontaneous speech processing, and to explore whether specific events or syntactic entities can be associated to them.

Acknowledgements. Authors would like to thank the CESAR project (funded under the ICT-PSP, Grant Agreement No. 271022), a partner of META-NET, for its support for the work done on the BEA corpus.

References

1. Balasko, B., Abonyi, J., Feil, B.: Fuzzy clustering and data analysis toolbox. Department of Process Engineering, University of Veszprém, Hungary
2. Bensaid, A.M., et al.: Validity-guided (Re)Clustering with applications to image segmentation. IEEE Transactions on Fuzzy Systems 4, 112–123 (1996)
3. Bruce, G., Touati, P.: On the analysis of prosody in spontaneous speech with exemplification from Swedish and French. Speech Communication 11, 453–458 (2003)
4. Clark, J.E., Yallop, C., Fletcher, J.: Introduction to Phonetics and Phonology, pp. 110, 116–118. Blackwell, Oxford (2007)
5. Gallwitz, F., Niemann, H., Nöth, E., Warnke, W.: Integrated recognition of words and prosodic phrase boundaries. Speech Communication 36, 81–95 (2002)
6. Gósy, M.: Magyar spontánbeszéd-adatbázis– BEA, In: Beszédkutatás 2008, pp. 194–207 (2008)
7. MacQueen, J.B.: Some Methods for classification and Analysis of Multivariate Observations. In: Proceedings of the 5th Berkeley Symposium on Mathematical Statistics and Probability, pp. 281–297. University of California Press (1967)
8. Moniz, H., et al.: Prosodically-based automatic segmentation and punctuation. In: Proc. of 5th International Conference on Speech Prosody, Chicago, Illinois (2010)
9. Mycock, L.: Prominence in Hungarian: the prosody–syntax connection. Transactions of the Philological Society 108(3), 265–297 (2010)
10. Selkirk, E.: The Syntax-Phonology Interface. In: Smelser, N.J., Baltes, P.B. (eds.) International Encyclopaedia of the Social and Behavioural Sciences, pp. 15407–15412. Pergamon, Oxford (2001)
11. Szaszák, Gy., Nagy, K., Beke, A.: Analysing the correspondence between automatic prosodic segmentation and syntactic structure. In: Proc. of Interspeech 2011, Florence, Italy, pp. 1057–1061 (2011)
12. Xie, X.L., Beni, G.A.: Validity measure for fuzzy clustering. IEEE Trans. PAMI 3(8), 841–846 (1991)
13. Wightman, C.W., Vielleux, N.M., Ostendorf, M.: Using Prosodic Phrasing in Syntactic Disambiguation: An Analysis-by-Synthesis Approach. In: Proceedings DARPA Speech and Natural Language Workshop, Asilomar, California (1991)

Czech Expressive Speech Synthesis in Limited Domain
Comparison of Unit Selection and HMM-Based Approaches*

Martin Grüber and Zdeněk Hanzlíček

Department of Cybernetics, Faculty of Applied Sciences
University of West Bohemia, Czech Republic
{gruber,zhanzlic}@kky.zcu.cz
http://www.kky.zcu.cz

Abstract. This paper deals with expressive speech synthesis in a limited domain restricted to conversations between humans and a computer on a given topic. Two different methods (unit selection and HMM-based speech synthesis) were employed to produce expressive synthetic speech, both with the same description of expressivity by so-called communicative functions. Such a discrete division is related to our limited domain and it is not intended to be a general solution for expressivity description. Resulting synthetic speech was presented to listeners within a web-based listening test to evaluate whether the expressivity is perceived as expected. The comparison of both methods is also shown.

Keywords: expressive speech synthesis, unit selection, HMM-based speech synthesis, communicative functions.

1 Introduction

Nowadays, production of synthetic expressive speech is a hot topic in the field of speech synthesis research. Modern text-to-speech (TTS) systems are able to produce high quality, intelligible and naturally sounding speech. The TTS systems can be used in our everyday life e.g. for reading emails and web pages, to make text documents available to vision impaired people [1] or to help hearing impaired people to better understand television programs [2]. Another application of TTS systems is in smart phones [3], navigation systems, dialog systems [4], or as a component of audiovisual speech synthesis systems [5]. However, for the human–computer interaction in a dialogue system, the naturalness and intelligibility are not a sufficient criterion to rate the quality of synthetic speech. In the direct contact with a human user, who expects to be served by another human, as presented e.g. in [6], the TTS system should express also some human's feeling and attitudes.

Thus, some kind of expressivity is necessary to be incorporated in the synthetic speech. That way the listeners could completely understand the information and its nature that is communicated with them. Various techniques incorporating expressivity into

---

* This research was supported by the Technology Agency of the Czech Republic, project No. TA01011264 and by the grant of the University of West Bohemia, project No. SGS-2010-054. The access to the MetaCentrum computing facilities provided under the programme "Projects of Large Infrastructure for Research, Development, and Innovations" LM2010005 funded by the Ministry of Education, Youth, and Sports of the Czech Republic is highly appreciated.

P. Sojka et al. (Eds.): TSD 2012, LNCS 7499, pp. 656–664, 2012.

synthetic speech have been introduced so far. Some of them perform the modification of acoustic parameters of synthesized speech [7] or voice conversion [8], others produce special speech or non-speech expressions or use emphasis [9] to evoke some expressivity. Methods using unit selection techniques are based on the creating of a unit inventory containing expressive speech units (or also other non-speech events) [10,11,12].

Since the general expressive speech synthesis is a very complex task, it is usually somehow limited (as well as limited domain speech synthesis systems are). In our work, the domain was restricted to conversations between seniors and a computer about their personal photographs. The work started as a part of a major project whose objective was to develop a virtual senior companion. The detailed motivation is described in [13,14].

To incorporate expressivity into our current Czech unit selection TTS system [15] and newly developed HMM-based system [16], two basic steps had to be performed.

First, an expressivity description was proposed. Many approaches have been suggested in the past, e.g. continuous description using a 2-dimensional space with two axis, one for positive/negative and one for active/passive determination of expressivity position in this space [17]. Another option is a discrete division, e.g. corresponding to the fundamental emotions like happiness, sadness, anger, joy, etc. Within our limited domain we decided to employ a slightly different approach, similar to the one described in [18], where so-called dialogue acts are proposed. A set of communicative functions was designed to fit our limited domain. This set is not a general solution for the problem of speech expressivity description. The communicative functions are assumed to describe more the part of a dialogue than the attitude of the speaker.

Next, the communicative functions need to be somehow incorporated into a speech corpus that is used in our speech synthesis systems. Thus, an expressive speech corpus was recorded. The process of the data preparation and the corpus recording is in more details described in [13]. In the resulting corpus, all utterances are labeled by a feature indicating the appropriate communicative function. This corpus is completed by merging with a part of neutral corpus (which is normally used for neutral speech synthesis) to cover all possible speech units that can occur in any input text. In the case of unit selection method, this corpus was used to build a speech unit inventory, in the case of HMM-based synthesis, corresponding statistical models were trained. These processes are described in Section 2. Thus, the systems were ready to be used for synthetic expressive speech generation.

The produced synthetic speech has to be evaluated to determine whether the expressivity was incorporated successfully or not, and whether the quality and intelligibility of the synthetic speech is still at an acceptable level. Both criterion were evaluated by using listening tests. The backgroud of those tests and the achieved results are described in Section 3. Finally, some conclusions and future work is suggested in Section 4.

2   Expressive Speech Synthesis

Expressive speech can be produced by using various synthesis methods. In our work, we focused on two of them that are currently widely used: the unit selection method and the HMM-based speech synthesis. For both approaches, expressive speech data need to be obtained and labelled with appropriate expressivity descriptors. The process of

Table 1. The set of the communicative functions and their occurrence rate in the expressive speech corpus

Communicative function (symbol)	Occurrence rate	Examples
directive (DIRECTIVE)	2.36%	Tell me that. Talk.
request (REQUEST)	4.36%	Let's get back to that later.
wait (WAIT)	0.73%	Wait a minute. Just a moment.
apology (APOLOGY)	0.59%	I'm sorry. Excuse me.
greeting (GREETING)	1.37%	Hello. Good morning.
goodbye (GOODBYE)	1.64%	Goodbye. See you later.
thanks (THANKS)	0.73%	Thank you. Thanks.
surprise (SURPRISE)	4.19%	Do you really have 10 siblings?
sad empathy (SAD-EMPATHY)	3.44%	I'm sorry to hear that. It's really terrible.
happy empathy (HAPPY-EMPTATHY)	8.62%	It's nice. Great. It had to be wonderful.
showing interest (SHOW-INTEREST)	34.88%	Can you tell me more about it?
confirmation (CONFIRM)	13.19%	Yes. Yeah. I see. Well. Hmm.
disconfirmation (DISCONFIRM)	0.23%	No. I don't understand.
encouragement (ENCOURAGE)	29.36%	Well. For example? And what about you?
not specified (NOT-SPECIFIED)	7.36%	Do you hear me well? My name is Paul.

recording natural expressive speech corpus and its annotation are presented in [13,19][1] where so-called communicative functions were used to describe the expressivity. The set of communicative functions is listed in Table 1 along with their occurrence rate in the expressive corpus.

Obviously, a lot of communicative functions occurred only sparsely in the corpus. For that reason, we decided to use only the most frequent ones in our experiments to obtain representative results – *SHOW-INTEREST, ENCOURAGE, CONFIRMATION, HAPPY-EMPATHY, SAD-EMPATHY* (that was chosen mainly to complete the set with supposedly contradictory pair of happy vs. sad empathy). We also used communicative function *NOT-SPECIFIED* that is related to the neutral speech.

The process of production of expressive speech using the aforementioned methods is described in the following subsections.

---

[1] Since the expressive corpus did not sufficiently cover the speech units that can occure in the Czech language, a part of the neutral corpus was merged with the expressive one. Only the sentences containing missing units were chosen and integrated into the expressive corpus (both corpora were recorded under the same conditions by the same female speaker with a relatively short time lag). This way a complete expressive speech corpus for employing in our TTS systems was created.

2.1 Unit Selection Method

Modifications of the standard unit selection algorithm are in details described in [14]. Thus, the process will be described only briefly.

The main issue of the unit selection method is to select the most appropriate sequence of speech units from the speech corpus (speech unit inventory). This sequence should form speech as smooth and natural as possible (in view of several prosodic and acoustic features). This is ensured by considering two criteria – target cost and concatenation cost.

The target cost reflects the level of approximation of a target unit by any of candidates from the speech unit inventory, in other words, how a candidate from the unit inventory fits the required target unit — a theoretical unit whose features are specified on the basis of the text to be synthesized.

The features affecting the target cost value are usually nominal, e.g. phonetic context, prosodic context, position in word, position in sentence, position in syllable, etc. Currently, these features are chosen manually, based on previous experience; in the future, an algorithm choosing them automatically on the basis of given data (or a speaker) nature might be suggested if it is possible [20]. The value of the target cost T_i for the unit candidate u_i is calculated as follows:

$$T_i = \frac{\sum_{j=1}^{n} w_j d_j}{\sum_{i=j}^{n} w_j} \tag{1}$$

where n is a number of features under consideration, w_j is a weigh of j-th feature and d_j is an enumerated difference between j-th features of a candidate for unit u_i and target unit t_i. The differences of particular features (d_j) will be further referred to as penalties.

In the case of expressive speech synthesis, the set of the features used within the target cost is extended with an additional feature, a communicative function. The penalty d_{cf} between candidate u_i and target unit t_i is calculated as follows:

$$d_{cf} = \begin{cases} 1 & \text{if } cf_t = cf_c \\ 0 & \text{otherwise} \end{cases} \tag{2}$$

where cf_t is the communicative function of target unit t_i and cf_c is the communicative function of the candidate for unit u_i.

Finally, the weigh of this penalty has to be specified since the target cost is calculated as a weighed sum of particular penalties. For our preliminary experiments, the weigh of the penalty for communicative function was determined ad-hoc; its value is one of the highest from all the weighs (e.g. it is 4x higher than the weigh for the phonetic context). It reflects our assumption that this feature should considerably influence the overall criterion.

2.2 HMM-Based Method

Nowadays, beside concatenative unit selection method, HMM-based speech synthesis is one of most researched synthesis methods [21]. In this method, statistical models (an extended type of HMMs) are trained from natural speech database. Spectral parameters, fundamental frequency and eventually some excitation parameters are modeled simultaneously by the corresponding multi-stream HMMs.

Table 2. A list of contextual factors and their values

Factors	Possible values
Previous, current and next phoneme	Czech phoneme set
Phone position in prosodic word (forward and backward)	1, 2, 3, 4, 5 …
Prosodic word position in clause (forward and backward)	
Prosodeme type	terminating satisfactorily/unsatisfactorily, non-terminating, formal null
Communicative function	see begin of Section 2

To model the variation of spectral and other speech parameters, so-called contextual factors are defined. They describe the general context of speech units within the utterance regarding the phonetic, prosodic and linguistics characteristics of speech for the given language, speaker, etc. Contextual factors utilized in our experimental TTS system are listed in Table 2. For a more thorough explanation see e.g. [16].

For a more robust estimation of model parameters, HMMs are clustered by using decision tree-based context-clustering algorithm, that analyzes the similarity of speech units in different contexts. As a result, similar units share one common model. The clustering trees created during that process are also employed for synthesis of speech units unseen within the training stage. In synthesis stage, trajectories of speech parameters are generated from these trained models in the maximum likelihood sense.

Within the HMM-based speech synthesis, several different methods for modeling of the expressivity or speaking styles have been introduced. The most simple approach is an independent training of HMMs for each expression (so-called style dependent model [22]). An evident disadvantage of that approach is a quite large amount of speech data needed for sufficient training of models for particular expressions.

A better solution is to train one set of HMMs for data for all expressions together, the particular expressions are distinguished by addition of a new contextual factor (so-called style mixed model [22]). In this approach, models corresponding to speech units, which are (almost) identical for more expressions, will be clustered and the common model will be trained from all corresponding data. On the other hand, in the case of substantial differences between speech units belonging to particular expressions, corresponding models remain independent to preserve the variability of those expressions.

Recently, methods based on model adaptation [23,24] are preferred because they allow to control the speech style or expression more precisely and require less training data. However, for our first experiments we decided to use the style mixed model with an additional contextual factor for communicative function. Advanced methods for modeling expression will be an objective of our future work.

3 Experiments and Results

To evaluate our modified TTS systems, two listening tests were performed – the first was focused on the perception of expressivity and the second assessed the quality of resulting synthetic speech. Both test were organized on the client-server basis using a specially developed web application. Thus, listeners were able to work on the test from their homes without any contact with the test organizers.

3.1 Comparison in Terms of Expressivity

The first listening test focused on the perception of expressivity in synthetic speech was carried out by 8 listeners. They listened to a big set of 170 synthetic utterances — 85 of them were synthesized by using the unit selection method (referred to as *USEL*) and 85 by using the HMM-based approach (referred to as *HMM*). The text contents of sentences were equal for both *USEL* and *HMM*. For each synthesis method, 35 sentences were of a neutral content and the remaining 50 sentences were of an expressive content. Moreover, disregarding the text content, 2 versions of both TTS-systems were evenly employed for synthesis of particular utterances: the default version (producing neutral speech) and the modified version employing communicative functions (producing expressive speech).

In this test, the listeners were asked to indicate whether they percieve any kind of expressivity or speaker's affective state in the presented utterance. In Table 3 and Table 4 the expressivity perception improvement is shown for both speech synthesis methods. Evidently, there is almost no improvement for the neutral text content whereas a slight improvement can be observed for the expressive text content.

Table 3. Comparison of used approaches in terms of expressivity perception considering neutral text content

approach	expressivity ratio using CFs	expressivity ratio w/out using CFs
HMM-based	4%	3%
Unit selection	10%	10%

Table 4. Comparison of used approaches in terms of expressivity perception considering expressive text content

approach	expressivity ratio using CFs	expressivity ratio w/out using CFs
HMM-based	15%	8%
Unit selection	45%	39%

Obviously, the difference between the perception of expressivity for the *USEL* and *HMM* is remarkable. Comparison of both synthesis methods disregarding the sentence content is presented in Table 5. We can conclude that expressivity is better perceived in speech synthesized by *USEL*. This can be also related to the result presented in Section 3.2, i.e. that the synthetic speech produced by unit selection method is of a better quality.

Table 5. Overall comparison of used approaches in terms of expressivity perception

approach	expressivity ratio using CFs
HMM-based	10%
Unit selection	28%

The overall improvement in the expressivity perception regardless of the used synthesis method achieved by employing communicative functions in the synthesis process is presented in Table 6. Considering neutral text content, no improvement was unfortunately achieved. However, for texts with an expressive content, the perception of the expressivity was more evident.

Table 6. Overall evaluation of improvement in perception of expressivity by listeners regardless of the approach

text content	expressivity ratio using CFs	expressivity ratio w/out using CFs
neutral	7%	7%
expressive	30%	23%

This result might suggest that not only the usage of communicative functions might influence the expressivity perception in the synthetic speech. The text content seems to be also important for the listeners when decide whether they feel any kind of expressivity or not.

3.2 Comparison in Terms of Quality

For the absolute evaluation of the speech quality, MOS (mean opinion score) listening test was organized. 12 participants of that test listened to 40 isolated utterances and rated them according to the standard MOS 5-point scale (1 = bad, 2 = poor, 3 = fair, 4 = good, 5 = excellent). Besides sentences synthesized by the *USEL* (18) and *HMM* (18), the test also contained several natural utterances (4). The results are presented in Table 7.

Table 7. Results of MOS test

Method	HMM-based synthesis	Unit selection	Natural speech
Score	2.71	3.51	4.44

4   Conclusion and Future Work

In this paper, our first experiments on expressive speech synthesis using HMM-based method were presented. We also compared the level of the perception of expressivity in synthetic speech produced by using that method and unit selection. The quality of synthetic expressive speech was also assessed for both methods. For the evaluation, web-based listening tests were performed.

From the achieved results we can conclude that the level of the expressivity perception increased when using expressive speech synthesis (employing the communicative functions), regardless of the used method. However, the improvement is apparent only for sentences with an expressive content. For those sentences, the expressivity perception rate increased from 23% to 30% while for sentences with a neutral content the rate remained almost unchanged at 7%. It is also remarkable that 23% of listeners perceived some kind of expressivity in utterances that were synthesized neutrally (i.e. without using communicative functions) but their content was expressive.

When comparing both synthesis methods, the expressivity is more perceivable for synthetic speech produced by unit selection algorithms (28% *USEL* vs. 10% *HMM*).

When comparing the quality of expressive speech, the *USEL* is preferred to *HMM*. This might be caused by the fact that our HMM-based speech synthesis system is only at an early stage of development and is supposed to be improved in the future. An overall improvement in the quality of synthetic expressive speech is also our future task. To improve the level of the expressivity perception using *USEL*, some modifications of the selection algorithm are planned to be done. For example, the similarity of speech features related to various communicative functions should be taken into account when a unit with required communicative function is not available.

References

1. Matoušek, J., Hanzlíček, Z., Campr, M., Krňoul, Z., Campr, P., Grüber, M.: Web-Based System for Automatic Reading of Technical Documents for Vision Impaired Students. In: Habernal, I., Matoušek, V. (eds.) TSD 2011. LNCS (LNAI), vol. 6836, pp. 364–371. Springer, Heidelberg (2011)
2. Matoušek, J., Vít, J.: Improving automatic dubbing with subtitle timing optimisation using video cut detection. In: Proceedings of ICASSP, Kyoto, Japan, pp. 2385–2388 (2012)
3. Tihelka, D., Stanislav, P.: ARTIC for assistive technologies: Transformation to resource-limited hardware. In: Proceedings of World Congress on Engineering and Computer Science 2011, San Francisco, USA, Newswood Limited, International Association of Engineers, pp. 581–584 (2011)
4. Švec, J., Šmídl, L.: Prototype of Czech Spoken Dialog System with Mixed Initiative for Railway Information Service. In: Sojka, P., Horák, A., Kopeček, I., Pala, K. (eds.) TSD 2010. LNCS, vol. 6231, pp. 568–575. Springer, Heidelberg (2010)
5. Krňoul, Z., Železný, M.: A development of Czech talking head. In: Proceedings of ICSPL 2008, pp. 2326–2329 (2008)
6. Ptáček, J., Ircing, P., Spousta, M., Romportl, J., Loose, Z., Cinková, S., Gil, J.R., Santos, R.: Integration of speech and text processing modules into a real-time dialogue system. In: Sojka, P., Horák, A., Kopeček, I., Pala, K. (eds.) TSD 2010. LNCS (LNAI), vol. 6231, pp. 552–559. Springer, Heidelberg (2010)
7. Přibilová, A., Přibil, J.: Harmonic model for female voice emotional synthesis. In: Fierrez, J., Ortega-Garcia, J., Esposito, A., Drygajlo, A., Faundez-Zanuy, M. (eds.) BioID MultiComm 2009. LNCS, vol. 5707, pp. 41–48. Springer, Heidelberg (2009)
8. Přibil, J., Přibilová, A.: Application of expressive speech in TTS system with cepstral description. In: Esposito, A., Bourbakis, N.G., Avouris, N., Hatzilygeroudis, I. (eds.) HH and HM Interaction. LNCS (LNAI), vol. 5042, pp. 200–212. Springer, Heidelberg (2008)

9. Hamza, W., Bakis, R., Eide, E.M., Picheny, M.A., Pitrelli, J.F.: The IBM expressive speech synthesis system. In: Proceedings of the 8th International Conference on Spoken Language Processing, ISCLP, Jeju, Korea, pp. 2577–2580 (2004)

10. Iida, A., Campbell, N., Higuchi, F., Yasumura, M.: A corpus-based speech synthesis system with emotion. Speech Communication 40, 161–187 (2003)

11. Hofer, G., Richmond, K., Clark, R.: Informed blending of databases for emotional speech. In: Proceedings of Interspeech, Lisbon, Portugal, International Speech Communication Association, pp. 501–504 (2005)

12. Bulut, M., Narayanan, S.S., Syrdal, A.K.: Expressive speech synthesis using a concatenative synthesiser. In: Proceedings of the 7th International Conference on Spoken Language Processing, ICSLP, Denver, CO, USA, pp. 1265–1268 (2002)

13. Grůber, M., Legát, M., Ircing, P., Romportl, J., Psutka, J.: Czech Senior COMPANION: Wizard of Oz Data Collection and Expressive Speech Corpus Recording and Annotation. In: Vetulani, Z. (ed.) LTC 2009. LNCS, vol. 6562, pp. 280–290. Springer, Heidelberg (2011)

14. Grůber, M., Tihelka, D.: Expressive speech synthesis for Czech limited domain dialogue system – basic experiments. In: 2010 IEEE 10th International Conference on Signal Processing Proceedings, vol. 1, pp. 561–564. Institute of Electrical and Electronics Engineers, Inc., Beijing (2010)

15. Tihelka, D., Kala, J., Matoušek, J.: Enhancements of Viterbi search for fast unit selection synthesis. In: Proceedings of Interspeech, Makuhari, Japan, pp. 174–177 (2010)

16. Hanzlíček, Z.: Czech HMM-Based Speech Synthesis. In: Sojka, P., Horák, A., Kopeček, I., Pala, K. (eds.) TSD 2010. LNCS, vol. 6231, pp. 291–298. Springer, Heidelberg (2010)

17. Russell, J.A.: A circumplex model of affect. Journal of Personality and Social Psychology 39, 1161–1178 (1980)

18. Syrdal, A.K., Kim, Y.J.: Dialog speech acts and prosody: Considerations for TTS. In: Proceedings of Speech Prosody, Campinas, Brazil, pp. 661–665 (2008)

19. Grůber, M., Matoušek, J.: Listening-Test-Based Annotation of Communicative Functions for Expressive Speech Synthesis. In: Sojka, P., Horák, A., Kopeček, I., Pala, K. (eds.) TSD 2010. LNCS, vol. 6231, pp. 283–290. Springer, Heidelberg (2010)

20. Tihelka, D., Romportl, J.: Exploring automatic similarity measures for unit selection tuning. In: Proceedings of Interspeech, Brighton, Great Britain, ISCA, pp. 736–739 (2009)

21. Zen, H., Tokuda, K., Black, A.W.: Statistical parametric speech synthesis. Speech Communication 51, 1039–1064 (2009)

22. Yamagishi, J., Onishi, K., Masuko, T., Kobayashi, T.: Modeling of various speaking styles and emotions for HMM-based speech synthesis. In: Proceedings of Eurospeech 2003, pp. 1829–1832 (2003)

23. Yamagishi, J., Onishi, K., Masuko, T., Kobayashi, T.: A style control technique for HMM-based speech synthesis. In: Proceedings of Interspeech 2004, pp. 1437–1440 (2004)

24. Nose, T., Kobayashi, Y.K.,, T.: A speaker adaptation technique for MRHSMM-based style control of synthetic speech. In: Proceedings of ICASSP 2007, pp. 833–836 (2007)

Aggression Detection in Speech
Using Sensor and Semantic Information

Iulia Lefter[1,2,3], Leon J.M. Rothkrantz[1,2], and Gertjan J. Burghouts[3]

[1] Delft University of Technology, The Netherlands
[2] The Netherlands Defence Academy
[3] TNO, The Netherlands

Abstract. By analyzing a multimodal (audio-visual) database with aggressive incidents in trains, we have observed that there are no trivial fusion algorithms to successfully predict multimodal aggression based on unimodal sensor inputs. We proposed a fusion framework that contains a set of intermediate level variables (meta-features) between the low level sensor features and the multimodal aggression detection [1]. In this paper we predict the multimodal level of aggression and two of the meta-features: Context and Semantics. We do this based on the audio stream, from which we extract both acoustic (nonverbal) and linguistic (verbal) information. Given the spontaneous nature of speech in the database, we rely on a keyword spotting approach in the case of verbal information. We have found the existence of 6 semantic groups of keywords that have a positive influence on the prediction of aggression and of the two meta-features.

Keywords: emotional words spotting, aggression detection, multimodal fusion.

1 Introduction

Our project addresses the problem of automatic audio-visual fusion in the context of aggression detection. In particular, we are analyzing a database with audio-visual recordings of aggressive and unwanted behavior in trains. The level of aggression in audio, video and both simultaneously has been annotated on a 3 point scale. Having these three types of annotation we explore methods of predicting the multimodal label given the audio and video labels. We find that there is a large diversity of combinations of the audio, video and combined sensor data, and no trivial fusion method based on simple rules or a classifier has a good performance in predicting the multimodal label given the unimodal ones. We observe that there are a number of concepts inherent in the multimodal data that are missed when only unimodal information is assessed. We propose to use an intermediate level in the fusion framework. This level contains the audio and video predictions and a set of five high level concepts (meta-features) that have an impact on the fusion process: Audio-Focus, Video-Focus, History, Context and Semantics [1].

In the context of this fusion framework, so far our predictions were based on low level sensor features from audio (acoustic) and video [2]. But the audio modality has two components: nonverbal (prosody – how things are said), and verbal (what is being said). In this paper we focus on both of these components to predict aggression. As

P. Sojka et al. (Eds.): TSD 2012, LNCS 7499, pp. 665–672, 2012.

opposed to our previous work, in this paper we add semantic information based on the linguistic content by extracting aggression related keywords.

This paper is organized as follows. In Section 2 we provide an overview of related work. We continue with a description of the database and how it was annotated. In Section 4 we present our fusion model and prove that it is beneficial. Details on the acoustic and linguistic features are given in Section 5. In Section 6 we describe the approach and results for multimodal aggression prediction based on the audio verbal and nonverbal features, and in Section 7 we use the same features to predict the Context and Semantics meta-features. The paper ends with conclusions and directions for future work.

2   Related Work

Research in the field of audio-visual fusion is getting more and more attention [3]. However, in the context of surveillance applications, the focus is mostly on video only, or on the acoustic component of sound, while linguistic information is mostly ignored. As we will show, the linguistic component provides a lot of relevant information.

In the field of emotion recognition, the linguistic component was successfully combined with prosodic information in [4]. Their approach was to use a Belief Network for spotting of emotional key-phrases, based on their frequency of appearance in emotional utterances. A string-based audio-visual fusion method for dimensional affect assessment is proposed in [5]. Besides head gestures, Action Units (AUs) and non-verbal acoustic events, they have used a keyword detection algorithm. The words relevant for each affect dimensions are detected by correlation based feature subset selection. The Dictionary of Affect in Language [6] contains valence and arousal scores for a large dataset of words. However, we could not use it because we observed that our database contains many expressions and bad words that are not in the database.

A different application of audio-visual fusion is event detection in team sports videos [7]. The use of semantic information from external sources like match reports or real-time game logs in detecting events proved to be a key contribution in their approach.

3   Dataset and Human Assessment

3.1   Database of Aggression in Trains and Its Annotation

We defined a set of rules that describe normal behavior in trains. A set of 21 scenarios were generated, each of them breaking one or more of the rules of normal behavior. The scenarios contain different abnormal behaviors like harassment, hooligans, theft, begging, football supporters, medical emergency and traveling without ticket. The total length of the audio-visual recordings is 43 minutes.

The level of aggression has been annotated on a three point scale (1-low, 2-medium, 3-high). The annotation has been done in three settings: audio only, video only and looking at and listening to the data simultaneously. For each of these settings, the data has been split into segments of homogeneous aggression level by two expert annotators. The data is unbalanced, dominated by neutral samples. More details about the database and its annotation can be found in [1].

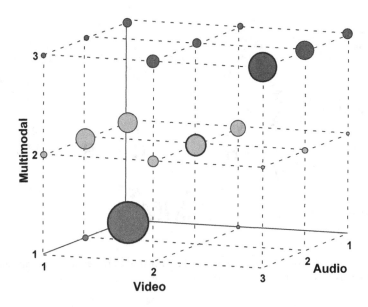

Fig. 1. Confusion cube of audio, video and multimodal annotations [1]

3.2 Analysis of Human Assessment

We want to understand how the audio, video and multimodal annotations relate to each other. Especially we are interested in those cases where these three labels do not agree. We have computed a 3D confusion matrix of the annotations, which we call a confusion cube. The axes of the cube correspond to the three types of annotation, audio-only (A), video-only (V) and multimodal (MM).

Given the difficulty of coming up with a multimodal label by having the audio and video labels, we divide the samples from confusion cube into two groups:

- On-diagonal – audio, video and multimodal are equal.
- Off-diagonal – for these points the multimodal label is not equal to at least one of the two unimodal labels. These are the samples that we consider challenging and on which we will base the following section. Note that 46% of the data falls in the off-diagonal case.

4 Fusion Model Based on Meta-information

In order to understand why there is no straightforward relation between the multimodal label and the unimodal labels we have carefully inspected the off-diagonal samples. It occurred that there are a number of intermediate factors that are not obvious from one modality only. We call these factors meta-features and we have identified a set of five that have an influence on fusion as follows [1]:

- **Audio-focus (AF)** – the rater was more influenced by the audio channel than by the video channel in his/her final assessment of the multimodal level of aggression.

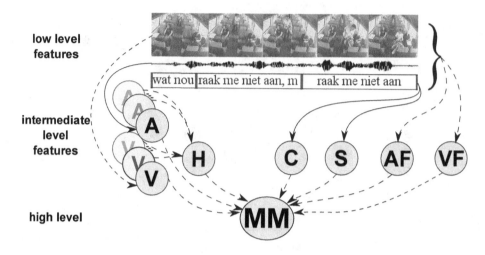

low level
features

intermediate
level
features

high level

Fig. 2. Fusion model based on meta-features. The dashed lines represent human annotations of the data streams. The solid lines are learned by classifiers.

– **Video-focus (VF)** – the video modality had the final impact on the multimodal assessment. The two meta-features are mutually exclusive. It can also be the case that none of them is active, when there is no dominant focus on one modality.
– **Context (C)** – the of aggression is strongly influenced by the region of interest or situations in which the actions is taking place and what is appropriate or not in that case.
– **History (H)** – illustrates the effect that negative events can have over time, even though there is no explicit aggression in the current segment.
– **Semantics (S)** – the semantic interpretation of the scene is pointing to abnormal behavior.

We have annotated the meta-features on the samples of the cube which were off-diagonal. More details about the annotation can be found in [1].

The proposed fusion model is depicted in Figure 2. Instead of trying to predict the multimodal label from the low level features, we propose to use an intermediate level. The picture shows the contributions of audio, video (depicted in green) as well as the meta-features (in orange) in constructing a multimodal output. It also shows that the previous audio and video assessments influence the History variable. The AF, VF and H meta-features are predicted based on low sensor features. For the prediction of Context and Semantics, we needed to add semantic information. The focus of the paper is to use the audio modality in terms of prosody and words to predict the level of aggression and the Context and Semantics meta-features. This is illustrated in the figure by solid lines.

To verify the effect of the meta-features on multimodal fusion we have trained a Random Forest (RF) classifier on the Audio and Video labels and predicted the Multimodal label. We then added the meta-features to the feature vector, and the performance improved from 86% to 92% weighted average. Class 3, which corresponds to the highest level of aggression and which is less represented in the data and harder to predict has a significant improvement of 18% absolute. More information can be found in [1]. In the next sections we continue with a semantic analysis of speech to compute Context, Semantics and their influence on the Multimodal label.

5   Acoustic and Linguistic Features

5.1   Acoustic Features

Vocal manifestations of aggression are dominated by negative emotions such as anger and fear, or stress. The audio feature set consists of features inspired from [8]: speech duration, mean, standard deviation, slope, range of pitch (F0) and intensity, mean and bandwidth of the first four formants F1–F4, jitter, shimmer, high frequency energy (HF500), harmonics to noise ration (HNR), Hammarberg index, center of gravity and skewness of the spectrum. These features are computed on segments of length equal to 2 seconds, because this resembles better what we can expect in real-time processing.

5.2   Linguistic Features

The used language has the characteristics of spontaneous speech, with a lot of interruptions, restarts, overlapping speech, slang, interjections and nonverbal utter-ances('um','eh','he'). The first step was to manually transcribe the speech from the database to text. Given the nature of the data no natural language processing approaches can be used successfully, since most utterances are not grammatically correct. A key-word spotting approach was used instead. To start with, we want to see what is the added values of linguistic analysis in perfect conditions, without being affected by tran-scription errors.

We define aggressive keywords as words or expressions that convey aggressive states or that are stimulating aggressive states. These keywords were selected manually from the transcription by three annotators. They have been clustered in 6 classes based on expert knowledge:

1. **positive emotions:** this class contains words that express positive emotions, e.g. 'nice', 'cool', 'helpful'.
2. **negative emotions:** this class contains words/expressions conveying negative emotions, e.g. 'irritated', 'don't want to', 'unfair', 'disturb', 'lousy'.
3. **actions:** this class contains words/ expression related to actions that relate to the fact that somebody is being disturbing, e.g. 'don't touch me', 'behave normally', 'leave me alone', 'stay still', 'go away', 'pay attention', 'stop'.
4. **context:** the words in this class are good indicators of special contexts. e.g. 'police', 'ambulance', 'thief', 'drugs', 'sniffing', 'dead', 'criminal', 'wallet'.
5. **cursing:** this class contains cursing and offensive words.
6. **nonverbal:** we have added semantic tags to a number of nonverbal sounds, e.g. singing, clapping, knocking, noise and repetitions. These can also be detected automatically but in this paper we first wanted to determine their added value.

In Figure 3 we show the number of occurrences per word class in the database. Given the five classes, we have created a 6 dimensional feature vector with binary values given the presence or absence of words from the class.

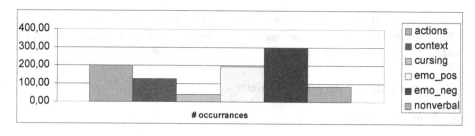

Fig. 3. Number of occurrences for each keyword class

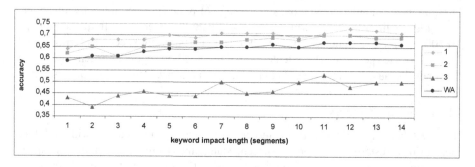

Fig. 4. Accuracies for prediction of the Multimodal label given different impact length of the keywords

6   Automatic Prediction of Multimodal Aggression

In this experiment we are predicting the Multimodal label of aggression based on the prosodic and linguistic features. As a first step, we use the prosodic features to predict the Audio label. For this we use a logistic regression classifier. The posteriors of this classifier concatenated with the linguistic features form the final feature vector for Multimodal aggression prediction. The results of all experiments are obtained using 10-fold cross-validation on the 2 seconds segments.

We have observed that the presence of our keywords does not have a strict local impact. Instead, the impact lasts for a longer time (in this case we consider a number of 2 seconds segments corresponding to the ones for which the prosodic features were computed). We test several configurations to find an optimal impact length.

With this setting, we compare two approaches. First, we use only the Audio posteriors to predict the Multimodal label. We then add the linguistic features to it, to illustrate their impact. In both cases we use a Random Forrest (RF) classifier.

A comparison of the performances of the RF classifier that predicts multimodal aggression using different impact length of the semantic words features is depicted in Figure 4. Because in a surveillance application we are interested in having the smallest miss rate for the aggressive cases, we choose the impact length of 11 segments as the most appropriate.

The results of using semantic keywords with an impact length of 11 segments shows an improvement of 10% absolute compared to using the Audio posteriors only.

Furthermore, the improvement of class 3 (the most aggressive cases) is of 16% absolute. We did an information gain feature ranking and the most relevant linguistic features were context, negative emotions and nonverbal. Even though the linguistic features provide an improvement in predicting the Multimodal label, the performance is not high. However, the video modality was not taken into consideration and the Multimodal labels is based on that as well.

7   Automatic Prediction of Context and Semantics Meta-features

This section describes the prediction from low level features of the Context and Semantics meta-features from the intermediate level of our fusion framework depicted in Figure 2. The successful prediction of the meta-features has a positive influence of the final multimodal aggression assessment. For predicting the Context and Semantics meta-features we use the Audio posteriors concatenated with the linguistic features with the 11 segments impact length. In both cases we use a Support Vector Machine classifier with a second order polynomial kernel. Note that in this experiment we only use the subset of the data for which the three modalities do not agree (the off-diagonal of the confusion cube), because these are the most interesting cases for multimodal fusion.

The prediction accuracies for Context and Semantics are presented in Table 1. For both cases we have used a feature ranker based on information gain. In the case of Semantics, the most useful features were the posteriors of the Audio aggressive classes (2 and 3), actions, nonverbal and cursing. In the case of Context, the Audio posteriors had no influence. The top ranked features were nonverbal, context and cursing.

Table 1. Prediction accuracies for Context and Semantics for class 0 (meta-feature is not activated), class 1 (activated) and their weighted average (WA)

Class	Accuracy	
	Context	Semantics
0	0.95	0.50
1	0.69	0.87
WA	0.91	0.73

8   Conclusion and Future Work

In this paper we have used prosodic and linguistic information in the form of semantically meaningful keywords to predict aggression. We have done the experiment in the framework of our project focused on multimodal fusion for aggression detection. Besides predicting the level of aggression, we have shown that the linguistic information is a rich source for predicting Context and Semantics, two of the meta-features of our fusion framework. In our future work we will combine the linguistic, prosodic and video modalities to predict all variables in the intermediate level of our fusion framework. Based on them the multimodal level of aggression will be automatically assessed.

References

1. Lefter, I., Burghouts, G., Rothkrantz, L.: Learning the fusion of audio and video aggression assessment by meta-information from human annotations. In: International Conference on Information Fusion, FUSION (in press, 2012)
2. Lefter, I., Burghouts, G., Rothkrantz, L.: Automatic audio-visual fusion for aggression detection using meta-information. In: IEEE Conference on Advanced Video and Signal Based Surveillance, AVSS (in press, 2012)
3. Atrey, P.K., Hossain, M.A., Saddik, A.E., Kankanhalli, M.S.: Multimodal fusion for multimedia analysis: A survey. Springer Multimedia Systems Journal, 345–379 (2010)
4. Schuller, B., Rigoll, G., Lang, M.: Speech emotion recognition combining acoustic features and linguistic information in a hybrid support vector machine-belief network architecture. In: Proceedings. IEEE International Conference on.Acoustics, Speech, and Signal Processing, ICASSP 2004, vol. 1, pp. I-577–I-580 (2004)
5. Eyben, F., Wöllmer, M., Valstar, M., Gunes, H., Schuller, B., Pantic, M.: String-based audiovisual fusion of behavioural events for the assessment of dimensional affect. In: 2011 IEEE International Conference on Automatic Face Gesture Recognition and Workshops, FG 2011, pp. 322–329 (2011)
6. Whissell, C.M.: The dictionary of affect in language, vol. 4, pp. 113–131. Academic Press (1989)
7. Xu, H., Chua, T.S.: The fusion of audio-visual features and external knowledge for event detection in team sports video. In: Proceedings of the 6th ACM SIGMM international workshop on Multimedia information retrieval, MIR 2004, pp. 127–134. ACM, New York (2004)
8. Lefter, I., Rothkrantz, L.J.M., Wiggers, P., van Leeuwen, D.A.: Emotion Recognition from Speech by Combining Databases and Fusion of Classifiers. In: Sojka, P., Horák, A., Kopeček, I., Pala, K. (eds.) TSD 2010. LNCS, vol. 6231, pp. 353–360. Springer, Heidelberg (2010)

Question Classification with Active Learning

Domen Marinčič, Tomaž Kompara, and Matjaž Gams

Jozef Stefan Institute, Jamova 39, 1000 Ljubljana, Slovenia
{domen.marincic,tomaz.kompara,matjaz.gams}@ijs.si

Abstract. In a question answering system, one of the most important components is the analysis of the question. One of the steps of the analysis is the classification of the question according to the type of the expected answer. In this paper, two approaches to the classification are compared: the passive learning approach with a random choice of training examples, and the active learning approach upgraded by a domain model where the learning algorithm proposes the most informative examples for the train set. The experiments performed on a set of questions in Slovene show that the active learning algorithm outperforms the passive learning algorithm by about ten percentage points.

Keywords: question classification, machine learning, active learning.

1 Introduction

When searching for information in the electronic form, keyword-based search engines are normally used. To determine the keywords requires a certain level of skill from the user. Furthermore, the result of the search is usually a list of documents, which the user has to go through to find the desired information. On the other hand, a question answering system allows the user to ask questions in natural language and returns the answer in a concise form. When searching for the answer, determining the type of information in the expected answer is one of the first steps to do. When an information source is searched for the answer, the pieces of information that conform to the specified type get higher score.

Consider the following toy example. The user enters a question expecting an answer about a location: *"Where is the office of professor John Smith?"*. The answer is searched for among the three sentences in the following text:

"(1) Prof. Smith is the dean of the Faculty of computer engineering. (2) His office is on the 2nd floor. (3) His office hours are on Mondays, 1 pm to 2 pm."

The first sentence contains a description of a person, the second sentence gives the information about a location, while the third sentence tells something about time. It is presumed that the coreference between the two pronouns "His" and the noun phrase "Prof. John Smith" is resolved. If the answer is searched for only by matching the keyword terms, the second and the third sentence are equally good answer candidates, because both sentences match the question in the two terms "Prof. John Smith" and "office". However, if the question is classified according to the expected answer type,

P. Sojka et al. (Eds.): TSD 2012, LNCS 7499, pp. 673–680, 2012.

the more likely answer candidate is the second sentence, which matches the type of the expected answer.

We studied active and passive learning upgraded by a domain model for question classification. Active learning chooses the most informative training examples, while during passive learning they are chosen randomly. The experiments performed on questions in Slovene show that with the upgraded active learning a 9.8 pp (percentage points) higher classification accuracy can be achieved than with the passive learning without the domain model.

2   Related Work

To our best knowledge, no literature about question classification using active learning for Slovene is available. We present some related work for English. In [1] a question classification experiment is briefly described, where an actively trained classifier achieves a 4.8 pp higher accuracy then a classifier built by co-training. Another type of active learning is described in [2] where a web-based QA engine relies on users' feedback to improve the question classification performance.

Further, a review of relevant publications about question classification in general is presented. In [3], an approach to question classification that extracts structural information using machine learning techniques is described. The patterns found are used to classify the questions. A report about question classification using several machine learning algorithms is available in [4]. The authors propose a special tree kernel function for the Support vector machine (SVM) classifier to improve the classification accuracy. In [5] the author shows that semantically enriched annotation together with the syntactic trees raises the accuracy of classification. The use of semantic features of head words with to achieve higher classification accuracy is presented in [6].

In most of the described related work about question classification, new features were invented to achieve better accuracy. These features require additional work on the annotation of questions. On the other hand, the active learning approach can improve the results only by choosing the most informative examples during the training process without adding further complex features.

3   Active Learning

In the pool-based active learning, the most informative train examples are chosen from a large pool of unannotated examples [7]. Usually the informativeness of an example is estimated by the confidence of a classifier about the class it assignes to the example. We upgraded this approach with a domain model and the TOPSIS technique to take into account specifics of the domain of the questions.

3.1   Example Informativeness

We describe how the domain model and the classification confidence are combined to estimate the informativeness of an example. We studied the Jozef Stefan Institute (JSI) domain, which is covered by the institute's question answering system[1]. The domain model has three features that consider the language model of the domain and the type of the questions asked; the bigger the value of a feature, the better a question fits the domain:

- *Fitting the domain vocabulary.* Here, a domain language model is used, where a word is represented by the value $f_w = f_{JSI} - f_{GF}$. f_{JSI} and f_{GF} are the respective frequencies of the word appearing on the JSI webpages and in the GigaFida corpus [8], a one-billion word reference corpus of Slovene language. The feature value: $F_1 = \sum_{w \in Q} f_w / n_{JSI}$, Q being the set of the words in the question and n_{JSI} the number of words in the question appearing on JSI web pages.
- *Presence of a wh-word.* $F_2 = 1$ if a wh-word is present, $F_2 = 0$ otherwise.
- *Number of words (n) in the question.* The pattern of common questions was followed: $F_3 = 0$ if $n = 0$; $F_3 = 1$ if $1 \geq n \leq 10$; $F_3 = 1/(l - 9)$ if $n \geq 11$.

To estimate the classification confidence, the weighed classifications of the J48 [9] base classifiers in the AdaBoostM1 algorithm [10] are used. The entropy of the set of the weighed predicted classes is calculated: the more even the class distribution, the bigger the entropy, the smaller the confidence.

Finally, the TOPSIS technique [11] is used to combine the domain model features and the classification confidence. TOPSIS identifies the ideal examples, described by various criteria, by simultaneous minimization of the distance from the ideal point and maximization of the distance from the negative ideal point. The distances incorporate the predetermined criterion importance by the relative weights, manually chosen in our case: 0.75 for the classification confidence and 0.15, 0.05, 0.05 for the respective domain model features F_1, F_2, F_3.

3.2 Training Algorithm

The active learning algorithm used in our work is outlined in Alg. 1. First, the seed set T_0 with a balanced distribution of the question classes is annotated. Second, the domain model features are calculated for all the questions in the pool L. Then, the algorithm enters a loop. The AdaBoostM1 classifier C_i with J48 decision trees is trained on the current train set T_i. In the next step, the exit condition is checked: if T_i has the desired size, the active learning algorithm terminates and returns the current classifier C_i as the result of learning. Otherwise, the algorithm continues to classify the examples of the pool L using the classifier C_i. Next, the informativeness of the examples is calculated using the TOPSIS technique. The three most informative examples are annotated and added to the set T_i to form the train set T_{i+1} used in the next iteration.

4   Preparing Data for Experiments

It is explained how a question is described with an attribute model. Further, the data sets used in the experiments are presented.

---

[1] available at http://www.ijs.si

Algorithm 1. Pool-based active learning

1: choose and annotate seed set T_0
2: calculate domain model features of questions in pool L
3: **while** true **do**
4:     C_i = train on T_i
5:     **if** T_i has desired size **then** return C_i
6:     classify L using C_i
7:     calculate the informativeness of examples by TOPSIS
8:     choose the three most informative examples from L without replacement
9:     annotate the examples and add them to T_{i+1}
10: **end while**

4.1   Attributes

The attribute description of a question consists of the class attribute and the attributes obtained from the words in the question. The class attribute values were chosen on the base of the expected answer type taxonomy described in [12]. Its first layer contains the types abbreviation, entity, description, human, location, quantity. We modified this set and used the following six values:

- Entity: object, non-human living creature, notion, etc.,
- Description: definition, reason, manner, etc.,
- Human: person, group of persons,
- Location: place, city, country, etc.,
- Quantity: size, time, money, etc.,
- Yes-no-answer: answered by "yes" or "no",

Further, there are four attributes with nominal values:

- Unigram on the first position. This attribute reveals important information about the expected answer, since this is usually a wh-word. It has nominal values – the 15 most frequent words appearing on the first position in all distinct questions of the GigaFida corpus, plus the value 'OTHER'.
- Bigram on the first position. The attribute is important when the question starts with a preposition, followed by a wh-word. It has nominal values – the 50 most frequent bigrams on the first position in all distinct questions of the GigaFida corpus, plus the value 'OTHER'.
- First wh-word on any non-first position. This attribute covers the cases, where the wh-word appears inside the sentence, not in the first position. If there are many wh-words, the attribute gets the value of the first one. The nominal values of the attribute were determined manually and comprise all the wh-words of Slovene plus the 'NONE' value.
- The fourth attribute is determined relatively to the second one and is analogous to the third one. If a bigram from the set of nominal values of the second attribute appears in a non-first position, this attribute gets its value. If there are many such bigrams, the attribute gets the value of the first one. This attribute has the same values as the second one.

The other attributes indicate the presence of a particular word in the question and have binary values:

- Six attributes, the presence of a noun in one of the six cases. The information about case can reveal the class of the question, e.g. nouns in expressions about a location are usually in the fifth case in Slovene.
- 12 attributes, the presence of a word with POS other than noun. Certain classes of questions may have specific distribution of the POS of the words.
- 1,000 attributes, representing the presence of one of the thousand most frequent words from the JSI web pages. The words are represented by the lemmas. The words characteristic for the domain carry important information about the class of the question.

4.2   Construction of Data Sets

The basis for the training datasets was the set Q_g containing all distinct questions of the GigaFida corpus (about 1 million). The questions were automatically lemmatized and assigned MSD-tags (MorphoSyntactic Description) by the a lemmatizer and MSD-tagger[2]. The following data sets[3] were derived from the set Q_g:

- Q_{ma}: the 100,000 questions with the biggest value of the domain model feature F_1 – the ones that best fit the language model.
- Q_{ra}: 100,000 randomly chosen questions.
- Q_{mp}: the 1,000 questions with the biggest value of the domain model feature F_1.
- Q_{rp}: 1,000 randomly chosen questions.
- Q_{sa}: the seed set for standard active learning, containing 10 randomly chosen questions from Q_{ra} with a manually balanced distribution of classes.
- Q_{sua}: the seed set for upgraded active learning, containing 10 most informative (described in Section 3.1) questions from Q_{ma} with a manually balanced distribution of classes.
- Q_t: 600 manually selected test questions obtained from the log of the questions posed by the users of the JSI's question answering system, leaving out the smalltalk questions, which are not a subject of this study.

The questions of the train and the test sets were paralelly annotated by two annotators. In the cases when the annotators disagreed, the conflicts was resolved mutually by both annotators.

5   Experiments

Four experiments were performed to compare the active and passive learning approach to question classification. In all experiments, the implementation of the machine learning algorithms in the WEKA software package was used [13] with the default parameter settings. The training phase of each of the four experiments proceeded as follows, producing 331 classifiers per experiment:

---

[2] The tagger and the lemmatizer are available on http://slovenscina.eu.
[3] The data sets are available upon request to the authors.

1. *Upgraded active learning (Upgraded AL)*: the set Q_{ma} was used as the pool of unannotated questions, the set Q_{sua} as the seed set. The training was performed as described in Alg. 1, each iteration producing one classifier.
2. *Standard active learning (Standard AL)*: the set Q_{ra} was used as the pool of unannotated questions, the set Q_{sa} as the seed set. Apart from the ranking of the examples according to their informativeness, the learning was performed as described in Alg. 1. To calculate the informativeness of the examples the domain model and the TOPSIS technique were excluded: the entropy was the only measure for the informativeness of the examples.
3. *Passive learning with the language model (Upgraded PL)*: the set Q_{mp} was annotated. 331 subsets of 10, 13, 16,..., 1,000 randomly chosen questions from Q_{mp} were created to match the size of the active learning train sets. One classifier was trained on each of the subsets.
4. *Random passive learning (Standard PL)*: the set Q_{rp} was annotated and 331 classifiers were built analogously to the third experiment.

In the testing phase all classifiers were evaluated on the test set Q_t. Although the set is rather small, we avoided cross-validation for two reasons:

- The examples of the test set reflect the real use of the question answering system since they are the questions actually posed by the users.
- Cross-validation may not give a proper estimate. The argumentation is that the difficulty distribution of the examples in the active learning train set is biased: by definition the train set contains difficult examples. Cross-validation thus gives a pessimistic accuracy estimate for active learning.

The results of some classification runs are presented in Tab. 1. The first row shows the size of the train sets, while the other rows show the accuracy of the classifiers. In the last column, the average accuracy of the classifiers trained on the sets larger than or equal to 619 examples is shown. In the graphs in Fig. 1 the accuracy of all classifiers runs is plotted.

Table 1. The accuracy in % measured in some of the classification runs

Train set size	22	37	52	76	100	202	400	601	802	1,000	Avg, size ≥ 619
Upgraded AL	64.1	70.2	72.2	72.7	72.9	73.2	73.0	73.9	75.2	75.5	74.6
Standard AL	63.2	72.5	78.2	74.9	76.5	72.5	74.2	74.2	74.4	74.4	74.4
Upgraded PL	57.6	60.9	70.2	59.9	63.6	63.6	65.2	70.0	70.2	66.7	68.8
Standard PL	55.6	69.1	68.9	61.6	60.2	52.6	64.2	59.1	64.4	59.9	64.8

The results show that the learning curve is very steep at the beginning for all the algorithms. The accuracy of the active learning classifiers not using the domain model "stalls" from the train set size of 292 examples on; actually only two distinct classifiers were built with the train sets larger than 292 questions. Despite the variation of the results of other experiments, which could be attributed to chance, we can observe that

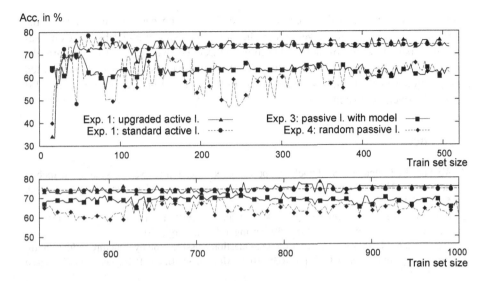

Fig. 1. The *upper graph* shows the results of the runs with the train set size up to 519 examples, the *lower graph* shows the rest of the results

the active learning classifiers give higher accuracy of about 10 pp from the train set size of about 350 up to about 500 examples, compared with the passive learning classifiers.

From the train set size of 619 examples on, no improvement can be observed in any of the experiments. By analyzing the average classification accuracy of the classifiers trained on the sets bigger than or equal to 619 questions we can conclude that the active learning algorithms clearly outperform the passive learning algorithms; the difference of the average accuracies measured in the first and fourth experiment 9.8 pp. Why active learning is better than passive learning could be explained by the fact that during active learning the essential examples are found that contribute to higher accuracy; for passive learning the chance to find these examples in the set of 100,000 questions by random choice is rather small. The average accuracy difference of 0.2 pp between both active learning algorithms is rather small, the domain model unfortunately does not help to raise the accuracy significantly if active learning is used. However, the domain model proves useful for passive learning, since the difference of the average accuracies measured in the third and fourth experiment is 4.0 pp.

6 Conclusion

Two approaches to the classification of questions according to the type of the expected answer were compared: the active and the passive learning approach. Furthermore, the application of a domain model describing the JSI domain was studied. The main findings of our study is that active learning outperforms the passive learning and that the use of the domain model with passive learning improves the classification accuracy. Unfortunately, no clear accuracy improvement can be observed comparing the plain active learning approach to the one upgraded by the domain model.

In the future, we intend to perform studies by varying certain parameters of the algorithms, e.g. the TOPSIS weights in the domain model, the number of examples annotated per iteration of active learning. Moreover, we intend to compare the complexity and accuracy of the automatically trained classifiers with hand crafted classification rules. Further, we plan to include the question classifier into the JSI question answering system.

References

1. Mishra, T., Bangalore, S.: Qme!: A speech-based question-answering system on mobile devices. In: Human Language Technologies: The 2010 Annual Conference of NAACL, pp. 55–63 (2010)
2. Song, W., Wenyin, L., Gu, N., Quan, X.: Advances in web and network technologies, and information management, pp. 148–159. Springer, Heidelberg (2009)
3. Van Zaanen, M., Pizzato, L.A., Mollá, D.: Question classification by structure induction. In: Proceedings of the 19th IJCAI, pp. 1638–1639. Morgan Kaufmann Publishers, San Francisco (2005)
4. Zhang, D., Lee, W.S.: Question classification using support vector machines. In: Proceedings of the 26th Annual International ACM SIGIR Conference on Research and Development in Information Retrieval, pp. 26–32. ACM, New York (2003)
5. Hermjakob, U.: Parsing and question classification for question answering. In: Proceedings of the Workshop on Open-domain Question Answering, pp. 1–6. Association for Computational Linguistics, Stroudsburg (2001)
6. Huang, Z., Thint, M., Qin, Z.: Question classification using head words and their hypernyms. In: Proceedings of the 2008 Conference on empirical methods in natural language processing, pp. 927–936. Association for Computational Linguistics (2008)
7. Olsson, F.: A literature survey of active machine learning in the context of natural language processing. Technical Report 3600, Swedish ICT (2009)
8. Logar Berginc, N., Iztok, K.: Gigafida–the new corpus of modern Slovene: what is really in there? In: The Second Conference on Slavic Corpora (2011)
9. Quinlan, J.R.: C4.5: programs for machine learning. Morgan Kaufmann Publishers Inc., San Francisco (1993)
10. Freund, Y., Schapire, R.E.: Experiments with a new boosting algorithm. In: Proceedings of the 13th ICML, pp. 148–156 (1996)
11. Hwang, C.L., Yoon, K.: Multiple Attributes Decision Making – Methods and Applications. Springer, Berlin (1981)
12. Li, X., Roth, D.: Learning question classifiers. In: Proceedings of the 19th international conference on Computational linguistics, vol. 1, pp. 1–7. Association for Computational Linguistics, Stroudsburg (2002)
13. Witten, I.H., Frank, E., Hall, M.A.: Data mining: practical machine learning tools and techniques. Morgan Kaufmann Publishers Inc., San Francisco (2011)

2B$ – Testing Past Algorithms in Nowadays Web

Hugo Rodrigues and Luísa Coheur

L^{2}F / INESC-ID Lisboa
Rua Alves Redol, 9, 1000-029 Lisboa, Portugal
{hugo.rodrigues,luisa.coheur}@l2f.inesc-id.pt

Abstract. In this paper we look into Who Wants to Be a Millionaire, a contest of multiple-answer questions, as an answer selection subproblem. Answer selection, in Question Answering systems, allows them to boost one or more correct candidate answers over a set of candidate answers. In this subproblem we look only to a set of four candidate answers, in which one is the correct answer. The built platform is language independent and supports other languages besides English with no effort. In this paper we compare some techniques for answer selection, employing them to both English and Portuguese in the context of Who Wants to Be a Millionaire. The results showed that the strategy may be applicable to more than a language without damaging its performance, getting accuracies around 73%.

Keywords: Question Answering, Who Wants to Be a Millionaire, Word Proximity, Web.

1 Introduction

Who Wants to Be Millionaire? (WWBM) is a famous TV contest, with national versions world-wide, where contestants answer multiple-answer questions (4 hypothesis) with the objective of taking home the maximum prize (usually 1 million of the local currency). Apart from a set of strategic decisions that a WWBM player has to take, his/her main problem is to be able to choose the correct answer between four given possible answers.

A solution to this problem is implemented by many Question Answering (QA) systems, as after extracting the answer candidates from the information sources these systems have to score them and choose one (or more) to be returned [1,2]. However, besides the fact that QA systems usually have to deal with more than 4 candidates, in WWBM candidate answers are not related, as opposite to candidate answers attained by QA systems. Thus, trying to solve WWBM can be seen as a subproblem of the QA scenario. In fact, it can be found in the literature attempts to solve specifically the WWBM challenge [3,4]. Both authors, by exploring simple techniques applied to the snippets returned by Web search engines or pre-defined knowledge sources, report accuracy results around 70%–75%.

In this paper, we evaluate how similar techniques perform, after more than 8 years, with the actual web, when dealing with the WWBM contest, both in English and in Portuguese (with exceptions of some stopwords filtering, presented techniques are

P. Sojka et al. (Eds.): TSD 2012, LNCS 7499, pp. 681–688, 2012.

language independent). Thus, we compare these techniques across time and also their performance when applied to different languages. This last comparison is particularly interesting because Portuguese is a morphologically richer language than English, and the World Wide Web (WWW) for Portuguese has a different magnitude from the English WWW.

The paper is organized as follows: in Section 2 we present some related work; in Section 3 we describe our system's architecture; in Section 4 we report our experiments; finally, in Section 5, we present the main conclusions and point to future work.

2   Related Work

To our knowledge, the first computational attempt to solve WWBM is the one described in [3], where the authors present an approach on the QA task (used at TREC [5]) and, later, on WWBM. Here, a query (a set of terms) is generated from the original question plus each one of the possible answers. The used knowledge sources are documents from pre-defined corpora or simply the top results returned by a search engine like *Google*. The first task described by the authors is to find the so called *extents*: passages from a document, where the answer may lay. These extents are classified and scored, considering the query terms appearing in it, as well as the self-information contained in each extent. This approach presupposes a probability associated with each extent, which is given by the probability of finding each term of the query in the extent[1]. A higher score is given to extents with lower probability of occurrence. Higher values do not imply a greater likelihood of the answer being present in the extent or its proximity, but the authors found empirically that this relation holds. However, in the context of WWBM an answer is one or more terms, not an whole extent; thus, by following a similar ranking strategy, each term is scored, based in a redundancy component, representing the number of occurrences of each term in the different passages. The accuracy accomplished for a 108-question test was 70%.

In 2003, [4] made another approach to WWBM. The strategy is simple, yet with good results. The web is used as the only resource, and three different approaches are reported, with ranging accuracies from 50 to 75%. The first technique (the baseline) comes out directly from the problem formulation: the correct answer is the one with more results retrieved by querying a search engine with a query in the form "answer . questionModified", where "questionModified" refers to the question filtered of stopwords. Also, for questions with negative constraints such as *Which of these plays is not written by Shakespeare? A: Hamlet B: Othello C: Romeo and Juliet D: Cats*, the chosen answer was the one with less results (an inversion due to the presence of *not*). Using *Google* the accuracy was set in about 50%. The authors also made some changes, as quoting the answers, which are sometimes multi-worded. Those changes, along with some other minor changes, increased the percentage of correct answers to 60%.

The second strategy cares about the position of the answer in a given document, when compared with other words present in the question. Therefore, authors implemented a word-proximity algorithm, which relies on the assumption that answers appear close to the terms contained by the question, giving this way more weight to them. The reason

---

[1] An assumption of independence of each term appearing in the snippet is made.

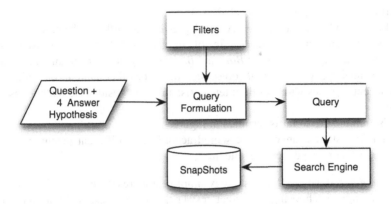

Fig. 1. Partial architecture overview of 2B$

behind this heuristic was the belief – as in [3] – that answers appear in documents close to question words. It is intended to weigh the distances, of a maximum radius, so that documents with too many references to an answer but not to the corresponding question words are worth less. The first ten pages returned by the queries were used in this process. Different values for radius were tested and authors concluded that the optimal value for radius is not trivial to get. Although, for higher values, it performs better than simple counting the results for each possible answer. When combining different parameterizations, accuracies reach 75%.

3   2B$ – Who Wants to Be a Millionaire

In this section we describe 2B$, our language independent system.

2B$ receives a question and a set of possible answers, from which one is the correct one. For each question and possible answer a query formulation is performed. Thus, for each question, four (in our case) distinct queries are envisaged, one for each hypothesis of answer. The process of query formulation involves different formats and filters (2B$ works with any combination of these):

- The answer can be quoted or not;
- The answer terms can appear before/after the question terms (AQ vs. QA format);
- Different filters can be applied to the question terms:
    - Wh-words–filter, where words such as *Where* and *Who* are eliminated;
    - Propositions–filter, where propositions like *At* are removed;
    - To Be–filter, that gets rid of different forms of the verb to be.

All together, these filters can be seen as the process of removing stopwords from the query as done in [4]. These filters have to be pre-defined for each language.

After the query formulation step, each query is submitted to a search engine . In what concerns the search engines 2B$ uses *Bing*[2] and *Blekko*[3]. This is due to the following

---

[2] http://www.bing.com
[3] http://blekko.com

facts: a) *Yahoo* API has been deprecated and is no longer available, as well as *Alta Vista*, which belongs to Yahoo!; b) *Google* API is deprecated as well, and the new one does not allow a significant number of queries free of charge; c) *AllTheWeb* also ceased to exist. In this way, the only survivors are *Bing* and *Blekko*. *Bing* is a traditional search engine, and works the same way as *Google* does. *Blekko*, on the other hand, has a limitation, as it only accepts queries with length of ten terms; any term besides the tenth is discarded blindly.

2B$'s architecture so far is resumed in Figure 1.

After this first step, answers are scored according to different techniques. In this paper, we focus on two different strategies:

- *Counting*: as done in [4], this strategy simply compares the total results returned by search engines, choosing as correct answer the query associated with the greatest value. The assumption behind this technique is that the correct answer often appears together with the question terms, thus hitting more results.
- *Word Proximity*: the idea is that answers usually appear near to terms present in the question. This way, consulting a set of documents one can search for the answers and, in case of finding them, check within a radius for the question terms. We use an algorithm suggested by [4], presented in Algorithm 1, and apply it to the top ten sites returned by the search engine.

Algorithm 1. Pseudocode for a scoring algorithm, giving more weight to answers near question words, within radius words

---

```
DistanceScore(documentSplit, questWords, ansWords, radius)
score, ansFoundWords = 0
for i = 1 to ||documentSplit|| do
    if documentSplit[i] ∈ ansWords then
        ansFountWords += 1
        for j = (i−radius) to (i+radius) do
            if documentSplit[j] ∈ questWords then
                score += (radius −|i − j|)/radius
            end if
        end for
    end if
end for
if ansFoundWords == 0 then
    return 0
else
    return score/ansFoundWords
end if
```

The algorithm receives an array representing the document split by its words, the words present in the question and answers, and a value for radius. Then it weighs the distances between terms and penalizes documents on which the answer is present often but not close to question terms. Each score corresponds to a single document (*i.e.* a page), so a final score needs to be computed based on the ten calculated scores. For this we implemented three different scoring methods: a) Mean, which does the average of the scores; b) Max, which returns the maximum value; c) and sMean, which ignores the maximum and minimum value (considering them outliers) and computes the average for the remaining values.

Is important to note that an inverse choice is made in cases where the question is negative, as in [4]. In these cases, we choose as correct answer the one with the least results.

4   Experiments

4.1   Experimental Setup

To test our system, we used a set of questions in English from some editions of the computer/console game 'Who Wants to Be a Millionaire?'. We use a subset of 100 questions, randomly picked, without caring about difficulty (in the human point of view the initial ones are the easiest)[4]. For the Portuguese set, we used a corpus of 100 questions from the local broadcast, manually transcribed, also randomly picked among a total of 180 questions.

The description of both corpus can be seen in Table 1.

Table 1. Used corpora in the experiments

Language	Number questions	Total words	Unique words	Longest question	Shortest question	Avg. Length
English	100	981	376	17	5	9
Portuguese	100	1155	565	18	3	11

As we can see, Portuguese is a much more rich and verbose language, using about 200 new words and in average two more words per question. This may be due to the verbs formulations, that in Portuguese have a different form for each person and number.

We should also note that we opted to not use a translation of the same corpus because some questions are local, that is, they only make sense in their mother language, such as questions about TV shows or important people of that country. For example, the Portuguese question *O que é uma pescadinha de rabo na boca?* refers to a traditional Portuguese dish. Translated it would be something like *What is a whitefish of tail in the mouth?*, which does not relate with any dish. Thus, although with different questions, we think it makes more sense to compare the behavior of the system within each language.

---

[4] The question set is available at *GameFAQs*, at http://www.gamefaqs.com/gba/
582399-who-wants-to-be-a-millionaire/faqs/37922

4.2 Results

As stated before, the system can work with a set of parameterizations. They include different features such as quoting the answers, the order of the answer in the query (first or last) or the filter used. Also, we found previously that our system performs slightly better when the answer comes first. With this in mind, we run a set of tests, for both English and Portuguese corpora.

For English, results are present in Table 2.

Table 2. Best results for English

Strategy	Scoring Method	Radius	Filters	Quoting	Bing	Blekko
Count			Wh,Prop,Be	Yes	50%	55%
			Wh,Prop	Yes	53%	**56%**
WordProx	Mean	40	Wh,Prop,Be	Yes	56%	52%
			Wh,Prop,Be	No	**63%**	48%
			Wh,Prop	Yes	57%	**57%**
			Wh,Prop	No	**64%**	51%
		30	Wh,Prop,Be	Yes	54%	56%
		50	Wh,Prop,Be	Yes	58%	**58%**
	Max	40	Wh,Prop,Be	Yes	43%	**60%**

We opted for a value of 40 for radius based on [4, Figure 1], which shows that the range 30–50 leads to the best results. Nevertheless, we ran some tests with 30 and 50 for radius. We can see that this feature does not have much impact, depending on the search engine.

When looking at the count strategy, we can see that *Blekko* performs better. This may have to do with *Blekko* being a 'spam-free' engine, as they describe it. With this, the developers mean that all web is not indexed. Instead, only a set of domains are allowed, in a attempt to favor data quality over quantity. This way, we believe that some queries that make no sense get no hits, while *Bing* tries at all cost to retrieve any data even if less or not related with it.

When comes to wordProximity, *Bing* has better results when not using quotes, as opposite to *Blekko*, which sees its performance decrease when removing them. Also, we can conclude that using Be-filter is useful: for *Bing* the accuracies increase slightly, while with *Blekko* they increase about 4%.

The last experiment tested a different scoring method: using the maximum value from the documents instead of the mean. The results for *Bing* were much worse. However, for *Blekko*, it was one of the best runs. Again, this may related to how *Blekko* was developed, being one of the top hits much more informative than the ones returned by *Bing*.

For Portuguese the results are showed in Table 3. We do not show results for *Blekko* because for almost all questions we got no hits. As we mentioned before, *Blekko* does not index all web, having instead a set of 'safe' domains. Thus, we believe that there are few Portuguese sites indexed.

Although overall results are slightly lower than for English, we can see that the behavior is similar. This decrease can be explained by the poorest data, in quantity

Table 3. Best results for Portuguese

Strategy	Scoring Method	Radius	Filters	Quoting	Bing
Count			Wh,Prop,Be	Yes	41%
			Wh,Prop	Yes	45%
WordProx	Mean	40	Wh,Prop,Be	Yes	54%
			Wh,Prop,Be	No	54%
			Wh,Prop	Yes	52%
			Wh,Prop	No	**58%**
		30	Wh,Prop,Be	Yes	**58%**
		50	Wh,Prop,Be	Yes	**56%**
	Max	40	Wh,Prop,Be	Yes	53%

and quality, for Portuguese language. For countings the difference is bigger, and it was expected to be so, due to the lack of information in great quantity.

4.3   Combining Methods

Because both search engines behave differently, we noticed that sometimes a configuration for *Bing* fails to answer some questions answered correctly by *Blekko* and vice-versa. This, allied with the assumption of greater precission for *Blekko* engine, makes another experiment come out naturally: the union of different runs, so that scores can be combined together, minimizing this way failures of each other. We combined different runs in a unique run, with different weights to each one. For this run we used three individual runs which are presented in Table 2: *Bing*, with WordProximity, mean as scoring method, 40 for radius and filters 'Wh,Prop'; *Blekko*'s counting; and *Blekko*, with WordProximity and max as scoring method. The weights were given manually, with values of 35%, 40% and 25% respectively. This run achieved 73% accuracy, a significant boost of almost 10% when compared with singular runs. Note that we did not use the best three results, but those we found to better complement each other. Also, we did not test all possible combinations, being possible that other runs and/or weights can achieve greater accuracies. Nevertheless, it is major improvement by just combining some of the best results.

5   Conclusions and Future Work

Although being a less complex problem than the one involved in the candidate answer selection step performed by QA systems, being able to choose the correct answer out of four in the context of the WWBM quiz is not a trivial task. In this paper we have shown how the simple counting returned by search engines is not enough to solve this problem, and how word proximity techniques can improve results. The best results are about 73%, a bit lower than previous results. However, our results cannot be compared straightforward with results obtained in the beginning of this century (around 70%–75% accuracy), as we are using different search engines and different corpora. The impressive growing of the WWW can be responsible for these results, but for even greater accuracies new techniques have to be envisaged.

For Portuguese language the results are promising, as the web contains much less redundancy, when compared with English. Although there is much room for improvement in the strategy, these are good signs that the approach is valid and, with some more work, can be adopted to any other language. Future experiments will include other languages to test our suspicions.

For instance, query expansion can be performed and search can be constrained to more reliable sources of information (for example, information from blogs can be ignored). In fact, some other possibilities offered by search engines can be tested. For instance, *Blekko* has an interesting feature that was not explored in this paper, but that can be useful to 2B$: the so called *slashtags*. Slashtags allow to parameterize a query, by restricting the search domain (slashtags can restrict to a domain, such as *health* or *sport*).

Also, the presented algorithm can be improved or substituted for another one to report more accurate results, as with this specification a question term closer to the answer has more weight than two or three question terms in the extremes of the considered snippet (a proximity versus quantity problem: which one is worth more?). It can also be substituted by latent models, such as LDA and LSA, which could understand if a given candidate answer belongs to the topic of the question or not.

Another point to explore is the combination of search engines. We accomplished the best results with this strategy, but improvements can still be made. The weights should be given in an automatic way, by learning techniques, avoiding over-fitting. Also, scoring techniques can be used to better aggregate results, given some features.

Acknowledgments. This work was supported by national funds through FCT – Fundação para a Ciência e a Tecnologia, under project PEst-OE/EEI/LA0021/2011.

References

1. Lin, J., Katz, B.: Question Answering from the Web Using Knowledge Annotation and Knowledge Mining Techniques (2003)
2. Schlaefer, N., Gieselmann, P., Sautter, G.: The Ephyra QA System at TREC (2006)
3. Clarke, C.L.A., Cormack, G.V., Lynam, T.R.: Exploiting redundancy in question answering. In: Proceedings of the 24th Annual International ACM SIGIR Conference on Research and Development in Information Retrieval, SIGIR 2001, pp. 358–365. ACM, New York (2001)
4. Lam, S.K., Pennock, D.M., Cosley, D., Lawrence, S.: 1 Billion Pages = 1 Million Dollars? Mining the Web to Play "Who Wants to be a Millionaire?", pp. 337–345 (2003)
5. Clarke, C.L.A., Cormack, G.V., Kisman, D.I.E., Lynam, T.R.: Question Answering by Passage Selection (2000)

Morphological Resources
for Precise Information Retrieval

Anne-Laure Ligozat[1,2], Brigitte Grau[1,2], and Delphine Tribout[3]

[1] LIMSI-CNRS,F-91403 Orsay Cedex, France
[2] ENSIIE, 1 square de la résistance, 91000 Evry
[3] LLF, 5 rue Thomas Mann, F-75205 Paris Cedex 13

Abstract. Question answering (QA) systems aim at providing a precise answer to a given user question. Their major difficulty lies in the lexical gap problem between question and answering passages. We present here the different types of morphological phenomena in question answering, the resources available for French, and in particular a resource that we built containing deverbal agent nouns. Then, we evaluate the results of a particular QA system, according to the morphological knowledge used.

Keywords: morphological resources, question answering.

1 Introduction

Question answering (QA) systems aim at providing a precise answer to a given user question. Their major difficulty lies in the lexical gap problem: the answering document may not contain the exact same words as the question. QA and IR systems must find a way of retrieving relevant documents without relying only on mere identity between words. Linguistic knowledge must thus be used, and integrating morphological knowledge has often been preferred over semantics because the integration of morphological knowledge often is more reliable.

Most of the research carried out so far made use of simple heuristic-based stemming techniques which cut off word endings (such as [1,2,3]). In most cases, the recall is slightly improved, but these techniques also produce some noise. Another way to use morphological knowledge is by extending the query, such as in [4], who significantly improve the results in most of the European languages for which they performed the experiment.

As we have shown, IR and QA applications mostly rely on partial or superficial morphological knowledge. However, some morphological resources are now able to provide detailed and precise knowledge about a large spectrum of morphological processes.

In this paper[1], we present the different types of morphological phenomena in QA, the resources available for French, and in particular a resource that we built containing deverbal agent nouns. Then, we evaluate the results of a particular QA system, according to the morphological knowledge used.

---

[1] This work was partly realized as part of the Quaero programme, funded by OSEO, French State agency for innovation.

P. Sojka et al. (Eds.): TSD 2012, LNCS 7499, pp. 689–696, 2012.
© Springer-Verlag Berlin Heidelberg 2012

2   Morphological Resources for QA

2.1   Morphological Phenomena in QA

[12] studied the most frequent derivational relations between questions and documents in French. An answering document can contain various types of variants of question words: a possible answer to the question *When was dynamite invented?* is *Alfred Nobel is the inventor of dynamite and patented it in 1867.* In order to detect the correspondance between the question and this document, the relation between *invented* and *inventor* must be recognized.

[12] manually annotated several corpora of questions and answering documents with the types of morphological relations between them. They showed that the most frequent relations in open domain are flexional and derivational relations. Concerning derivations, the most common types are denominal adjectives (*région-régional* for *region-regional*), and nominalizations, in particular action nouns (*inaugurer-inauguration* for *inaugurate-inauguration*) and agent nouns (*réaliser-réalisateur* for *direct-director*).

In this work, we used these observations to integrate morphological knowledge into a QA system. To this end, we first considered the existing morphological resources for French. Two French morphological resources exist, for deverbal action nouns and relational adjectives, which we will now present.

2.2   Derivational Resources for French

Verbaction[2] is a lexical resource containing action names derived from a verb [5,6]. It contains 9,393 noun-verb pairs, such as *renouveler-renouvellement* (*renew-renewal*).

Prolexbase [3] is a multilingual dictionary of proper nouns [7,8]. Although it does not contain explicit morphological knowledge, it gives information about relational nouns and adjectives derived from proper nouns. For example, the noun *Français* and the adjective *français* (*French*) are related to the entry *France*. Prolex contains 76,118 lemmas and 20,614 derivational relations.

Derivational resources thus exist for French, but concerning QA needs, a resource for deverbal agent nouns was lacking, so we built one semi-automatically, which we called VerbAgent, in reference to Verbaction.

3   Construction and Validation of a Resource for Deverbal Agent Nouns

As a first step, we automatically derived verbs from nouns, using formal properties of nouns. In French, some suffixes are frequently used to form deverbal agent nouns, such as *-eur*, as in *danseur* (*dancer*) derived from the verb *danser* (*dance*). Nine such suffixes were identified:

---

[2] http://redac.univ-tlse2.fr/lexicons/verbaction.html
[3] http://www.cnrtl.fr/lexiques/prolex/

1. *-eur* (*danser* > *danseur*)
2. *-euse* (*chanter* > *chanteuse*)
3. *-rice* (*inspecter* > *inspectrice*)
4. *-eresse* (*défendre* > *défenderesse*)
5. *-aire* (*signer* > *signataire*)
6. *-ant* (*attaquer* > *attaquant*)
7. *-ante* (*diriger* > *dirigeante*)
8. *-ent* (*adhérer* > *adhérent*)
9. *-ente* (*présider* > *présidente*)

We then used a lexicon of inflected forms, Morphalou[4], to extract all nouns ending with one of the suffixes, and then checked if a corresponding verb existed in Morphalou. This verification was based on some twenty rules, such as: suppress the agent suffix *-eur* and replace it with the infitive suffix *-er*.

4,067 noun-verb pairs were generated. Yet, a formal resemblance between a noun and a verb does not guarantee that they are morphologically related: for example the pair *accentuer-accentueur* is extracted, although these two words are not morphologically related. Moreover, the noun and verb can belong to the same derivational family but without the noun being derived from the verb, such as in *rougir-rougeur*. A validation is thus necessary.

3.1 Manual Validation

Manual validation consisted in checking that, for each considered pair, the noun was derived from the verb, and corresponded to an agent noun. We verified the semantic link or definitions in the TLFi[5] if necessary, for example when the noun was rare. 363 pairs were examined, among which 76% were correct i.e. contained a verb and the corresponding agent noun and 24% incorrect.

Errors mostly come from nouns with *-ant* or *-aire* suffixes, which can denote agents, but also commonly denote non agent nouns (such as *adoucissant*). Other errors come for example from nouns with an *-eur* suffix, which are frequently associated with an instrument (such as in *aspirateur*).

As this manual validation is very time-consuming and delicate, we defined several methods for an automatic validation.

3.2 Automatic Validation

We experimented several methods for an automatic validation of pairs.

Definition Extraction from a Dictionary. First, we extracted the definition from the dictionary Littré[6]. Indeed, agent noun definitions usually begin with a phrase such as *Celui qui* (*The one who...*) followed by the corresponding verb. For example, the

---

[4] http://www.cnrtl.fr/lexiques/morphalou/
[5] http://atilf.atilf.fr/: on-line French dictionary
[6] XMLittré is an electronic version of the French dictionary Littré.

definition of the agent noun *chanteur (singer)* is *Celui, celle qui chante, qui fait métier de chanter (The one who sings, whose profession is to sing)*.

Using the pattern *Celui, (celle)? qui* for detecting such definitions, we extracted 2,944 nouns. Yet, using only a pattern does not guarantee that the noun is derived from the verb. For example, it extracts the definition *Celui, celle qui joue du piano (A person who plays the piano)*. Thus, we added a simple constraint on the verb form: the first two characters of the noun and the verb must be the same. This way, we extracted 1,121 nouns, which we hope to be more accurate, although less complete (for example the noun *agresseur* is not extracted because its definition is *Celui qui attaque le premier*, and does not contain the corresponding verb *agresser*).

We compared these lists of nouns to the manually annotated part of the resource. Using the definition pattern only, 92 annotated nouns are extracted, among which 87 were considered as actual deverbal agent nouns. Using the additional constraint on the verb, 60 nouns are extracted, which were all manually annotated as correct.

As 275 noun-verb pairs were manually annotated as correct, this automatic validation method does not present a very good recall (22% with the verb constraint); yet it is very precise. The low recall can be explained by the existence of different definition patterns (such as for *agresseur*), and by the absence of some rare words from the dictionary (such as *avaliseur*).

Cooccurrents. In order to evaluate if two words are semantically related, it is also possible to rely on their contexts in corpus. Thus, we exploited a cooccurrence network [9], extracted from a French corpus of articles[7]. Our hypothesis is that if a verb-noun pair shares cooccurrents, words in the pair are semantically related and more likely to be the result of a derivation.

We extracted, for each verb-noun pair, their closest cooccurrents[8], and considered that a pair was correct if it had at least one common cooccurrent. The main disadvantage of this method is the absence of many words, due to the limited size of the corpus: only 869 pairs are present in the network (both the noun and the verb exist) on the 4,067 pairs of the resource.

In order to test the relevance of this method, we compared the pairs presenting at least one common cooccurrent with the manually annotated part of VerbAgent. 85 annotated pairs are found in the cooccurrent network, among which 56 have a common cooccurrent, 45 of which being annotated as correct, and 11 as incorrect. Errors are usually semantically related pairs, but with a noun that does not correspond to an agent, such as *accablant-accabler* or *accélérateur-accélérer*. This method thus seems to give a clue on the relation between the noun and the verb, but a larger corpus would be needed to get more complete results.

---

[7] This corpus was constructed based on 24 months of articles from Le Monde newspaper, using a 20 word window, and without taking order into account. Only cooccurrents with a frequency higher than 5 were kept. This network contains 31,000 words. Cohesion between words is based on mutual information estimation.

[8] All nouns were lemmatized according to the TreeTagger lemmatization usage, since this tagger was used to build the cooccurrence network.

N-grams. In order to validate an equivalence of meaning of the two words, we exploited the distributional idea which states that related words share a same context. We considered a context made of one word, and defined rewriting rules for defining the usage of verb *vs* the usage of an agent noun. Our purpose is to recognize such rewritings: *chanteur d'opéra* (*opera singer*) vs *chanter un opéra* (*to sing an opera*) to validate the pair *chanteur-chanter* (*sing-singer*).

We used the Google Books Ngrams resource, which contains n-grams of words computed on digitized books. We collected n-grams which contain the nouns and verbs of VerbAgent, followed by either a determinant or a preposition plus a word. We considered that a pair is valid if both words share at least a same context, i.e. are used in relation with a same word. We extracted 1,795 n-grams corresponding to 231 pairs. Their evaluation on the reference set shows that within the 19 pairs found, 1 is not valid. This method seems to be precise; however it has a low recall.

Combination. Table 1 presents the number of noun-verb pairs validated by each method, as well as by their combination. The second column indicates for example that 170 pairs were validated by all three methods, among which 15 were manually annotated as correct, and none was manually annotated as incorrect.

Table 1. # noun-verb pairs validated by each method

littré	*		*	*	*			
coocs.	*	*		*		*		
n-grams	*	*	*				*	
# found pairs	170	191	163	79	790	161	223	2290
correct	15	9	11	7	54	14	16	179
incorrect	0	10	0	0	5	1	12	61

The Littré seems to have the best recall, with a good precision. Yet, 2,290 pairs are found by none of the methods, mostly because the verb or noun frequencies are two low. A possible improvement could be to use larger or more adapted corpora (for example results of Web queries using these words), or to use additional resources (such as other dictionaries).

4   Contribution of Morphological Knowledge to the Answering Process

4.1   General Description of QAVAL

QAVAL [10] is a QA system for French. Lucene is used to select shorts passages (instead of documents) which are then analyzed by a shallow terminological parser, Fastr [11], for recognition of terms and their variants. Best passages are selected according to the presence of question terms. Candidate answers are then extracted from these passages and a machine learning validation system applies several criteria to rank the candidates.

4.2   Use of Morphological Knowledge in QAVAL

Morphological variants are handled at two stages in QAVAL: at passage retrieval and at passage selection as said before. At passage retrieval, the collection indexation and interrogation use stemming, which contributes to the presence of morphological variants of question terms.

4.3   Experiments

We conducted tests on two kinds of documents, Web documents and newspaper articles. We used 147 factual questions on a Web collection and 479 questions from CLEF and EQUER campaigns. To evaluate the impact of morphological resources, we calculated the MRR[9] on the first 10 passages selected for these questions, with and without morphological variants. We focused on the 10 best passages because after this rank, it is very difficult to extract a correct answer which would be proposed in the first ranks. Terms are searched in 150 passages retrieved by the search engine. After their annotation by terms which allows their weighting, we only keep the 50 best passages.

Results are presented in Table 2. The first two columns indicate the collection which is searched, and the total number of questions studied. Column *#q SS* gives the total number of questions that can be answered without taking into account variations and *#q VAR* with variations. A question can be answered if it is associated with at least one passage containing the expected answer.

In order to compare the QA system performances under the same conditions, we determined the questions which can be answered (column *#q OK* Table 2). Then, we kept among these questions those such as at least one passage contains variations of the question terms (column ExistVAR, subset of the column *#q OK*). Columns *MRR SS* and *MRR VAR* give the MRR computed on the passages associated to this last set of questions, annotated without and with variants respectively.

Table 2. QAVAL results when selecting passages, with and without morphology

collection	#quest.	#q OK	#q SS	#q VAR	#q Ex-istVAR	MRR SS	MRR VAR
clef05	197	187	175	174	125	0.6298	0.6486
clef07	156	92	86	82	49	0.5269	0.5484
equer	126	117	105	105	96	0.6782	0.7039
quæro	147	125	106	113	76	0.3984	0.4347
total	626	521	472	474	346	0.5778	0.6027

The overall number of questions which can be answered does not vary in the two cases. However, differences come from the impact of variations in the ranking process. Adding morphological knowledge systematically improves the MRR, for each collection, and each question set. On all collections, the MRR without any morphological

---

[9] Mean Reciprocal Rank: average of the reciprocal ranks of the first correct answers.

knowledge is 0.5785 and 0.6096 when taking into account morphological knowledge. This kind of improvement, even small, is important for a QA system. Extraction of answers is based on the capacity of a system to compare a question and an answering passage, especially when the wordings are different. A good matching will rely on resources having a good coverage.

QAVAL searches for all kinds of morphological variations. Thus, we evaluated which kinds of variants are found in all the passages retrieved by Lucene, and in the passages which contains a correct answer (see Figure 1). Following the study conducted in [12], we categorized variations in verb-action variants (*action* in the table), verb-agent (*agent*), location noun-adjective (*relat*) and others, as for example adjective-adverb etc. Percentages of each kind of variations are computed for questions coming from different QA campaigns and we can see that if *others* variations are numerous in all the passages, they are less important in the correct passages. On the other hand, *relat* and *action* variants are well represented in correct passages. We can also see that *agent* variants are less frequent; however these results have to be confirmed on a larger corpus. The cumulate results are given in the last two columns.

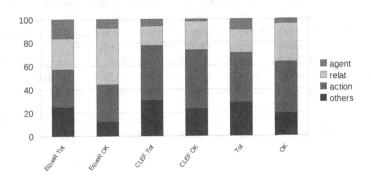

Fig. 1. Repartition of kinds of variants in passages (all passages: tot, correct ones: OK)

5   Conclusion

We presented in this paper a method for constructing a precise terminological resource containing morphological relations. This method leads to some errors and requires validation. Variant occurrences in corpora or resources were used to validate some relations automatically.

We also experimented using morphological resources in a question answering system, QAVAL, and found that such a knowledge leads to a better selection of passages, and that some kinds of morphological derivations are more frequent in passages which contain the correct answer than in other passages.

References

1. Lennon, M., Pierce, D.S., Tarry, B.D., Willett, P.: An evaluation of some conflation algorithms for information retrieval. Journal of Information Science 3(4) (1988)
2. Harman, D.: How effective is suffixing? Journal of the American Society of Information Science 42(1) (1991)
3. Fuller, M., Zobel, J.: Conflation-based comparison of stemming algorithms. In: Proceedings of the Third Australian Document Computing Symposium (1998)
4. Moreau, F., Claveau, V.: Extension de requêtes par relations morphologiques acquises automatiquement. In: Actes de la Troisième Conférence en Recherche d'Informations et Applications, CORIA 2006 (2006)
5. Hathout, N., Tanguy, L.: Webaffix: Discovering Morphological Links on the WWW. In: Proceedings of the Third International Conference on Language Resources and Evaluation, Las Palmas de Gran Canaria, Espagne, pp. 1799–1804 (2002)
6. Hathout, N., Namer, F., Dal, G.: An Experimental Constructional Database: The MorTAL Project. In: Many Morphologies, pp. 178–209. Cascadilla Press (2002)
7. Tran, M., Maurel, D.: Prolexbase: un dictionnaire relationnel multilingue de noms propres. Traitement Automatique des Langues 47, 115–139 (2006)
8. Bouchou, B., Maurel, D.: Prolexbase et LMF: vers un standard pour les ressources lexicales sur les noms propres. Traitement Automatique des Langues 49, 61–88 (2008)
9. Ferret, O.: ANTHAPSI: un système d'analyse thématique et d'apprentissage de connaissances pragmatiques fondé sur l'amorçage. Ph.D. thesis, Paris Sud (1998)
10. Grappy, A., Grau, B., Falco, M.H., Ligozat, A.L., Robba, I., Vilnat, A.: Selecting Answers to Questions from Web Documents by a Robust Validation Process. In: Web Intelligence (2011)
11. Jacquemin, C.: Syntagmatic and paradigmatic representations of term variation. In: Proceedings of the 37th Annual Meeting of ACL (1999)
12. Bernhard, D., Cartoni, B., Tribout, D.: A Task-based Evaluation of French Morphological Resources and Tools. Linguistic Issues in Language Technology 5(2) (2011)

Author Index